**ADAPTATIONAL
BIOLOGY
Molecules
to Organisms**

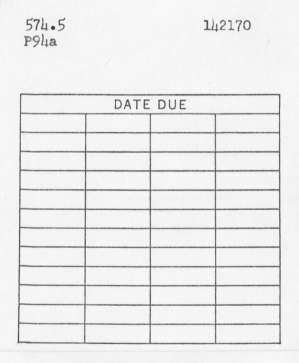

ADAPTATIONAL BIOLOGY
Molecules to Organisms

C. LADD PROSSER

Department of Physiology and Biophysics
University of Illinois
Urbana, Illinois

A WILEY-INTERSCIENCE PUBLICATION

JOHN WILEY & SONS

New York • Chichester • Brisbane • Toronto • Singapore

Library of Congress Cataloging in Publication Data:

Prosser, C. Ladd (Clifford Ladd), 1907–
 Adaptational biology.

 (Environmental science and technology)
 "A Wiley-Interscience publication."
 Bibliography: p.
 Includes index.
 1. Adaptation (Biology)   2. Adaptation (Physiology)
3. Physiology, Comparative.   I. Title.   II. Series.
QH546.P697   1986   574.5   86-1563
ISBN 0-471-89485-0
ISBN 0-471-89486-9 (pbk.)

Printed in the United States of America

10 9 8 7 6 5 4 3 2 1

PREFACE

This book is intended as a broad look at general biological concepts from the viewpoint of a comparative physiologist. It is neither a textbook nor a source or reference book for specialists. It is my hope that specialists will find this book useful as an overview of other specialties and for relating one area of biology to another. Every chapter has been read by several experts, each of whom has noted the incomplete coverage of his own field—which is appropriate for chapters that present viewpoints and unresolved questions.

Adaptational Biology: Molecules to Organisms was begun some eight years ago in an effort to provide intellectual bridges between molecular and whole-organism biology. The scope broadened during the writing of the book, and the evolutionary theme is pervasive. There are many recent books on evolution but no others from the viewpoint of general and comparative physiology. I consider it important for physiologists to be aware of the implications of evolutionary theory and equally important for evolutionists to know some physiology.

Not since the post-Darwinian period have there been such diverse positions in biology. Reductionists see little to be gained from holistic studies, and whole-organism biologists do not recognize the value of molecular analysis. Within each discipline there is extreme specialization, and consequently our journals, symposia, and societies become increasingly narrow. Very few generalists remain. Narrow specialization is encouraged by granting agencies.

There is a conflict as to whether evolution occurs by the accumulation of small changes (gradualism) or by abrupt major changes (saltation). In part, the discordant interpretations are due to the differences in time scale of population genetics and paleontology. It is my view that evolution results in biological diversity at all levels, from populations and subspecies to phyla and kingdoms. One objective of this book is to describe in some detail the functional diversity of organisms. Much more genic diversity is shown by molecular measurements than by observations of phenotypes.

There is disagreement as to whether natural selection can be recognized at the molecular level. Selection acts on phenotypes—whole organisms—not on genotypes per se. Most phylogenetic descriptions are necessarily based on morphological characters. The basic biochemical patterns of life processes were established very early in evolution, before there were fossil-forming organisms. Many biochemical characters are highly conserved, and for recent evolution, for example, the evolution of fishes, anatomy gives more clues than biochemistry. Variations in primary sequences of proteins are relevant to cladistic analysis, but amino acid sequence differences in the midregion of a protein may have little effect. Functionally critical regions and higher-order structure of proteins are highly conserved and have survived selection.

There is conflict about the species concept. The term *species* has several meanings. The concept of reproductive isolation, which serves well for animal classification, is not fully applicable to higher plants, in which hybridization and ecological isolation lead to species formation; reproductive isolation is also not relevant to asexually reproducing organisms. Another viewpoint is cladistic phylogeny. I propose in this book a physiological definition of species which emphasizes the uniqueness of physiological adaptation to ecological niche and geographic range.

A related question is whether most genetic changes are neutral or adaptive. Chapter 5 argues that measurements of kinetic properties reveal subtle adaptive differences between proteins that might otherwise be considered neutral.

There is uncertainty as to why there is so much more genetic variation than is indicated by proteins and by morphology. Is the excess DNA superfluous? Why are there such extensive noncoding regions in eukaryotic genomes? Why are many proteins encoded as long precursors and then cleaved to smaller molecules before they come into function?

There is a philosophical gap and a lack of communication between ethologists and membrane biophysicists. Some ethologists take the position that animal behavior can be studied meaningfully only in nature; the brain is a black box. The opposite standpoint is taken by biophysicists, who have made great strides in understanding nerve conduction and synaptic transmission. These two views of brain function are rarely brought together. To what extent is the neurobiology of relatively simple invertebrates similar to that of vertebrates, especially mammals?

One of the most controversial aspects of evolutionary biology concerns the relevance of neurobiology to human nature. Sociobiologists emphasize the long evolutionary history of *Homo* and the genetic basis for human behavior. Social scientists disagree and emphasize that culture plays a far greater role then heredity in determining human characters. Chapter 10 deals with the coevolution of genetically transmitted and culturally transmitted characters.

In this book I point out the validity of each of the conflicting positions and aim to provide some balance between them. In the following pages controversies are not avoided. Unifying themes are the temporal continuity of life forms and functions, the dynamic nature of life, biological diversity at all levels of organization, and physiological adaptation to the total environment—physical and biotic—at every stage of biological history. This book presents the point of view of a generalist, a naturalist in the classical sense.

References at the ends of chapters have been selected as examples and guides to detailed literature. Titles are abbreviated.

Chapters were read by the following persons and I am grateful to each of them for suggestions: Theodore Bullock, Philip Best, Peter Carras, Thomas Ebrey, Martin Feder, Albert Feng, Rhanor Gillette, William Greenough, Harlan Halvorsen, Eric Jakobsson, Leonard Kirschner, Michael Klein, F. Treville Leger, Robert Metcalf, Jr., David Nanny, Dennis Powers, John Roberts, David Shapiro, Carl Woese, and Colin Wraight.

I am grateful to the Department of Physiology and Biophysics, J. E. Heath and D. Buetow, chairmen, at the University of Illinois for facilities and logistic support.

I express special thanks to my wife, Hazel Blanchard Prosser, who critically and constructively read several versions of every chapter. Without her positive comments and encouragement the book would not have been completed.

C. LADD PROSSER

Urbana, Illinois
March 1986

CONTENTS

ADAPTATIONAL BIOLOGY
Molecules to Organisms

1

THEORY OF ADAPTATION

The term *biological adaptation* has several meanings according to context and usage. Physiological adaptation refers to all those properties—morphological and functional—which favor normal life processes in a delineated environment. Every organism is impinged upon by physical properties of the environment—gravity, temperature, water, ions, oxygen, nutrients, various chemicals, and light, as well as mechanical, magnetic, and electrical stimuli. One environmental parameter does not act alone and it may have different effects according to accompanying conditions; for example, the action of temperature on an aquatic organism differs with the salinity. Every organism is also influenced by its biotic environment—predators, prey, kin, competitive and cooperative conspecific and heterospecific individuals. No organisms, or very few, live in a constant environment; most organisms experience a fluctuating environment, cyclic or random.

All cells interact with their surroundings and are dependent on them. Unicellular organisms are chemically unlike their environment and the cell composition of multicellular organisms differs from the surrounding fluid. The first sign of death is a breakdown of isolating mechanisms and loss of cellular integrity.

Criteria of physiological adaptation can be identified at all levels of biological organization. Populations and societies are adapted in their capacity to reproduce; populations of some animals exhibit social behavior. An organism is adapted in the integration of vital processes and in the unified functioning of tissues and cells. A subcellular organelle is adapted in the activity of its contained proteins, lipids, and nucleic acids. Organisms must be adapted as integrated wholes, not as separate parts. However, analysis of the properties of components elucidates adaptive mechanisms.

ADAPTATION: DETERMINANTS AND STRATEGIES

The determinants of an adaptation are of three kinds, (1) genetically specified, (2) environmentally induced, and (3) developmental. Environmentally induced and developmental variations are possible only within genetically fixed limits. Ontogeny, canalization, determines the possibilities open to

1

later stages. A phenotype is not completely described even when both genotype and environment have been specified. Unless an organism has passed through essential developmental stages and experiences, adaptive properties are not realized.

The frequency of occurrence of a given character in a population follows a distribution curve, normal or skewed. Two populations from different environments may have the same distribution of an adaptive character, for example, tolerance of heat or cold. To ascertain whether the populations are similar genetically for that character they can be cross-acclimated, individuals can be transplanted from one environment to another, or they can be acclimatized to an environment beyond the normal limits of either population. Genetic differences may appear in the capacity of one population to extend its range. Extension of range is more usual for a population living near its limits than for one living in its midrange (Fig. 1-1).

The adaptive procedures followed by a given organism determine where the organism can live and reflect also where its ancestors probably lived. The capacity to live in an environment not occupied by forebears indicates that adaptive evolution has occurred. The essence of evolution is the production and replication of adaptive diversity. Adaptive diversity can be viewed either temporally or spatially. The fossil record provides a limited view (primarily morphological) of the diversity of organisms in time. Ecophysiology provides a picture of variation in present-day organisms. Evolutionists tend to emphasize either the temporal or spatial view. In this book, contemporary diversity will be examined at all taxonomic levels—phyla, classes and orders, genera and species, subspecies, and local populations. Our goal is to give a picture of adaptation as it occurs at the present instant in history.

It is my view that excessive attention has been devoted to the origin of species and that adaptive mechanisms at other levels are equally important in describing biological diversity. Populations of a species in different parts

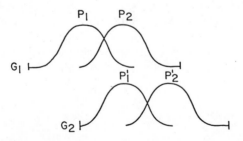

Figure 1-1. Two genotypes G_1 and G_2 each with phenotypic expressions P_1, P'_1, P_2, and P'_2 at different states of acclimatization. P_1 and P_2 can be interconverted, as can P'_1 and P'_2. Reprinted with permission from C. L. Prosser in *The Species Problem*, E. Mayer, Ed., Copyright 1957, American Association for the Advancement of Science.

of a geographic range are adapted to the sum of local conditions, physical and biotic. Populations at the two ends of a cline, for example, those from northern and southern latitudes, show little or no gene exchange with each other; in midregions, however, there may be evidence of incomplete speciation in the form of intermediates and/or hybrids. Adaptive properties of species, subspecies, and local populations may be analyzed in terms of molecular mechanisms. Higher taxonomic levels of plants and animals (orders, classes, phyla) show adaptive molecular properties, just as species and subspecies do, identification of these properties in present-day organisms may provide insight concerning the environments occupied by their ancestors and answer such questions as: Why are all echinoderms and cephalopods marine? Why are there no marine insects and amphibians (except secondarily)? Why are numerous orders of mammals restricted to the tropics?

ADAPTATIONS: LIMITATIONS—SEMANTIC AND STATISTICAL

The concept of adaptation is used in this book in two ways: (1) natural selection of adaptive characters is considered as a determining mechanism of evolution; (2) adaptation to the environment—both physical and biotic—accounts for the way of life of present-day organisms. The concept of adaptation in evolution is based largely on inferences regarding the function of observed structures, for any of which there may be several interpretations. Where motile organisms live is determined not only by adaptive structures and functioning molecules, but also by the organisms' behavior and the impinging biotic influences.

Genetic changes may be biologically neutral, negative, or advantageous (Ch. 4); fixation of a character in a population may result from natural selection by random means, such as genetic drift. Evolution by natural selection builds on preexisting structures and chemical compounds; frequently a selected character is secondary in that it evolved initially for another function. Feathers may have evolved in birdlike reptiles where they served for temperature regulation and later become adapted by birds for flying. Many mechanisms serving originally for respiration and vasomotion came to be used for temperature regulation in birds and mammals. The term *preadaptation* implies some anticipation of the later function, an unacceptable concept, and Gould has suggested *exaptation* as more general and objective (6,7).

A goal of this book is to give reductionist explanations of holistic phenomena in present-day organisms. Molecular descriptions of physiological properties throw light on the evolution of organisms. Comparative physiology describes aspects of ecology and behavior in terms of adaptation at all levels of organization and taxonomy. Extrapolation from molecular to organism-population levels is based on assumptions regarding causal relations. The present account will use the word *adaptation* while recognizing the underlying assumptions and multiple meanings of the word.

A major difficulty in describing either ancestral or present-day organisms lies in inferring causation from correlation. When event A is followed by event B, what is the probability that A is the cause of B? Statistically, no number of replications of a correlation can *prove* causation, but several procedures permit reasonable conclusions of causal relations based on correlations. One method is to omit event A and see whether B fails to occur. Another procedure is to alter conditions or to make substitutions for event A. Additionally, multiple observations of individual cases strengthen confidence in a conclusion. An example is the amino acid sequence in hemoglobins of fishes; certain sequences are correlated with differences in affinity for oxygen at different pH levels, and these are correlated with the O_2 content of the water in which the fish lives. The probability of a causal relation between the hemoglobin structure and the ecology of fish is strengthened by observing numerous species and by maintaining one kind of fish in containers of water of different O_2 content. The chapters that follow give examples of causal relations inferred from correlations, but it is implicit that proof is usually incomplete and the conclusions are probabilistic.

GENOMES AS DYNAMIC CONTROLLERS

Genes exert control of phenotypes by coding for proteins; the order of transcription from genes to RNA, followed by translation to proteins, was established early in evolution. Genetic control of a phenotype is not exerted simply by one gene determining one protein. In a few organisms (viruses), the genetic messages are carried by RNA. The orderly expression of specific genes during development is controlled by feedback mechanisms that are not yet well known. The regulation of genetic action is at the stages of transcription, translation, and posttranslation. Genes and the proteins encoded by them evolved in families in which macromutations that occur rarely led to the modern diversity of proteins. Some families of proteins evolved very slowly; others showed frequent gene duplications. Examples of macromutational evolution within protein families are vertebrate hemoglobins and calcium-binding proteins. The primordial genes of most protein families are not well known; the relationships among the hemoglobins of invertebrates, microorganisms, and vertebrates are speculated upon. Within single branches of a family of gene-proteins, micromutations have occurred that have resulted in differentiation of species and varieties. These genic duplications and rearrangements occur in shorter times than macromutations. For example, many isotypes of α and β hemoglobins occur in different fishes. Major genetic changes code for active (catalytic or binding) sites on proteins. These sites are relatively stable. Micromutations code for structural sequences of amino acids that may be at some distance from active sites and are less stable and less conserved than macromutations.

Genomes show considerable plasticity over many generations (9). Genomes of eukaryotes contain far more DNA than is adequate to code for the known complements of proteins (Ch. 3). The noncoding DNA has been considered as reserve, as "parasitic," or as possibly serving regulatory functions. Coding regions within a genome are exons; these may occur singly or in duplicated or repeated DNA sequences. In most eukaryotic genes, exons are interrupted by introns that are noncoding. Some noncoding regions are initiation and termination segments, others are pseudogenes that may have had a genetic function in ancestral genomes. The sequence of exons, introns, and pseudogenes for a given family may not be the same in related taxa; for example, the alpha hemoglobin genes of various mammals (10). Genes for a given family of proteins may be clustered or linked on one chromosome, or they may occur on separate chromosomes. In a few organisms the entire genome is replicated (polyploidy). Many proteins, probably all that are to be secreted for use outside the cell, are formed in long sequences comprised of several polypeptides that become cleaved away posttranslationally. An example is the formation of polypeptide neurotransmitters (Ch. 12). How the genes came to code for long precursors with peptides that are discarded before a final protein results is not known.

In conclusion, genomes are not static. Some portions of a genome may change relatively rapidly, within a few hundreds or thousands of generations, while other parts of genomes are conservative and change very little over millions of generations. Environmental influences, such as high temperature or chemical mutagens, can increase mutation rates; some regions of a genome are more susceptible than others. However, genomic changes are random and undirected except by natural selection of organisms.

SPECIES

Species may be defined in several ways, according to the use to be made of the concept (20). The species of a taxonomist represents the judgment of specialists as to how a group of plants or animals can most logically be classified. Some systematists emphasize similarities—these are the "lumpers"—while others base the groupings on differences and classify into many species—these are the "splitters." Consequently, what is named a species differs according to the taxonomist. One attempt to deal with subjective judgment has been made by numerical taxonomists who feed the information on the distinguishing characters into computer programs and arrange plants and animals into species according to quantitative similarities. This procedure fails to take account of subjective judgments and does not take into account many adaptive physiological characters.

Biological species are populations of like organisms which exchange genes. Reproductive isolation may be established by several mechanisms: (1) cytological incompatibility, such as polyploidy or failure of chromosome pairing; (2) physiological incompatibility, for example, hormonal; (3) gross morphological mating differences; (4) incompatibility of mating because of photoperiod, circannual rhythms or nutrition; (5) behavioral isolation by courtship and mating differences; (6) geographic separation; and (7) ecological separation.

A variant of the biological species is the physiological species. This is based on the concept that no two species can occupy the same ecological niche and geographic range throughout the life cycle. Each species can most accurately be described in terms of its physiological adaptations to its ecological niche and geographic range. The concept of physiological species is more consonant with processes of natural selection than the traditional concept of biological species. However, physiological species are more difficult to characterize.

Neo-Darwinian theory states that evolution occurs by species, not by local populations of individuals (1). Speciation may occur as the result of the accumulation of small genetic changes, called micromutations (gradualism), or by macromutations followed by periods of no change (punctuated equilibrium) (2,5). The processes are different for asexually reproducing organisms and for those in which hybridization is associated with colonization (Ch. 4). Gradualism is well known for physiological characters; major changes in animal behavior may serve the same purpose as macromutations.

Regardless of the nature of the biological unit of evolution, adaptive changes in physiology provide the basis on which natural selection operates. Molecular analyses of genetic composition in populations show that there is much more reserve DNA than is estimated from phenotype measurements. Molecular properties are highly conserved and small variations in genes and proteins are useful for classification.

However, physiological adaptations are present at all taxonomic levels— local populations, subspecies, species, genera, families, orders, classes, and phyla. The essence of evolution is the emergence of new organisms that survive and replicate; the understanding of the physiology of adaptive diversity contributes to our knowledge of how evolution has occurred and of where present-day organisms live.

CLASSIFICATION OF PHYSIOLOGICAL ADAPTATION

There are two broad classes of adaptive physiological responses. One refers to internal states as a function of environment. Constancy of *internal state* for some parameter is called homeostasis. A second class of adaptation includes all responses concerned with normal *activities* as a function of the environment; constancy of an activity is called homeokinesis (15).

Internal State; Homeostasis

Ever since the first living organisms appeared, certain cellular properties have been maintained: high intracellular concentration of potassium; consistent levels of redox potentials; constancy of pH; and retention of organic metabolites. The homeostasis within cells is similar for unicellular and multicellular organisms; the latter show in addition some extracellular fluid regulation.

Conformity

At the cellular or whole-organism level, the internal state may change to *conform* with the environment (Fig. 1-2a). The prefix *poikilo-* refers to a varying internal state. Internal temperature of poikilothermic organisms is essentially the same as that of their environment; poikilosmotic organisms have internal osmotic concentration varying with and conforming to that in the environment. In some animals, particularly marine invertebrates, oxidative metabolism varies in proportion to the partial pressure of oxygen in the environment; this is metabolic conformity. The term *ectothermy* refers to the warming or cooling of poikilotherms via external sources of heat or cold; ectothermy is not the equivalent of poikilothermy.

Regulation

The second pattern of internal state as a function of variation in the environment is regulation (Fig. 1-2b). The prefix *homeo-* refers to constancy of internal state. Homeothermic animals maintain relatively constant body temperature over a wide environmental temperature range. *Endothermy* refers to the use of internal sources of heat for raising body temperature; endothermy is not synonymous with homeothermy. Homeoosmotic animals maintain constant internal osmotic concentration in altered environmental

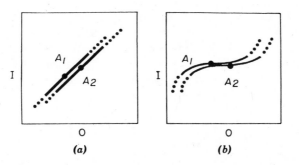

Figure 1-2.   Internal state (*I*) as function of the same parameter outside the organism (*O*). Patterns for (*a*) conformers and (*b*) regulators. A_1 and A_2 refer to two states of acclimatization. From *Comparative Animal Physiology,* Third Edition, C. Ladd Prosser, Ed., Copyright © 1973 by Saunders College Publishing/Holt, Rinehart and Winston. Reprinted by permission of CBS College Publishing.

salinity. Regulation of ions within living cells is universal, even in osmoconformers. For example, in some marine invertebrates the concentration of potassium in hemolymph (extracellular fluid) may vary with the potassium in the medium, but the intracellular potassium concentration is held relatively constant. Most animals are metabolic regulators over wide ranges of oxygen supply.

When an environmental parameter changes, regulation maintains internal constancy until a limit is reached beyond which regulation fails. The mechanisms of failure of regulation at low or high limits are by integrated chains of events, molecular examination of which is an important part of adaptational physiology. Many variations in the two patterns of internal state—conformity and regulation—are known. An animal may regulate to environmental change in one direction of a parameter and conform in the opposite direction. For example, many estuarine and intertidal invertebrates are osmoconformers in normal sea water and in above-normal salinities, but are regulators in dilute media.

The prefix *hetero* is sometimes used to refer to an intermediate condition or to limited conformity and regulation; heterothermy may be temporal or spatial. In a temporal heterotherm the animal's body temperature falls in dormancy of hibernation seasonally or is lowered diurnally (hummingbirds and bats). Some heterotherms regulate temperature in the body core more than in skin and extremities; this is spatial heterothermy.

Either conformity or regulation can be adaptive. Conformity permits life over a wide range of internal states with small expenditure of energy. Regulation permits life over a wide range of environmental variations but at high energetic cost (Fig. 1-2). *Acclimation* under controlled conditions or *acclimatization* in nature can shift the midpoint and the limits for change of internal state in either conformers or regulators. A conformer can function over a wider range of internal state, usually within narrower environmental limits; a regulator can function over a narrower internal range but a wider environmental range.

The physiological characterization of an internal state as conformity or regulation is to be distinguished from the ecological characterizations *steno-* and *eury-*. Stenohaline organisms are limited to narrow ranges of salinity, while euryhaline organisms tolerate wide salinity ranges. Stenothermal animals live in a limited temperature range, while eurythermal ones function over a wide range of environmental temperature.

Biological Activity; Homeokinesis

Two general categories of adaptation favoring constancy of biological activity (homeokinesis) in different environments are recognized: capacity and resistance adaptation.

Capacity Adaptations

Capacity adaptations are functional properties which permit relative constancy of biological activity over a "normally" varying environment. Many capacity adaptations are homeokinetic: positive metabolic adaptations of enzymes and lipids maintain constancy of energy output, membrane alterations maintain ionic and electrical gradients, and neural modifications subserve constancy of behavioral activity. Negative metabolic adaptations conserve food reserves during dormancy. Capacity adaptations occur at all levels of biological organization in both conformers and regulators and often are measured as rate functions. Capacity adaptations function especially in cycling environments and in natural acclimatization.

Several patterns of capacity adaptation with respect to rate functions have been described in detail for temperature, and they apply to other parameters as well. Figure 1-3 represents a rate function of a poikilotherm, the activity of an energy-yielding enzyme, or rate of O_2 consumption, measured over a temperature range T for animals acclimated at several temperatures T_1, T_2, T_3. The rate measured at the acclimation temperature, AT, is relatively constant, indicating compensation. Figure 1-4 shows direct and acclimated rate functions. The rate at an intermediate temperature (T_2) is designated as unity; when the temperature at measurement is reduced the rate declines and when temperature is raised, the rate increases. These are normal direct responses, that is, Q_{10} effects. Figures 1-4 and 1-5 diagram steady states attained some time after transfer from T_2 to a lower temperature T_1 or higher temperature T_3 (Fig. 1-6). If positive acclimation occurs over a period of days, the rate rises in cold and declines in warmth, compensating for the altered thermal regime. If compensation is complete, at steady state the rate is the same at all three temperatures (pattern 2 of Fig. 1-4). Overcompensation occurs in some systems (pattern 1), but partial compensation is most common (pattern 3). No compensation occurs when the rate remains the same after some days as it was directly after transfer (pattern 4); some systems (enzymes or organisms) show undercompensation or inverse acclima-

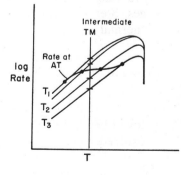

Figure 1-3. Rates of a biological function (e.g., enzyme activity) measured at different temperatures (TM) for organisms acclimated at low (T_3), medium (T_2), or high (T_1) temperatures. Rates measured at acclimation temperatures (●) indicate compensation. Reprinted with permission from C. L. Prosser in *Biological Adaptation*, H. Hildebrandt, Ed., Copyright 1982, Thieme-Stratton, Inc.

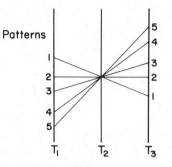

Figure 1-4. Patterns of rate function as changed by temperature acclimation, according to nomenclature of Precht. T_2 is intermediate temperature, T_1 is lower, and T_3 is higher temperature. Pattern 4 shows no compensation, merely Q_{10} effect; pattern 2 perfect compensation, rate the same at all temperatures after acclimation; pattern 3 shows partial acclimation, the most common pattern; pattern 1 shows overcompensation; pattern 5 shows inverse compensation. Reprinted with permission from C. L. Prosser in *Biological Adaptation,* H. Hildebrant, Ed., Copyright 1982, Thieme-Stratton, Inc.

tion (pattern 5). Another representation (Fig. 1-5) compares entire rate/temperature curves for a given function. These curves can show translation to right or left, with the same Q_{10} at different temperatures of acclimation. The curves may show rotation with a change in Q_{10} according to acclimation; the rotation may be about a midtemperature or it may be about a low or high temperature, in which case there is acclimation only in the range in which the curves diverge. Frequently there is a combination of translation with rotation. Analogous patterns of salinity acclimation have been described for metabolism after conforming organisms have been transferred from one salinity to another.

Various tissues differ in ability to show capacity acclimation, and in a given tissue one enzyme may compensate while another does not (18). For example, in many fishes acclimation in the cold shows positive rate compensation (patterns 2 or 3 of Fig. 1-4) in oxidative enzymes of the TCA cycle and electron transport chain, but hardly at all for glycolysis. This may relate to the greater dependence on oxidation associated with high dissolved O_2 at low temperatures. In general, hydrolytic and degradative enzymes show little or no thermal acclimation and may exhibit inverse effects. There are differences between tissues and species. Instead of maintaining homeokinesis, some kinds of conformers become dormant in cold. Many amphibians and reptiles show metabolic compensation at high temperatures, but become dormant in cold. Some species of fishes and insects, and desert reptiles and amphibians, estivate during periods of scarce water or high temperature. The adaptive significance of inverse acclimation (pattern 5 of Fig. 1-4) is to conserve energy reserves during periods of inactivity.

In thermoregulating animals metabolic patterns are different from those of thermoconformers (21). Within a certain ambient temperature range, the

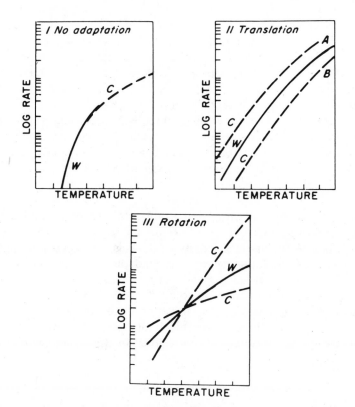

Figure 1-5. Three patterns of rate functions in cold (*c*) and warm (*w*) acclimated organisms. Rate functions measured at different temperatures. From *Comparative Animal Physiology, Third Edition*, C. Ladd Prosser, Ed., Copyright © 1973 by Saunders College Publishing/ Holt, Rinehart and Winston. Reprinted by permission of CBS Publishing.

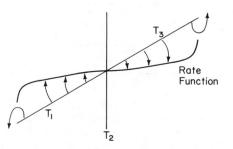

Figure 1-6. Diagramatic representation of a reaction rate measured at different temperatures for animals acclimated at three temperatures T_1, T_2, T_3. Reprinted with permission from C. L. Prosser in *Biological Adaptation*, H. Hildebrandt, Ed., Copyright 1982, Thieme-Stratton, Inc.

thermoneutral zone, the metabolism of a bird or mammal is minimal; metabolism increases as environmental temperature falls below a critical temperature T_c (Fig. 1-7). Elevated metabolism in cold maintains internal temperature in the face of increased heat loss. Below some low T_a, metabolism no longer maintains constant T_b. At temperatures above the thermoneutral zone, insulative mechanisms tend to keep the internal temperature constant; however, if body temperature rises, metabolism increases. The compensatory changes which tend toward internal regulation involve cellular activities which are hormonally triggered. Besides biochemical changes, morphological and behavioral responses counteract effects of cold or heat.

Resistance Adaptations

Resistance adaptations occur at environmental extremes and limit geographic and seasonal distribution of plants and animals. Resistance or tolerance adaptations may permit extension of range, within genetically determined limits; acclimatization of resistance adaptations confers tolerance of diurnal and seasonal extremes. For example, many plants and animals are narrowly limited by temperature at some stage in their life cycle; resistance adaptations allow them to withstand lower or higher temperatures than those at which activity is optimal. Tolerable limits may vary with stage in life cycle; for example, some crustaceans tolerate dilute sea water or fresh

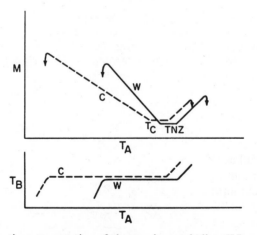

Figure 1-7. Schematic representation of changes in metabolism (M) and body temperature (T_B) of homeothermic animals at different ambient temperatures (T_A). TNZ is thermoneutral zone. Below lower critical temperature (T_C) metabolism increases. Above TNZ when T_B rises, metabolism also increases. Metabolism-temperature slope lower for cold (C) acclimated than for warm (W) acclimated animals. Arrows indicate failure of regulation at low or high T_A's. Body temperature (T_B) constant to lower T_A for cold- than for warm-acclimated and converse at high ambient temperature. Reprinted with permission from C. L. Prosser in *Biological Adaptation,* H. Hildebrandt, Ed., Copyright 1982, Thieme-Stratton, Inc.

water, except at certain developmental stages when they can live only in full-strength sea water.

Resistance adaptations can be examined at all levels of biological organization. One measure of resistance adaptation in populations is the capacity to reproduce. An essential feature of evolution is reproduction of populations. For individuals, resistance adaptation is commonly measured by survival. Different criteria of lethality apply for various kinds of organisms and at extremes of environmental stress. Actuarial statistics are used for time-stress measures of mortality. The LD_{50} (survival limit for 50 percent of a population in a specified time) for a given time-stress combination is used as a measure of both acclimatory and genetic differences. In general, the lethal limits for whole organisms (plants or animals) are relatively narrow and represent failure of integrated function. An isolated tissue or cell can survive at greater extremes. For example, muscle responses to electrical stimulation are used to test heat resistance; isolated muscle tolerates higher heat than the intact animal. Resistance at the molecular level is measured by protein denaturation or by lipid melting; these changes occur at wider thermal extremes than the changes tolerated by tissues and cells. Another measure of cellular resistance adaptation is maintenance of ionic and electrical gradients across cell membranes when the cells are in extreme environments.

In general, thermal tolerance of enzymes from animals and plants correlates with tissue or whole organism tolerance; the enzymes, however, can function in conditions more extreme than those tolerated by the tissues from which they came. This suggests genetic correlates of properties of enzymes with survival of intact organisms even though the organisms may not die because of failure of the specific enzymes. A general conclusion is that the more highly integrated a biological system, the narrower its range of tolerance. The measures of resistance adaptation—reproduction, survival, tissue function, membrane selectivity, protein denaturation—can be made with several stresses—temperature, salinity, oxygen, hydrostatic pressure. Acclimation can shift the tolerance limits within a narrow range which is genetically determined; acclimation can also change the critical temperature or salinity beyond which regulation of the internal state fails. Physiological functions which fail at the limits of regulation may not be the same as those which fail at the point of lethality.

Resistance acclimations to cold, heat, salinity, and oxygen may involve extensive biochemical changes. For example, some polar fishes synthesize antifreeze glycoproteins, other fishes make antifreeze polypeptides, and some arctic insects in winter contain large amounts of glycerol. These compounds lower the freezing point of body fluids to below temperatures at which unprotected fluids would freeze. Cold-hardening in plants results from biochemical changes which provide for tolerance of temperatures in the fall which could not have been tolerated during the growing season. Many invertebrates make resistance adaptations to hypoxia by shifting metabolic pathways from oxidation to glycolysis.

A useful measure of resistance in animals is coma—behavioral or respiratory paralysis. Coma can occur at a fairly sharp endpoint and is reversible. Also, as environmental extremes are approached, many animals show behavioral deficits. In fishes and many invertebrates the first sign of stress at extremes of temperature—heat or cold—is hyperactivity and reflex hyperexcitability; this may be followed by motor incoordination, then loss of righting ability, and finally coma (15a). A similar sequence is seen on exposure to toxic substances, such as heavy metals or pesticides. The behavioral sequence is similar in insects and fishes. Mammals made hypothermic or hyperthermic show increased activity, escape behavior, uncoordination, and finally respiratory coma. Since coma and death occur with less cooling or heating than will denature their proteins, and since integrated functions and behavior fail before metabolic functions, the nervous systems of animals appear to be more readily disrupted than the other body systems.

In many organisms—animals, plants, and microorganisms—an early sign of failure at an environmental extreme is loss of selective permeability of cell membranes; potassium leaks out and sodium enters. Leaky membranes are indicative of functional failure and ultimately of cell death. Another measure of resistance is reduction of energy metabolism below the point needed for maintenance of normal selective permeability and active ion pumps. In extreme cold, failure to tolerate freezing, particularly of bound water, is critical. Freezing tolerance varies with genetic strains of plants and animals and can be altered somewhat by acclimatization. Salinity tolerance in organisms which survive in high osmotic concentrations, for example, halophilic bacteria, requires enzymes genetically adapted to high ionic strength.

TIME COURSE OF ADAPTIVE CHANGES

The adaptive changes in internal state or in rate functions occur over a wide time course which can be divided into three parts: direct or immediate responses, acclimation or acclimatization occurring during days or weeks, and genetic alterations which may persist during many generations. Chapters 6 to 9 deal with molecular and whole-organism adaptations which occur at different times after environmental change.

The response of an organism to an altered environment depends on the rate and intensity at which the organism is subjected to the change. Intensity-time curves define the responses. If the environmental change is made very slowly, there may be no direct visible response, but slow acclimation can occur (Fig. 1-6). In a cycling environment, such as diurnally changing ambient temperature, the responses of animals and plants are different from what they are in a constant environment.

For every organism there are "optimal" states for metabolic and other activities—optima of cell temperature, osmotic and ionic concentration, O_2

partial pressure, light intensity, and circadian period. Deviations from the optima in either direction result either in reduced enzyme activity or in requirement of increased energy for cellular and organismic regulation (8a). Changes in reaction rates, as shown by Q_{10} values, are adaptive in that eurythermal animals tend to have lower Q_{10}s than stenothermal ones. Kinetic constants which define enzyme catalysis, for example apparent K_m and k_{cat} values, are sensitive measures of adaptive direct changes which are genetically determined (Ch. 5). Physical properties of cells and body fluids may alter directly with changes in the environment. The pH of aqueous solutions, such as protoplasm, decreases as temperature rises. Phospholipids of membranes at high temperatures become more fluid and probably more permeable. Much behavior of animals is direct response to external stimuli causing movement into "preferred" environments (Ch. 10).

During the enusing period, the second time period of adaptation, acclimatory changes can occur in an individual animal or plant, changes which may or may not compensate for the environmental alteration. The amount of change is limited by the genome. When a single environmental parameter is altered in the laboratory in a controlled fashion the adaptive process is acclimation. Under natural conditions where several variables act together, as during seasonal changes, the process is acclimatization. Acclimatory changes usually occur within days or weeks and consist of either metabolic reorganization or long-term modification in behavior. Not all organisms acclimatize; some enter a state of dormancy—in prolonged cold, drought, or other extremes. Some animals escape adverse conditions by migration. Capacity for acclimatization may vary with the stage in life history. Organisms which live in a cycling environment—diurnal, annual—usually have more capacity for acclimatization than those living in a relatively uniform or constant environment, such as the tropics, deep sea, or polar sea. Alterations in substrates, such as nutrients, may induce synthesis of specific enzymes (16a). The molecular mechanisms of acclimation constitute a major concern of adaptational physiology.

A partial list of the observed biochemical changes in acclimation follows (8a):

1. Changes in concentration of an enzymatic protein may result from increased (or decreased) synthesis or from changes in relative rates of synthesis and degradation. For energy metabolism there may be changes in number and size of mitochondria, hence of membrane surface for enzyme binding. Hypertrophy or atrophy of an organ may result in alteration of total enzyme activity without change in specific activity (18a).
2. Acclimation may bring about changes in the relative amounts of isozymes or allozymes; examples are proteins such as acetylcholine esterase and hemoglobin, enzymal regulators, or coenzymes.
3. Alterations in fatty acid composition; fatty acids formed at low temperatures are more unsaturated than those formed at high temperatures.

4.   Gross shifts in metabolic pathways.

5.   Changes in synaptic membranes such that central nervous function and resulting behavior may be altered; "preferred" temperature may change with acclimation.

The third time period for adaptive changes extends from generation to generation over geologic time. The long-term changes in organisms may be similar in nature to those in individuals during acclimation, but the changes differ in that they arise through genetic alterations—mutation, inversion, gene exchange, and other chromosomal events. When a change in coding for a protein occurs, allelic forms (allotypes) may appear; those which are enzymes are called allozymes. When gene duplications occur, multiple forms of a protein (isotypes or isozymes) may result. An allozyme or an isozyme may have different kinetics from its predecessor and this may open up new adaptive possibilities. Under favorable natural conditions some changes become fixed and lead to the formation of varieties and subspecies, these are the basis for clines in environmental gradients. Many genetic changes are neutral or harmful, but the dynamic genome is so variable that there can be some positive changes that become fixed by selection (Ch. 4).

Genetic adaptations, like acclimatory adaptations, may be complete, incomplete, or noncompensatory. Behavioral and metabolic activity of poikilothermic animals from Arctic or Antarctic regions are not the same as of animals from temperate zones; genetic compensation is not complete. However, motor activity and metabolism of cold-living animals at subzero temperatures are much greater than of temperate-zone animals at near-zero degrees. Some functions compensate more completely than others. Genetically determined resistance adaptations may be more compensatory than capacity adaptations.

Interactions of the three determinants of adaptation—genotype, environment, and developmental pattern—are essential for the adaptations of whole organisms. Developmental causation occurs within limits set by genotype and environment. If a normal sequence of development fails to occur, heredity and environment cannot substitute. Many genes are transcribed only at certain stages in a life cycle. Early stages differ from the adult stage in sensitivity to environmental stress. Development may, therefore, influence a phenotype at all three periods—immediate, during acclimation, and after generations of selection.

HIERARCHICAL ORGANIZATION

Adaptive properties of whole organisms differ from those of isolated parts. The range of tolerance of environmental extremes is narrower for intact animals than for tissues and cells, which in turn have lower tolerance than their constituent proteins and lipids. Organisms are adapted as unified wholes and it would be disadvantageous for component parts to be adapted by different amounts or at different rates. Natural selection acts on whole phenotypes rather than on genotypes (with a few exceptions such as "parasitic" genes).

Levels of Organization

Biological systems can be arranged by level of organization as follows:

Communities
Independent individuals
Dependent (colonial and parasitic) organisms
Organs
Tissues
Cells
Organelles
Molecules
Atoms

Some kinds of organisms are relatively independent of the biotic environment; others are very dependent. At each level there are several types of analyses and one finds properties that are qualitatively different from those at other levels. One aim of adaptational biology is to describe a more complex level in terms of one nearer the molecular. In general, complexity is determined by geometry or spatial interactions and differences in temporal sequences of actions. Complexity is a statistical concept; it is a state with very low probability that it will occur as a result of random events (2a).

Examples of differences between levels are numerous. Social behavior can hardly be described in terms of synaptic interactions in the brain. Methods for describing the regulation of blood pressure are very different from those for describing the contraction of vascular smooth muscle fibers. A mitochondrion carries out an organized sequence of electron transfer reactions; the linkage between steps has not been fully simulated by measurements on isolated enzymes. A hemoglobin molecule in solution has an affinity for oxygen different from its affinity when it is in a red blood cell.

Emergent Evolution

Several important generalizations can be made regarding the hierarchy of biological organization. A primary principle is that of emergent properties. This was enunciated by Morgan (11e). This principle states that at each more complex level of organization some properties are qualitatively different from those at simpler levels. Quantitative differences may lead to qualitative changes (16). The properties of the more complex organization could not have been inferred from the less organized components. The principle of emergence is useful for a given category of properties. It is a truism that detailed knowledge of all the nucleotides and proteins in an organism would not make it possible to synthesize the organism. The concept of emergence has been useful in psychology and philosophy insofar as it is from the neuroendocrine organization of humans that properties of thought and abstraction emerge. Sensation, intelligence, personality, esthetics are not predictable from studies of the nervous system in isolation but rather are emergent from its organization. No neuroendocrine analysis can quantitate such human qualities as love or hate. Yet these properties are as real and as valid for study as the regulation of blood pressure or hormonal control of reproduction.

The photochemical reactions in retinal cones are essential for color vision, but the total sensations of color which permit appreciation of a painting require a different level of integration. The responses of the hair cells in the cochlea permit the hearing of tones but cannot account for the enjoyment of a symphony. Psychological reactions depend on the balance of synaptic transmission, and behavior can be influenced by drugs, but attitudes and motivation are more than the sum of synaptic transmissions.

Each level of organization has its own principles, is described by its own vocabulary, and provides its own description of life; and at any given level, the vocabulary characterizes the kinds of measurement and interpretation. Study of animal behavior is very different from study of cell biology and from molecular biochemistry. Each level provides for a relatively independent discipline. One's view of nature depends on the viewpont of the level of observation. Some theoreticians have argued that it is not possible to extrapolate from one level to another; each organizational level can only be studied on its own. Analysis, or breakdown from complex to simple, is easier than synthesis, or putting parts together. A goal of this book is to show that extrapolation between levels is possible, and indeed desirable, to provide bridges between whole-organism and molecular levels, though some of the bridges are tenuous and based on correlation rather than causation.

Information Content

Information content clarifies hierarchical analysis. This has led to the aphorism that "the whole is greater than the sum of its parts."

One use of information theory concerns the information content of the whole (I_{whole}) and of parts (I_{parts}). The description of each part includes all possible connections with other parts plus the internal or nonconnecting information: $I_{\text{parts}} = I_{\text{connections}} + I_{\text{internal}}$. The concept of information is closely related to the concept of entropy because both are related to the degrees of ordering, or the constraints, of a system. The maximum amount of information contained in a system corresponds to the number of different ways in which the system may be ordered. The amount of information that may be conveyed to an observer from a system in a given situation depends on the observer's ability to discern ways in which the system is ordered. The ordering may be either spatial or temporal; it may be in the structure of the system (enzymes, nucleic acids, etc.), or in the dynamic response of the system (pattern of action potentials produced by neurons, cyclic release of hormones, etc.). Another mode of ordering is in connections between parts of the system, for example, synaptic connections in neural networks. Since information is contained in these connections, information is always lost in the process of taking the system apart to examine the parts, hence $I_{whole} > I_{parts}$.

Mutual dependency of parts organized into a whole causes reduction of uncertainty. The amount of nonredundant information associated with the whole is less than the sum of the information content of the separate parts, and the difference is the information content of the constraints. A protein molecule in solution has many possible states and has high information (I) content. The same molecule in a mitochondrion has fewer possible states because of the constraints of the organization, hence it has lower I content. In organisms, as cellular division of labor is established, I_{parts} decreases; for example, a unicellular organism, such as a protozoon, has higher I content than a cell of a metazoan but less than the total metazoan. It is impossible to reassemble a total living system after breaking it into components. After homogenization of a tissue and lysis of mitochondria the component enzymes can be measured separately and in coupled reactions, but they cannot be reassembled. On disruption, information of interaction, of interdependence, and of different reactive components is lost. In this respect, a living system differs from a machine, such as a watch, the components of which in isolation have not lost information and can be reassembled. Complementarity in a biological system means that information content increases exponentially as connections increase within the system.

Biological Indeterminacy

Another aspect of information theory which concerns hierarchical organization is that of biological indeterminacy (12). In a computer the number of information states goes up exponentially (or by some other nonlinear function) as the number of interacting elements increases. In a binary or linear system each element $A, B, \ldots$, may have two states—plus or minus, yes or no. Combined, elements AB can have four states, ABC eight states, and so on. In nonlinear systems the output of two or more interacting units is not the sum of the inputs but is some other function. Output information (P)

may be greater or less than input: $P > (A + B)$ or $P < (A + B)$. A random system of interactions yields unpredictable general patterns. Alternative pathways may be stabilized in different proportions. In systems in which elements have more than two states, for example, network neurons onto which multiple signals converge, the total information of the system increases disproportionately with more elements than when elements occur in only two states.

Many examples of randomness or quasi-randomness (randomness within limits) in biological systems are known. The unit behaviors within a system may be random but in each there is some monitoring for control, often by feedback from products of the system. The monitor freezes the output at some point (12). Genetic changes (mutations and gene rearrangements) are undirected, but rates of mutation can vary with such environmental factors as temperature, and some mutagens have higher probability of causing certain mutations than others. Whether a particular genetic change is retained depends on developmental factors. A DNA molecule may show spontaneous random changes in nucleotides, but in the transcription of the molecule the inherent proofreading mechanisms tend to prevent gross errors. In small populations of animals or plants the properties of a given individual are indeterminate and in accordance with Gaussian distribution, but in a large population or over a geographic range, natural selection opposes randomness. In a brain, randomness of single neurons may provide a basis for neural plasticity, with the net output leading to a given behavioral act monitored according to instructions contained in experience and in the immediate neuroendocrine milieu. In a cellular organelle, such as a mitochondrion, the reactions of a single molecule may be indeterminate, but the average net output may be controlled by feedback of products and environmental parameters. A distinction is made between randomness and noise or jitter. Some fluctuations in complex systems are adaptive in providing for alternative actions. In a nerve network, fluctuation in excitability or neuronal jitter prevents excessive phase-linking with input. Physiological jitter permits selection of significant inputs and rejection of nonsignificant inputs.

In evolution, direction or order is imposed on random variation by selection. The flow of qualitative information is necessarily from genotype to phenotype, although the phenotype can have quantitative effects on expression of genotypic information. Direction in evolution is given by natural selection; evolutionary change is more horizontal than vertical. In this book, the use of terms "higher" or "lower" is avoided in describing levels of complexity. Randomness occurs at many points in single elements in living systems, while feedback controls and monitors provide for adaptive plasticity.

WHAT IS LIFE?

The distinction between living and nonliving depends on the context and the use. An essential feature of life in whole organisms is adaptation to the environment, and the essence of evolution is the production of adaptive diversity. A legal definition of life and death in humans is relatively simple—heartbeat, electrical brain waves. However, many cells, skin for example, continue to metabolize for several days after cessation of the legally recognized criteria. A motile sperm and a receptive ovum are as alive as the zygote they form; however *when* a developing human being becomes a social individual is a legal, not a biological, definition.

Before a molecular description of vital adaptedness can be made, the nature of life can be considered in reductionist terms. Several cellular criteria of life are self-replication, capacity to synthesize proteins, energy liberation, and selective membrane permeability. Cell replication is essential for reproduction and growth, yet many cells, such as nerve cells in animals and wall cells of the vascular systems of plants, do not divide during the adult life span of the organism. A process of genic replication was essential in the first organisms. Replication of a RNA genome of Q_B virus has been achieved *in vitro* (19). Proteins do not self-replicate, but protein synthesis can occur in a cell-free system in the presence of appropriate nucleotide templates. Protein synthesis implies transduction from genetic templates; RNA encodes the sequential alignment of amino acids (Ch. 3). Selectively permeable membranes enable intracellular composition to be different from the extracellular environment; synthetic membranes can provide boundaries similar to those in cells; vesicles prepared from lipoprotein membranes permit *in vitro* study of permeability and electrical properties of membranes. Production of energy, and synthesis and degradation of macromolecular intermediates, characterize living cells. Biochemists have isolated the component steps of intermediary metabolism, and long metabolic pathways have been made to function *in vitro*.

No single reductionist criterion of life is possible. A holistic conclusion is that, at the cellular level, the integration of all component processes is essential. This does not imply any mysterious force—an *elan vital* in living cells—but rather that the properties of integrated organisms amount to more than the sum of their components. One goal of adaptational biology is understanding the complex interactions among chemical components.

Thermodynamics of Living Systems

The laws of thermodynamics are the basis of physical chemistry. Biophysicists and biochemists have considered the extent to which thermodynamic laws may be applied to living systems. The first law of thermodynamics states that the total energy of a system (plus its interacting surround) remains contant; however, there may be transformation of one form of energy to another: chemical $\leftrightarrows$ mechanical $\leftrightarrows$ heat. The second law of thermodynamics states that the entropy of a system (including the surround) increases toward equilibrium at constant temperature (T), pressure (P), and volume (V); this is sometimes stated as the tendency toward randomness, toward the running down of energy in a system. Free energy is that energy capable of doing work; according to the second law, this decreases toward a minimum. $\Delta G\ddagger = \Delta H\ddagger - T\Delta S\ddagger$ where G is free energy, H is enthalpy, S is entropy, and T absolute temperature. The formulation of the principles of thermodynamics in rigorously quantitative terms by Gibbs and other physical chemists required that limits be imposed, that is, that the system be a closed one. In an equilibrium state, net fluxes across boundaries are zero. Enzyme kinetics will be discussed in thermodynamic terms in Chapter 5.

Living organisms are open systems, and there is continuous exchange of matter and energy between organisms and their surround. Strict boundaries do not exist and net fluxes of energy and matter are not zero. For living organisms a nonequilibrium or irreversible thermodynamics has been developed (11). Katchalsky noted that "life is a constant struggle against the tendency toward entropy"—a struggle "to produce entropy at a minimum rate by maintaining a steady state" rather than an equilibriuim. Schrödinger commented that "life evades decay toward equilibrium." In life the second law applies to the system as a whole—organism plus environment—hence the interaction between organism and environment becomes paramount. The boundary between environment and organism is not sharp but must be arbitrarily defined. In an equilibrium state, net fluxes across boundaries are zero; in a steady state, fluxes across boundaries are not zero. However, at any instant input equals output. Life converts energy from a closed homogeneous equilibrium to a nonhomogeneous open system; a biological entity removes energy from a stable closed system. Evolution is usually an irreversible increase in organization.

In steady state, the system must have a nonzero flux of energy across boundaries. The biosphere takes up radiant energy from the sun and reradiates the energy to space. Living cells take up energy in the form of chemical bonds in macromolecules, and release heat and smaller molecules (CO_2, water) across the surface membrane. In both examples, there are net fluxes across the surface of the system but the system is in a steady state. A steady state is a first-order approximation. The examples, biosphere and cells, are oscillatory, but they are in steady state over an intermediate range.

The change in entropy in a living system is given by: $dS = d_eS + d_iS$, where d_eS is entropy inflow from the surround and d_iS is entropy production from irreversible processes in an isolated system. In a nonequilibrium system there is competition between binding or activation energy (E) and kinetic energy (proportional to temperature (T)) in that probability of transition per unit time is proportional to exp (E/KT), where K is the Boltzmann gas constant. E tends to hold the system together and T tends to disrupt it. This competition is amplified at macroscopic levels.

In an open nonlinear system the kinetics of synthesis and degradation are exponential or according to some other nonlinear function. In an open system with exchange of energy and matter, a steady nonequilibrium state may be achieved, but in a system with many components it is much more likely that the system will oscillate rather than maintain a constant steady state. This is a general finding of systems theory and is relevant to the ubiquity of biorhythms. Oscillations amplify reactions and become controlled by feedback from products.

The response of any system to a small perturbation is the sum of exponential functions. There is one exponential term for every degree of freedom that the system has. If the real part of any one of these exponential terms is positive, the system will not be quiescent at a steady state, but will oscillate or be unstable. Since the more complex a system and the more exponential terms there are, there will likely be at least one with a positive real component. It follows that the more complex a system is (the more interconnected components it has), the more likely it is to oscillate.

Before life appeared, nonequilibrium thermodynamics operated because the earth was receiving radiant energy from the sun and re-radiating it into space. Numerous authors have pointed out that it is most unlikely that random events could have yielded the macromolecules of life. But these arguments have been based on equilibrium thermodynamics; with the establishment of boundaries, exchange of energy and matter, and exponential multiplicative processes, nonlinear steady state conditions made possible self-organization, self-maintenance, and self-replication. "Mutations" from one steady state to another could occur (3,4).

Eigen has presented a plausible scenario (4,17). In a prebiotic period on our planet, single-stranded RNA molecules were self-replicating (as inferred from Q_β virus) in the absence of enzymes (but possibly catalyzed by metal ions). Molecules of RNA may have consisted of some 4000 nucleotides (Ch. 2). Errors (mutations) occurred; some mutants of these RNA molecules may have been better adapted to the immediate environment than others, and these adapted forms constituted relatively stable "quasi-species." Selection for growth of a particular RNA would have occurred with a certain probability. A particular sequence of nucleotides would survive only if copying errors did not accumulate. The next step is postulated to have been RNA-instructed synthesis of proteins and formation of hypercycles in which coupling between two or more RNAs and their coded proteins occurred. Figure 1-8 diagrams an autocatalytic cycle; the gene I or RNA_1 codes for enzymatic protein E_1 that catalyzes replication of RNA_2 which is template for replicase E_2 which in turn catalyzes replication of RNA_1. Thus template and enzyme are coupled and a number of multiplicative couples constitutes a closed loop which is conserved as a self-replicating system or hypercycle. Hypercycles may have competed with one another in early prebiotic evolution and selection was based on optimization of growth which was in accordance with nonlinear kinetics. The appearance somewhat later of double-stranded DNA provided for fewer errors in replication and DNA became coupled to RNA-protein hypercycles. Natural selection, largely dictated by feedback from proteins, ensued. Details of Eigen's hypothesis are given in Chapter 3.

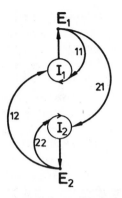

Figure 1-8.   Diagram of simple autocatalytic cycle of two RNAs I_1 and I_2 coupled both directly and reciprocally by the proteins E_1 and E_2 which they encode. An information-carrying gene RNA (I_1) codes for a primitive enzyme (E_1) that catalyzes formation of another gene RNA (I_2) which helps to replicate I_1 through its translation product E_2. Cooperation between two self-replicative elements I_1 and I_2 requires that reciprocal catalysis is more efficient than autocatalysis. Reprinted with permission from M. Eigen, *Naturwissenschaften*, Vol. 65, p. 354. Copyright 1978, Springer-Verlag, New York.

SUMMARY

This chapter provides a theoretical background for subsequent detailed chapters. Animals and plants are adapted as wholes to physical and biotic environments. Holistic descriptions of physiological adaptation account for distribution in different milieus. Reductionism analyses molecular adaptations to different life styles. Extrapolation from molecular to whole-organism or from holistic to reductionist properties requires attention to: (1) differences in language and concepts according to the level in the hierarchy of biological organization; (2) the emergence of unpredicted properties in the transition from simple to complex levels; and (3) differences in information content of wholes and parts. This book presents many examples of information bridges between molecular and whole-organism adaptations.

The essence of evolution is to provide biological diversity at all levels—populations and ecosystems, species, higher taxonomic categories. Description of species in terms of physiological adaptation accounts for niche and range selection and adds a dimension to morphological and reproductive descriptions.

An understanding of the physiological adaptation of present-day organisms helps to elucidate evolutionary history. The origin of life, of metabolic pathways, mechanisms of replication, and protein synthesis, can be inferred from physiological adaptations of present-day organisms. Patterns of adaptation are homeostatic and homeokinetic and include capacity and resistance (tolerance) adaptations. The determinants of adaptation are genetic, developmental, and environmental. Physiological measurements on several kinds of organisms and recognition of physiological adaptations are essential for understanding both the evolutionary history and the present distribution of organisms.

A final section deals briefly with the nature of life and gives a warning that living systems follow nonequilibrium thermodynamics in which net flux of energy and matter is not zero, but entropy production is kept at a minimum by steady states rather than by equilibria.

Subsequent chapters or groups of chapters will present holistic problems—ecological, physiological, evolutionary—and offer explanations in terms of molecular adaptations.

REFERENCES

1. Allen, G. E. Ch. 5, pp. 81–102 in *Evolution from Molecules to Man,* Ed. D. S. Bendall, Cambridge University Press, New York, 1983. Updating evolution since Darwin.

2. Ayala, F. J. Ch. 19, pp. 387–402 in *Evolution from Molecules to Man,* Ed. D. S. Bendall, Cambridge University Press, New York, 1983. Macroevolution and microevolution.

2a. Dawkins, R. Ch. 20, pp. 403–428 in *Evolution from Molecules to Man,* Ed. D. S. Bendall, Cambridge University Press, New York, 1983. Universal Darwinism.

3. Eigen, M. *Naturwissenschaften* **58**:465–573, 1971; **65**:7–41, 341–364, 1978. Molecular self-organization and early stages of evolution. *Q. Rev. Biophys.* **4**:149–212, 1971.

4. Eigen, M., Gardiner, W., Schuster, P., and Winkler-Oswatish, R. *Sci. Am.* April: 88–118, 1981. The origin of genetic information.

5. Eldredge, N. and S. J. Gould. pp. 82–115 in *Models in Paleobiology,* Ed. T. J. M. Schopf, Freeman, San Francisco, 1972. Punctuated equilibrium.

6. Gould, S. J. and E. S. Urba. *Paleobiology* **8**:4–15, 1982. Exaptation.

7. Gould, S. J. and R. C. Lewontin. *Proc. R. Soc. London B* **205**:581–598, 1979. Critique of adaptationist program.

8. Hazel, J. R. and C. L. Prosser. *Physiol. Rev.* **54**:620–677, 1974. Molecular mechanisms of temperature compensation in poikilotherms.

8a. Hochachka, P. and G. N. Somero. *Biochemical Adaptation,* Princeton University Press, Princeton, NJ.

9. Jacob, F. Ch. 7, pp. 131–141 in *Evolution from Molecules to Man,* Ed. D. S. Bendall, Cambridge University Press, New York, 1983. Molecular tinkering in evolution.

10. Jeffrys, A. J. et al. Ch. 9, pp. 176–195 in *Evolution from Molecules to Man,* Ed. D. S. Bendall, Cambridge University Press, New York, 1983. Evolution of gene families.

11. Katchalsky, A. and P. Curran. *Nonequilibrium Thermodynamics in Biophysics.* Harvard University Press, Cambridge, MA, 1965.

12. Prigogine, I. and G. Nicholis. *Q. Rev. Biophys.* **4**:107–148, 1971. Biological order, structure and instabilities.

13. Prosser, C. L. pp. 339–369 in *The Species Problem,* Ed. E. Mayr. AAAS, Washington, DC, 1957. Molecular mechanisms of temperature adaptation.

14. Prosser, C. L., Ed. *Comparative Animal Physiology,* 3rd ed. Saunders, Philadelphia, 1973; 2nd ed. 1961.

15. Prosser, C. L. pp. 2–22 in *Biological Adaptation. Marburg/Lahn,* Ed. G. Hildebrandt and H. Hensel, Thieme, Stuttgart, 1982. Theory of adaptation.

15a. Prosser, C. L. and D. O. Nelson. *An. Rev. Physiol.* **43**:281–300, 1981. Role of neuron systems in temperature adaptation.

16. Quastler, H. *The Emergence of Biological Organization.* Yale University Press, New Haven, CT, 1964.

16a. Reisenauer, A. M. and G. M. Gray. *Science* **227**:70–72, 1980. Enzyme induction.

17. Schuster, P. pp. 50–87 in *Biochemical Evolution,* Ed. H. Gutfreund, Cambridge University Press, New York, 1981. Prebiotic evolution.

17a. Morgan, C. L. *Emergent Evolution.* Henry Holt, New York, 1923.

18. Shaklee, J. B., Christiansen, J. A., Sidell, B. D., Prosser, C. L., and Whitt, G. S. *J. Exper. Zool.* **201**:1–20, 1977. Molecular aspects of temperature acclimation in fish.

18a. Sidell, B. D. Ch. 1, pp. 103–120 in Cossins, A. R. and P. Shetline, *Cellular Acclimation to Environmental Change.* SEB Symposia 17, Cambridge, 1983. Thermal acclimation in fishes.

19. Spiegelman, S. with I. Haruna. *Proc. Nat. Acad. Sci.* **54**:579–587, 919–927, 1965. Experimental analysis of precellular evolution. *Q. Rev. Biophys.* **4**:213–253, 1971.

20. Stebbins, G. L. and F. J. Ayala. *Science* **213**:967–971, 1981. Is a new evolutionary synthesis necessary?

21. Wang, L. C. H. and J. W. Hudson, Eds. *Strategies in Cold.* Academic Press, New York, 1978.

2

ORIGIN OF LIFE AND OF METABOLIC PATHWAYS

PREBIOTIC EVOLUTION

More biochemical innovations were made before there were organisms than have been made since. Prebiotic chemical evolution occurred for approximately a billion years. Many "decisions" were made as to which prebiotically formed molecules were to be retained or discarded. We can only speculate on the events of prebiotic evolution, but there is no doubt that some molecules were selected over others. Prebiotic events can be pieced together from geochemical evidence, from properties of present-day compounds, and from simulation experiments.

The age of the earth is estimated at 4.65×10^9 years (4.65 billion years, byr). This value is obtained from ratios of chemical elements and from radioisotope dating of the earth, planets, the moon, and meteorites. Geophysical evidence indicates that during the earth's molten state gaseous components, including CO_2, N_2, and H_2O, became dissolved. As the molten earth cooled and a crust formed, an atmosphere and hydrosphere were gradually established by outgassing of volatiles from the crust. Such volatiles continue to be emitted from volcanoes. Crust rocks, such as barites, contain CO_2, H_2O, and H_2S (9). The atmosphere of the primitive earth is conjectured to have contained much H_2O, CO_2, some N_2, H_2, CO, and SO_2, and argon and small amounts of CH_4, NH_3, and H_2S; the atmosphere was weakly reducing (9). Some geophysicists maintain, by analogy with the present atmosphere of Venus, that CO_2 was present in large quantity. The oldest sedimentary deposits yet found are in Greenland and Australia and are dated at 3.5 to 3.9 billion years (byr) (Table 2-1). These and slightly more recent deposits are the earliest carbonaceous rocks. The Archean age is dated from that time (3.5 byr) to the lower Proterozoic (2.6 byr). The oldest sedimentary rocks are mainly siliceous and contain little limestone.

During the period of less than one billion years after formation of the earth's atmosphere and hydrosphere, chemical reactions occurred that led to formation of many kinds of organic molecules. The sources of energy

Table 2-1.   Estimates of Time of Events and Appearance of Life Forms[a]

Events and Life Forms	Age in 10^9 Years before Present
Chordates	0.4–0.5
Metazoans	0.5–0.7
Late glaciation	0.7
Eukaryotes	1.2–1.4
Early glaciation	2.0
Cyanobacteria	2.2–3.0
Stromatolites	2.8–3.4
Cellular fossils	3.0
Sulfate photosynthesizing bacteria	3.1–3.3
Oldest sedimentary rocks	3.4–3.8
Chemical synthesis	3.5–4.0
Origin of earth	4.65

[a]From Cloud (9) and others.

available for spontaneous organic reactions were: sunlight of short wavelengths (UV) absorbed by methane, CO_2, CO, and H_2O; sunlight, both long and short wavelengths absorbed by NH_3 and H_2S; electric energy of lightning and corona discharge; radioactive decay; volcanic heat; heat from deep ocean vents; and atmospheric pressure waves (43). Organic molecules were formed, probably more kinds than are found today; these compounds probably decomposed very slowly in the absence of organisms. Shallow saline pools, often transitory, may have provided favorable milieus for interactions between molecules. Organic reactions could also have occurred on fog droplets in the atmosphere. Recent experiments with clays in suspension show on their surfaces oriented catalysis and some stereoisomerisms. The hypothesis that early biosynthesis occurred on clay particles is becoming increasingly accepted. It is probable that organic molecules accumulated in the oceans, but it is doubtful that much synthesis occurred there.

Extraterrestrial Synthesis: Time Limits

Recently there has been revival of the idea of the extraterrestrial origin of life—that life compounds reached earth by meteorite bombardment (29,10). This hypothesis has been considered along with the contention that in the time postulated for organic synthesis on earth, the probability is infinitesimal (1 in 10^{400} or lower) that complex molecules, such as DNA, could have been formed randomly.

Regardless of whether the earliest syntheses occurred in space or on earth, if the biological complexity could not be arrived at by chance, then some miracle or catastrophe must have been involved. The assumption that random associations would be inadequate is based on classical equilibrium thermodynamics. Life is, however, based on nonequilibrium, steady-state thermodynamics. The kinetics of synthesis are exponential, multiplicative, and regulated by product feedback, and tend to be oscillatory (Ch. 1). Possibly the first replicating genetic material was an RNA, probably tRNA, which became coupled in hypercycles to proteins (14). *In vitro* replication of an RNA (53,57,58), and synthesis of a protein coded by an RNA (43), have been achieved. Polymerization of oligonucleotides of more than ten monomers has been achieved in the absence of proteins but with complementary purines or pyrimidines as templates and with metal catalysts (15a,29a,44,46a,62). This is strong evidence that nucleic acids may have replicated prior to the synthesis of proteins (46a).

There is much evidence from galactic space probe collections and analysis of newly landed meteors that some extraterrestrial synthesis of biomolecules does occur. Its importance for the origin of life on earth is, however, questionable. The absence of biomolecules on Mars supports the hypothesis of the earth origin of life. Any organic molecules that might have been present on Mars would have been decomposed by the high ultraviolent radiation and superoxides in the environment. Evidence from geological dating and examination of craters of the moon shows that the earth was heavily bombarded by large and small planetesimals or meteorites during the period from 4.6 to 3.8 byr ago. It was during this period that the earth was cooling and its crust solidifying. The hydrosphere and molten lithosphere on the surface of the earth prior to 4.0 byr ago would have been most inhospitable to the establishment of any extraterrestrial organic molecules (9).

Superficial layers of some recently fallen meteorites have been found to contain organic molecules, especially amino acids. These organic molecules may have been acquired as the meteorite passed through the earth's atmosphere or after it landed. However, a study of amino acids from the interior of a recently arrived meteorite (the Murchison in Australia) showed that it contained 14 of the 20 commonly occurring amino acids. These include some familiar amino acids (Glu, Asp, Pro, Ala), some unusual ones (α-aminoisobutyric, ornithine, isovaline, pseudoleucine), minor amounts of Ser and Thr, and no Tyr, Met, or Phe. Some L and D amino acids are present, but they are not fully racemized (present in equal amounts); therefore it is suggested that some stereoselective synthesis may occur in space (15).

The contention that complex biomolecules could not have been formed by chance in the time available ignores the type of amplification provided by nonlinear thermodynamics. Although extraterrestrial synthesis of biomolecules, such as amino acids, does occur in space, this source has probably not contributed to the origin of life on earth. The earlier Haldane-Oparin hypothesis is apparently valid if modified to take account of probable catalytic sites other than the oceans.

Evidence for Spontaneous Synthesis of Life

Evidence for chemical syntheses on the primitive earth has been obtained by experiments with closed systems in which solutions containing presumed primitive atmospheric components were subjected to electric discharge or to ultraviolet radiation at the energy levels postulated for the prebiotic earth. In laboratory experiments, after relatively short times (days) organic molecules were formed—aldehydes, phosphides, amides, followed by sugars, amino acids, and short-chain fatty acids (43). Evidence for the synthesis of organic molecules at primitive radiation levels comes from analyses of carbonaceous chondrites in meteorites and from samples of particulates collected in interstellar space.

Geochemical data provide evidence by which extrapolation can be made from the present to the early atmosphere and oceans. If there was much CO_2 in the primitive environment, why is there so little at present? What is the source of O_2 in the atmosphere and oceans?

Carbon dioxide balance is closely correlated with the presence of calcium. At low CO_2 levels, Ca silicates are deposited, and at higher CO_2 levels $CaCO_3$ and $MgCO_3$. CO_2 combines with Ca to form $CaCO_3$, which precipitates in the oceans at their present pH (8.1) whenever atmospheric concentrations of CO_2 are greater than $10^{-3.5}$ atm (bar). The CO_2 from the atmosphere is buffered by Ca in rivers. Calcium becomes buried as $CaCO_3$ in marine sediments at the rate of some 0.5×10^{14} g/yr (27). Cycling between the biosphere and hydrosphere-atmosphere is continuous and results in relatively constant partial pressure of CO_2 in the atmosphere. The degassing of the earth contributes 0.9×10^{14} g/yr of CO_2. Human activities are estimated to liberate CO_2 into the atmosphere at some 4.2×10^{15} g/yr.

Atmospheric CO_2 is increasing as a result of combustion; the concentration measured in Antarctica increased from 3.2 ppm in 1957 to 3.4 ppm in 1971; in metropolitan areas the concentrations are greater. If the CO_2 should increase to several times the present level there is danger of a greenhouse effect, such as the one responsible for the high temperature on Venus (9). Initially, melting of polar icecaps could occur, with resultant raising of ocean level; and if warming became extreme, oceans could evaporate.

Oxygen concentration in the primitive atmosphere was very low. Volcanic gases contain virtually no oxygen. Two methods of eventual production of O_2 on Earth are recognized: photolysis of water by solar irradiation and photosynthesis by green plants, algae, and bacteria. Evidence regarding the early presence of O_2 comes from some of the oldest sedimentary rocks (3.76 byr) from Greenland, in which there are bands of iron oxide, alternately Fe-rich and Fe-poor (but Si-rich) layers (6,9). The banded iron formations (BIF) contain ferric iron, which indicates the presence of O_2. One view is that oxygen was generated by primitive photosynthesizing microorganisms; that these were (as some are at present) poisoned by free O_2; and that the banded iron formations created sinks that kept the O_2 concentration low. The earliest fossils of photosynthesizing microorganisms until recently were considered to be 2.8 byr old; recently stromatolites (algae mats) 3.4 byr of age were found in Australia (37). Modern cyanobacteria are similar in metabolic properties to the constituents of ancient algal mats (61,66).

Another and less favored view of the origin of oxidation on Earth is that the early oxidation of iron may have been by oxygen produced by photolysis of water (6). Probably oxygen was derived from both processes, photolysis and photosynthesis, in different amounts. No more banded Fe was formed after the ozone screen was established. Neither hypothesis, photosynthesis or photolysis, explains the periodicity in deposition of iron oxide. The earliest organisms were certainly anaerobic; levels of O_2 in the atmosphere and oceans must have been extremely low. Whatever the sources, O_2 appears to have increased to about one to three percent of present concentrations at 2 byr. After the increase in atmospheric oxygen, ozone was formed in a layer that shielded the earth from strong ultraviolet radiation so that it could no longer effect synthesis of organic molecules. All of the O_2 in the present atmosphere (1.2×10^{21} g) could be generated by photosynthesis in 4×10^6 years (27). The rate of production of O_2 equals the rate of consumption. In the Archean atmosphere some O_2 probably went to oxidize material from volcanic gases: CO to CO_2, SO_2 to SO_3, SO_3 to SO_4^{2-}. At the present time some O_2 is consumed by weathering and oxidation of igneous rocks. Oxygen at the surface of the oceans is at 6.5 ml O_2/kg and is at equilibrium with oxygen in the atmosphere; the concentration of oxygen declines gradually at increasing depth in the ocean to less than 1 ml/kg at 1000 m below the surface (the O_2 minimum layer) (Ch. 6); and lower yet the oxygen concentration increases so that from about 2000 m to 4000 m depth oxygen is present at about 3 ml/kg H_2O.

Of other biologically important elements, igneous rocks contain a small percentage of Na, K, and Mg; sedimentary rocks are richest in Si and contain less Ca and Al. Acids formed from volcanic emissions weathered the rocks and released Ca, Al, and Si to rivers. Sulfate was fixed as gypsum; Mg was fixed partly as silicate and partly as carbonate. Na became the most abundant cation in seawater and Cl the most abundant anion. SO_4 occurs at much higher concentrations than Ca in seawater. Some phosphorus was buried in organic deposits along with carbonates and silicates. It is calculated that the runoff from rivers sufficiently balances deposits of salts so that the composition of ocean water has been relatively constant for 10^9 years (27).

A wide variety of organic compounds can be produced with electric discharge, and slightly different compounds with ultraviolet radiation. These compounds include fatty acids—formic, glycolic, lactic, acetic, propionic, succinic, and others; amino acids—glycine, alanine, glutamic, aspartic (both D and L acids); amino acids not used in proteins—α-aminobutyric, norvaline, diaminobutyric; and other N compounds—sarcosine, urea, methylurea, methylalanine. In interstellar space occur cyanide, formaldehyde, cyanamide, methanol, methylamine, acetaldehyde, and other compounds (9). Carbonaceous chondrites contain D and L amino acids, D-isobutyric acid, and others. Under the action of ultraviolet, HCN in solution goes to a dimer, this to a tetramer, and then to adenine; pyruvate goes to cytosine; formaldehyde in the presence of $Ca(OH)_2$ can go to ribose sugar.

Early in the reaction with electric discharge in a simulated primitive atmosphere the concentration of NH_3 decreases and HCN and C_2N_2 increase, as do simple aldehydes. Formaldehyde reacts with HCN and water to form amino acids (41). A number of biologically essential compounds could have been formed under the postulated primitive conditions: (1) formaldehyde condenses to the pentose that is present in nucleic acids; (2) HCN becomes converted via several steps to adenine; (3) adenine combines with ribose to form adenosine; and (4) adenosine combines with triphosphate to form adenosine triphosphate (ATP). Purine rings are made from glycine, formate, glutamine, aspartic, and CO_2. Pyrimidine rings are formed from aspartate, CO_2, and glutamine. The pyrimidine-purine components of nucleic acids could have been formed as soon as glycine, formate, aspartate, and glutamine appeared in the biosphere (25). The extent to which these and other reactions occurred during the prebiotic period cannot be known, and the environmental conditions during prebiotic synthesis can only be postulated. However, the reactions do occur under simulated primitive conditions, and many of these compounds are evidently formed also in extraterrestrial space. It is likely that the components of protoplasm were synthesized in quantity.

In the experimental model systems and in chondrites both L and D amino acids are formed, yet organisms use only L amino acids in protein synthesis. However, organisms can metabolize D amino acids, and D amino acids occur in bacterial cell walls. In model systems, for example with electric discharge, both D and L amino acids are formed and both D and L forms of ribosides are produced. In the synthesis of nucleic acids only D ribosides are used. Present-day proteins are formed of L amino acids and the helix of DNA contains D pentoses (35). Why these configurations were selected is not known. Another "decision" was the selection of a few—20—amino acids for protein composition. The same 20 amino acids are used in protein synthesis by all organisms. Some organisms require certain amino acids in the diet, while others can make the acids they use whenever these are not available in the medium. Each essential amino acid is coded by base sequences in DNA and tRNA (Ch. 3). Other and very different amino acids are synthesized in the experimental model system and have been found in meteorites and in some organisms; the amino acids that are not incorporated in proteins may have other functions, for example, ornithine in protein catabolism. Sufficient codons are available for 64 amino acids. Selection of the coding for the 20 essential amino acids must have resulted from some adaptive advantages. Some 300 naturally occurring amino acids are known, thousands are possible, yet only 20 are regularly incorporated into proteins. Suitable amino acids from each class have been selected: acidic, basic, hydrophobic, imidazole, hydroxy, sulfur, aromatic. Which alternative within each class of amino acids prevailed was determined by the ease of synthesis prebiotically, the probable abundance in primitive seas, the sensitivity to degradation, and the reactivity (66a). The most abundant amino acids made in model systems are alanine and glycine; glycine is a structural amino acid in polypeptides. Proline is important at the kinks in secondarily folded peptides. Amino acids which are bulky and occur in relatively large amounts in the interior of proteins are valine, isoleucine, and leucine. Histidine is important in proton transfer reactions and cysteine provides disulfide bridges.

Another "decision" in prebiotic evolution was the use of ATP as the universal energy currency. Other triphosphates—UTP and GTP—could serve the same function, and some are used for special purposes in present-day organisms, but ATP came into general use. A small amount of energy is derived from hydrolysis of pyrophosphate PP to inorganic phosphate (P_i), and this may have preceded the functioning of nucleotide phosphates as sources of energy. The selection of certain purines and pyrimidines as determinants of the genetic code was another "decision." The adaptive implications of these decisions can only be inferred.

Formation of Macromolecules

The next step in prebiotic synthesis was the formation of large molecules—polypeptides and proteins, large carbohydrates, fats, nucleic acids. From the purines, porphyrins were synthesized which complexed with metals, especially iron, to form hemes. Amino acids polymerized as polypeptides. In a model system, complexes of amino acids can be joined with adenylates, adsorbed onto clays and then polymerized to polypeptide chains (19a,50a). Also, when mixtures of amino acids are heated for a few hours at temperatures between 100°C and 150°C, water is removed and the amino acids polymerize to make "proteinoids." Some sequences are formed more frequently than others. Chains of approximately 200 amino acids have been made by heat polymerization (19a,50a). However these are not self-replicating molecules and the reactions do not occur in solution. Protein synthesis requires nucleic acid coding. Replication of nucleic acids probably preceded protein synthesis.

Several requirements must have been met in the transition to what are usually recognized as true living organisms. Bounding membranes with selective permeability were essential. Such membranes can be made synthetically from macromolecules, and if the molecules are the proper distance apart and have suitable surface charges, some ions, such as K^+, are accumulated, while others, such as Na^+, are excluded. As soon as the concentration of organic molecules within cells increased and they became separated from the medium, a selectively permeable membrane with some means of preventing osmotic swelling was essential. A clear sign of death in modern cells is for the membrane to become leaky so that intracellular and extra-cellular ion concentrations become similar.

Early cells presumably leaked protons, which would have been formed by breakdown of organic molecules. Specific channels for proton efflux may have prevented volume increase. An early enzyme to appear was probably a membrane ATPase, which served as a proton pump, to increase proton transport and to maintain intracellular pH (24b).

Besides the membrane, another requirement for the evolution of cells was the use of proteins as catalysts to promote reactions in which the proteins themselves were not destroyed. The synthesis of amino acids and of purine-pyrimidine bases, pentoses, and organophosphates provided the building blocks for self-replicating and protein-synthesizing molecules. The origin of the genetic code is obscure. One-letter and two-letter codes for amino acids have been proposed (Ch. 3), but the three-letter code using four bases is universal and was probably selected because of steric advantages and the number of possible combinations: 4^3, or 64.

The next critical step in evolution was the origin of macromolecules that became organized as self-replicating and protein-coding quasi-species. A detailed hypothesis for the evolution of self-replicating macromolecular complexes of nucleic acids and proteins has been presented by Eigen (14). It is postulated that the first "genes" were transfer RNA (tRNA) molecules. In all cells of present-day organisms, tRNAs function to transfer amino acids to ribosomes, where they are coupled to form proteins. Each amino acid is the substrate carried by one to three different tRNAs. The tRNAs of all organisms are very similar, and are single chains with 76 homologous positions, with some of the same bases at corresponding positions. tRNAs occur in a cloverleaf form with the anticodon in the middle of the folded molecule (Fig. 2-1).

The structure of tRNA is highly conserved; for example, the tRNAs for phenylalanine of *Drosophila* and of man differ in one nucleotide pair. Phylogenetic trees of the different tRNAs do not correspond to trees based on other characters. Phylogenetic trees for several tRNAs—for example, tRNA for Phe with anticodon GAP, or Met with anticodon CAU—differ within chloroplasts from those within mitochondria, and both of these from cytoplasmic tRNA. These three tRNAs diverged from what was probably a common ancestor. The probable ancestral ratio of bases was GC:AU of 3:1. The GC content is high in primitive tRNAs, lower in the derived tRNAs in mitochondria. The first amino acids to be synthesized were probably Gly and Ala (as in model syntheses); then came Asp, Glu, and Val, and later Ser, Ileu, Thr, and Leu (66a). The first tRNAs were complementary for the early amino acids; they were GGC, GCC, GAC, and GUC, which code for Gly, Ala, Asp, and Val. It is postulated that the first tRNAs served a dual function—to attach to the coded amino acid and to transfer it to a peptide chain. In the process of transfer the tRNA changes its configuration. The key to the evolution of codons is in the evolution of translation. Nonpolar aliphatic

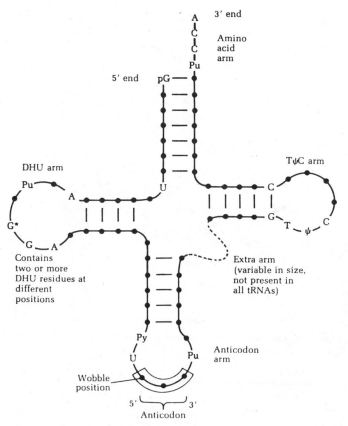

Figure 2-1.   Generalized structure of tRNAs, Pu-purine nucleoside, Py-pyrimidine nucleoside, ψ-pseudouridine, G-Guanosine. From A. L. Lehninger, *Principles of Biochemistry,* Worth Publishers, New York, 1982, p. 876.

amino acids use a codon with U as middle term; Asp and Glu are coded by GAX, Ileu and Met use AUX, Phe and Leu use UUX, where X is a variable base (66a). One reason for concluding that tRNAs may have been the first macromolecules is their high degree of conservatism.

Ribosomes are organelles on which essential steps of protein synthesis occur. They are composed of subunits of ribosomal (r) RNA (Figs. 2-2, 2-3). Messenger (m) RNAs vary according to the proteins which they encode.

On the basis of the *in vitro* replication of RNA of the Q_B virus (58) it is postulated that the first enzymes to be selected were replicases. The simplest replicating feedback cycle is between an RNA and its coded protein (Fig. 1-8). The next step is the coupling of several of these into hypercycles (Fig. 2-4) (14,53). A hypercycle with translation of a polynucleotide serves (1) as its own template for its reproduction, and (2) as codon for translation

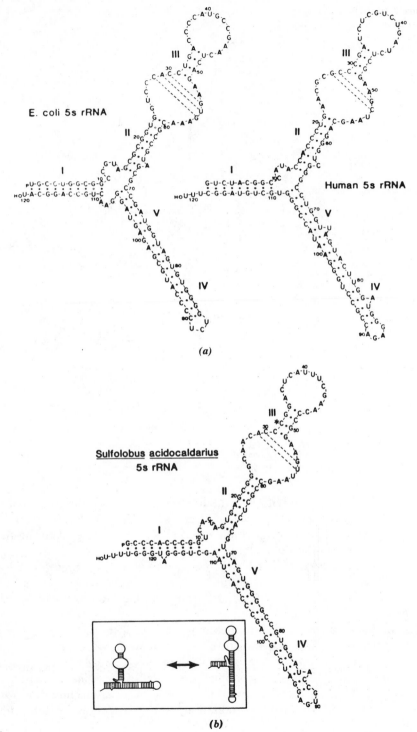

(a)

(b)

36

Figure 2-2. Schematic comparison of models of secondary structure of 5 S RNA from (*a*) eubacterium *Escherichia coli,* eukaryote *Homo* and (*b*) archaebacterium *Sulfolobus acido-caldarius.* Reprinted with permission from D. Stahl et al., *Nucleic Acid Research,* Vol. 9, No. 22, p. 6133. Copyright 1981, IRL Press, Oxford.

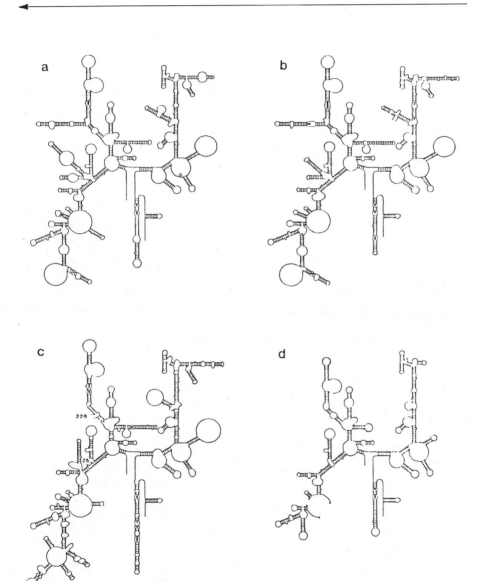

Figure 2-3. Schematic comparison of models of secondary structure of 16 S-like rRNAs of (*a*) eubacterium, *E. coli,* (*b*) archaebacterium, *H. volcanii,* (*c*) eukaryote, *S. cervisiae,* and (*d*) "minimal" or reduced unit of 16 S rRNA. Reprinted with permission from C. R. Woese et al., *Microbiology Reviews,* Vol. 47, No. 4, p. 664. Copyright 1983, American Society for Microbiology, Washington, D.C.

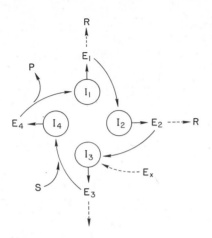

Figure 2-4. Schema of replication and information transfer by a RNA-protein hypercycle. I's indicate RNAs, E's represent proteins. Other symbols given in text. Reprinted with permission from P. Schuster in *Biochemical Evolution*, H. Gutfreund, Ed., Copyright 1981, Cambridge University Press, London.

to a protein. Cooperation between two self-replicative elements is more efficient than catalytic self-replication. A four-member hypercycle with translation is shown in Figure 2-4. I's represent complementary strands of RNA acting as templates, E's proteins encoded by corresponding I's and serving two functions: as replicases for the successive RNA and as catalysts for products R. S represents substrate mononucleotides entering the cycle, P product which is replicated RNA. R, S, and P are indicated for single steps in the four-member cycle (14,53).

Two requirements for a primitive organism were self-replication and membrane structure. Rigid cell walls probably appeared very early. The first organisms were probably heterotrophic; that is, they derived energy from molecules in the milieu. As these molecules became depleted, advantage accrued to autotrophic organisms, those which could synthesize sugars and other compounds and store energy as adenylates. The first processing by heterotrophs of metabolites for energy was anaerobic—by hydrolysis of polyphosphates, then by fermentation or glycolysis of sugars and fatty acids. Several energy-yielding pathways evolved, many of which persist in present-day anaerobes. Many anaerobic pathways remained functional after aerobic respiration had evolved.

An early step was differentiation into cells. Some organisms—myxomycetes and many fungi—are multinucleate but acellular. Organization into cells provides optimum surface area per unit protoplasmic volume and limits the volume within which protein synthesis and nuclear control take place. Possibly distribution of metabolites and O_2 from surface to interior occurred by diffusion before cytoplasmic streaming evolved. Cell size, the distance of metabolic organelle from the surface, was limited by the distance for diffusion.

ORIGIN OF METABOLIC PATHWAYS

Prokaryotes are organisms that are separate cells but do not have discrete nuclei, condensed chromatin, or nucleoli; the metabolic diversity of the different kinds of prokaryotes exceeds that of more complex organisms. Prokaryotes are not a coherent taxonomic group. The first prokaryote cells were bacteria; they have had at least 3.8 byr of evolution. Genetic processes and protein synthesis in prokaryotes and eukaryotes are similar (Ch. 3). The genetic systems of phage and of *Escherichia coli* are better understood than those of any other organisms. Knowledge of the physiology of bacteria living in different environments may provide clues to early evolution. However, extrapolation from existing bacteria to the earliest organisms is difficult because there has been continuing evolution among bacteria, and present-day forms are probably unlike ancestral species. There has been much biochemical convergence, in that similar but not identical compounds and enzyme pathways have evolved several times. There has been so much parallel evolution that it is more useful to describe diversity than to seek direct lines of phylogeny.

Evidence regarding the phylogeny of several prokaryotic organisms has come from comparison of nucleotide sequences in DNAs and RNAs (11). Measurements of metabolic patterns of bacteria living in different environments have been made mainly on organisms whose genetics is not well known. Figure 2-5 is an attempt to combine biochemical and physiological evidence of the relationships of bacteria (13). Much less is known about the phylogeny of prokaryotes than that of eukaryotes. Consequently Figure 2-5 is a representation of diversity more than of phylogeny. It is based on metabolic pathways of present-day bacteria, on nucleic acid sequences, particularly in 16 S rRNAs, and on amino acid sequences of cytochrome c (a highly conserved protein). Much structural, biochemical, and genetic evidence indicates that very early in cellular evolution primitive prokaryotes (progenotes) diverged as archaebacteria and eubacteria.

True bacteria, eubacteria, are of many metabolic types; for one of these, *E. coli,* some of the DNA and RNA structures have been ascertained. The second group of prokaryotes, the archaebacteria, live in unusual environments: methanogens synthesize methane from H_2 and CO_2; thermoacidophiles live in water at high temperatures, often of high acidity; halophiles live in hypersaline media. One group of halophiles photosynthesize by a pigment, bacteriorhodopsin, that is unlike chlorophyll. *Thermoplasma acidophilum,* isolated from a hot acid mine, is an archaebacterium with isoprenoid cell membrane and without cell wall. Thermoplasma has a high protein to DNA ratio; its DNA is stabilized by proteins which resemble the histones of eukaryotes; its 5 S rRNA resembles eukaryotic 5 S rRNA. Possibly *Thermoplasma* has properties of the postulated progenote (55).

The ribosomal rRNAs in three groups of archaebacteria are more similar

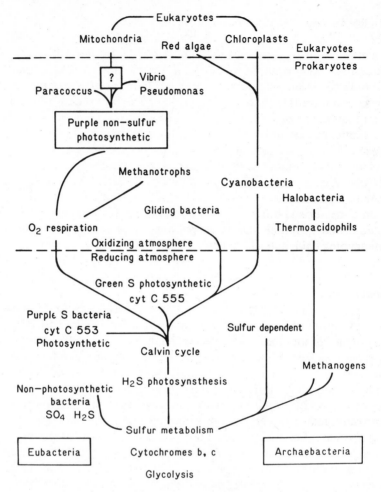

Figure 2-5.   Schema to indicate possible biochemical relationships during evolution of prokaryotes. Modified from References 12, 65, and others.

to those of eukaryotes than to rRNAs of eubacteria, and rRNAs of three different groups of eubacteria are four times more similar to each other than rRNAs of eubacteria are to rRNAs of archaebacteria. Archaebacteria have 16 S and 5 S rRNA but not 18 S RNA such as is found in eukaryotes (18).

Ribosomes, the organelles at which protein synthesis occurs, consist of two particles, 30–40 S and 50–60 S; subunits of the smaller of these are 16 S or 18 S and of the larger are 5 S and 23 S.

Ribosomal 16 S subunits occur in archaebacteria, eubacteria, and in eukaryotic chloroplasts and mitochondria; there are 18 S subunits in yeast, slime mold, and animals (*Xenopus*). Structural domains are common to different rRNAs (Fig. 2-2), and the early origin of 16 S rRNA (Fig. 2-3) is probable (65). The structures of 5 S rRNA from the archaebacterium *Sulfolobus acidocaldarius,* from a eubacterium *Escherichia coli,* and from a eukaryote *Homo* are compared in Figure 2-2. Three regions are similar; a molecular stalk (I), a tuned helix (II), and a common arm base (III); *E. coli* has a prokaryotic loop (IV) and in the archaebacterium a segment V between regions IV and I, as in human 5 S rRNA (67).

The sequences of tRNAs are different in archaebacteria from the sequences in eubacteria or eukaryotes. Some of the base modifications in archaebacteria tRNAs resemble those in yeast more than those in eubacteria. The tRNA of archaebacteria has methylpseudouridine instead of ribosylthymine (49,67). Cell walls of archaebacteria lack muramic acid, which is present in all eubacteria. Phospholipids of cell membranes of archaebacteria are branched chains (phytanyls) with ether linkages; phospholipids of eubacteria and eukaryotes have ester linkages.

Chemoautotrophy

Gradual depletion of organic molecules in the medium may have made chemoautotrophy, photoautotrophy, and anaerobic oxidation advantageous. These processes were important for long periods before chlorophyll evolved; oxygen levels were then raised by water-reducing photosynthesis. Oxidation occurs whenever an electron is removed from a higher energy source to a lower one—that is, from a molecule of more negative redox potential to a more positive molecule. The midpoint of an oxidation-reduction "titration" curve is the potential corresponding to 50 percent oxidized molecules and is designated as E_m. Organic substrates are highly negative, glucose has E_m of -0.45 V. Metabolic intermediates and coenzymes carry H atoms (protons plus electrons); cytochromes carry electrons. An energy decrease of approximately 53 kcal/mol occurs in the transport of a pair of electrons from NADH to O_2. The standard free energy of formation of ATP is 7.3 kcal, hence several ATPs are formed per electron pair transported. E_m for 1/2 O_2/ $2H^+$ is $+0.82$ V at pH 7.0, E_m of cytochrome c, is $+0.23$ V and of NADH is -0.32 v (35).

Chemotrophic bacteria obtain energy by oxidation of inorganic compounds, such as sulfides, or by oxidation of simple organic compounds, such as methane and alcohols. The most general oxidations are of organic metabolites—sugars, short-chain fatty acids, amino acids, fats. Oxidations in the sense of production of reducing units occur in steps of the glycolytic-fermentative chain. Only after the atmosphere became oxidative was there evolution of the complete tricarboxylic acid (TCA, Krebs) cycle and the transport of electrons via cytochromes to oxygen.

Photosynthesis is a process that is the reverse of respiration. Photosynthesis uses the energy of light to raise a molecule of chlorophyll to an excited state and to displace an electron from the chlorophyll. The electron acceptor is of high energy-negative redox potential. Electrons are then transferred along a chain of carriers, iron sulfur compounds, quinones, and cytochromes. ATP and NADPH are generated and a low-energy electron is returned to chlorophyll which can again be excited by light. Protons are derived from H_2S or H_2O; CO_2 is reduced by a series of reductive steps either by a chain from C_3 to C_4 acids or by the formation of a 5-carbon sugar (ribulose) via the Calvin cycle (Fig. 2-9) (8). Several types of photosynthesis evolved before oxygen-producing photosynthesis by cyanobacteria. Bacterial photosynthesis uses a single photocenter; cyanobacteria and plants use two photocenters.

Anaerobic Sulfur-Metabolizing Bacteria

Some kinds of nonphotosynthetic sulfur bacteria reduce sulfate to sulfide or thiosulfate. *Desulfovibrio* and *Desulfotomaculum* reduce SO_4^{2-} in the presence of ATP to adenyl sulfite (APS) and pyrophosphate (PP). The pyrophosphate is then hydrolyzed to inorganic phosphate (P_i). In *Desulfotomaculum* the reactions are

$$ATP + SO_4^{2-} \rightarrow APS + PP_i$$

$$PP_i + H_2O \rightarrow 2 P_i \text{ with liberation of 6.9 kcal/mol}$$

This reaction is an electron-coupled phosphorylation. Substrate-level phosphorylation is proportional to the amount of sulfate reduced to sulfite (1,36).

Sulfur bacteria can grow on a medium containing H_2, CO_2, SO_4, and acetate or lactate. They make ATP by the reactions

$$acetate + PP_i \rightarrow acetyl P + P_i$$

$$ADP + acetyl P \rightarrow acetate + ATP$$

The overall reaction with lactate is

$$2 \text{ lactate} + ADP + P_i + SO_4^{2-} \rightarrow acetate$$
$$+ 2 CO_2 + S^{2-} + ATP + 3 H_2O$$

Sulfate reduction uses a flavonucleotide hydrogen carrier, one (or two) cytochromes as electron carriers (36).

Other kinds of sulfur-metabolizing bacteria oxidize H_2S; some of them oxidize thiosulfates to sulfates or polythionates. The best known live in sulfide-rich muds or sulfide effluents. Several types of sulfide oxidizing chemoautotrophic bacteria live in sulfide-containing vents in marine trenches (some as deep as 2500 m) (59). Some of these bacteria are symbionts within and provide the nutrients for worms which live near the hot vents where O_2 is available. Some sulfur bacteria live in gills of mussels in sewage outfall (16). Energy of chemoautotrophy is derived from inorganic sources, not from the sun. Some sulfur-metabolizing bacteria are anaerobic; others are aerobic, facultatively anaerobic. By energy obtained from oxidation of H_2S these bacteria fix CO_2 and make malate, which is part of the TCA cycle; glucose is an end product (24a,52,56a,30). Chemolithoautotrophic synthesis of carbohydrate is as follows:

$$CO_2 + H_2S + O_2 + H_2O \rightarrow [CH_2O] + H_2SO_4$$

or

$$CO_2 + 4H_2S + O_2 \rightarrow [CH_2O] + 3 H_2O + S^-$$

These bacteria also make ATP and NADH which are used in organic synthesis.

Another mode of chemoautotrophy occurs in an anaerobic group of archaebacteria—the methanogens (3,19,69). These bacteria oxidize H_2 or alcohols or organic acids and reduce CO_2 to make methane as follows:

$$4 H_2 + HCO_3^- + H^+ \rightarrow CH_4 + 3 H_2O$$

or

$$4 CH_3OH + H_2O \rightarrow 3 CH_4 + CO_2$$

Methanogenic bacteria which occur in the stomach of present-day ruminants may use NH_4 as a source of nitrogen and acetate as a carbon source. Methanogens derive energy by fermentation of alcohols or organic acids such as acetate; they have no cytochromes or quinones but do have a unique coenzyme which serves in methyl transfer (70). The cell walls of methanogens contain phytanyl, a branched-chain ether-linked lipid, not peptidoglycan like eubacteria. The ribosomal RNA resembles that of other archaebacteria (3,65,67). Some methanogens from deep-sea vents are thermophilic and have a generation time of 37–65 minutes at 100°C, but do not grow at 70–75°C (5).

One kind of aerobic eubacteria which is chemoautotrophic oxidizes methane as a source of energy. Methanotrophic bacteria use the methane produced by methanogens in hypoxic muds or in rumen of cattle. Both methanogens and methanotrophs occur in deep-sea vents. Some methanotrophs use O_2 as a hydrogen acceptor and use the carbon of methane for synthesis of amino acids. Some methanotrophic bacteria can oxidize propane, ethane, or butane to alcohols, aldehydes, and fatty acids. One kind of methanotroph incorporates carbon from methane in oxidation of formaldehyde via ribulose monophosphate cyclase; another kind of methanotroph uses a serine pathway (26). Methanotrophic bacteria have broad-spectrum oxygenases or dismutases which detoxify oxygen; they may be an ancient group which has evolved to an aerobic metabolism, or they may have evolved in the ecological niche provided by methane-producing bacteria (65).

Photosynthetic Bacteria

Photoautotrophy, the use of light energy for organic synthesis, evolved in anaerobic bacteria. At present there is considerable diversity in patterns of bacterial photosynthesis; some photosynthetic bacteria are anaerobic, some are capable of respiration when O_2 is available, and some are obligately aerobic. All photosynthesis requires a chlorophyll which traps light energy. Chlorophylls are magnesium-containing porphyrins. Metalloporphyrins probably evolved very early prebiotically; four pyrrol rings form a porphyrin which is reactive with a metal. Iron-containing porphyrins, such as heme compounds, are found in the cytochromes of both anaerobic and aerobic bacteria. Besides having chlorophyll, photosynthetic organisms have electron carriers of several kinds—ferredoxins, flavoproteins, quinones, iron-sulfur proteins, and several cytochromes. The Cyanobacteria release oxygen. Some kinds of bacteria use H_2S as a source of H_2, some use H_2O. Some bacteria fix CO_2 by use of certain steps of the tricarboxylic acid cycle run in the reductive direction; other bacteria use 5-carbon sugars in a process similar to the Calvin cycle of green plants (33).

Green sulfur bacteria, Chlorobiaceae, and purple sulfur bacteria, Chromatiaceae, are obligately anaerobic. They oxidize H_2S (or thiosulfate) and are capable of anaerobic glycolysis. Their reaction center, bacteriochlorophyll, absorbs light at 840 nm and is thereby activated to a high energy state. High energy electrons are carried by Fe-S proteins and by several cytochromes (Fig. 2-6). Sulfur bacteria derive electrons indirectly from NAD-$NADH_2$ for recharging their chlorophyll, and use H_2 from H_2S and fix CO_2 in sugar synthesis (32).

The overall photosynthetic reaction is

$$2\ CO_2 + H_2S + 3\ H_2O \xrightarrow{h\nu} 2[CH_2O] + H_2\ SO_4 + H_2O$$

They can also oxidize elemental sulfur:

$$2S + 3CO_2 + 5H_2O \xrightarrow{h\nu} 2H_2SO_4 + 3[CH_2O]$$

The Chromatiaceae can also use thiosulfate:

$$2\ Na_2S_2O_3 + 4\ CO_2 + 6\ H_2O \xrightarrow{h\nu} 2\ H_2SO_4 + 2\ Na_2SO_4 + 4[CH_2O]$$

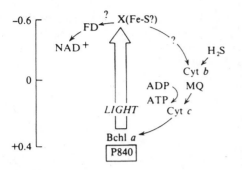

Figure 2-6. Photosynthetic electron transport in green sulfur bacteria, *Chlorobium.* Modified from K. Roa, D. Hall, and R. Carmack in *Biochemical Evolution,* H. Gutfreund, Ed., Copyright 1981, Cambridge University Press, London, and C. Wraight in *Photosynthesis I,* Govindjee, Ed., Copyright 1982, Academic Press, New York.

Purple sulfur bacteria metabolize glucose anaerobically and have enzymes of the glyoxylate cycle which is characteristic of higher plants. Most Chromatiaceae use bacteriochlorophyll a or b, which has a single photocenter. Some purple sulfur bacteria use a chlorophyll with absorbance maximum of 900 nm as the reaction center (21).

Purple sulfur bacteria have ribulose diphosphate carboxylase, a cyclic electron transport enzyme used in the Calvin cycle for CO_2 reduction (Fig. 2-9). Green sulfur bacteria lack enzymes of the Calvin cycle but fix CO_2 into acids (malate) of the TCA cycle.

Gliding filamentous bacteria Chloroflexaceae are facultative aerobes, have one photocenter, bacteriochlorophyll c, and can use either H_2S or organic molecules as H_2 source. These bacteria are capable of aerobic metabolism and can grow in darkness. They carry out anoxygenic photosynthesis.

One kind of archaebacteria, the Halobacteria, make use of light to generate ATP. They grow in brine, in salterns with NaCL concentrations higher than 12–15 percent. Halobacteria cell walls consist of a glycoprotein and the plasma membrane is 60 percent phospholipid. The enzymes of Halobacteria require high salt concentrations; their rRNAs and tRNAs differ significantly from those of eubacteria (65,67). Halobacteria can grow anaerobically without light if supplied with sufficient amino acids. In nature Halobacteria appear red due to carotenoids, which possibly protect the cells from solar radiation. When in the stationary phase of the life cycle, these cells produce a purple membrane composed of a compound, bacteriorhodopsin, which is very similar to that in the rods of vertebrate eyes (Ch. 14). Purple-membrane Halobacteria normally form ATP in the presence of O_2. In the absence of oxygen, their ATP levels decline but are immediately restored if the cells are illuminated. Restoration of ATP after addition of O_2 can be blocked by cyanide, but the recovery by illumination is not so blocked. When intact cells or preparations of the purple membranes are illuminated the medium becomes acid due to liberation of H^+. In intact cells in light, a large transmembrane potential is generated. The extrusion of protons provides a gradient that permits intracellular potassium concentration to reach $3M$ in a medium of $0.5\ M\ K^+$; ATP is generated by proton pumping activity of the bacteriorhodopsin and functions in ionic control (Ch. 9). Halobacteria have cytochromes which differ from the cytochromes of aerobic eubacteria and eukaryotes. Halobacteria can grow and synthesize ATP without oxygen but in metabolism of amino acid substrates they normally use an oxidative pathway. When they evolved in relation to the transition from low to high oxygen atmosphere is unknown.

Rhodospirillaceae and Derivative Bacteria

The purple nonsulfur photosynthetic bacteria, the Rhodospirillaceae, occupy a key position in phylogeny (Fig. 2-5). They have apparently given rise to several kinds of bacteria, photosynthetic and nonphotosynthetic, and to obligate and facultative aerobes. On the basis of the amino acid sequences of their cytochromes c, the nucleic acid patterns of their 16 S rRNA, and the nature of their photosynthetic membranes, some 14 species have been arranged in three groups. The ancestors of one group, six species with medium-size cytochromes c, probably gave rise to endosymbionts which became mitochondria in eukaryotes (1,13,51). Some photosynthetic Rhodospirillaceae can use organic acids as a H_2 source; they have only one photocenter, bacteriochlorophyll, with absorption at 870 nm. In photosynthesis, light energy is trapped, as shown in Figure 2-7, and electrons are transferred via quinone to a ubiquinol: cyt c_2 oxidoreductase complex. Additional energy is provided by a cycle between high- and low-potential cytochrome b, and additional reducing capacity is provided by electrons from a substrate couple. From the ubiquinone complex, electrons pass to cytochome c_2 and from this the BChl is recharged.

Cyanobacteria

An important advance was made by Cyanobacteria or their ancestors. Cyanobacteria were formerly called blue-green algae but they are prokaryotes, not true algae, which are eukaryotes. Cyanobacteria (Cyanophyceae) differ from most other photosynthesizing bacteria in having two photocenters like higher plants and in liberating free oxygen from H_2O (Fig. 2-8).

System I functions at the reducing end of the scale, system II at more oxidizing potentials. CO_2 is the ultimate electron acceptor. The overall reaction is

$$CO_2 + 2 H_2O \rightarrow CH_2O + H_2O + O_2$$

In Cyanobacteria and higher plants photocenter I (P-700) at E_m of $+400$ mV is excited by light and goes to an excited state P-700 of $E_m < +700$ mV. The excited electron is then passed down a chain of Fe-S carriers to ferredoxin (Fd), and oxidoreductase to reduce $NADP^+$ to NADPH, E_m of -320 mV.

$$2 \text{ Fd}_{red} + 2H^+ + NADP^+ \rightarrow 2 \text{ Fd}_{ox} + NADPH + H^+$$

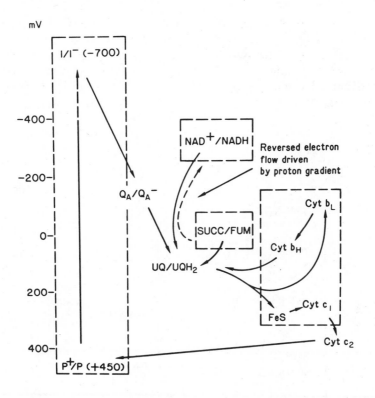

Component	$E_m(mV)$
P 870 (P$^+$/P)—bacteriochlorophyll a	$+450$
^{1}/I$^-$—bacterial pheophytin (electron acceptor)	-700
^{Q}A/^{Q}A$^-$—quinone intermediate	-180
UQ/UQH$_2$—uboquinone	$+90$
cyt b$_H$—cytochrome b high potential	$+60$
cyt b$_L$—cytochrome b low potential	-90
cyt c$_1$—cytochrome c$_1$	$+260$
cyt c$_2$—cytochrome c$_2$	$+340$

Figure 2-7. Schema of photosynthesis in purple non-sulfur bacteria, *Rhodospirillaceae*. Modified from K. Roa, D. Hall, and R. Carmack in *Biochemical Evolution,* H. Gutfreund, Ed., Copyright 1981, Cambridge University Press, London, and C. Wraight in *Photosynthesis I,* Govindjee, Ed., Copyright 1982, Academic Press, New York.

Photosystem II (P-680) of E_m + 1100 mV is also energized by a photon and the high-energy electron goes to an activated molecule of E_m −620 mV. An electron from this center is then carried via a plastoquinone porter to the cytochrome b and f complex and then via plastocyanin to fill the hole caused in photocenter I when this system was excited. The hole in photocenter II is filled by an electron arising from water and O_2 is released:

$$2 H_2O + 2 NADP^+ \rightarrow O_2 + 2 NADPH + 2H^+$$

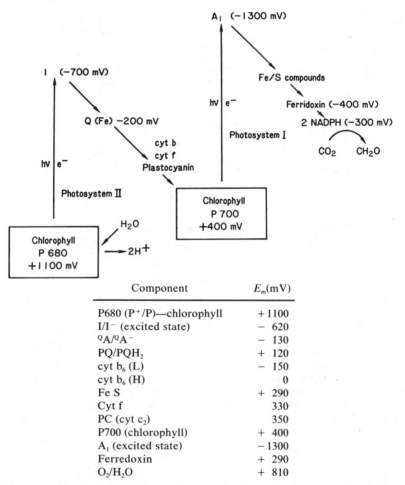

Component	E_m(mV)
P680 (P$^+$/P)—chlorophyll	+ 1100
I/I$^-$ (excited state)	− 620
QA/QA$^-$	− 130
PQ/PQH$_2$	+ 120
cyt b$_6$ (L)	− 150
cyt b$_6$ (H)	0
Fe S	+ 290
Cyt f	330
PC (cyt c$_2$)	350
P700 (chlorophyll)	+ 400
A$_1$ (excited state)	−1300
Ferredoxin	+ 290
O_2/H_2O	+ 810

Figure 2-8.   Schema of two-photo center photosynthesis in *Cyanobacteria* and green plants. Modified from K. Roa, D. Hall, and R. Carmack in *Biochemical Evolution,* H. Gutfreund, Ed., Copyright 1981, Cambridge University Press, London; C. Wraight in *Photosynthesis I,* Govindjee, Ed., Copyright 1982, Academic Press, New York; and A. L. Lehninger, *Principles of Biochemistry,* Worth Publishers, New York, 1982, p. 657.

Photosystem I alone can drive a cyclic flow of electrons via the cytochrome b-f complex to generate additional ATP. The schema of Figure 2-8 is similar in higher plants and Cyanobacteria. Chlorophyll a occurs in both reaction centers for high energy transfer; chlorophyll b functions in light harvesting rather than in energy transfer. Chlorophyll b occurs only in photosynthesizing eukaryotes.

Capture of CO_2 is by reactions of the Calvin cycle (Fig. 2-9). CO_2 enters at the step between the 5-carbon sugar ribulose diphosphate to form two molecules of phosphoglycerate. By several steps the 3-carbon compounds, energized by ATP, go to fructose-6-P and to glucose-6-P and finally to glucose. By a series of reactions the cycle is completed by synthesis of ribulose diphosphate. The total reaction is

$$6 \text{ ribulose, } 1,5\text{-diphosphate} + 6 \text{ } CO_2 + 18 \text{ ATP} + 12 \text{ } H_2O + 12 \text{ NADPH}$$
$$+ 12 \text{ } H^+ \rightarrow 6 \text{ ribulose } 1,5\text{-diphosphate} + \text{glucose} + 18 \text{ } P_i + 18 \text{ ADP}$$
$$+ 12 \text{ NADP}^+$$

It has been argued (9) that the bulk of atmospheric oxygen in the early Cambrian or Precambrian came from Cyanobacteria. Stromatolites were formed as early as 3.4 byr ago in Western Australia, 3.0 byr ago in Canada. The possibility that early stromatolites might have been formed by other bacteria, possibly anoxygenic photosynthetic bacteria similar to Chloroflexaceae has not been considered. In some hot springs, for example in Yellowstone Park, Cyanobacteria are now forming stromatolites (66). Most Cyanobacteria have aerobic oxidative metabolism in which O_2 is the terminal

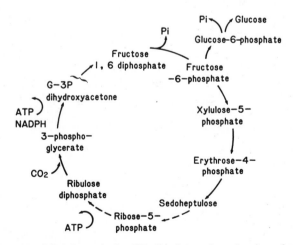

Figure 2-9. Diagram of Calvin cycle for CO_2 fixation and production of glucose in photosynthesis. Modified from K. Roa, D. Hall, and R. Carmack in *Biochemical Evolution,* H. Gutfreund, Ed., Copyright 1981, Cambridge University Press, London, and A. L. Lehninger, *Principles of Biochemistry,* Worth Publishing, New York, 1982, p. 665.

electron acceptor and aerobic metabolism is independent of light. Cyanobacteria can fix atmospheric nitrogen. They lack the peptidoglycan cell wall
which is characteristic of other eubacteria. The blue-green *Anabaena* fixes
CO_2 at the beginning of a light period; later it fixes N_2 and in the dark liberates O_2 (46).

Summary of Metabolism of Bacteria

The adaptive diversity of bacteria results from many changes during the 3.8
billion years of their evolution; some bacteria brought about changes which
altered the environment and afforded new ecological situations. The origins
of prokaryotes are not known and may have been multiple; prokaryotes certainly originated from anaerobic ancestors. The fermentative metabolic patterns of the ancestors for making ATP persist to the present in virtually all
organisms. Heterotrophy in the sense of the utilization of organic substances
already present was lost temporarily, to be revived millions of years later by
bacteria which depend on organic substances formed by other bacteria. The
use of nonorganic energy sources—H_2S, SO_4^{2-}, CH_4 and H_2^- for ATP production—has been retained in some bacteria.

   The archaebacteria and eubacteria diverged from primitive prokaryotes
and differ in the nucleotide sequences of rRNAs and tRNAs, and in cell
walls and lipids. The archaebacteria colonized extreme environments; one
kind uses H_2 as an energy source to produce methane, another kind, perhaps
descended from inhabitants of saline waters, has a pigment, bacteriorhodopsin, which uses sunlight to product ATP. Two kinds of eubacteria now use
bacteriochlorophyll, which closely resembles the chlorophyll of eukaryotes;
instead of H_2O these photosynthesizing bacteria derive electrons from H_2S
and use ATP formed in the light reaction from H_2S for biological work; O_2 is
poisonous to them.

   Several anaerobic methods for ATP formation have evolved, and at least
two kinds of photosynthetic pigments. Anaerobic oxidation with iron-containing cytochromes and iron-sulfur proteins as electron carriers was present
long before the atmosphere contained significant oxygen. Probably aerobic
oxidation and tolerance of O_2 were secondarily derived by halophiles, methanotrophs, and nonsulfur photosynthesizing bacteria. The Cyanobacteria
developed a two-step photosynthesis that opened the way for using H_2O
instead of H_2S as a proton source (Fig. 2-8). Later, oxygen was present in
quantity and several preexisting kinds of organisms, as well as new ones,
developed pathways that used oxygen in energy liberation.

Origin of Eukaryotes

Organisms are classified in two superkingdoms, prokaroyotes and eukaryotes. Prokaryotes include two kingdoms, eubacteria and archaebacteria. Eu-

karyotes include four kingdoms—protists, animals, plants, and fungi (40,9a). Table 2-2 contrasts prokaryotes and eukaryotes in respect to a number of characters. The terms prokaryote and eukaryote are better used for description of structural and functional characters than for tracing phylogeny.

A eukaryote is an organism with a true nucleus; a prokaryote is an organism without a discrete nucleus. In a prokaryote (including *Cyanobacteria*) the DNA is in a single circular chromosome associated with the cell membrane. In eukaryotes the DNA is condensed into discrete chromosomes within a nuclear membrane. Presence of a nucleus provides for separation of transcription from translation in protein synthesis. The DNA of eukaryotes is supported by the histones and some other basic proteins. Histones are very similar in all eukaryotes. Prokaryotes lack histones, but have other basic proteins as support for DNA.

In a eukaryote the rRNA is stored in a nucleolus before moving out to a ribosome; a prokaryote lacks a nucleolus and the rRNA enters the cytoplasm directly as it is synthesized. In eukaryotes, division of chromosomes is via mitotic spindles, usually with centrioles and tubulin filaments. Cell cleavage in animals makes use of the contractile proteins myosin-actin. Cells of "higher" plants divide by laying down a plate between the two cells. Eukaryote genes have coding segments (exons) interspersed with noncoding segments (introns); there may be separate exons that code for different segments of one protein. A DNA codon may be repeated in multicopies in a eukaryote, but usually not in a prokaryote (2). Formylated methionine is a code initiator in both prokaryotes and eukaryotes; after initiation the formyl group is removed in most eukaryotes, but not in prokaryotes. Cell walls of prokaryotes have amino-sugars with peptide tails; cell walls of eubacteria have peptidoglycans, most of them containing muramic acid. Eukaryotes lack such cell walls. Eukaryotes make numerous fatty acids, especially polyunsaturated; prokaryotes make fatty acids with only 2, 3, or 4 double bonds. Many bacteria locomote by movement of a flagellum consisting of a single fibril of a protein, unlike the proteins of motile eukaryotes. Most eukaryotes have cilia with compound fibrils of 9 + 2 filaments of tubulin. Many prokaryote and probably all eukaryote cells have actin as a basic protein of cell movement. In eukaryotes, chromosome reduction occurs in the sequence of meiotic divisions during the production of germ cells. Sex in the sense of unequal germ cells is present only in eukaryotes. Where interchange of genetic material between organisms occurs in prokaryotes, it is by similar but not always equivalent cells. The development of sex led to the greater diversity of eukaryotes.

The first eukaryotes appeared 1.8–2.0 byr ago; their origin is obscure. Red algae are primitive eukaryotes that lack bacteria-type flagella; in red algae the 16 S chloroplast RNA is similar to Cyanobacteria rRNA. *Euglena*, a green dinoflagellate, is a primitive eukaryote. The basic proteins which

Table 2-2. General Characteristics of Prokaryotes and Eukaryotes[a]

Prokaryotes	Eukaryotes
[Characteristics of archaebacteria ("A") and eubacteria ("E")]	
Loop of double-stranded DNA near cell membrane. No nuclear membrane	Double-stranded DNA condensed into chromosomes. Nuclear membrane. Nucleolus. Histones.
Asexual reproduction. Fertilization can occur between equivalent strains of some prokaryotes	Unequal fertilization; sex Mitosis; meiosis

Cell walls
  Various (A)
  Muramic acid (E)
	No cell walls in animals; cellulose in plants

Membrane lipids
  Ether-linked branched aliphatics (A)
  Ester-linked straight chains (E)
  Sterols usually absent, many have terpenes or tetrapyrenoids
	Ester-linked aliphatic chains
	Steroids in all cell membranes

Transfer RNAs
  Ribothymidine in common arm (E)
  Pseudouridine in common arm (A)
  Initiator tRNA carries methionine (A)
  Initiator carries formyl-methionine (E)
	Ribothymidine in common arm of most tRNAs
	Methionine in initiator tRNAs

Ribosomal subunits
  30 S, 50 S
  Insensitive to chloramphenicol (A)
  Sensitive to chloramphenicol (E)
  16 S (18 S) rRNA of 1400–1600 nucleotides
  mRNA binding site AUCACCUC at 3' end of 16 S RNA (E and A)
	40 S, 60S
	Insensitive to chloramphenicol
	1800 nucleotides
	AUCACCUCC site absent

Metabolism—variable anaerobic, some are facultative aerobic	Mostly obligate aerobic; combined oxidation and glycolysis
Unsaturated fatty acids 2, 3, or 4 double bonds	Polyunsaturated fatty acids

Electron transport variable (E)	Mitochondrial electron transport
	  Lysosomes
	  Golgi apparatus

Actins rare or absent	Actins; tubulins

[a]From References 67, 40, and others.

support *Euglena* DNA in the discrete nucleus are transient, are present during the period of gene activation, and the ratio of histone to DNA is 0.1, compared to 1.0 in more complex eukaryotes.

It is probable that mitochondria in eukaryotes originated from bacteria, the descendants of which later became symbiotic. Chloroplasts in eukaryotes descended from the ancestors of modern Cyanobacteria and became symbiotic (34,40,54). Mitochondria, both plant and animal, contain DNA which codes for some of the mitochondrial components of the respiratory chain. Mitochondrial DNA, like bacterial DNA, is circular. The gene for cytochrome c is in the nucleus; the mitochondria code for cytochrome c oxidase. Inheritance of mitochondrial DNA is through the ovum—uniparental, not Mendelian. The initiator and terminator sequences of mitochondrial DNA are different from those of the nucleus—UGA terminal in the nucleus, T or A in mitochondria. The largest mitochondrial genome is in higher plants. Yeast has nine mitochondrial genes, including split genes, one which codes for ATPase, two genes for cytochrome oxidase, a single gene for each cyt d, 14 S rRNA, and 5 S rRNA. Differences in mitochondrial rRNAs indicate separate origins of mitochondria in fungi from those in complex plants and animals (34).

Chloroplasts of algae and higher plants contain DNA; chloroplast genes are uniparental by the female line. Some proteins of photosynthesis, RUB-P-carboxylase, are coded in both nucleus and chloroplast. The 16 S rRNA of chloroplasts is different in its oligonucleotides from the 18 S rRNA of cytoplasm. DNA sequences of the chloroplast of *Euglena* resemble those of Cyanobacteria in 32–47 percent of sequences.

Evidently mitochondria and chloroplasts arose by symbiosis between prokaryotes and primitive eukaryotes, the chloroplasts coming from ancestors of Cyanobacteria and the mitochondria from nonphotosynthetic but aerobic bacteria, probably ancestors of Rhodospirillaceae. In time some of the genes may have been transferred to the nuclei of the host; membranes eventually came to bound the chloroplasts and mitochondria (40,54). A contrary hypothesis is that chloroplasts and mitochondria differentiated like other organelles in eukaryotes and that some genes of eukaryotes became partitioned between nucleus and cytoplasmic organelles.

It is probable that eukaryotes first arose at least 700 million years (myr) ago, during the Precambrian, and that the earliest eukaryotes did not leave a fossil record. One group of soft metazoan eukaryotes, the ediacarians, of wide occurrence, left fossils that have been dated to more than 650 myr ago. These resembled and may have given rise to medusoid cnidarians and bilateral wormlike organisms that may have evolved to annelids (21a). In the early Cambrian beginning at about 500 myr ago animal phyla expanded extensively and apparently all of the animal phyla were established within a geologically short period. Some animals have not evolved into new groups (e.g., Porifera) since the Cambrian. In some phyla certain classes are either extinct or are now represented only by relicts—brachiopods, crinoids, trilobites, ammonites. Judged by the fossil record and by biochemical characters, fungi evolved in the Precambrian before plants—horsetails (Equisetales) and club mosses (Lycopodales) in the Cambrian (420–430 myr), conifers and ferns at 350–345 myr, and angiosperms in the Mesozoic (190 myr).

Measurements on the ciliate protozoans of the species complex *Tetrahymena* (more than 13 recognized species) suggest that this genus diverged very early in eukaryote evolution, possibly one byr ago (45). The highly conserved protein histone 4 of *Tetrahymena* differs by more than 15 sequences from histone 4 of other eukaryotes. The structure of *Tetrahymena* calmodulin and cytochrome c is not typical of eukaryotes. *Tetrahymena* 5 S rRNA has an oligonucleotide sequence intermediate between eubacteria and eukaryote sequences. *Tetrahymena* is a living fossil in its biochemistry (45).

Metabolic Cycles and Enzyme Sequences of Energy Liberation

Enzymatic pathways that function today in eukaryotes had their origin in prokaryotes; as evolution proceeded, there were changes in function. Diverse pathways were used in different kinds of anaerobic metabolism; some of these metabolic paths persisted in aerobiosis. Besides liberating energy and synthesizing organic compounds, cells maintain redox and acid-base balance. Important components of metabolic evolution were (1) the utilization of energy in small packets, (2) the transfer of specific groups from metabolites to protein receptors, and (3) the coupling of separate reactions into cycles and chains. Some coupled sequences of reactions probably developed in precellular evolution. Kinetic properties of some metabolic enzymes are presented in Chapter 5; data on the ecological significance of these pathways are given in Chapter 6. The following is an outline of general energy-yielding pathways and suggestions as to their evolution.

A bioenergetic reaction that was established early is substrate-level phosphorylation and dephosphorylation. Interconversions of AMP to ADP and of ADP to ATP generate high-energy bonds. The most used energy-transfer unit is the high-energy terminal phosphate of ATP. The standard free energy of hydrolysis of the terminal phosphate under physiological conditions is 7.3 kcal/mol; the calculated free energy ΔG is 12.4 kcal/mol (35). ADP is phosphorylated to ATP with transfer of high-energy phosphate to ATP at numerous steps during a metabolic sequence. Pyrophosphate (PP) yields considerable energy (7.1 kcal/mol) on hydrolysis to inorganic phosphate (P_i). Hydrolysis of PP is used in a few energy-requiring reactions and possibly provided energy for chemical work prior to the use of ATP (33,35).

Very early in cell evolution, agents for transfer of H_2 or protons and electrons provided appropriate oxidation-reduction reactions resulting in redox balance. In all organic oxidation-reduction sequences H_2 is transferred from a negative substrate toward a receptor (e.g., O_2) of more positive redox potential. Transfer agents are usually coenzymes that are capable of reversible oxidation-reduction. The origin of coenzymes has not been much considered by evolutionary biochemists. One suggestion is that the active transfer groups on coenzymes may have evolved before the protein binding parts of the molecules. In vertebrates, some coenzymes at present are modified from vitamins, but earlier organisms may have synthesized coenzymes. Anaerobic prokaryotes have a variety of transfer substance, some of which occur also in aerobic eukaryotes as cofactors of enzyme reactions. In photosynthesis several carriers transfer high-energy electrons from an excited state of a photosystem to electron receptors. A few examples of transfer molecules and suggestions regarding their evolution follow.

The most widely used transfer agent is nicotinic adenine nucleotide, which is alternately reduced and oxidized, $NADH_2 \leftrightharpoons NAD^+$; this coenzyme consists of nicotinamide and adenine coupled by ribose (Fig. 5-12); the reactive part of the molecule is at carbon one on the amide range. A related agent for proton transfer is $NADP \leftrightharpoons NADPH_2$, in which one sugar is esterified with phosphate. Some enzymes of glycolysis and of oxidation use NAD, others use NADP, and it is probable that NAD^+ preceded $NADP^+$ in early evolution.

Another agent for transfer of H_2 is ubiquinone or coenzyme Q. This is a fat soluble quinone with a long isoprenoid side chain. CoQ collects reducing equivalents from NADH dehydrogenase and from flavin-linked dehydrogenases. Its reaction is reversible: $CoQ \leftrightharpoons CoQH_2$ (35).

Agents for transfer of electrons probably evolved very early. The best known of these are metalloporphyrins. Several heme-containing cytochromes are present in anaerobic bacteria; cytochromes serve in electron transport in photosynthesis (Fig. 2-7). Cytochromes are heme derivatives, metalloporphyrins, which probably appeared early in evolution (42). Cytochrome c_2 is present in purple nonsulfur bacteria that can photosynthesize anaerobically but respire aerobically. Cytochrome c_6 occurs in Cyanobacteria and in chloroplasts of plants where it transfers electrons between photosystems I and II (11). Cytochrome c is a central carrier in aerobic energy metabolism by the TCA cycle.

Flavins, such as flavin mononucleotide (FMN) and flavin adenine nucleotide (FAD), are derived in mammals from vitamin B_2 (riboflavin). The reactive group consists of two adjacent carbons in an alloxazine ring; these are alternately oxidized and reduced by a shift between two and one double bonds. Flavins probably occurred in primitive prokaryotes.

Transfer agents for acetyl groups function in initiation of the TCA cycle. Coenzyme A has as its reactive group a sulfhydryl at one end of the long molecule.

Pyridoxal phosphate (or pyridoxamine phosphate), in mammals derived from vitamin B_6, is the amino group transfer agent in nitrogen acid metabolism. One of its functions is as a cofactor in transaminations. A bacterial amino acid decarboxylase has been found in which pyruvate substitutes for pyridoxal-5-phosphate as the cofactor (4).

Biotin is a carrier in a number of enzymatic carboxylation reactions. A derivative of folic acid, tetrahydrofolic acid (FH_4) is a transfer agent in transport of 1-carbon groups ($-CH_3$, $-CHO$, and others). One of these folic acid derivatives is a coenzyme in synthesis of thymidylic acid, a nucleotide building block of DNA.

In summary, numerous transfer molecules may have functioned prior to the evolution of specific enzymes. The reactive groups of the transfer agents can occur in other molecules and it is probable that considerable transformation of these agents has occurred in the course of evolution of metabolism.

Metabolic Sequences

The most universal of metabolic pathways is glycolysis, the conversion of glucose to pyruvate. The overall reaction is

$$C_6H_{12}O_6 + 2\ ADP + 2\ P_i + 2\ NAD + 2\ ATp \rightarrow 2\ CH_3COCOOH$$
$$+ 4\ ATP + 2\ NADH_2$$

There is a net gain of 2 ATP and two reducing equivalents (35). A flow chart of the entire pathway is given in Figure 2-10. Glycolysis to pyruvate requires 10 enzymes and can be considered as consisting of two phases: glucose to glutaraldehyde-3-phosphate and glyceraldehyde-3-phosphate to pyruvate. Pyruvate can then be reduced to lactic acid with oxidation of $NADH_2$:

$$CH_3COCOOH + NADH_2 \rightarrow CH_3CHOHCOOH + NAD^+$$

Pyruvate can also be decarboxylated to acetaldehyde which is then reduced to ethanol:

$$CH_3COCOOH \rightarrow CH_3CHO + CO_2$$

$$CH_3CHO + NADH + H^+ \rightarrow CH_3CH_2OH + NAD^+$$

The entire sequence occurs anaerobically, and the generation of two ATPs is the same whether products are lactate or ethanol (Fig. 2-10). The origin of the 10-enzyme sequence has not been much considered by evolutionary biochemists. Some individual steps have multiple functions: (1) glucose-6-phosphate to 6-phosphogluconate and to the pentose shunt; (2) entry of fatty acids at the step between PEP (phosphoenolpyruvate) and pyruvic acid; and (3) the role of isomerase between dihydroxyacetone phosphate and glyceraldehyde phosphate in fatty acid metabolism. However, the complete gly-

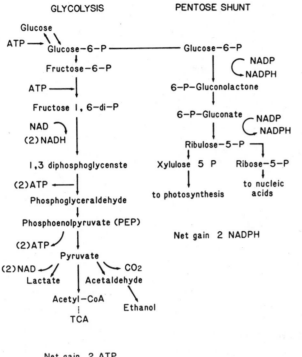

GLYCOLYSIS   PENTOSE SHUNT

Figure 2-10. Flowchart of anaerobic metabolic pathways, glycolysis, and parallel pentose shunt. Modified from A. L. Lehninger, *Principles of Biochemistry,* Worth Publishers, New York, 1982, pp. 404, 408, 457.

colytic chain occurs intact in both anaerobic and aerobic organisms and the origin of individual steps is obscure.

A probable early anaerobic pathway was carboxylation of the 3-carbon pyruvate to give a 4-carbon oxaloacetate (OAA):

$$\text{pyruvate} + CO_2 + ATP \rightarrow \text{oxalacetate} + ADP + P_i + 2H^+.$$

OAA then reacted with $NADH_2$ to give NAD and malate; this lost a H_2O to form fumarate which in turn went to succinate. Thus fumarate and succinate are electron acceptors; that is, the sequence from pyruvate to succinate is reductive. Some of these reactions occur in CO_2 fixation (reduction) by photosynthesizing sulfur bacteria.

Pyruvate is a key substance as the terminal of the glycolytic chain and for entry into the tricarboxylic acid (TCA) cycle (Fig. 2-11) (4). A cluster of enzymes and coenzymes react with pyruvate; how these evolved and in what reactions pyruvate participated before the TCA cycle are not known.

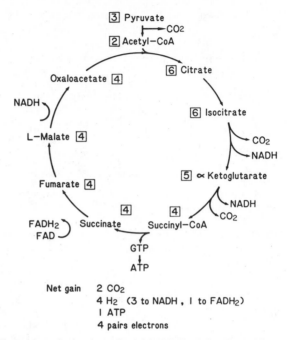

Figure 2-11. Flowchart of tricarboxylic acid (TCA) cycle of aerobic metabolism (number of C atoms in boxes). Modified from A. L. Lehninger, *Principles of Biochemistry,* Worth Publishers, New York, 1982, p. 441.

The key reaction for entry into the TCA cycle is a decarboxylation of pyruvate followed by a condensation with acetyl CoA. Coenzyme A consists of adenine, ribose phosphate, pantothenic acid, and mercaptoethanolamine. The reactive group is a sulfhydryl on the mercaptoethanolamine. It is probable that other SH-compounds may have served in reactions of pyruvate in anaerobic metabolism. In some bacteria, pyruvate can react anaerobically with CoA to form acetyl-CoA plus hydrogen gas. Another way to form acetyl-CoA anaerobically is by a reaction with ferredoxin. Once acetylCoA was formed by oxidative decarboxylation of pyruvate a reaction with OAA could occur to form citrate:

$$OAA + AcCoA \rightarrow citrate + CoA$$

Several of the reactions that later became incorporated into the TCA cycle had probably been used in anaerobic metabolism of prokaryotes. Transketolation and transaldolation combined with the early steps of glycolysis in what is known as the pentose shunt. Synthesis of pentoses was needed for nucleic acids. The "shunt" provides reducing capacity, NADP as H^+ acceptor.

At about this stage in biochemical evolution, CO_2 fixation by the Calvin-

Benson cycle appeared. This cycle (Fig. 2-9) uses 3 ATP for each CO_2 reduced; 3-phosphoglycerate feeds into the sequence of glycolytic enzymes acting in reverse to its usual direction and for six turns of the cycle one glucose is synthesized.

Over a long period some oxygen was being formed by photolysis by the action of ultraviolet light on water. Anaerobes were poisoned by O_2. Nonspecific oxygenases evolved as protection against oxygen. Anaerobic photosynthesis used bacteriochlorophyll with one photocenter and made carbohydrate but did not liberate oxygen.

Cyanobacteria appeared having the photosynthetic pathway now known in algae and higher plants (Fig. 2-5). Cyanobacteria had two photocenters, produced oxygen in quantity, and were much more efficient in organic synthesis than the bacteria with anaerobic photosynthesis. Photosynthesis by Cyanobacteria probably began about 3 byr ago as judged by the presence of Cyanobacteria in stromatolites. Liberation of oxygen and use of O_2 as an electron acceptor in energy metabolism led to accumulation of O_2 in the atmosphere (47,48). It is probable that at about the time of appearance of aerobic photosynthesizing organisms, the glyoxylate shunt evolved. This shunt feeds two molecules of acetate into synthetic pathways and forms succinate. The glyoxylate shunt is used by plants in production of 4-carbon acids, CO_2, and reduced NADH.

With aerobic conditions and O_2 as an electron acceptor, probably with a linking of TCA cycle enzymes, an important change occurred. It is postulated that α-ketoglutarate dehydrogenase catalyzed the reaction from α-ketoglutarate (α-KG) to succinate. The sequence from OAA to succinate was then less efficient and functioned in the oxidative rather than in the reductive mode. Direction of the sequence then reversed to be oxidative from succinate to fumarate to malate to oxaloacetate (Fig. 2-12) (20). The TCA cycle was then complete.

Cytochrome c, ubiquinone, and cytochrome oxidase came to serve as electron transporters to O_2. Other cytochromes had been functional in anaerobic pathways, but with aerobic respiration, oxygen accepted four electrons per molecule and the efficiency of energy liberation increased.

Modifications of the TCA cycle permitted oxidation of fats and amino acids. Some metabolism of these substances had occurred anaerobically, but efficiency increased with aerobic pathways. The ornithine cycle for disposing of products of deamination probably evolved once OAA and α-ketoglutarate were available.

GENERAL SUMMARY

The beginnings of life on earth are more speculated upon than known. There is some evidence from (1) geochemical data regarding age and composition of rocks, (2) characteristics of early fossils, (3) experimental models in which

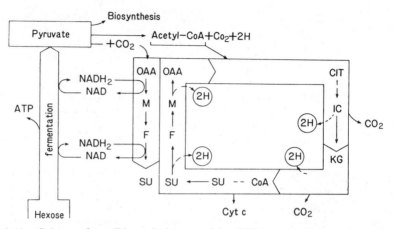

Figure 2-12. Schema of possible evolutionary origin of TCA cycle. Three major routes are: (1) Anaerobic reductive sequence, Oxaloacetate → malate → fumarate → succinate. (2) Conversion of pyruvate to acetyl CoA plus CO_2; this led to pathway for biosynthesis of glutamate via oxaloacetate plus acetyl CoA, citrate-isocitrate, α-ketoglutarate to glutamate (not shown). (3) Oxidative modification by ketoglutarate, plus CoA + NAD to give succinate CoA + CO_2 + NADH; the direction of proton flow was reversed in aerobic condition to succinate → oxaloacetate. Modified from H. Gest, *FEMS Microbiology Letters,* Vol. 12, p. 214. Copyright 1981, Elsevier Scientific Press, Amsterdam, The Netherlands.

atmospheric conditions probable on early earth are simulated, and (4) biochemical studies of present-day organisms to ascertain reactions that are so universal that they must have been present in early life forms. The basic constituents of organisms were formed during the period of some 1.0 to 1.2 byr before there were discrete cellular organisms. The organization of organic compounds into metabolic pathways may have occurred during the next billion years in primitive cells which left no fossils. The Oparin-Haldane hypothesis of spontaneous synthesis of life constituents under the influence of solar and other sources of energy appears valid, modified to take account of probable sites other than oceans.

Some universal biochemical properties can be arranged in a probable scenario of the following steps: selection of L amino acids as building blocks for proteins; formation of nucleotides, some coupled to sugar-phosphates as energy stores, some replicating to make polymers which became nucleic acids; arrangement of phospholipids and proteins as bounding membranes with selective permeability; selection of nucleotides as the basis for a self-replicating genetic code (the three-letter code was early established); replication of nucleic acids (probably tRNAs at first) in association with amino acid assembly; synthesis of some proteins, several of which had a feedback interaction with the tRNA; complexing of metals with porphyrin rings which became versatile electron-carrier molecules; inorganic ions used in some chemical reactions and as intracellular solutes. All cells concentrated potassium and had higher H^+ than the milieu.

Possibly the first "organisms" were hypercycles of transfer RNAs and proteins which facilitated replication of a hypercycle; many hypercycles coupled together to form "quasi-species." The earliest organisms were undoubtedly heterotrophic. An energy-yielding pathway which is so universal as to have evolved very early is anaerobic glycolysis. Before glycolysis or fermentation could be established, the evolution of numerous transfer agents which maintain redox balance must have taken place; these persisted in coenzymes. There may have been separate functions of individual steps in anaerobic glycolysis before the entire pathway was established. As living material was differentiated into cells, membrane mechanisms for electrical, osmotic, and volume regulation were developed.

Various types of chemoautotrophy appeared in organisms, particularly the use of sulfur compounds, possibly also of molecular hydrogen. The first photoautotrophic organisms were anaerobic and synthesized sugars by pathways of single photocenters without liberating oxygen. Some modern bacteria carry out photosynthesis anaerobically but may metabolize carbohydrates aerobically. Diversity of photosynthetic pathways in bacteria indicates considerable testing before the two-activation center mode of photosynthesis evolved in ancestors of Cyanobacteria. However, the Cyanobacteria appeared some 3.0–3.5 byr ago, indicated by the presence of fossil stromatolites. The transition of the atmosphere from reducing to oxygen-containing occurred at this time.

Very early the archaebacteria evolved—bacteria with characteristic nucleic acid composition, unique cell walls, and metabolic capabilities which permitted invasion of environmental extremes. Modern eubacteria are the result of long and divergent evolution and differ substantially from their probable ancestors.

Eukaryotic algae are estimated to have been present on earth for some 1.5 to 2.0 byr, higher plants and animals for about 0.7 to 1.0 byr. More is known from fossils about the evolutions of eukaryotes than of prokaryotes. Further back in time the record is obscure. Throughout evolution there have been many "trials" of modes of life and many parallel lines. Some biochemical patterns that were established early in evolution have changed; enzymes have taken on new functions. There have been separate origins of similar compounds, for example the cytochromes. The environment has changed, especially with respect to oxygen, hence what was adaptive at one period was no longer so at a later time. In general, the trend has been toward greater complexity and biochemical divergence. Most striking from our present perspective are the biochemical similarities in all forms of life.

REFERENCES

1. Almassy, R. J. and R. E. Dickerson. *Proc. Nat. Acad. Sci.* **75**:2674–2678, 1978. Evolution of purple nonsulfur photosynthetic bacteria.

2. Angerer, R. and B. Hough-Evans. Ch. 1, pp. 1–30 in *Receptors and Hormone Action,* vol. 1, Ed. B. W. O'Malley and L. Birnbaumer, Academic, New York, 1977. Repetitive sequences of genes in eukaryotes.

3. Balch, W. E. and R. Wolfe. *Microbiol. Rev.* **43:**260–296, 1979. Methanogenic archaebacteria.

4. Baldwin, J. E. and H. Krebs. *Nature* **291:**381–382, 1982. Evolution of metabolic cycles.

5. Baross, J. A. et al. *Nature* **298:**366–368, 1982. Thermophilic bacteria from submarine hydrothermal vents.

6. Cairns-Smith, A. G. *Nature* **276:**807–808, 1978. Precambrian solution chemistry; banded iron formation.

7. Chappe, B. et al. *Science* **217:**65–66, 1978. Polar lipids of archaebacteria in sediments and petroleum.

8. Clayton, R. K. and W. R. Sistrom. The photosynthetic bacteria. Cambridge University Press, Cambridge, 1978.

9. Cloud, P. *Paleobiology* **2:**351–387, 1976. *Science* **196:**729–736, 1968. *Nature* **296:**198–199, 1982. Nature of prebiotic atmosphere and hydrosphere; origin of life forms.

9a. Corliss, J. O. *Biosystems* **17:**87–126, 1984. Kingdoms of eukaryotes.

10. Crick, F. *Life itself, its origin and nature.* Simon and Schuster, New York, 1981.

11. Dayhoff, M. O. and R. M. Schwartz. in *Origin of Life,* Ed. H. Noda., Academic Publishers of Japan, Tokyo, 1978, pp. 547–560. Evolution inferred from RNA sequences.

12. Dickerson, R. E. *J. Mol. Biol.* **100:**473–491, 1976. *Proc. Nat. Acad. Sci.* **75:**2674–2678, 1978. *Sci. Am.,* Sept. 1978.

13. Dickerson, R. *Nature* **283:**210–212, 1980. Evolution and gene transfer in purple photosynthetic bacteria.

14. Eigen, M. et al. *Sci. Am.* **244:**88–119, 1981. *Naturwissenschaften* **65:**7–41, 1978. *Naturwissenschaften* **68:**217–228, 282–292, 1981. Ch. 6, pp. 105–130 in *Evolution from molecules to men,* Ed. D. S. Bendall, Cambridge University Press, New York, 1983. Models of self-replicating hypercycles and quasi-species.

15. Engle, M. and B. Nagy. *Nature* **296:**837–840, 1982. Amino acids in meteorite from Australia.

15a. Fakhrai, H., L. E. Orgel, and J. H. VanKoode. *J. Mol. Evol.* **17:**295–306, 1982. Synthesis of oliogonucleotides with metal catalysts on simple templates.

16. Felbeck, H. and G. Somero. *Nature* **293:**291–293, 1981. *T.I.B.S.* **7:**201–204, 1982. Sulfur oxidizing bacteria from deep-sea vents.

17. Fischer, F. et al. *Nature* **301:**511–513, 1983. Sulfur metabolism of thermophilic bacteria.

18. Fox, G. E. et al. *Science* **209:**457–463, 1980. Phylogeny of prokaryotes.

19. Fox, G. E. et al. *Proc. Nat. Acad. Sci.* **74:**4537–4541, 1977. Methanogenic bacteria.

19a. Fox, S. W. and K. Dose. *Molecular Evolution and the Origin of Life.* Freeman, San Francisco, 1972, Ch. 6–7, pp. 196–270. Self assembly of polyamino acids.

20. Gest, H. *Microbiol. Lett.* **12:**209–215, 1981. Evolution of citric acid cycles.

21. Gibson, J. et al. *Current Microbiol* **3**:59–64, 1979. Phylogeny of purple photosynthetic bacteria.

21a. Glaessner, M. F. *The Dawn of Animal Life*. Cambridge University Press, New York, 1984.

22. Granick, S. and S. I. Beale. *Adv. Enzymol.* **46**:37–203, 1978. Porphyrin metabolism.

23. Gupta, R. and C. R. Woese. *Current Microbiol.* **4**:245–249, 1980. tRNA of archaebacteria.

24. Hall, J. B. *J. Theoret. Biol.* **30**:429–454, 1971. Evolution of prokaryotes.

24a. Hand, S. C. and G. N. Somero. *Biol. Bull.* **165**:167–181, 1983. Energy metabolism of animals from hydrothermal vents and deep sea.

24b. Harris, D. A. *Bio-Systems* **14**:113–121, 1981. ATPase in evolution of cells.

25. Hartman, H. *J. Mol. Evol.* **4**:359–370, 1975. Origin and evolution of metabolism.

26. Higgins, I. J. et al. *Nature* **286**:561–564, 1982. Methanotrophic bacteria.

27. Holland, H. D. *Acta Geochimica et Cosmochimica* **36**:637–651, 1972. Geological history of sea water.

28. Hori, H. and S. Osawa. *Proc. Nat. Acad. Sci.* **76**:381–385, 1979. Evolution of 5 S RNAs and phylogenetic tree of 5 S RNA species.

29. Hoyle, F. *Evolution from Space*. Dent, London, 1981.

29a. Inoue, T. and L. E. Orgel. *J. Mol. Biol.* **162**:201–217, 1982. *Science* **219**:859–862, 1983. Models for RNA polymerization.

30. Jannasch, H. W. and C. O. Wirson. *Biosci.* **29**:592–598, 1979. *J. Bact.* **149**:161–165, 1982. *Oceans* **27**:73–78, 1984. Chemolithotrophic bacteria. pp. 677–709 in *Hydrothermal Processes at Seafloor Spreading Centers,* Ed. P. Rona et al., Plenum, New York, 1984.

31. Jukes, T. H. *Nature* **246**:22–26, 1973. Origin of genetic code.

32. Knaff, D. B. Ch. 32, pp. 629–640 in *Photosynthetic Bacteria,* Ed. R. K. Clayton and W. Sistram. Plenum, New York, 1979. Patterns of bacterial photosynthetic cycles.

33. Kobayashi, K. et al. pp. 421–426 in *Origin of Life,* Ed. H. Noda, Academic Publishers of Japan, Tokyo, 1978. Bacterial metabolism.

34. Küntzel, H. and H. G. Köchel. *Nature* **293**:751–755, 1981. Origin of mitochondria.

35. Lehninger, A. L. *Principles of Biochemistry*. Worth, New York, 1982.

35a. Lewin, B. *Gene Expression,* vol. 1, Wiley, New York, 1974. Models of tRNAs and rRNAs.

36. Liu, C. L., N. Hart, and H. Peck. *Science* **217**:363–364, 1982. *J. Bacteriol.* **145**:966–973, 1981. Sulfur metabolism of nonphotosynthetic bacteria.

37. Lowe, D. R. *Nature* **284**:441;–443, 1980. Australian stromatolites.

38. Luehrsen, K. R. and C. Woese. *Current Microbiol.* **4**:123–126, 1980. rRNA of *Tetrahymena thermophila*.

39. Mah, R. A. et al. *Annu. Rev. Microbiol.* **31**:309–341, 1977. Biogenesis of methane. *Origin of Eukaryotic Cells,* Yale University Press, New Haven, CT, 1970. *Symbiosis in Cell Evolution,* Freeman, San Francisco, 1981.

40. Margulis, L. *Evol. Biol.* **7**:45–78, 1974. *Origin of Eukaryotic Cells,* Yale University Press, New Haven, CT, 1970. *Symbiosis in Cell Evolution,* Freeman, San Francisco, 1981.

41. Matthews, C. et al. *Science* **198**:622–624, 1977. Evidence for HCN polymers as protein ancestors.

42. Mauzerall, D. *Ann. New York Acad. Sci.* **206**:484–494, 1973. Origin of porphyrins in chlorophyll.

43. Miller, S. L. and L. E. Orgel. *Origins of Life on Earth.* Prentice-Hall, Englewood Cliffs, NJ, 1974.

44. Mills, D. R. and S. Spiegelman. *Science* **180**:916–927. *Proc. Nat. Acad. Sci.* **58**:216–224, 1967. Phage RNA as template for synthesis with replicase.

45. Nanney, D. L. Molecular diversity and evolutionary antiquity of the *Tetrahymena pyriformis* species complex. In press.

46. Olson, J. M. pp. 1–37 in *Evolutionary Biology,* Ed. M. K. Hecht and W. Steere, Rockefeller University Press, New York, 1978. Precambrian evolution of photosynthesis. Lewis, B. *Gene Expression,* vol. 1, Wiley, New York, 1974. Models of tRNAs and rRNAs.

46a. Orgel, L. E. et al. *J. Mol. Evol.* **17**:295–306, 1982. Synthesis of oligonucleotides with nucleotide template.

47. Padan, E. *Annu. Rev. Plant Physiol.* **30**:27–40, 1979. *Cyanobacteria.*

48. Paere, H. W. and P. E. Kellar. *Science* **204**:620–622, 1979. Blue green alga fixes N_2 and CO_2 and produces oxygen.

49. Pang, H. et al. *J. Biol. Chem.* **257**:3589–3592, 1982. Transfer RNA in archaebacteria.

50. Pfennig, N. *Annu. Rev. Microbiol.* **31**:275–290, 1977. Phototrophic bacteria.

50a. Ponnamperuma, C., Ed. *Exobiology,* North-Holland, Amsterdam, 1972.

51. Rao, K. K. et al. Ch. 5, pp. 150–202 in *Biochemical Evolution,* Ed. H. Gutfreund, Cambridge University Press, New York, 1981. Patterns of photosynthesis in bacteria.

52. Ruby, E. G. and H. W. Jannasch. *J. Bacteriol.* **149**:161–165, 1982. *Appl. Environ. Microbiol.* **42**:317–324, 1981. Chemolithotrophic sulfur-oxidizing bacteria.

52a. Schopf, W., Ed. *Earth's Earliest Biosphere.* Princeton University Press, Princeton, NJ, 1983.

53. Schuster, P. Ch. 2, pp. 15–87 in *Biochemical Evolution,* Ed. H. Gutfreund, Cambridge University Press, New York, 1981. Prebiotic evolution.

54. Schwartz, R. M. and M. O. Dayhoff. *Science* **199**:295–403, 1978. Origin of mitochondria from prokaryotes.

55. Searcy, D. G. *Trends Biochem. Sci.* **7**:183–185, 1982. Thermoplasma, a primordial cell.

56. Snell, E. E. *T.I.B.S.* **2**:131–135, 1977. Pyruvate containing enzymes.

56a. Somero, G. N., J. Siebenaller, and P. Hochachka. Ch 7, pp. 261–330 in *The Sea,* vol. 8, Ed. G. Rowe, Wiley, New York, 1983. Metabolism of hydrothermal vent organisms.

57. Spiegelman, S. and I. Haruna. *Proc. Nat. Acad. Sci.* **55**:1539–1554, 1966. *Q. Rev. Biophys.* **4**:213–253, 1971. A rationale for an analysis of RNA replication.

58. Spiegelman, S. et al. *Proc. Nat. Acad. Sci.* **58:**217–224, 1967; **59:**139–144, 1968. *Science* **150:**884, 1965; **180:**916–927, 1973. *In vitro* synthesis of viral synthesis of RNA in presence of replicase.

59. Southward, A. J. et al. *Nature* **293:**616–619, 1981. Bacteria from deep-sea worms.

60. Stahl, D. et al. *Nucl. Acids Res.* **9:**6129–6137, 1981. Structure of 5 S rRNA from *Sulfolobus, Escherichia,* and *Homo.*

61. Stanier, R. Y. and G. Cohen-Bazire. *Annu. Rev. Microbiol.* **31:**225–275, 1977. Biology of *Cyanobacteria.*

62. Sumper, M. and R. Luce. *Proc. Nat. Acad. Sci.* **72:**162–166, 1973. De novo production of self-replicating RNA.

63. Towe, K. M. *Nature* **247:**657–661, 1978. Evolution of oxygen in atmosphere.

64. Truker, H. G. Ch. 35, pp. 677–690 in *Photosynthetic Bacteria,* Ed. R. K. Clayton and W. R. Sistron, Plenum, New York, 1979. Sulfur metabolism in bacteria.

65. Van Valen, L. M. and V. C. Maiorana. *Nature* **287:**248–250, 1980. Origins of archaebacteria, eubacteria, and eukaryotes.

66. Walter, M. R. *Am. Sci.* **65:**563–571, 1977. *Science* **178:**402–405, 1972. *Stromatolites,* Ed. M. R. Walter, Elsevier, New York, 1976.

66a. Weber, A. L. and S. L. Miller. *J. Mol. Evol.* **17:**273–284, 1981. Reasons for 20 amino acids in proteins.

66b. Woese, C. R. *Naturwissenschaften* **60:**447–459, 1973. Evolution of genetic code.

67. Woese, C. R. et al. *Nature* **254:**83–86, 1975. *Sci. Am.* **244:**98–122, 1981. *J. Mol. Evol.* **10:**1–6, 1977. *Microbiol. Rev.* **47:**596–644, 1983. Ch. 11, pp. 209–233 in *Evolution from Molecules to Man,* Ed. D. S. Bendall, Cambridge University Press, New York, 1983. Archaebacteria and evolution of prokaryotes. Ribosomal RNAs.

68. Woese, C. et al. *Nature* **283:**212–214, 1980. *Nucl. Acid Res.* **9:**6129–6137, 1981. 16 S rRNA in purple photosynthetic bacteria.

69. Wolfe, R. S. et al. *Microbiol. Rev.* **43:**260–296, 1979. *Adv. Microbiol. Physiol.* **6:**107–176, 1972. Methanogenic bacteria.

69a. Wraight, C. A. Ch. 2, pp. 17–61 in *Photosynthesis* I, Ed. Govindjee, Academic, New York, 1982.

70. Zeikus, J. G. *Bact. Rev.* **41:**514–541, 1977. *Annu. Rev. Microbiol.* **34:**423–464, 1980. Biology of methanogenic bacteria.

3

ADAPTATIONAL GENETICS

Self-replication is characteristic of all living organisms. Replication and the genetic code, selective membranes, and the enzymes for anabolism and catabolism all probably evolved during the same period. The evolution of sexual reproduction in eukaryotes extended the genetic range of metabolic patterns. The nucleolus and the condensation of DNA with histones into chromosomes accompanied the increase in genetic capability. The evolution of the genetic code, of the transcription process from DNA to RNA, of the sequence of reactions in protein synthesis, and of developmental and tissue differences in gene expression all remain to be fully elucidated. What determined the α-helix structure of DNA? How and in what order did the several kinds of RNA become established? What was the origin of mitochondrial and chloroplast genomes? The presence of these mechanisms in nearly all organisms indicates that the total pattern which evolved was so versatile that, with relatively few changes, it served the needs of most organisms, both prokaryotes and eukaryotes, in diverse environments.

The principles of genetics are based on the Mendelian laws of inheritance and on the linear arrangement of heritable units on chromosomes. Genetic plasticity resides in point mutations and in structural changes, such as gene transpositions, crossings-over, deletions, and insertions. The genome, according to this traditional view, is quasi-static, and genetic change is slow and unresponsive to the environment. Each gene, especially in eukaryotes, consists of long stretches of nucleic acids (exons) that encode for the messenger RNA which specifies amino acid sequences of proteins. Each gene also contains introns that do not encode RNAs. The genome is much more dynamic than was formerly perceived. The techniques of recombinant DNA and cloning will lead to changes in interpretation of much of the data presented in this book. This chapter summarizes background information about genetic coding and protein synthesis, then presents some of the principles of molecular genetics and suggests their bearing on physiological adaptation.

NUCLEIC ACIDS

DNA

The genetic material is DNA, a double-stranded helix consisting of chains of alternating deoxyribose and phosphate. The origin of double stranding of DNA and RNA is very ancient; in some viruses in which genetic material is in single strands of RNA, the strands are doubled during replication (69). Connecting the two chains are pairs of nucleotides, five for each turn of the helix, like rungs of a ladder (Fig. 3-1) (58). Each rung consists of two of the four nucleoside bases: a purine (either adenine, A, or guanine, G), hydrogen-bonded to a pyrimidine (either cytosine, C, or thymine, T). Each nucleotide base is attached to one sugar-phosphate; the distance between the two sides of the helix is 20 Å. Internucleotide distance is 3.4 Å, which is near the interpeptide bond distance in a protein (3.6 Å). The distance per turn of the right-handed helix is 34 Å or 10 residues. The ratios of nucleotides are 1:1, guanine to cytosine, and adenine to thymine. DNAs rich in G and C are more dense than those with much A and T. During replication the chains become separated by an unzipping process; enzymes, one of them a DNA polymerase, with energy from ATP provide for production of a complementary chain. In diploid cells each chromosome consists of two pairs of double-helical DNA chains (chromatids); replication of the chains occurs prior to mitosis.

The alpha-helix is a right-handed double strand. A left-handed Z-DNA has been discovered in crystals of tetranucleotide alternating d(G-C) and d(C-G) residues with antiparallel sugar-phosphates, 12 base pairs per turn. Histone proteins bind to the Z-DNA (80,89) as well as to right-handed strands. Evidence that Z-DNA is present in natural DNA is the fact that antibodies for left-handed Z-DNA bind to interband regions of Drosophila polytene chromosomes. Z-DNA may be a conformational switch in control of transcription (64). Other nonstandard DNA structures may occur in equilibrium with the Z-DNA (89).

RNA differs from DNA in having ribose rather than deoxyribose as the sugar moiety and uridine for thymine. The code which is transcribed to messenger RNA (mRNA) for different proteins is contained in the nucleotides of the DNA chain. Twenty "essential" amino acids are used in proteins and four nucleotides are used in RNAs of all organisms. One nucleotide could code for only four amino acids. Two nucleotides per amino acid (aa) could code for 4^2, or 16 amino acids. The generalized code consists of three nucleotide bases, which can give 4^3, or 64 possible combinations. Some triplets are nonsense codes and do not designate an amino acid. Three sequences are used for termination and one for initiation of protein synthesis (48). In transcription to RNA, the nucleotides of mRNA are paired to complementary nucleotides of a strand of DNA. The mRNA is a complementary copy of the DNA sequence from which it is transcribed. Each triplet specifies a

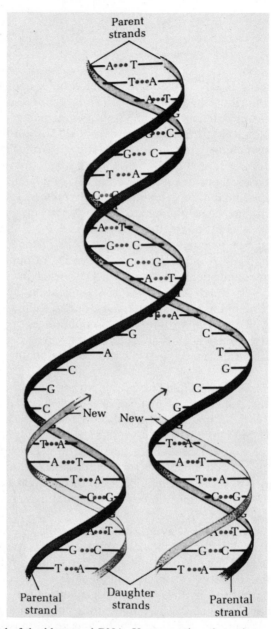

Figure 3-1. Model of double strand DNA. Upper portion shows intact molecule, middle portion separation of strands, and lower portion replicated complementary strands. From A. L. Lehninger, *Principles of Biochemistry,* Worth Publishers, New York, 1982, p. 808.

certain amino acid, and the first two nucleotides are usually similar or the same (except for Leu, Val, and Ser); the third nucleotide is more variable (Fig. 3-2). The third base by its wobble can allow recognition of other anti-codons. An archetype code may have had 16 anticodons for 15 amino acids (54a). Only 16 amino acids could have been coded by a doublet code; the third letter permitted additional amino acids and termination sequences. Proteins may have been formed by polymerization before nucleic acids appeared, but protein synthesis as it occurs in prokaryotes and eukaryotes could hardly have preceded DNA and RNA. There is redundancy in that some amino acids are encoded by 6, 4, 3, 2, or 1 triplet as follows:

1 codon	Met, Trp
2 codons	Glu, Gln, Asp, Asn, Cys, Lys, His, Tyr, Phe
3 codons	Ile
4 codons	Val, Pro, Thr, Ala, Gly
6 codons	Leu, Ser, Val, Arg

All codons with U as the second letter are for amino acids with nonpolar or hydrophobic side chains—Phe, Leu, Ile, Met, Val. Codons with the first two letters G and A code for negatively charged amino acids—Asp, Glu. DNA high in GC is more resistant to ultraviolet light and is in general more stable than DNA rich in AU. The DNA of bacteria is generally high in GC; DNA of eukaryotes is lower in GC content (69).

SECOND LETTER

FIRST LETTER	U	C	A	G	THIRD LETTER
U	UUU UUC } Phe UUA UUG } Leu	UCU UCC UCA UCG } Ser	UAU UAC } Tyr UAA Chain End UAG Chain End	UGU UGC } Cys UGA Chain End UGG Trp	U C A G
C	CUU CUC CUA CUG } Leu	CCU CCC CCA CCG } Pro	CAU CAC } His CAA CAG } Gln	CGU CGC CGA CGG } Arg	U C A G
A	AUU AUC AUA } Ile AUG Met	ACU ACC ACA ACG } Thr	AAU AAC } Asn AAA AAG } Lys	AGU AGC } Ser AGA AGG } Arg	U C A G
G	GUU GUC GUA GUG } Val	GCU GCC GCA GCG } Ala	GAU GAC } Asp GAA GAG } Glu	GGU GGC GGA GGG } Gly	U C A G

Figure 3-2. Representation of polynucleotide sequences for 1-, 2-, or 3-letter codes.

Transcription to RNA is accomplished by a number of DNA-dependent RNA polymerases. The 5' end of RNA is aligned opposite the 3' end of the DNA template and bases are paired as follows:

DNA	mRNA
thymine	adenine
adenine	uracil
cytosine	guanine
guanine	cytosine

Prokaryotes have only one type of RNA polymerase; eukaryotes have at least three types of nuclear RNA polymerase; mitochondria and chloroplasts have one each. In transcription the polymerase moves along a DNA molecule from the 3' to 5' end and only one strand of DNA is transcribed. The rate of transcription is high; 50 nucleotides are incorporated per second in multiplying bacteria. In eukaryotes, polymerase I transcribes the 18 S and 28 S rRNA, polymerase II transcribes mRNA, and polymerase III transcribes tRNA and 5 S rRNA (48).

Messenger RNA

Messenger RNAs vary in number according to coded proteins in different cell types; there may be many thousand different mRNAs in a multicellular organism (68). mRNAs differ in length approximately according to the size of coded proteins; they are large molecules with Svedberg sedimentation (S) values (a measure of size) in the range from 6 S to 30 S and they average 1200 nucleotide pairs per molecule. The turnover rate of mRNA is high; in eukaryotes the $t_{1/2}$ usually averages about 20 hours, in bacteria a few minutes. In *E. coli* the mRNA makes up 3 percent of the total RNA; the remainder is tRNA and rRNA. In one polysomal mRNA from sea urchin *Strongylocentrotus* blastula there are some 13,000 mRNAs corresponding to 2 percent of the total genome. The ratio $(A + U)/(G + C)$ in mRNA is similar to that of $(T + A)/(C + G)$ in DNA but is higher than the ratio in ribosomal RNA (68,69). The AU content of mRNAs is higher in eukaryotes than in prokaryotes and species differences are considerable. Messenger RNAs may be retained briefly in the nucleus and show little or no post-transcriptional modification of base sequences, but they are extensively processed, for example, by addition of 3' poly-A tails, 5'7-methyl-G cap, methylation, elimination of introns, and splicing. Not only is mRNA used for amino acid coding in protein synthesis, but mRNA strands contribute to the structure of polyribosomes during protein synthesis, since several ribosomes can be attached to the mRNA at one time.

Ribosomal RNA

Ribosomal RNAs (rRNAs) are mostly transcribed as large molecules; these are cleaved and complexed with protein to form ribosomes on which protein synthesis occurs (69). In eukaryotes a 5 S rRNA passes from nucleus to cytoplasm, apparently without being processed; larger rRNAs are processed from a single large precursor synthesized in the nucleolus. In some insects, fishes, amphibians, reptiles, and plants the primary rRNA transcript is 34–40 S, in birds and mammals it is 40–45 S. In the nucleolus many nucleotides become methylated and in some nucleotides U is replaced by pseudo-U. The ribosomes (rRNA + protein) of eukaryotes are approximately 80 S (mammals) and dissociate to 60 S and 40 S particles. In prokaryotes the ribosomes are 70 S and dissociate to 50 S and 30 S particles.

The number of ribosomes in a cell varies according to the amount of protein being synthesized. A single *E. coli* may contain 15,000 ribosomes during its active growth phase. Ribosomal RNAs constitute as much as 90 percent of total RNA in actively metabolizing cells. Genes for the different rRNA subunits are highly redundant—some 200 for each. In *E. coli* the 30 S subunit contains a 16 S rRNA (1542 nucleotides) plus a copy of each of 21 different proteins. The 50 S subunit contains a 5 S rRNA (120 nucleotides), a 23 S rRNA (2904 nucleotides) plus a single copy of 13 different proteins and 4 copies of one unidentified protein (114). The rRNA genes are highly conserved, and there are homologies in archaebacteria, eubacteria, and eukaryotes (Fig. 2-2) (Ch. 2). Ribosomal proteins may facilitate protein synthesis and may stabilize the helices. Translation can continue after some ribosomal proteins are lacking (79).

Transfer RNA

Transfer RNAs (tRNAs) contain 73–93 (average 80) nucleotides; thus tRNAs are small molecules, with sedimentation coefficient 4 S (69). There are 30 to 100 different tRNAs; a minimum of 20 transcribe the essential amino acids, several tRNAs for each of most amino acids, three tRNAs for termination of polypeptide sequences, and one tRNA for initiation. The number of tRNAs is slightly different in different organisms. tRNAs undergo posttranscriptional modification, such as methylation, pseudo- and dihydro-uracylation. In eukaryotes, the tRNA genes are transcribed via RNA polymerase III (a large protein of 10 subunits) to form the various tRNA precursors. The tRNA precursors are large molecules which are cleaved by nucleases and modified to become mature 4 S tRNAs. The tRNAs differ not only in nucleotide sequence but also in tertiary structure. tRNAs are translocated from the nucleus by a carrier-mediated process. In *Xenopus* oocytes the maximum rate of transfer is 19×10^8 mol/min (117). Many tRNAs may have a common carrier (114).

Molecules of tRNA are folded in cloverleaf fashion (Fig. 2-1) (Ch. 2). In folding there is pairing of nucleotides, purines opposite pyrimidines. The 3′ end has CCA, and below this is a region rich in AT; the adenine becomes amino-acylated for attachment of the amino acid. At the 5′ end of the molecule is a guanine. Below the 3′ end is the T-C arm where a synthetase can attach. The third arm is the anticodon region where three nucleotides attach to the codon of mRNA. Between arms II and III some tRNAs have an extra arm. At the beginning of protein synthesis, specific amino acids (aa) are activated by formation of an adenylated amino acid complex with a synthetase:

$$ATP + aa + synthetase \rightarrow synthetase\text{-}aa\text{-}AMP + 2\ P_i$$

$$synthetase\text{-}aa\text{-}AMP + tRNA \rightarrow aminoacyl\text{-}tRNA + enzyme + AMP$$

Not only are there specific tRNAs for most amino acids, but also there is an activating aminoacyl synthetase for each amino acid. In some eukaryotes there are five tRNAs each for Leu and Ser. However, the same aminoacyl-tRNA synthetase serves all tRNAs for any one amino acid.

RNAs: Nuclear-Heterogeneous, Mitochondrial, and Chloroplast

In the nucleus of eukaryotes there are heterogeneous RNAs (hnRNAs), the function of which is not agreed upon. hnRNAs vary in size; in mammals the number of bases ranges from 1000 to 50,000 (69). For example, globulin hnRNA has 1500 bases. Some 90 percent of hnRNAs are retained in the nucleus and do not enter the cytoplasm.

Mitochondria and chloroplasts have messenger rRNAs (mtmRNAs and cmRNAs) which are coded by mitochondrial and chloroplast DNAs. Mitochondrial and chloroplast DNAs are circular. Yeast mitochondrial DNA has 78,000 base pairs (bps), human has 16,569, and mouse 16,295 bps (11). Mitochondria and chloroplasts also have their own RNAs and tRNAs. In yeast the circular mtDNA has been mapped; it encodes for 15 S rRNA and 21 S rRNA (Fig. 3-3). Yeast mitochondria have 22 tRNA genes, *Euglena* chloroplasts have 26 tRNA genes. In the alga *Chlamydomonas* chloroplast circular DNA is 7 percent of the total DNA. The yeast mitochondrial genome has many introns; for example, the gene for one subunit of cytochrome oxidase has 7 introns. In complex eukaryotes, cytochrome oxidase consists of 7 subunits, 3 coded by mitochondrial and 4 by nuclear DNA (8,17,18). The mitochondrial genome encodes for large and small mRNAs, for 22 tRNAs (30 percent fewer than in nuclei), and 15 mRNAs. Of the latter, 5 are not translated, 6 are for unidentified proteins, 3 for 3 subunits of the cytochome oxidase complex, a mtATPase and cytochrome (16a).

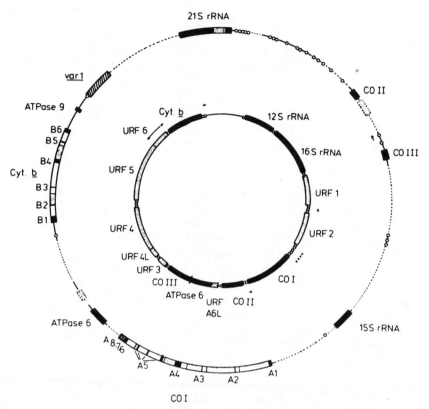

Figure 3-3. Organization of yeast (inner) and human (outer) mitochondrial genomes. Human genome drawn to twice "normal" size relative to yeast. DNA sequences encoding known gene products indicated by solid bars, unassigned reading frames by stippled areas, tRNAs by open circles. Reprinted with permission from P. Borst and L. Grivell, *Nature*, Vol. 290, p. 443. Copyright 1981, MacMillan Journals Ltd., London.

Animal mtDNAs are of fairly constant size—approximately 15 kilobase pairs (11,16a). Nucleotide sequences of human, bovine, and mouse mtDNA indicate a compact organization with no intervening sequences. The mitochondrial ribosomes of *Neurospora* are 73 S or 80 S in size, varying with the preparation procedure. Mitochondrial ribosomes are less elongated than cytoplasmic ribosomes. The mitochondrial ribosomal RNAs (mtrRNAs) have lower G and C content than those of cytoplasm. Yeast mtrRNAs are lower in GC than cytoplasmic rRNAs and lower also than those of *E. coli* (11). The size of mtDNA is similar for sea urchin, flatworm, Drosophila, and man; in yeast it is five times as large. Yeast mitochondrial mRNA for cytochrome B has a leader sequence of 100 nucleotides. Mammalian mtRNAs differ markedly with species. Human and bovine mtRNAs show 74 percent homology (8).

The mitochondrial genomes of higher plants are large, and nuclear and mitochondrial codes may differ; tryptophan, for example, in maize is coded in nuclei by UGG, in mitochondria by CGG. Fungal and mammalian mtrRNA use UGA to code for Tyr rather than as a termination signal. Yeast mitochondria use CUA (the usual nuclear code for Leu) to code for Thr. Mitochondrial tRNAs use AGA and AGG as stop signals rather than for coding Arg; they use AUA for Met rather than for Ileu. Thus, the genetic code is not universal; mitochondria have features like prokaryotes and eukaryotes as well as unique ones. The early symbionts that became mitochondria may have been very different from present-day mitochondria in their DNAs and RNAs. Exceptions to the universal genetic code occur also in ciliate protozoans (42a).

Chloroplasts contain a set of genomic components. Maize chloroplasts code for the ribosome sequence for ribulose-bi-p-carboxylase (77). This enzyme, Ru-b-P-case fixes CO_2 and is the most abundant protein in corn leaves, probably the most abundant protein in all living matter. The protein consists of 8 small units (13,000 da) and 8 larger units (52,000 da). The small units are coded by the nucleus, the large ones by chloroplasts. Chloroplasts use UGG to code for Try, while mitochondria use UGA (77) (4).

Mitochondria are transmitted in ova, not in sperm; mitochondrial inheritance is maternal. There are many more mitochondria per cell than nuclei, hence there is much redundancy in the mtRNA. In mammals (*Homo, Peromyscus*) mitochondrial DNA variants are more frequent and occur in fewer generations than nuclear DNA variants. Phylogeny based on mitochondrial DNA is unlike that based on nuclear DNA. Mitochondrial DNA crosses species boundaries because it is not linked to reproductive isolation. Primate mtDNAs differ between individuals more than nuclear DNAs. Genetic changes in mtDNA are more frequent than in nuclear DNA and it is estimated that mtDNA evolves some 8 to 16 times faster than the nuclear genome (4a).

DNA fragments from both mitochondria (mt) and chloroplasts (cht) of mung bean, pea, spinach, and corn were cloned. All chloroplast DNA clones could hybridize to one mtDNA fragment. Chloroplast and mitochondrial DNA sequences are more similar for closely related species than for distantly related species. Segments of mtDNA can hybridize to segments of chtDNA. It is probable that transposition has occurred between the two kinds of organelles relatively recently (107a).

PROTEIN SYNTHESIS

The first step in protein synthesis is the activation of an amino acid to form an aminoacyl bond to a tRNA. The subsequent sequence is initiation, elongation, and termination (Fig. 3-4). These events are best known for prokaryotes such as *E. coli* (66).

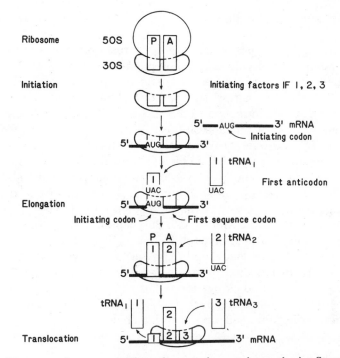

Figure 3-4. Diagrammatic representation of events in protein synthesis. Steps 1–3 formation of initiation complex; step 4 elongation; step 5 translocation. Modified from A. L. Lehninger, *Principles of Biochemistry,* Worth Publishers, New York, 1982, pp. 884, 886, and 889.

Initiation

Protein synthesis begins with an initiation complex which includes methionyl-tRNA, a ribosome subunit, an mRNA, and initiation factors (proteins). In prokaryotes, initiation is by a formyl Met tRNA. Mitochondria and chloroplasts also use formyl Met tRNA. Eukaryotes use a similar tRNA for initiation but their Met is not formylated. The 70 S ribosome dissociates to 50 S and 30 S rRNA components; an initiating factor (IF3) combines with the 30 S fragment; this binds to the appropriate mRNA and a second initiating protein (IF1) is added. Then the tRNA with Met plus a third initiating factor (IF2) and GTP (for energy) are added to the complex, and the anticodon of the tRNA is attached to mRNA at the code AUG, which is the codon for Met. Then the 50 S ribosomal fragment combines with the complex, the three IFs leave, and translation from the mRNA starts.

Elongation and Translocation

A ribosome has two types of active sites for reaction with tRNAs: (1) P-sites where a peptidyl tRNA donates a polypeptide chain to the next aminoacyl tRNA; and (2) A-sites where a transfer complex positions an incoming aminoacyl tRNA. Elongation is by addition of amino acids in series, and the resulting dipeptide is attached to the second tRNA in the A-site, while the deacylated Met tRNA remains on the P-site. With the Met-tRNA on the P-site of the 70 S ribosome, the next tRNA with its activated amino acid binds with an elongation factor (EF-T) plus GTP to the 70 S ribosome at the A-site. The anticodon of the tRNA (with its aminoacyl attached) becomes positioned to the proper codon on the mRNA. For formation of the first peptide linkage the amino group of the second aminoacyl tRNA attaches to the esterified carboxyl of the Met-tRNA. Polypeptide chain formation begins at the amino terminal of the polypeptide.

In translocation the deacylated tRNA in the P-site is released and methionyl tRNA is translocated from the A-site to the vacant P-site, a process which requires a second elongation factor EF-2 and GTP. As the process is repeated for successive aminoacyl-tRNAs, the mRNA tracks through a groove in the ribosome and a polypeptide chain is constructed.

Termination

Termination occurs when a particular signal in the mRNA is reached. Termination signals are UAA, UAG, or UGA. In prokaryotes three releasing proteins bind to the ribosome, the polypeptide chain is released, the mRNA and the last tRNA are also released, and the 70 S ribosome is ready to dissociate into its 50 S and 30 S subunits to start a new initiation process.

DYNAMIC GENETICS

Genetic change is by (1) point mutations, (2) duplications, and (3) chromosome rearrangements. These three kinds of genetic change determine protein polymorphism. Structural genes fix amino acid sequences and contribute to microevolution; chromosomal rearrangements lead to the macromutations of major genetic change. In geological time domains some protein families evolve very slowly (histones), others rapidly (globins).

Genomes are dynamic and change in many ways, sometimes rapidly. From amino acid sequencing of proteins, corresponding mRNA maps have been constructed from which DNA sequences can be deduced. However, ambiguity of the genetic code makes this imprecise. Instead of starting with proteins, it is now possible to determine for a number of genes the coding sequences which are used to designate primary patterns in proteins. Most eukaryotic genes have much longer noncoding than coding regions. As a result of DNA analyses, far more genomic diversity is recognized than was envisioned from earlier genetic methods.

ISOLATION AND SEQUENCING OF mRNA AND GENES

For preparation of messenger RNA and DNA, a tissue which synthesizes a given protein in quantity is used as source, for example, parathyroid glands for parathyroid hormone RNA, pancreatic islet cells for insulin RNA, reticulocytes for hemoglobin RNA, gastric mucosa for gastrin RNA (76). To prepare mRNA (Fig. 3-5), tissue is homogenized and total RNA is isolated from protein by protease digestion or by extraction with denaturing agents, such as phenol, guanidinium salts, or cesium chloride, or by a combination of these methods. For most eukaryote proteins mRNA can be separated from the ribosomal RNA by affinity chromatography on gel columns of oligo(dt)cellulose or poly(U)sepharose. The mRNA can be fractionated further by size in sucrose gradient centrifugation or gel electrophoresis. To assay for specific mRNAs, the mRNA may be translated in a cell-free protein synthesis system, such as a wheat germ extract, which has tRNAs, ribosomes, initiation and elongation factors and releasing factors. Addition of appropriate mRNA plus radiolabelled amino acids results in labelled protein, which may be identified by immunoprecipitation with antibody raised against the protein and/or by gel electrophoresis; the protein serves to identify the mRNA.

Once a single-stranded RNA (3′ at one end, 5′ at the other end) is isolated, treatment with a reverse transcriptase in the presence of a primer, usually oligo(dT) which hybridizes to the poly(A) region of the mRNA, produces a complementary DNA (cDNA) with reversed 3′ and 5′ ends. The single-stranded DNA is then made double-stranded by DNA polymerase which uses a hairpin loop at the 3′ end of the cDNA as primer. Double-stranded cDNA may be produced by this method in very small quantities and is not completely pure. To obtain large quantities of essentially pure sequence the double-stranded cDNA is amplified and isolated by recombinant DNA technique (molecular cloning) (Fig. 3-5). This involves insertion of the cDNA into a plasmid DNA to form a recombinant molecule which is then used to transform bacteria. Transformed bacteria express and replicate only one plasmid; thus selection of individual colonies of transformed bacteria allows the "cloning" of individual DNA molecules. The pure cDNA may then be used as a hybridization probe to assay for mRNA or for the nuclear genes (1). This is useful in studies of gene regulation involving changing levels of mRNA, the numbers of genes, the structure of the gene, and as an assay to isolate the gene from the total genomic DNA. The cDNA may also be used to determine the nucleotide sequence of its complementary mRNA. At present this is generally done by a chemical method (76) or by a chain termina-

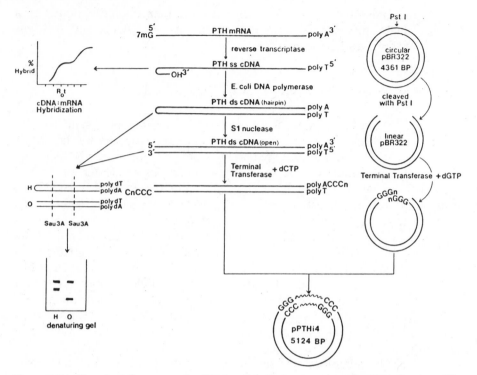

Figure 3-5. Flowchart for sequencing DNA and deducing sequence of mRNA for a specific protein. Example of parathormone, PTH. Procedure starting with PTH mRNA shown at left. The steps are: (1) conversion of mRNA by action of reverse transcriptase to cDNA; (2) action of DNA polymerase to form double strand cDNA with hairpin loop; (3) breaking loop to give two strands of cDNA; (4) adding terminal nucleotides (CC). Preparation of plasmid shown at right. (1) Cut circular plasmid with enzyme to give linear strands. (2) Add terminal nucleotides (GG) by aid of transferase. Final step (lower middle) is annealing the single strands of cDNA with CCC ends attaching to complementary strands of plasmid with GGG ends. Plasmid is then grown in E. coli and DNA harvested, the bacterial DNA removed, and cloned DNA treated with restriction enzymes which cleave at different points. The resulting fractions are identified by electrophoresis. From personal communication with B. Kemper.

tion enzymatic method (94,95). In the chemical method specific fragments of DNA are obtained after digestion with restriction endonucleases. The DNA then undergoes a limited chemical cleavage by reactions that are specific for each of the four bases. For each reaction a set of fragments of differing length are produced that are terminated by a specific base. The sizes of the resulting fragments are ascertained by gel electrophoresis and autoradiography of labelled fragments, from which a nucleotide sequence can be deduced. Some 350 restriction endonucleases with 85 recognition specificities are available (87a).

In the Sanger method, *E. coli* DNA polymerase I is used to synthesize a complementary sequence of a single-stranded DNA sequence by extension of a small fragment used as a primer. In each of four reactions the deoxynucleotide triphosphate analog of one of the bases is added, which causes an occasional termination of replication and thus produces a set of fragments terminated at a given base similar to that produced by the chemical method. As in the chemical method the sequence can be deduced from the sizes of the terminated fragments.

Isolation of genes poses a formidable technical problem. In mammals, which contain about 3×10^9 bp (base pairs) a unique copy gene is about 1 part in 3×10^6, making isolation by using separation techniques based on physical properties an impossible task. This problem has been solved by using the informational content of the gene as a selection device. In essence this is done by using a specially constructed λ bacteriophage containing DNA, in which a middle segment is not needed for replication of the phage. The middle DNA may be replaced by any DNA, for example a mammalian DNA. Populations of λ phage produced in this way carry DNA which is representative of essentially a complete genome of a given species, and are called "libraries." The DNA fragment corresponding to a specific gene can be selected by hybridizing DNA from phage plaques to a radioactive DNA probe. Libraries for numerous gene species have been constructed.

PROPERTIES OF DYNAMIC GENOMES

Nucleotide sequences have been ascertained for genes of a number of prokaryotes and eukaryotes. In prokaryotes, the sequences in genes are colinear with their mRNA and protein products. In eukaryotes the genes are often noncolinear. In eukaryotes, but not in prokaryotes, coding regions of DNA (exons) are frequently interrupted by noncoding regions (introns or intervening sequences). The sequences in the middle of introns may be comparatively variable, sequences at the ends more conserved (Fig. 3-6). Introns may prevent crossing-over, and appear to be required for expression of mature mRNA for coding some proteins, such as globin mRNA, though not for other mRNAs. Exons occur in domains; rearrangement of domains by introns makes for imprecise recombinations. Intron sequences are found in DNA, not in mRNA. Introns may be eliminated as exons are spliced together during transcription. Introns may be transcribed during the initial stages of production of messenger RNA but are processed out at some stage before the mRNA reaches a ribosome.

At the 5' end of a gene sequence is an initiation sequence, the TATA (thymidine-adenine) box, sometimes also a CCAT box. In *E. coli* the initiation codon at the 5' end is AUG, and the termination sequence at the 3' end is usually UGA. The noncoding or intervening sequences often begin with GT or GU and end with AG. The different coding regions may correspond

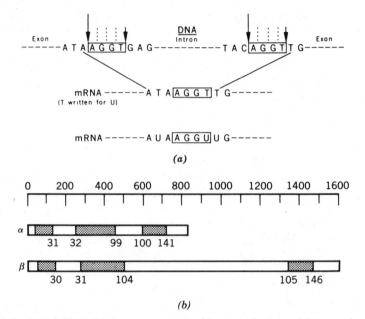

Figure 3-6. (*a*) Model indicating arrangement of intron and exons with transcripts, excision of intron, and splicing of exons. Reprinted with permission from F. Crick, *Science,* Vol. 204, p. 267. Copyright 1979 by the AAAs. (*b*) Fine structure of α and β globin genes. Solid and open boxes represent exon and intron sequences. The α-like globin genes contain introns of approximately 95 and 125 bp between codons 31 and 32, and 99 and 160. The β-like globin genes contain introns of approximately 122–130 and 850–900 bp located between codons 30 and 31, and 104 and 105. Reprinted with permission from N. J. Proudfoot, *Science,* Vol. 209, p. 1330. Copyright 1980 by the AAAS.

with specific regions of a protein. The exact mechanisms of removal of intervening sequences and splicing of coding sequences are not well understood. Splicing requires ATP and a transcriptase (20).

The noncoding introns account for genomes that are much longer than is necessary for mRNAs. Histone and interferon genes have no introns, whereas collagen genes have approximately 55 introns (4). The gene for human β tubulin lacks introns. The β-globin gene of mouse Hb has 2 introns of 116 and 646 base pairs (bps) and a coding sequence of 432 bps. In the globin family the number and location of introns is conserved; the size and sequences are not. All globin genes have 2 introns, and it is probable that these evolved before the α and β duplication (Ch. 4), some 500 myr ago. The dihydrofolate reductase gene of mouse has five intervening intron sequences; the entire gene is 32,000 bps (base pairs) long but only 568 bps serve to code mRNA (1). The protein ovalbumin (chicken) has 386 amino acids, mRNA of 1872 bps, of which 1158 bps code for amino acids; the ovalbumin gene has 7700 bps—four times as long as the mRNA and seven times as long as is required for the amino acids. There are two other ovalbumin-like genes, each with 8 exons and 7 introns; the exons are of the same length but the introns differ (21). The gene for alcohol dehydrogenase (ADH) of Drosophila has an intron of 65 bps between coding nucleotides 32 and 33 and an intron of 70 bps between codons 167 and 168 (85). Rats have two insulin genes, one with two introns, the other with one intron (69a).

Introns evolve faster than exons; insertions and deletions are common in introns. Splicing during evolution could provide new genes by fusion of duplicated transcription units. For example, a chicken gene for ovotransferrin which consists of 17 exons may have evolved through duplication by fusion of ancestral genes 7 or 8 exons long. Intragenic crossing-over between misaligned intron sequences could result in duplication of an exon, and this could lead to gene amplification or to generation of homologous exons (12).

In the dynamic genome there may be genes that move from one location to another—jumping or nomadic genes. These genes occur in multiple copies. Splicing occurs when one or more internal stretches become deleted and the remaining ends are joined. Genes called "copia" in *Drosophila melanogaster* occur in many copies, and there may be 20–30 families of these. Copias may constitute as much as 5 percent of the total Drosophila DNA and may occur in 20–40 sites. Copia occurs as short sequences—5000 bps—which are nomadic and repetitive; the first 276 bps at one end are repeated in inverted orientation at the other end (29,84). Another Drosophila nomadic gene codes for a heat shock protein (67). Some mutations in Drosophila appear to result from insertion of segments, such as copia, into a gene rather than from point mutation. In a yeast the codon for protein which synthesizes histidine—the his-4 locus—is a movable element. Movable genes occur also in bacteria. The shuffling of exons may result in assembly of new proteins by novel reassortments. When or how genes move from one location to another, in or between chromosomes, is not known; such transposition can occur in cells maintained in culture.

A repetitive element of yeast called sigma is located within 16–18 bps of the 5' end of several tRNA genes. Sigma occurs in different locations in related yeast strains (28). An example of dispersed repetitive sequences occurs in humans and rodents, the Alu family, so named because it is cleaved by the bacterial endonuclease Alu. In human DNA the Alu family constitutes 3 to 6 percent of the total DNA—a large fraction of the noncoding introns. It is estimated that the human genome has 500,000 copies of Alu sequences. Each is a short sequence, some 300 bps occurring in clusters. Alu DNA sequences of the green monkey can hybridize with human sequences. In several rodents the repetitive segments are 130 bps long. In human HeLa cells as much as 5 percent of hnRNA can be transcripts of the Alu family; the only function ascribed thus far to the Alu family is that some members are templates for an RNA polymerase (96).

Satellite DNAs are highly repetitive sequences for which no mRNA is complementary. Mouse satellite DNA amounts to 6 percent of nuclear DNA and occurs in condensed regions of every chromosome. Satellite DNAs are identified by use of restriction enzymes and may be relicts (16, 103).

Some movable genes are pseudogenes. A pseudogene resembles a coding gene but may have lost an intron, may have an extra intron, or may be inserted at a new location and be unable to code mRNA (75). Pseudogenes have been referred to as dead genes which are evolutionary relicts. Pseudogenes are frequent in eukaryotes, not in prokaryotes (43). Globin genes have pseudogenes, especially between the genes for embryonic and adult globins. In a mutant mouse, one pseudogene resembles an α-globin gene which lacks introns; functional genes are on chromosome 11, pseudogenes on chromosomes 15 and 17 (70).

Transforming genes can enter receptive cells and modify the genome. A gene from a human bladder tumor can be introduced into mouse fibroblasts in culture and transform the fibroblasts into cancer cells. Retroviruses use host tRNA as a primer for transcription of viral DNA. Reverse transcription is the transfer of genetic information from RNA to DNA; it occurs normally in viruses, especially of plants. It has been suggested that Alu genes of mammals may be processed genes in which self-copying RNA is put back into DNA.

Somatic mutations occur during development of an organism but do not enter the genome. Somatic mutations in immune systems can provide substitutions so that an immunized mammal can keep pace with continuous antigenic drift, for example, of influenza hemagglutinins.

PRO-PROTEINS AND PRE-PRO-PROTEINS

Several posttranslational or cotranslational events modify the structure of proteins. These are described for secreted proteins and are also observed in polypeptide sequences coded by multiple genes. Precursor proteins (proproteins) are coded by mRNA, and may be preceded by pre-pro-proteins. The initial translation may be of a very long polypeptide chain which is cleaved by intracellular proteolytic enzymes.

Pro-proteins were first discovered for processing of secreted proteins in eukaryotes. In some bacteria similar pro-protein precursors are cleaved at the cell membrane (30). Examples of secreted proteins follow.

Insulin

One of the proteins for which pro- and then pre-pro- forms were first discovered was insulin. Cell-free translation measurements have been used in combination with nucleotide sequence measurements on cloned insulin

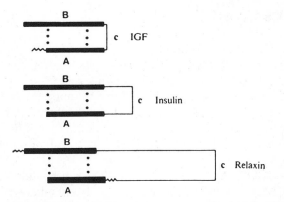

Figure 3-7. Schematic comparison of prohormone structures for insulin-like growth factor (IGF), insulin, and relaxin. Reprinted with permission from P. Hudson, *Nature,* Vol. 291, p. 130. Copyright 1981, MacMillan Journals Ltd., London.

genes. Insulin consists of two chains, A and B, connected by two disulfide bridges (Fig. 3-7). The protein is synthesized in islet cells of the pancreas. In mouse, rat, and several fishes there are two insulins, I and II, which are products of nonallelic genes; each insulin has A and B chains. Pre-pro-insulin is a chain of 110 amino acids. Processing consists initially of elimination of the first 24 amino acids in the pre-region; these serve as a hydrophobic signal sequence for transfer through the microsomal membranes. There remains pro-insulin which then folds and splits into A and B chains, disulfide bridges are formed, and by proteolysis a C-peptide is cleaved away (2). In rat the gene for insulin I has one small intron; the other gene (II) has two introns, one of 119 bps, the other of 499 bps. The smaller intron in each encodes the 5' noncoding region of mRNA (74). It is proposed that the two pre-pro-insulin genes are products of a recent duplication. Precise elimination of an intron suggests an RNA-processed intermediate. The two insulin genes (I and II) in rat differ in sequences by 4 percent from those in rabbit. From amino acid sequences it is estimated that the rat genes I and II diverged some 20–35 myr ago (74). In chicken the gene for pre-pro-insulin has 324 nucleotides (83); there is a 3.5 kb (kilobase) intron which interrupts the coding region connecting the A and B peptides and there is also a 119 bp intron which interrupts the sequence of the 5' noncoding region of mRNA. Sequences for the C peptide of insulin are established for rat, human, chicken; the first part consisting of 14 residues is well conserved; the second part of 10 residues shows much divergence. Calculations of times of divergence of the genes for several species of animals indicate that replacement sites (where one amino acid is replaced by another) alter very slowly, while silent sites (where nucleotide changes cause no amino acid replacements) diverge rapidly.

Gastrin and Cholecystokinin

Gastrin is a small polypeptide of 17 amino acids. Progastrin as coded by mRNA is 104 amino acids long (Fig. 3-8). The cDNA has 602 nucleotides, of which 312 are in the entire mRNA coding region; of these, 61 are in the 5' untranslated region, 56 in the 3' region and 86 in the poly A tail (10,50a,116). Figure 3-8 shows the amino acid sequences from the mRNA, beginning at the 5' end, and the cDNA sequences from the 3' end. Nucleotides in untranslated regions are indicated by negative numbers. The initiation codon (AUG), the termination codon (UGA, UAG), and the polyadenylation site (AAUAAA) are underlined. The amino acid sequence derived from mRNA begins with Met at position -75 and ends with Pro at position $+29$. The final protein product (gastrin) begins at Gln at position 1 and ends with Phe at position 17 (116). Pro-cholecystokinin (CCK) consists of 39 amino acids; it is first cleaved to CCK of 33 amino acids, then this to three secreted peptides of 12, 8, and 4 amino acids (31).

Parathormone (PTH)

Several properties of synthesis are illustrated by the coding and synthesis of parathormone (PTH) (60). This protein consists of 84 amino acids (8 in human PTH are different from bovine). The pro-PTH has 6 additional amino acids and pre-pro-PTH has 25 additional ones. mRNA for pre-pro-PTH prepared from beef parathyroid glands has 700 nucleotides, of which 345 are essential for coding (Fig. 3-9). At the 3' terminus is a poly A sequence, at the 5' terminus a 7-methylguanosine. There is, therefore, a long untranslated sequence. By use of reverse transcriptase and DNA polymerase, double-stranded cDNA complementary to PTH mRNA was synthesized and characterized by specific restriction endonucleases. The cDNA was cloned in bacterial plasmids and the sequence determined for 12 overlapping restriction fragments. PTH illustrates the extensive modification which occurs before the final production of a relatively small protein (60). Besides the two cleavages of the precursors, the N-terminal methionine is probably removed; later, PTH itself is cleaved in the circulating blood to two fragments,

---

Figure 3-8. Primary structure of porcine preprogastrin mRNA. Nucleotide numbers above the mRNA sequences beginning with initiation codon. Nucleotides in untranslated region given by negative numbers. Major restriction sites underlined below the cDNA sequences. Initiation codon (AUG), termination codon (UGA, UAG), and polyadenylate site (AAUAAA) underlined. Amino acid sequence derived from mRNA given in upper line and begins with Met at position -75 and ends with Pro at position 29. The sequence of secreted gastrin begins with Gln at position 1 and ends with Phe at position 17. Reprinted with permission from O. J. Yoo et al., *Proc. Natl. Acad. Sci. USA*, Vol. 79, p. 1052. Copyright 1982, National Academy of Sciences, Washington, D.C.

Protein (NH₂-)

mRNA (5'-)

cDNA (3'-) (C)₁₈(A)₁₆

```
                                                                    -75
                                                          Met--Gln--Arg--Leu--Cys
  -61  -58        -49                                       -1
   AUG GAG AAC    UGA ... GGC ACC AGG CCA ACA GCA GCA CAC CUG CCU CCC AGC UCU GCA GCA CAC CUG     AUG CAG CGA CUC UGC
 G                                                                        -40            -20                        -1
 C TAC CTC TTG    ACT ... CCG TGG TCC GGT TGT CGT CGT GTG GAC GGA GGG TCG AGA CGT CGT GTG GAC     TAC GTC GCT GAG ACG
                                                                        Pst I                                    Hinf
```

```
 -70
 Ala--Tyr--Val--Leu--Ile--His--Val--Leu--Ala--Leu--Ala--Ala--Cys--Ser--Glu--Ala--Ser--Trp--Lys--Pro--Gly--Phe--Leu--Gln
                            30                              60                                                      90
 GCC UAU GUC CUG AUC CAU GUG CUG GCU CGG GCC UGC UCU ACG      GAA GCU UCU UGG AAG CCU GGC UUC CAG CUG CAA
 CGG ATA CAG GAC TAG GTA CAC GAC CGA GCC CGG ACG AGA TGC      CTT CGA AGA ACC TTC GGA CCG AAG GTC GAC GTT
 Leader Sequence                                             Hind III                              Pvu II
```

```
                          -40                                                   -30
 Asp--Ala--Ser--Ser--Gly--Pro--Gly--Ala--Asn--Arg--Gly--Lys--Glu--Asp--Arg--Leu--Asp--Arg--Leu--Ala--Ser--His
                              120                              150
 GAU GCG UCC UCA GGA GCC AAC AGG GGC AAA GAG GAU CUG CGG GAU CAG CCA GCC GCC UCU CAC
 CTA CGC AGG AGT CCT CGG TTG TCC CCG TTT CTC CTA GAC GCC CTA GTC GGT CGG AGA GTG
```

```
 -20                                                                  -10                                         I
 His--Arg--Arg--Gln--Leu--Gly--Leu--Gln--Gly--Pro--His--Leu--Val--Ala--Asp--Leu--Ala--Lys--Lys--Gln--Gly--Pro--Trp--Met
                              180                              210                                             240
 CAC CGA AGG CAG CUG GGG CUC CAG GGG CCC CCU CAC CUG GUG GCA GAC CUG GAC CGG AAG AAG CAG GGG CCA UGG AUG
 GTG GCT TCC GTC GAC CCC GAG GTC CCC GGG GGA GTG GAC CAC CGT CTG GAC CTG GCC TTC TTC GTC CCC GGT ACC TAC
 Pvu II
```

```
          10                      17  Phe--Gly     20  Arg--Arg    Ser--Ala--Glu--Gly--Asp--Gln--Arg--Pro--  29  End.
 Glu--Glu--Glu--Glu--Ala--Tyr--Gly--Trp--Met--Asp
                              270                              300                                             315
 GAG GAG GAA GAA GCA UAU GGA UGG AUG GAC UUC GGC      CGC GCG GCG UCA CGA CUC CUU GUC UCA GGA GAC CAG CGU CCC     UAG
 CTC CTC CTT CTT CGT ATA CCT ACC TAC CTG AAG CCG      GCG CGC CGC AGU GCU GAG GAA CAG AGU CCT CTG GTC GCA GGG     ATC
```

```
      330                                330                  360                  369        382   387      390          401
 AACC GAGCUC CAGAGCCCAGCCACCUCCUAGGCCAUC CCAGUCCAGCCACA     AAGGCCAAGUCCC UGA      ACT AAUAAA CUAGCUUCCAACGG UAG   (A)₈₆  C₂₀
 TTGG CTCGAG GTCTCGGGTCGGTGGAGGATCCGGTAG GGTCAGGTCGGTGT     TTCGGTTCAGGG          ACT TTATTT GATCGAAGGTTGCC      (T)₈₆  G₂₀
      Sst I
```

```
5'  G GGG GGG GGG GGG GTT TTA TCA GCC TTC TCA GGT TTA CTC AAC TTT GAG AAA GCA GCT AAT ACA TYT
          10        20        30        40        50        60        70


                           met met ser ala lys asp val lys val ile val met leu ala
GAA AGA AGA TTG TAT CCT AAG ACG TGT GTT AAT ATG ATG TCT GCA AAA GAC ATG GTT AAG GTC ATG ATT GTC CTT GCC
  80        90       100       110       120       130       140       150       160


ile cys phe leu ala arg ser asp gly lys lys ser val lys lys arg ala val ser glu ile gln phe met his asn leu gly
ATC TGT TTT CTT GCA AGA TCA GAT GGG AAG AAG TCA GTT AAG AAG AGA GCT GTG AGT ATA CAG TTT ATG CAT AAC CTG GGC
       170       180       190       200       210       220       230       240


lys his leu ser met gly arg val glu trp leu arg lys lys leu gln asp val his asn phe val ala leu gly ala
AAA CAT CTG AGC TCC ATG GAA AGA GTG GAA TGG CTG CGG AAA AAG CTA CAG GAT GTG CAC AAC TTT GCC CTT GGA GCT
       250       260       270       280       290       300       310       320


ser ile ala tyr arg asp gly ser gly gln arg pro arg lys lys glu asn val leu val glu his gln lys ser
TCT ATA GCT TAC AGA GAT GGT AGT GGT CAG CGA CCT CGA AAA AAG GAA AAT GTC CTG GTT GAG AGC CAT CAG AAA AGT
       330       340       350       360       370       380       390       400


leu gly glu ala asp lys ala asp val leu ile ile ala lys pro gln stop
CTT GGA GAA GCA GAC AAA GCT GTA TTA ATT ATT AAA GCT AAA CCC CAG TGA AAA CAG ATA TGA TCA GAT CAC TGT
       410       420       430       440       450       460       470       480


TGT AGA CAG CAT AGG GCA ACA ATA TTA CAT GCT GCT AAT GTG TTC ACC TTC TAT TAA GTG CCA GTA TGA CCA ACC
       490       500       510       520       530       540       550       560


TTT ATT GCT AGC TGT GAT ACC TAC AAT TTT AAT TGA GTA TTT TGA TTC TAC TTT CAT CTA AGA GCT CTT TTA ATA ATT
       570       580       590       600       610       620       630       640


CTA TTT CTA TTG ATT CCA AAT AAA TGA AGT AGT TTA GTA TTA AAA AAA AAA AAA AAA AAA AAA AAA AAA AAA
       650       660       670       680       690       700       710       720


AAA AAA AAA CCC CCC CCC CCC CCC CCC CCC CCC 3'
       730       740       750       760
```

86

one of which is active. Thus modification involves three and possibly four specific proteolytic cleavages. The adaptive function of these many steps and how they evolved remain unknown.

Collagen

Collagen consists of three α chains in a rodlike helix with 338 repeats of an amino acid triplet, Gly-X-Y, where X and Y are usually proline or hydroxyl-proline. Each chain is synthesized as a pro-peptide with nonhelical regions and amino and carboxyl terminal ends. There are nine different types of collagen coded by nine genes. Chicken genes for collagen have been cloned; a sequence of 54 kbp contains the 38 kbp of pro-collagen. The gene has about 50 introns and 14 exons, all multiples of the 9 bp code for the repeating triplet Gly-X-Y. The collagen genes appear to have more introns than any other genes which have been cloned (102,115).

Neuropeptides

Several families of peptides have components which function as neurotransmitters in animals (Ch. 12). Some of the same polypeptides occur widely, for example, in *Tetrahymena, Neurospora,* and *E. coli.* It is probable that the genes for polypeptide neurohormones have been subject to change and that the coded peptides have had different functions during animal evolution; the present functions as neurotransmitters may be secondary. Coding and synthesis are known for a few neuropeptides which are genetically linked to very dissimilar polypeptides (55). One polypeptide family includes the peptide hormones ACTH, MSH_1, MSH_2 (melanocyte-stimulating hormones), and β endorphin; MSH and αEnd are cleavage products of βLPH (α-lipotropin) which may have other functions than as precursor of MSH and endorphins. The gene complex occurs in hypothalamus, anterior and intermediate pituitary, and adrenal cortex, with minor differences in each. The mRNA for the complex but not the DNA has been isolated. Sequences of smaller molecules (MSH) occur within the ACTH and βLPH to give MSH and endorphin (Fig. 12-8). A precursor consisting of 260 amino acids has an active end of 130 amino acids which include ACTH at sequences 1–39, MSH at sequences 41–58, and endorphin at 61–91. ProACTH-endorphin is cleaved to

Figure 3-9. Nucleotide sequence of bovine parathormone cDNA. Amino acid of prepro-parathyroid hormone indicated above the respective triplet codons. Personal communication from B. Kemper.

ACTH intermediate plus LPH; the former then goes to a 16K fragment plus ACTH, the latter to LPH and endorphin (23,35,90).

The enkephalins, Met-enkephalin and Leu-enkephalin, have opiate properties resembling endorphins; enkephalins are of different origin and are much smaller molecules. The primary structure of precursor sequences for mRNA codes for human Met-enk and Leu-enk are given in Chapter 12, Figure 12-19 (31). One mRNA encodes 5 to 7 enkephalin molecules. Pre-pro-enkephalin mRNA has long leader and terminal noncoding regions.

Globins

The best known proteins for genetic patterning and history are the α and β globins of vertebrates. Globins probably evolved from a single locus by gene duplication, gene migration (even between chromosomes), and structural mutations. Hemoglobins likewise show a sequence of developmental expression (53). The evolution of hemoglobins and relations between structure and function are considered in Chapters 5 and 6. In man the α globins are encoded by four functional genes and one pseudogene, the β globins by five functional genes and two pseudogenes. The α-like polypeptides include embryonic ζ, fetal α, and adult α (gene duplicate); the β-like globins are embryonic ε, fetal γ, and adult β and δ. Two forms of fetal γ globin are Gγ and Aγ, in which position 136 is glycine or alanine. At birth the ratio of Gγ to Aγ is 3:1 (111).

Adult hemoglobin (Hb A$_1$) consists mainly of two α and two β ($\alpha_2\beta_2$); Hb A$_2$ constitutes 2 percent of total Hb and is $\alpha_2\delta_2$. Human embryos prior to eight weeks of age have $\zeta_2\varepsilon_2$. From the eighth week to term there is increasing fetal hemoglobin (HbF) $\alpha_2\gamma_2$; at birth, the γ chains are replaced by β and δ (83a,111). The transitions from ζ to α and ε to γ are asynchronous.

Genes for α globins are on chromosone 16, for β globins on chromosome 11 (86). DNA fragments containing globin genes have been isolated by use of endonucleases; in a 40 kb segment the sequence is: Gγ-Aγ-δ-β (6a). Complete sequences of Gγ, Aγ, and adult δ and β are available. The α and β loci diverged by duplication some 85 myr ago. Linkage arrangements are given in Figure 5-9 (Ch. 5). The genes for β-type globin consist of 2280 nucleotides with 30 replacements between γ and ε, 36 between ε and β (33). Introns may have been fixed before the α-β division.

Three β globins of goat (fetal, preadult, adult) show 90 percent homologies and differ mainly in introns; in addition, two β pseudogenes occur. It is concluded that in goat the three functional β globins diverged recently and that the ancestors of the pseudogenes and of functional genes diverged earlier (73,97). In mouse, major and minor β genes differ by 9 amino acids out of 146. Mouse has three α globin genes and two pseudogenes (73).

Histones

Histones, which form the structural backbone of chromosomes to support DNA, are among the most conserved of all proteins. Five histone genes have been identified. Of 132 positions in H_4 and H_3, only three amino acids in the sea urchin are different from those in a calf (108).

NUMBER OF GENES

Correlations have been made of the complexity of organisms with (1) total amount of DNA, (2) number of genes, (3) organization of genetic material, and (4) regulatory genes which modulate action of structural genes. The only correlations of biological diversity with number of chromosomes are for polyploids. A bacterium has a single circular chromosome; the diploid number of chromosomes in eukaryote plants and animals ranges from 4 to 78. Polyploidy is relatively common in plants, less common in animals. Polyploidy is often associated with large cell size and may permit greater diversity of specific proteins. For example, salmonid fishes are tetraploid and in them there are more isozymal forms of enzymes, such as lactate dehydrogenase, than in diploid fishes.

In a nematode worm *Caenorhabditis* the haploid genome contains 0.8×10^8 base pairs, 20 times the amount of DNA in *E. coli*. In general, somatic cells have 30 percent less DNA than germ cells (13).

The total amount of DNA per cell tends to be greater in more complex organisms. This is shown in Figure 3-10 (14) and by the following data:

Organism	Total Amount of DNA in Picograms/ Cell
bacteriophage T_4	0.00024
various bacteria	0.002–0.06
yeast	0.005
sponge	0.1
mollusc	1.2
plant (*Fritillaria*)	100
lungfish	100
amphibian	7.0
chicken	2.5
human	5.6

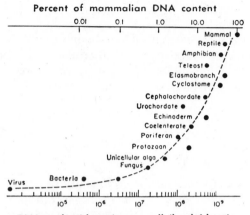

Percent of mammalian DNA content

DNA nucleotide pairs per cell (haploid set)

Figure 3-10. Minimum haploid DNA content in species of various levels of organization. Reprinted with permission from R. Britten and E. Davidson, *Science,* Vol. 165, p. 352. Copyright 1969 by the AAAS.

Related organisms differ greatly in content of DNA; the range in unicellular algae is 500-fold. The DNA content per cell is 30 times greater in a mammal than in a sponge. The number of nucleotide pairs of DNA per organism (prokaryote) or per haploid cell (eukaryote) is as follows (68):

Organism	Nucleotide Pairs	Number of Chromosomes
virus SV-40	5.4×10^3	
virus T$_4$	1.8×10^5	
bacteria (*E. coli*)	4.5×10^6	
mold (*Neurospora*)	8.6×10^9	
alga (*Euglena*)	5.8×10^9	
alga (*Chlamydomonas*)	1.2×10^8	
slime mold	3.0×10^7	
nematode	0.8×10^8	
plant (*Lilium*)	1.8×10^{11}	6
sponge	1.1×10^8	
Drosophila	1.4×10^8	4
sea urchin	1.8×10^9	18
amphioxus	1.0×10^9	12
salamander (*Amphiuma*)	1.9×10^{11}	
human	3.2×10^9	24

Specific correlations are lacking. Two species of buttercups (Ranuncula-ceae) differ 180-fold in haploid DNA, two flagellates (*Euglena* and *Chla-*

mydomonas) 25-fold. A lily or a salamander has 20–30 times as much haploid DNA content as a human cell.

A more meaningful correlation with biological adaptation than the number of chromosomes or the quantity of DNA may be the number of coding and noncoding nucleotide pairs. A widely used method to determine coding and noncoding nucleotide pairs is annealing and hybridization. When a double strand of DNA is heated, the strands separate, and on cooling they reassociate, with matching purines and pyrimidines pairing at different rates. The annealing is given by the ratio of number of nucleotides associated at a given time to the initial number (percent association), and it is plotted against the number of moles of nucleotides per liter per second, the C_0t measurement. The value for 50 percent association ($C_0t_{1/2}$) is a measure of number of pairing nucleotides. By another method, when excess mRNA is hybridized with DNA, the number of structural genes can be estimated from the kinetics of the hybridization, since each polypeptide chain is coded by its mRNA. The percentage of nucleotide pairs hybridized is measured (48).

In both prokaryotes and eukaryotes most proteins and hence most mRNAs are coded by some 1000 to 2000 DNA nucleotide pairs. In viruses and bacteria the number of coding genes corresponds closely to the number of types of RNA. A bacteriophage (X 174) has 5385 nucleotides for 10 proteins (101). The number of structural genes in *E. coli* is about 3000, which corresponds to the estimated total number of kinds of RNA. There may be many overlapping frames of coding DNA. In eukaryotes (except some yeasts) the number of nucleotide pairs is far greater than the number of polypeptides coded by structural genes. In humans 48 chromosomes have enough DNA for 3×10^6 proteins of average size, but the actual number of different proteins in the body is estimated at 30,000 to 150,000.

Sequences of nucleotide pairs in DNA may be repetitive or nonrepetitive (Table 3-1). Prokaryotes have little or no repetitive DNA; their DNA sequences are mostly unique and nonrepetitive; regulatory genes are relatively few and are close to structural genes. Eukaryotes have many repetitive DNA sequences, which are interspersed among nonrepetitive sequences. In general, mRNAs are transcribed by nonrepetitive DNAs (68). The genes coding

Table 3-1. Percent of Repetitiveness in DNAs as Measured by c_0t Analysis (68)

	Nonrepetitive	Low and Middle Repetitive	High Repetitive
E. coli	100		
Slime mold	70	30	
Drosophila	74	13	13
Snail	38	>12	18
Cow	60	38	2

for ribosomal RNA are much smaller in number than those coding for all the mRNA, even though in eukaryotes rRNA may be 90 percent of cytoplasmic RNA.

In summary, the number of genes and total DNA per genome have increased during evolution but there is no close correlation between amounts of DNA and taxonomic level.

EXCESS DNA

Several attempts have been made to explain the great amount of DNA that is present but not needed to code for the mRNA. (Coding genes account for only 10–15 percent of total DNA.) One view is that much of the DNA is "junk" or incidental and of no phenotypic function (32). It is argued that natural selection operates on some portions of the genotype, that is, on a gene rather than on an organism or population. DNAs lacking benefit or harm to the phenotype are of no selective advantage to the phenotype, and the nontransposable DNA may favor nonphenotypic selection. Such DNA would not be readily eliminated and is replicated without any effect on the phenotype. Much of this selected DNA could be self-maintaining in the sense of a parasite—hence the expression "selfish gene" (32), or facultative gene (68,80,81). The proportion of noncoding DNA is greater in organisms with slow development and long life cycle. Facultative DNA elements are replicative but not transposable, and selection would not delete them. A function of noncoding DNA may be merely its own survival within the genome. In the absence of phenotypic selection, excess DNA would require a long time to be eliminated (32,82).

Another proposed explanation of the excess DNA is that much of it has a nucleoskeletal function (19). There is a 40,000-fold range in haploid DNA content in eukaryotic organisms. The amount of noncoding DNA correlates with cell and nuclear volume, cell cycle time, and minimum generation time. It is postulated that cell growth is determined by the nuclear volume and area of the nuclear envelope, which would influence transport of RNA out of the nucleus. In some kinds of organisms, r selection (Ch. 4) favors small cells, rapid growth rate, and low DNA content. In other kinds of organisms, K selection (Ch. 4) favors large cells, slow growth rate, and high DNA content. Structural proteins in chromosomes, both lysine-rich and arginine-rich histones, may repress some gene expression. The amount of DNA per nucleus is large in organisms in which generation time is long and duration of meiosis is prolonged. For example, in annual plants, cell cycle times are shorter than in perennials; nuclear DNA quantity is lower in annuals. In 24 species of plants the duration of meiosis ranged from 21 to 274 hours; the DNA content ranged from 83 to 120 pg/cell (5).

Large nucleus and cell size are characteristic of polyploidy. Polyploid lampbrush chromosomes occur in large germ cells of amphibians.

The excess DNA may code for nuclear RNA, some of which may be regulatory to DNA (26). In a sea urchin, some 30 percent of the single-copy DNA codes for hnRNA. Most of the hnRNA remains in the nucleus but a small amount of it is the precursor for cytoplasmic RNA. The means by which hnRNA regulates DNA function is not known.

Another hypothesis is that the property of DNA to undergo change—by mutation, chromosomal rearrangement, replication and amplification—is favored by selection. Mutability ensures variability and capacity for phenotypic change. Control by DNA of chromosomal functions is an inherent property guaranteeing mutability that may be favored by selection. There must be a pool of DNA to provide for mutability (preadaptation). There are known mutants for variation in meiosis—chromosome pairing, chromosome condensation, segregation, recombination (24). Mutants which affect meiosis are not detected by coded proteins. Deletions, insertions and frameshifts constitute some 90 percent of mutations in the lac I gene of *E. coli* (36).

It is probable that each of these proposed explanations for the excess DNA has some validity. The absence of much noncoding DNA in prokaryotes and the large excesses in eukaryotes probably relate to the evolutionary diversity of eukaryotes.

EVOLUTION OF GENETIC MECHANISMS

Genetic mechanisms are so universal that they must have evolved very early, probably before there were organisms. Universal genetic mechanisms are the genetic code of DNA, the synthesis and functioning of the several RNAs, and the sequence of events in protein synthesis. Some enzyme proteins function in nucleotide replication, other proteins function in transduction and translation, with minor differences with kind of organism. Numerous biochemical "decisions" must have been made very early—the use of the 20 amino acids in proteins, the sequences of purines and pyrimidines, the types of RNAs—transfer, ribosomal, messenger. Somewhat later there may have evolved "start and stop" base sequences, long noncoding sequences, and the production of pro- and pre-pro-proteins that are later cleaved away. Both cooperation and competition took part in the molecular evolution of genetic systems. Description of the molecular genetics of modern organisms gives a basis for speculation concerning the origin and evolution of genetic systems.

Proteins can be synthesized *de novo* from amino acids under special conditions, such as high temperatures, but proteins thus formed are not self-replicating. The property of self-replication required nucleic acids.

It is probable that the first genes were RNAs (34,104). Some present-day RNAs are self-replicating in the presence of appropriate catalytic proteins and can fold to give a three-dimensional structure with molecular stability. Some viruses use a single-stranded RNA as the genome. The virus Q_B which can infect *E. coli* has a single-stranded RNA of about 4500 nucleotides. Only part of the molecule contains a genetic message; the remainder recognizes protein. A portion of Q_B RNA consisting of 220 nucleotides can replicate itself in the presence of a specific replicase enzyme (101). Some replication of viral RNA can take place in the presence of a sufficient number of nucleotides (base + sugar + appropriate phosphate) without an enzyme but with RNA as a template. In the absence of an RNA template but in the presence of Q_B replicase plus nucleosides, several products, some of them 60 nucleotides in length, were obtained. Short polymers were formed from monomers in the absence of the replicase (34). In these experiments mutant RNA polymers survive; these are called quasi-species. In such RNA replication many errors occur.

It is argued (34) that RNA replication, the production of quasi-species and hypercycles in homogeneous solution, could not have provided sufficient testing of products for selection to operate. Organization into cells was necessary for testing which mutants would survive better than others; compartmentalizing to organelles within cells increased the probability of certain reactions and of error correction.

Metal ions, Pb and Zn, catalyze condensation of nucleoside 5'-phosphorimidazolides (Imp A or Imp G) to oligonucleotide chains. From Imp A on poly (U) template or Imp G on poly (C) template, chains of 40 nucleotides are formed. In absence of metal catalyst shorter chains are produced (82a).

The best candidate for a primitive RNA is tRNA; tRNAs are the smallest RNAs and are limited in number; each tRNA binds to one essential amino acid. tRNAs fold in a cloverleaf pattern which has properties of a double-stranded coil; a central region, the anticodon, matches an mRNA and could have served primitively in initiating polypeptide association (Fig. 3-4). tRNAs of archaebacteria resemble those of some eukaryotes (114). Some genes for tRNAs in archaebacteria are split with short introns as in eukaryotes; no such RNA splicing is known in eubacteria (114).

New properties appeared with the evolution of DNA; it is double-stranded. Errors in eukaryote DNA are 10^{-8} to 10^{-11} per nucleotide replication; DNA is associated with proofreading exonucleases (46). Errors in RNA genomes (as in RNA viruses) are 10^6 more frequent than in DNA; Q_B virus errors are 10^{-3} to 10^{-4} per nucleotide doubling. Error rates in the first RNA genes may have been 10^{-1} to 10^{-2} (87a). The third position in the base sequence in RNA is more variable by some 4–5 times than the first and second positions. Replication of DNA requires some 20 different enzymes—more than for any replicating RNA. DNA replicates in both strands of the helix and from 3' toward the 5' ends. All tRNAs have CCA at the 3' end and all mRNAs translate their message from the 5' toward the 3' end (converse to DNA) (68).

The base code for amino acids may have been initially fixed by translation of classes of amino acids. Codons for similar amino acids occur near one another—negatively charged together, positively charged together, aliphatic amino acids together.

Several hypotheses for the selection of the genetic code have been proposed. One is that the interaction between amino acids and bases is fixed by steric properties resulting from the water shell surrounding an amino acid (14). Evidence for a stereochemical explanation has come from models of DNA; second codons were removed, thus creating cavities. The 20 L-amino acids fit into cavities and their position is fixed by H-bonding and by steric constraints. Side chains of the 20 amino acids resemble the purine and pyrimidine bases with α amino groups at N-9 of a purine or N-1 of a pyrimidine. The *stop* codons TAA and TAG do not accommodate any amino acid (45).

Understanding the evolution of genetic mechanisms can be aided by model systems. Many of the decisive events probably occurred before organisms appeared, but refinement and acceleration were promoted by organization into cells. Nucleotides and proteins which are adapted to biotic conditions became replicated.

Another kind of genetic evolution took place via transposable or jumping genes. By transposition from one site to another, new gene sequences were made possible (22a). In bacteria the transfer of genes between species can result in transduction, transformation, and sexduction. Viruses can activate certain genes of cells of eukaryotes and cause them to become tumors; fibroblasts in particular are subject to transformation of the genome. Examples of probable transfer of eukaryotic genes into symbionts are known. There are suggestions of possible transfer of a gene from one species to another. Leghemoglobin of the root nodules of leguminous plants closely resembles the globins of vertebrates. One suggestion is that the gene was transferred by a virus from an animal to the legume. The possibility of gene transfer between species has important evolutionary implications.

GENE REGULATION

In prokaryotes, regulatory genes control the transcription of structural genes. Most regulatory genes are transcribed to mRNA, which codes proteins that either activate or repress structural genes. A single regulatory gene may modulate a series of structural genes; the set of contiguous regulatory and structural genes in a prokaryote is called an operon.

The organization of an operon is best known for *E. coli*. The lac operon (Fig. 3-11) consists of three structural genes and two regulatory genes (the structural genes code mRNAs for a β-galactosidase, galactose permease, and galactoside acetylase); two regulators, which code for activator and repressor proteins, respectively; and three promoter-operator genes. Synthesis of the three enzymes is controlled by the concentration in the medium of a sugar, such as lactose. One of the regulators, repressor gene I, encodes

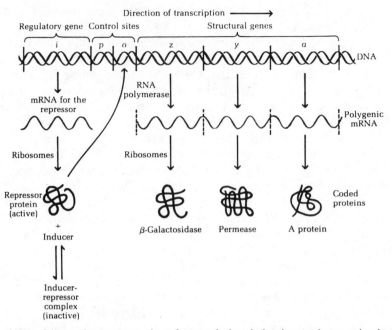

Figure 3-11. Schematic representation of transcription, induction, and repression in the lac operon. *p* and *o* are control sites, *i* is regulatory, *z, y,* and *a* are structural genes which encode mRNAs. In absence of inducer the product of *i* gene represses transcription of structural genes. In presence of inducer an inactive repressor complex is activated to allow transcription. From A. L. Lehninger, *Principles of Biochemistry,* Worth Publishers, New York, 1982, p. 905.

the mRNA for repressor protein, which acts on the operator gene to repress the transcription of mRNA by the three structural genes. The lac repressor gene has 1080 bases; the lac repressor protein has 360 amino acids (36). In the presence of an inducer sugar, the repressor protein is inactivated and the operator and promoter via their mRNAs and respective proteins signal RNA polymerase to start transcription at a structural gene. Mutations of any of the regulatory genes can affect the rate of synthesis of enzymes coded by the three structural genes without affecting the amino acid sequences of their enzyme products. The depression caused by an appropriate sugar may involve a second mediator, cyclic-AMP, which increases in concentration when the repressor (glucose) is present.

In the operon for tryptophan there are five structural genes, each coding for one step in tryptophan metabolism. The repressor gene encodes for repressor mRNA, and its protein product is inactive until tryptophan is present. Thus the product tryptophan represses transduction of the genes which would lead to further production of tryptophan (7). In *E. coli* nine structural genes encode for synthesis of histidine; when His is omitted from the growth medium, His mRNA increases by 10 times and the enzyme for His synthesis increases by 25 times. The corepressor for histidine synthesis is not the amino acid, as for tryptophan, but His-tRNA (68). The adaptive function of an operon is that a single regulatory gene can control the action of numerous structural genes which may be located separately in the genome. Transduction starts at the operator gene which by its encoded protein begins transcription of the first structural gene (Fig. 3-4).

The prokaryote *Salmonella* has two genes for synthesis of the motility protein flagellin. These genes are labeled H_1 and H_2. H_2 has an adjacent repressor rh_1; H_1 has a repressor rh_2. A promoter is needed for transcription of H_2; if the controller is inverted, the promoter is disconnected (98).

The function of regulatory genes is less clear in eukaryotes than in prokaryotes, but products and reactants affect the synthesis of proteins. Agents such as hormones, nutrients, and possibly physical factors in the environment may regulate transcription of structural genes. A model for eukaryotes similar to that for prokaryotes was developed by Britten and Davidson (14). It is postulated that all structural genes require activation. Sensor proteins respond to such agents as hormones and to some substrates; sensor proteins act on sensor genes which control the transcription of integrator genes. The integrator genes code for secondary regulatory proteins which, as activator proteins, bind to DNA receptor sites; transcription of structural genes is then stimulated. A full set of structural genes can be activated by a stimulating agent acting via appropriate regulator genes. A commonly accepted view is that the regulated turning on of a gene in a cell type requires (1) the gene to be located in a transcriptionally active (nuclease sensitive) chromosome domain (this confers potential for expression); and (2) cell-specific signals to be present—protein, methylation pattern, receptors.

The number of regulatory genes is difficult to estimate, but in the few operons known there are fewer regulator than structural genes. The influence of one regulatory gene may be wide. It is established for maize that regulators may be located at some distance from the structural genes which they control. There may be overlap of two or more sets of structural genes which are controlled by one regulatory gene.

Regulation of gene expression in eukaryotes is more complex than in prokaryotes. Brown (15) listed some of the control mechanisms in eukaryotes:

1. Some chromatin may be lost in early stages of development; this loss renders the cells incapable of becoming germ cells.
2. Gene amplification occurs as an increase in the DNA coding for particular RNA products, for example, the genes for rRNA in oocytes. Specific genes are replicated at given stages of a life cycle.
3. Methylation of cytosine at the 5-position; this may alter the activity of a gene.
4. Transcriptional control is by selective action of various RNA polymerases; posttranscriptional control is by processing of RNA before it

leaves the nucleus. Cleavage of long mRNAs is by specific nuclear enzymes.

5. Translational regulation occurs when pro- and pre-pro- proteins are cleaved during activation. Translational control is important when some proteins are made in large amounts at one stage of development or are made for secretion. The gene codes for corresponding families of proteins are related to one another.

ADAPTIVE PROPERTIES OF GENOMES AND PROTEIN SYNTHESIS

Environmental factors can influence the rates but not the kinds of mutations. A few natural antibiotics do have qualitative effects on certain regions of the genome. Spontaneous mutations or errors of DNA replication are much more frequent than induced mutations. Mutation occurs at varying rates in different kinds of organisms. The basal mutation rate for a single nucleotide has been estimated for eukaryotes at 10^{-9} to 10^{-11}/gene/DNA replication (88). Error rate in RNA genomes (viruses) is much higher than in DNA (46) and evolution of RNA viruses is rapid. The probability of a two-base substitution is 3×10^{-10}/nucleotide/DNA replication. Rate of mutation can be increased by high temperature (*Drosophila*), by ionizing or ultraviolet radiation, or by mutagenic chemicals. Enriched nutrition can increase the mutation rate in *E. coli* 10^3-fold. DNA that is rich in GC is more resistant to ultraviolet radiation than DNA that is rich in AU. CG in mammalian DNA is selected against because methylation decreases stability. The compound 2-aminopurine effectively increases mutation rate in *E. coli*. Not all sites are equally susceptible to mutation; some regions are "hot spots." In the lac I gene 5-methylcytosine is associated with high mutability (36).

Changes or errors in one base of the triplet code may be undetected for those amino acids which are coded by several triplets. Such changes take place in a few out of several thousand nucleotides and are silent mutations.

Metabolic pathways involving families of enzymes evolve together (113). Examples are the utilization of lower fatty acids instead of carbohydrates in ruminants; increases in pancreatic endonucleases in animals with populations of intestinal microorganisms; enzymes leading to production of end-products of anaerobic metabolism; and enzyme systems for use of new nutrient resources (e.g., various sugars by parasitic animals and bacteria) (68). Growth of mutant organisms utilizing new nutrients can outstrip growth of wild type organisms. Groups of genes are controlled and evolve together, for example, the electron transport system, the TCA cycle, glycolytic path, and protein degradation. Genes for components of a system need not be linked on one chromosome; genes for α and β Hb are separate, genes for 6PGDH and 6-P-gluconate DH are on separate chromosomes.

Many genes are expressed only at certain stages during development, for example, the different α and β globins. Temporal changes in enzyme patterns may be controlled by the cellular environment—chemical and spatial or geometric factors. A sea urchin blastula makes use of mRNA deposited by the parent in the egg, but by 10 hours after fertilization the embryo makes new mRNA; the proportions of different histones, too, change with developmental stage.

Some cells stop dividing at a certain stage but continue to make proteins, for example, vertebrate neurons. Transplants of nuclei in various animals indicate that DNA is not lost during development. Each oocyte of an amphibian *Xenopus* makes several thousand times more ribosomes during germ cell development than does a single somatic cell. In *Xenopus* there are two classes of 5 S rRNA genes—an oocyte type and a somatic cell type. Injection of somatic nuclei into oocytes causes oocyte genes to remain inactive; oocyte nuclei injected into liver cells remain inactive but if returned to oocytes are reactivated (63). Gene expression for two chorion proteins in *Drosophila* is amplified tenfold during a five-hour period in late stages of production of ovarian follicle cells (101).

MODULATION DURING TRANSCRIPTION AND TRANSLATION

An example of modulation at the level of transcription is alteration of production of calcitonin by a "mutant" RNA. Calcitonin is a peptide of 32 amino acids, normally synthesized in thyroid tissue. Calicitonin is formed from a precursor containing 136 amino acids, coded by a gene of 545 nucleotides (51). Thyroid tissue of some mutant rat lines shows smaller than normal amounts of calcitonin synthesized (92). Conversion of cells from high to low calcitonin content is due to production in the mutant of a new mRNA. The second form of mRNA is normally predominant in the hypothalamus where a related peptide (CGRP) is synthesized. Apparently two linked genomic regions alternate in expression according to tissue, and control of splicing is by transcription (3).

Examples of modulation by cell environment are viruses, which are incapable of synthesizing protein when outside a host cell. Some viruses, for example, most bacteriophages, inject DNA into the host; the host (bacterium) under the influence of viral DNA produces RNA which is complementary to the viral DNA; the host then produces specific proteins for replication and transcription. Other viruses (many on plants) have no DNA and use RNA as their genetic material. Inside a host cell the viral RNA is translated directly to protein. Infected leaves of a plant may make one polymerase for replication of viral RNA and another polymerase normal to uninfected plants.

A silkworm *Bombyx,* in the last four days of its fifth instar, makes in its silk glands quantities of the protein fibroin. The cells of a silk gland become polyploid and enlarge, the DNA which encodes for fibroin increases in proportion to other (total) DNA, and in a few days each fibroin gene gives rise to 10^4 molecules of mRNA which are very stable and produce 10^9 molecules of protein. The tremendous increase in protein synthesis is the result of polyploidy and of amplification during translation (109).

As animals age, errors in protein sequences may be more frequent than in young animals. Specific activity of some enzymes, such as aldolase and superoxide dismutase, in old rats is less than in young adults because of the smaller amount synthesized; kinetic properties are unaltered. Some other enzymes show no changes with aging (57). Tyrosine aminotransferase of rat liver is induced by glycocorticoids and shows rapid turnover. In old rats the molecules are more easily degraded by heat or proteases; the effect is posttranslational, and is not due to errors in DNA.

ENVIRONMENTAL EFFECTS

The modes of action on the genome of physical factors in the environment have been little investigated. Many changes in the biochemistry of plants and animals occur with seasonal adaptation. Some Arctic fishes synthesize an antifreeze polypeptide; Arctic insects produce quantitites of glycerol as winter approaches (Ch. 7). In early winter, 70–75 percent of the mRNA in the liver of flounders may be that which codes for antifreeze polypeptides; in spring and summer virtually no antifreeze mRNA is made (71). The synthesis of antifreezes in fish may be controlled by reproductive hormones that vary in level with season and water temperature. Another mechanism for enhanced protein synthesis may be an increase in the elongation factor I. Compensatory acclimation to cold by fishes results in part from enhanced synthesis of energy-yielding enzymes. Acclimatory responses occur in intact fish and in isolated liver cells; increased synthesis plus decreased degradation result in high enzyme content.

Synthesis of some enzymes in bacteria is induced by the substrate and repressed by products of the enzyme system. Constitutive enzymes may be synthesized in bacteria in the absence of specific agents.

Examples of induction and repression are also known for eukaryotes. In rats the synthesis of appropriate hydrolytic enzymes (amylase or trypsin) of pancreas and intestinal mucosa is triggered by diets high in carbohydrate or protein (36a). In mammals, liver ornithine aminotransferase and cytosolic MDH can be induced by cortisol; G-6-Pase is induced by a high protein diet. In rat liver a high carbohydrate diet induces an increase in mRNA for 6-P-gluconate dehydrogenase (50). Whether these increases are by direct action of the nutrients or by hydrolytic products or by hormones is not clear.

In barley seeds, gibberellic acid ($10^{-10}M$) induces the synthesis of α-amylase and thus releases sugar; ethylene induces synthesis of peroxidase; light acts to increase synthesis of anthocyanin; nitrate induces production of nitrate reductase. In *Euglena,* inorganic phosphate (P_i) represses synthesis of acid phosphatase.

In mammals and birds, steroid hormones initiate transcription of mRNA for synthesis of specific proteins. In rat uterus, estradiol combines in the cytoplasm of uterine cells with a receptor protein to form a 5 S complex which passes to the nucleus where it initiates transcription of DNA to several RNAs, which then leave the nucleus as messenger, ribosomal, or transfer RNA. Synthesis of RNA polymerases II and III or elongation factors of several cytoplasmic proteins is stimulated; some increase in DNA synthesis also occurs. In female birds estrogen and progestin act on the oviduct to stimulate synthesis of ovalbumin and conalbumin. The concentration of mRNA is rate-limiting for the protein synthesis. Actinomycin D blocks transcription of the new mRNA in these systems (59). Prolactin and thyroid hormones also act on transcriptional and translational processes. Some catecholamines act via cyclic AMP on translational events in the cytoplasm, but not on nuclear events.

Analysis of the nucleotide genetic sequence is aided by use of certain blocking drugs. In eukaryotes, cycloheximide inhibits amino peptidyl transferase and actinomycin D blocks RNA transcription. In prokaryotes chloramphenicol inhibits peptide transferase; streptomycin and tetracycline block protein synthesis by binding to the 30 S RNA subunit. Catecholamines act via cyclic AMP on translational events in the cytoplasm, but not on nuclear events.

CONCLUSIONS

Genomes provide the substrate on which environmental and developmental influences act and increase biological diversity. Phenotypic variation of individuals and populations occurs within genetically fixed limits. An individual organism shows adaptive plasticity—biochemical changes in response to physical environmental stress, behavioral changes according to stage in life cycle and biotic environment. In populations there is variation—between individuals, local populations, races, subspecies. The quantitative analysis of variation in groups of organisms constitutes population genetics. Diversity of populations is largely determined by properties of genomes. Species originate and higher taxonomic groups evolve by continuing genetic reorganization.

A broad historical view of the genetic basis for biological variation and evolution is as follows: Darwin pointed out the selective interaction between environment and organism. It remained for classical genetics during the first third of this century, using chromosomal studies, to give a structural and functional interpretation of heredity. Then followed an emphasis on speciation by accumulation of micromutations in geographic or ecological isolation. During the 1950s and 1960s an explosive advance occurred in the identification of genes as base pairs in DNA, the sequences of coding by RNA, and the steps in protein synthesis. A major advance of the 1970s was the development of methods for isolation and rapid sequencing of specific RNAs and DNAs and for estimating numbers of genes. It became apparent that genomes are not static, but are highly dynamic. Genes contain many non-coding regions, their coding sequences (exons) are interrupted by long introns; genes may migrate; DNA polymorphism greatly exceeds that of RNA and only a part of an mRNA molecule may be translated to protein.

A central problem of modern biology is the explanation of evolution and adaptation to the environment in terms of genetic replication and protein synthesis. The essential processes are (1) DNA replication, (2) synthesis of tRNA, rRNA, mRNA, and nuclear and mitochondrial chloroplast RNA and DNA, (3) transcription of a code to mRNA, (4) initiation, elongation, and translocation of amino acids on ribosomes, and (5) the translation of the code from mRNA to a polypeptide. Each step requires one to three specific proteins, and energy is provided by either ATP or GTP. The entire sequence is similar in prokaryotes and eukaryotes, with some minor diffferences. By the processes of replication and protein synthesis, genotypes are made manifest in phenotypes and phenotypes are reproduced. Why and how should so many steps have evolved? The many steps have some redundancy and provide a guarantee of uniformity of each kind of protein molecule; errors, although they occur, are minimal. If the processes of DNA replication and protein synthesis are similar in all organisms, how did the observed diversity of living organisms arise? How much of the sequence evolved before there were organisms? To what extent can the primitive genetic code and its evolution be inferred from properties of present-day genomes?

The genetic code is expressed in proteins that are synthesized as specified by nucleotide codons; a protein may be different in different tissues of an organism. Each cell of a plant or animal contains the same complement of genes (except that, in a few kinds of organisms, some chromatin is eliminated during development). Yet in a mammal, reticulocytes make mainly hemoglobin, liver cells make serum proteins, brain cells make neither of these. The specificity of a hormone-producing cell is exact. A major unknown is the mechanism of differentiation of a fertilized egg into cells and organs. Partial solutions of these problems may be found in the hormonal and environmental regulation of nuclear function and protein synthesis. Besides the genes controlling protein synthesis, genes are known which regulate meiosis and cell division. Other actions of genes have not been identified but they may occur. For example, there is much genetic control of animal behavior (Ch. 10), but how this is coded and how the code is expressed remains to be ascertained.

Many aspects of the relationship of molecular genetics to evolution remain unexplained. The terminology of classical genetics, words such as *locus* and *allele,* can be used only in a general sense. A few of the genetic problems which must be taken into account in any molecular theory of speciation and evolution are listed as follows:

Regulation at the level of transcription.   Regulatory genes have been described in prokaryotes, and postulated but not clearly established in eukaryotes. Regulation of transcription may be of sequences upstream in a gene from the coding region; regulation could be by feedback influences from RNA (as in retroviruses). Transcriptional amplification can occur by gene cooperation.

Quantitative control of protein synthesis at the level of translation. Polyploidy of organisms or of single tissues, with stability of mRNA, permits enhanced protein synthesis. Specialized cells periodically make large quantities of a protein.

Extracellular control of synthesis.   Induction and repression are well known in prokaryotes, not so well in eukaryotes. Environmental modulation of the genome could be by direct effects on transduction and translation; modulation may be by way of intracellular messengers. Amount and kind of protein present in a cell can be determined by the balance between synthesis and degradation.

Split genes and splicing in the formation of new DNA sequences. Posttranscriptional splicing may lead to formation of new duplicative units. Sections of a divided gene correspond to regions of the coded protein. "Jumping" and transposable genes make new combinations and contribute to speciation.

Long regions of noncoding DNA sequences.   Silent areas and introns in a codon are often longer than coding sequences. Why noncoding sequences are retained is not understood. One suggestion is that these sequences provide a reserve of DNA which may come to be tapped in evolutionary change.

Conformation or folding of DNA.   Three-dimensional views of DNAs and RNAs suggest functional properties which could be independent of sequence changes. Functional differences between right-handed and left-handed DNA are not understood.

Long sequences of DNA, of mRNA, and of precursor proteins which are cleaved away at different stages of protein synthesis.   Amino acid sequences of secreted proteins are usually less than one-tenth the length indicated by the coded sequence.

Relationship between genic changes and natural selection.   Changes in a genotype may be neutral, negative, or positive for selection. Much genetic change is not detected by selective forces, for example, changes in noncoding sequences, or amino acid substitutions which function equally well. Micromutations of structural genes may permit polymorphism and make possible the extension of ecological niches and ranges. Larger genic changes, especially rearrangements, may provide for saltatory evolution.

The dynamic nature of genomes is the basis for population diversity and evolutionary plasticity. To find out how the dynamic genome responds to environmental parameters or to the influences of development, aging, and population pressure is a major challenge to biologists.

REFERENCES

1. Abelson, J. *Science* **209:**1319–1321, 1980. Gene cloning.
2. Albert, A. G. and M. A. Permutt. *J. Biol. Chem.* **254:**3483–3492, 1979. Proinsulin in catfish pancreas.
3. Amara, S. G. et al. *Nature* **298:**240–244, 1982. Structure of calcitonin gene.
4. Anziano, P. G. et al. *Cell* **30:**925–932, 1982. Functional domains in introns.
4a. Avise, J. C. and R. A. Lansman. Ch. 8, pp. 147–164 in *Evolution of Genes and Proteins,* Ed. R. Koehn and M. Lei, Sinauer, Sunderland, MA, 1983. Polymorphism of mitochondrial DNAs.
5. Bennett, M. D. and J. B. Smith. *Proc. R. Soc. Lond. B.* **181:**81–135, 1972. Polyploidy in plants.
6. Bernard, S. A. et al. *Cell* **27:**497–505, 1981. Promoter genes.
6a. Bernards, R. et al. *Proc. Natl. Acad. Sci.* **76:**4827–4831, 1979. Structure of the human G_δ-A_δ-δ-β-globin gene locus.
7. Bertrand, K. et al. *Science* **189:**22–26, 1973. Regulation of tryptophan operon.
8. Bibb, M. J. et al. *Cell* **26:**167–180, 1981. Gene organization in mouse mtDNA.
9. Blanchetot, A. et al. *Nature* **301:**732–733, 1983. Seal myoglobin gene.
10. Boel, E. et al. *Proc. Natl. Acad. Sci.* **80:**2866–2869, 1983. Cloning of gastrin cDNA.
11. Borst, P. and L. Grivell. *Nature* **290:**443–444, 1981. Mitochondrial genomes.
12. Breathnach, R. and P. Chambon. *Annu. Rev. Biochem.* **50:**349–383, 1981. Split genes in eukaryotes.
13. Brenner, S. *Genetics* **77:**71–104, 1974. Genetics of *Caenorhabditis* behavior.
14. Britton, N. and E. Davidson. *Science* **165:**350–357, 1969. *Q. Rev. Biol.* **48:**365, 1973. Regulatory genes in eukaryotes.
15. Brown, D. D. *Science* **211:**667–674, 1981. Gene expression in eukaryotes.
16. Brown, S. D. and G. A. Dover. *Nature* **285:**47–48, 1980. Satellite DNA.
16a. Brown, W. M. Ch. 4, pp. 62–88, in *Evolution of Genes and Proteins,* Ed. R. Koehn and M. Lei, Sinauer, Sunderland, MA, 1983. Evolution of mitochondrial DNA.
17. Buetow, D. and W. M. Wood. Ch. 1, pp. 1–85 in *Intracellular Biochemistry,* vol. 5, Ed. D. B. Roadyn, Plenum, New York, 1978. The mitochondrial translation system.
18. Burton, N. and J. S. Jones. *Nature* **306:**317–318, 1983. Mitochondrial DNA.
19. Cavalier-Smith, T. *Nature* **256:**463–468, 1975. *J. Cell. Sci.* **34:**247–278, 1978. Nuclear volume control by nucleoskeletal DNA.
20. Cech, T. et al. *Cell* **31:**147–157, 1982. *Science* **218:**872–874, 1982. Introns and exons; *Tetrahymena.*
21. Chambon, P. *Sci. Am.* **244:**60–71, 1981. Split genes.
22. Childs, G. et al. *Cell* **23:**651–663, 1981. Pseudogenes-orphons.
22a. Cohen, S. W. and J. A. Shapiro. *Sci. Am.* **242:**40–49, 1980. Transposable genes.
23. Comb, M. et al. *Nature* **295:**663–666, 1982. Human Met and Leu enkephalin.

24. Cox, E. C. *Annu. Rev. Genet.* **10:**135–156, 1976.

25. Crick, F. *Science* **204:**264–271, 1979. Split genes and RNA splicing.

26. Davidson, E. H. et al. *Devel. Biol.* **55:**69–84, 1977. *J. Molec. Biol.* **56:**491–506, 1971. *Q. Rev. Biol.* **48:**565–613, 1973. *Science* **204:**1052–1059, 1979. Regulation of gene expression in eukaryotes.

27. Dayhof, M. O. *Atlas of Protein Sequence and Structure,* National Biomedical Research Foundation, Washington, DC, 5:Suppl. 3, 1978.

28. del Rey, F. J. et al. *Proc. Natl. Acad. Sci.* **79:**4138–4142, 1982. Repetitive elements in genes of yeast.

29. Dickerson, R. E. *J. Molec. Evol.* **1:**26–45, 1971. Structure and evolution of cytochrome c.

31. Docherty, H. and D. T. Steiner. *Annu. Rev. Phys.* **44:**625–638, 1982. Post-translational proteolysis in polypeptide hormone synthesis.

32. Doolittle, W. F. and C. Sapienza. *Nature* **284:**601–603, 1980. Selfish genes.

33. Efstratiadis, A. et al. *Cell* **21:**653–668, 1980. Evolution of human beta globin genes.

34. Eigen, M. et al. *Sci. Am.* April 1981, p. 88. *Naturwissenschaften* **65:**7–41; **68:**217–228, 1980. Origin of genetic information.

35. Eppler, B. A. and R. E. Mains. *Endocrin. Rev.* **1:**1–27, 1980. Cleavage of pro-ACTH/endorphin.

36. Fambaugh, P. J. et al. *Nature* **274:**765–780, 1978. Methods of gene sequencing; lac I gene.

36a. Felber, J. P. pp. 220–232 in *Biological Adaptation,* Ed. G. Hildebrandt and H. Hensel. Thieme, Stuttgart, 1982. Adaptations of pancreatic exocrine functions in altered diets.

37. Felsenfeld, G. and J. McGhee. *Nature* **296:**602–603, 1982. DNA methylation.

38. Filner, P. et al. *Science* **165:**358–367, 1969. Enzyme induction in plants.

39. Finnegan, D. J. *Nature* **292:**800–801, 1981. Transposable elements in eukaryotes.

40. Fisher, S. E. and G. S. Whitt. *J. Molec. Evol.* **12:**25–55, 1978. Evolution of isozyme loci for creatine kinase.

41. Flanagan, J. G. and T. H. Rabbitts. *Nature* **300:**709–713, 1982. Human immunoglobulin heavy chain genes.

42. Fox, C. E. et al. *Proc. Natl. Acad. Sci.* **74:**4537–4541, 1977. *Science* **209:**357–463, 1980. Classification of methanogenic bacteria and phylogeny of prokaryotes.

42a. Fox, T. *Nature* **314:**132–133, 1985. Exceptions to universal genetic code in protozoans.

43. Gilbert, W. *Nature* **271:**501–502, 1978. *Science* **214:**1305–1312, 1981. Function of pseudogenes; procedures for sequencing.

44. Hartl, D. L. *Principles of Population Genetics,* Sinauer, Sunderland, MA, 1980.

45. Hendry, L. B. *Proc. Natl. Acad. Sci.* **78:**7440–7444, 1981. Stereochemical models of DNA.

46. Holland, J. et al. *Science* **215:**1577–1585, 1982. Errors in speciation of DNA and RNA.

47. Haschemeyer, A. E. *Phys. Zool.* **50:**11–42, 1977. Protein synthesis in cold acclimated toadfish.

48. Hood, L. et al. *Molecular Biology of Eukaryotic Cells,* Benjamin, Menlo Park, CA, 1974.

49. Hudson, P. et al. *Nature* **291:**127–131, 1981. Sequence coding of cDNA for rat relaxin.

50. Hutchison, J. S. and D. Holten. *J. Biol. Chem.* **253:**52–57, 1978. Induction of 6-P-gluconate DH in liver.

50a. Ito, R. et al. *Proc. Natl. Acad. Sci.* **81:**4662–4666, 1984. Gene encoding human gastrin.

51. Jacobs, J. W., et al. *Science* **213:**457–459, 1981. Calcitonin mRNA encodes multiple polypeptides.

52. Jain, H. K. *Nature* **288:**647, 1980. Stereochemistry of DNA.

53. Jeffreys, A. J. and S. Hains. *Nature* **296:**9–10, 1982. Gene duplication.

54. John, B. and K. R. Lewis. *Chromosome Hierarchy,* Clarendon, Oxford, 1975.

54a. Jukes, T. H. *J. Molec. Evol.* **19:**219–225, 1983. The archetypal codon.

54b. Kaine, B. P. and C. Woese. *Proc. Natl. Acad. Sci.* **80:**3309, 1983. Split genes in archaebacteria.

55. Kakidani, H. et al. *Nature* **298:**245–249, 1982. Sequence analysis of cDNA for porcine endorphin.

56. Kanopky, C. *Proc. Natl. Acad. Sci.* **55:**274–281, 1966. Mutagenicity in *E. coli.*

57. Kanungo, M. S. *Biochemistry of Aging.* Academic Press, New York, 1980.

58. Karp, G. *Cell Biology,* McGraw-Hill, New York, 1979. Ch. 10. Nature of genes.

59. Katzenellenbogen, B. *Annu. Rev. Physiol.* **42:**17–35, 1980. Steroid hormone receptors.

60. Kemper, B. and D. F. Gordon. *Nucleic Acid Res.* **8:**5669–5683, 1980. *Proc. Natl. Acad. Sci.* **78:**4073–4077, 1981. Cloning of cDNA for parathyroid hormone.

61. Konkel, D. et al. *Cell* **18:**868–873, 1979. Cloned β-globin genes, mouse.

62. Korn, J. et al. *Proc. Natl. Acad. Sci.* **80:**4253–4257, 1983. Protein structural domains in a nematode.

63. Korn, L. J. and J. B. Gurdon. *Nature* **289:**461–465, 1981. Transplantation of nuclei in Xenopus.

64. Lafer, E. M. et al. *Proc. Natl. Acad. Sci.* **78:**3546–3550, 1981. Z-DNA.

65. Leder, P. et al. *Science* **209:**1336–1339, 1980. *Nature* **293:**196–200, 1981. Globin genes of mouse.

66. Lehninger, A. L. *Principles of Biochemistry,* Worth, New York, 1982.

67. Levis, R. et al. *Cell* **21:**581–588, 1980. Transposable genes in copia *Drosophila.*

68. Lewin, B. *Cell* **4:**77–93, 1975. Units of transcription and translation.

69. Lewin, B. *Gene Expression. I. Bacterial Genomes,* 1978. *II. Eukaryotes,* 1980.

III. Plasmids and Phages. Wiley, New York.

69a. Lewin, R. *Science* **217**:921, 1982. Origin of introns.

70. Liebhaber, S. A. et al. *Nature* **290**:26–29, 1981. Genes of human alpha globin.

71. Lin, Y. and J. K. Gross. *Proc. Natl. Acad. Sci.* **78**:2825–2829, 1981. Cloning of winter flounder antifreeze cDNA.

73. Little, P. F. *Cell* **28**:683–684, 1982. Globin pseudogenes.

74. Lomedico, P. *Cell* **18**:545–558, 1979. Rat pre-pro-insulin genes.

75. Marx, J. *Science* **197**:853–927, 1977; **216**:969–970, 1982. Viral messengers in animal cells; retroviruses.

76. Maxam, A. M. and W. Gilbert. *Proc. Natl. Acad. Sci.* **74**:560–564, 1977. Method for sequencing DNA.

76a. McClintock, B. *Science* **226**:792–801, 1984. Responses of genomes to challenges.

77. McIntosh, L. et al. *Nature* **288**:556–560, 1980. Chloroplast gene sequences in maize.

78. Miller, J. H. et al. *J. Molec. Biol.* **109**:275–301, 1977. Silent mutation.

79. Noller, H. F. and C. Woese. *Science* **212**:403, 1981. Speciation of 16 S rRNA.

80. Nordheim, et al. *Nature* **294**:417–422, 1981. Left-handed Z-DNA.

81. Orgel, L. E. and F. H. C. Erich. *Nature* **284**:604–607, 1980. Selfish genes not contributory to phenotype. Excess DNA.

82. Orgel, L. E., F. H. C. Crick, and C. Sapienza. *Nature* **288**:645–646, 1980. Selfish genes.

82a. Orgel, L. E. et al. *J. Molec. Biol.* **144**:567–577, 1980. *J. Molec. Evol.* **17**:295–302, 302–306, 1981. Synthesis of oligonucleotides without replicase but with RNA templates and metal catalysts.

83. Perier, F. et al. *Cell* **20**:555–566, 1980. Chicken pre-pro-insulin gene.

83a. Peschle, C. et al. *Nature* **313**:235–238, 1985. Switching of hemoglobin genes in human embryos.

84. Potter, S. et al. *Cell* **20**:647, 1980. Inverted gene sequences in *Drosophila*.

85. Powers, D. et al. *Proc. Natl. Acad. Sci.* **78**:2717–2721, 1981. ADH gene of *Drosophila*.

85a. Proudfoot, N. J. *Nature* **286**:840–841, 1980. Pseudogenes.

86. Proudfoot, N. J. et al. *Science* **209**:1329–1335, 1980. Structure and transcription of human globin genes.

87. Razin, A. and A. D. Riggs. *Science* **210**:604–610, 1980. DNA methylation and gene function.

88. Reanney, D. *Nature* **307**:318–319, 1984. Genetic noise in evolution.

89. Rich, A. *Science* **214**:1108–1110, 171–176, 1981; **222**:495–496, 1983. Z-DNA. **80**:1921–1923, 1983. *Sci. Am.* **238**:52–62, 1978. Three-dimensional structure of tRNA.

89a. Roberts, R. B. *Proc. Natl. Acad. Sci.* **48**:897–900, 1962. Specific endonucleases; alternative codes and templates.

90. Roberts, J. L. and E. Herbert. *Proc. Natl. Acad. Sci.* **74**:5300–5304, 1977. Sequence of precursor of ACTH-endorphin.

91. Roeder, G. S. and G. R. Fink. *Cell* **21:**239–249, 1980. Transposable elements in yeast.

92. Rosenfeld, M. G. et al. *Nature* **290:**63–65, 1981. Altered expression of calcitonin genes associated with mRNA polymorphism.

93. Sakano, H. et al. *Nature* **277:**627–633, 1979. Introns and coding of DNA segments of immunoglobulin.

94. Sanger, F. *Science* **214:**1205–1210, 1981. Nucleotide sequence determination in DNA.

95. Sanger, F. et al. *Nature* **265:**687–695, 1977. Nucleotide sequences of bacteriophage.

96. Schmid, C. W. and W. R. Jelinek. *Science* **216:**1065–1070, 1982. Dispersed repetitive sequences in Alu family of genes.

97. Schon, E. A. et al. *Cell* **27:**359–369, 1981. Structure and evolution of hemoglobins of goat.

98. Schuster, P. pp. 15–87 in *Biochemical Evolution,* Ed. H. Gutfreund, Cambridge University Press, New York, 1982. Prebiotic evolution.

99. Shen, L. P. et al. *Proc. Natl. Acad. Sci.* **79:**4575–4579, 1982. DNA sequences of human somatostatin II.

100. Simon, M. et al. *Science* **209:**1370–1374, 1980. Phase variation in genes of bacterial flagella.

101. Smith, M. *Amer. Sci.* **67:**57–67, 1979. Nucleotide sequence of a bacterial virus.

102. Solomon, E. and K. Cheah. *Nature* **291:**450–451, 1981. Collagen evolution.

103. Southern, E. M. *J. Molec. Biol.* **94:**51–69, 1975. Long range periodicities in mouse satellite DNA.

104. Spiegelman, S. *Ann. Rev. Biophys.* **4:**213–253, 1971. Replicating of viral RNA.

105. Spradling, A. C. and D. Mahowald. *Proc. Natl. Acad. Sci.* **77:**1096–1100, 1980. Gene amplifications in oogenesis in Drosophila.

106. Springe, M. et al. *Nucl. Acid Res.* **8:**r1–r22, 1980. Compilation of tRNA sequences.

107. Stein, J. P. *Proc. Natl. Acad. Sci.* **80:**6485–6489, 1983. Tissue-specific expression of chicken calmodulin pseudogenes.

107a. Stern, D. B. and J. D. Palmer. *Proc. Natl. Acad. Sci.* **81:**1945–1950, 1984. Homologies between mitochondrial and chloroplast DNAS.

108. Sures, I. et al. *Cell* **15:**1033–1044, 1978. Histone genes.

109. Suzuki, Y. J. et al. *J. Molec. Biol.* **70:**637–649, 1972. Fibroin genes in *Bombyx.*

110. Warren, T. and D. Shields. *Proc. Natl. Acad. Sci.* **79:**3729–3733, 1982. Somatostatin precursors.

111. Weatherall, D. J. and J. B. Clegg. *Cell* **16:**467–479, 1979. Genetics of human hemoglobin.

112. Wilcken-Bergmann, G. et al. *Proc. Natl. Acad. Sci.* **79:**2427–2431, 1982. Repressor genes in *E. coli.*

113. Wilson, A. C. *Annu. Rev. Biochem.* **46:**73–639, 1977. Biochemical evolution. Stadler Symposium, U. MO., 1975.

114. Woese, C. et al. *Naturwissenschaften* **60:**441–446, 1973. *Proc. Natl. Acad. Sci.* **74:**5088–5090, 1977. *J. Molec. Evol.* **11:**245–252, 1978. Phylogeny of prokaryotes. Ch. 11, pp. 201–233, in *Evolution from Molecules to Man,* Ed. D. S. Berdall, Cambridge University Press, New York, 1983.

115. Wozney, J. et al. *Nature* **294:**129–135, 1981. Structure of collagen genes.

116. Yoo, O. J. et al. *Proc. Natl. Acad. Sci.* **79:**1049–1053, 1982. Sequence analysis of cDNA of gastrin.

117. Zasloff, M. *Proc. Natl. Acad. Sci.* **80:**6436–6440, 1983. Transport of tRNA from nucleus to cytoplasm oocytes.

4

BIOLOGICAL VARIATION, MOLECULAR CLOCKS, AND SPECIATION

BIOLOGICAL VARIATION

Variation is characteristic of life. Variation results from genetic or environmental influences acting within genetic limits. This chapter is concerned primarily with variation that is genetically determined. The term *polymorphism* refers to the presence of differences between individuals in a species; *diversity* refers to the presence of differences between species. In prokaryotes and some asexually reproducing eukaryotes, individuals within a clone are alike, but one clone may differ from another. In eukaryotes variation is evident between species, populations, and, in complex animals, between individuals. Variation is the basis of population genetics and provides the material that may or may not be selected. Experimentalists attempt to minimize individual variation by maintaining constancy of heredity and environment, by inbreeding plants and animals, and by statistical analyses that factor out variations between individuals. However, for naturalists, variation has much biological meaning, and experimentalists could well give more attention to variation, recognizing that not all biological properties follow Gaussian distribution.

Most structural and physiological characters are determined by multiple genes; assortment may not be entirely according to Mendelian ratios. A single gene may affect more than one character, that is, it may be pleiotropic. Virtually all plants and animals are diploid, a few are polyploid; an individual can be homozygous or heterozygous at a given gene locus. In rigorous experiments that test for mutation rates, efforts are made to obtain stocks that are homozygous for a certain chromosome, that is, ones that are isochromosomal. Each cell in one diploid organism carries the same complement of genes, except for somatic mutations or chromosomal alterations or for developmental rearrangements caused by antigenic stimuli. Differences in expression of genes may be temporal—at stages of development—or spatial—in tissues and organs. Expression of a genotype is affected by environmental conditions. Multiple genetic forms of organisms of the same taxonomic status—intraspecific and intrageneric variation—were first described from morphological criteria. Many structure variations are adaptive—animal coloration matching background, plant root structure and leaf pattern conserving water supply.

Variation in Proteins

Many genetic variations are known for proteins, structural and enzymatic. For observing variations, the most widely employed methods make use of charge and size differences of the proteins—electrophoresis, isoelectric focusing, and differential centrifugation; amino acid sequencing is also used. Many functional proteins consist of single polypeptide chains—that is, they are monomeric—and on the assumption of one gene for one polypeptide, variation results from mutations. Other proteins are dimeric or tetrameric and the component chains may be alike or unlike and coded at different genetic loci. Mutant proteins are allotypes (allelic proteins, allozymes). When several loci (sometimes on more than one chromosome) provide the coding, the polymeric proteins are multilocus isotypes, isozymes. Allozymes and isozymes can be identified by bands on an electrophoretic gel; genetic polymorphism is detected within a species by comparing the bands of a given protein in many individuals.

In related individuals, differences in protein patterns supplement morphological differences as the basis for distinguishing populations and species. Both protein differences and morphological differences are descriptive and do not indicate any selective advantage or disadvantage. Some sibling species of animals cannot be readily distinguished morphologically, but their proteins may be distinguishable from the corresponding proteins of other species (53) in electrophoretic patterns. Animal species that are alike in morphology and electrophoretic pattern but differ in behavior are called ethospecies.

A second level of protein analysis is the sequencing of amino acids in a polypeptide and the examination of its primary structure. Further analysis is of folding—secondary and tertiary structure, often done by X-ray diffraction. Sequence analysis requires breaking a long chain into short chains by means of specific proteases and then running the fragments on an amino acid sequencer. From the amino acid sequence it is possible, since each amino acid is coded by a known nucleotide triplet, to estimate the mutations in the DNA locus. Differences in DNA are reflected in mRNA patterns which specify the pattern of amino acids. Comparisons of amino acid sequences are much used for ascertaining lineages or clades. As assayed by presently available methods, the number of DNA sequences in a genome exceeds that of RNA, and there are more variations of mRNA than of amino acid sequences in a protein. Not all DNA is transcribed; some mRNA specifies portions of proteins that are not finally translated or are cleaved away, and some amino acids are not critical for protein function. Nucleotide sequences in DNA and protein structure and sequences have shown that there is much greater variation within populations than had been previously supposed on structural grounds; there are vast reserves of genetic material in most eukaryotes.

Another indicator of variation in animal populations is the number or diversity of species. Several definitions of species are used (Ch. 1), but for the purposes of this discussion, species are distinct populations of a given kind of animal which, when living sympatrically, make little or no gene exchange.

For higher plants, reproductive isolation is not as critical a criterion of speciation as for animals. There is much hybridization between plant "species," and local populations show considerable variation according to their microenvironments. Two populations of *Achillea millefolium,* one from an Aleutian island, the other from the San Joaquin valley of California, cross readily, and F_1 and F_2 generations are fully fertile; yet in their native environments the plants are markedly different in stem length, root habit, and other characters (45). Plants tend to have more alleles per locus than animals, and the incidence of a given allele varies with habitat. In *Pinus ponderosa,* for example, one allele of needle peroxidase has lower frequency, and heterozygosity is greater, when the trees are on a slope facing south than when on a north slope.

For ascertaining diversity, besides electrophoresis and sequencing of amino acids and nucleotides, immunological methods are used. Antibodies are prepared against a pure protein and the degree of reaction with its related proteins is measured and compared. Immunological compatibility is especially informative for estimating the relatedness of organisms from different taxa.

Cytochrome c is a single protein chain which is present with some variations in all aerobic and in some anaerobic organisms. Cytochrome c has undergone many changes during evolution and several cytochromes c occur in bacteria. Hemoglobin (Hb) and lactate dehydrogenase (LDH) are tetrameric proteins. Adult vertebrate Hb consists of two isozymal chains, α and β, either of which can occur as several allotypes. LDH consists normally of two isozymes, A and B, which vary in proportion according to tissue and organism; each of these can vary allelically. Hb and LDH may each have more than two forms, each coded at a different locus and expressed at a different developmental stage (fetal Hb), or in a different tissue (retinal LDH in fishes) (87).

Polymorphism and Diversity   .

Polymorphism is usually considered to be an indication of genetic differences within a species. However, a specific gene may be manifest only in a given environment, at a certain stage in development, or in a given hormonal state. Polymorphism (P) in a large population of an organism is expressed as percentage variation at a given locus. Environmental pressures may favor one phenotype over others. For example, populations of land snails and of moths in Britain have been selected in nature by color patterns corresponding to background, with reduced visibility to predator birds.

Heterozygosity (h) is genic diversity and is given by $1 - x_i^2$, where x_i is the frequency of the ith allele. Average heterozygosity (H) is the mean of hs over all loci examined (101).

One way to measure species diversity is to count the number of species present in a given habitat or community. A measure of variation within a species is the estimation of heterozygosity for a single character, and of average heterozygosity for a number of loci (101).

Multiple alleles at one locus, multiple loci for a single character, and interactions between regulatory and structural genes make for diversity of single characters. The probability of fixation of adaptive mutations in yeasts is greater in diploid than in haploid populations. Evolutionary advantages of diploidy may be: (1) diploidy generates genetic variability; (2) diploidy gives alternatives to deleterious recessive mutations; and (3) diploids evolve faster than haploids in new environments. Polymorphism is greater for polyploid organisms than for diploid organisms; in many plants and some animals (e.g., salmonid fishes) high ploidy provides for greater biochemical potential. Parthenogenic animals, however, may have higher polymorphism than was formerly believed (115).

In the haploid bacterium *E. coli* 829 clones showed genetic heterozygosity of 0.23 for 4 or 5 enzymes (78). In another study on *E. coli,* 20 enzymes examined from 109 clones showed in 18 enzymes 4–5 times more diversity than most eukaryotes show. If the different electromorphs were randomly distributed, the number of combinations would be astronomical, yet the number of types (combinations of enzyme forms) is limited; evidently recombinations are rare (100).

Many of the analyses of mobility of proteins from different populations do not relate the diversity to the adaptive properties of the allotypes and isotypes (for example, nonspecific esterases and phosphatases of unknown function). Electrophoretic patterns are relevant in population analyses, just as morphological characters are in genetics; for either protein or morphological polymorphism it must be emphasized that natural selection acts on the phenotype, not on the genotype *per se*. An alternative view is that some noncoding DNA (selfish genes) may be selected and thus serve like phenotypes (26).

We shall discuss the biological meaning of polymorphism before discussing molecular mechanisms. Polymorphism may be considered at different taxonomic levels: polymorphism consists of variation in individuals or populations within a species, or it may occur as diversity of species within a genus, family, or order. The concept of polymorphism within taxa is so broad that no single ecological or physiological description is adequate.

Balanced polymorphism refers to the maintenance of some proportion of alleles at a single genetic locus; this implies an advantage for heterosis for a particular gene (allozyme). Balanced polymorphism could result from selection in a heterogeneous environment. When multiple loci interact, such as those that encode isozymes, there may be interlocus balance, a kind of polymorphism which is important physiologically. One view of balanced polymorphism is that such diversity has adaptive value; another view is that many neutral genes may be maintained by genetic drift. Probably each view has some validity.

Some hypotheses or interpretations of polymorphism of single and multiple interactive loci are the following:

1. Polymorphism preadapts organisms to heterogeneous environments, to changed or numerous niches; entry into disturbed environments is thus more feasible.

2. Genetic heterosis is sometimes advantageous in hybrids between strains, for example in domesticated animals and plants. Interspecies hybrids may be larger and faster in growth but not fertile (mule, sunfish), or they may be less fit and slower in growth (carp, catfish). In higher plants hybridization is often advantageous for occupation of new niches (95).

3. Enzyme balance may be favorable; multiple forms of an enzyme widen the range of possible metabolic function and buffer against environmental perturbations.

4. A single character, such as a protein allozyme variant, may be carried by linkage to other characters that are adaptive.

5. Metabolic coordination favors homeokinesis (energetic constancy). Energy-yielding pathways may make use of multiple forms of given enzymes.

6. Enzyme diversity may permit entry into new environments.

7. Behavioral diversity may permit exploitation of different biotic environments.

8. Heterozygosity of the total genome favors metabolism, growth, reproduction, and genetic stability.

Evolutionary, Ecological, and Physiological Correlates of Polymorphism and Diversity

Biological correlations of variation with age of species, ecology, environmental heterogeneity, geography, position in a range, and life history have been attempted. Such correlations may apply to some groups, but it has proven impossible to make generalizations. Correlations have been attempted between protein polymorphism and many biological variables.

Attempts to correlate enzyme polymorphism with rates of evolution and speciation have not been successful. The rate of evolution differs greatly in various kinds of plants and animals and during different times in evolutionary history. Mammals during the past 1–2 myr have shown much speciation, amphibians much less, yet protein polymorphism among amphibians is as great as among mammals (125). Some groups of invertebrates have shown little speciation (Limulus), whereas others are more diverse (some insect orders). Polymorphism as measured by protein diversity is high in some very ancient species—Limulus, Tridacna—hence a low level of speciation is not necessarily due to lack of protein diversity (5,102). In Limulus 24 proteins (16 enzymatic and 8 nonenzymatic) were measured in 64 individuals from several localities; the populations were on average polymorphic at 25 percent of loci; these values are not significantly different from human populations with 36 percent, or from *Drosophila similis* with 25 percent polymorphism (101,102). We may conclude that polymorphism is as much maintained in old, slowly evolving species as in recent species.

A survey of heterozygosity of many proteins in a large number of animals indicated that H in invertebrates is greater than in vertebrates from similar environments (89):

| | Heterozygosity | |
Environment	Invertebrates	Vertebrates
tropical	0.11	0.047
temperate	0.13	0.05
marine	0.098	0.05
terrestrial	0.157	0.039

Heterozygosity of proteins does not necessarily correlate with species, morphological or reproductive. Five species of desert pupfish (*Cyprinodon*) from hot springs in southern California show almost no differences in electrophoretic patterns of 20 proteins or in temperature tolerance (16). Of two families of freshwater fishes the Cyprinidae contain more than 200 species, the Centrarchidae 30 species; protein diversity is low but speciation high in Cyprinids, the reverse in Centrarchids. There is also nonagreement between enzyme polymorphism and speciation in sunfish genus *Lepomis*. Eighteen loci were examined in 10 species and on the basis of enzyme patterns four groups of species were identified which do not agree with classifications based on morphology. Some enzymes are monomorphic for a single allele in all 10 sunfish species; in two groups of sunfish species there are 2–4 alleles in common; other sunfish show 5–7 alleles in common. In three species of sunfish, heterozygosity was high—11 percent per individual; in two species it was low—less than 3 percent (3).

Polymorphism is correlated with life histories. *Capitella capita* (polychaete worm) is a complex of at least six sibling species that are similar morphologically but do not interbreed in the laboratory; in nature the larvae of different species disperse at different times, and their breeding season may last from two weeks to three or four months according to species. At least 12 loci for proteins in *Capitella* are polymorphic (46). In general, animals that disperse eggs (ophiuroids) show more polymorphism than species that carry their eggs (many crustaceans) (46).

Two general life patterns have been described by the logistic growth model. Plants or animals of one pattern are continual colonizers, occupy temporary habitats, have short generation times, tend to be small individuals, and to allocate small amounts of food energy to reproduction. Annual plants are examples. Organisms with the other pattern have longer generation times, live where environmental stresses are minimal, tend to be of larger size, and to put more resources into reproduction. Perennial plants, trees and shrubs are examples (71).

Evidence has accumulated that total genomes are homeostatic—that is, that a genome which is altered by a disturbance or stress compensates by self-repair. When a chromosome is broken, the ends reattach, sometimes to other chromosomes, and new linkages are established (76a). Structural genes are influenced by regulatory ones, and the expression of one gene may be influenced by genes at other loci.

In several species of estuarine molluscs correlations have been established between genetic heterozygosity and general fitness. Many individual bivalves (*Munidia*) were grown in separate compartments, and physical measurements were made over an extended period; then each individual was extracted for isozymal and allozymal determinations of some 6 to 9 polymorphic enzymes. In general, heterozygotes showed higher growth rate and food uptake, lower oxygen consumption rate and scope for growth (amount of food to produce a given amount of tissue). The enzymes assayed were for different functions, and the conclusion is that total heterozygosity made for superior growth, survival, and reproductive capacity (38a,b,60a). Similar correlations have been made for plants. The composite plant *Liatris cylindracea* was tested electrophoretically for 27 loci; individual plants high in heterozygosity flowered earlier and more profusely than the less heterozygous ones. General biological fitness is determined by a total genome rather than by single loci. The total genome has a homeostatic function (65a).

A limitation of the usefulness of electrophoretic measurements on proteins is that only a few proteins are examined. Enzymes are more easily studied than structural proteins and polymorphism of some kinds of enzymes is critical. Comparison of polymorphism of enzymes in *Drosophila* and in the butterfly *Colias* indicates in regulatory enzymes such as adenyl kinase and hexokinase a polymorphism greater than in nonregulatory enzymes, such as fumarase, malic dehydrogenase, and triose isomerase; higher heterogeneity was found for enzymes acting on multiple substrates—esterases, phosphatases, peptidases (53).

The physiological advantage of a particular allozyme is shown in a glutamate-pyruvate transaminase (GPT) of an intertidal copepod *Tigriopus californicus*. Adults with a fast migrating allozyme (homozygous GPT[f/f] or heterozygous GPT[f/s]) accumulate free amino acids, especially alanine, in a hyperosmotic medium. A change in salinity evokes a rapid response; also GPT[f/f] is more heat tolerant than GPT[s/s] (homozygous for slow form). Larvae with GPT[s/s] show higher mortality under hyperosmotic stress than GPT[f/f] larvae (18a).

In *Drosophila melanogaster* a fast migrating allozyme of 6P-gluconate dehydrogenase has lower activity than a slow migrating allozyme; cells with the slow form have higher pentose shunt activity and produce more NADPH for lipid synthesis than other allotypes (21a).

Two allozymes of phosphoglucose isomerase (PGI) of a sea anemone from the northeastern coast of the United States differ in kinetics. An electrophoretically slow form is more heat stable; K_m for G6P is higher for the slow allozyme and K_m-temperature curves are U-shaped. The fast allozyme has higher V_{max} values than the slow one at all temperatures. The greatest difference in velocity between allozymes is at 25°C; therefore the fast form has an advantage over the slow allotype at relatively high environmental temperatures. The frequency of occurrence of the fast form is high south of Cape Cod, but only 0.5 in northern Maine (49). These sea anemones show no clinal variation in two loci (2 alleles each) of leu-aminopeptidase and no clinal variation in 3 alleles of phosphoglucomutase. In *Drosophila, Colias,* and *Homo,* heterogeneity is greatest for enzymes that utilize several substrates—phosphatases, esterases, peptidases. Moderately high heterogeneity is found for the regulatory enzymes: adenyl kinase, ADH, G-3PDH, glycerol 6PDH, hexokinase. Heterogeneity is least for nonregulatory enzymes: α-glyc-PDH, aldolase, fumarase, MDH, triose isomerase. No polymorphism was found in P-glucomutase, saliva amylase, pepsin, or carbonic anhydrase. Enzymes that utilize several substrates and that occur at branch-points in metabolism, for example, PFK, have high variability (53,89).

Some ecological correlations with protein polymorphism have been noted. Many kind of plants show local polymorphism according to the space or water available. Five species of goldenrod in the same pasture differ in abundance with local supply of moisture. The goldenrod species differ in general form, seed size, and number of seeds per stem (79,122). Clones of cattail *Typha latifolia* taken from cold and from warm locations in California were grown on the same temperature regime. Apparent activation energy (E_a) for glycolate oxidase was three times greater for the population from the warm locale (77). Genetic variation in self-pollinating, highly homozygous plants is exemplified in oats *Hordeum spontaneum* that showed polymorphism at 28 gene loci and in spikelet structure; structure was correlated with local humidity and soil type (83).

Anolid lizards have dewlaps of pattern, color, and size used in territorial advertisement, courtship, and aggression. The dewlap diversity is greater in dark forests than in light, open areas. Twenty populations of the shore snail

Littornia angulifer from islands of mangrove trees in the Gulf of Mexico were examined for frequency of five alleles of esterase II; differences were noted for populations on islands only 300 m apart; no correlates with physical factors were discerned.

A series of populations in a cline varies over a geographic range. Clinal variation is often correlated with temperature. In the killifish *Fundulus heteroclitus* two alleles are known for one isozyme of lactate dehydrogenase (LDH-B); at the northern end of the range, Halifax, the fish are homozygous for one allele B^b, at the southern end, Florida, populations are homozygous for the allele B^a; the two genotypes occur in latitudinally-related proportions off the mid-Atlantic states (87d,90) (Fig. 4-1). Concentration of ATP in red blood cells of *Fundulus* is proportional to swimming performance in the B^b allotype; maximum swimming speed (4.3 body lengths/second) is greater in fish with B^b than in those with B^a (3.6 body lengths/second). When fish are exposed to low O_2 or to high temperature, Hb:ATP ratio in red blood cells decreases; this alters Hb function and sets swimming speed. O_2 stress triggers hatching and fish with B^aB^a hatch before those with B^bB^b (87d,90). The temperature of the rivers in which steelhead trout live correlate with the gradient for allozymes of LDH and IDH (96).

Clines for allozymes MDH-A, LDH-B, and GPI-B in *Fundulus* are shown in Figure 4-1. Clines for fishes with several allozymes and isozymes of some 13 enzymes have been established for largemouth bass. Figure 4-2 shows the geographic distribution of two alleles of cytosolic MDH-B. Black areas represent frequency of the B^2 allele. The northern and Florida populations differ also in allelic frequency of MDH, IDH, superoxide dismutase, creatine kinase, aspartate transferase, glucose-P-isomerase, and other enzymes. Average total heterozygosity for 8 loci in northern populations is 0.024, for Florida populations 0.041; in regions of intergrades the heterozygosity is 0.51 (87b). The adaptive correlations with allozymes are not clear, but differences are sufficient to conclude that the northern and southern populations are subspecies. Probably temperature at some stage in development is selective, since enzyme-substrate affinity is greater (lower K_m) for the allozyme of MDH predominant in Florida fish (47b).

Enzymes coded by 14 loci showed clinal polymorphism in populations of the butterfly *Colias* from low, intermediate, and high altitudes. Montane populations of *Colias* show polymorphism as high as 35 percent; lowland populations show less heterozygosity—7 percent (53).

Clinal variations in nitrogen metabolism are correlated with salinity in the

Figure 4-1. Distribution of one of two allotypes for three enzymes: LDH B^a, MDG A^b, and GPI B^b in killifish (*Fundulus heteroclitus*) from northern to southern ends of its range, Halifax to Florida. Each enzyme shows a shift in predominant allotype in mid-Atlantic states. Reprinted with permission from D. Powers and A. Place, *Biochem. Genetics*, Vol. 16, pp. 597, 598. Copyright 1978, Plenum Publishing, New York.

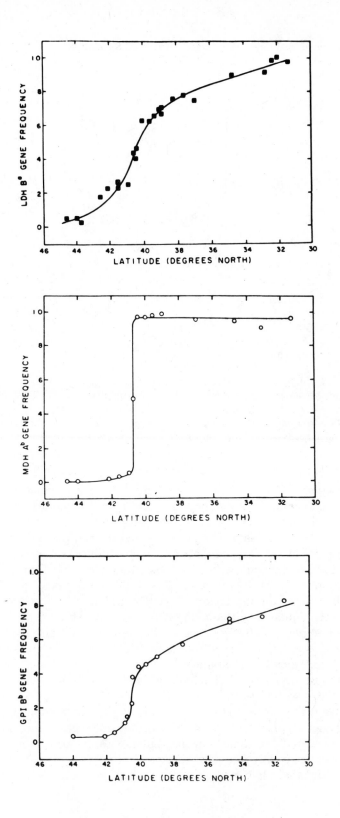

Figure 4-2. Distribution of alloptypes of cytosolic malate dehydrogenase (MDH) in large-mouth bass. Solid symbols represent frequency of the B[1] allele and open symbols represent frequency of the B[2] allele. Reprinted with permission from D. P. Phillip et al., *Canadian Journal of Fisheries and Aquatic Sciences,* Vol. 38, p. 1719. Copyright 1981, Department of Fisheries and Oceans—Scientific Information and Publications Branch, Ottawa, Ont., Canada.

bivalve *Mytilus edulis*. Leucine aminopeptidase (Lap), which removes N-terminal amino acids from polypeptides occurs in three alleles, named from electrophoretic mobility 98, 96, 94. One allotype (Lap-94) occurs with a frequency of 0.55 in specimens from ocean water at the tip of Long Island and at a frequency of 0.15 in specimens from the dilute water of inner Long Island Sound. Specimens with Lap 94 accumulate more free amino acids (FAAs) in their tissues than specimens with other allotypes. In a dilute medium such accumulation is disadvantageous, in that excretion of amines and of some NH_3 regulates cell volume (47a). Juveniles show high mortality of Lap-94 genotypes in dilute sea water, hence the clinal distribution of the three allotypes results from differential mortality during migration from the ocean to inner Long Island Sound (47a,60).

A clinal variation in alcohol dehydrogenase in *Drosophila* in southeastern Asia correlates with local rainfall.

It has been suggested that heterogenous and unstable environments lead to greater phenotypic variety than uniform and stable environments; this has been corroborated for some animals, but not for others. A heterogenous environment, especially one with high productivity, provides many ecological niches and the number of species is often great—for example in coral reefs. Polar environments are less varied, and species diversity is less than in tropic or temperate zones; H (average heterozygosity) for euphausids (crustaceans) in the tropics was 0.21, in temperate waters and the Antarctic 0.058 (120). In contrast to marine invertebrates, *Drosophila* species diversity in the tropics is less than in temperate climates; H for GPDH in the tropics was 0.02, in the temperate zone 0.22 (53).

For 9 species of lizards on islands in the Gulf of California 20 loci were examined (14 populations). Heterozygosity was greater on larger islands, possibly correlated with time since divergence of populations (109).

Clones of leguminous plants *Lathyrus japonicus* collected at four locations of different mean summer temperature showed genetically different malate dehydrogenases. The enzymes of warm-climate clones in New Jersey were more thermostable, less sensitive to high temperatures, and had higher activation energies than the enzymes in cold-climate clones from Hudson Bay (105).

The deep sea provides a uniform environment where nutrients are obtained as they fall from upper regions of the ocean; low temperature, high hydrostatic pressure, and low nutrient availability make for low metabolism and great longevity of animals living there. In this constant environment, species diversity is high (98) (Fig. 4-3). Protein diversity in animals in the deep sea is as great as it is in related tropical species. For 20 enzymes in 13 species the H for deep benthic species was 0.110, for tropical reef species 0.017, for inshore tropical species 0.071, for pelagic species 0.036, and for Antarctic species 0.05–0.025 (108). In 8 species of deep-sea invertebrates 74 loci for enzyme coding were examined; 30–44 percent were polymorphic. Species diversity for fish is high in communities from the depths of Lake Baikal; in shallow African lakes relatively few fish species occur (98). A general correlation exists between constancy of environment and high diversity.

Most electrophoretic measurements have been made under relatively standard conditions. The use of sequential electrophoresis—different gels, different pHs, along with heat denaturation and electrofocusing—have revealed much greater protein polymorphism than had previously been shown by linear gradients. In *Drosophila persimilis* 60 isochromosomal lines from three geographic populations were examined for allozymal variation of the multisubstrate enzyme xanthine dehydrogenase. Sequential electrophoresis showed 23 alleles; only 5 had been distinguished by previous methods (23). In 146 lines from 12 populations of *D. pseudoobscura* 37 allelic classes were obtained, with average heterogeneity of 0.72. In 60 lines of *D. persimilis* xanthine DH showed 47 alleles, ADH 18, and ODH 11. In local inbred human populations hemoglobin variants occur without apparent physiological significance. Most fish have multiple hemoglobins—both isotypes and allotypes. For example, eelpouts (zoarcids) have three different Hb chains—one α- and two β-like—with variants in each (48). Histones show little variability—for histone-4, heterozygosity (H) is only 0.03 between peas and cows.

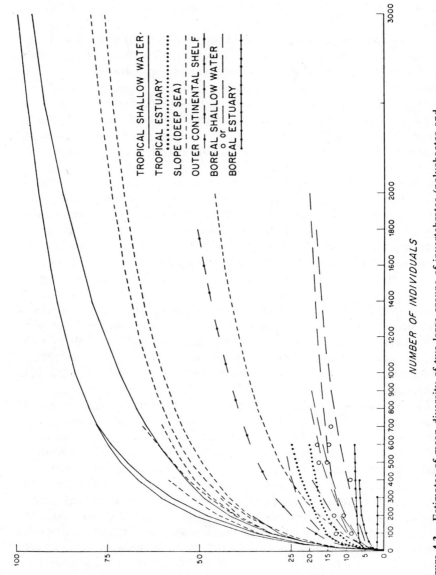

Figure 4-3. Estimates of mean diversity of two large groups of invertebrates (polychaetes and bivalves) as given by number of different species as a function of total numbers of individuals of each phylum from different marine environments. Personal communication from Howard Sanders.

The invasion of new habitats opened up by disturbance makes opportunities for new biological characters. In central California, populations of wild oats *Avena* in the foothills are twice as polymorphic for five loci as populations in the cultivated central valley where habitats are less varied (22).

Genetic changes that confer resistance to insecticides, especially organophosphates, have appeared since exposure of insects to intensive spraying (88). Some of the resistance to insecticides, for example, in red spider mite and green rice leafhopper, is by modification in sensitivity of acetylcholinesterase to organophosphates. Genetic evidence shows the changes to be in single genes. In the fly *Musca domestica,* the poison-resistant effects include changes in rate constants for the activity of the hydrolytic enzyme. Another effect is induction of a nonspecific oxidase that is resistant to the toxins; this change in *Musca* is carried in three genes. Comparable genetic changes have taken place in strains of bacteria that have become resistant to antibiotics. Constantly developing resistance to specific lethal agents leads to a continuing search for new pesticides and antibiotics.

An example of polymorphism resulting from agricultural practice is altered choice of food plants. Maggots of a fly *Rhagoletis pomonoma* formerly fed on hawthorn, then shifted to apples and in some localities to cherry. A sibling species *R. mendax* prefers blueberries and differs from *R. pomonoma* in fumarase electrophoretic mobility (11).

Summary of Polymorphism

Related organisms were first classified on anatomical criteria. Some variants are adaptive—cryptic coloration, courtship behavior in animals, root structure, leaf pattern in plants. The polymorphism of nucleotides in DNA and RNA is greater than is sufficient to code for the differences in sequences in proteins. Neutral or near-neutral nucleotide substitutions are accumulated by genetic drift. There is a great deal of genetic noise that is not translated into proteins, and there is also much protein diversity that may be neutral and not adaptive. Nucleotide sequencing of mRNAs and DNAs shows much greater variation of individuals, strains, and species than was apparent before DNA hybridization and cloning became possible. Proteins are highly polymorphic, as shown by measurements based on charge differences. Proposed correlations of protein variation with age of species, geography, or uniformity of environment have not been substantiated. Many of the variations in protein patterns, hence in coding nucleotides, appear nonadaptive, but subtle physiological measurements may eventually show that they are useful (Ch. 5). Clines of allotypes and isotypes are correlated with temperature, salinity, aridity, and availability of nutrients; populations at ends of clines may be separate species, subspecies, or full species. Genetic homeostasis is indicated by the correlation of biological fitness with mean heterozygosity.

MUTATIONS: NEUTRAL AND ADAPTIVE

Adaptation to the environment is the basis of natural selection and evolution. Selection of adapted phenotypes determines continuance and fixation of gene patterns. Much of adaptational biology is devoted to identifying the properties of proteins, hence of the nucleotides that code their synthesis. There is no evidence that mutations are directed; substitutions in DNA sequences appear to be initiated randomly under normal conditions; some mutagens may act selectively. Frequency of mutation can be altered, for example, by temperature, radiation or mutagenic chemicals.

A few quantitative relations should be stated; the following formulation is by Kimura (57,69):

$$N = \text{effective reproducing population size}$$
$$\mu = \text{mutation rate per gene per generation}$$
$$V = \text{probability of fixation of a mutation}$$
$$k = \text{rate constant for a mutant substitution}$$
$$S = \text{selective advantage}$$

In diploid organisms the number of new mutants per generation $= 2N\mu$ and

$$k = (2N\mu)/2N$$

$$\text{since } V = 1/(2N)$$

$$\text{therefore } k = 2N\mu$$

The number of new alleles per generation $= 2\ Nu$, which for a neutral gene $= 1$. If mutations are neutral $k = \mu$, and is independent of population size. If a mutation is selected, $k = 4NS\mu$, since the time to fix an allele is $4N$ generations.

Two theoretical measure of diversity are heterozygosity H and genetic distance D. Heterozygosity describes variation within a population. Genetic distance measures variation between populations, that is, accumulated number of codon substitutions per locus since time of divergence of two populations.

$$H = 4N\mu/(4N\mu + 1)$$

Heterozygosity depends on both population size and on mutation rate, as shown in Figure 4-4.

Genetic distance D between two populations for a character is as follows (81): Let X and Y be the two populations and X_i and $Y_i =$ frequencies of the

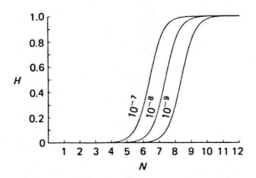

Figure 4-4. Neutral mutation equilibrium relationship between heterozygosity (*H*) and population size (*N*) shown logarithmically. Three different mutation rates. From M. Soule, in *Molecular Evolution*, F. J. Ayala, Ed., Sinauer, Sunderland, MA, 1976.

*i*th alleles in *X* and *Y*. If mutation is random the probability *j* of a particular allele in either population *X* or *Y* is

$$j_X = \Sigma X_i^2 \quad \text{and} \quad j_Y = \Sigma Y_i^2$$

Probability of identity of a gene in both *X* and *Y* is

$$j_{XY} = \Sigma X_i\, Y_i$$

and normalized identity $I = J_{XY}\sqrt{J_X J_Y}$ where *J* is the mean of all *j*'s. Thus genetic distance $D = -\ln I$.

The Hardy-Weinberg relation describes the variability in a population at equilibrium in the absence of forces that affect instability—decrease (by selection) or increase (by mutation, gene flow) when mating is random (67). Exceptions could occur in balanced selection. Let two alleles in the same organism be designated as *A* and *B*; the frequency of *A* is *p* and of *B* is *q*, and if there is no selection of a mutation, $p + q = 1$ or $p = 1 - q$ and $q = 1 - p$. Probability *m* of a homozygote *AA* is p^2 and of *BB* is q^2; *m* of a heterozygote is $2\,pq$. Since gene frequencies do not change at equilibrium,

$$\text{frequency of } A = p = (1 - q)$$
$$\text{frequency of } B = q = (1 - p)$$

The frequency of *A* and *B* alleles after generations of random mating without selection or mutation is the same as that in the gametes that gave rise to it—that is, the initial ratio is maintained. Hardy-Weinberg equilibrium occurs in the absence of selection.

According to the preceding equations, if the distribution of two mutants follows Hardy-Weinberg equilibrium and there is no change in frequency, the selection of either one is unlikely. It follows from the Kimura equation that if the rate of substitution is equal to the mutation rate and indifferent to the size of the effective population, the mutation is neutral. Deviation of frequency of one of two alleles from Hardy-Weinberg equilibrium or from the Kimura equation may indicate selective advantage. The probability that an allele continues or that it is eliminated is determined by its frequency, which depends on selection, mutation, population size, and gene flow. A population is a derivative of ancestral populations. Random fluctuations can occur and can be important, especially in small populations. This is genetic drift. In an isolated situation a few pairs or one pair may establish a population. This is the founder effect (67). Balanced polymorphism is the maintenance of the alleles of a single gene in constant proportion. This usually implies selective advantage of heterozygotes and advantage of continuation of both alleles in a fixed proportion. Physiologically, balance between different loci for an isotypic protein may be important but difficult to analyze genetically. When 23 outbreeding and 7 inbreeding species of plants were compared. Heterozygosity was lower in outbreeders than in inbreeders. Neither pattern corresponded to the Hardy-Weinberg distribution. Inbreeding species showed more multilocus association and more geographic variation than outbreeding species (14).

Selectionists contend that each amino acid substitution that is retained has been selected. Neutralists contend that most changes in amino acid sequences are nonfunctional substitutions. Dynamic genetics demonstrates long noncoding regions in genomes and many nucleotide substitutions that are not reflected in proteins. Noncoding sequences of nucleotides have been called parasitic or facultative and may be retained because there has not been negative selection. For example, a mouse globin pseudogene of no apparent function evolves faster than a functional α gene; the pseudogene has a high rate of nucleotide substitution (101). It has long been recognized by geneticists that most mutations, spontaneous or induced, are deleterious or negative. It is rare that a mutation detected in a genetic experiment is beneficial. It is clear that the number of mutations is far greater than is detected by electrophoretic methods, and that the number of neutral base substitutions in nucleotides of DNA is even greater than the number of changes in protein structure. Many mutants of protein structure are trivial and one amino acid may be as good as another for the protein function. This is especially true for amino acids in the core of a large protein molecule. One form of a protein out of many can become established in an organism by genetic drift. Extensive silent nucleotide sequences may represent selection against sequences that would encode for deleterious amino acid patterns. An example of hidden polymorphism of noncoding nucleotides is alcohol dehydrogenase (ADH) of *Drosophila melanogaster.* Two allozymes, (ADH-fast) and (ADH-slow), differ by a single amino acid substitution. Sequencing of clones shows some 43 allelomorphs; all but one are cryptic and consist of silent or noncoding sequences in both exons and introns. Four of the non-coding regions are highly conserved (62). Some amino acid substitutions may result from errors of translation; these would not be detectable in measurements of nucleotide replacements.

Some proteins evolve at relatively constant rates for millions of years; the rate of evolution of a given protein may be similar in unrelated kinds of organisms (125). The rates of evolution are different for different proteins (42,43). The rate of evolution in number of amino acid substitutions per locus per 10^9 years in vertebrates is, for fibrinopeptides 9.0, pancreatic ribonuclease 3.3, hemoglobin chains 1.4, lysozyme 1.0, insulin 0.5, cytochrome c 0.3, histone IV 0.006 (125,126). The rate of evolution of proteins is proportional to elapsed time, not to number of generations. This suggests that changes at given loci may not be determined by environmental influences.

The amino acid sequences in proteins can differ without measurable functional differences, thus sequence mutation may be neutral; however, tertiary and quaternary structure and composition of critical regions of a protein may be conserved. For example, human and horse α-hemoglobins are different in 42 amino acids, yet the hemoglobins (Hbs) are similar according to several functional measurements (87a).

Selectionists note that gene variations may be adaptive in the aggregate even when individual nucleotide changes are neutral. Changes in families of proteins, sequential enzymes, and balanced parallel pathways may have contributed more to the process of evolution than changes in single proteins. Analysis of single mutations is supplemented by observations on pleiotropisms and linkages. Genomes evolve as integrated units. Examples are cited of amino acid substitutions at a distance from binding and catalytic sites but having effects on enzyme function. The exact sequences at distant sites are not as critical as the sequences at binding sites. Tertiary and quaternary structure may determine the function of a protein. Polymorphism guarantees the retention of characters of dissimilar importance. All phenotypes are capable of some environmentally induced variations, the limits of which are genetically fixed. A phenotypic change that permits extension of ecological range is more likely to become fixed in a population at the environmental limit than in the center of the range. The effect of phenotypic variation on genotypic fixation is called the Baldwin effect (see Fig. 2-1).

When the frequencies of several mutants or heterozygotes (H values) are calculated, many do not conform to the Hardy-Weinberg equilibrium. This means that the distribution of many alleles is not random.

The effect of selection pressure is corroborated by kinetic measurements of enzyme function (Ch. 5). Gross determination of such parameters as O_2 consumption, maximum enzyme velocities, lethality, reduced fecundity, give little information as to what may have been selected. The adapative values of small changes in protein structure may be demonstrated by measurements of kinetic properties, effects of ionic strength, entropic and enthalpic energies, and selective synaptic failure.

In summary, there is much genetic "noise"; not all nucleotide substitutions that are retained for generations are demonstrably adaptive. However, it is certain that in many structural genes small changes code for protein alterations that correlate with fitness in certain environments.

MOLECULAR CLOCKS

One objective of protein analysis is to ascertain the degree of relationship between populations within a species, and between closely related species, by measurements of amino acid sequences, charge, molecular size, and conformation. A second objective of protein analysis is to learn, by comparing related organisms, the age of taxonomic groups and the time of divergence from ancestors. Times for attainment of protein diversity are matched against ages deduced from fossils for calibration of the "protein clock." A third objective of the study of protein structure is to ascertain the evolution of families of proteins—that is, the relations not between taxa of organisms but between proteins that have similar general functions but differ in details. Analysis of the relations within families of proteins and nucleotides may help to account for the excess DNA in eukaryote genomes, for the long precursors in protein synthesis, and for the extreme polymorphism of proteins.

Methods

Several methods are used for estimating the times of divergence of given protein sequences and, from these, the degree of relationship and genetic distance between taxa. Each method uses the number of amino acid (aa) substitutions or of base pair (bp) replacements in DNA and mRNA codons. One procedure is to prepare a matrix for a sequence in a given protein or nucleic acid derived from each of several organisms that are to be compared. Abbreviations for amino acids or nucleotides are used in alignment in the matrix (50), as in the following example: amino acids (aa) are abbreviated by first letter, and a bracket indicates unidentified amino acids. Codons for mRNAs corresponding to specific amino acids are listed in Chapter 2. X refers to any base, Y to either pyrimidine, and Z to either purine.

								Amino Acids	mRNA Codon
amino acids	1	2	3	4	5	6	T	= threonine	ACX
organism	a	[ ]	I	L	M	T	Q	Q	= glutamine	CAZ
	b	D	I	L	M	S	Q	K	= lysine	AAZ
	c	D	V	E	M	T	K	S	= serine	UCX, AGY
	d	D	V	E	[ ]	T	Q			

I	=	isoleucine	AUY, AUA
L	=	leucine	CUX, UUX
E	=	glutamine	GAZ
D	=	aspartate	GAY
M	=	methionine	AUG
V	=	valine	GUX

The number of *differences* between organisms a, b, c, d may be shown in two ways:

By aa alignment:

	a	b	c	d
a	—			
b	2	—		
c	4	4	—	
d	4	4	2	—

By nucleotide codons:

	a	b	c	d
a	—			
b	2	—		
c	5	5	—	
d	5	5	2	—

The matrix is used to derive presumed branch points or nodes that represent ancestors (Fig. 4-5).

Another procedure is to prepare branching diagrams with legs corresponding in length to mutation distance obtained from the number of substitutions. For example, three animals A, B, C show distances for a protein indicated in Figure 4-6.

Matrix analysis has been made of proteins, DNAs, and RNAs. Trees (cladograms) are constructed in which the length of branches corresponds to the number of amino acid or nucleotide replacements and each branch point is the average sum of mutations (9,80). By the use of computers it is

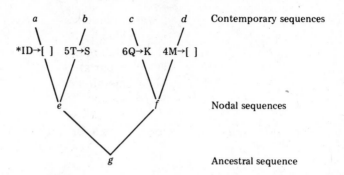

Figure 4-5. A genealogical tree of sequences *a, b, c,* and *d* in text table. Reprinted with permission from L. Hood et al., *Molecular Biology of Eukaryotic Cells,* Copyright 1975, Benjamin Inc., Menlo Park, CA.

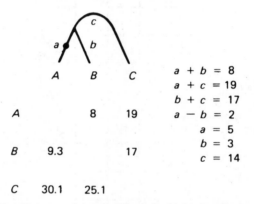

	A	B	C	
A		8	19	$a + b = 8$
				$a + c = 19$
				$b + c = 17$
B	9.3		17	$a - b = 2$
				$a = 5$
				$b = 3$
C	30.1	25.1		$c = 14$

Figure 4-6. Branching diagrams of data in Fig. 4–5 with leg length corresponding to mutation distance. From M. Soule, in *Molecular Evolution,* F. J. Ayala, Ed., Sinauer, Sunderland, MA, 1976.

now possible to compare full sequences in proteins, for example, the 110 amino acids in homologous cytochromes c.

Examples

Immunological differences agree with the number of amino acid substitutions per 100 residues. If two species have diverged from a common ancestor, the number of sequence differences or immunological differences divided by divergence time based on the fossil record gives the rate of macromolecular evolution in each lineage (125) (Fig. 4-7).

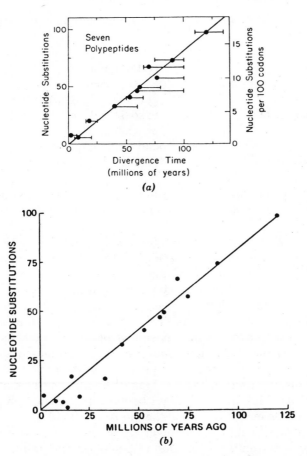

Figure 4-7. (*a*) Estimates of number of nucleotide substitutions for seven proteins in mammals as a function of time since divergence in millions of years based on fossils. Peptides are cytochrome c, myoglobin, α-Hb, β-Hb, fibrinopeptides A and B, insulin C. Reproduced, with permission, from the *Annual Review of Biochemistry*, Volume 46, © 1977 by Annual Reviews, Inc. (*b*) Nucleotide substitutions as a function of paleontological time; 7 proteins and 17 mammals; outermost point represents marsupial-placental divergence. From W. Fitch in *Molecular Evolution*. F. J. Ayala, Ed., Sinauer, Sunderland, MA, 1976.

Unit evolutionary period (time in 10^6 years for 1% difference in aa sequences between two lineages):

histone 4	400	prolactin	5
collagen	36	carbonic anhydrase B	4
GlutDH	55	Hb α	3.7
LDH-A$_4$	19	Hb β	3.3
LDH B$_4$	13	carbonic anhydrase C	2.1
cytochrome c	15	fibrinopeptide B	1.1
insulin	14	immunoglobulin	0.7

Rates of protein mutations in PAMs (accepted point mutations) per 100 residues per 100 myr (27):

histone (H4)	0.1	lactalbumin	27
glucagon	1.2	kappa casein	33
cytochrome c	2.2	glutamate dehydrogenase	0.9
lactate dehydrogenase	3.4	vasoactive intestinal peptide	2.6
insulin	4.4	parvalbumin	7.0
myoglobin	8.9	carbonic anhydrase	16
hemoglobins α and β	12	pancreatic ribonuclease	21

Evolution is slow in conservative proteins that are relatively indispensable and in proteins in which genetic change would be deleterious to the organisms.

Histones show the least variation throughout evolution. The function of histones in providing the structural skeleton of chromosomes has not changed in 10^9 years. Sea urchin histone genes are exceptional in containing more silent nucleotides than are known in any other organisms (116). Histones of ciliates (such as *Tetrahymena*) are so different from those of other eukaryotes as to indicate an early divergence of ciliates from the main eukaryotic stem (39). Cytochrome c is similar in all aerobic cells. In rapidly evolving proteins there are many sites where several amino acid substitutions may occur without change in the function of the protein; examples are serum albumins and fibrinopeptides. Hemoglobin is intermediate in rate of evolution. In insulin, region C bridges between the A and B regions; more sequence variation in C is observed than in A and B (125).

When immunological distance (diversity of amino acid sequences) for a given protein is plotted as a function of geological age, a relatively linear relation appears (6,36,126). Serum albumin of 580 residues shows immunological distance corresponding to a divergence time of mammals and birds of 300 myr. Albumin evolved rather steadily, 1.7 immunological units per million years in several mammals (36). Cytochrome c nucleotide replacements per 10 codons per 10^8 yr are estimated at 6.6 in the transition from invertebrates to vertebrate ancestors; 5.3 from vertebrate ancestors to amniotes (500 myr), 3.0 from amniote ancestors to birds, and 1.8 from primate ancestors to old world monkeys (35 myr) (41).

An atlas of members of protein families (27) lists 106 cytochromes c, 43 fibrinopeptides A, and 35 fibrinopeptides B. Within protein families resemblances in sequences are noted for avian lysozyme to mammalian lactalbumin, insulin to relaxin, parvalbumin to troponin C. In proteins of a single family the sequences are so similar as to indicate a common ancestor.

The proteins of a number of primates have been compared (125). In *Homo* (man) and *Pan* (chimpanzee) α Hb differs by one amino acid, serum albumin differs by six, and carbonic anhydrase by three amino acids; of a total of 2633 amino acid sites in man and chimpanzee, 19 are different. These differences are, by immunological methods, 7.2/1000, and by electrophoresis 8.2/1000 (36,59). The times of divergence of Homo from Pan are estimated on the basis of differences in (1) Hb 1–1.5 myr ago, (2) immune differences 4.2–5 myr, (3) fossils 3–4 myr. The short time for Hb difference has been interpreted as decelerated evolution (41,43). Gorilla Hb differs by two amino acids from Pan and Homo.

Evolution of primates has been claimed to provide evidence for both saltation and gradualism. Several primate genera persisted for millions of years and then died out. Two lines, the hominids and pongids, diverged from the line that gave rise to old world monkeys. The hominid line includes chimpanzee, gorilla, and human, which are very similar biochemically, and *Australopithecus* which is known only from fossils. The four hominid genera apparently separated from a forebear about 405 myr ago; *Homo* is known from 2.2 to 1.8 myr ago. The extant members of the hominid clade have been compared as to morphology, immunology, DNA hybridization, chromosome banding, and electrophoresis of some 23 proteins (18). *Homo* and *Pan* have nearly identical sequences in fibrinopeptides, cytochromes c, a, b, and α and β hemoglobin; *Homo* and *Pan* differ by only three amino acids in carbonic anhydrase and by six amino acids in serum albumin (125). In humans, mitochondrial DNAs differ by only 0.36 percent in nucleotide sequences; intraspecific differences in nonprimate mammals are 3 to 30 times greater (34). There is no correlation of genetic differentiation (determined by electrophoresis) with speciation; differences *between* species may be less than *within* species (24). Divergence of pongids from hominids apparently occurred about 10 myr ago; the surviving pongid is the orangutan; *Sivapithecus* and *Ramapithecus* became extinct (2).

Divergence at the time of the hominid-pongid transition indicates transilience—a few genes brought about large changes. The differences in brain structure, especially in the speech area, and the resulting behavior are described in Chapter 10. Behavioral changes were probably critical in the evolution of primates; stasis of punctuated equilibrium was minimal (40). Changes in regulatory loci may have been more critical than point mutations of structural genes.

Proteins from 36 pairs of bird species that are capable of hybridization differ in immunological distance from each other for albumin by 12 units and transferrin by 25 units. From known rates of evolution of these proteins it is estimated that the average sibling species bird pair diverged from a common ancestor about 22 myr ago. Corresponding periods for hybridizable species of frogs are 21 myr and for placental mammals 2 to 3 myr (91).

Evidence from sequences and from immunological distances for numerous proteins (especially serum albumin lysozymes and ribonucleases) indicates that specific proteins have evolved at characteristic rates and that each protein has its own rate of evolution. The range of rates of evolution of proteins is greater than 100-fold (27).

Acceleration of protein evolution at the time of radiation of jawed fishes and tetrapods is indicated by cladistic analyses of gamma globins, cyto-chromes c, hemoglobins, and Ca-binding proteins. During the emergence of jawed fishes and tetrapods, evolution was five times faster than during the following 210×10^6 years. Acceleration occurred during evolution from early mammals to primates, and the rate slowed from early primates to man (44). Periods of deceleration bring about evolutionary stabilization (homeo-stasis) and provide buffering against environmental changes (40). Differ-ences in rates of evolution of proteins indicate that the changes are nonran-dom. Probability of ancestral common proteins is indicated by homology between pairs of proteins from the same animal as follows: βHb and αHb 93.2 percent, LDH A and LDH B 75.2 percent, CA-b and CA-C (carbonic anhydrase) 50.6 percent; more distant ancestry is shown in the similarity of parvalbumin and troponin C, 22 percent. Midge (insect) Hb differs from hag-fish Hb by 26.4 percent (30). Two mouse β-globin genes differ by 17 nucleo-tides, of which 7 are noncoding, and 10 produce 9 amino acid replacements; these genes may have separated from a common ancestor 50 myr ago (61).

Vertebrate Hemoglobins

Vertebrate hemoglobins have been examined for (1) amino acid substitutions at corresponding positions in the different Hb chains, (2) substitutions in corresponding positions in Hbs from different species, and (3) substitutions in critical functional regions of the molecules compared with distant sites. The phylogeny of different globins in vertebrates is indicated in Figure 4-8. The most primitive chordate Hb was probably a monomeric molecule similar to the Hb of cyclostomes (*Agnatha*); myoglobin persists as a monomer in other vertebrates. Some association of the monomers may have occurred in primitive jawed vertebrates. Many functional changes resulted from the transition from monomeric to tetrameric Hb, the most important being coop-erativity among the four chains. Probably the first such Hb was a homotet-ramer which is designated as $\alpha_2\alpha_2$. A gene duplication in early jawed fishes at about 500 myr ago resulted in formation of two Hb lines—α-like and β-like (Fig. 4-8c). Some 27 of 30 subunit sites for contact between chains occur in similar positions in α and β chains. Once α and β loci were separated, tetrameric Hb consisted of heterotetramers ($\alpha_2\beta_2$). Many changes within each chain occurred in fishes; these include α-β contacts which are respon-sible for cooperativity in O_2 binding, sites associated with effects of pH on O_2 loading and unloading (the Bohr effect, Ch. 6), and sites at which organo-phosphates bind (ATP in fishes, 2,3 diphosphoglycerate in mammals).

Soon after the α-β separation, duplications resulted in the formation of embryonic Hbs. The evolution of both α and β chains was rapid in fishes; evolution of positions for critical functions was four times as fast as for ex-ternal positions. The rate of evolution of hemoglobin slowed in early tetra-pod divergence. Goodman (42) estimates that between an amniote (reptile) and a eutherian (mammal) ancestor, the period from 300 to 90 myr ago, the rate of β and α evolution was only 8–9 NR/100 codons/10^8 yr. Evolution was faster in marsupials than in other mammals (41,42,43). From mammalian ancestors to primates, the period from 90 to 65 myr ago, there was a four- to fivefold acceleration. From a primate ancestor to man (35 myr to the present) the rate decelerated and α chain substitutions were only 2 NR/100 codons/10^9 yr.

In early mammals a duplication of the β locus occurred with the appearance of the Hb gene; β is similar to γ in 73 percent of sequences (30); γ replaces β in fetal Hb. Several duplications occurred in different groups of fishes so that in many fishes there are several different β loci (isotypes) (43). In mammals there are introns within α and β loci and α and β genes are located on different chromosomes. In very early embryos of primates, εHb precedes γ. The β chain has duplicated in adult primates to form a δ chain. Evolution of the α chain has been slow—there are no differences between *Pan* and *Homo* (31). In birds, of six residues that bind inositol pentaphosphate (IPP), four are the same as for DPG in mammals, two are different. No changes occurred in these sites in the avian lineage; in birds a mean of 5.7 nucleotide replacements occurred in the β lineages, 22.8 in α lineages. Dates in myr before the present for the appearance of divergence computed from Hb evolution are compared with paleontological dates as follows (40):

	Protein Clock Date (myr)	Paleontological Date (myr)
gnathostome (first jawed fish)	485	395–425
teleost-tetrapod	387	370–400
amphibian-amniote	260	340–350
bird-mammal	180	290–320
eutherian	90	70–90

From the Hb data, acceleration appears to have occurred after the emergence of jawed vertebrates, followed by deceleration in the amniote lineage. The fastest early evolution was for sites used in O_2 unloading; later evolution was for sites of contact between α and β chains, for the Bohr effect, and for organophosphate binding.

Cytochromes c

Cytochromes c evolved in early prokaryotes and may have provided a structural frame on which monomeric myoglobin-hemoglobin later developed. Some features of cytochrome c (cyt c) are conserved and are critical as binding sites for cytochrome oxidase (cytox) and for cytochrome c reductase (cyt c red) (29). Binding sites of high and low affinity for cytox occur in eukaryotes (73). A cladogram for cyt c is presented in Figure 4-9, which indicates similarities of cytochromes c of fungi, higher plants, invertebrates, birds, and mammals. The rate of change was steady throughout eukaryote evolution at 8–9 NR/100 codons/10^8 yr; the rate slowed to 3 NR/100 codons/10^8 yr

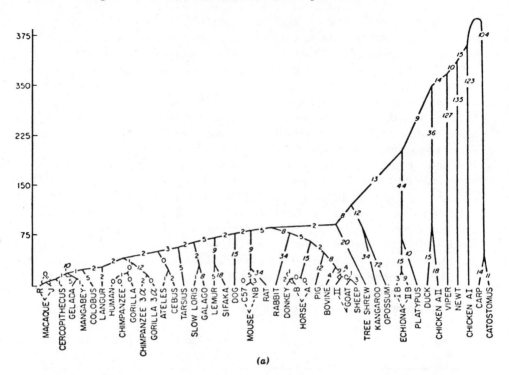

(a)

Figure 4-8. (*a*) Genealogy of 45α hemologin sequences and 726 nucleotide replacements. Ordinate scale in millions of years. (*b*) Genealogy of 28 β hemoglobins requiring 514 nucleotide replacements. (*c*) Evolutionary tree of human globulins. (*a, b*) Reprinted with permission from M. Goodman, *Molecular Anthropology,* Copyright 1976, Plenum Publishing, New York. (*c*) Reprinted with permission from A. J. Jeffreys, in *Evolution from Molecules to Man,* D. S. Bendall, Ed., Copyright 1983, Cambridge University Press, Cambridge, UK.

from early vertebrates to amniotes and then accelerated from therian ancestors to primitive primates but decelerated in higher primates. In lower primates a surge of evolution occurred and changes in oxidase and reductase interactions evolved 13 times faster than the amino acid sequences in external positions (42). Cytochrome c of *Tetrahymena* is the most generalized eukaryotic cytochrome c thus far sequenced.

Ca-Binding Proteins

Calcium-binding proteins constitute a family for which sequence data suggest a series of evolutionary divergences (Fig. 4-10). Calcium-binding proteins are known only in eukaryotes and function in the transfer of calcium to catalytic proteins that require Ca^{2+} to regulate their activity. Many enzymes are modulated by calcium and have high affinity for it when activated, but release Ca when inactive. The Ca modulator family consists of seven members; well characterized are: (1) calmodulin, which probably occurs in

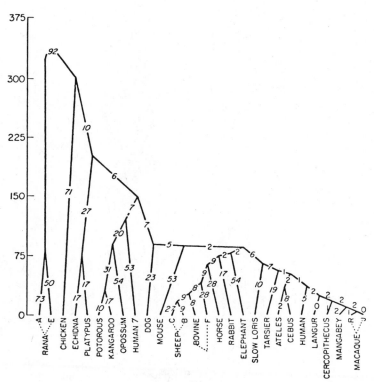

(b)

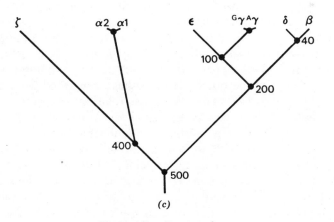

(c)

Figure 4-8. *Continued*

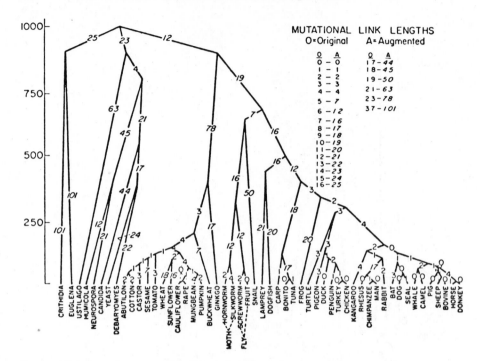

Figure 4-9. Genealogy requiring 521 nucleotide replacements for 53 cytochrome c amino acid sequences. Ordinate in million years. Reprinted with permission from M. Goodman, *Molecular Anthropology*, Copyright 1976, Plenum Publishing, New York.

most eukaryotic cells, both plant and animal; (2) troponin c of striated skeletal muscle; (3) myosin regulatory light chains (mollusc); (4) alkaline light chains of myosin; (5) parvalbumin from muscle of fishes and turtles (small amounts also in mammals, birds, and *Limulus*); (6) an intestinal calcium-binding protein, and (7) a phenylalanine-rich acidic Ca-binding protein. The latter two differ from the others and may have separated from an ancestral family line very early (44). There are close relationships between parvalbumin and regulatory light chains, and between troponin c, calmodulin, and the alkaline light chain (Fig. 4-10). Calmodulin binds four calcium ions; it has four domains, each with 30 to 40 amino acid residues and each binding one Ca^{2+}. Calmodulin is an acidic protein of molecular weight 16,700 that modulates Ca^{2+} binding to numerous enzymic proteins (kinases). Spinach calmodulin is very similar to that of beef brain; calmodulin of the coelenterate *Renilla* differs from that of beef in only six residues. Clearly, calmodulin is highly conserved and has evolved very slowly. Later in evolution parvalbumin and light chain myosin lost Ca-binding in domain II (7,44). In each lineage of the Ca-binding proteins, early evolution was six to ten times faster than recent evolution. Calmodulin of *Tetrahymena* differs from all others;

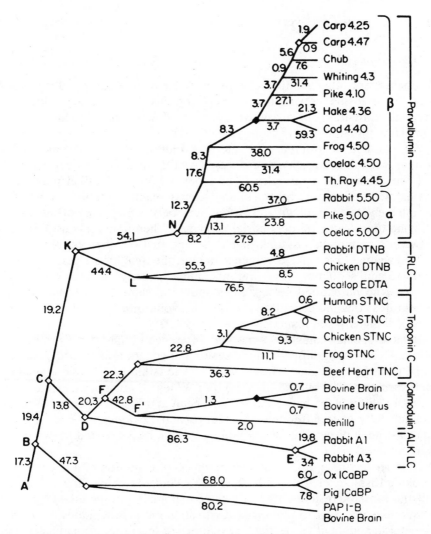

Figure 4-10. Genealogical tree for family of calcium-binding proteins. Numbers on branches indicate genetic distances from ancestral branch points. Reprinted with permission from M. Goodman et al., *J. Molec. Evol.,* Vol. 13, p. 336. Copyright 1979, Springer-Verlag, New York.

this corroborates the very early divergence of *Tetrahymena* from other eukaryotes.

Other Proteins

Cladograms of other families of proteins have been constructed (9): neuropeptides, myosins, several neurotoxins from invertebrates, and glucagons. Carbonic anhydrases occur as two isozymes; CA II appears to have evolved more slowly than CA I in recent years (72,117).

An example of difference in rate of evolution of two isozymes of one protein is alanine-aspartate-transferase (AAT); one isozyme occurs in cytosol, the other in mitochondria. The isozymes have identical polypeptide chains (400 amino acid residues) and similar reaction features, but micro-complement tests show greater differences between species of the cytosol isozyme than of the mitochondrial isozyme. When immunological distance between a mammalian cytoplasmic AAT and that of other vertebrates was plotted against time since divergence in myr, the time for a 1 percent change in amino acid sequence was 6 myr in mammals, 12 myr in other vertebrates, whereas the mitochondrial isozymes show the same rate of change in all vertebrates; this indicates that the rate of evolution of cytoplasmic protein (AAT) changed when mammals emerged (97b).

Related species can show a marked difference in the rate of protein evolution. This is exemplified in three sea urchin genera in the Caribbean compared with the same genera in the Pacific. The Isthmus of Panama separated each species pair about 2.5 myr ago (66). When 18 nucleotide loci are compared for differences between the Caribbean and Pacific species, *Diadema* urchins show little difference, *Euchidoris* and *Echinometra* each show 20 times as much difference. Much more mutation has occurred in two genera than in the other.

Another example of the inconstant tempo of evolution of a protein is superoxide dismutase, for which there are calculated to be 30.9 amino acid substitutions/100 residues/10^6 yr for three mammals, compared with 5.8 substitutions from yeast to mammals. Critical amino acid residues for binding to metal ligands and for tertiary structure are the same in mammals, *Drosophila*, and yeast (65a). A comparison of coding regions of 11 genes suggests faster evolution in rodents than in humans. More changes occur in 5' and 3' untranslated regions than in other gene regions (127a).

Mitochondrial DNA is transmitted through maternal inheritance. Mitochondrial DNA nucleotide sequences were compared with DNA sequences in nuclei of cells of four primates. The rate of mutation for the tRNA site was 0.02 substitutions per base pair per 10^6 years in mitochondria, 10 times more than in the nuclear DNA. Also, mitochondrial genes for rRNA showed more rapid change than nuclear genes for rRNA (17,17a). Some nucleotides in mtDNA may be free from evolutionary constraints. It has been estimated from sequence analysis of the gene for cytochrome oxidase II that 94 percent

of the nucleotide substitutions in mtDNA of *Rattus norvegicus* are silent
(17a). Thus substitutions for genes that encode for amino acids are fewer
than substitutions in the total genome. Wild Danish mice have mtDNA like
Mus domesticus from southern and western Europe, and have nuclear DNA
like *Mus musculus* from Scandinavia and eastern Europe. The mtDNA in
Danish mice may have come from one *M. domesticus* female (34a). In plants
(maize) amino acid sequences show that chloroplasts and mitochondria
evolved separately. In plants the mitochondrial genome has four to five times
as many base pairs as there are in animals.

Similarity of structure of 5 S rRNAs was noted in Chapter 2 (Fig. 2-2).
All of these RNAs are 118 to 122 nucleotides in size. Comparison of nucleotide sequences in 5 S rRNAs indicates an early separation of two groups of
eukaryotes, (1) the ancestors of vascular plants and green algae and (2) the
ancestors of metazoa, protozoa, and slime molds. A phylogenetic tree based
on 5 S rRNAs in 54 metazoans and protozoans is given in Figure 4-11. There
is remarkable agreement with cladograms based on other criteria (85a).

Phylogenetic Conclusions

Amino acid sequences in proteins reflect nucleotide sequences in mRNA,
and these correspond to coding in DNA. Probably more than half of nucleic
acid substitutions are silent changes not translated into protein structure.
Noncoding sequences may become fixed by genetic drift, and the evolution
of coding sequences by selection of proteins may protect silent regions of
DNA from extinction. Some noncoding regions are essential for transcription of specific genes. Differences in mitochondrial DNA sequences indicate
more rapid changes than in nuclear DNAs. Nucleotide sequences of DNAs
indicate faster evolution in noncoding regions.

Analyses of the same proteins of different plants or animals, of different
members of a family of proteins, and of regions of related proteins elucidate
molecular mechanisms of adaptation. Some indispensable proteins, such as
histones, have changed little or not at all since the time of primitive ancestors. Histones fit into helical grooves of DNA, hence the entire histone molecule is essential. Some proteins, such as cytochromes c, have persisted
throughout evolution with few substitutions since early prokaryotes. Other
proteins, such as hemoglobin, may have arisen independently several
times—in N-fixing bacteria of legumes, in protozoans, in animals of several
invertebrate phyla, and in ancestral vertebrates; hemoglobins in vertebrates
have changed by gene duplications and by a relatively small number of point
mutations. Other proteins have diverged as branches and new members of a
family. Still other proteins, particularly small ones, such as hormonal polypeptides and fibrinopeptides, have been subject to many substitutions, especially in regions distant from active sites. In many proteins long regions

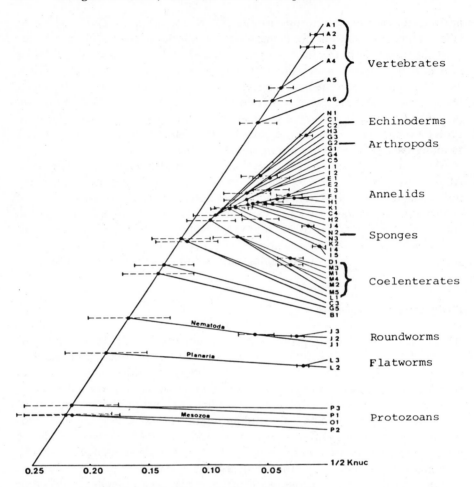

Figure 4-11. Phylogenetic tree of animals based on 5 S rRNAs; 73 sequences from 54 multicellular and 11 unicellular animals. Relationships are given in terms of thousands of nucleotides (Knuc). Reprinted with permission from T. Ohama and T. Kumazaki, *Nucleic Acid Research*, Vol. 12, No. 12, p. 5104. Copyright 1984, IRL Press, Washington, D.C.

that are not critical for function have evolved more than regions of active binding.

Cladograms of five multilocus isozymes in chordates show divergence at different times. Some separations occurred very early, some enzymes moved to other chromosomes—LDH-A and LDH-B. Other enzymes arose more recently—LDH-C from LDH-B, creatine kinase-B from CK-A, G3PDH-A. Some isozymes are expressed in certain tissues: muscle has LDH-A, G3PHD-A, CK-A. Some isozymes occur only in certain classes of animals—LDH-C in teleosts (35).

Proteins have evolved at different rates. For some, the rate of change appears to have been relatively constant; for many others, the rate has accelerated and decelerated. Acceleration has often been associated with radiation of species and of higher taxa; deceleration has often been associated with stability of structure. An explosive phase in evolution is often followed by a slowed rate of change—stabilized selection. In microevolution, some mutations may be neutral; in macroevolution, mutations leading to new proteins in a family and to structural changes adaptive to new functions have occurred. Not all regions of one protein molecule are subject to the same rate of change. In enzymes, the critical sites of catalysis and coenzyme binding are more conserved than distant sites. Albumin lysozymes of chicken and of bobwhite quail show four amino acid differences, of which three are in buried positions, one in a region of antigenic determinant (91).

Data for comparing proteins of distantly related taxa are less reliable than for comparing closely related taxa. The degree of kinship between living organisms estimated from several different proteins usually shows agreement. Genealogical comparisons of families and orders are likely to be less reliable than comparisons of species. Paleontological data provide means of calibration for rates of biochemical change. Times of divergence estimated from fossils are subject to some differences of interpretation. For many closely related species, a few proteins (cytochrome c; α and β hemoglobin) have been sequenced. For recently diverged species more detailed cladograms are available, and the fossil record is better than for more ancient divergences. The more extensive data indicate differences in the rate of evolution of various proteins and in the rate of appearance of new members of protein families. For some proteins it is possible to measure rates of evolution of different domains and linkages. Most proteins which have been examined have enzymic or transport functions in metabolism. Little is known of the evolution of the proteins that take part in cell membrane functions, differentiation and development, and replication or transcription of gene messengers.

SPECIATION

Definitions of Species and Speciation

Diversity characterizes all forms of life, and a recognition of diversity is essential for basic and applied biology. One measure of diversity is the number of species. Several definitions of species and modes of speciation are recognized (Ch. 1). The taxonomic definition of species is a group of organisms distinguished by "key" characters, usually of structure. Taxonomic species represent the best judgment of specialists. Some systematists, the lumpers, classify organisms into few species; others, the splitters, classify plants and animals into many species. The systematist's classification is useful to experimentalists for identifying organisms and may, but need not, correspond to phylogenetic relationship. Numerical taxonomy puts many characters on a computer program and makes a quantitative classification based neither on the opinions of systematists nor on whether key characters are adaptive. Phylogenetic trees, cladograms, show plant and animal relationships without reference to adaptation.

Biological species are defined as reproductively isolated population units; separate species are organisms of the same kind that do not exchange genes. This definition of species has limitations and exceptions. Interspecific hybridization is possible for many kinds of organisms (some fishes, plants); but the hybrids (F_1s) between biological species are often infertile (e.g., between species of sunfishes, between horse and donkey). Sterility may be in F_2 generations rather than in F_1s. Local populations within a cline may interbreed freely, while the populations at the outer extremes are incapable of interbreeding with one another; examples are some circumpolar and island birds and sibling species of *Rana pipiens* (76,79a). Reproductive isolation may be by immunological or cytological incompatibility, failure of development of hybrid embryos, differences in time of sexual maturity, and behavior that favors mating only with a conspecific (74).

In higher plants, many hybrids are fertile; some hybrid offspring can exploit habitats unfavorable for parent species. Plant population biologists consider species as complexes of populations rather than as reproductively isolated units (51,95). Local populations of plants may be highly adapted. Most plant species have limited distribution; a local population may develop distinctive characters that persist as genetic entities. Evolution of plant species is by geographic and ecologic differentiation, hybridization, polyploidy, asexual reproduction, and autogamy (self-pollination) rather than by reproductive isolation, as in animals.

No two species can occupy throughout their life cycles the same ecological niche or geographic range. Each species is physiologically adapted to its niche and range; an account of adaptive properties is necessary for a complete species description. Physiological species are described by the functional characters on which natural selection acts. Description of physiological species is more difficult than of reproductive isolation or morphological characters of taxonomy. For asexually reproducing organisms, classification is necessarily based largely on physiological characters. Unique adaptations include not only the physiological properties associated with the physical parameters of an environment but also the characters that limit reproductive continuity.

Sibling species are those that are distinguishable with difficulty or not at all by morphological criteria but are physiologically isolated. Examples are species of a polychaete *Capitella* distinguished electrophoretically, sibling species of mosquitoes distinguished by malaria-carrying, species of crickets identified by chirp and of birds by song.

A procedure for measuring diversity at the genetic level is the determination of homology of DNA sequences in genomes, both nuclear and mitochondrial. Restriction maps are constructed and homologous regions identified.

Mitochondrial DNA is maternally inherited, and the mitochondrial genome consists of relatively few genes (more in plants than in animals). Analysis of mtDNAs by use of endonucleases is particularly useful for tracing the phylogeny of parthenogenetic animals. One genus of lizards contains two parthenogenetic and four diploid species. Sequencing of mtDNAs shows that one bisexual species is the maternal parent for both a second diploid species of lizard and a parthenogenetic species (17a). Comparison of mtDNA of several species and populations of frogs in eastern Europe indicates the direction of hybridization and resulting gene transfer (introgression) (111a).

Sympatric speciation is proposed to occur when isolation may be by way of food plant selection, host-parasite relationship, feeding or reproductive behavior, or difference of breeding season. Two species of neuropterans, lacewings (*Chrysopa*) that hybridize in the laboratory and produce fertile F_1 and F_2 progeny, do not hybridize in nature. One species (*C. carnea*) lives in deciduous forests, the other (*C. downesi*) in coniferous woods. The two lacewing species differ in time of entering diapause so that one species (*carnea*) has two generations per year, the other (*downesi*) one generation. *C. downesi* has a 2°C higher threshold for development. Apparently habitat selection by the lacewings preceded changes in reproduction and life cycle (118).

Speciation is not proportional to protein heterogeneity. There are 3050 named species of frogs, 4600 of mammals; the amount of protein diversity in anurans and mammals is similar, yet the geologic age of origin of anurans is estimated at 150 myr ago and of mammals at 75 myr ago. Two species of *Rana* differ by 29 amino acid substitutions in the β chain of Hb; this is more difference than there is between β chains of Hb in *orders* of mammals. Some species of frogs hybridize in nature, although they diverged more than 30 myr ago. Mammals do not hybridize after 10 myr of separation (125). Antigenic differences of albumins evolved no faster among frogs than among mammals. Change in chromosome number has come 20 times faster in mammals than in frogs—3.5 myr in mammals and 70 myr in frogs for similar changes (74). Species of hylid frogs were compared; for the accumulation of 1.7 immunological distance per myr it was calculated that Australian frogs diverged from American frogs 75 myr ago and North American from South American 65 myr ago (74).

Homo and *Pan* are clearly very different in behavior, anatomy, and reproductive isolation, yet *Homo* and *Pan* are similar in most protein sequences. *Homo* and *Pan* differ in chromosome number and by nine pericentric chromosomal inversions (24). The temperatures of annealing of their DNAs differ by 1.1°C. Melting temperatures for DNAs of sibling species of *Drosophila* may differ by 3.0°C (125). Genetic distance between genera is greater than between species and between species is much greater than between populations (24). It is concluded that protein differences may not correspond to other criteria of species.

Speciation by Gradual Change

The classical (neo-Darwinian) view of the origin of species is as follows. Mutations occur in individuals and become fixed by balanced polymorphism or by genetic drift in populations that are interbreeding; accumulation of population differences leads to varieties, races, and subspecies. Spatial separation permits reproductive barriers to form; spatial separation varies with the mobility of the organism and may be geographic, ecological, or by feeding or mating habits. Biological isolation becomes established by secondary characters. In this view, speciation results from the accumulation of micromutations accompanying some form of spatial separation. Population differences may occur progressively over a geographic range, for example, in clines. Micromutations are changes in structural genes, those which code for sequences in proteins; these changes may be transitory on a geological time scale, and may be deleterious, neutral, or adaptive. According to this view speciation is the outcome of adaptation to the environment and natural selection acts on species (6).

Examples of the development of reproductive isolation in different environments are seen in small populations of *Drosophila*. *Drosophila* were reared in cages variously maintained at high and low temperatures and humidities. Differences, mainly behavioral, that resulted in reproductive isolation evolved within a hundred generations (56). *Drosophila melanogaster* have been selected for bristle number (important in mating); after fewer than 50 generations the flies had 32 bristles in one group, 85 bristles in another. A line of *D. melanogaster* selected for high locomotor activity of males accomplished more matings in a given time (121). Mice have been selected in breeding experiments for body size; progeny of a group of large mice were selected to an average of 32 g in weight, small mice to 13 g average. Corn has been selectively bred for oil content, which reached 5 percent in one line, 19 percent in another.

Diversification by small changes is illustrated in the many species of cichlid fish that live in the varied habitats of lakes. Comparison of these fishes shows a change from generalized to specialized structures. Changes in the jaw apparatus as slight as a shift of insertion of one muscle and development of a basopharyngeal joint implemented a change from swallowing to masticating. The same muscle had changed in its attachments when the fish evolved an altered mode of feeding. A few genetic changes in this fish jaw led to behavioral and ecological isolation (70).

A modification of the neo-Darwinian view is that small groups of similar organisms—varieties within a population—may result from changes in "open systems," whereas the major changes of speciation result from genetic reorganization in "closed systems" (20,21). Each genome has potential open and closed systems. Open systems involve protein sequences and are sensitive to selection pressures in clines and subspecies. Closed systems involve blocks of genes, perhaps linked genes, and codons for protein families not readily open to environmental influence.

One pattern of speciation by clustered genes or by major genetic changes is the founder-flush effect (20,21,67,74). A population may grow slowly or maintain a stable level for a long period. Rather suddenly, at a major environmental change, a crash in the population occurs; the few surviving individuals may become founders of a new species. Examples are (1) geological or climatic changes, (2) island formation, and (3) changes in agricultural practice. Examples of founder effects are *Drosophila* in the Hawaiian Islands, where there are 650–700 species; a species may be limited to one valley (21,119). One or two founder flies may have been blown or carried from Maui to Hawaii.

Genetic drift is change in the probability of frequency of a variation brought about by random processes. An initial frequency of expression of characters in a state of balanced polymorphism can change stochastically. Neutral or maladaptive traits can reach a high frequency, especially if linked to adaptive genes; change in gene frequency can occur by chance without advantage. Genetic drift is most effective in small populations (37).

Examples of incipient microevolution are responses by insect species to the expansion of range with agriculture—after vegetation changes from forest to cultivated fields, application of insecticides, and so on. Linking by land bridges, for example at Panama 2 to 3.5 million years ago, made invasion of new environments possible. Bottlenecks, partial separation of occupied land areas, resulted from continental glaciation in the Pleistocene some 10 to 30×10^4 years ago. Protein variation between members of a species is greatest in the middle of a range and least where selection and drift occur at the margins (20). *Drosophila willistoni* reared in cages at two temperatures developed more heterozygosity in 22 proteins when reared in heterogeneous (cycling) than in constant temperature (1a).

A different view of the origin of species is that the many small genetic changes that are retained—micromutations—provide for considerable polymorphism, but that speciation requires rare and large genetic changes—macromutations, genetic changes which have effects of greater magnitude than alterations in amino acid sequences of proteins. These macrogenetic changes may involve genetic rearrangements, translocations, inversions, exchanges between blocks of DNA, or changes in chromosome number.

Macromutations may be changes in regulatory genes that control the expression of structural genes. Posttranscriptional and translational events regulate the expression of structural genes; regulation is of protein production, which may include blocks of structural genes. Some actions regarded as macrogenetic changes include the following:

1. Gene rearrangements, translocations, inversions, conversions.
2. Mutations in regulatory genes, any one of which may modulate many structural changes.
3. Changes that alter a family of related proteins so as to permit new functions, changes in rate-limiting enzymes, changes in a metabolic sequence; alterations in balance of a gene complex, transilience.
4. Changes in receptor molecules for hormones or neurotransmitters.
5. Transport proteins that make new substrates or nutrients available for use.
6. Genetic determinants of behavior, especially reproductive and feeding behavior.

It is reasonable to conclude that microevolution could account for major evolutionary changes (6).

An example of action of regulatory genes in bacteria is the facultative shift to utilization of new kinds of carbon sources.

Macrogenetic differences between mammalian genotypes are shown in changes in major metabolic pathways, such as in ruminant digestion. Macroevolutionary changes in the action of a hormone take place, such as the changed functions of prolactin during the course of vertebrate evolution. Other examples of macroevolution are phylogenies of families of proteins. Macroevolution of many kinds of plants results from the occurrence of polyploidy. Behavioral changes, for example, in signs for species and mate recognition, and in sites for oviposition, have effects for species separation comparable to the effects of metabolic regulatory genes. The nature of genetic control of critical changes in behavior is not well known; it could involve changes in synaptic membranes or neural connections.

Punctuated Equilibrium; Saltation

The preceding account pictures evolution as gradual change, regardless of whether speciation takes place by accumulated micromutations or by macromutations. The gradualist view of evolution is held in various forms by most geneticists, ecologists, systematists, biochemists, and physiologists. Population geneticists have demonstrated changes sufficient for speciation within relatively few generations—hundreds or thousands. A different view of evolution has been proposed by some paleontologists—punctuated equilibrium, saltation, evolutionary change occurring in steps after long periods of no discernible change. In this view, single species remain constant during geological periods that are very long compared with the time for genetic changes observed in modern populations. Strata of fossil-bearing rocks often show extensive radiation of a plant or animal group during one period followed by a long period of stasis. The evolution of a taxon is proposed to be not continuous but saltatory. Species are distinguished by paleontologists by relatively large morphological differences; in fossils, biochemical differences are not detected. The fossil records are discontinuous even when many strata are well preserved. The rise and decline of species of plants or animals from one level to another is called *punctuated equilibrium*. The causes of the sudden appearance or disappearance of species are not clear (32,45,94,112). Some taxa show explosive radiation whenever niches become available.

The fossil record adds distribution in time to the distribution in space that is apparent to ecologists. Correlations of major evolutionary transitions with geological changes are not as close as some paleontologists hoped. The times for appearance of new species are thousands of years, and once formed, many species remain for several million years.

The intrinsic rate of speciation is considered as exponential: $N = N_0 e^{RT}$, where R is the intrinsic rate constant and the N_0 number of species at the beginning of a series, N the number of species after time T. For bivalves, R is 0.06 myr^{-1} and for mammals $R = 0.13$ to 0.29 myr^{-1} (125). The life span of a species is estimated at some 65 myr for bivalve molluscs, 2 myr for mammals (112). Speciation of eukaryotes has been rapid whenever new trophic levels have become established.

In some animal groups that were once considered as showing discontinuities, intermediates have been found, for example North American horses. The rate of evolution of a group is not constant; species of some mammals (rodents, cats) have evolved faster than others (hominids, dogs).

Many groups of animals, plants, and bacteria have evolved, flourished, and then disappeared or greatly diminished. In the Precambrian, algae were dominant; at some 600–500 myr ago protist herbivores appeared then declined, persisting in only a few forms (32). Equisetales, cycads, and pteridophytes were dominant during long periods. Large reptiles, including the dinosaurs, were dominant for some 150 million years.

Slow major geological events are continental drift, formation of land bridges, and shifts in oceans. Fast geological events are catastrophes and climatic change. Either slow or fast events may account for some transitions in flora and fauna. Continental masses approached each other during the Silurian and Devonian. A single large supercontinent, Pangea, existed about 225 myr ago and persisted during the Permian. Separation into continents began in the Cretaceous, leading to the present continents in the Cenozoic. Glaciation has occurred several times during the earth's history; important cooling occurred between the Permian and Triassic, the most recent glaciation ending 10,000 years ago.

A major transition occurred at the Cretacean-Tertiary boundary approximately 100 myr ago (Table 4-1). Reptiles declined and some classes became extinct; birds and mammals radiated extensively (10). A layer of iridium-enriched clay in sediments deposited between the Cretaceous and Teritary has been interpreted as indicating that a large meteorite collided with the earth (1). Clouds of dust presumably resulted in marked cooling and may have caused the extinction of many plants, marine invertebrates, and possibly of the dinosaurs. Some 96 percent of marine species of vertebrates and invertebrates died out at the Permian–Triassic transition (94a). A contrary view is that dinosaurs reached a peak in the Cretaceous, declined gradually, and became extinct at 100 to 75 myr ago (97a). A statistical examination of some 200,000 known species extinctions reveals periods of high extinction rates between the Permian and the present, with a periodicity of 26×10^6 years. The causes of periodic extinctions are unknown; however, two of the periods apparently correspond with possible meteorite collisions (94a).

Major extinctions occurred at the end of the Permian (225 myr ago) when many shallow-water invertebrates died out. Some classes, for example, bivalve molluscs, recovered and radiated into many species; others—brachiopods and trilobites—diminished or became extinct. The rate of speciation in mammals in recent times is much greater than the rate in invertebrates, such as bivalves. Fossil evidence indicates that in the Cambrian, evolution of bivalves was fast—30 genera of cockles in the Pontian Sea in 5 myr (112). Some invertebrates (ciliates and *Limulus*) and plants (*Acacia*) show virtually no recent speciation but do show considerable protein polymorphism. This is in contrast to some other organisms—desert pupfish *Cyprinodon* of several species are very similar in protein patterns. The rate of appearance of new species of invertebrates reached a maximum in the post-Cambrian, was low during the Triassic, and increased to a peak in the Tertiary. The rise and fall of species has not occurred at a constant rate for any plants or animals. Fossil records show that many species are long-lived. Nine lineages of mammals show that many species have continued for 2 myr; species of marine bivalves, gastropods, and foraminiferans have continued for more than 10 myr. Pollen of plants shows that of species extant 4 myr ago, about half are still living (112). Breaking of land bridges, for example, between Asia and North America, and establishment of bridges, as in Panama, are recent events that have altered the distribution of plants and animals.

Many extant species are now the same in structure as they were in the Pleistocene—beetles from 3 myr ago, mammals 2–3 myr, mosses and liverworts 4–5 myr (112). There was a sudden appearance of flowering plants in the Cretaceous. At times when one group of organisms declined or disappeared, a group of new species evolved rapidly (8). When the dinosaurs vanished, niches occupied by them were taken over by mammals in an explosive radiation of species. Mammals radiated in the Cenozoic, some 66 myr ago. Some vertebrate groups—horses, rabbits, teleosts—are fast-evolving, with 1.6 to 2.8 new species/lineage/myr, and large chromosome changes—0.4 to 1.4 chromosome alterations/lineage/myr. Other vertebrate groups—whales, frogs, lizards are slow-evolving, with 0.3 new species/lineage/myr and very

Table 4-1. Life Forms in Geological Ages (Compiled)

Geological Era	Period		Beginning of Periods Age (millions of years bp)	Characteristic Biota
Cenozoic	Quaternary	Recent	0.05	
		Pleistocene	2.5	Humans
	Tertiary	Pliocene	10	
		Miocene	27	Primates
		Oligocene	38	
		Eocene	55	Large carnivores
		Paleocene	70	Early placental mammals
Mesozoic	Cretaceous		130	Flowering plants; last of dinosaurs, primitive birds
	Jurassic		180	Flying reptiles, primitive mammals
	Triassic		225	Ammonites, cycads
Paleozoic	Permian		270	Reptiles, last of trilobites
	Carboniferous	Pennsylvanian	315	Amphibians Sporophytes
		Mississippian	350	Crinoids
	Devonian		405	Ferns; amphibians
	Silurian		440	Land plants; brachiopods
	Ordovician		490	Trilobites; cephalopods
	Cambrian		575	Most invertebrate phyla
	Precambrian		>650	Ediacarian fauna Eukaryotes Divergence of eubacteria and archaebacteria Progenotes

low rates of chromosome change—0.03 changes/lineage/myr (19). Crocodiles have persisted for at least 130 myr.

Summary and Interpretation of Speciation

Probably speciation has occurred both by accumulation of micromutations in clines and by macromutations in steps of changes in body form and reproductive behavior. It is clear that small genetic differences—in allozymes and isozymes—provide for phenotypic variations, some of which are adapted to the environment. Macromutations account for evolution of species, not of populations; the rates of speciation have differed greatly in various kinds of organisms at different geological times. Saltation (transilience) cannot be induced by natural selection of micromutations but results from genetic discontinuities (macromutations) (119). During geological intervals when little morphological change occurred, there could have been extensive physiological change.

There is no close correlation between genetic differentiation (electrophoretic diversity) and speciation. Single genes have had far-reaching effects in population differentiation; transilience is indicated where polygenic systems have undergone major changes.

Some of the disagreements between proponents of gradualism and of saltation may be resolved by the accumulating evidence for surplus DNA for which no coding function is apparent and for the extreme protein polymorphism with neutral replacements of amino acids. One function of the excess DNA and the protein variants may be to provide a pool of preadaptations. These could be of little value in speciation or in adaptations of single proteins. However, changes in DNA, particularly in regulatory genes that provide for alterations in a family of proteins or in the enzymes of a sequence of reactions, may lead to major change in the organism. In protein families there is often overlap of function. Evolution of new regions or sequences in a molecule may make possible a major extension of function and the synthesizing of new members of a protein family. Thus the explanation of saltation in evolution may be found in the large reserves of DNA and amino acid substitutions in critical regions of proteins; these are preadaptations which become functional only in rare combinations which then make possible a biochemical or behavioral change.

The definitions of a species as reproductively isolated and the processes of speciation as accumulation of micromutations or as macromolecular changes are based largely on observations on animals. The debate regarding evolution by gradual accumulation of small changes as opposed to large steps—punctuated equilibrium—is based largely on animal fossils. The modes of speciation are different in asexually reproducing organisms and in higher plants from the mode in sexually reproducing animals. In prokaryotes and cellular eukaryotes which produce asexually, and in self-fertilizing plants, species are set apart by morphological, physiological, and ecological differences, not by reproductive isolation. In many asexually reproducing organisms, body form and nutritional requirements are altered according to environment. There are mating types in microorganisms where equal nuclear exchange occurs, but this is unlike the fusion of haploid gametes in sexual reproduction.

Parthenogenetic animals and self-pollinating plants may be less variable and may evolve more slowly than diploid, sexually reproducing organisms. However, allelic frequencies may be nearly as high as in diploid species; parthenogenetic weevils (*Curculionidae*) have high genetic variation which does not conform to the Hardy-Weinberg equilibrium (115).

In higher plants, no general definition of species applies. Species of plants are complexes of ecologically adapted populations; reproductive isolation does not have the same importance as in animals. Each plant species contains a complex of interacting genes. Many closely related plant species interbreed, and the hybrids are fertile. Many of the products of hybridization are adapted to changing environments; interspecific recombinations provide for widened ecological ranges. Many kinds of plants are reproductively separated by the mode of pollination or the season of flowering.

Early fossil plants show high rates of formation of new species and short life spans of species; fossil plants in later strata show less species diversity. Early vascular plants became abundant in the Silurian-Devonian, pteridophytes in the Carboniferous, gymnosperms in the Jurassic; angiosperms have been most numerous since the Tertiary (84).

In most plants the biological unit is the local population or clone; population differences are suited to habitats. Most plant populations live in very limited areas, even those with wind dispersal of seeds. Eucalyptus stands only a few hundred meters apart differ in growth habit and drought resistance (95). Two subspecies of a sunflower *Stephanomeria exigua* differ in so many morphological and physiological characters that they are reproductively isolated. One sunflower appears to be the recent progenitor of the other (45). Trees and most perennials show much hybridization, annuals show little hybridization. Annuals are *r*-strategists with little hybridization. Polyploidy is frequent in herbaceous perennials capable of some vegetative reproduction. Chromosome numbers are low in annuals, especially in colonizers. Variation occurs to a surprising extent in self-fertilizing plants, for example *Avena* (barley) (22,113,114). Two populations of barley from different sources were crossed and frequency of four allozymes were scored; after 41 generations locus pairs showed the same frequencies in each of the two lines as at the start of the breeding experiment. This is an example of coadaptation or of inheritance as a correlated gene complex (22). Many plants show steep clines, ecotypes. The structure of a plant provides for meeting its needs in the environment; a tree in full sun has leaves throughout; a tree in the shade grows leaves around the periphery. Island populations become separate species, because of both isolation and ecological differences. Polyploidy provides for enzyme multiplicity and genetic diversity with resulting possible entry into new niches.

Some plant geneticists propose to abandon the biological species definition for plants and to recognize only natural populations. The plant population ecologist P. H. Raven states: "Selection for better adapted populations in a mosaic environment continually provides a supply of evolutionary novelties. Reproductive isolation is neither necessary nor an inevitable end point for plant species, and interspecific hybridization itself may be a highly adaptive and nearly universal feature of population systems in plants, especially those with longer generations. Ecotypes, races, subspecies, and semispecies cannot be regarded as stages in the evolution of species. Rather they are seen as taxonomic evaluations of particular patterns of variation in nature which are themselves the product of the interplay between the genetic diversity of populations and the environment" (95). The concept of saltation applies to plant speciation wherever marginal populations are examined; variation is greater where some physical factors are challenging.

In summary, plant species are useful for estimates of diversity; the plant population units conform to the definition of physiological species in which adaptation is the essential feature. As local populations are examined for diversity of DNA sequences, isolation by methods other than reproduction, are recognized in animals as well as plants.

GENERAL CONCLUSIONS

In no other field of biology is there so much ferment as there is concerning the mechanisms of evolution. Throughout much of evolutionary literature there has been the tacit assumption that simimlar mechanisms operate at all taxonomic levels. Historically, evidence for diversity and phylogeny came from systematics. Classification of organisms has long been a way of arranging living things in ways useful for the understanding of natural variation. All taxonomies, from Aristotle, Linnaeus, and Darwin to the modern systematists, are descriptions of diverse organisms in structural terms. Paleontological descriptions are necessarily based on morphological characters. The concept of relatedness, of lineages, was a natural consequence of taxonomy. Classification is the ordering by specialists of kinds of organisms. Classification need not correspond to lineage. The recognition that one kind of organism evolved from a preceding kind led naturalists to construct phylogenetic trees and to present relationships in cladograms. Relating fossil forms to extinct plants and animals led to temporal considerations, and the history of life was pursued in several directions—the paleontological record, the biochemical assignment of times for divergence of related organisms.

With the description of physiological and biochemical properties of organisms and with the classification of microorganisms in metabolic terms, comparisons of function complemented those of structure. Physiological and behavioral characters were added to morphological descriptions of species. Early in the development of comparative physiology, functions were correlated with life histories and ecology. During the first half of the twentieth century classical genetics gave an explanation of variation in terms of mutations, gene rearrangements, and structural changes in chromosomes. Genetics came to be used as the basis for variation, physiological and structure, and as proteins and nucleic acids were sequenced, it became evident that there is far more biological diversity than had been supposed on morphological grounds. Variation occurs at all levels of biological organization—between individuals, local populations, and between all taxonomic categories. The classifications of organisms were compared with evidence from physiological variation and with sequences of amino acids in proteins and of nucleotides in DNA. Relatedness could be given a functional and genetic basis. Similarly, from differences in primary structures—sequences of amino acids and nucleotides—distances of phylogeny were estimated. By the use of fossils for calibration, proteins and DNA have been used as clocks to estimate times of divergence. Accuracy decreases as older divergences are tested. Phylogenetic trees have been constructed for many proteins, for example, hemoglobins, cytochromes c, albumins, and Ca-binding proteins. In general, the relationships estimated from proteins and nucleic acids confirm those based on morphology. Some proteins change more than others; albumins evolve rapidly, histones slowly. Evolution of single proteins has not been at a uniform rate: hemoglobins in vertebrates showed periods of acceleration and deceleration.

Much biological relatedness is inferred from electrophoretic patterns and from primary structure. Noncritical regions of a protein can undergo much change during evolution, whereas more critical regions—catalytic sites, ligand binding sites—change very little. Only recently, variations in higher order conformations have been considered in cladistics. Proteins occur in families and the relations of the histories of members of a family can be estimated, for example, the sequence of evolution of α and β hemoglobins in vertebrates. A phylogenetic tree shows more sequence differences between distantly related than between closely related organisms. These differences reflect time of divergence, for example, between cytochrome c of yeast, mollusc, fish, and mammal. A few examples of the use of kinetics in describing relationships are given in Chapter 5. Evolutionists have been more active in estimating relationships and times of divergence than in giving adaptive explanations of divergence. Regulatory genes may be more important than structural genes in the evolution of higher taxonomic categories.

This book presents many examples of the conservative nature of protein and nucleotide evolution. Anaerobic pathways evolved early in prokaryotes and have been retained with minor changes (including reversal of some enzymes, Ch. 2) in aerobic eukaryotes. Biochemical conservatism is shown by heme-linked enzymes, energy transfer by electron transport, and enzymes of the Calvin cycle and of the pentose shunt. Other examples are the proteins of movement, for example, actins in bacteria, a few plants, and all animals. The bilayer of phospholipid is universal in plasma membranes and cell organelles. Insulin is similar in its three-dimensional structure in all vertebrates and has recently been found in invertebrates (insects). Vasotocin-like peptides occur in all vertebrates and in molluscs and insects. Actins differ more from tissue to tissue in a mammal than between classes and phyla. The neuropepetide substance P is found in all classes of vertebrates. Some hormones have a different function in one class of vertebrates from what they have in another class; an example is prolactin. Even at the genetic level protein sequences may not be very different, for example, heme proteins in chimpanzee and human. We conclude that families of similar proteins have long evolutionary histories; and that while between higher taxonomic categories—phyla, classes—sequences do differ, especially in noncritical regions of proteins, the adaptive significance of these differences is not well known. In the evolution of higher taxonomic categories, the differences in families of proteins and DNAs. Major morphological changes correlate with differences in life habits between classes and orders. Frequently a structure which served a given function becomes altered and serves another function; for example, several kinds of fishes have made the transition from water breathing to air breathing: in the most successful, a swim bladder has become a lung. In evolution of temperature regulation in reptiles, birds, and mammals, structures for locomotion, respiration, and water balance were changed.

In contrast with higher taxonomic categories, this book presents many examples of correlations of protein structure with life histories, and with geographic distribution and interactions between lower taxonomic categories and the environment. Chapter 5 presents measurements of kinetic properties of enzymes and conformations of proteins that correlate with habitat and with classification at lower taxonomic levels. For microevolutionary adaptations to the environment, physiological and biochemical characters are critical. Small genetic changes are selected as organisms extend their range or niche. Functional characters may change over relatively short times and provide the basis for gradual evolution. Chapters 6–9 present many examples of genetically based protein differences adaptive to the environment. Physiological adaptation to the environment—physical and biotic—is necessary for natural selection.

Evolutionary theory has emphasized speciation, especially as it occurs in animals. Reproductive isolation of species is of less importance in the evolution of plants and asexual organisms, and in parthenogenetic and autogamous organisms.

In animals, reproductive behavior contributes to isolation. Relatively small changes in the nervous system may cause drastic changes in behavior which, for speciation, may be functionally comparable to macromutations. Reproductive behavior has a genetic basis. Whether speciation can occur sympatrically—that is, in populations which are in breeding contact—has been much debated; it now appears that ecological separation may be as contributory as geographic. Behavioral differences in food preferences may be analogous to changes in regulatory genes. Behavior and anatomy may evolve independently of changes in protein composition, and may reflect macromutations or nucleotide rearrangements which cannot be detected by available methods.

The history of life is complex; myriad mechanisms have contributed. Many genetic alterations—mutations in structural genes—are not directly adaptive and may not be selected. Some genetic changes—preadaptations—remain latent and can be effective only in combination with others. Some small changes that are important for adaptation of populations may be cumulative. However, for evolution of larger taxonomic groups, especially animals, macromutations or genetic changes in gene complexes and chromosome rearrangements and duplications are essential. The rate of evolution of proteins, species, and phyla is variable with time. Sometimes evolution is very slow, at other times rapid. Both gradual change and saltation have taken place. The same kinds of differences occur at all biological levels. Polyploidy, hybridization, changes in whole metabolic pathways vary in importance in different kinds of organisms. Not all evolutionary change is by direct natural selection at local levels, but stochastic changes in genotypes may, by genetic drift, lead to organismic variation (74,106,126). In the long run, however, evolution depends on organism-environment interactions.

It is futile to search for a single mechanism to account for the panorama of life. Population geneticists observe adaptive changes that can occur in a few generations; paleontologists observe the fossil record with periodic changes evident over millions of years. The explanation of evolution also depends on the kinds of organisms observed—animals, higher plants, asexual eukaryotes, or prokaryotes. Physiological and genetic experiments are not possible with fossils; within static groups of fossils designated as species, adaptive physiological variation may have occurred without changes in structure. An understanding of the history of life requires a breadth of approach that has not previously been possible. Each discipline emphasizes its way of looking at evolution; an integration of approaches is the direction for the future.

REFERENCES

1. Alvarez, W. et al. *Science* **23**:1135–1141, 1984. Impact theory of mass extinction; invertebrate fossil record.

1a. Anderson, W. W. *Evolution* **27**:278–284, 1973. Genetics, body size, *Drosophila*, temperatures.

2. Andrews, P. and J. E. Cronin. *Nature* **297**:541–546, 1982. Evolution of primates.

3. Avise, J. C. and M. H. Smith. *Am. Nat.* **108:**458–472, 1974. Species clusters of sunfish *Lepomis*.

4. Avise, J. C. et al. *Proc. Natl. Acad. Sci.* **76:**6694–6698, 1979. Mitochondrial DNA clones and phylogeny of pocket gophers *Geomys*.

5. Ayala, F. J. et al. *Evolution* **27:**177–191, 1973. Genetic variation in *Tridacna*.

5a. Ayala, F. J. *Evol. Biol.* **8:**1–78, 1975. Genetic differentiation during speciation.

6. Ayala, F. J., Ed. *Molecular Evolution.* Sinauer, Sunderland, MA, 1976. pp. 387–402 in *Evolution from Molecules to Men,* Ed. D. S. Bendall, Cambridge University Press, New York, 1983.

7. Baba, M. L. et al. *J. Molec. Ecol.* **17:**197–213, 1981. Protein cladistics.

8. Bakker, R. T. *Nature* **274:**661–663, 1978. Evolution of dinosaurs and origin of flowering plant.

9. Barker, W. C. and N. O. Dayhoff. *Compar. Biochem. Physiol.* **62B:**1–5, 1979. Evolution of homologous families of proteins.

10. Benton, M. J. *Nature* **302:**16–17, 1983. Correlations of evolution with geological events.

11. Berlocher, S. H. and G. L. Bush. *Syst. Zool.* **31:**136–155, 1982. *J. Hered.* **71:**63–67, 1980. *Ann. Ent. Soc. Am.* **73:**131–137. Electrophoretic analysis of *Rhagoletis* (*Dipteran*) phylogeny.

12. Bosson, W. et al. *Proc. Natl. Acad. Sci.* **77:**5784–5788, 1980. Isozymes of human liver ADH.

13. Bowen, S. T. *Biochem. Genet.* **15:**409–437, 1971. Hemoglobins of Artemia.

14. Brown, A. H. D. *Theor. Pop. Biol.* **15:**1–42, 1979. Enzyme polymorphism in plant populations.

14a. Brown, G. D. and M. V. Simpson. *Proc. Natl. Acad. Sci.* **79:**3246–3250, 1980. Mitochondrial DNA evolution in two species of rat.

15. Brown, J. R. *Fed. Proc.* **35:**2141–2144, 1976. Structural origins of mammalian albumin.

16. Brown, J. H. and C. Feldmeth. *Evolution* **25:**390–398, 1971. Speciation in desert pupfish *Cyprinodon*.

17. Brown, W. M. Ch. 4., pp. 62–88 in *Evolution of Genes and Proteins,* Ed. M. Nei and R. K. Koehn, Sinauer, Sunderland, MA, 1983. Evolution of mitochondrial DNA.

17a. Brown, W. M. and J. W. Wright. *Science* **203:**1247–1249, 1979. *Proc. Natl. Acad. Sci.* **76:**1967–1971, 1979. *J. Molec. Evol.* **18:**225–239, 1982. Divergence of DNA sequences in mitochondrial genomes, primates.

18. Bruce, E. J. and J. F. Ayala. *Evolution* **33:**1040–1056, 1979. Phylogenetic relation between man and ape.

18a. Burton, R. S. and M. W. Feldman. *Compar. Biochem. Physiol.* **73A:**441–445. *Biochem. Genetics* **21:**239–251, 1983. Allozyme polymorphism in hyperosmotic stress in copetod *Tignopus*.

19. Bush, G. L. *Annu. Rev. Ecol. Syst.* **6:**339–364, 1974. *Proc. Natl. Acad. Sci.* **74:**3942–3946, 1977. Modes of animal speciation.

20. Carson, H. L. *Cold Spring Harbor Symposium on Quantitative Biology* **24:**87–

105, 1959. *Stadler Symposium* **3**:51–70, 1971. *Proc. Natl. Acad. Sci.* **71**:3517–3521, 1974. *Am. Nat.* **109**:83–92, 1975. Genetics of speciation in *Drosophila*.

21. Carson, H. L. et al. *Annu. Rev. Ecol. Syst.* **7**:311–345, 1976. *Proc. Natl. Acad. Sci.* **76**:1929–1932, 1979. *Evolution* **36**:132–140, 1982. Incipient speciation in *Drosophila* in Hawaii.

21a. Cavener, D. R. and M. R. Clegg. *Proc. Natl. Acad. Sci.* **78**:4444–4447, 1981. Polymorphism of enzymes of pentose shunt and glycolysis in *Drosophila*.

22. Clegg, M. T. et al. *Proc. Natl. Acad. Sci.* **69**:2474–2478, 1972. Species are not the evolutionary unit of plants.

23. Coyne, J. A. et al. *Genetics* **84**:593–607, 1976; **87**:285–304, 1977. *Proc. Natl. Acad. Sci.* **75**:5090–5093, 1978. Genetic heterogeneity of enzymes in two species of *Drosophila*.

24. Cronin, J. E. et al. *Nature* **292**:113–122, 1981. Tempo and mode in hominid evolution.

25. Czelusniak, J. et al. *Nature* **298**:297–300, 1982. Phylogeny of avian and mammalian hemoglobin.

26. Dawkins, R. *The Selfish Gene,* Oxford University Press, New York, 1976. *The Extended Phenotype*. Oxford University Press, New York, 1982.

27. Dayhoff, M. O. *Atlas of Protein Sequence and Structure*. vol. 4, 1969; vol. 5, 1978.

28. De Lange, R. J. and E. L. Smith. *Annu. Rev. Biochem.* **40**:279–314, 1971. Structure and function of histones.

29. Dickerson, R. E. J. *J. Molec. Evol.* **1**:26–45, 1971. *Enzymes* XI, Ed. O. Beyer, 397–547, Academic, New York, 1975. Structure and function of cytochrome c.

30. Doolittle, R. F. *Science* **214**:149–159, 1981. Similar amino acid sequences—chance of common ancestor.

31. Efstratiadis, A. et al. *Cell* **21**:653–668, 1980. Evolution of β-globulins.

32. Eldredge, N. and S. J. Gould, pp. 82–115 in *Models in Paleobiology,* Ed. T. J. M. Schopf, Freeman, San Francisco, 1972. Punctuated equilibrium: an alternative to phyletic gradualism.

32a. Endler, J. A. *Geographic Variation, Speciation, and Clines*. Princeton University Press, Princeton, NJ, 1977.

33. Eventoff, W. and M. G. Rossman. *Crit. Rev. Biochem.* **3**:111–140, 1975. Cladogram of nucleotide binding enzyme.

34. Ferris, S. D. et al. *Proc. Natl. Acad. Sci.* **78**:6319–6323, 1981. Polymorphism in mitochondrial DNA of apes.

34a. Ferris, S. D. et al. *Proc. Natl. Acad. Sci.* **80**:2290–2294, 1983. Interspecific transfer of mtDNA between two species of mice.

36. Fisher, S. E. et al. *Genetica* **52**:73–85, 1980. Evolution of multilocus isozyme systems in chordates.

36. Fitch, W. M. *Science* **155**:279–284, 1967. *Syst. Zool.* **20**:406–416, 1971. Methodology for constructing molecular clocks. pp. 160–178 in *Molecular Evolution,* Ed. F. Ayala, Sinauer, Sunderland, MA, 1976.

37. Futuyama, D. *Evolutionary Biology,* Sinauer, Sunderland, MA, 1979.

38. Fyhn, U. E. H. and B. Sullivan. *Biochem. Genetics* **11**:373–385, 1974. Hemoglobin polymorphism in *Opsanus*.

38a. Garton, D. W. *Physiol. Zool.* **57**:530–543, 1984. Multiple locus heterozygosity and physiological energetics of growth in estuarine snail *Thais*.

38b. Garton, D. W., R. K. Koehn, and T. M. Scott. *Genetics* **108**:445–455, 1984. Heterozygosity and physiological energetics of growth in clam *Mulina*.

39. Glover, C. V. *Proc. Natl. Acad. Sci.* **76**:585–589, 1979. Histones in *Tetrahymena*.

40. Goodman, M. et al. Ch. 9, pp. 149–159 in *Molecular Evolution*, Ed. F. Ayala, Sinauer, Sunderland, MA, 1976. Protein sequences in phylogeny.

41. Goodman, M. et al. *Nature* **303**:546–548, 1983. Human origins inferred from hemoglobins of African apes.

42. Goodman, M. et al. *Nature* **253**:603–608, 1975. pp. 321–353 in *Molecular Anthropology*, Ed. M. Goodman et al., Plenum, New York, 1976. *J. Molec. Evol.* **17**:114–120, 1981. *Prog. Biophy. Molec. Biol.* **38**:105–164, 1981. Molecular clocks based on sequences in proteins, hemoglobins, and cytochrome c.

43. Goodman, M. et al. *Syst. Zool.* **31**:376–399, 1982. *Acta Zool. Fennica* **169**:19–35, 1982. *Nature* **303**:546–548, 1983. Molecular evolution above the species level.

44. Goodman, M. et al. *J. Molec. Evol.* **13**:331–352, 1979. pp. 347–354 in *Calcium-Binding Protein*, Ed. F. L. Siegel et al., Elsevier, New York, 1980. Evolution of calcium-binding protein.

45. Gottlieb, L. D. Ch. 11, pp. 265–286 in *Topics in Plant Population Biology*, Ed. A. Solbrig et al., Columbia University Press, New York, 1979. Ch. 8, pp. 123–140 in *Molecular Evolution*, Ed. F. Ayala, Sinauer, Sunderland, MA, 1976. Recent evolution of two species of plants.

46. Grassle, J. F. and J. P. Grassle. *Science* **192**:567–569, 1976. pp. 347–364 in *Marine Organisms*, Ed. B. Battaglia, Plenum, New York, 1978. Sibling species in polychaete *Capitella*.

47. Graves, J. et al. *Evolution* **37**:30–37, 1983. LDH differences in species pairs of fishes.

47a. Hilbish, T. J. et al. *Nature* **298**:688–689, 1982. Allozyme polymorphism and cell volume regulation in molluscs.

47b. Hines, S. A. et al. *Biochem. Genetics* **21**:1143–1151, 1983. Allozyme distribution of MDH in large-mouth bass.

48. Hjorth, J. P. *Biochem. Genetics* **13**:379–391, 1975. Molecular and genetic structure of hemoglobin in eelpout *Zoarces*.

49. Hoffman, R. J. *Biochem. Genetics,* **19**:129–144, 1981. Geographic variation of allozymes of *Metridium*.

50. Hood, L. E. et al. *Molecular Biology of Eukaryotic Cells*, Benjamin, Menlo Park, CA, 1974.

51. Jain, S. Ch. 7, pp. 160–187 in *Topics in Plant Population Biology*, Ed. A. Solbrig, Columbia University Press, New York, 1979. Adaptive strategy in plant evolution.

52. Jeffreys, A. J. et al., Ch. 9, pp. 175–195 in *Evolution from Molecules to Men,*

Ed. D. S. Bendall, Cambridge University Press, New York, 1983. Evolution of globin genes.

53. Johnson, G. B. *Science* **184**:28–37, 1974. *Carnegie Inst. Annu. Rep. Pl. Biol.* **75**:440–449, 1975. *Biochem. Genetics* **14**:403–426, 1976. Ch. 3, pp. 46–59 in *Molecular Evolution*, Ed. F. Ayala, Sinauer, Sunderland, MA, 1976. Enzyme polymorphism, *Drosophila,* and butterfly *Colias.*

54. Jones, J. S. *Nature* **286**:757–758, 1980; **289**:743–744, 1981; **293**:427–428, 1981. Models of speciation.

55. Jukes, T. H. *Science* **210**:973–978, 1980. Silent nucleotide substitutions and molecular clock.

56. Kilian, D. et al. *Evolution* **34**:730–737, 1980. Speciation in *Drosophila.*

57. Kimura, M. J. *Nature* **229**:467–469, 1971; **267**:275–276, 1977. *Sci. Am.* **241**:98–122, 1979. *Proc. Natl. Acad. Sci.* **76**:2858–2861, 1979. Neutral genes in molecular evolution.

58. King, J. and T. Jukes. *Science* **164**:788–797, 1969. Non-Darwinian evolution.

59. King, M. C. and A. C. Wilson. *Science* **188**:107–116, 1975. Protein evolution in human and chimpanzee.

60. Koehn, R. K. et al. *Proc. Natl. Acad. Sci.* **77**:5385–5389, 1980. Aminopeptidase cline in *Mytilus.*

60a. Koehn, R. K. and P. M. Gaffney. *Marine Biol.* **82**:1–7, 1984. Genetic heterozygosity and growth rate in *Mytilus edulis.*

60b. Koehn, R. K. et al. Ch. 6, pp. 115–136 in *Evolution of Genes and Proteins,* Ed. M. Nei and R. K. Koehn, Sinauer, Sunderland, MA, 1983. *Marine Biol.* **82**:1–7, 1984. Heterozygosity and biological fitness.

61. Konkel, D. A. et al. *Cell* **18**:865–873, 1979. Evolution of recent globin genes in mouse.

62. Kreitman, M. *Nature* **304**:412–417, 1983. Nucleotide polymorphism in alcohol dehydrogenase.

63. Kretsinger, R. H. *Int. Rev. Cytol.* **46**:323–393, 1976. *Crit. Rev. Biochem.* **8**:119–174, 1980. Evolution of Ca-binding protein.

64. Lakovaara, S. and A. Saura. *Genetics* **69**:377–384, 1971. Genetic variation in population of *Drosophila obscura.*

65. Langley, C. H. and U. M. Fitch. *J. Molec. Evol.* **3**:161–179, 1976. Methods for molecular clock.

65a. Lee, Y. M., D. Friedman, and F. Ayala. *Proc. Natl. Acad. Sci.* **82**:824–828, 1985. Sueproxide dismutase as molecular clock.

65b. Lerner, I. M. *Genetic Homeostasis,* Oliver and Boyd, London, 1954.

66. Lessios, H. A. *Nature* **280**:599–601, 1979. Phylogency of Panama sea-urchins.

67. Lewontin, R. C. *Genetic Basis of Evolutionary Change,* Columbia University Press, New York, 1974.

68. Levinton, J. S. *Biol. Bull.* **165**:686–698, 1983. Latitudinal variation in growth of polychaete worm.

69. Li, W. H. et al. *Nature* **292**:237–238, 1981. Neutral genes and negative selection.

70. Liem, K. F. *Syst. Zool.* **22**:425–441, 1973. Evolution, cichlid fish jaw muscles.

71. MacArthur, R. H. and E. O. Wilson. *Theory of Island Biogeography,* Princeton University Press, Princeton, NJ, 1967.

72. Maren, T. H. et al. *Compar. Biochem. Physiol.* **67B:**69–74, 1980. Evolution of carbonic anhydrases.

73. Margoliash, E. *Federation Proceedings, Societies for Experimental Biology* **35:**2125–2130, 1976. Evolution of cytochrome c.

74. Maxson, L. R. and A. C. Wilson. *Syst. Zool.* **24:**1–15, 1975. *Nature* **225:**397–400, 1975. *Proc. Natl. Acad. Sci.* **71:**2843–2847, 3028–3030, 1974. Evolution of serum albumins in amphibians.

75. Mayr, E. *Populations, Species, and Evolution,* Harvard University Press, Cambridge, MA, 1969. With W. B. Provine, *The Evolutionary Synthesis,* Harvard University Press, Cambridge, MA, 1980.

76. Mayr, E. *Science* **214:**510–516, 1981. Principles of biological classification.

76a. McClintock, B. *Science* **226:**792–801, 1984. Response of genomes to stresses.

77. McNaughton, S. J. *Am. Nat.* **106:**165–172, 1972. Enzyme thermal adaptations in plants.

78. Milkman, R. *Science* **182:**1024–1026, 1973. Electrophoretic variations in clones of E. coli.

79. Mooney, H. A. and S. L. Gulmon. Ch. 13, pp. 316–337 in *Topics in Plant Population Biology,* Ed. A. Solbrig et al., Columbia University Press, New York, 1979. Environmental and evolutionary constraints on photosynthesis by higher plants.

79a. Moore, J. A. *Amer. Zoologist* **15:**837–849, 1975. Species complex of Rana pipiens.

80. Moore, G. W. et al. *J. Molec. Biol.* **105:**15–37, 1976. Evolution of proteins, cytochrome c.

81. Nei, M. *Am. Natl.* **106:**283–292, 1972. Formulas for genetic distance.

82. Nevo, E. et al. *Evolution* **28:**1–23, 1974. Genetic variation and speciation in pocket gophers.

83. Nevo, E. *Evolution* **33:**815–833, 1979. Genetic diversity in wild barley.

84. Niklas, K. J. et al. *Nature* **303:**614–616, 1983. Patterns in plant diversification.

85. Oakeshott, J. G. et al. *Evolution* **36:**86–96, 1982. Alcohol dehydrogenase clines in *Drosophila,* different continents.

85a. Okama, T. et al., *Nucl. Acids. Res.* **12:**5101–5106, 1984. Evolution of animals deduced from 5 S rRNAs.

86. Paquin, C. and J. Adams. *Nature* **302:**495–500, 1983. Fixation of adaptive mutations in yeast populations.

87. Pasdar, M. et al. *Biochem. Genetics* **22:**931–956, 1984. Expressions of enzyme activities in interspecific sunfish hybrids and backcross progeny.

87a. Perutz, M. F. *Molec. Biol. Evol.* **1:**1–8, 1983. Species adaptation in hemoglobin.

87b. Philip, D. P., W. F. Childers, and G. Whitt. *Trans. Am. Fisheries Sco.* **112:**1–20, 1983. Biochemical evaluation of northern and Florida subspecies of largemouth bass.

87c. Philip, D. P., H. P. Parker, and G. S. Whitt in *Isozymes: Current Topics in Biology and Medicine* **10:**193–237, 1983. Isozymic analysis of patterns of gene expression during development of hybrid fish.

87d. Peace, A. R. and D. A. Powers. *Proc. Natl. Acad. Sci.* **76**:2354–2358, 1979. Clinal distribution of LDH-B allozyme.

88. Plapp, F. W. *Annu. Rev. Entomol.* **21**:179–197, 1976. Biochemical genetics of insecticide resistance.

89. Powell, J. R. *Evol. Biol.* **8**:79–119, 1975. Protein variations in natural populations of animals.

90. Powers, D. et al. *Biochem. Genetics* **16**:593–607, 1978. *Science* **216**:1014–1016, 1982. Protein differences in a cline, 7 Fundulus.

91. Prager, E. M. and A. C. Wilson. *J. Biol. Chem.* **246**:5978–5989, 1971; **247**:2905–2916, 1972. *Proc. Natl. Acad. Sci.* **71**:200–204, 1975. Protein evolution in vertebrates.

92. Pryor, S. C. et al. *Compar. Biochem. Physiol.* **65B**:663–668, 1980. Biochemical genetics of *Culex pipiens* complex.

93. Ramshaw, J. A. et al. *Genetics* **93**:1019–1037, 1979. Variations in sequences in human hemoglobin.

94. Raup, D. and S. Gould. *Syst. Zool.* **23**:305–322, 1974. Models for morphological evolution.

94a. Raup, D. M. *Science* **206**;217–218, 1979. *Proc. Natl. Acad. Sci.* **81**:801–805, 1984. Periodic extinctions of species from late Permian to present.

95. Raven, P. H. *Syst. Bot.* **1**:284–316, 1977. Ch. 19, pp. 467–481 in *Topics in Plant Population Biology,* Ed. A. Solbrig, Columbia University Press, New York, 1979. pp. 3–10 in *Canad. Bot. Assoc. Bull. Suppl.,* vol. 13, 1980. Hybridization and speciation in higher plants.

96. Redding, J. M., and C. B. Schreck. *J. Fish. Res. Bd. Canad.* **36**:544–551, 1979. Enzyme polymorphism in steelhead trout.

97. Rossmann, M. G. *Federations Proceedings Societies for Experimental Biology* **35**:2112–2114, 1976. Evolution of heme binding and NAD⁺ binding protein. *Deep Sea Res.* **14**:65–78, 1967.

97a. Russell, D. *Nature* **307**:360–361, 1984. Gradual decline of dinosaurs.

97b. Sandregger, P. et al. *Nature* **275**:157–159, 1978. Evolution of alanine-aspartate transferase.

98. Sanders, H. *Am. Nat.* **102**:243–282, 1968. *Deep-Sea Res.* **14**:65–78, 1967. Diversity of benthic populations.

99. Scanlon, B. E., L. Maxson, and W. E. Duellman. *Evolution* **34**:222–229, 1980. Albumin evolution in marsupial frogs.

100. Selander, R. K. and B. R. Levin. *Science* **210**:545–547, 1980. Genetic diversity in populations of *E. coli.*

101. Selander, R. K. Ch. 2, pp. 21–45 in *Molecular Evolution,* Ed. F. Ayala, Sinauer, Sunderland, MA, 1976. Genetic variation in natural populations.

102. Selander, R. K. et al. *Evolution* **24**:402–414, 1970. Variation in limulus proteins.

103. Sepkoski, J. J. et al. *Nature* **293**:435–437, 1981. Diversity in plants throughout evolution.

104. Siebenaller, J. and G. N. Somero. *Science* **201**:255–256, 1978. Pressure adaptation of enzymes from two species of Sebastolobus.

105.   Simon, J. P. *Oecologia* **39**:273–278, 1979. Latitudinal variation of thermal properties of MDH in a legume.

106.   Simpson, G. G. *Tempo and Mode in Evolution,* Columbia University Press, New York, 1944. *The Major Features of Evolution,* Columbia University Press, New York, 1953.

107.   Snyder, L. R. G. *Genetics* **89**:511–530, 1978. Genetics of hemoglobin in *Peromyscus*.

108.   Somero, G. and M. Soule. *Nature* **249**:670–672, 1974. Variation in deep-sea fishes.

109.   Soule, M. *Am. Nat.* **106**:429–446, 1972. *Evolution* **27**:593–600, 1973. Genetic variation in population of lizards.

110.   Soule, M. Ch. 4, pp. 60–77 in *Molecular Evolution,* Ed. F. Ayala, Sinauer, Sunderland, MA, 1976. Determinants of allozyme variation.

111.   Spiess, E. B. and H. L. Carson. *Proc. Natl. Acad. Sci.* **78**:3088–3092, 1981. Sexual selection in *Drosophila silvestris* of Hawaii.

111a.   Spolsky, C. and T. Uzzell. *Proc. Natl. Acad. Sci.* **81**:5802–5805, 1984. Interspecies mitochondrial DNA in amphibians.

112.   Stanley, S. M. *Paleobiology* **4**:26–40, 1978. *Macroevolution, Pattern and Process,* Freeman, San Francisco, 1979. *Evolution* **36**:460–473, 1982. Macroevolution and the punctuation model from fossil records.

113.   Stebbins, G. L. *Canad. J. Genet. Cytol.* **14**:453–462, 1972. Evolution of higher plants.

114.   Stebbins, G. L. and F. J. Ayala. *Science* **213**:967–971, 1981. Is a new evolutionary synthesis necessary?

115.   Suomalainen, E. *Genetics* **74**:489–508, 1973. Evolution in parthenogenetic animals.

116.   Sures, I. et al. *Cell* **15**:1033–1044, 1978. Histone variation in sea urchins.

117.   Tashian, R. E. et al. *Adv. Human Genetics* **7**:1–57, 1970. *Biochem. Genetics* **15**:885–896, 1977. Genetics of carbonic anhydrase.

118.   Tauber, C. A. and M. S. Tauber. *Nature* **268**:702–705, 1977. Sympatric speciation in neuropterans.

119.   Templeton, A. R. *Evolution* **33**:513–517, 1979; **34**:719–729, 1980. *Genetics* **92**:1265–1282, 1979. *Annu. Rev. Ecol. Syst.* **12**:23–48, 1981. Modes of speciation; founder effects, to *Drosophila* and others.

120.   Valentine, J. W. Ch. 5, pp. 78–94 in *Molecular Evolution,* Ed. F. Ayala, Sinauer, Sunderland, MA, 1976. Genetic strategies of adaptation.

121.   van Dijken, F. R. and W. Scharloo. *Behavior Genetics* **9**:555–561, 1980. Selection for locomotor activity in *Drosophila*.

122.   Werner, P. A. Ch. 12, pp. 287–310 in *Topics in Plant Population Biology,* Ed. A. Solbrig, Columbia University Press, New York, 1979. Competition between species of goldenrod.

123.   Whitt, G. S. *Am. Zool.* **21**:549–572, 1981. pp. 271–289 in *Evolution Today,* Ed. G. G. Scudder, Hunt Inst. Carnegie-Mellon University, Pittsburgh, 1981. *Copeia* 721–725, 1982. Developmental genetics and evolution of fishes.

124. Williamson, P. G. *Nature* **293:**437–443, 1981. Palaeontology of speciation in Cenozoic molluscs from a fossil bed in Africa.

125. Wilson, A. C. et al. *Proc. Natl. Acad. Sci.* **63:**1088–1092, 1969. Ch. 13, pp. 225–234 in *Molecular Evolution,* Ed. F. Ayala, Columbia University Press, New York, 1976. *Annu. Rev. Biochem.* **46:**573–639, 1977. Biochemical evolution.

126. Wilson, A. C. et al. *Proc. Natl. Acad. Sci.* **71:**2843–2847, 1974; 3028–3030, 1974. Immunological methods, animal relationships.

126a. Wilson, A. C. et al. *Proc. Natl. Acad. Sci.* **72:**5061–5065, 1975. *Science* **188:**107–116, 1975. Biochemical analyses of evolutionary genetics.

126b. Wright, J. W. et al. *Herpetologica* **39:**410–416, 1983. Use of mtDNA to assess relations of parthenogenetic and sexual lizards.

127. Wright, S. *Evolution and Genetics of Populations,* University of Chicago Press, Chicago, 1968.

127a. Wu, C.-I. and W. H. Li. *Proc. Natl. Acad. Sci.* **82:**1741–1745, 1985. Higher rates of nucleotide substitutions in rodents than in man.

128. Yunis, J. J. and O. Prakash. *Science* **215:**1525–1538, 1982. Chromosomal history of man.

5

ENZYME KINETICS AND PROTEIN STRUCTURE

Genomes contain extensive noncoding regions of DNA, many mutations are neutral, and some protein variants do not appear to be adaptive. Commonly used measures of variation in protein structure and function are crude approximations—patterns of gel electrophoresis, O_2 consumption, and maximum velocity (V_{max}) for enzyme activity. Measurements are often made at nonphysiological pH, temperature, and substrate concentration. With more critical measurements, however, adaptive properties due to small changes in protein sequence and conformation are observed. Kinetic analyses of enzymes from similar organisms living in different environments have shown differences not made evident by grosser methods. Kinetic measurements are (1) V_{max} (maximum velocity V_m at physiological pH and temperature; (2) Michaelis constant, K_m (ratio of rate constants); (3) S_{50} (substrate concentration for half maximum velocity); (4) k_{cat} (turnover number of an enzyme); (5) E_a (Arrhenius activation energy); (6) $\Delta G\ddagger$ (activation free energy); (7) $\Delta S\ddagger$ (estimate of entropy); (8) $\Delta H\ddagger$ (enthalpy of activation); and (9) binding constants for substrates and cofactors. E_a and S_{50} are frequently measured with crude preparations, such as tissue homogenates, whereas $\Delta G\ddagger$ and K_m are measured with purified enzymes (28). Kinetic properties are related to (1) amino acid sequences and thus to the nucleotide sequences that code them, (2) secondary, tertiary, and quaternary structures, and (3) distribution of hydrophobic and hydrophilic bonds and charged groups at sites of cofactor and of substrate binding, as well as at distant sites on an enzyme.

Some kinetic measurements together with a few examples of structure–function relations in proteins will be given in this chapter; applications to adaptation in different environments will be presented in later chapters.

SUMMARY OF KINETIC EQUATIONS

The following principles of enzyme kinetics are presented in more detail in texts of biochemistry (20,32,36). A reaction in which substrate [S] is catalyzed by enzyme [E] to go to product [P] with recovery of enzyme is described by the Michaelis-Menton equation:

$$[E] + [S] \underset{k_2}{\overset{k_1}{\rightleftharpoons}} [ES] \underset{k_4}{\overset{k_3}{\rightleftharpoons}} P + [E]$$

[ES] is enzyme-substrate complex; four velocity constants are indicated. If P is considered to be zero, as when products are removed when formed, k_4 can be disregarded.

The equilibrium constant of enzyme-substrate binding is K_S.

$$K_S = \frac{k_1}{k_2} = \frac{[E][S]}{[ES]}$$

At equilibrium, K_S is the concentration of substrate at which half of the substrate is bound to the enzyme and half is free.

The equilibrium constant for the catalyzed reaction is the Michaelis constant K_m, which is the ratio of velocity constants (when k_4 is disregarded):

$$K_m = \frac{k_2 + k_3}{k_1}$$

A quantity which can be obtained without measuring velocity constants is k_{cat}, the turnover number for E. Velocity (v) is measured at different concentrations of substrates and k_{cat} calculated:

$$v = \frac{k_{cat}[E][S]}{K_m + [S]}$$

k_{cat} is a first-order rate constant

$$[ES] \xrightarrow{k_{cat}} P + [E]$$

When S is high so that E is limiting, such that $S >>> K_m$ the velocity V $= V_m$:

$$V = V_m = k_3[E]$$

When substrate concentration is low ($S <<< K_m$) then

$$v = \frac{k_{cat}}{K_m}[E][S]$$

The ratio k_{cat}/K_m is an apparent second-order rate constant for the reaction of $[E]$ with $[S]$.

If $k_2 \gg k_3$, k_3 can be neglected and $K_m = k_2/k_1$. If $k_3 \gg k_2$, $K_m = k_3/k_1$. For measurements of true K_m it is necessary to have a pure enzyme of known or measurable concentration as well as known substrate concentrations. Each enzyme has its characteristic K_m for a given pH and temperature. Qualitatively, K_m is inversely related to enzyme-substrate binding, and enzyme-substrate affinity is reciprocally proportional to K_m; affinity is proportional to $1/K_m$.

A simple representation of reaction kinetics is to plot velocity as a function of $[S]$ (Fig. 5-1a). Such plots are hyperbolic for monomeric enzymes and for some multimeric ones; plots of v versus $[S]$ are sigmoidal for multimeric enzymes. From such a plot K_{50} or K_S can be obtained as the concentration $[S]$ where $v = \frac{1}{2}(V_m)$.

A method for measuring K_m is the Lineweaver-Burke equation,

$$\frac{1}{v} = \frac{K_m}{V_m} \cdot \frac{1}{[S]} + \frac{1}{V_m}$$

which is plotted as $1/v$ versus $1/[S]$ (Fig. 5-1b).

A statistical method of measuring K_m is a fit of the equation

$$v = \frac{V_m}{1 + K_m/S}$$

K_m is obtained from a plot of S/v against S. This plot varies less from linearity than a plot of $1/v$ (67). Computer programs are now available for calculating K_m and k_{cat}.

An approximation that can be made with crude enzyme preparations, such as homogenates, rather than with pure enzymes is K_m apparent ($K_m$app), which is the concentration of substrate for half-maximum velocity. Instead of $K_m$app the term $K_{0.5}$ is preferable.

EXAMPLES OF KINETICS MEASUREMENTS

Many evolutionary changes in K_m for single enzymes have occurred under different environmental conditions. For constant substrate values, genetically controlled differences in K_m maintain constant enzyme efficiency.

Examples of the usefulness of K_m are given for temperature adaptation; as temperature is lowered, the binding of E to S increases, which counteracts the reduction in kinetic activity due to cooling. Organisms from warm en-

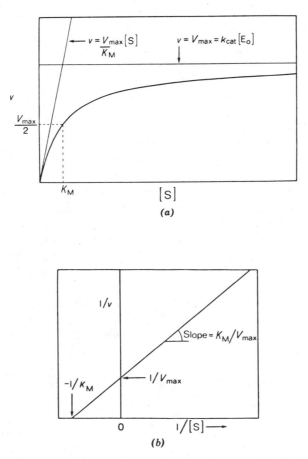

$$v = V_{max}\frac{[S]}{K_M}$$ $$v = V_{max} = k_{cat}[E_o]$$

v

$\frac{V_{max}}{2}$

K_M $[S]$

(a)

$1/v$

Slope $= K_M/V_{max}$

$-1/K_M$ $1/V_{max}$

0 $1/[S] \longrightarrow$

(b)

Figure 5-1. (*a*) Linear plot of an enzyme-catalyzed reaction; velocity plotted as a function of substrate concentration. (*b*) Lineweaver-Burk plot of Michaelis-Menton relation. From A. Fersht, *Enzyme Structure and Mechanisms.* Copyright 1977 by W. H. Freeman and Company. All rights reserved.

vironments tend to have K_m-temperature curves displaced toward high temperature, thus maintaining a relatively constant K_m under physiological conditions (Fig. 5-2). Evolutionary pressures tend to make k_{cat} large and K_m appropriate to physiological concentrations of substrate. Muscle LDH (A_4) from cold-water fish has higher k_{cat} and catalyzes the production of lactate from pyruvate several times faster than LDH-A_4 from warm-water fish or from mammals when both are measured at low or intermediate temperature. K_m and k_{cat} for LDH-A_4 for NADH or pyruvate are as follows (25):

	Normal Cell Temperature	K_m (NADH) (mM)	k_{cat} (NADH) (s^{-1})	t of Measurement (°C)
rabbit	37	8.5	155	20
halibut	10–20	26.75	542	20
rabbit	37	22.7		10
halibut	12	37.4		10
Trematomus	−2	55.3		10
Fundulus	10–20(LDHB[b]B[b])	0.056(Pyr)	179(Pyr)	10 (45)

In halibut muscle K_m is independent of temperature over the range 10–25°C (Fig. 5-2a). For many energy-yielding reactions K_m approaches physiological [S]; for hydrolytic (digestive) reactions K_m is lower than physiological [S], thus providing constant products over a wide range of [S] or without constant feeding (11). K_m for CO_2 of photosynthesis by phytoplankton is lower than for photosynthesis by terrestrial or emergent aquatic plants since light intensity is less with depth in water. The higher K_m of photosynthesis by canopy trees than of understory trees and shrubs correlates with higher light intensity in the canopy. In the marine algae in which growth is limited by nutrients, K_m is low for nitrates and ammonia (12).

For nonregulatory glycolytic enzymes K_m values are higher than measured substrate concentrations—that is, $K_m/S > 1$. In some metabolic pathways the K_m of the first step may be low and the initial enzyme functions at its maximum velocity; hence this step controls the pathway. For example, hexokinase, the first enzyme in glycolysis, has a K_m for glucose of 0.1 mM while glucose concentration in a red blood cell is about 5 mM. K_m values of later steps in glycolysis are higher. A low K_m for the first step in a metabolic pathway prevents later steps from being overloaded and accumulating reactive intermediates. Activities of control enzymes in metabolic pathways are often regulated by varying K_ms of critical substrates by allosteric effects (20). Values of k_{cat}, K_m, and enzyme efficiency (k_{cat}/K_m) for several mammalian enzymes are given in the Tables 5-1 and 5-2 (20).

The kinetic treatment for multisubstrate reactions involves two or more additional rate constants. For example, as shown later, reactions of LDH or ADH require binding of cofactor NAD or NADH before binding of substrate, that is, the formation of a ternary complex and release of cofactor after products are released.

Aqueous solutions, extracellular and intracellular fluids, decrease in pH as temperature is raised. For the most common biological buffers (imidazole groups of proteins) the change is 0.016 pH unit per degree C. When enzymes of poikilotherms are assayed at a constant pH but a different temperatures, the conditions are nonphysiological and K_m values show a steep decline on cooling (68). However, when pH is adjusted at each temperature the K_m variation with temperature is less (Fig. 5-2b) (25). pH optima for several

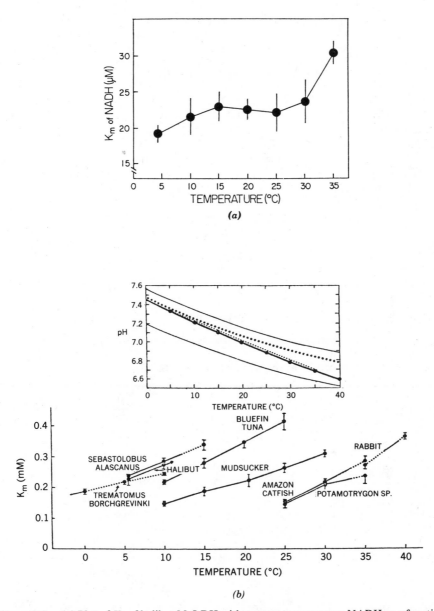

Figure 5-2. (*a*) Plot of K_m of halibut M_4 LDH with respect to coenzyme NADH as a function of temperature of assay. Reprinted with permission from G. W. Somero and G. S. Greaney, *Journal of Comparative Physiology,* Vol. 137, p. 118. Copyright 1980, Springer-Verlag, New York. (*b*) Upper figure: pH as function of temperature, upper and lower lines, limits of intracellular values; dotted line, water; middle solid line, imidazole buffer. Lower figure: Plot of K_m for M_4LDH with pyruvate as substrate as a function of temperature for species living in Antarctic, temperate, and tropical waters. All measurements made at pHs corrected for temperature. Reprinted with permission from G. W. Somero and P. H. Yancey, *Journal of Comparative Physiology* Vol. 125, p. 131. Copyright 1980, Springer-Verlag, New York.

Table 5-1. Selected Values of k_{cat} and K_m (20)

Enzyme and Substrate	k_{cat} (s^{-1})	K_m (mM)	k_{cat}/K_m (s^{-1}/M)
Chymotropsin			
Ac-Tyr-NH$_2$	0.17	32	5
Ac-Pro-Tyr-Gly-NH$_2$	4.4	32	140
Ac-Phe-NH$_2$	.07	31	2
Pepsin			
-Phe-Gly	0.5	0.3	1.7×10^3
-Phe-Gly-Gly	6	0.6	1×10^4
Acetylcholinesterase			
Acetycholine	1.4×10^4	9×10^{-5}	1.6×10^8
Carbonic anhydrase			
HCO$_3$	4×10^5	2.6×10^{-2}	1.5×10^7
Catalase			
H$_2$O$_2$	4×10^7	1.1	4×10^7
Fumarase			
Malate	9×10^2	2.5×10^{-5}	3.6×10^7
Triosephosphate isomerase			
Glyceraldehyde-3-phosphate	4.3×10^3	4.7×10^{-4}	2.1×10^8

Table 5-2. Substrate Concentrations and Enzyme K_m

Enzyme, Substrate, Tissue	Substrate Concentration [S] (mM)	Enzyme K_m (mM)
Lactate dehydrogenase		
Pyruvate (red blood cells)	51	59
Lactate (red blood cells)	2900	8400
Pyruvate (brain)	116	140
Pyruvate kinase		
PEP (red blood cells)	23	200
Glyceraldehyde phosphate dehydrogenase		
G3P (muscle)	3	70

enzymes of trout shift with temperature; acetyl CoA carboxylase has $pH_{opt}/\Delta T$ of -0.27, while fatty acid synthesis has $pH_{opt}/\Delta T$ of -0.43 (26).

For many enzymes and transport proteins, hemoglobin for example, the binding to ligands (substrate, cofactors, transported substances) follows a sigmoid curve rather than a hyperbolic relation. The sigmoid curve results from cooperativity between binding sites or from interactions with allosteric effectors. Binding at one site causes an increase in affinity at another site.

Cooperativity can be negative or positive. Cooperativity is described for a protein by the relation

$$\log \frac{v}{V_{max}-v} = n \log [S]$$

where n is the Hill constant (Fig. 5-3) and is given by the slope of $\log v/V_{max}-v$ versus $\log [S]$.

The best known example of cooperativity is for the interactions of the four subunits of hemoglobin with oxygen. When the log of percentage of hemoglobin combined with O_2 is plotted against $\log P_{0_2}$, the measured Hill constant values for tetrameric hemoglobin are 2.6–2.8; for monomeric Hb n = 1. Enzymes with four binding sites, for example, pyruvate kinase, have cooperativity constants of 2.3–2.8 (20,38).

ENERGETICS

Energy is required for all chemical reactions. Some driving energy for catalyzed reactions comes from the binding energy of $E + S$, some from external heat. Total activation energy is given by the Arrhenius constant E_a:

$$E_a = RT \ln \frac{k_2}{k_1} (T_1 - T_2)$$

where R is the universal gas constant, T_1 and T_2 are two temperatures in degrees Kelvin, and k_1 and k_2 are rate constants at the two temperatures.

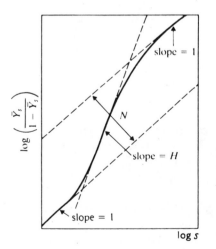

Figure 5-3. Diagram of Hill plot as a measure of cooperativity between subunits of a protein. Y is degree of binding and can be replaced by v. The slope approaches one at very high and very low substrate concentrations. Reprinted with permission from M. Dixon and E. Webb, *The Enzymes*, Copyright 1979, Academic Press, New York.

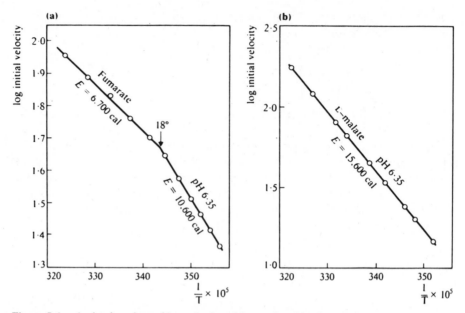

Figure 5-4. Arrhenius plots of log velocity of forward and backward reactions of fumarate hydrase (malate ⇌ fumarate) as function of reciprocal of temperature (K). Reprinted with permission from M. Dixon and E. Webb, *The Enzymes,* Copyright 1979, Academic Press, New York.

V_{max} is proportional to the difference between the two velocity constants. Strictly, activation parameters should be estimated from theoretical or calculated V_{max} values so that temperature-dependent changes in K_m do not invalidate the rate versus temperature plots. E_a is given by the slope of the plot of log $(k_2 - k_1)$ or log V_{max} versus $1/T$ (Fig. 5-4). For enzyme reactions, E_a is in kcal/mol.

Arrhenius plots are extensively used for measuring activation energies for many biological reactions. The rate-temperature relation is usually linear for poikilotherms, but sometimes the plots show a break, with E_a larger at low than at high temperatures. Arrhenius plots for enzymes of homeotherms often have a break or are curvilinear.

The equilibrium constant between the activated complex ES‡ and E + S is proportional to the activation energy ΔG‡. The activation energy is composed of two terms, the activation energy of bond making and the algebraically negative energy of bond breaking. The relation is given by the energy plot in Figure 5-5 (20).

The free energy associated with transitions from ground state to activated state is of two sorts, entropic (ΔS‡) and enthalpic (ΔH‡):

$$\Delta G‡ = \Delta H‡ - T\Delta S‡$$

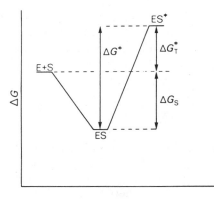

Figure 5-5. Diagram of energy barrier for enzyme-catalyzed reactions. Activation energy $\Delta G_T\ddagger$, energy of binding ΔG_s, $\Delta G\ddagger$ difference between energy of activated complex ES$\ddagger$ and ground state of ES. From A. Fersht, *Enzyme Structure and Mechanisms.* Copyright 1977 by W. H. Freeman and Company. All rights reserved.

where $\Delta G\ddagger$ is free energy change in calories per mol, T is temperature (Kelvin); $\Delta S\ddagger$ is in entropy units (or cal/mol/K); $\Delta H\ddagger$ is in cal/mol.

Entropy ($\Delta S\ddagger$) is, in simple terms, a dimensionless measure of energetic randomness; the more disordered a system, the higher its entropy. Enthalpic energy is a measure of heat content $\Delta H\ddagger$ (38). When two molecules combine as in enzyme-substrate interaction, there is an increase in order and loss of entropy; translational and rotational entropies decrease; this may be partially compensated by a gain of internal entropy. Living systems are open thermodynamically, and their energetics opposes the nondynamic tendency toward increased entropy. Formation of an H bond is exothermic and results in decreased entropy. Whether $\Delta G\ddagger$ is positive or negative depends on relative contributions of $\Delta H\ddagger$ and $\Delta S\ddagger$; if $T\Delta S\ddagger$ is large relative to $\Delta H\ddagger$, then $\Delta G\ddagger$ is negative. The reference point of entropy is the entropy of water, which is zero at $0°K$. $\Delta H\ddagger$ can be calculated from the Arrhenius plot for a purified enzyme: $\Delta H\ddagger = E_a - RT$. To obtain $\Delta S\ddagger$ the specific velocity is required; that is, true [E] (active enzyme concentration) must be known.

The equation for calculation of $\Delta S\ddagger$ as developed by Eyring is

$$\ln \frac{K}{T} = \ln \frac{k_B}{h} - \frac{\Delta H\ddagger}{RT} + \frac{\Delta S\ddagger}{R}$$

where k is the rate constant. k_B is the Boltzmann constant and h is Planck's constant. When appropriate values are inserted, the equation becomes (36a,38)

$$\frac{\Delta S\ddagger}{4.58h} = \log k - 10.75 - \log T \frac{E_a}{4.58T}$$

A method for obtaining approximate $\Delta H\ddagger$ and $\Delta S\ddagger$ for impure enzyme preparations is to assume a constant relation between these quantities and $\Delta G\ddagger$ for a given enzyme. $\Delta H\ddagger$ is plotted as a function of $\Delta S\ddagger$ for pure preparations of the same enzyme from a number of species, and the slope is in degrees

Kelvin (Fig. 5-6). Then for the impure system E_a is obtained from an Arrhenius plot and $\Delta H\ddagger$ is calculated; $\Delta S\ddagger$ is obtained from the plot for $\Delta H/\Delta S$ for the enzyme. $\Delta G\ddagger$ is then computed for the system at a given temperature.

$\Delta H\ddagger$ and $\Delta S\ddagger$ covary. The energy barrier is overcome by increase in enthalpy (heat content) or decrease in entropy (increased order). Endothermic animals have higher heat content than ectotherms. For a constant $\Delta G\ddagger$, changes in $T\Delta S\ddagger$ are relatively greater in enzymes of ectotherms, whereas in endotherms changes in $\Delta H\ddagger$ are greater. Low enthalpy reduces temperature dependence. The energy of activation ($\Delta G\ddagger$) is less in ectotherms and this corresponds to lower values of E_a than in endotherms (6,60). Values of $\Delta G\ddagger$ for a given reaction are similar at physiological temperatures but at low temperatures $\Delta G\ddagger$ is greater for ectotherms than endotherms and ΔH is less negative for warm-bodied, more negative for cold-bodied animals. The following data for lactate dehydrogenase show relatively constant $\Delta G\ddagger$ at physiological temperatures, lower enthalpy, and higher entropy in cold- than in warm-bodied animals (58a).

Enzyme and Substrate	Source	Normal Cell Temperature (°C)	$\Delta G\ddagger$ (kcal/ mol)	$\Delta H\ddagger$ (kcal/ mol)	$\Delta S\ddagger$ (kcal/ mol/°K)
LDH-A$_4$	chicken	39	12.9	10.5	8.7
(pyruvate)	rabbit	37	13.2	12.5	2.5
	tuna	10–25	12.5	8.77	13.4
	halibut	8–15	12.5	8.77	13.7

LDH from fish from different depths were compared with rabbit LDH; when assayed at one atmosphere $\Delta G\ddagger$ was similar, but in fish, ΔH increased and $\Delta S\ddagger$ decreased with depth of habitat (Ch. 8) (58a). These data suggest some compensation for high pressure.

Animal	$\Delta G\ddagger$	$\Delta H\ddagger$	$\Delta S\ddagger$
surface fish	14.0	10.5	12.7
midwater fish	14.2	11.8	8.7
deepwater fish	14.3	12.5	6.4
rabbit	14.3	12.5	6.4

Figure 5-6. Plots of activation enthalpy ΔH‡ versus activation entropy $\Delta S\ddagger$. (*a*) plot for pyruvate kinase from several species. Reprinted with permission from G. W. Somero and P. S. Low, *Journal of Experimental Zoology,* Vol. 198, p. 6. Copyright 1976, Alan R. Liss, Inc., New York. (*b*) Plot for myofibrillar ATPase from named species. Reprinted by permission from *Nature,* Vol. 257, p. 622. Copyright © 1975, MacMillan Journals Limited.

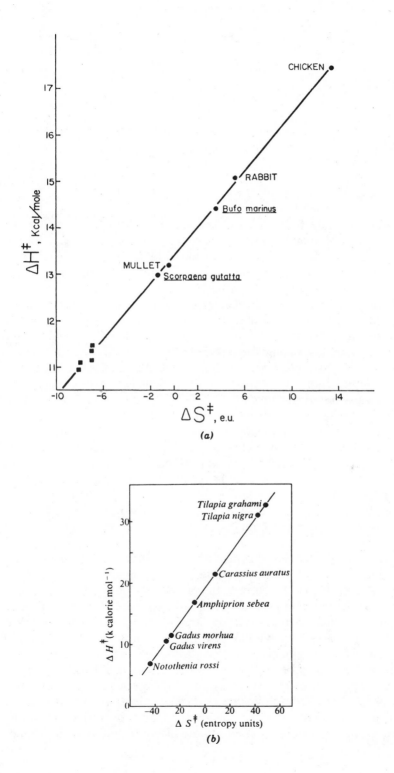

(a)

(b)

177

Mg-Ca myosin ATPase was compared for fish from different temperatures (69):

Fish	Habitat Temperature (°C)	$\Delta G\ddagger$	$\Delta H\ddagger$	$\Delta S\ddagger$
Pomocentrus	18–26	17.6	26.5	32.0
Cottus	3–12	16.3	13.5	−9.8
Notothenia	0–3	16.1	11.3	−17.7
Champsouffalos	−1 to 2	15.9	7.4	−31.4

$\Delta G\ddagger$ was lowest for fish from cold water; $\Delta H\ddagger$ and $\Delta S\ddagger$ decreased in parallel.

Other critical adaptive properties of enzymes include the rigidity of tertiary and quaternary structure. Molecular folding can be estimated from volume changes measured by effects of hydrostatic pressure. Molecular conformation, size, and shape, can be estimated from X-ray diffraction, from NMR (nuclear magnetic resonance), from fluorescence polarization, and from other physical measurements (52). The covalent bonds that link the amino residues within a protein or the bases within DNA and RNA molecules are relatively strong. However, hydrogen bonds, electrostatic (ionic) bonds, and van der Waals interactions are weak bonds. The secondary, tertiary, and quaternary structures of proteins depend on weak chemical bonds, and it is these bonds that give inherent flexibility to proteins. Conformational changes are essential for precise alignment of reactive centers between enzyme and substrate. Some weak bonds can be formed with positive enthalpy, others with negative enthalpic changes. The higher order conformation of proteins is dictated by primary structure, which is genetically coded. The folding due to weak bonds is sensitive to environmental parameters, such as temperature, hydration, pressure, and charge distributions. Inactivation of enzymes at extremes of temperature or pressure may result from the environmental effects on the flexible structure of enzyme proteins (20,28,52).

Binding sites between enzymes and substrates can be characterized by using analog cofactors and substrates. One property which may be influenced by the primary structure of the protein is charge distribution; replacement of an amino acid at a distance can influence a binding site. Binding sites may be separate from catalytic sites. Allosteric effects result from modification of catalysis by binding of a ligand at a noncatalytic site. Binding frequently is accompanied by removal of water or shifts in charges within an enzyme, and may be determined by distribution of hydrophobic and hydrophilic amino acids. Lower $\Delta G\ddagger$ and $\Delta H\ddagger$ are achieved in part by exergonic exposure of hydrophilic groups during activation; in cold-adapted fish, more efficient catalysis occurs at low temperature by more hydrated enzymes (60). Electrostatic interactions are also affected by ionic strength; the presence of some organic solutes may protect against denaturation by inorganic ions (Ch. 9).

Cell membranes, both plasma membranes and those of intracellular organelles, contain lipids as well as proteins, usually in a ratio of 9 to 1 on a molar basis, lipid to protein. Proteins are, therefore, imbedded in a lipid surround. The lipids in eukaryotes are mostly phospholipids plus small amounts of sterols. The phospholipids have fatty acid chains which vary in percent saturated fatty acids according to environmental conditions. The increased unsaturation and resulting greater fluidity when membranes are made at low temperatures are discussed in Chapter 7. Degree of unsaturation is determined by desaturases and by enzymes for *de novo* synthesis. Both genetic and environmental control of membrane lipids are demonstrated by analyses of membranes from organisms living at different temperatures and pressures. Homeoviscous adaptation provides constancy of membrane fluidity and thus of the environment for membrane proteins. Membrane-bound enzymes, such as those of mitochondria, vary in activity according to degree of saturation of fatty acids in the membranes; permeability of plasma membranes is also much influenced by membrane fluidity (58b).

Thermal stability of a protein is correlated with the cell temperature at which the protein normally functions (Fig. 5-7). Temperatures for heat dena-

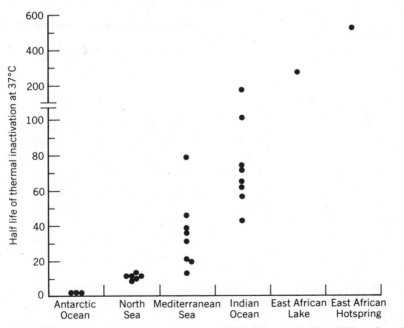

Figure 5-7. Half-time for thermal inactivation at 37°C for myofibrillar ATPase from fish living in the different environments stated on abscissa. Reprinted with permission from I. A. Johnston and W. J. Ublesby, *Journal of Comparative Physiology,* Vol. 119, p. 200. Copyright 1977, Springer-Verlag, New York.

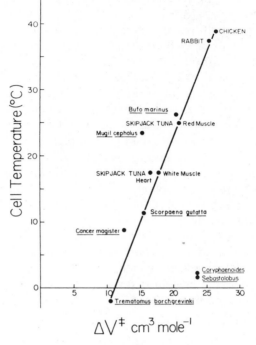

Figure 5-8.   Relation between normal cell temperature and activation volume $\Delta V\ddagger$ for pyruvate kinase of different species. Reprinted with permission from G. W. Somero and P. S. Low, *Journal of Experimental Zoology*, Vol. 198, p. 9. Copyright 1976, Alan R. Liss, Inc., New York.

turation or for 50 percent inactivation of pyruvate kinase from different animals that function within designated temperature ranges follow: rabbit (from 30–40°C) 62°C; mullet (18–30°C) 58°C; *Bufo* (25–32°C) 57°C; shark (8–17°C) 52°C; abalone (8–15°C) 45°C; and ice fish (0–−2°C) 42°C (59,60). Conformational changes often occur when ligands bind; an enzyme must be sufficiently flexible to permit these changes at biological temperatures. Since protein structure may become more rigid at low temperatures, it may be necessary for the inherent rigidity to be low for enzymes that function in the cold. This is reflected by lower heat inactivation temperatures. There is a compromise between rigidity and flexibility.

Another measure of the capacity for conformational change is the change in volume (ΔV), which can be measured by effects of hydrostatic pressure. This is given by

$$\Delta V = \frac{-RT(\ln k_2 - \ln k_1)}{P_2 - P_1}$$

where k_1 and k_2 are rate constants at pressures P_1 and P_2. ΔV reflects the molecular flexibility of a protein. Enzymes from mammals and fish which live at one atmosphere are more sensitive to pressure than enzymes from deepwater fish, as shown by ΔV (cm^3/m) as follows for M_4 LDH (28):

Enzyme Source	ΔV
beef	36.0
dogfish	5.4
sculpin	10.0
Antimora	0.0

Other examples of adaptive properties with respect to volume changes under pressure are given in Figure 5-8 and in Chapter 8.

In conclusion, most of the correlations made in the past of amino acid composition with enzyme function have been based on V_{max} measurements. More subtle measurements are needed to reveal the adaptive significance of small differences in amino acid sequences and conformations. Kinetic and structural properties of proteins are adaptive for life patterns. The following sections present several examples of adaptive kinetic properties of proteins. Further examples are given for specific environmental parameters in Chapters 7 and 8.

ADAPTIVE PROPERTIES OF SPECIFIC PROTEINS

Iron Porphyrin Proteins; Hemoglobins and Cytochromes

Porphyrins consist of rings of fused tetrapyrroles. They probably evolved during the period of prebiotic evolution. Porphyrins complex with iron to form hemes, which then become bonded to specific proteins; an example is cytochrome c. Heme proteins (cytochromes) evolved in anaerobic bacteria (Ch. 2); several cytochromes now occur in anaerobic and aerobic bacteria (14). Cytochromes c function in all types of photosynthesis and in electron transport in all aerobic cells. Chlorophyll contains a magnesium-porphyrin. With increased body size of animals, there was need for oxygen-transporting pigments; the transition from cytochrome to hemoglobin must have been relatively direct. Metalloproteins probably evolved during the prebiotic period of chemical evolution. In some metalloproteins, the metal is attached directly to the protein, iron in the O_2-transporting pigment hemerythrin, copper in hemocyanin. The best known transport pigment is hemoglobin, in which iron is attached to a porphyrin ring which is bonded to a globin. No molecular similarities have been noted between hemoglobins and other O_2-transporting proteins (hemerythrin, hemocyanin) (Ch. 6).

Globins vary considerably in size, amino acid composition, charge, solubility, and other physical properties which are adapted to oxygen transport. Hemoglobins occur intracellularly in some protozoans (Paramecium), in some neurons of molluscs, and in many muscle cells as myoglobins. Probably in all of these the hemoglobins act as facilitators of O_2 transfer to oxidative enzymes. Rhizomes of some legumes contain nitrogen-fixing bacteria; the cytoplasm of legume root cells has leghemoglobin (leghb), which probably facilitates transfer of O_2 to the bacteria. Transfer hemoglobins occur as large molecules in solution in the plasma in many invertebrates, (annelids, a few insects), and in special cells (blood corpuscles or coelomocytes) of some polychaetes, echinoderms, and vertebrates. Hemoglobin molecules are usually smaller in transporting cells than as free molecules in extracellular fluids. Since hemoglobins occur in legume roots, in some ciliates and in several unrelated animals, a few genera of molluscs, echinoderms, insects, and in all vertebrates, it is likely that hemoglobins (Hbs) have evolved several times.

Invertebrate Hbs vary in size from 34,000 Da in cells to several million Da for those occurring as free molecules in solution. Most invertebrate Hbs are monomeric and show hyperbolic O_2 equilibrium curves which are only slightly affected by pH. Invertebrate Hbs differ in peptide sequences from vertebrate Hbs. A direct evolution from invertebrate Hbs to those of vertebrates has not been demonstrated; functional similarities do occur. By packaging small Hbs in cells the total Hb concentration in the blood can be elevated; equivalent concentrations of large Hb molecules in solution would make the blood excessively viscous.

Hemoglobins of Vertebrates

This section deals with structure and relationships of vertebrate hemoglobins, and with some adaptive correlations with structure. Genetic and developmental properties are given in Chapter 3, phylogenetic relations in Chapter 4, and ecological correlations of hemoglobin function with O_2 transport in Chapter 6.

Vertebrate hemoglobins (Hbs) are tetrameric except in cyclostomes and occur in red blood cells; myoglobins (Mbs) are monomeric and occur mainly in cells of red muscle (12a). A molecule of myoglobin has heme in a crevice below the surface of the coiled protein (Fig. 5-9). The molecular weight of the monomer myoglobin is about 17,000 Da. In adult mammals each tetramer consists of two sets of each chain, α-type and β-type. In adult human Hb each α chain contains 141 amino acid residues and each β chain 146 amino acid residues; there are 75 differences in sequences between α and β chains and 65 similarities. The α and β chains of vertebrate Hbs differ in the amino acid sequence at the COOH (carboxy-) terminal. In fishes the carboxyl terminals of α chains have -Tyr-Arg; the terminals of β chains have -

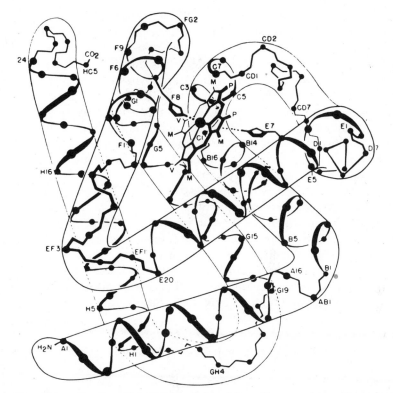

Figure 5-9. Reconstruction of myoglobin deduced from X-ray analyses. Heme located on upper middle loop. From A. Fersht, *Enzyme Structure and Mechanisms*. Copyright 1977 by W. H. Freeman and Company. All rights reserved.

Tyr-His when the O_2 affinity is pH sensitive, or -Tyr-Phe when the Hb is pH insensitive. Numerous mutants, particularly of the β chain, are known; these are allotypes (allohemoglobins). In addition, hemoglobins occur in several isotypic forms (isohemoglobins) coded at different loci (Ch. 3). Fishes have more different Hbs (isotypes) in circulating blood than mammals, each Hb tetramer consisting of amino acid chains which are homologous to the mammalian α and β chains. A single electrophoretic type may be composed of more than one structurally distinct hemoglobin, as demonstrated by ion exchange chromatography on DEAE cellulose, by amino acid sequencing, or by functional measurements. In fish hemoglobins more than one type of α chain and β chain associate in different proportions as hybrid tetramers.

From amino acid sequences and corresponding nucleotide replacements (NR) the relationship between Hbs of various vertebrates has been calculated (23a). About 500 myr ago a primitive heme protein apparently gave rise to myoglobin (Mb), and this in turn to monomeric Hb such as occurs in cyclostomes. Gene duplication leading to separate α and β lines of Hb occurred in jawed fishes about 400 myr ago (12b,23a). Both α and β lines evolved rapidly at first, then slowly. In the β line a duplication in primitive mammals (marsupials) led to formation of a γ chain; this chain replaces β in fetal Hb (HbF). The loci for γ and β are on the same chromosome and they differ in sequences by 9 amino acids. HbF has higher O_2 affinity and lower potentiation of O_2 binding by DPG (diphosphoglycerate) than adult HbA_1. A recent gene duplication in primate β led to δ Hb. The number of amino acid sequence differences between β and δ in human is 11, in chimpanzee 10 (23a). Early human embryos have another substitution for β, a mutant called ε (10), and a substitution for α called ζ. The principal human adult Hb is Hb A_1 which consists of 2α, 2β; fetal Hb is 2α 2γ. A second adult form HbA_2 in primates consists of 2α and 2δ chains and constitutes 2–3 percent of the circulating Hb molecules. Many mutants of Hb are known, more in β loci than in α loci. An example of an allotype in humans is the hemoglobin of sickle cell anemia in which the β chain is altered to βS by substitution of a valine for a glutamate. In primates, α and β Hb genes are on different chromosomes and each is a split gene with introns between the active nucleotide sequences. In the normal β chain the number of sequence differences from human hemoglobin are: for cebus monkey 6, corresponding to divergence 20 myr ago; for kangaroo 38 amino acid substitutions in 130 myr; for frog 67 residues in 320 myr (23a).

In nonprimate mammals and most other vertebrates the blood contains several Hbs (isotypes and allotypes). For example, in the rodent *Peromyscus* Hbs c, d, and e are polymorphic with multiple alleles. In cattle two allotypes of the β chain differ in three positions. In sheep, two allotypes of the β chain differ in seven residues and the tetramers have been designated as HbA and HbB. Unfortunately, a uniform terminology has not been adopted for multiple Hbs.

Tetrameric Hb is roughly spherical, 65 Å × 55 Å × 50 Å, with four heme pockets in clefts between the monomers. The β and α pairs are related on a twofold axis. To identify the relative positions of each member of a pair, subscripts are used; these subscripts do not imply chemical differences between members of a pair. Figure 5-9 is a stereo view of the molecule from one surface. There are two kinds of interlocking contacts of side groups between chains and with neighboring hemes. The $\alpha_1\beta_1$ contact has 34 side chains, the $\alpha_1\beta_2$ contact 19. The α-α and β-β interactions change on oxygenation. On deoxygenation the two α hemes move 1 Å closer to each other and the two β hemes separate by 6.5 Å, with distances in Å as follows (13):

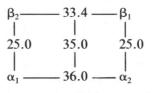

Oxyhemoglobin

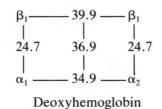

Deoxyhemoglobin

Bony fishes have numerous isotypes and allotypes of Hb. There may be more than one type of α chain and/or β chain; hybrids or various combinations of the chains are common, each with 2 α type and 2 β type chains. One group of fish—carp, cichlids, plaice, flounder, *Fundulus,* and coelacanth— have Hbs all of which are sensitive to pH and temperature. A second kind of bony fish—salmonids, trout, salmon, catastomids (suckers), and eels— have some Hbs which are pH and temperature sensitive, and have other Hbs that are pH and temperature insensitive. A third kind of fish—tuna and porbeagle shark—have Hbs which are pH sensitive and temperature insensitive (5). Fish hemoglobins in which O_2 binding is independent of pH have the carboxyl-terminal histidyl residue of the β chain replaced by a phenylalanyl residue and the amino terminal of the α chain acetylated.

In many hemoglobins, affinity for O_2 is reduced when the pH is lowered (Bohr effect). A normal Bohr effect is a shift in an oxygen equilibrium curve to the right, with acidification such that unloading is facilitated when CO_2 is added and loading is enhanced when CO_2 is lost (Ch. 6). O_2 affinity is also modulated by organophosphates: 2-3 diphosphoglycerate (DPG) in mammals, inosinephyrophosphate (IPP) in birds, adenosine triphosphate (ATP) and guanosine triphosphate (GTP) in fishes. The positions affected by pH and by organophosphate are on the β chains. The Bohr effect depends on the imidazole groups of C-terminal histidines of β-chains and on α-amino groups of α chains; β chains with a phenylalanine substituted for the terminal histidyl residue show no or reverse Bohr effect. 2,3-Diphosphoglycerate (2-3 DPG) lies in a cleft between the two β chains and binds between protonated α-NH_2 groups of the two β chains. Related to the Bohr effect is the Root effect, according to which the Hb becomes only partly saturated with O_2 when pH drops; this is a property of many fish hemoglobins. Fetal hemoglobin has a γ chain which differs from the β chain in 19 amino acid substitutions, three of these at DPG binding sites. Low DPG binding capacity favors ability of fetal Hb to compete for O_2 (8,23a).

In the killifish (*Fundulus*) four distinct Hbs are separable by electrophoresis (Hb I, Hb II, Hb III, and Hb IV) (42a). Each of the four shows essentially the same O_2 affinity and each is sensitive to pH, temperature, and organophosphate. Ion exchange chromatography and amino acid end-group analysis show that there are two different subunits for Hbs I and IV, four different subunits for each Hb II and III. Two different α and two different β chains are coded at separate loci (α^a and α^b) and (β^a and β^b). Each individual fish has loci for both types of α and β chains. While all possible Hb combinations are formed, some are less stable than others. The stable Hbs of *Fundulus* are shown in Figure 5-10. Although all these hemoglobins have similar oxygen affinities, temperature responses, and Bohr effects; some fish have red cells that are more capable of unloading oxygen. Fish with LDH-B^bB^b have elevated ATP content in red cells. When fish were swum to exhaustion at 10°C, those with LDH-B^bB^b swam longer (46).

The American eel has five hemoglobins in seven isotypes. Type I has both fast (F) and slow (S) isotypes as distinguished by electrophoresis; S is cathodal, F is anodal. Type II has fast, medium (M), and slow isotype, all of which are anodal. In type II F, M, and S have different β chains; F and M have the same α chain but S has a different α—that is, Hb II has three different β and two different α chains. Hb type I has one α chain similar to II β, the other α chain different; its β chains are different from those in II (22).

A stingray (*Dasyatis*) has five hemoglobins, one of which (Hb III) comprises two-thirds of the total. The amino terminal end of the β chain differs in its amino acids from this region in teleosts. The carboxyl terminal end has Tyr and His, and there is a large Bohr effect.

Rainbow trout have four Hbs numbered I to IV according to their electrophoretic mobility from cathode to anode. Two (Hb I and Hb IV) constitute

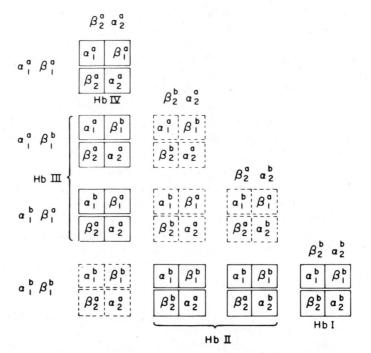

Figure 5-10. Composition of stable hemoglobins of *Fundulus*. Reprinted with permission from P. A. Mied and D. A. Powers, *Journal of Biological Chemistry,* Vol. 253, p. 3526. Copyright 1978, American Society of Biological Chemists, Inc., Bethesda, MD.

most of the circulating Hb; in Hb I (anionic) the carboxyl terminal of the β chain is Tyr-Phe whereas in IV it is Tyr-His. Hb I shows no effect of pH or temperature on the P_{50}; Hb IV is very sensitive to pH, temperature, and ATP (3).

Amia, a primitive bony fish, has seven Hbs (5). An eelpout (*Zoarces viviparus*) has six major Hb components (26a).

Hb structure has been correlated with physiological adaptation to the environment of catostomid fish (47). These fish have six Hb chain loci—two of α and four of β chains. One subgenus (*Pantosteus*) has one or more cathodal Hb components, the other (*Catostomus*) has no cathodal component. The β chain of the cathodal Hb has a carboxy terminal of -Tyr-Phe while the anodal β chain ends in -Tyr-His. The anodal Hbs are sensitive to both pH and ATP, while the cathodal Hb is insensitive to pH and is negligibly affected by ATP. In the anodal Hb the Bohr effect increases as ATP is added. *Pantosteus* occurs in fast streams; its cathodal Hb has high O_2 affinity, no Bohr effect, and is probably used for O_2 transport in emergency exertion. *Catostomus* occurs in pools and sluggish water; all of its Hbs are anodal and sensitive to pH (Ch. 6).

Genetic variations which are associated functionally with amino acid sequences may be identified with respect to either defects or to altered function of different Hbs. Many mutants (some 250) of human Hb have been detected electrophoretically or by electrofocusing; only a few of these have been correlated with altered function (49). A tabulation of a selected few of these follows (7):

Mutant Hb (human)	Chain, Amino Acid Position	Substitution	Effect
S	β 136	glu → val	Solubility of oxyHb decreased by 100×, sickle cell anemia, cells more resistant to malaria than with Hb A
C	β 136	glu → lys	
I	α 16	lys → asp	
Norfolk	α 57	gly → asp	
Kansas	β 103	ans → thr	Low O_2 affinity
Zurich	β 63	his → arg	Low O_2 affinity
Hammersmith	β 42	phe → ser	Low O_2 affinity, altered heme attachment
Tarrant	α 126	asp → asn	Change in α-β contact, bridge to arg 141 of other α, high O_2 affinity

Mutant Hb (human)	Chain, Amino Acid Position	Substitution	Effect
Fort de France	α 45	his → arg	Binding to propionic of heme altered, increased O_2 affinity
Titusville	α 94	asp → asn	Altered β-α contact, low O_2 affinity

Functional correlates of regions of Hb molecules provide clues to why there are so many Hbs (43). It is possible to prepare tetramers of the same chain, homotetramers, *in vitro*. A homotetramer of β chains lacks cooperativity in the binding of O_2, hence the reversible binding of O_2 needs heterotetrameric structure. The most slowly evolving positions are those for heme contact; however, these evolved seven times faster in preamniotes than in aminotes (23a).

A tabulation of positions serving different functions and estimated nucleotide replacements/100 codons/10^6 yr follows (23a):

Function	β Chain	α Chain	NA Replacements/ 100 codons/ 10^6 yr β	α
α-β contacts	10	10	0.02	0.01
heme contacts	21	19	0.02	0.02
nonsalt α-β	16	14	0.16	0.13
DPG contacts	4		0.1	
interior contacts	21	19	0.09	0.09
exterior aa's	70	72	0.2	0.15

The main binding site for 2,3-diphosphoglycerate (DPG) is in a cleft between the two β subunits which is formed by at least four positively charged groups from each chain: the $α-NH_3$ group of the terminal valine, the imidazole at position 2, the lysyl residue at 82, and histidyl at 143 (48). The Bohr effect ($\Delta \log P_{50}/\Delta pH$) depends on release of protons when the Hb binds to oxygen. Almost 75 percent of this effect of pH depends on the α chains and on the carboxyl-terminal histidyl residues of the β chains.

Hemoglobins differ in tertiary and quaternary structures which correlate with function (43a). An α chain has 7 and a β chain has 8 helical segments. During uptake of O_2, tertiary change occurs around the heme and a quaternary rotation of one α-β dimer relative to the other so that the cavity between β chains can bind an organophosphate. Deoxy Hb has low affinity for O_2, high affinity for H^+, Cl^-, organophosphate, CO_2. Human and horse Hb differ in 42 amino acid sequences, but the two Hbs are similar in higher order structure and in functional properties. Perutz (43a) concludes that many differences in amino acid sequences are not significant for O_2 transport, that they originated by random genetic drift, but that structure, charge, and binding properties at specific sites account for the physiological differences between Hbs.

The phylogenetic relations between hemoglobins from different animals, as inferred from amino acid sequences, were discussed previously. What are the adaptive values of the extreme diversity of Hbs? Tetrameric structure provides for sigmoid O_2 binding and cooperativity between different chains. In general, α chains are more positively charged than β chains. Adaptations for binding of O_2 in different ranges of P_{O_2}, that is, different O_2 affinities, are determined by structure. Differences in effects of pH, temperature, and organophosphates are highly adaptive. There are many isotypes which in a given animal show similar respiratory properties; some isotypes occur in small proportions in circulating blood. Many allotypes actually represent neutral mutations, for example, in the many allotypes of human Hb. Subtle measurements may eventually reveal the adaptive functions of some of them. The family of hemoglobins provides clear evidence for both saltatory and continuous evolution via molecular changes, some of which relate to ecology and life history.

Cytochromes c

Cytochromes are heme-containing proteins that function in electron transport. The family cytochromes c is extremely conservative biochemically in that the protein's structure and functions are similar in primitive bacteria and advanced eukaryotes (Fig. 5-11). From similarities in the fold into which the heme binds, from some invariant amino acid sequences, and from functions in electron transport it has been concluded (14) that cytochromes c evolved very early, probably in ancestors of anaerobic photosynthesizing bacteria (Ch. 2). Anaerobic sulfur bacteria, green or purple, use cytochromes in photosynthesis. Bacteria that use sulfate produced by sulfur photosynthesizers or from other sources for anaerobic respiration contain a cytochrome; an example is *Desulfovibrio,* which has a cytochrome c (absorbing at 553 nm). Nonsulfur purple bacteria are capable of photosynthesis under anaerobic conditions. These bacteria can also respire using NADH generated from organic substrates by TCA cycle enzymes for reducing substance in the presence of oxygen and in darkness. Apparently nonsulfur purple bacteria use the same cytochrome sequence—b and c_2—in photosynthesis and respiration. The cyanobacteria and all eukaryotic algae and plants

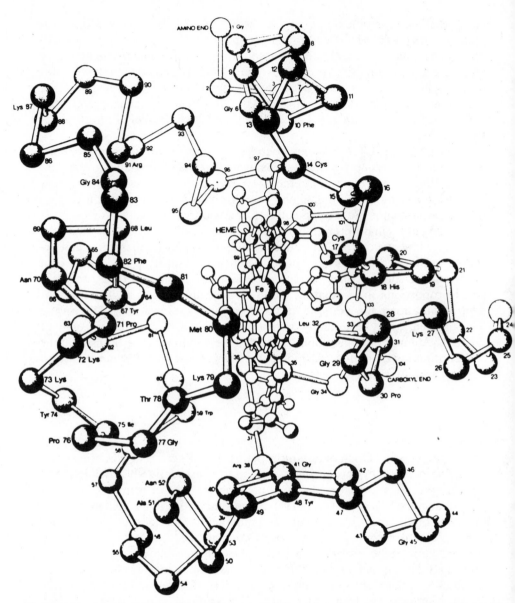

Figure 5-11. Model diagram of cytochrome c with heme in center of molecule. Reprinted with permission from R. E. Dickerson and R. Timkovich, *The Enzymes,* Copyright 1975, Academic Press, New York.

use H_2O as a reductant, and have two photocenters (Fig. 2-8). They use a cytochrome series for electron transport between the two photocenters. The aerobic *Paracoccus* and the mitochondria of all eukaryotes use a common cytochrome chain which transports electrons from a reducing substance to coenzyme Q, to cytochromes b, to cytochrome c and cytochrome oxidase (cytochrome a-a_3), and to oxygen (Fig. 6-1). Mitochondrial cytochromes c have maximum absorption at 550 nm; bacterial cytochromes c may absorb at slightly longer wavelengths (555 nm in the anaerobic photosynthetic green sulfur *Chlorobium*). Bacteria have cytochrome c in the cell membrane, eukaryotes have it in mitochondria. It is concluded that cytochromes c evolved in electron transport, first in anaerobic metabolism and photosynthesis, then in combined aerobic photosynthesis and oxidation, and that cytochromes c have persisted in eukaryote photosynthesis and respiration.

Cytochrome P-450 occurs in both prokaryotes and eukaryotes. In mammalian liver it is found in microsomes where it functions in nonspecific oxidation. P-450 can be induced by various toxic materials—drugs and poisons—and it reacts with molecular O_2.

Cytochromes c have been classified as short (*Pseudomonas,* with 82 amino acid residues), medium (tuna and most vertebrates with 103 residues), and long (the purple nonsulfur bacterium *Rhodospirillium,* with 112 residues) (14). In more than 50 different eukaryotes the redox potential of cyt-c approximates $+254$ mV and maximum absorption is at 550 nm. There are 27 invariant amino acid residues in eukaryotic cytochromes c—in fungi, insects, fish, plants, and mammals (39a).

In all eukaryotes cyt c is contained in the inner membrane of mitochondria. Amino acid composition and positions have been examined in cytochromes c of many eukaryotes; comparisons between different organisms show that the amino acid sequences, particularly of hydrophobic and aromatic chains, are highly conserved. Phenylalanine at position 10 is invariant. In cytochrome c heme is bonded to two cysteines (positions 14 and 17) on one side, and to a methionine (position 80) and histidine (18) on the other side (Fig. 5-11). A loop between the amino acid sequence 57-74 interacts with cyt c reductase, and a helix at sequences 92-102 interacts with cyt oxidase (42b). The unit evolutionary period, that is, the time for 1 percent change in amino acid composition, is 20 myr for cyt c, compared with 5.8 myr for Hb, 1.1 myr for fibrinopeptides, and 500 myr for histone IV (14). The average number of residue differences from mammals are: to birds 10.6, to reptiles 14.5, to several invertebrates 22.1, to fungi 43.1 (39a). Cladograms for cytochrome c agree reasonably well with the relation indicated by paleontology (14) (Chapter 3).

Interpretations of possible adaptive functions of structural differences are found in interactions with cyt oxidase and cyt reductase (39a). For beef and horse heart, cyt c has two binding sites for cyt oxidase; the K_m of high binding affinity of cytochrome oxidase is 10^{-7} to 10^{-8} $M,$ low affinity K_m is about 10^{-6} M. The high affinity site is sensitive to ions and inhibited by ATP. Yeast cyt c has a single binding site of low affinity, K_m 4 × 10^{-3} m. *Euglena* has a single site with K_m 4 × 10^{-7} M (39a). Bacterial cyts c_2 react well with mitochondrial cyt c reductase, poorly with mitochondrial cyt oxidase. This may indicate that the oxidases have diverged farther than the reductase.

PURINE AND PYRIDINE NUCLEOTIDE-ACTIVATED ENZYMES

The principal purine cofactor is adenine and the principal pyridine cofactor is nicotinamide (Fig. 5-12*a*) (36). An early step in evolution was the binding of adenine to a ribose; adenosine is adenine-ribose. The next step was phosphorylation of adenosine to AMP, ADP, ATP (Fig. 5-12*b*). These are energy carriers, discussed in Chapter 2.

As energy-transferring enzymes evolved, first anaerobically, then aerobically, purine-pyridine derivatives were used as cofactors. The best known of these is NAD or nicotinamide adenine dinucleotide (Fig. 5-12*a*). Many enzymes use diphosphopyridine nucleotide (NAD^+) as coenzyme: malate DH, lactate DH, alcohol DH, 3-phosphoglyceraldehyde DH, α-glycerophosphate DH, and β-hydroxybutyrate DH. Some energy-transferring enzymes use flavin nucleotides, such as flavin adenine dinucleotides (FAD), as prosthetic groups: *d*-amino acid oxidase, xanthine oxidase, glycine oxidase, and succinate dehydrogenase. Some enzymes use triphosphopyridine nucleotides (NADP) as cofactor: isocitrate dehydrogenase (IDH), malate DH, glucose -6-P-DH, and 6-P-gluconate DH (36).

Several of the NAD linked enzymes are dimers—ADH, MDH; others are tetramers—LDH, GAPDH. These proteins have similar functions with different substrates, and have similar tertiary structure. In each the nicotinamide attaches in a cavity which has hydrophobic amino acids on one side (NAD^+ binding) and hydrophilic groups on the other (substrate binding) side. Crystallographic analysis is available for several of these enzymes. In LDH the substrate (pyruvate or lactate) lies between nicotinamide and histidine 195. Orientation of the monomers within tetrameric LDH is inverted relative to that in GAPDH; in LDH the coenzyme sites are well separated on the outside of the molecule, in GAPDH they are close to the subunit interfaces; thus NAD^+ sites on the monomers can interact (52,53,54).

Lactate Dehydrogenase

A highly conserved and much studied enzyme that uses NAD^+ as cofactor is lactate dehydrogenase (LDH), which catalyzes the reaction

$$\text{lactate} + NAD^+ \rightleftharpoons \text{pyruvate} + NADH + H^+$$

LDH is composed of subunits of molecular weight 35,000 which function

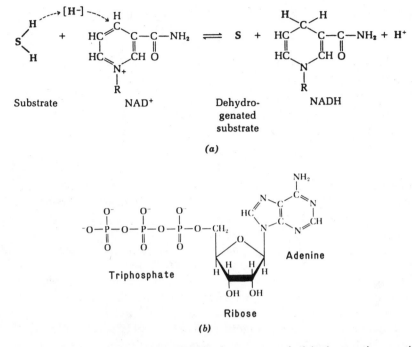

Figure 5-12. (*a*) Nicotinamide portion of NAD$^+$ in an enzymatic dehydrogenation reaction; S is substrate. (*b*) Structural formula of adenosine triphosphate (ATP). From A. L. Lehninger, *Principles of Biochemistry,* Worth Publishers, New York, 1982, pp. 256 and 9.

as dimers in some fungi and arthropods, more usually as tetramers—in *E. coli,* invertebrates such as *Balanus,* lobster, and in all vertebrates. LDH occurs as several isozymes; the primitive vertebrate form is A (or M for muscle) LDH; this gave rise by duplication in fishes to B (or H for heart) LDH. The two isozymes can occur in proportions A_4, A_3B, A_2B_2, AB_3, and B_4. In general, the B enzyme is predominant in tissues which are highly aerobic—heart, brain—and it has high affinity for NAD. The A form is predominant in tissues low in O_2, such as white muscle, and tissues which produce but do not utilize lactate; LDH A has high affinity for NADH and competes with the TCA cycle for NADH. The A form resists inhibition by lactate. Data for trout follow (2):

	B_4	A_4
K_m pyruvate (mM)	0.037	0.65
K_m lactate (mM)	5.8	30
Percent inhibition by pyruvate	40 percent	11 percent
K_i pyruvate (mM)	0.075	0.187
K_i lactate (mM)	29	103

Comparison of LDH K_m for pyruvate in several species follows (56):

Species	A_4 (mM)	B_4 (mM)
hagfish	0.53	0.45
mackerel	0.56	0.069
man	0.83	0.08
Fundulus	0.40	0.14

The proportions of the A and B forms vary with species and tissue. Liver contains 90 percent B LDH in sheep and deer, 5 percent in horse, man, and mouse. In red blood cells the LDH consists of 100 percent B in rabbit, 25 percent in mouse (34a). In cyclostomes, lampreys have only A_4, but in hagfish the heart has a B_4 which resembles A_4 more than it does the heart isozyme of other vertebrates, in that hagfish B_4 is inhibited less by pyruvate and the hagfish heart can function at low O_2 (56).

A third isozyme of LDH, C, occurs in spermatocytes of birds and mammals. In the retina and brain of many fishes another form of LDH, the E-type, has been found; E_4 is more sensitive to lactate inhibition than B_4 and may prevent accumulation of lactic acid in the highly aerobic nervous tissues (58,66). Hybrids between different isozymes can be made *in vitro*.

A number of allozymes, especially of LDH-B, have been found, and the terminology for these is not consistent. In rainbow trout there are two alleles (37). In the killifish *Fundulus* the allozyme more abundant in populations in cold water (*Halifax*) is B^b, and is B^a in warm water populations (Florida) (45) (Fig. 4-1). Adaptive functional differences are described in Chapter 4. In a minnow at 25°C there is more AA, in cold water more A'A' (42). In trout, the presence of one allozyme of LDH-B provides more stamina for swimming at low P_{O_2} than the alternate allozyme (35). In trout the proportions of A and B allozymes may change with temperature of acclimation. Where less O_2 is available at high temperature, LDH activity is high in several kinds of fish (60).

The kinetics of LDH differs whether the reaction is from pyruvate to lactate or the reverse. The enzyme binds to either substrate only if first bound to coenzyme NAD^+ in going from lactate to pyruvate, NADH for the reaction pyruvate to lactate. High-speed measurements show that the NAD^+ binds rapidly and the reaction to pyruvate is fast. As pyruvate dissociates, all four bound NAD^+ molecules are reduced. The rate-limiting step is the slow dissociation of NADH from enzyme (20):

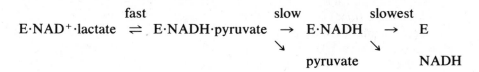

Properties of the ternary complexes are critical:

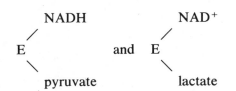

Kinetic measurements have been made on LDHs purified by affinity chromatography from muscles of a wide variety of vertebrates. Constants have been obtained for reactions from pyruvate to lactate, from lactate to pyruvate, for binding of NADH (6); some of the available kinetic constants are given in Table 5-3. Differences between A_4 LDHs are adaptive for the life patterns and habitats occupied; kinetic variations reflect a few genetically determined amino acid substitutions. Adaptive functions of allotypes of LDH are given in Chapter 4. Binding and catalytic sites are highly conserved (18).

A few general conclusions can be drawn regarding A_4 LDHs. Activation energies are lower for enzymes which normally function over a wide range of temperature (in poikilotherms) and for animals living at low temperatures than for enzymes of high temperature animals (homeotherms). Enzymes from cold-living species are less temperature-sensitive and have lower Q_{10}'s than enzymes from animals that maintain warm body temperature. When measured at body temperatures the maximum velocities of reactions are higher for homeotherms, but in poikilotherms velocities are reduced less by cold. When measured at 5°C the reaction rate of A_4 for pyruvate to lactate is three times faster in halibut than in rabbit enzyme (60). Arrhenius activation energy E_a for halibut enzyme is 6.7 kcal/M in the high temperature range, 14.5 in the low range; however, K_m for NADH is unaffected by temperature over the range of 10–25°C (25). Figure 5-13 compares Arrhenius plots for LDH for pyruvate to lactate and the reverse for enzyme from flounder muscle FM, beef muscle BM, and beef heart BH (6). Values of k_{cat} are several times greater for halibut than for rabbit enzyme at low temperature. The differences between $\Delta G\ddagger$ values for different species are relatively small; this is because both $\Delta H\ddagger$ and $\Delta S\ddagger$ decline together, resulting in relative constancy of $\Delta G\ddagger$. Within the normal range, however, $\Delta H\ddagger$ values change more than $\Delta S\ddagger$ values for poikilotherms.

The ratio k_{cat}/K_m is conserved at physiological temperatures. The turnover numbers k_{cat} are high for enzymes of poikilotherms at their normal temperatures. K_m's are modified by the cellular milieu—ionic composition, pH, and temperature. Since pH decreases in aqueous solutions with warming (68), measurements made at constant regulated pH but at different temperatures are nonphysiological (25). When reactions are measured at pHs adjusted for temperature, such that the OH^-/H^+ ratio is held relatively constant, K_m values are compensatory over a wide range of temperatures (Fig. 5-2b). Ge-

Table 5-3. Kinetic Data for A$_4$ (Muscle) LDH (6,18,25,27,52)

Substrate, Animal, Reference	Normal Temperature (°C)	K_m (μM)	$\Delta G^\ddagger$	$\Delta H^\ddagger$	$\Delta S^\ddagger$	$\Delta V^\ddagger$
Oxamate as Pyruvate Analog (27)						
Beef	37		5.7	15.4	32.3	36
Dogfish	15		5.1	11.0	19.8	5.4
Sculpin	5–15		5.1	8.2	11.1	10.0
Antimora	2		4.7	6.1	4.9	0.0
Pyruvate (27)						
Chicken	40		13.18	10.5	8.7	
Rabbit	37	6.7	13.3	12.52	2.5	4.2
Tuna	15		12.9	8.77	13.8	
Halibut	12	14.8	12.9	8.77	12.7	0.23
Antimora	2			6.1		
Trematomus	–2	17.4				0.43
Trout (35)	15	4				
Pyruvate (38)						
Rabbit	37		13.3	12.5	2.5	4.2
Coryphenoides	2–10					3.9
Halibut	8–15		12.9	8.77	13.7	.23
Trematomus	–2		6.1			.43
Pyruvate (58a)						
Rabbit	37		14.36	12.5	6.7	
Tuna	15–30		14.15	11.38	10.0	
Coryphaenoides	2–10		14.22	12.55	6.4	2.6
Antimora	2–5		14.34	11.81	8.7	0.3
Sebastoldus altivellus	4–12	0.395	14.25	11.98	8.1	8.1
Sebastoldus alascanus	4–12	0.399	14.0	10.51	12.6	12.8

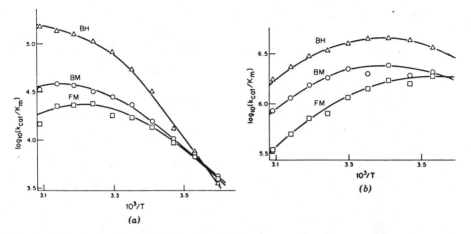

Figure 5-13. Arrhenius plots of log (k_{cat}/K_m) for pyruvate to lactate and from lactate to pyruvate by M_4 LDH purified from flounder muscle (FM), beef muscle (BM), and beef heart (BH). (*a*) Forward direction at saturating NAD^+ (2 m*M*) concentrations. The K_m is for lactate. (*b*) Reverse direction at saturating NADH (0.1 m*M*) concentrations. The K_m is for pyruvate. Reproduced by permission of the National Research Council of Canada from the *Canadian Journal of Biochemistry*, Vol. 53, 1975.

netically determined compensation is indicated for purified A_4 LDH from three species of Pacific barracuda, *Sphyraema* (58a):

	S. argentea	*S. lacasna*	*S. ensis*
normal temperature mid range (°C)	18	23	26
K_m at 25°C (m*M*)	0.34	0.26	0.20
k_{cat} at 25° (s^{-1})	893	730	658
K_m at mid-temperature	0.24	0.24	0.23
k_{cat} at mid-temperature	667	682	700

At physiological temperature the ratio k_{cat}/K_m is relatively constant (24).

Differences in $\Delta V\ddagger$ are greatest for LDHs from homeotherm muscles, less for muscle from shallow-water fishes, and negligible for deep-water fishes; benthic species are less affected by hydrostatic pressure. For NADH binding, H-bonds are stabilized by cold, while hydrophobic bonds are destabilized by pressure (27). Sensitivity to high temperature, heat denaturation, is less in enzymes from warm tissues; enzymes of homeotherms function at much higher temperatures than enzymes of low-temperature animals. In each kind of animal the enzymes function at temperatures above lethal. Hence, denaturation of these enzymes is not the cause of death.

The three principal isoenzymes of LDH differ in concentration of pyruvate for maximum activity (Fig. 5-14). Explanations of kinetic differences in

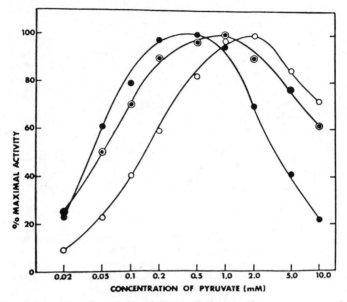

Figure 5-14.   Comparison of enzyme activity in relation to substrate concentration for three isozymes of LDH, A$_4$, B$_4$, and C$_4$. Pyruvate as substrate. $\bigcirc$ = A$_4$, $\bullet$ = B$_4$, $\circledcirc$ = C$_4$. Reprinted with permission from C. L. Markert, *Advances in Biosciences,* Vol. 6, p. 523. Copyright 1970, Pergamon Press, Oxford, UK.

terms of amino acid sequences are relatively meager. The A isozyme (A$_4$) of LDH contains more Lys and Arg, and is cathodal; the B isozyme (B$_4$) has more Asp and Glu and is anodal. LDHs from cold-habitat animals depend on hydrogen bonds and electrostatic interactions in cofactor binding. In contrast, LDHs of warm-body animals show more reliance on hydrophobic stability (28). Amino acid substitutions at a distance from binding sites may provide sufficient charge differences to affect kinetic constants. It is concluded that molecular adaptations in structure make for high catalytic efficiency, for relative constancy of K_m under physiological conditions, and for reduced temperature coefficients of activation energy (18).

Correlations of function with molecular structure have been noted. Most LDHs use L-lactate, a few (*Limulus* and *E. coli*) use D-lactate (29). The L-specificity is determined by Arg at position 171, His at 195, and by carbon-4 of NAD. All LDH molecules are similar in structure; an amino terminal arm serves for association of the dimers; coenzyme binds in a hydrophobic pocket. NAD$^+$ binds at Arg 101 to the phosphate, at Arg 109 to carboxyl, and at Asp 53 to ribose by a hydrogen bond. Tyr 83 and Asp 53 form a hydrophilic pocket in which adenine attaches. Valines at positions 32 and 247 support the nicotinamide ring (29). The catalytic binding site for substrate is between NAD$^+$ and His 195 and Arg 171 (Fig. 5-15) (29).

Several substitutions have been noted for functionally similar amino

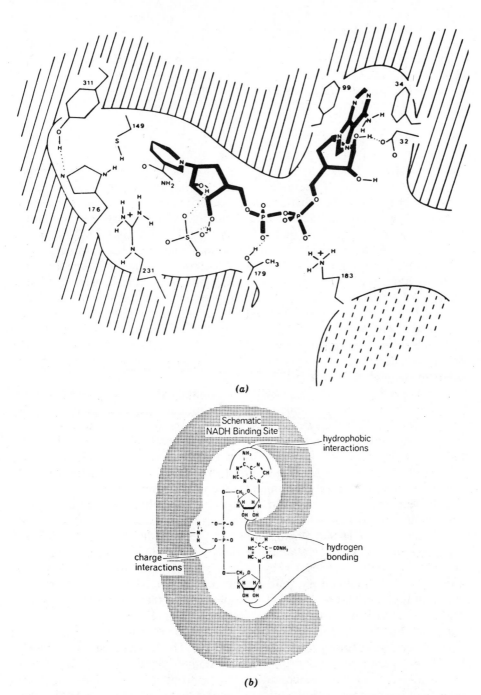

Figure 5-15. (*a*) Diagrammatic representation of NAD⁺ binding to GAPDH. Reprinted with permission from M. G. Rossman, *The Enzymes,* Copyright 1975, Academic Press, New York. (*b*) Diagrammatic representation of NADH binding to M₄LDH. Reprinted with permission from P. W. Hochachka, *Comparative Biochemistry and Physiology,* Vol. 52B, p. 25. Copyright 1975, Pergamon Press, Oxford, UK.

acids. In the vicinity of the site for coenzyme binding, an alanine in B-LDH replaces glutamate in A-LDH. Fewer differences occur between the same isozyme in different species than between different isozymes of one species.

Alcohol Dehydrogenase (ADH)

Alcohol dehydrogenases require NAD^+ as cofactor. In yeast, ADH catalyzes the conversion of acetaldehyde to ethanol. In mammals, liver ADH acts on a broad range of substrates and primarily converts ethanol from food or formed by intestinal bacteria to acetaldehyde. Yeast ADH is tetrameric, with molecular weight 140,000. *Drosophila* ADH consists of 255 amino acids; its gene has two introns (4). Liver (mammalian) ADH is dimeric and each unit has two Zn atoms bound to loci rich in cysteine. Specific sequences have been recognized for binding Zn, NAD, and substrate. Mammalian ADH has 374 amino acids; isozymes are coded at three autosomal loci—ADH_1, ADH_2, ADH_3. Several alleles are known at the ADH_2 and ADH_3 loci. The composition of ADH is highly conserved; human and horse ADH are identical for 90 percent of their amino acid sequences. As with LDH, the coenzyme binding is with adenine in a hydrophobic pocket; ribose binds to the carboxyl of Asp 223.

Alcohol dehydrogenase is essential in animals such as fruit flies, which consume alcohols in decomposing fruit. In humans the levels of liver ADH vary with age and diet. In *Drosophila melanogaster* two principal allozymes, fast (F) and slow (S), can be separated electrophoretically. In Miami populations the frequency of the slow form was 0.9, in New York State 0.5 (65). A cline of *Drosophila* in Mexico occurs with ADH-S at low frequency (0.51) in mountain populations and at high frequency (0.99) in the tropics. ADH-F may be selected at low temperatures. However, the clinal correlations could be with diet rather than with temperature. In North America, Australia, and Southeast Asia ADH-S in *Drosophila* decreases with distance from the equator. The clines in Asia correlate more closely with rainfall than with temperature (42c). ADH is coded in *Drosophila* by a structural gene on chromosome II and by a regulatory gene on chromosome III. The fast ADH is less stable in heat (40°C) and less stable to pH change than the slow form (55). The slow ADH is most active on ethanol (EtOH), the fast on propanol; propanol and butanol inhibit activity competitively on ethanol. ADH may comprise 1 percent of total protein in an adult fly. The enzyme can be induced by adding alcohol to the diet (41). The gene that encodes *Drosophila* ADH consists of 255 nucleotides containing two introns, and the amino acid sequence of the protein in *Drosophila* resembles that in yeast and in horse liver ADH (4). *Drosophila* reared on high EtOH show increase in the fast allele; low humidity also favors the fast form (42c). Eggs of homozygotes for the slow allele show less survival in alcohol than eggs of homozygous fast alleles

(4a). An alcohol dehydrogenase can be induced in maize by methanesulfate (57).

In a nematode *Paragrellius,* addition of 7 percent alcohol to the diet resulted in a three-fold increase in ADH activity. This was not due to stimulation of transcription (actinomycin had no effect on the induction), but appeared to be an increase in translation of mRNA; it was blocked by actidine or pyrrazole.

OTHER PROTEINS

Carbonic Anhydrase

Carbonic anhydrase (CA) catalyzes the reaction $CO_2 + H_2O \rightleftharpoons H_2CO_3$. Carbonic anhydrase is of widespread occurrence in animals and plants. It occurs in both monocotyledenous and dicotyledonous plants, especially in the chloroplasts, which contain two forms of carbonic anhydrase. In vertebrates the enzyme occurs as two or more isozymes. CA is abundant in red blood cells of terrestrial vertebrates, gills of fishes, rectal glands of elasmobranchs, lenses of vertebrates (but not of marine fishes), and kidneys of mammals, where it may aid acid-base balance. CA has a molecular weight of 30,000 in most vertebrates, 37,000 in sharks, 40,000 or 45,000 in plants. In mammals two isozymes (I and II) have been characterized and numerous allozymes are known, especially of CA I. Most mammalian red blood cells contain both I and II; in mouse the proportion is 46 II to 31 I; sheep red cells have CA II only, and in mammalian kidneys isozyme I is the principal CA. CA II is present in more kinds of tissues than CA I (4).

CA has been sequenced for a number of mammals, and of the 260 amino acids I and II are identical in 60 percent of sequences; the two isozymes are similar in tertiary structure. Correlations of structure with function are few; sequences of histidine at positions 94, 96, and 119 provide for binding of zinc. Functional differences between CA in three tissues in dogfish (*Squalus acanthias*) indicate three isozymal forms (39).

Source	k_{cat} (s^{-1})	K_m (mM)	Enzyme Activity
dogfish red blood cells	2.5	5.0	2000
dogfish ciliary fold	8.0	14.0	600
dogfish rectal gland	12.0	14.0	3200
goosefish red blood cells	50.0	10.0	25000
human CA II	25	9	23000
human CA I	3.0	4.0	23000

CA activity is low in cyclostomes and elasmobranchs and high in teleosts and most terrestrial vertebrates. Turtles have both high- and low-activity enzymes. From a cladogram for mammalian CAs it is deduced that CA I diverged from CA II some 75 million years ago, and that evolution of CA II has been somewhat slower than of CA I (64).

Fibrinogens, G-3-P-DH, Influenza Virus

Correlation of structure with function is found in proteins which form chains, often helices (16). Vertebrate fibrinogen is a six-chain protein of three pairs of nonidentical polypeptide chains (Fig. 5-16). The chains $\alpha^2\ \beta^2$ are connected by disulfide bonds. The α chains are low in tyrosine, high in Arg and Ser; α has less Pro, β more Met and Cys. In a central domain of the fibrinogen six amino terminals are gathered together; terminal domains are formed by two three-stranded ropes. When fibrinogen is converted to fibrin under the action of thrombin, fibrin monomers are formed. Thrombin-sensitive linkages occur between α and β chains. In human, bovine, and lamprey fibrinogen these regions have similar amino acid sequences, and COO$^-$ termini are similar in β and α chains. It is deduced that all three chains descended from a common ancestor (16). The similarity in sequences and tertiary structure throughout vertebrates may be related to the function of fibrinogen—to clot blood.

Structure-function correlations have been described for families of other proteins. For example, Ca-binding proteins have identifiable structural domains (Ch. 4) (23b).

The three-dimensional structure of glyceraldehyde 3-P-dehydrogenase (G3PDH) from two thermophilic bacteria, *Stearothermophilus* and *Thermus,* resembles that for the same enzyme from lobster, yeast, and pig. Amino acid sequences in the two thermophilic bacteria are 70 percent similar. The enzyme is tetrameric; each chain has a NAD$^+$ binding site and a catalytic domain. Cysteine is needed for catalysis, but the thermophiles have less cysteine than is in the enzyme from mesophiles. The thermophiles have more hydrophobic interactions in the so-called S-loop. Peripheral hydrophobic amino acids prevent access of water to the enzyme core; buried ionic bonds also contribute to thermal stability. It is concluded that the thermophilic and mesophilic enzymes have similar functional domains and general structure but that substitutions in specific regions are responsible for heat tolerance in the thermophiles (65a).

An example of considerable functional change with substitutions of single amino acids is given by the large glycoprotein molecules of influenza virus. Twelve strains of influenza virus associated with outbreaks of the disease differ in virulence and sensitivity to antibodies. The virus is a trimeric glycoprotein of molecular weight 224,640 daltons; it consists of a hydrophilic carbohydrate-containing domain on the external surface, small hydrophobic

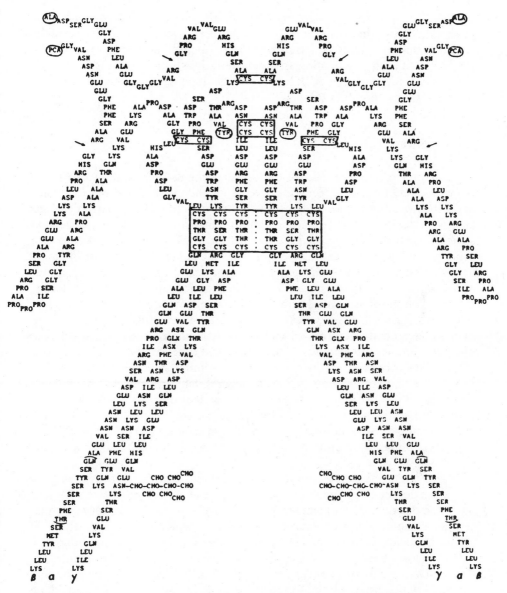

Figure 5-16. Molecular model of human fibrinogen showing all six chains "in register" at sequences comprising a disulfide girdle. Reprinted with permission from R. F. Doolittle, *The Plasma Proteins,* F. W. Putnam, Ed., Copyright 1975, Academic Press, New York.

peptides spanning the host cell membrane, and a hydrophilic domain on the inner side. The domain for carbohydrate has five conserved residues and the amino terminus which penetrates the host membrane is also conserved. Four sites of antibody binding are identified. Substitution of at least a single amino acid, for example, ASP for Gly on the binding side-chain alters binding of antibody threefold (66a).

CONCLUSIONS

This chapter marshals evidence that measurements of kinetic properties of enzymes and of transport proteins reveal more adaptive properties than have been shown by measurements of maximum velocities or by patterns of gel electrophoresis. Examples are given of correlations between molecular structure and kinetic properties. Similar correlations will be presented for proteins taking part in neural transmission in Chapter 12. Changes in amino acid sequences coded by the genetic system can alter binding of substrates and cofactors and can change charge distribution and hydration. Clearly, micromutations of structural genes can alter enzyme proteins in subtle adaptive ways. Some key enzymes interact with several substrates and coenzymes. Genetically determined variants of enzyme proteins are adapted to life style, temperature, ionic strength, pH, and other modifying parameters.

Families of related enzyme proteins have similar structural domains. The NAD^+ binding domains are similar in several dehydrogenases. The nucleotide binding domain of hexokinase resembles that of lactate dehydrogenase and of some kinases (15). Folds for heme are similar in hemoglobins and cytochromes c. Several enzymes which contain zinc (carbonic anhydrase, carboxypeptidase, alcohol dehydrogenase) have similar bonds with the metal.

Many amino acid substitutions in noncritical regions of a protein may make no detectable change in function. However, proteins with different amino acid sequences but similar function may have similar tertiary and quaternary structure with comparable domains in which critical sequences of amino acids are conserved (53). Binding sites for nucleotide cofactors and substrates are highly conserved, and two binding sites may be connected. The genes of globins and of dehydrogenases are divided by introns into functional domains, for example, for coenzymes and substrates. Enzymes which use flavin nucleotides as cofactors resemble those which use NAD^+ or NADP.

We conclude that the comparative study of related proteins and of the same functional proteins (enzymes or O_2 transporters) from different organisms can show the adaptive functions of genetic variants. Kinetic analysis provides links between micromutations, biological diversity, and evolution.

REFERENCES

1. Atkinson, M. R. et al. *Biochem. J.* **80**:318–323, 1961. *Cellular Energy Metabolism and Its Regulation,* Academic Press, New York, 1977. Kinetics of reactions involving pyridine nucleotide coenzymes.

2. Bailey, G. and S. Lim. In *Isozymes,* Ed. C. Markert, Academic Press, New York, vol. 4, pp. 29–54, 1975. Kinetics of lactate dehydrogenase.

3. Barra, D. et al. *Febs Lett.* **35**:151–154, 1973. *Nature* **293**:587–589, 1981. Properties of fish hemoglobin.

4. Benyajati, C. et al. *Proc. Natl. Acad. Sci.* **78**:2717–2721, 1981. The gene for alcohol dehydrogenase in *Drosophila.*

4a. Bijlsoma-Meels, G. *Heredity* **42**:79–89, 1979. Sensitivity of *Drosophila* to alcohol.

5. Bonaventura, J. et al. *J. Exper. Zool.* **194**:155–174, 1975. Survey of fish hemoglobins.

6. Borgmann, V. et al. *Biochemistry* **13**:5152–5158, 1974. *Canad. J. Biochem.* **53**:998–1004, 1196–1206, 1975. Kinetics of heart LDH.

7. Braunitzer, G. et al. *Adv. Protein Chem.* **19**:1–73, 1974. Structure of embryonic hemoglobin.

8. Brunori, M. et al. *Proc. Natl. Acad. Sci.* **78**:6076–6080, 1981. Mutations of human hemoglobin.

9. Chandrasekhar, K. et al. *J. Molec. Biol.* **76**:503–518, 1978. Conformational changes in LDH associated with binding to substrate.

10. Clegg, J. B. and J. Gagnon. *Proc. Natl. Acad. Sci.* **78**:6076–6080, 1981. Structure of chain of human embryonic Hb.

11. Cornish-Bowden, A. *J. Molec. Biol.* **101**:1–9, 1976. Effects of natural selection on enzyme catalysis.

12. Crowley, P. H. *J. Theoret. Biol.* **50**:461–475, 1975. Natural selection and K_ms.

12a. Cullis, A. F., M. F. Perutz, et al. *Proc. R. Soc. Lond.* **A265**:161–187, 1962; **B173**:113–140, 1969. X-ray structure of myoglobins.

12b. Czlusniak, J. M. et al. *Nature* **298**:297–300, 1982. Adaptive evolution of avian and mammalian Hbs.

13. Dickerson, R. E. and I. Geis. Ch. 3, pp. 1–119 in *Structure and Action of Proteins,* Harper and Row, New York, 1969. Structure of hemoglobin.

14. Dickerson, R. E. *Sci. Am.* 1981. 137–149. *with* R. Timkovich, pp. 397–547 in vol. XI, part A, *The Enzymes,* Ed. P. D. Boyer, Academic Press, New York, 1975. Structure, evolution, and functions of cytochromes c.

15. DiMichele, L. and D. A. Powers. *Nature* **296**:563–564, 1982. Function of LDH genotypes in *Fundulus* embryo.

15a. Dixon, M. and E. Webb. *Enzymes,* Academic Press, New York, 1979, p. 425, Fig. VIII. 37. Hill plot.

16. Doolittle, R. F. *Fed. Proc.* **35**:2145–2149, 1976. Structure of fibrinogen.

17. Eaton, U. A. *Nature* **284**:183–185, 1980. Coding and function of hemoglobin.

18.   Eventoff, W. et al. *Proc. Natl. Acad. Sci.* **74**:2677–2681, 1977. Structural adaptation of LDH.

19.   Everse, J. and N. O. Kaplan. pp. 29–43 in *Isozymes* II, Ed. C. Markert, Academic, New York, 1975. Biological functions of dehydrogenases.

20.   Fersht, A. R. *Proc. R. Soc. Lond.* **B187**:397–407, 1974. *Enzyme Structure and Mechanisms,* Freeman, San Francisco, 1977. Enzyme structure and mechanisms.

21.   Gilbert, W. *Nature* **271**:501–502, 1978. Meaning of introns in genetic codes.

22.   Gillen, R. G. and A. Riggs. *J. Biol. Chem.* **248**:1961–1969, 1973. Structure and function of eel hemoglobin.

23.   Goldberg, A. *Proc. Natl. Acad. Sci.* **77**:5794–5798, 1980. coding of the gene for ADH in *Drosophila*.

23a. Goodman, M. et al. *Nature* **298**:297–300, 1982. *J. Molec. Evol.* **17**:14–120, 1981. *Prog. Biophys. Molec. Biol.* **38**:105–164, 1981. Evolution of vertebrate hemoglobins.

23b. Goodman, M. et al. pp. 347–354 in *Ca-Binding Proteins,* Ed. F. L. Siegel et al. Elsevier, New York, 1980. Domains in calmodulin family of proteins.

24.   Graves, J. E. and G. N. Somero. *Evolution* **36**:96–106, 1982; **37**:30–37, 1983. Ecological differences in LDH of marine fishes.

25.   Greaney, G. S. and G. N. Somero. *J. Compar. Physiol.* **137**:115–122, 1980. *Biochemistry* **24**:5322–5332, 1979. Binding and catalytic rate constants for LDH, hydration change during catalysis.

26.   Hazel, J. R. et al. *J. Compar. Physiol.* **123**:97–104, 1978. Effect of assay temperature on pH optima of trout enzymes.

26a. Hjorth, P. P. *Biochem. Genetics* **13**:379–391, 1975. Hb structure in *Zoarces*.

27.   Hochachka, P. *Compar. Biochem. Physiol.* **52B**:25–31, 1975. *Int. J. Biochem.* **8**:183–186, 1977. Active sites on LDH.

28.   Hochachka, P. and G. Somero. *Strategies of Biochemical Adaptation,* Saunders, Philadelphia, 1973. *Biochemical Adaptation,* Princeton University Press, Princeton, NJ, 1983.

29.   Holbrook, J. J. et al. pp. 191–292 in vol. 11, *The Enzymes,* Ed. P. D. Boyer, Academic Press, New York, 1975. Review of structure and function of LDH.

30.   Hsu, S. S. and R. E. Tashian. *Biochem. Genetics* **15**:885–896, 1977. Genetic control of carbonic anhydrase isozymes.

31.   Ibsen, K. H. et al. *Compar. Biochem. Physiol.* **65B**:473–480, 1980. Pyruvate kinase isozymes.

32.   Jencks, W. P. *Adv. Enzymol.* **43**:219–410, 1975. Binding energy, specificity and enzyme catalysis.

33.   Johnston, I. A. *J. Compar. Physiol.* **119**:195–206, 1977. Adaptive properties of myofibrillar ATPase.

34.   Kao, Y. and T. M. Farley. *Compar. Biochem. Physiol.* **60B**:153–155, 1978. Thermal modulation of LDH.

34a. Kaplan, N. O. et al. *Science* **136**:962–969, 1962. Isozymes of LDH.

35.   Klar, G. T. et al. *Compar. Biochem. Physiol.* **63A**:229–235, 1979. LDH alleles in trout.

36. Lehninger, A. L. *Principles of Biochemistry,* Worth, New York, 1982.

36a. Lehrer, G. M. and R. Barker. *Biochemistry* **7:**1533–1539, 1970. Conformational changes in rabbit muscle aldolase.

37. Lim, S. T. and G. Bailey. *Biochem. Genetics* **15:**707–721, 1977. LDH isozymes in salmonids.

38. Low, P. S. and G. N. Somero. *Proc. Natl. Acad. Sci.* **70:**430–432, 1973. *J. Exper. Zool.* **198:**1–12, 1976. Roles of free energy, enthalpy and entropy of activation of enzymes.

39. Maren, T. H. et al. *Compar. Biochem. Physiol.* **67B:**69–74, 1980. Carbonic anhydrase in fishes.

39a. Margoliash, E. et al. *Fed. Proc.* **35:**2125, 1976. Kinetic properties of cytochrome c.

39b. Markert, C. L. *Advances in Bioscience* **6:**523, Fig. 10, Pergamon Press, Oxford, 1970.

40. Markert, K. and F. Molher. *Proc. Natl. Acad. Sci.* **45:**753–763, 1959. Developmental changes in LDH.

41. McDonald, J. F. and F. J. Ayala. *Genetics* **89:**371–388, 1978. Genetics and biochemistry of ADH, *Drosophila,* induction by dietary alcohol.

42. Merritt, R. B. *A. Nat.* **196:**173–184, 1972. Geographic distribution of LDH allozymes in minnow.

42a. Mied, P. A. and D. A. Powers. *J. Biol. Chem.* **253:**3521–3528, 1978. Subunit composition of *Fundulus* Hb.

42b. Moore, G. W. et al. *J. Molec. Biol.* **105:**15–37, 1976. Cytochrome c sequences.

42c. Oakeshott, J. *Evolution* **36:**86–96, 1982. Clines of *Drosophila melanogaster* for ADH and Gluc 3PDH.

43. Perutz, M. F. *Nature* **28:**726–739, 1970; **299:**421–427, 1982. Stereochemistry of hemoglobin.

43a. Perutz, M. F. *Molec. Biol. Evol.* **1:**1–28, 1983. Species adaptation in a protein molecule, hemoglobin.

44. Philipp, D. P. and G. S. Whitt. *Devel. Biol.* **59:**183–197, 1977. Gene expression during teleost embryogenesis.

45. Place, A. R. and D. A. Powers. *Proc. Natl. Acad. Sci.* **76:**2354–2358, 1979. Genetic variation and catalytic efficiency of LDH. *Biochem. Genetics* **16:**593–607, 1978.

46. Powers, D. A. *Science* **216:**1014–1016, 1982. *Nature* **296:**563–564, 1981. Correlations of LDH allotype with swimming and hatching in *Fundulus*. pp. 63–84 in *Isozymes* IV Ed. C. Markert, Academic Press, New York, 1975. LDH gene frequencies.

47. Powers, D. A. and A. B. Edmundson. *J. Biol. Chem.* **247:**6686–6693, 6694–6707, 1972. Hemoglobins of catastomid fishes.

48. Riggs, A. *Federation Proceedings* **35:**2115–2118, 1976. Structure-function correlations of hemoglobins.

49. Riggs, A. and Q. Gibson. *Proc. Natl. Acad. Sci.* **70:**1718–1720, 1973. Properties of mutants of human hemoglobin.

51. Rossman, I. B. and V. M. Ingram. *Proc. Natl. Acad. Sci.* **78:**4782–4785, 1981.

Embryonic hemoglobin of chickens.

52. Rossman, M. G. et al. pp. 61–102 in vol. XI, part A, *The Enzymes,* Ed. P. D. Boyer, Academic, New York, 1975. *J. Molec. Biol.* **109:**99–129, 1977. Evolutionary and structural relationships of dehydrogenases.

53. Rossman, M. G. and P. Argos. *J. Biol. Chem.* **250:**7525–7532, 1975. *J. Molec. Biol.* **105:**75–96, 1976. Comparison of heme binding pockets in globin and cytochrome B_5.

54. Rossman, M. G. and P. Argos. *Annu. Rev. Biochem.* **50:**497–532, 1981. Protein folding.

55. Sampsell, B. *Biochem. Genetics* **15:**971–988, 1977. (With S. Sims) *Nature* **296:**853–855, 1982. Genetics of heat stability of alcohol dehydrogenase of *Drosophila.*

56. Sidell, D. *Science* **207:**769–770, 1980. Cyclostome LDH.

57. Schwartz, D. *Proc. Natl. Acad. Sci.* **73:**582–584, 1976. ADH genes in maize.

58. Shaklee, J. B. and G. S. Whitt. *Copeia* 563–578, 1981. Gene expression for LDH in 35 species of fishes.

58a. Siebenaller, J. F. and G. N. Somero. *J. Compar. Physiol.* **129:**295–300, 1979. *Science* **201:**255–257, 1978. Kinetics of enzymes from fish from different depths.

58b. Sinenski, M. *Proc. Natl. Acad. Sci.* **71:**522–525, 1974. Homeoviscous adaptation of membrane lipids.

59. Somero, G. N. pp. 221–234 in *Isozymes,* vol. II, 1975. (with P. S. Low) *Nature* **266:**276–278, 1977. Energy relations for LDH.

60. Somero, G. N. *J. Exper. Zool.* **194:**175–188, 1975; **198:**1–12, 1976. *Annu. Rev. Ecol. Syst.* **9:**1–29, 1978. Adaptations of kinetic properties of enzymes from poikilotherms.

62. Takano, T. and R. E. Dickerson. *Proc. Natl. Acad. Sci.* **77:**6371–6375, 1980. Cytochrome c structure.

63. Tan, A. I. and R. Noble. *J. Biochem.* **248:**2880–2888, 7412–7416, 1973. Modulation of hemoglobin by organophosphates.

64. Tashian, R. E. and N. D. Carter. pp. 1–56 in *Advances in Human Genetics 7,* Ed. H. Harris and K. Hirschner, Plenum, New York, 1976. Biochemical genetics of carbonic anhydrase.

64a. Van Delden, W. et al. *Genetics* **90:**161–191, 1978. ADH polymorphism in *Drosophila.*

65. Vigue, C. L. and F. M. Johnson *Biochem. Genetics,* **9:**213–227, 1973. Isozyme variability of ADH in *Drosophila.*

65a. Walker, J. E. et al. *Europ. J. Biochem.* **108:**549–565, 1980; **108:**567–579, 1980; **108:**587–597, 1980; **108:**581–586, 1980. Structure of D-glyceraldehyde-6-P DH in relation to thermal stability in thermophilic bacteria and mesophilic eukaryotes.

66. Whitt, G. et al. in *Isozymes,* IV, 381–400, 1975. *J. Exper. Zool.* **175:**1–36, 1970. Isozymes of LDH teleosts. *Science* **166:**1156–1158, 1969.

66a. Wiley, D. C. et al. *Nature* **289:**366–373; 373–378, 1981; **304:**459–462, 1983; **311:**678–680, 1984. Antigenic reactions and amino acid structure of strains of influenza virus.

67. Wilkinson, G. N. *Biochem. J.* **80:**324–332, 1961. Statistical estimations of enzyme kinetics.

68. Wilson, T. L. *Arch. Biochem. Biophys.* **179:**378–390, 1977; **182:**409–419, 1977. Interrelations of temperature and pH for enzyme function.

69. Yancey, P. H. and G. N. Somero. *J. Compar. Physiol.* **125:**129–134, 135–141, 1978. Urea-requiring LDH of marine elasmobranch fishes; temperature dependence of intracellular pH.

6

OXYGEN

The most abundant elements in the cosmos are hydrogen (86.3 percent) and helium (13.2 percent); oxygen comprises only 0.09 percent (54). Before organic molecules were formed on earth, H_2O and CO_2 came into the atmosphere by out-gassing. Water condensed to form the oceans, CO_2 reacted with water and silicon to form silicon carbonates of limestone. The atomic abundance of the principal elements in the earth's crust relative to silicon is Si—1.0, H—0.14, C—0.0018, N—0.00014, and O—2.9 (54,60). Hydrogen of the primordial atmosphere was mainly free or bound in water; oxygen was in H_2O, carbon in CH_4 and CO_2, nitrogen was free and in NH_3 (Ch. 2). At present the atomic abundance of oxygen in the earth's crust is 58 percent.

Oxygen reserves in the earth's crust are as follows:

	Emols O_2 (E = erda = 10^{18})	Percent of Total Oxygen
lithosphere	371,000	90.5
hydrosphere	38,500	9.4
atmosphere	37.7	0.009
biosphere	0.0812	0.00002
total	410,000	100

Free O_2 was formed initially by photodissociation of H_2O under action of solar radiation. Early in the Precambrian, atmospheric O_2 in larger quantities was formed by photosynthesis. Virtually all of the earth's O_2 is now formed by photosynthesis, and the turnover is such that photosynthesis can replace all of the atmospheric oxygen in 2500 years. It is estimated that 99 percent of the O_2 in the biosphere is contained in plants. The rate of O_2 production by photosynthesis equals the sum of the rates of O_2 removal by respiration, organic decay, methane oxidation, and weathering.

Figure 6-1 summarizes the major oxygen reserves and turnover routes on earth. Reserves are in Emols (1 erda mol = $10^{18}M$); rates are in Emols O_2/myr (10^6 yr). Oxygen is the element best qualified for storage and utilization of energy because of its abundance, accessibility, high thermody-

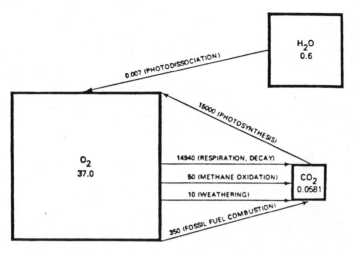

RESERVOIRS ARE IN EMOLES O₂ AND RATES ARE IN EMOLES O₂/MYR

Figure 6-1. Oxygen cycle, reservoirs, and turnover rates of oxygen. Reprinted with permission from D. L. Gilbert, *Oxygen and Living Processes,* Copyright 1982, Springer-Verlag, New York.

namic potential, and low reaction rate. Oxygen has a very positive redox potential; organic metabolites are negative.

Evidence that the primordial atmosphere was reducing, that it was virtually devoid of oxygen, and that primitive organisms were anaerobic was given in Chapter 2. Early prokaryotes were poisoned by oxygen; photosynthesis by purple or green sulfur bacteria and by halobacteria did not produce oxygen. Oxygen, a product of photosynthesis of Cyanobacteria, changed the atmosphere from anaerobic to aerobic. The function of metabolism, either anaerobic or aerobic, is to liberate energy for biological work. Biochemical synthesis of proteins, nucleic acids, and intermediate compounds requires energy. As cells became organized, energy was required for many processes other than synthesis—maintenance of cell membranes, movement, generation of bioelectric potentials, and other functions. In both anaerobic and aerobic metabolism, energy is delivered as adenosine triphosphate (ATP), the "universal currency," in packets of some 7.6 kcal/mol. Anaerobic metabolism occurs via glycolysis or fermentation, aerobic metabolism via oxidation of substrate.

Oxidation is a more efficient means than glycolysis or fermentation for providing energy from organic molecules; from one mol of glucose there are formed 36 mols of ATP by oxidation to CO_2 and H_2O, or 2 ATPs by glycolysis to lactic acid or by fermentation to ethanol. Photosynthesis transforms solar energy into a chemical energy reserve in the form of molecular oxygen and carbohydrate. In photosynthesis, photodissociation of water to oxygen and hydrogen requires 680 kcal/mol; the subsequent combination of the hydrogen with CO_2 to form water and sugar requires an additional 8 kcal for a total of 688 kcal to form 1 mol of sugar from 12 mols of H_2O and 6 mols of CO_2. In oxidative respiration, electrons are transported by steps from substrate of negative oxidation-reduction potential toward receptor molecules of more positive redox potentials. In photosynthesis the net flow of electrons is toward lower or more negative potentials; this flow against a normal redox gradient is made possible by energy from sunlight (Ch. 2). The net reactions are:

Photosynthesis:

$$2\ H_2O + CO_2 \xrightarrow{h\nu} (CH_2O) + H_2O + O_2$$

Oxidative respiration:

$$(CH_2O) + O_2 + H_2O \rightarrow 2\ H_2O + CO_2 + \text{released energy}$$

Oxidative respiration maintains the appropriate redox balance by reversibly reduced and oxidized coenzymes. The most used of these is $NAD^+/NADH$ (Ch. 2). The general equation for redox ratio is

$$K_{eq} = \frac{[NADH][S][H^+]}{[NAD^+][P]}$$

where [S] is metabolite and [P] products. The metabolite is negative to products and the coenzyme (NADH) transfers electrons down the redox gradient (Fig. 6-2). Redox balance must be maintained for normal energy flow.

Metabolic intermediates can be "titrated" for their oxidation-reduction

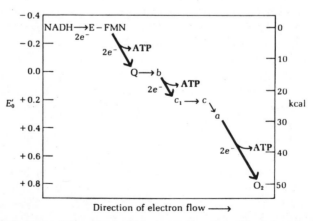

Figure 6-2. Redox potential gradient which shows direction of flow electrons in respiratory chain of mitochondria. E—FMN represents NADH dehydrogenase, Q is ubiquinone, b, c_1, c, and a are cytochromes. From A. L. Lehninger, *Principles of Biochemistry*, Worth Publishers, New York, 1982, p. 475.

potential in relation to a standard hydrogen electrode much as a proton donor (acid) can be titrated. The midpoint of the curve relating potential to
percent oxidation is E_0 or standard redox potential for each compound. A
tabulation of E_0 values in volts follows (76):

	E_0
acetate $+ 2H^+ + 2\ e^- \rightarrow$ acetaldehyde	-0.580
$NAD^+ + 2H^+ + 2e^- \rightarrow NADH + H^+$	-0.320
pyruvate $+ 2H^+ + 2e^- \rightarrow$ lactate	-0.185
2 cyt c (reduced) $+ 2e^- \rightarrow$ 2 cytochrome c (oxidized)	$+0.254$
$\frac{1}{2} O_2 + 2H^+ + 2e^- \rightarrow H_2O$	$+0.816$

Redox enzymes or e^- transferring proteins are pyridine-linked dehydrogenases which use NAD^+ or $NADP^+$, flavin-linked dehydrogenases which use
FAD^+ or FMN^+, Fe-S compounds, cytochromes, and ubiquinone (CoQ)
(Ch. 5).

At the time of the origin of life, the sources of energy for synthesis were
solar ultraviolet, lightning, and geothermal heat. Early life forms used energy derived anaerobically from the degradation of preformed organic molecules (Ch. 2). Many of the enzymes of anaerobic metabolism persist in present-day organisms. As protoplasm became organized, certain compounds,
such as ATP, must have been selected very early for transferring energy and
for maintaining redox balance. In anaerobic metabolism oxygen was a waste
product; oxygen is in fact toxic to anaerobic cells. The first oxidative enzymes probably served to rid cells of toxic oxygen. In the transition from
fermentative (anaerobic) to oxidative metabolism some critical enzyme pathways reversed (Fig. 2-12). Some oxygenases persist in anaerobic organisms,
in which they protect against free radicals, peroxides, and ozone. Some
plants have two general oxidases, (1) a CN-sensitive cytochrome system
similar to that in other eukaryotes, and (2) a CN-resistant oxidase for oxidation of ubiquinone and flavoprotein by a pathway which leads to superoxide formation (111a).

Sufficient oxygen for metabolism diffuses into unicellular and small multicellular aerobic organisms. There are no multicellular obligate anaerobes.
Some animals can remain active without oxygen for long periods. As organisms' requirements for oxygen increased, diffusion to their cells became insufficient to provide the P_{O_2} necessary for interior enzymes. As the size of
organisms increased, means of ventilation and transport pigments evolved.
Photosynthesis and aerobic metabolism became coupled in higher plants;
vascular systems evolved and served to transport O_2 from leaves to other
tissues. In air-breathing animals, oxygen transport by respiratory pigments
became coupled with the excretion of CO_2 and H_2O. In air-breathing animals
ventilation and transport are closely related; in aquatic animals which respire
by gas exchange to water, rather than to air, diffusion removes CO_2, and the

coupling of oxygen uptake with CO_2 excretion is less close than in terrestrial animals.

ECOLOGICAL AVAILABILITY OF O_2

The availability of oxygen in the air is fixed by the concentration of atmospheric O_2, which comprises one-fifth of the atmosphere. Oxygen availability in water is fixed by the solubility of O_2 and the atmospheric pressure. The partial pressure of oxygen (P_{O_2}) in the atmosphere decreases with higher altitude. At sea level, the oxygen pressure equals 159 torr (211 mbar) in the gas phase. The pressure of one atmosphere is 760 mm Hg at $O°C$; the equivalent is 760 torr or 770.11 bar. At higher altitudes the barometric pressure decreases nonlinearly by 11–12 percent per kilometer and the amount of O_2 decreases proportionately. Oxygen concentration in water decreases with higher temperature and with increased salinity. Sea water contains only three-fourths as much O_2/ml as fresh water. The relevant parameter for diffusion of oxygen is partial pressure, not concentration. Diverse adaptations of organisms permit them to occupy ecological niches ranging from those saturated with oxygen at atmospheric levels to niches with zero oxygen. Hypoxia occurs in some tissues after high levels of activity.

The only habitats that are supersaturated with oxygen are warm salt marshes and shallow lakes, where photosynthesis outstrips utilization of O_2 during daylight hours. Shallow-water plants in warm, sunny ponds exude bubbles of oxygen. In the light, sea anemones with symbiotic green algae (Zooxanthellae) produce superoxide radicals (O_2^-) and the anemones have twice the superoxide dismutase activity of anemones lacking the symbionts (41). Swim bladders of many kinds of fishes contain gas which is predominantly oxygen; this comes from oxyhemoglobin, which releases its O_2 in the presence of gas gland secretory cells which produce much lactic acid. Permeability of swim bladder epithelium to oxygen is low, and countercurrent exchange prevents backflow of O_2.

In oceans the absolute P_{O_2} is the same at depths as at the surface (except for an O_2 minimum at 100–300 m due to O_2 removal by organisms); but, because of increase in hydrostatic pressure of 1 atm per 10 m depth, the fraction of pressure due to oxygen decreases. Many muds are essentially devoid of O_2 and are rich in anaerobic organisms, such as sulfide-producing bacteria. Parasites live where O_2 levels are near zero in the intestinal contents of some host animals. Aquatic animals that live intertidally are periodically subjected to O_2-supplying water and to air from which they have little means to extract oxygen. Animals that are normally water breathers in streams that dry up seasonally either have some means of breathing air or go into estivation when the streams become dry. Animals that live where the oxygen supply is intermittent have muscles that function anaerobically during periods of hypoxia.

Numerous structures for external respiration have evolved. Direct diffusion is adequate for unicellular organisms and for multicellular organisms in which water is carried by channels directly to tissues, as it is in sponges. Animals with permeable skin, such as some aquatic animals and many that live in moist air, may take in oxygen by cutaneous routes. Gills in which the water flows opposite to the blood flow in countercurrent exchange are more efficient than gills lacking countercurrent.

Many kinds of animals have made the transition from water to land. In some amphibious animals gas exchange occurs by modified swim bladder, branchial chamber, skin, or stomach. Life in air required new gas exchange structures. In insects, tracheal tubes transport air directly to tissues; their tracheoles penetrate into muscle fibers. Birds and mammals have lungs and bronchi that develop as outpouchings of the digestive tract. In birds the air passes through lungs to air sacs and back through parabronchi, and is exposed a second time to blood. In mammals and reptiles air passes into alveoli and the O$_2$ exchange is from a ventilated pool of air.

O$_2$ CONSUMPTION MODIFIERS—GROSS METABOLISM AND RESPIRATION

Many adaptations of whole organisms permit an output of energy appropriate to the life style of the plant or animal (39). A few of these will be mentioned.

Size

When animals or plants of a given kind are compared, small ones have a higher mass-specific metabolic rate than large ones. This is true of individuals of different sizes in a species and of adults in different species. Absolute metabolic rate (M) is given by ml O$_2$ consumed per minute by an entire organism; mass specific metabolism $\dot{V}_{O_2}$ is ml O$_2$/g organism weight/minute. It has been shown that the relationship between metabolism and mass is allometric for various invertebrates, poikilothermic vertebrates, birds, mammals, trees, and shrubs. The relationship is also evident in unicellular organisms—bacteria have higher metabolism than protozoans, which are larger. The relation is given by $M = KW^b$, where the exponent b varies from strict proportionality ($b = 1$) to $b = 0.5$. M = metabolism in O$_2$ consumed in a given time, W = mass in grams. In the majority of animals b is 0.67 to 0.75. The value of K describes the absolute level of metabolism, that is, the vertical intercept on a plot of log M as a function of log W (Fig. 6-3). The mass coefficient K varies according to body composition and is constant within a species. Covariance analysis shows that the exponent b is influenced by K

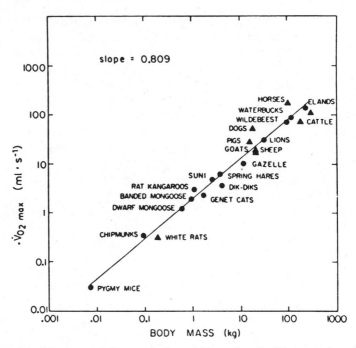

Figure 6-3. Maximum or activity metabolism of 14 species of wild mammals and 7 species of domestic mammals as a function of mass; logarithmic coordinates. Reprinted with permission from C. R. Taylor, K. Johansen, and L. Bolis, *A Companion to Animal Physiology*, Copyright 1982, Cambridge University Press, London.

as well as by other modifiers of metabolism. In mature mammals the standard metabolism $\dot{V}_{O_2 std} = 0.188\ W^{0.75}$ (30,114). The rate of metabolism ($\dot{V}_{O_2}$ varies such that in very small mammals (approximately 1-2 g shrews) the $\dot{V}_{O_2}$ increases to an asymptote. Because $\dot{V}_{O_2}$ can be equated with heat production, a very high $\dot{V}_{O_2}$ may pose thermal problems. If $\dot{V}_{O_2}$ of a very large animal, such as a whale or a horse, were at the metabolic rate of a mouse, the temperature of the internal organs would be intolerably high. Metabolism is high not only for whole small animals but also for each of their tissues. Oxidative activity is greater per mitochondrial protein and the number of mitochondria per cell and of cristae per mitochondrion is greater in small animals than in large ones. How organism size controls enzyme activity is far from understood. There must be cellular sensing of level of energy production and feedback to protein synthesis and enzyme activity.

In homeotherms different functions scale to body mass by different exponents: heart mass and stroke volume 1.0, metabolism 0.73, heart rate and breathing rate 0.25, and life span 0.25 (30). One gram of muscle of a 30 g mouse consumes O$_2$ at 20 times the rate of a gram of muscle from a 300,000 g horse.

For birds and mammals the higher mass-specific metabolism was explained as compensation for the heat loss as surface/volume ratio increased. This was disputed by observations that the values of b are similar but not identical to values relating metabolism to body surface. For unicellular organisms it is postulated that more energy is needed for maintaining ionic gradients and active transport when the surface membrane area per volume of protoplasm is larger. The explanation of b values similar to those of birds and mammals on the basis of thermoregulation cannot apply to most organisms which are multicellular poikilothermic. In multicellular organisms a large proportion of metabolic energy is used to maintain ionic gradients of all body cells. Tissue cells of most animals and plants are similar in size, at least for a given tissue type, but the number of mitochondria per cell is higher in small animals. Small animals of a motile species tend to move faster, hence they use more energy per unit weight. Probably for a given kind of animal there is an optimum size and an optimum energy production for its life style. The adaptive functions of metabolism-size relations remain to be elucidated.

Activity

Another type of scaling is the energetic cost of locomotion as a function of body size (Fig. 6-4). It is practically impossible to measure standard or rest metabolism; rather, $\dot{V}_{O_2}$ is best measured at several rates of exercise, with extrapolation to zero activity to give the standard value. For animals of similar mass the cost of locomotion, that is, the higher $\dot{V}_{O_2}$ above resting metabolism, is least for swimmers, more for fliers, and most for runners. The energetic cost of running for a land mammal may be 20 times the cost of flying for a bird (118). Efficiency of modes of locomotion differs with the resistance of water, air, substrate, and with body form (Fig. 6-5). The cost of locomotion increases as the 0.6 power of body mass in a number of mammals, 0.76 in fishes (13). In some moths the value of b at rest is 0.77, in hovering flight 0.83, during warm-up 0.90. (Fig. 6-6) (5,6). A bumblebee at rest consumes 2 ml O$_2$/g/hr at 20°C; in free flight it consumes 166–188 ml O$_2$/g/hr (71). A 1 kg iguana at T_B of 35.6°C consumes at rest 2 ml O$_2$/min, in full activity 9 ml O$_2$/min. A 1 kg mammal at rest consumes 9 ml O$_2$/min and in high activity 54 ml O$_2$/min (13).

In mammals trained to run at a rate corresponding to maximum metabolism, $\dot{V}_{O_2max}$ is 10 times as great during running as at rest (standard) metabolism. The relation to mass is $\dot{V}_{O_2max} = 1.92\ W^{0.81}$ for mammals ranging in size from pigmy mice and chipmunks to cattle, horses, and elands (114). Work *rate* in running is independent of body size.

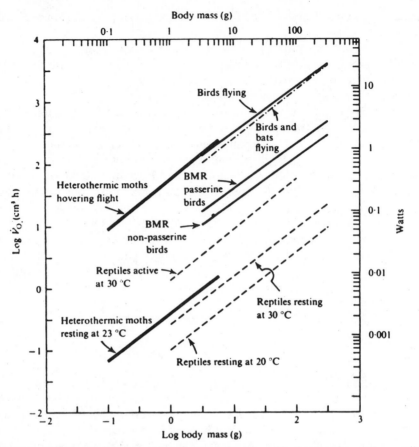

Figure 6-4.   Standard and active metabolism as a function of body mass for moths, reptiles, birds and bats, passerine birds, and nonpasserine birds. Reprinted with permission from G. A. Bartholomew and T. M. Casey, *Journal of Experimental Biology,* Vol. 76, p. 19. Copyright 1978, Cambridge University Press, London.

Homeotherms and Poikilotherms

Metabolic scope is the difference between the metabolism of rest and of activity; this can be compared in different animals by measuring ATP production (in $\mu M/g$) during periods of rest and of maximum activity. For animals of similar body size (mass), homeotherms need 5–10 times as much energy for maintenance as poikilotherms when measured at the same body temperature. Furthermore, active homeotherms require at least 50 times as much energy as active poikilotherms of similar size. Homeotherms have greater aerobic scope; most poikilotherms depend predominantly on increased anaerobic metabolism in activity. Examples for two mammals and two reptiles follow (13):

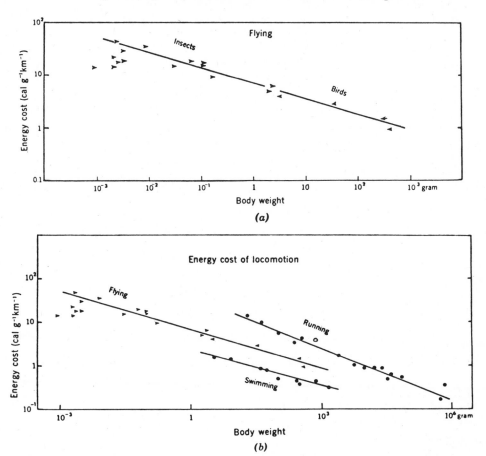

Figure 6-5. (*a*) Energy cost of flying in insects and birds. (*b*) Cost of locomotion in many animals—running, flying, and swimming. Reprinted with permission from K. Schmidt-Nielsen, *Science,* Vol. 177, pp. 224, 226. Copyright 1972 by the AAAS.

ATP Production in 5 Minutes of Maximum Activity (μ*M*/g)

	Mass (g)	Aerobic	Anaerobic	Total
mountain vole	25	40	13	53
kangaroo rat	35	98	9	107
lizard	13	24	22	46
snake	262	23	29	52

The total aerobic power is 50–100 percent greater in birds than in mammals, and 10–20 times greater in mammals than in fishes or amphibians (13,20).

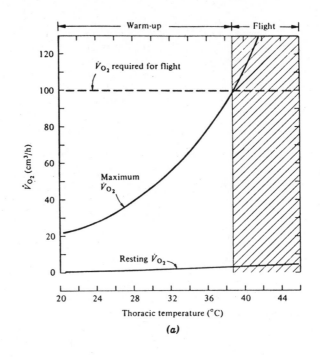

(a)

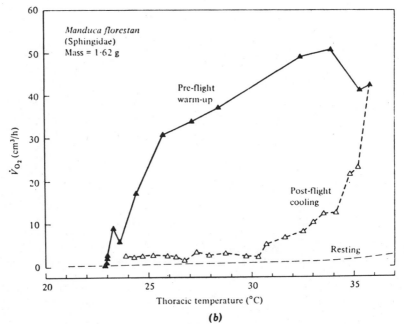

(b)

For example, a 1 kg reptile has a standard $\dot{V}_{O_2}$ of 0.07, and active metabolism of 0.4 ml O_2/g/hr, compared to a bird's values of 0.79 and 9.5. Invertebrates that are capable of active flight (insects) or running (crabs) have much greater metabolic scope than vertebrates of similar body mass. Many poikilotherms have greater capacity than homeotherms for obtaining energy by anaerobic processes, and can survive hypoxia longer.

Metabolic rates of active mammals, such as cats, are greater than rates of sluggish mammals, such as sloths—a genetic difference that may be expressed in numbers of mitochondria per tissue. An active fish, salmon, has a $\dot{V}_{O_2}$ of 0.37, a sluggish cave fish a $\dot{V}_{O_2}$ of 0.06–0.16 ml O_2/g/hr. The aerobic scope is greater for birds than for fish (13). There is a balance of metabolic rate between cost and benefit for size, amount, and speed of activity, temperature, nutrition, and way of life (104).

Oxygen Debt

An important adaptation of animals is the capacity to accumulate an O_2 deficit in exercise. Products of anaerobic metabolism accumulate during exercise; postexercise O_2 consumption is increased in proportion to the products that are reconverted to energy-rich stores. In large mammals, lactic acid is produced by muscles, and is carried to the liver where 75–80 percent of it is reconverted by gluconeogenesis to carbohydrate. This is known as payment of the oxygen debt. Many poikilothermic vertebrates accumulate lactic acid in the blood during exercise; lactate in blood of a lizard increases from 0.5–2.0 to 9–20 mM during vigorous exercise. However, some poikilothermic animals (amphibians) are predominantly aerobic in exercise. Payment of an O_2 debt may be complete before the lactate returns to preexercise levels; in salmon $\dot{V}_{O_2}$ is recovered in three hours, lactate in eight hours. In a salamander, recovery of $\dot{V}_{O_2}$ is proportional to P_{O_2} in the air and is faster than the decline of blood lactate (45). Some animals with limited gluconeogenesis do not accumulate lactate but excrete it. In many reptiles and amphibians, anaerobic metabolism provides energy for only the first few minutes or burst of activity; prolonged activity is mainly aerobic. Many animals with large aerobic scope (*Bufo, Scaphiopus*) are incapable of bursts of activity. Animals with high anaerobic scope (*Rana, Hyla*) rely on rapid movements for escape or foraging (13). A fast-moving frog fatigues rapidly, has low aerobic scope, forms lactate in exercise; a sluggish toad is mainly aerobic in exer-

---

Figure 6-6. (*a*) Diagram of relation between thoracic temperature, rate of O_2 consumption in flight, rate of O_2 consumption, $\dot{V}_{O_2}$, during warm-up, and flight temperature in a sphinx moth. (*b*) Relation of O_2 consumption to thoracic temperature using preflight warm-up, postflight cooling, and rest in a sphinx moth. Reprinted with permission from G. A. Bartholomew and D. Vleck, *Journal of Experimental Biology*, Vol. 90, pp. 29, 26. Copyright 1981, Cambridge University Press, London.

cise. Rapid activity by anurans is possible only by those frogs with high anaerobic scope (13). Behavioral adaptations may be more important than metabolism in determining feeding and escape patterns.

When diving mammals, such as seals, are forcibly submerged they show bradycardia (slowing of heart rate) and accumulate much lactic acid; a seal can hold its breath for as long as one hour. When they dive voluntarily there is no change in heart rate and no accumulation of lactate for 20–25 minutes; they can feed at depths of 400 m. In longer voluntary dives, some shift to glycolysis occurs. The normal O_2 stores of seals appear adequate for their usual dives to be supplied aerobically (75). In summary, several means have evolved for supplying the energy needs of exercise (63,64).

Tissue Diversity

A further gross metabolic adaptation to activity in vertebrates is the proportion of red to white muscles. White muscles generally fatigue rapidly and are highly glycolytic; red muscles, containing myoglobin, are capable of prolonged slow contraction and are rich in oxidative enzymes. Vertebrates that are capable of bursts of vigorous activity have more white muscle than red and are more capable of glycolysis. In large fishes, red muscles are used in slow swimming and cruising, white muscles in high-velocity swimming. In chickens the specific oxygen consumption of red muscle is 2.4 times greater than of white muscle; when newly hatched chickens are chilled, the cytochrome oxidase activity of red muscle doubles but oxidative activity of white muscle is not increased (3).

Brown adipose tissue (BAT), brown fat, occurs in many neonate mammals, in cold-acclimated adult mammals, and in most hibernating mammals, in suprascapular pads. When a hibernating ground squirrel arouses, its body temperature rises, the brown fat warms faster than any other tissue and yields heat to the thorax. The stimulus for arousal is norepinephrine (NE), which can stimulate O_2 consumption of BAT as much as 30-fold *in vivo* or *in vitro*. BAT accounts for 5.5–6.0 percent of total $\dot{V}_{O_2}$ during arousal; the lipid content of BAT may decrease by 12.5 percent. During arousal fatty acids are released from triglycerides, phosphorylation is uncoupled from oxidation, and heat is liberated by oxidation of fatty acids (78).

The uses of metabolic energy differ according to tissue function. Oxygen consumption by tissues rich in ion transporting cells, such as kidney, is reduced by 70 percent if Na-K ATPase is poisoned by ouabain. The utilization of energy for Na pumping is high in brain, lower in other organs. Cells that are actively engaged in protein synthesis, for example, exocrine and endocrine glands, reticulocytes, and dividing and growing tissues, use more energy for protein synthesis than resting cells. Motile cells, muscles, and cilia, use much energy for the interaction of the proteins used in cell movement.

Nutrition markedly influences energy metabolism. When an alligator feeds, its metabolism goes up by 300 percent (37). Animals consuming a high protein diet differ from herbivores in enzyme pathways and water requirement. Dietary protein may stimulate oxidation by specific dynamic action. In many physiological experiments the nutritional level may influence efforts to measure effects of other environmental parameters on metabolism.

In conclusion, oxygen consumption is influenced by so many internal and environmental factors that measurement as a function of a single variable may lead to incomplete or erroneous conclusions. The molecular controls by which cellular metabolism in different cell types is regulated are not well known.

Ventilation

$\dot{V}_{O_2max}$ of tissues is directly proportional to the product of tissue volume × mitochondrial volume. Muscle metabolism is determined by capillary density within the muscle, and by capillary permeability to O_2 and metabolites. Metabolism is scaled also to alveolar surface and to the diffusing capacity of the boundary between air and blood (131).

In air-breathing vertebrates O_2 uptake and CO_2 excretion are coupled and take place in the lungs; acid is excreted by kidneys. In water-breathing animals the blood P_{CO_2} is much lower than in air breathers and is regulated by HCO_3^- in the blood. Evolution from water to air breathing was a major transition because CO_2 is 50 times more soluble than O_2 in water. If extraction of O_2 is similar by gills and lungs to provide the O_2 necessary for comparable metabolism, some 27 times the volume of water must pass over gills as must pass as air into lungs. Given the same CO_2 gradient from venous blood to air or to water, most CO_2 goes out from air breathers via the lungs; but from aquatic animals CO_2 goes out to the water through any permeable surface. In air breathers external respiration and P_{O_2} are primarily regulated by CO_2 concentration (less by P_{O_2}). CO_2 diffuses freely across gills to water and blood pH is not regulated by ventilation as it is in air breathers (Fig. 6-7). In frogs, O_2 uptake is mainly by lungs during activity by the skin during hibernation and under water; CO_2 elimination is mainly cutaneous at all times. Lungfish use gills, lungs, or skin according to environmental conditions; an increase in CO_2 in the water brings about a shift from gill breathing to lung breathing. A garfish in aerated water uses the lung for 42 percent of its oxygen consumption, in hypoxic water for 100 percent; the CO_2 output is via gills. In hypoxia, gill circulation is reduced and pulmonary ventilation is increased. Air breathing by fish is more closely linked to O_2 than to CO_2 (110). In a Panama catfish, air breathing gradually decreases during hypoxic acclimation as gills take over CO_2 excretion (58). Crabs that spend time in air use gills for aerial respiration; the gill size is smaller than in aquatic crabs (115). In general, those crabs and molluscs that can live in both media have lower metabolism in air than in water. CO_2 transport in air-breathing verte-

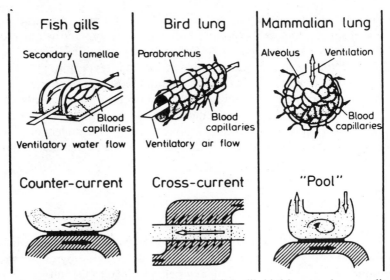

Figure 6-7. Scheme of respiratory apparatus of fish gill, bird lung, and mammalian lung. Reprinted with permission from Peter Scheid in *A Companion to Animal Physiology*, C. R. Taylor et al., Eds. Copyright 1982, Cambridge University Press, London.

brates is mediated by red blood cells, in which carbonic anhydrase (CA) catalyzes hydration of CO_2 to H_2CO_3, and Cl^- exchanges for HCO_3^-. In fishes the CA in red blood cells is low compared to mammals; the gills of many fishes contain CA of high activity.

Most fishes that swim at slow speeds ventilate by rhythmic branchial breathing. In high-speed swimming the mouth may remain open in ram-jet gill ventilation. In many pelagic fishes, ram gill ventilation is made at swimming speeds of 2.7–4.7 body lengths per second. Some pelagic fish, for example, mullet, do not ram ventilate. Scombrids—mackerel, tuna—are obligatory ram ventilators and survive only under conditions in which they can swim faster than 1 km/hr. Ram gill ventilation reduces the cost of breathing below active branchial breathing cost and improves swimming efficiency (51,100).

Environmental Oxygen

When subjected to a range of P_{O_2} levels, animals show one of two kinds of response or an intermediate pattern. In *metabolic regulators* O_2 consumption (M or $\dot{V}_{O_2}$) is relatively constant down to a critical P_{O_2} but in *metabolic conformers* the $\dot{V}_{O_2}$ varies in proportion to the P_{O_2} in the environment (Fig. 6-8). The distinction between regulation and conformity is not sharp. Most animals are metabolic regulators; many of the conformers are invertebrates. Of two related crabs, *Pachygrapsus* is a conformer, *Carcinus* is a regulator

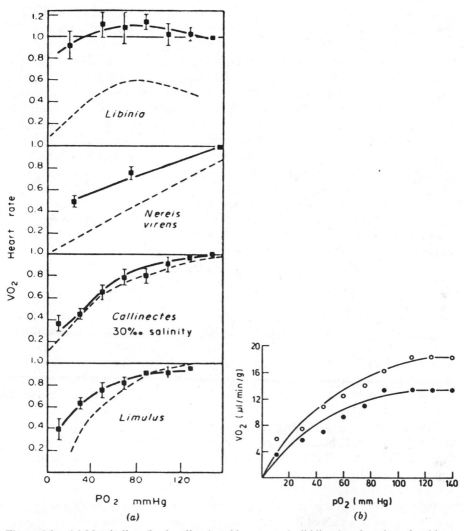

Figure 6-8. (*a*) Metabolism (broken lines) and heart rate (solid lines) as function of ambient P_{O_2} in four species of invertebrate animals. *Libinia,* a metabolic regulator, and *Nereis, Callinectes,* and *Limulus,* metabolic conformers. Reprinted with permission from P. L. de Fur and C. P. Mangum, *Comparative Biochemistry and Physiology,* Vol. 62A, p. 288. Copyright 1979, Pergamon Press, Ltd, Oxford, UK. (*b*) Effect of $\dot{V}_{O_2}$ on O_2 uptake by *Ophelia.* Solid circles normal respiration, open circles after four hrs anaerobiosis. Reprinted with permission from J. Beis et al., *Comparative Biochemistry and Physiology,* Vol. 67A, p. 304. Copyright 1980, Pergamon Press, Ltd., Oxford, UK.

down to 50 torr (0.066 bar) (27). A barnacle is a metabolic conformer and takes up more O_2 in air than in water (91). Newborn mammals survive periods of hypoxia for longer periods than adults (1).

After water-breathing invertebrates sense reduced P_{O_2} they increase ventilation. In mammals, ventilation is mediated mainly by plasma CO_2 via its effect on the respiratory center, and to a lesser degree by reduced P_{O_2} sensed by the carotid sinus. In hypoxia, ventilation increases. There is synergism in that CO_2 stimulates ventilation most in low P_{O_2}, and low O_2 stimulates most at high CO_2. Humans acclimatized at high altitude have low alveolar CO_2 and the respiratory center is more sensitive to CO_2 than it is in humans who dwell at sea level. Many aquatic animals are effective metabolic regulators; in them (fish, crustaceans) ventilation is stimulated by reduced O_2 in water, probably via O_2 receptors on the gills.

Some acclimation to O_2 level is possible; fish maintained at low P_{O_2} have lower $\dot{V}_{O_2}$ and lower critical P_{O_2} levels than normoxic fish. In a flounder, acute hypoxia resulted in doubling of the ventilation volume, but $\dot{V}_{O_2}$ nevertheless decreased from the normal 0.45 ml O_2/kg/m to 0.12; after acclimation to reduced O_2, the $\dot{V}_{O_2}$ partially compensated by rising to 0.24 ml/kg. Crayfish compensate for hypoxia initially by faster breathing and heart rate; alkalosis develops which increases affinity of the blood hemocyanin for O_2 (133). *Artemia* after acclimation to low O_2 shows reduced metabolism and lower critical P_{O_2} (124). A polychaete worm *Ophelia* after hypoxic exposure shows reduced $\dot{V}_{O_2}$, but critical pressure remains the same (11) (Fig. 6-8b). In general, on exposure to reduced P_{O_2} there is initial reduction in $\dot{V}_{O_2}$; this is followed during acclimation by gradual recovery of $\dot{V}_{O_2}$ and decrease in critical P_{O_2}.

Any aerobic biological process does best in an optimum oxygen concentration. In a yeast the maximum enzyme activities for three enzyme pathways occur at the following P_{O_2} levels: palmitoyl-CoA desaturase at 0.03 to 3 mbar, anaerobic acetyl-CoA synthetase at 3-10 mbar, and aerobic acetyl-CoA synthetase at 210 mbar.

Invertebrate animals show respiratory patterns that correlate with life habits. Of two species of holothurians (sea cucumbers), one (*Sclerodactyla*) is a metabolic regulator when P_{O_2} is reduced; in low oxygen it shows little decline in ventilatory frequency and a large increase in ventilatory volume. Another holothurian (*Cucumaria*) shows less metabolic regulation. *Sclerodactyla* lives subtidally in mud which is often hypoxic, while *Cucumaria* lives under rocks in pools where hypoxia is rare (24). A mud shrimp *Callianassa* can survive in air for two weeks but its $\dot{V}_{O_2}$ in air is 40 percent less than in water (48). If oxygen is low many invertebrates shift to anaerobic metabolism. A survey of 31 species of invertebrates in gradually reduced oxygen supply showed that some of them ceased withdrawing O_2 before the environmental O_2 was exhausted. Shutdown of aerobic metabolism occurs in some jellyfish at 15–55 mm Hg (torr) O_2, in polychaete annelids at 0–25 torr (according to life habits), in clams and snails at 4–13 torr, in a sea urchin at 21–31, and in a burrowing holothurian at 0–2 torr (79). Some snails (*Nassarius*) and some aquatic insects carry an air bubble under water and utilize the contained O_2. Many snails are metabolic conformers with *b* values of 0.59 (2). In *Mytilus*, metabolism acclimates when P_{O_2} is reduced to 55 mm Hg; no aerobic shutdown occurs (10). In some invertebrates the O_2 stores in blood provide for brief periods of hypoxia at low tide. Blood P_{CO_2} and HCO_3^- increase in an intertidal crab *Uca* when temperature rises; blood pH in a subtidal *Cancer* and *Callinectes* is adjusted by excretion of HCO_3^- (82). The metabolic pathway used in exercise may be different from that in hypoxia. A swimming leech accumulates lactate and metabolizes glycogen; in hypoxia a resting leech initially obtains reducing capacity by converting malate to succinate (139).

A balance exists between reactions that produce ATP and those in which ATP is required. Tolerance of hypoxia resembles tolerance of hypothermia in that, as activity of ATP requiring ion pumps declines, membrane leakage is reduced. Homeotherms are less capable of such membrane homeostasis than are poikilotherms and are consequently more sensitive to both hypoxia and hypothermia.

Summary

In whole organisms many adaptive mechanisms have evolved that permit continued activity despite reduced O_2. Most of the adaptive changes in whole plants and animals are related to body form and life style. Body size, locomotor activity, habitat, and nutritional pattern each is correlated with an adaptive structure. The need to deliver O_2 to tissues led to the development of ventilatory and transport organs; among the most efficient means are the tracheas of insects. In many kinds of animals, mechanisms for increasing ventilation and O_2 extraction evolved, usually as neural reflexes. Survival in hypoxia led to new anaerobic pathways, and to behaviors that brought the animals into environments with more oxygen. Parasitic animals make use of anaerobic methods of energy liberation that are low in efficiency. A holistic analysis considers the costs and benefits of each mechanism that modifies exchange of O_2 and CO_2 with the environment.

O₂ TRANSPORT

The shape and dimensions of some organisms allow them to use diffusion for delivery of O_2. Diffusion was no longer adequate as aerobic eukaryotes became larger and the distance from an O_2 source to cell interior was greater than 1 mm. Plant leaves are usually flat and have stomata opening to extracellular channels. In some animals, oxygen is transported directly to tissues—via water in sponges and coelenterates, via air in tracheate animals, such as insects. In other animals, oxygen-carrying pigments evolved; metalloproteins are Fe pigments—hemoglobins, chlorocruorins, hemerythrin, and Cu pigments—hemocyanins. Transport pigments serve several functions: (1) they combine with O_2 at respiratory surfaces, then circulate and release O_2 at tissues; (2) they provide a store of O_2 for brief periods of hypoxia; (3) they buffer pH changes; (4) as carbamino compounds they combine with CO_2 to a limited degree; and (5) they serve as colloidal anions (96a).

Heme compounds must have evolved early (Ch. 2); they are present as cytochromes in all aerobic cells and in some anaerobic bacteria. From cytochromes there evolved intracellular hemoglobin (myoglobin) which facilitates O_2 diffusion. Transport hemoglobins (Hbs) may occur either in solution in plasma or in corpuscles. Presumably Hbs evolved separately in unrelated animals, the heme moiety remaining constant, the globins differing. One variant of heme is the chlorocruorin of serpulid and sabellid polychaetes in which in one pyrrole ring a vinyl group is replaced by a formyl. The globins of invertebrate Hbs are extremely varied—in size, charge, and primary structure. No direct evolutionary link to vertebrate Hb is apparent. Some invertebrate Hbs of large molecular weight are in solution in hemolymph; smaller invertebrate Hbs are packaged in cells. The large size of Hbs in solution may prevent their loss from an animal by diffusive filtration. An advantage of Hb containment in red blood cells is that the chemical environment inside the red cell can differ from that of the plasma. Further, the viscosity of whole blood is three times as great as plasma and 50 percent greater than hemolyzed blood (105).

Vertebrate Hemoglobins

The evolution of the Hbs of vertebrates and the adaptive properties of Hbs as proteins were mentioned in Chapter 4. Many fishes and amphibians have multiple Hbs (isotypes and allotypes), and how these different Hbs compare to the α and β chains of Hbs of mammals is not clear. In all vertebrates (except cyclostomes) the circulating Hbs are tetrameric and each tetramer contains at least two different chains, α and β, according to their terminal amino acids. Most genetic variants are in β chains; a few are in α chains.

The affinity of Hb for O_2 is described by a dissociation curve of percent saturation as a function of partial pressure of O_2 (P_{O_2}). The curve is sigmoid for tetrameric Hb and hyperbolic for monomeric Hb, such as myoglobin and lamprey Hb, as well as for most invertebrate Hbs. Another measure of O_2 transport is oxygen capacity, the amount of O_2 that can be combined with blood Hb at saturation with air. The position of the O_2 equilibrium curve on the oxygen axis is given by the percent P_{O_2} for half-saturation (P_{50}), which is the inflection point on a sigmoidal equilibrium curve. Adaptation of hemo-

globins to the oxygen need of the organism are of two kinds: (1) the genetically determined structure of the Hb molecule provides for differences in O_2
affinity, for changes in P_{50}, in P_{O_2}'s of loading and unloading, and in sensitivity of the protein to pH and temperature; and (2) induced adaptations to
reduced O_2 result in an increase in the amount of transport Hb and increased
concentrations of modulators, such as organophosphates.

The shape and position of the O_2 equilibrium curves of purified Hb are
different from those of Hb contained in red blood cells because of the effect
of intracellular components. The properties of the equilibrium curves are
determined partly by the primary structure of Hb (genetically coded) and
partly by external modulators. When pH is lowered, the O_2 affinity decreases and the Hb saturation-P_{O_2} curve shifts to the right; this favors unloading of O_2 in the tissues. The converse effect, shifting to the left as pH
rises, favors O_2 loading at low P_{O_2}; the pH effect is known as the Bohr effect.
An example of adaptive value of the Bohr effect is in exercising animals;
liberation of lactic acid lowers blood pH, and by the shift in position of the
O_2 equilibrium curve, delivery of oxygen to the tissues is favored.

In some Hbs, particularly in fishes, the percent saturation of arterial
blood at equilibrium with air decreases with declining pH; this is the Root
effect, which is important in O_2 secretion into swim bladders. Addition of
CO_2 to a closed vessel containing trout blood may cause the appearance of
a bubble of O_2. The Root effect is also functional in the secretion of O_2 by
the choroid rete of the eye of fishes (135).

The position and shape of the O_2 equilibrium curve is influenced by cellular organophosphates—DPG (diphosphoglycerate) in mammals, IHP (inositol hexaphosphate) in birds and some reptiles. In most fish ATP (adenosine triphosphate) is the modulating organophosphate, in some fish GTP
(guanosine triphosphate); in catfish and lungfish it is ITP (inosine
triphosphate).

Hemoglobins of fishes show more genetic variation than Hbs of any other
animals. Diversity of structure of α and β chains in fish Hbs was described
in Chapter 3. Hbs of sluggish fish that can live in water low in O_2 have low
P_{50}'s and show little Bohr effect—catfish, carp, and tarpon. Hb of fish that
are active swimmers and live in well-oxygenated cold water—suckers,
trout—have low O_2 affinity, large Bohr effect (Figs. 6-9, 6-10). Fish that live
in relatively stagnant water have low P_{50}s and the Bohr effect, while it may
be considerable as a percentage of the O_2 equilibrium curve, does not put
the curve out of the useful range. By contrast, in fish living in well-oxygenated water and with high P_{50} or low O_2 affinity, an increase in CO_2 from 2 to
10 torr may move the O_2 equilibrium curve so far to the right that the fish
suffocate even in ample O_2. Thus CO_2, which favors unloading in the tissues,
prevents loading in the gills. For example, when P_{O_2} was kept at 160 mm Hg
an active fish, the shiner (*Notropus*) died at 0.88 mm P_{CO_2}, while a sluggish
fish, a bullhead from low O_2 water, survived until the P_{CO_2} reached 338 mm

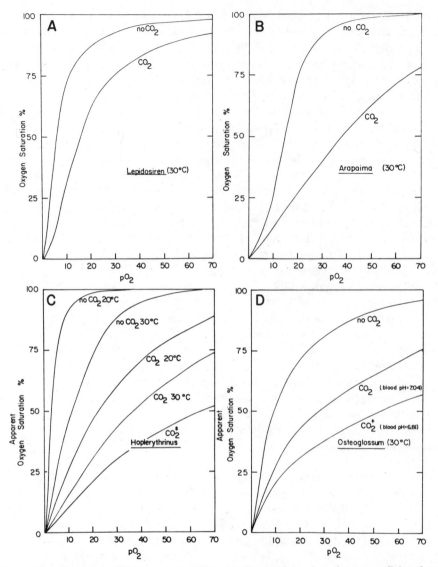

Figure 6-9. Oxygen equilibrium curves of blood of several species of Amazon fishes from different habitats. Measurements with and without CO_2. (*a*) Lungfish *Lepidosiren,* high O_2 affinity; (*b*) air-breathing fish *Arapaima,* high O_2 affinity, large Bohr effect; (*c*) facultative air-breather *Hoplerythrinus,* high O_2 affinity, high CO_2 and temperature effects; (*d*) shallow-water *Osteoglossum,* low O_2 affinity; (*e*) fast-water, highly aerobic fish *Schizodon;* (*f*) fish from hypoxic water, *Sternopygus,* high O_2 affinity; (*g*) temperature-sensitive *Potomorhaphis;* (*h*) fish with large Root effect *Mylossoma.* Reprinted with permission from D. Powers, *Comparative Biochemistry and Physiology,* Vol. 62A, pp. 75, 76, 77, 79. Copyright 1978, Pergamon Press, Ltd., Oxford, UK.

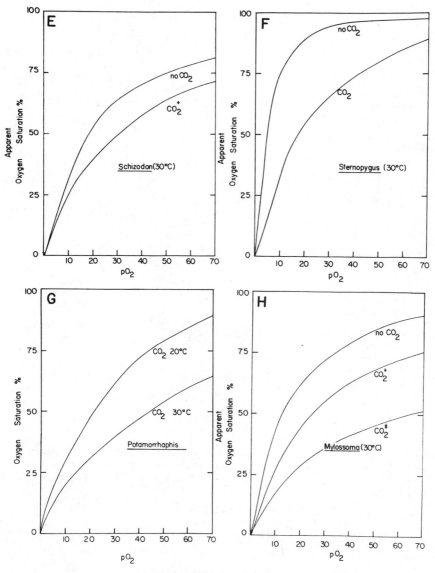

Figure 6-9. *Continued*

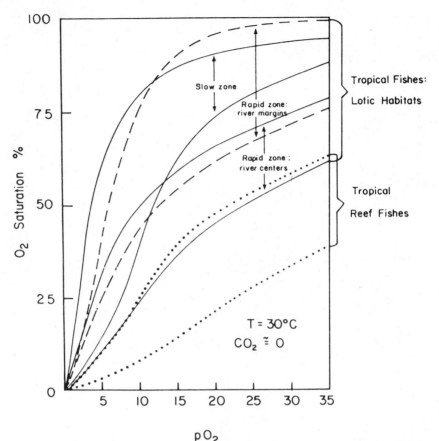

Figure 6-10. Comparison of O_2-Hb affinity for water-breathers of rapid water motion zones and reefs compared with slow zone species. Reprinted with permission from D. Powers et al., *Comparative Biochemistry and Physiology,* Vol. 62A, p. 84. Copyright 1978, Pergamon Press, Ltd., Oxford, UK.

Hg. The Hb of active fish, such as mackerel and sea robin, is more CO_2-sensitive than the blood of sluggish elasmobranchs. In trout at least four hemoglobins occur: one is insensitive to pH, temperature, and ATP, while another Hb is sensitive to these factors and has a large Root effect (16,59,17). The effect of ATP is greater at low than at high pH, irrespective of whether there is a large Bohr effect.

Air-breathing fishes tend to have high O_2 capacity; some have high affinity and considerable Bohr effect (69). O_2 equilibrium curves of 42 species of Amazon fishes show in correlation with habitat that those from slow or stagnant water have Hb of high O_2 affinity, while those from fast water and stream riffles have low O_2 affinity (Figs. 6-9, 6-10) (95).

Structure of Hb chains correlates with function. In general, anodal Hbs

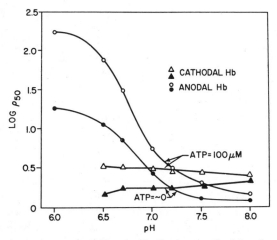

Figure 6-11. Interaction between ATP and pH effects on cathodal and anodal hemoglobins of catostomid fish *Pantosteus*. Cathodal Hb insensitive to pH but slightly sensitive to ATP at low pH. Anodal Hb sensitive to both pH and ATP, more sensitive to ATP at low pH. Reprinted with permission from D. Powers, *Annals of the New York Academy of Sciences*, Vol. 241, p. 482. Copyright 1974, New York Academy of Sciences.

have normal Bohr effect and are sensitive to ATP, while cathodal Hbs show little or no Bohr effect and are relatively insensitive to ATP. ATP stabilizes the structure of deoxy Hb, especially at low pH. *Fundulus,* a shallow-water fish from brackish water, has four different Hbs. In acclimation of *Fundulus* to low O_2, the ATP/Hb ratio decreases and hematocrit increases, resulting in increase in O_2 affinity (94).

One genus of suckers (Catostomids) has six globin loci—two of α and four of β chains. One subgenus (*Pantosteus*) has one or more Hb components that migrate cathodally in electrophoresis; the other subgenus (*Catostomus*) has no cathodal component. The β chain of the cathodal Hb has a carboxyl terminal of -Tyr-Phe whereas the anodal migrating β chain ends in -Tyr-His. The anodal Hbs are sensitive to pH and ATP, while the cathodal Hb is insensitive to pH and is negligibly affected by ATP (Fig. 6-11). In the anodal Hb the Bohr effect increases as the ATP increases. *Pantosteus* occurs in fast streams and its cathodal Hb has high O_2 affinity, no Bohr effect, and is probably used for O_2 transport only in emergency exertion. *Catostomus* occurs in pools and sluggish water; all of its Hbs are anodal and sensitive to pH (93).

The Hb of some fishes is very sensitive to temperature. In three species of trout, the P_{50} rises about 1 mm Hg of O_2 with each rise of 1°C. At higher temperatures the amount of oxygen dissolved in water is diminished, so that the combined effect on the equilibrium curve and on dissolved O_2 forces the fish to seek cool, oxygenated water. Sensitivity of blood of a trout, *Salmo gairdneri,* to both high temperature and to CO_2 is shown by the following values of P_{50} (31):

Temperature (°C)	P_{50} at Different P_{CO_2}		
	No CO_2	3 mm CO_2	7–8 mm CO_2
10	9	18.5	
15	14	20	38
20	18.5	27	38

Antarctic icefish (Chaenichthyidae) lack red blood cells, yet their oxygen consumption is nearly as high as in fish having Hb; however, the $\dot{V}_{O_2}$ of both kinds of Antarctic fish is lower than the $\dot{V}_{O_2}$ of temperate water fish. Icefish skin is well vascularized, and cutaneous respiration is high relative to gill respiration. $\dot{V}_{O_2}$ at 1°C is 0.02 ml O_2/g/hr, the critical P_{O_2} is 50 mm Hg, and the blood volume is high. These fish obtain sufficient O_2 in solution in their pale blood (97). In temperate-water fishes that are maintained under hypoxic conditions the concentrations of ATP and of GTP decrease.

Among mammals, species that normally move rapidly—sprinters, pouncers, and dashers—have Hbs with high P_{50}, and moderate Bohr effect, and their RBCs (red blood cells) are low in DPG. Mammals which are endurance runners, divers, or hibernators have moderate to high Bohr effects, low and moderate P_{50}'s, and high DPG content (72). At high altitude and in hibernacula of mammals P_{O_2} is low, and in mammals living in those environments the equilibrium curve moves to the left; affinity is increased because of a fall in the DPG content of the cells. Genetic differences occur between mountain and sea-level mammals, between Himalayan sherpas and sea-level-dwelling humans.

Adaptations of Hbs have been noted for birds. The P_{50} for a distance runner, the *Rhea,* is 20.7 torr; for a large, moderate flier, the pigeon, 29.5 torr; for fast passerines 41.3 torr; diving birds have Hb of higher O_2 affinity, greater Bohr effect, and higher Hb content in red cells than in cells of nondivers. Land turtles, such as *Testudo,* have lower P_{50} (17.2 torr) and larger Bohr effect than those turtle species capable of prolonged dives; *Caretta* has a P_{50} of 29 torr. Skin breathing amphibians have lower O_2 affinity (P_{50} for *Desmognathus* is 26 torr, for *Rana* 39 torr) than water breathers (*Necturus* P_{50} is 15 torr).

Hbs synthesized during embryonic development, are different from the adult form (HbA). In mammalian embryos fetal Hb (HbF) is formed; this contains a γ chain in place of the β chain of adult Hb, and the HbF is made in liver rather than in bone marrow. In very early embryos a third beta form occurs—Hbε; an embryonic alpha type is δ chain (Ch. 3). Genes for the α and β chains are at different loci on the same chromosome. The affinity for O_2 is greater in HbF than in HbA. For example, in humans there is a P_{50} of 20 torr for HbF, 28 torr for HbA. Tadpoles have Hb with greater O_2 affinity than frogs; the tadpole Hb shows no Bohr effect. Salamander larvae have five Hbs, adults have three (50). Hemoglobin of salmon fry has high O$_2$ affinity (P_{50} = 3.9 mm Hg) and Bohr effect -1.73; adult Hb has P_{50} = 14 mm Hg and Bohr effect $-.18$ (55,99). An embryonic Hb also occurs in ovoviviparous dogfish. Adult eels in fresh water have 2 Hb bands, while elvers from sea water have 9 Hb bands, of which 3 are cathodic, 3 anodic (greater in amount) and 3 intermediate. During several months of adaptation to fresh water, the eels' cathodic Hb increases in amount relative to the anodic (57, 99).

Chickens have five Hbs, which can be determined electrophoretically. One of these is the major Hb during the first week of incubation, the second is predominant shortly before hatching. Synthesis of the adult Hbs begins on the sixth day of incubation. The developmental sequence of Hbs is an example of the activation of different genes at stages in development. Changes in DPG concentration also occur during development; in pig the concentration increases from 2 to 10 μM/g heme and the P_{50} shifts from 22 to 35 torr at birth, while in chickens and ducks DPG peaks just before hatching. The adaptive function of the developmental changes is that a pigment of high affinity facilitates O$_2$ transfer in mammals from maternal blood to tissues in utero, in birds from allantois to embryo, and in amphibians provides for O$_2$ transfer from water.

In the killifish *Fundulus heteroclitus* two genotypes homozygous for the B form of lactate dehydrogenase (LDH) occur, LDH B^{a}B^a and LDH B^{b}B^b. The red blood cells of the aa genotype have a low ratio of ATP to Hb and an O$_2$ saturation curve to the left of that of the bb form (94) (Fig. 6-12).

An adaptive response of O$_2$ transport to reduced environmental P_{O_2} is synthesis of additional Hb. A direct response of mammals to altitude hypoxia is release of stored red blood cells. Production of reticulocytes and synthesis of hemoglobin increase within a few hours of hypoxia. Dogs reared at 15,000 feet had 40 percent more Hb in blood and 66 percent more myoglobin in muscle than dogs at sea level. Similar increases in Hb occur in rats maintained in the laboratory at reduced P_{O_2}. Stimulation of Hb synthesis is initiated by a hormone erythropoietin from the kidney; juxtaglomerular cells of the kidney may sense the reduction of P_{O_2}. Some animals that live at altitude, for example, llamas, transferred to sea level show little reduction in blood Hb.

Some invertebrates respond to reduced P_{O_2} as vertebrates do by synthesizing more Hb (127,128). The crustaceans *Daphnia* and *Artemia* are red in water low in O$_2$, and pale in water high in O$_2$ (66).

Summary

The genes for globin synthesis are best known for mammals. The history of β and α Hbs is well established. A number of gene duplications have given

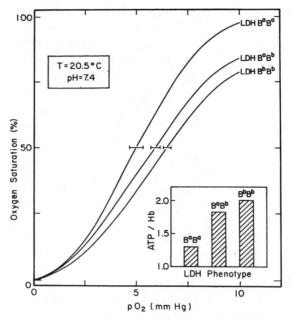

Figure 6-12. Oxygen saturation curves and ATP/Hb ratios for *Fundulus* of two genotypes for LDH and the heterozygote. Reprinted by permission from *Nature*, Vol. 277, p. 241. Copyright © 1979 Macmillan Journals Limited.

rise to different mutant chains, more in β than in α chains. Several Hb chains function in transport of O_2 during developmental stages; embryonic Hbs have higher affinity for O_2 than adult Hbs in both amniotes and proamniotes. The genes that code for different Hbs are active sequentially during development. The diversity of isohemoglobins and allohemoglobins is greatest in animals that live in a wide range of P_{O_2}, P_{CO_2}, and/or temperature. Functionally, O_2 affinity is adaptively modified by pH and by temperature. Fishes have multiple hemoglobins, some of which function independently of acidity and temperature, others with adaptive changes in O_2 affinity that enable the fish to function in diverse environments and at different levels of activity. Besides differences in the Hbs, intracellular modulators (organophosphates) act on O_2 affinity so that Hb in cells functions differently from Hb in solution. The amount of organophosphates can vary according to O_2 supply, for instance, at altitude. Two adaptations for temperature compensation in fishes are: (1) Hbs may have low sensitivity of O_2 binding to temperature (genetic adaptation); or (2) changes in organophosphate may shift the O_2 dissociation curve so that the temperature effect is reduced (acclimation). Transport pigments show correlations of molecular structure with life habits and milieu.

METABOLIC PATHWAYS

The enzymatic pathways for energy production in organisms vary with life style and habitat. Anaerobic pathways of metabolism evolved before aerobic pathways (Ch. 2). Anaerobic fermentation to ethanol and glycolysis to lactate use essentially the same enzymes until the terminal steps and yield two mols of ATP per mol glucose. Oxidation of glucose to CO_2 and H_2O yields 36 mols ATP; the early steps are in common with anaerobic routes.

Glycolysis

A generalized diagram of glycolysis is given in Figure 6-13. There are minor differences between the steps in bacteria and eukaryotes, and between different tissues of the same animal. Most of the enzymes of glycolysis are in solution in the cytosol, and are not bound to membranes. The universality of the glycolytic-fermentative sequence suggests that the enzymes evolved very early (Ch. 2). The intermediates between glucose and pyruvate are phosphorylated compounds; phosphates make the compounds highly polar, negatively charged, impermeant through cell membranes, and structurally specific for enzyme binding. Two general stages in glycolysis are: (1) phosphorylation of hexose followed by cleavage to a three-carbon sugar, glyceraldehyde-3-phosphate; and (2) oxidoreduction of the glycerophosphate to pyruvate. In the first stage two ATP molecules are used, in the second stage four molecules of ATP are formed, so that the net yield is two mols of ATP per mol of glucose degraded. Hexokinase occurs in three isozymes which differ in K_m for various hexoses. Phosphofructokinase (PFK) acts at a control point and has several allosteric modulators—ATP, citrate, and long-chain fatty acids. Fructose diphosphate aldolases of bacteria, yeasts, and fungi differ from those of plants and animals in metal requirements. The nucleotide NAD^+ accepts electrons from phosphoglyceraldehyde and forms NADH. Glyceraldehyde-3-phosphate dehydrogenase has four binding sites for the reduced coenzyme NAD^+. Phosphoenopyruvate (PEP) is a compound that has several allosteric modulators. The concentration of pyruvate is determined by a balance between PEPCK (PEPcarboxy kinase) and PK (pyruvate kinase) and LDH. In aerobic cells, pyruvate may not go to lactic acid but to acetyl-CoA, which then feeds into the TCA (tricarboxylic acid) cycle. Other metabolites—fatty acids and amino acids—also convert to acetyl CoA via aerobic pathways.

TCA Cycle and Electron Transport

The aerobic system is of two parts: the TCA cycle (Figs. 2-11 and 6-14) and the electron transport chain (Fig. 6-2). Some of the enzymes of the TCA

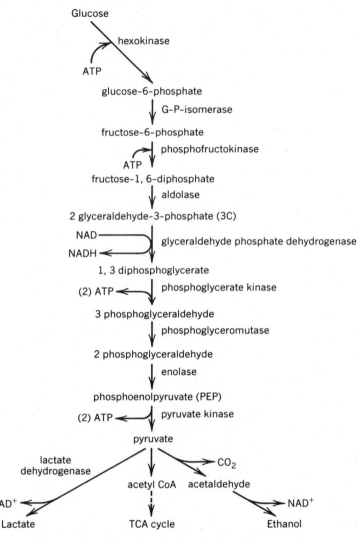

Figure 6-13. Flowchart for glycolysis and fermentation.

cycle evolved in bacteria, where they serve other functions. In eukaryotes the TCA enzymes are in mitochondria; in the cytosol of some tissues there are isozymes of some TCA cycle enzymes (mitochondrial MDH and cytosolic MDH). The overall result of TCA cycle activity is generation by dehydrogenation of four pairs of H atoms and two molecules of CO_2. Three pairs of protons reduce NAD^+ and one pair reduces the flavin adenine dinucleotide (FAD) of succinate dehydrogenase. These four pairs of H atoms become H^+ ions and the corresponding electrons pass along the respiratory chain. Plants and some bacteria have a modified TCA cycle, the glyoxylate

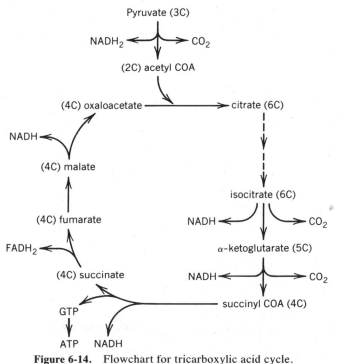

Figure 6-14. Flowchart for tricarboxylic acid cycle.

cycle, in addition to the TCA cycle; the glyoxylate cycle is used for metabolism of fatty acids and for synthesis of glucose. For example, seeds convert acetate derived from fat into carbohydrate.

A variant in the pathway of glycolysis that must have evolved very early is the pentose shunt of the phosphogluconate pathway (Figs. 2-10, 6-15) (117). Glucose-6-phosphate, instead of going directly to fructose-6-phosphate, is converted to 6-P-gluconate and then via ribulose-5-P to xylulose-5-P and pentose-5-P. One function of this shunt is to produce reducing power outside mitochondria in the form of NADPH. This is important in tissues that synthesize fatty acids and steroids. A second function is the synthesis of pentose (ribose) for formation of nucleic acids. In higher plants the P-gluconate path is modified to function in the dark reaction of photosynthesis. In lactobacilli, the PG shunt is used for fermentative degradation of pentoses. The PG shunt uses sugar labeled only in the 1-position, whereas the normal glycolytic pathway oxidizes both 6- and 1-positions. Use of the pentose shunt increases in cold acclimated poikilotherms, perhaps because of high fat metabolism in the cold. The pentose shunt must be very ancient, as may be deduced from its function in the synthesis of ribose and in photosynthesis.

The electron transport chain is present in all aerobic cells and transports

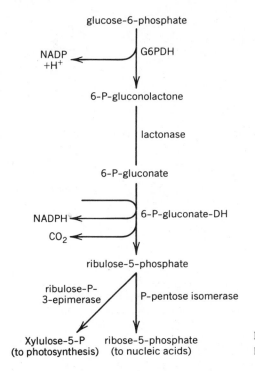

glucose-6-phosphate

NADP +H$^+$ G6PDH

6-P-gluconolactone

lactonase

6-P-gluconate

NADPH   6-P-gluconate-DH

CO$_2$

ribulose-5-phosphate

ribulose-P-3-epimerase   P-pentose isomerase

Xylulose-5-P   ribose-5-phosphate
(to photosynthesis)   (to nucleic acids)

Figure 6-15. Diagram of pentose phosphate pathway.

electrons (with accompanying protons) to oxygen. The TCA cycle and electron transport chain together make 15 mols of ATP per mol of pyruvate and twice that per mol of glucose:

$$2 \text{ pyruvate } + 5 \text{ O}_2 + 30 \text{ ADP} + 30 \text{ P}_i \rightarrow 6 \text{ CO}_2 + 30 \text{ ATP} + 34 \text{ H}_2\text{O}$$

Energy charge is ½ (ADP + 2 ATP)/[(AMP) + (ADP) + (ATP)]. An additional two molecules of ATP may be generated outside mitochondria per pair of electrons transferred by oxidation of NADH; this makes a total of 36 mols of ATP formed in oxidation of 1 mol of glucose. The energy available in hydrolysis of the first phosphate of ATP is 7.3 kcal/mol. Released energy varies according to many factors—pH, substrate concentration, temperature, and cofactors. The total free energy in ATP is 12.5 kcal. The theoretical $\Delta G\ddagger$ for oxidation of glucose is 686 kcal/mol, and the generated ATPs provide for recovery of approximately 40 percent of this.

Properties and evolution of cytochromes were described in Chapter 2. Many cytochromes are present in both anaerobic and aerobic bacteria. Electron transport by cytochromes—iron-porphyrin-containing proteins—evolved prior to aerobic conditions; cytochromes have been retained with modification throughout biochemical evolution. The role of cytochromes in the electron transport pathway is given in Figure 6-2.

Virtually all of the oxygen in the atmosphere has come from photosynthesis. Three modified pathways of photosynthesis are known in plants, each adaptive to a kind of environment. The most common path for CO_2 assimilation is the C-3 reductive path in which the first products of CO_2 fixation are 3-carbon acids and the first enzyme used is ribulose-biphosphate carboxylase (RuBPCase). CO_2 combines with ribulose diphosphate (RuDP) and water to form phosphoglycerate which is then converted to pyruvate (Ch. 2). The C-3 path makes use of the Calvin cycle and six turns of the cycle yield one glucose (Fig. 2-9). This pathway requires continuous inflow of CO_2 via stomata; plants using the C-3 path tend to lose water through stomata. In C-4 plants, the first enzyme is phosphoenolopyruvate carboxylase (PEPCase) and 4-carbon dicarboxylic acids are formed—oxaloacetate, malate. C-4 plants may use the C-3 cycle in bundle sheath cells and C-4 enzymes in mesophyll. A third pattern is the CAM or crassulacean acid metabolic path. CMA plants have a temporal rather than a spatial separation of C-3 and C-4 enzymes. They assimilate CO_2 at night, then close the stomata and use the stored CO_2 for C-3 photosynthesis by day (14,90).

The CAM pathway is especially adapted to arid environments. C-4 and CAM photosynthesis are also adaptive to high temperature and high light intensities. The rate of CO_2 uptake in rising leaf temperatures increases less steeply in C-3 than in C-4 plants. Quantum yield, amount of CO_2 fixed per quantum of light absorbed, is relatively independent of temperature in C_4 plants, but in C-3 plants it decreases as temperature is raised (14).

Adaptations to Differences in O_2 Supply

Animals differ in their need for oxygen according to activity and environment. Several biochemical adaptations tend to provide for energy needs at different O_2 levels. One means is the generation of an oxygen debt, usually in the form of accumulated lactate, during muscular activity, and the payment of the debt by increased O_2 consumption afterward and by resynthesis of glycogen. Another adaptation is the presence of separate glycolytic and oxidative muscle fibers. A third adaptation to O_2 need is the presence of isozymes of multifunctional enzymes. There are two isozymes of LDH, one predominantly in high oxygen tissues, such as the heart and brain, the other in low oxygen tissues, such as skeletal muscle, particularly white fibers (Ch. 5).

The most highly aerobic animals are fast-flying insects. The tracheal system brings oxygen directly into the interior of muscle fibers that contain large mitochondria (101,137). Fast insect muscles are incapable of anaerobic work; they fail to accumulate glycolytic products and lack lactate dehydrogenase. In many insects the blood sugar is a nonreducing disaccharide, trehalose, which reaches high concentrations and is cleaved to glucose as needed. Flight muscles of flies have two additional pathways not found in vertebrate muscles. In the utilization of sugar, G-3P goes to DHAP which is converted to α-GP; this cycles in and out of mitochondria (Fig. 6-16) and yields four protons to the oxidative chain with the production of four ATPs.

Proline is present in the cytosol in high concentrations; it penetrates mitochondrial membranes and in a series of steps some 13 ATP molecules are generated. In addition, proline is converted to α-ketoglutarate, which combines with acetyl CoA (from pyruvate) to activate the TCA cycle. Once the cycle is fully activated it runs at high level during flight and all the pyruvate formed goes to CoA, none to lactate. The combination of the α-GP cycle and incorporation of proline produces 21 mols of ATP per mol G6P and proline prior to the TCA cycle; this is seven times the three ATPs formed in vertebrate muscle. Overall, the fast insect muscle can produce an additional 18 mols of ATP above the 36 mols of ATP that are gained per mol of glucose by oxidation in other organisms. In addition, in some insect muscles the OAA that is formed from proline is decarboxylated to pyruvate, which feeds into the TCA cycle and provides for complete oxidation of proline. New enzymes were not evolved for this highly energetic system; rather, new uses were made of preexisting enzymes to make the most energy-productive system known in animals. The system is highly adaptive for small, fast-flying animals, which have limited capacity for storage of reserve metabolites (28,62,63,70). Insect brains have very high oxidative metabolism; moth brains use predominantly fatty acids, and bees use carbohydrates.

Another widely occurring biochemical adaptation for efficient aerobic metabolism is the oxidation of fat in combination with sugar. Fat is favored over glycogen for storage of energy sources since it has only one-eighth the mass of glycogen; fat provides twice as much energy per gram oxidized as carbohydrate or protein. Oxidation of fat requires approximately twice as much oxygen per gram as carbohydrate oxidation. Fat, when completely oxidized, yields twice as much metabolic water as glycogen, hence its metabolism is favored in desert animals. Finches that overwinter in temperate regions burn predominantly unsaturated fats on cold nights (32). Hibernating mammals utilize fat during their long periods without food, and on arousal their brown fat deposits (BAT) produce heat. The primary metabolites of mammalian hearts are fatty acids, of fish hearts carbohydrates and fats, of elasmobranchs ketone bodies (138). The complete oxidation of fatty acid (e.g., palmitate) yields 130 mols of ATP per mol palmitate, compared to 36 mols ATP for a mol of glucose (9.3 kcal/g compared with 4.1 for carbohydrate). However, the oxidation of fat can occur only when carbohydrate is also being glycolyzed, or when an amino acid is deaminated and dehydrogenated or decarboxylated. Carbon skeletons can come either from carbo-

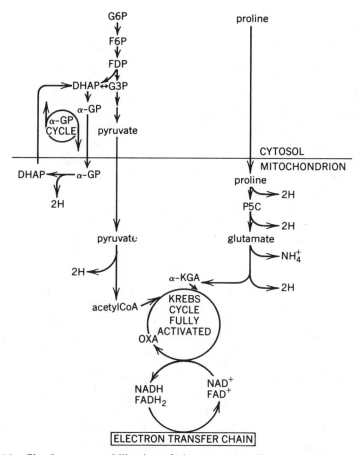

Figure 6-16. Simultaneous mobilization of glucose and proline during flight initiation in blowfly. Modified from P. W. Hochachka and G. N. Somero, *Strategies of Biochemical Adaptation* (Saunders, Philadelphia, 1973).

hydrate intermediates or from amino acids. Oxaloacetate can be formed from aspartate or from malate, and irrespective of the source it keeps the TCA cycle supplied with 4-carbon intermediates. Phenylalanine and tyrosine enter the 4-carbon path via acetoacetate. The amino acids Arg, His, Gln, and Pro enter the TCA cycle via α-ketoglutarate. Ala, Thr, Gly, and Ser enter the TCA cycle via pyruvate and acetyl-CoA; amino acids account for only 10 percent of the animal body's energy (76).

Long-distance fliers (birds) have large fat stores used in migratory flight. A migratory insect (locust) uses carbohydrate during the first 20–30 minutes of flight, then undergoes a shift to fat.

Facultative Anaerobiosis

Many animals live in a precarious balance for oxygen. They are aerobic whenever oxygen is available, but are able to live anaerobically in reduced oxygen and usually excrete the products of anaerobiosis without acquiring much oxygen debt. Life without oxygen or at very low P_{O_2} is made possible by pathways that lead to products which can be metabolized oxidatively or can be excreted; small amounts of energy are thus liberated anaerobically. There is considerable diversity in the products formed by facultative anaerobes. Parasitic helminths and some aquatic molluscs show facultative anaerobiosis and make several different products of glycolysis (44,88).

The nematode *Ascaris* appears not to possess a cytochrome system, and the TCA cycle is not important in this parasitic worm. Enzymes in the cytosol of *Ascaris* muscle form PEP by glycolysis, and the PEP is carboxylated by PEPCK (phosphoenol pyruvate carboxykinase) to OAA (oxaloacetic acid), which then goes to malate. In mitochondria of *Ascaris,* acetyl CoA and OAA condense to form citrate; this does not enter a cycle but goes to succinate, which is excreted. Both PEPCK and PK (pyruvate kinase) are active in *Ascaris* and in related parasitic worms and the ratio V_{max}/K_m is similar for the two enzymes. Succinate is a sink for two metabolic paths; it is formed from both OAA and from α-KGA (α-ketoglutaric acid) (Fig. 6-19) (85). *Ascaris* can also utilize O_2, when it is present, by a flavin system that yields H_2O_2 (103).

Many parasitic helminths form succinate via OAA, malate, and fumarate; eight ATPs are formed per mol of G6P used. In addition, glutamate is used via α-KGA to form succinyl CoA, which goes to succinate as an endproduct. Alternatively, succinyl CoA can be converted via malonyl and propionyl CoA to propionate, which is excreted. Some helminths have active β-oxidation of palmitate to produce acetyl CoA. The trematode *Fasciola* has no PK, low levels of LDH, and fixes CO_2 as indicated in Figure 6-19. The flatworm *Moniezia* forms some lactate as well as succinate (25). Diversity of metabolism of parasitic helminths is related to habitat and life cycle. The free cercariae of schistosomes are aerobic; the adult worms are anaerobic. The alternative products produced by helminths are acetate, succinate, propionate (Fig. 6-17), isobutyrate, and isovalerate, and, in some worms, lactate. It is unclear why some facultative anaerobes form mainly one endproduct, others a different endproduct. The pathways of facultative anaerobiosis are less energy-productive per mol of substrate than aerobic pathways; potentially useful substrates are excreted, but these parasitic animals live in a sea of nutrients and very low oxygen, to which they are well adapted.

Mud-burrowing polychaetes tolerate periods of hypoxia at low tide and form succinate and propionate but no lactate (139). Muscle of the polychaete *Arenicola* forms propionate from succinate under anaerobic conditions; this is enhanced by malate (106).

Intertidal molluscs are subjected to periodic hypoxia during ebb tides. In

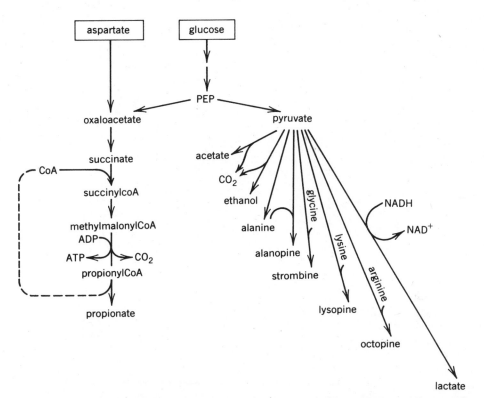

Figure 6-17. Pathways for production of products of glycolysis in bivalve molluscs and facultative anaerobic invertebrates, PEP as branchpoint. Fig. 5-3, p. 157, in *Biochemical Adaptation* by P. W. Hochachka and G. N. Somero, Princeton University Press, 1984.

air the pallial fluid of an oyster drops in pH from 7.6 to 6.7 in one hour, then holds steady for several weeks, after which the pH falls to 5.4. Mollusc muscles form phosphoenol pyruvate (PEP) by glycolysis. At neutral or alkaline pH (when the tide is in and the shell is open) PK converts PEP to form pyruvate plus ATP and alanine, but in an acid mediium (when the tide is out and the shell is closed) PEP is carboxylated by PEPCK to form OAA which then goes to malate and succinate plus fumarate (Fig. 6-18). Both pathways are functional at neutral pH. LDH is very low or absent and no lactate is formed (77). Oyster heart produces alanine and succinate in a 2:1 ratio (36).

Molluscan muscles can form several unusual imino compounds (conjugated secondary amino or imino acids); when pyruvate combines with alanine, alanopine [HN (CHCH$_3$CO$_2$H)$_2$] is formed and NAD is generated (Fig. 6-19). Octopine results from interaction of pyruvate with arginine, lysopine from pyruvate plus lysine, and strombine from pyruvate with arginine, (Fig. 6-21). Lysopine and octopine also occur in crowngall tumors of plants (90a). For each of these 1 mol of NAD$^+$ is formed from NADH per mol of amino acid used (49). The formation of reduced coenzyme occurs without lowering

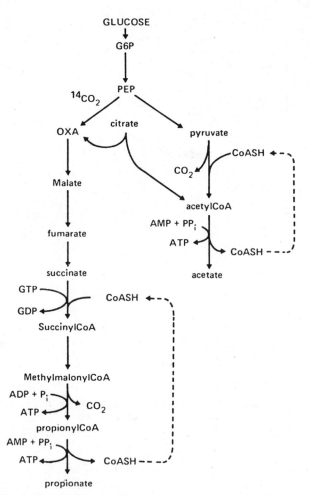

Figure 6-18. Metabolic map of probable origins of acetate and propionate in helminths. Reprinted with permission from P. Hochachka, *Living without Oxygen,* Copyright 1980, Harvard University Press, Cambridge, MA.

of the pH of body fluid. Octopine accumulates in working molluscan muscle when phosphoarginine is used as a source of high energy phosphate, analogous to phosphocreatine in vertebrate muscle. The adductor of a scallop *Pecten* is in two parts—a fast catch muscle and a slow holding muscle (Ch. 16). Both muscles are rich in octopine dehydrogenase (ODH) but have no LDH or PEPCK. During the holding function of the catch muscle glycolysis increases by 75 times and octopine is produced (4,40). Foot muscle of *Cardium* generates ATP glycolytically by production of octopine in exercise, of lactate in hypoxia (52). Land snails are more tolerant of hypoxia than inter-

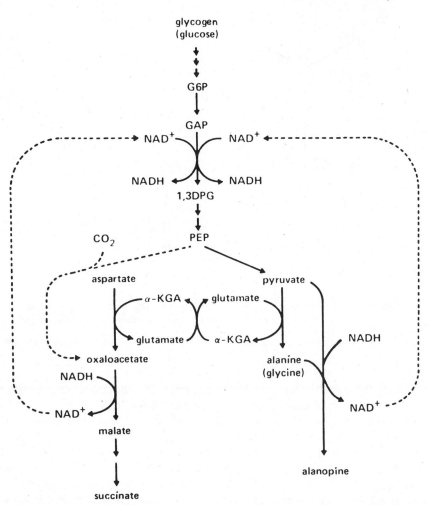

Figure 6-19. Pathways of anaerobiosis in molluscs and some parasites to succinate or alanopine. Reprinted by permission from P. Hochachka, *Living without Oxygen,* Copyright 1980, Harvard University Press, Cambridge, MA.

tidal molluscs; land snails produce lactate during hypoxia. Red muscle of a marine snail *Busycotypus* has aerobic and anaerobic isozymes of PK; S_{05} for pyruvate is 0.047 mM in the aerobic form, 0.85 mM in the anaerobic isozyme (91a).

A bivalve from the high intertidal zone gapes when exposed to air, and its adductor muscle is aerobic. A species from the subtidal zone keeps its valves closed when in air, and its muscle is highly anaerobic (89). In general, bivalve molluscs subjected to environmental anoxia accumulate alanine,

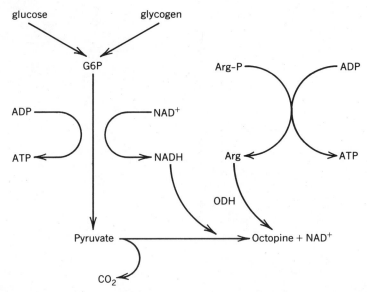

Figure 6-20. Pathway for anaerobic production of octopine in molluscs. Reprinted with permission from K. B. Storey and J. M. Storey, *The Mollusca,* Vol. 1, Copyright 1983, Academic Press, New York.

succinate, or propionate; in work anaerobosis they accumulate imino acids—octopine, alanopine, strombine (4,113) (Fig. 6-21).

The mantle muscle of squid is highly aerobic and dependent on a continuous supply of oxygen via the blood. In exercise, lactate does not accumulate, but arginine phosphate may decrease from 10 to 1 μm/g; proline decreases, and alanine increases. The mantle muscle may use the α-KGP cycle like insects (113,113a). In squid muscle, octopine is formed during fast activity and this is reconverted to pyruvate during recovery. Squid brain has LDH which competes with ODH (49). Cephalopod mantle muscle produces octopine, not lactate, in exercise and anoxia; alanopine and strombine are produced in bivalve and gastropod molluscs, not in cephalopods. Sepia muscle has four isotypes of octopine dehydrogenase (113a).

Goldfish and carp can survive hypoxia for many days, especially at low temperatures under the ice in winter. They survive by converting acetyl CoA to acetate, then to acetaldehyde and ethanol. This pathway provides a gain of one ATP and two reducing equivalents of NAD (63).

Diving mammals use aerobic glycolysis and do not accumulate lactic acid when they make short or voluntary dives. When they undergo forced dives or prolonged voluntary dives they show increased blood concentrations of lactate and pyruvate (from glycolysis) and of alanine (from amino acid metabolism). Species of aquatic mammals which are capable of prolonged dives have more LDH-B and more PK in heart and brain than nondivers (63). They

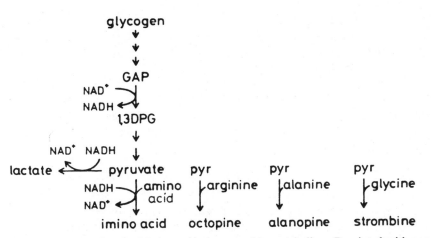

Figure 6-21. Origin of several imino acids in anaerobic metabolism. Reprinted with permission from K. B. Storey and J. M. Storey, *The Mollusca,* Vol. 1, Copyright 1983, Academic Press, New York.

are able to couple amino acid utilization with glycolysis, like the facultative anaerobic invertebrates.

CONCLUSIONS

Life originated in an aqueous environment lacking oxygen, or low in oxygen (Ch. 2). The earliest organisms obtained energy anaerobically. After oxygen in the atmosphere increased (some from photodissociation of water, more from photosynthesis) the anaerobic organisms were poisoned by the oxygen. Detoxifying enzymes—oxygenases—evolved, and some of these have persisted. A few kinds of bacteria continue at present as obligate anaerobes. Present-day animals that are mainly anaerobic (parasites) appear to have lost oxidative capacity. The persistence in eukaryotes of anaerobic pathways— glycolysis and fermentation—up to the present is evidence for the usefulness of the capacity for life without oxygen.

The advantages of aerobiosis over anaerobiosis include the much greater yield of high-energy phosphates, the stepwise sequence in oxidation-reduction from negative substrate to positive oxygen, and the alternate paths which make possible the release of energy from several substrates. The presence of both oxidative and glycolytic paths in organisms provides versatility, redundancy, and safety factors in energy liberation. Some oxidative enzymes—cytochromes and dehydrogenases—evolved in early prokaryotes and have persisted to the present with minor changes. Green plants that produce oxygen have oxidative enzymes similar to those of animals.

With the appearance of aerobic enzymes, many new structures evolved that enabled organisms to obtain oxygen from environments differing in oxygen availability. Oxygen uptake by way of skin limited body size; gills of many kinds appeared and digestive epithelia developed the capacity for respiration in water. The transition to terrestrial life made air breathing essential; some present-day fishes and crustaceans are able to breathe in either water or air. Lungs, with accompanying circulatory devices, are of many kinds. In many kinds of small animals, for example, insects, air goes directly to tissues through ramifying tubes.

With the evolution of oxidative enzymes and increase in body size and tissue diversity, there appeared circulatory systems and oxygen carriers. Pigments, especially hemoglobins, became adapted by structure and by effects of pH and of modifying organophosphates to transporting O_2 in various environments. Many animals have several hemoglobins in their blood that differ in O_2 affinity and that function at different temperatures, salinities, and environmental P_{O_2}. Both allotypes and isotypes of Hbs are adaptive. Several genes in one animal code for hemoglobins of different O_2 affinities. Hemoglobin genes may become active at different stages in development.

Aerobic organisms had to contend with an O_2 supply that varied from time to time. One set of adaptations to varying O_2 consists of changes in ventilation via gills, lungs, or tracheae. Most animals regulate ventilation according to the level of P_{O_2} or P_{CO_2}; the regulation is largely via the central nervous system. Another kind of adaptation to low oxygen is the shift from oxidative to glycolytic metabolism. Glycolysis gives only one-eighteenth as much energy as oxidation of glucose; but in extreme hypoxia this may be enough for survival. Capacity to make this shift varies greatly in animals; several products of anaerobiosis are produced by different animals, especially invertebrates. The temporal change from aerobic to anaerobic pathways permits periods of activity (e.g., exercise) longer than can be supplied by O_2 and allows for survival for limited periods of time without ventilation (for example, bivalve molluscs in air).

The rate of mass-specific oxygen consumption is influenced by many internal and external factors. Small organisms have higher metabolic rates than large ones. In each cell of a smaller animal there are more mitochondria and more oxidative enzymes. The nature of sensing mechanisms and feedback to enzyme synthesis is not well known. Modifiers of O_2 consumption are activity, hormonal and reproductive state, sex, temperature, salinity, altitude, and life habit.

From the distribution of specific enzymes, metabolites, and products it is possible to speculate concerning the evolution of energy-yielding pathways and to relate these pathways to the physiological demands on organisms. The use of high-energy phosphate bonds (terminal phosphate of ATP) for energy transfer probably evolved very early, and this led to selection of proteins for phosphorylation of adenylates. Why, in all energy-requiring reactions, is the energy delivered in packets or quanta of ATP? Two suggestions have been made. One is that complete oxidation of glucose, which has a $\Delta G\ddagger$ of 686 kcal/mol, in one massive amount would be wasteful in heat generated. Another suggestion is that since energy use in biosynthesis, in muscle contraction, and in ion pumping is in small amounts by steps, delivery in quanta is efficient.

Once the energy currency was established, reactions for providing phosphorylation of ADP to ATP appeared. Probably in early metabolic reactions substrates were incompletely broken down. This was appropriate, since carbon compounds were present in quantity and biochemical pathways were limited. In the anaerobic environment the first energy-yielding pathways were fermentative (to alcohols) or glycolytic (to 3- and 2-carbon acids). These processes persist in both anaerobic and aerobic cells, but the net yield of ATPs is small—2 per mol glucose. The pentose shunt (HMP) probably evolved in anaerobic conditions. This provided additional reducing power (NADPH) and pentoses which were used in the synthesis of nucleic acids. Somewhat later, the pentose shunt became essential for the dark reaction of photosynthesis. Anaerobic utilization of fats probably came later; fats provide more energy but require concurrent catabolism of carbohydrate.

Anaerobic photosynthesis yielded high-energy organic compounds without liberating oxygen (Ch. 2). More efficient photosynthesis came with a process in which oxygen was released. In bacteria, heme compounds, cytochromes, presumably were used in electron transfer under anaerobic conditions; there are cytochromes in bacterial cell membranes. Cytochromes appeared before the evolution of mitochondria in eukaryotes. Prokaryotes use enzymes of the TCA cycle for different purposes than eukaryotes; presumably these enzymes were taken over for the TCA cycle by eukaryotes. The liberation of oxygen by photosynthesis created a problem for prokaryotes to which O_2 was toxic; nonspecific oxygenases evolved as means of getting rid of O_2.

The development of two-photocenter photosynthesis in Cyanobacteria set the stage for eukaryote evolution. Some bacteria became incorporated as chloroplasts with the packaging of chlorophyll. Other kinds of symbionts became incorporated as mitochondria, in which the tricarboxylic acid cycle and electron transport agents such as flavoproteins and cytochromes gave much more complete use of substrates than was possible by glycolysis. Oxygen was then the proton acceptor. The evolution of the mitochondrial pathways started an evolutionary explosion that led to diverse plants and animals that occupied a wide range of ecological niches. Two modifications of CO_2 reduction were added to the primitive path to 3-carbon acids; each of these (C-4 and CAM paths) is adaptive to a particular type of environment.

More recently, animals have become adapted to continuing activity when in reduced oxygen. Adaptive pathways provide energy anaerobically in exercise, excrete glycolytic products rather than burn them, and use proteins and fats for energy under special metabolic conditions. The combined oxidation of amino acids together with glycolysis led to the production of end-products which now appear in facultatively anaerobic animals, parasites, and intertidal invertebrates. Anaerobic energy production in vertebrates yields principally lactic acid. Several facultatively anaerobic invertebrates make and may excrete lower fatty acids; other invertebrates produce imino acids. The products of anaerobiosis may be different in exercise hypoxia from the products of environmental hypoxia. The most energy-effective oxidative pathway is in insects; it consists of three processes—the α-ketoglutarate shunt, glycolysis, and the oxidative chain. Some 54 ATPs may be generated in contrast with the 36 by direct glucose oxidation. Cephalopods, like insects, are highly efficient oxidative animals.

In aerobic metabolism large quantities of protons are generated; this process could cause marked intracellular acidification. Anaerobic pathways utilized by animals in reduced oxygen yield various end products—lactate, alanopine, octopine, succinate, propionate. Each glycolytic pathway produces 2 mols of protons per mol of substrate fermented. There is normally a close balance between rate of H^+ production and (1) removal of protons by oxidative phosphorylation (ATP synthesis) and (2) reoxidation by reduced coenzymes and cytochromes.

The protons liberated are utilized in the synthesis of ATP in the absence of oxidation. ATP is hydrolyzed to ADP, a proton and P_i by ATPases in various energy-utilizing processes, for example, in muscle contraction. Increase of H^+ in a cell increases glycolysis and reduces ATPase activity. Thus production of protons in each mode of glycolysis is balanced by ATP hydrolysis, and intracellular pH is maintained relatively constant (33).

There are more diverse adaptations to oxygen and carbon dioxide than to other environmental factors. Morphological, physiological, and biochemical adaptations have allowed occupancy of ecological niches of different O_2 availability. Environments with P_{O_2} above 1 atm of air are rare (supersaturated ponds), but environments with reduced P_{O_2} are common—sluggish fresh water, contaminated waters, high altitudes, gut of vertebrates. The general patterns of metabolic pathways are similar in all aerobic organisms; they liberate energy in small packets in a stepwise fashion. Many enzymes of intermediary metabolism function in a coordinated sequence. The sequence, coded genetically, is modifiable by energy needs and environmental conditions. The integrated sequence is also adapted to the utilization of substrates—carbohydrates, proteins, fats—under changed conditions. Enzymes and products formed are characteristic of the organism or tissue—low energy requirements in some molluscs, high energy requirements in insects. Facultative anaerobiosis, temporally or environmentally induced, results from the fine-tuning of sequence of many enzymes.

Mechanisms evolved for providing energy when the atmosphere changed, new niches were occupied, and organisms adapted to new life styles made possible the diversity of life forms.

REFERENCES

1. Adolph, E. F. *Respir. Physiol.* **7**:356–368, 1969. Survival without oxygen in infant mammals.

2. Akerlund, G. *Compar. Biochem. Physiol.* **47A**:1065–1075, 1974. Oxygen consumption in snails.

3. Aulie, A. and H. J. Grav. *Compar. Biochem. Physiol.* **62A**:335–338, 1979. Oxidative capacity of muscle and liver of chickens.

4. Baldwin, J. and A. M. Opie. *Compar. Biochem. Physiol.* **61B**:85–92, 1978. Octopine dehydrogenase in bivalve adductors.

5. Bartholomew, G. A. and T. M. Casey. *J. Exper. Biol.* **76**:11–25, 1978. O_2 consumption of moths.

6. Bartholomew, G. A. and D. Vleck. *J. Exper. Biol.* **90**:17–32, 1981. $\dot{V}_{O_2}$ in warm-up of moths.

7. Bartholomew, G. A. and D. Vleck. *J. Compar. Physiol.* **B132**:285–288, 1979. O_2 consumption in iguana.

8. Bashamohideen, M. and H. Kunnemann. *Zool. Anz. Jena.* **202**:163–171, 1979. Energy-yielding reactions in fish Idus.

9. Bauer, C. et al. *J. Compar. Physiol.* **B136**:67–70, 1980. Modulation of Hb in llama and camel.

10. Bayne, B. L. and D. R. Livingstone. *J. Compar. Physiol.* **114**:129–142, 1977. Response of *Mytilus* to low oxygen.

11. Beis, J. et al. *Compar. Biochem. Physiol.* **67A**:303–305, 1980. O_2 consumption of polychaete worm.

12. Belman, B. W. and A. C. Giese. *Biol. Bull.* **146**:157–164, 1974. O_2 consumption by echinoderms.

12a. Belmer, R. T. and A. D. Strobusch. *J. Appl. Physiol.* **42**:571–577, 1977. Critical size of newborn homeotherms.

13. Bennett, A. F. et al. *J. Compar. Physiol.* **87**:351–360, 1973. *Am. J. Physiol.* **226**:1149–1151, 1974. *Compar. Biochem. Physiol.* **48A**:319–327, 1974. *Annu. Rev. Physiol.* **400**:447–469, 1978. *Science* **206**:649–654, 1979. Metabolism in relation to activity in amphibians, reptiles, and fishes.

14. Berry, J. and J. Dorinton. pp. 263–343 in *Photosynthesis,* vol. 2, Ed. Govindjee, Academic Press, New York, 1982. Environmental effects on photosynthesis.

15. Bjorkman, O. *Brittonia* **18**:214–224, 1966. Photosynthesis and respiration in goldenrod.

16. Black, E. C. et al. *J. Fish Res. Board Canada* **23**:1–13, 1966. O_2 transport by fish hemoglobins.

17. Bonaventura, J. et al. *Biochem. Biophys. Acta* **371**:147–154, 1974. Hb of *Torpedo.*

18. Bonaventura, J. et al. *J. Exper. Zool.* **194**:155–174, 1975. Structure and function of hemocyanin.

19. Bowen, S. T. et al. *Biol. Bull.* **155**:273–287, 1978. *Artemia* hemoglobin.

20. Brett, R. *Am. Zool.* **11**:99–113, 1971. Activity metabolism in salmon.

21. Brett, S. S. and G. Shelton. *J. Exper. Biol.* **80**:251–269, 1979. Ventilation in *Xenopus*.

22. Bridges, C. R. et al. *Compar. Biochem. Physiol.* **62A**:457–462, 1979. O₂ content of *Sepia* blood.

23. Brix, O. et al. *J. Compar. Physiol.* **129**:97–103, 1979. Effects of pH on O₂ transport in *Buccinium*.

24. Brown, W. I. and J. M. Schick. *Biol. Bull.* **156**:272–288, 1979. Gas exchange in holothurians.

25. Bryant, C. pp. 35–69 in *Advances in Parasitology,* vol. 13, Ed. B. Dawes, Academic, New York, 1975. pp. 89–94 in *Biochemistry of Parasites,* Ed. Van Den Bossche, 1976. Metabolism in parasitic worms. North Holland, New York.

26. Burggren, W. N. and G. Shelton. *J. Exper. Biol.* **82**:75–92, 1979. Gas exchange in turtles.

27. Burke, E. M. *Biol. Bull.* **156**:157–168, 1979. Metabolism of intertidal crabs.

28. Bursell, E. *Compar. Biochem. Physiol.* **52B**:235–238, 1975. Oxidative metabolism in muscle of flies.

29. Bushnell, P. G. *J. Compar. Physiol.* **147**:41–47, 1982. Arterial blood of lemon shark in exercise.

30. Calder, W. A. *Annu. Rev. Physiol.* 43:301–322, 1981. *Q. Rev. Biol.* **56**:1–16, 1981. Scaling of physiological processes in homeotherms. *Size, Functions, and Life History,* Harvard University Press, Cambridge, MA, 1984.

31. Cameron, J. *Compar. Biochem. Physiol.* **38A**:699–704, 1971. O₂ equilibrium curve of trout Hb.

32. Carey, C. et al. *J. Compar. Physiol.* **125**:101–113, 1978. Seasonal acclimatization in finches.

33. Castellini, I. and G. Somero. *J. Compar. Physiol.* **143**:191–198, 1981. Buffering by vertebrate muscle.

34. Chefurka, W. et al. *Compar. Biochem. Physiol.* **37**:143–165, 1970. Pentose cycle in insects.

35. Christiansen, J. and D. Penney. *J. Compar. Physiol.* **87**:237–245, 1973. Anaerobic metabolism in frogs.

36. Collicutt, J. M. and P. W. Hochachka. *J. Compar. Physiol.* **115**:147–157, 1977. Anaerobic metabolism of oyster heart.

37. Coulson, R. A. and T. Hernandez. *Compar. Biochem. Physiol.* **65A**:453–457, 1980; **69A**:1–13, 1981. O₂ debt in reptiles; comparisons with mammals.

38. de Fur, P. L. and C. P. Mangum. *Compar. Biochem. Physiol.* **62A**:283–294, 1979. Heart rates of invertebrates in relation to O₂ consumption.

39. Dejours, P. *Principles of Comparative Respiratory Physiology,* Elsevier, New York, 1981.

40. de Zwaan, A. et al. *Biochem. Soc. Symp.* **41**:133–168, 1976. *J. Compar. Physiol.* **B137**:97–104, 105–114, 1980. Enzymes in anaerobic metabolism in molluscs.

41. Dykens, J. A. and M. Schick. *Nature* **297**:579–580, 1982. O₂ production by endosymbiotic algae in sea anemones.

42. Ellington, W. R. and C. S. Hammen. *J. Compar. Physiol.* **122:**347–358, 1977. Metabolic compensation to low oxygen in sea cucumber.

43. Emilio, M. G. and G. Shelton. *J. Exper. Biol.* **85:**253–262, 1980. Gas exchange in *Xenopus.*

44. Fairbairn, D. *Biol. Rev.* **45:**29–72, 1970. Biochemical adaptation in parasites.

45. Feder, M. *Compar. Biochem. Physiol.* **70A:**497–508, 1981. *J. Exper. Zool.* **220:**33–42, 1982. O_2 consumption by anuran larvae.

46. Felbeck, H. *J. Compar. Physiol.* **137:**183–192, 1980. Role of amino acids in anaerobic metabolism of *Arenicola.*

47. Felbeck H. and G. Somero. *Thermobiology* **37:**201–204, 1982. Sulfur oxidizing bacteria of hydrothermal vents.

48. Felder, D. L. *Biol. Bull.* **157:**125–139, 1979. Respiration in mud shrimp.

49. Fields, J. H. A. et al. *Canad. J. Zool.* **54:**871–878, 1976. *Archiv. Biochem. Biophys.* **201:**110–114, 1980. *J. Thermal Biol.* **88:**35–45, 1981. Enzyme pathways in anaerobic metabolism of marine invertebrates.

50. Flavin, M. et al. *Compar. Biochem. Physiol.* **61B:**533–537, 1978. Hb in metamorphosing salamander.

51. Freadman, M. A. *J. Exper. Biol.* **83:**217–230, 1979. Swimming energetics of fishes. U. E. Fyhn et al. *Compar. Biochem. Physiol.* **62A:**39–66, 1979. Hb in Amazon fishes.

52. Gade, G. *J. Compar. Physiol.* **137:**177–182, 1980. Energy metabolism of foot muscle of Cardium.

53. Garlick, R. L. et al. *Compar. Biochem. Physiol.* **62A:**219–226, 1979. Hb in air-breathing catfish.

54. Gilbert, D. L. *Respir. Physiol.* **5:**68–77, 1968. Ch. 5, pp. 73–101 in *Oxygen and Living Processes,* Ed. D. L. Gilbert, Springer, New York, 1981. Oxygen in life.

55. Giles, M. A. and D. J. Randall. *Compar. Biochem. Physiol.* **65A:**265–271, 1980. Hb in developing salmon.

57. Gillen, R. G. and A. Riggs. *J. Biol. Chem.* **248:**1961–1969, 1973. Structure and function of Hb of *Anguilla.*

58. Graham, J. B. and T. A. Baird. *J. Exper. Biol.* **96:**53–67, 1982. Air breathing fishes.

59. Greaney, G. S. and D. A. Powers. *J. Exper. Zool.* **203:**339–350, 1978. Allosteric modifiers of fish Hb.

60. Herberg, G. H. Ch. 4, pp. 65–72 in *Oxygen and Living Processes,* Ed. D. L. Gilbert, Springer, New York, 1981. Origins and history of terrestrial O_2.

61. Hernandez, T. and R. A. Coulson. *Compar. Biochem. Physiol.* **67A:**283–286, 1980. Anaerobic metabolism in alligator.

62. Hochachka, P. W. et al. *Science* **178:**1056–1060, 1972. *Am. Zool.* **13:**543–555, 1973. *Compar. Biochem. Physiol.* **50B:**17–22, 1975. *Biochem. Soc. Symp.* **41:**3–31, 1976. Metabolic pathways of anaerobiosis in invertebrates and diving mammals.

63. Hochachka, P. W. and G. N. Somero. *Strategies of Biochemical Adaptation,*

Saunders, Philadelphia, 1973. Ch. 2, Oxygen, Ch. 3, Carbon dioxide, in *Biochemical Adaptation,* Princeton University Press, Princeton, NJ, 1984.

64. Hochachka, P. W. *Living without Oxygen,* Harvard University Press, Cambridge, MA 1980.

65. Hoffmann, R. J. and C. P. Mangum. *Compar. Biochem. Physiol.* **36:**211:228, 1970. Hb of *Glycera.*

66. Hoshi, T. et al. *Sci. Rep. Niigata Univ.* **10:**79–86, 1973. Synthesis of Hb in *Daphnia.*

67. Imamura, T. et al. *J. Biol. Chem.* **247:**2785–2797, 1982. Amino acid sequences in Hb of *Glycera.*

68. Jackson, D. C. and B. A. Braun. *J. Compar. Physiol.* **129:**339–342, 1979. Respiratory control in frogs.

69. Johansen, K. et al. *Canad. J. Zool.* **56:**898–906, 1978. Respiratory properties of blood of Amazon fishes.

70. Jutsum, A. R. and G. J. Goldsworthy. *J. Insect Physiol.* **22:**243–249, 1976. Fuel for flight in *Locusta.*

71. Kammer, A. E. and B. Heinrich. *J. Exper. Biol.* **61:**219–227, 1974. Metabolic rates in active bumblebee.

72. Kay, F. R. *Compar. Biochem. Physiol.* **57A:**309–316, 1977. Comparative properties of mammalian blood.

73. Khan, M. A. and C. A. de Kort. *Compar. Biochem. Physiol.* **60B:**407–411, 1978. Proline as substrate in insect muscle.

74. Kinney, M. and F. White. *Respir. Physiol.* **31:**309–325, 1977. O_2 consumption in turtles.

75. Kooyman, G. L. et al. *J. Compar. Physiol.* **138:**355–346, 1980. Metabolism in voluntary diving in seals.

76. Lehninger, A. L. *Principles of Biochemistry,* Worth, New York, 1982.

77. Livingstone, D. R. and B. L. Bayne. *Compar. Biochem. Physiol.* **48B:**481–497, 1974. Enzymes from mantle of *Mytilus.*

78. Lyman, C. and R. C. O'Brien. *Ann. Acad. Sci. Fenn. A. IV* **71/72:**213–320, 1964. Effects of autonomic drugs on hibernating ground squirrels.

79. Mangum, C. P. *Am. Zool.* **13:**529–541, 1973. *J. Exper. Mar. Biol. Ecol.* **27:**125–140, 1977. Oxygen consumption by invertebrates in different oxygen environments.

80. Mangum, C. P. and G. Polites. *Biol. Bull,* **158:**72–90, 118–128, 1980. O_2 transport in *Busycon.*

81. Markl, J. et al. *J. Compar. Physiol.* **133:**167–175, 1979. Crustacean hemocyanin.

82. McMahon, B. R. et al. *J. Exper. Biol.* **60:**195–206, 1974. *J. Compar. Physiol.* **128:**109–116, 1978. *J. Exper. Biol.* **80:**271–285, 1979. O_2 consumption and ventilation in crustaceans.

83. Meints, R. H. and C. Forehand. *Compar. Biochem. Physiol.* **58A:**265–268, 1977. Hb switching in frogs.

84. Miller, K. I. and K. Van Holde. *J. Compar. Physiol.* **115:**171–184, 1977; **143:**261–267, 1981. Structure and function of crustacean hemocyanin.

85.   Moon, T. W. et al. *Compar. Biochem. Physiol.* **56B**:240–254, 1977. *J. Exper. Zool.* **200**:325–336, 1977. Metabolism of a parasitic worm.

86.   Mooney, H. A. and W. D. Billings. *Ecol. Monographs* **31**:1–28, 1961. Ecology of alpine and arctic plants.

87.   Moore, P. D. *Nature* **280**:193–194, 1979; **304**:310, 1983. Ecology of photosynthesis.

88.   Mustafa, T. and P. W. Hochachka. *Compar. Biochem. Physiol.* **45B**:639–655, 656–667, 1973. Enzymes of glycolysis in molluscs.

89.   Nicchitta, C. V. and W. R. Ellington. *Biol. Bull.* **165**:708–722, 1983. Metabolism of high intertidal and subtidal molluscs.

90.   Ogren, W. L. and R. Chollet. Ch. 7, pp. 191–230 in *Photosynthesis,* vol. 2, Ed. Govindjee, Academic Press, New York, 1982. Photorespiration in plants.

90a. Otten, L., et al. *Biochem. Biophys.* **A485**:268–277, 1977. Amino acids from crown gall tumors.

91.   Petersen, J. A. et al. *J. Exper. Biol.* **61**:309–320, 1974. Respiration of a barnacle.

91a. Plaxton, W. C. and K. B. Storey. *European J. Biochem.* **143**:257–265, 1984. Aerobic and anaerobic forms of PK from red muscle of whelk.

92.   Portner, H. et al. *J. Compar. Physiol.* **133**:227Z–231, 1979. Recovery from anaerobiosis in *Arenicola*.

93.   Powers, D. A. et al. *J. Biol. Chem.* **247**:6686–6693, 6694–6707, 1972. *Science* **177**:360–362, 1972. *Am. Zool.* **20**:139–162, 1980. *Ann. New York Acad. Sci.* **241**:472–490, 1974. Hb of fish from fast and slow water.

94.   Powers, D. A. et al. *J. Biol. Chem.* **253**:3521–3528, 1978. *Nature* **277**:240–241, 1979. Hb function in Fundulus.

95.   Powers, D. A. et al. *Compar. Biochem. Physiol.* **62A**:67–85, 1978. Oxygen equilibrium of blood from Amazon fish.

96.   Prichard, R. K. and P. J. Schofield. *Compar. Biochem. Physiol.* **24**:697–710, 1968. Glycolytic pathway in liver fluke.

96a. Prosser, C. L. Ch. 5 in *Comparative Animal Physiology,* Ed. C. L. Prosser, Saunders, Philadelphia, 1973.

97.   Ralph, R. and I. Everson. *Compar. Biochem. Physiol.* **27**:299–307, 1968. Metabolism of Hb-less Antarctic fish.

98.   Reischl, E. *Compar. Biochem. Physiol.* **58A**:217–221, 1977. O_2 equilibrium of Hb from catfish.

99.   Riggs, A. et al. *Fed. Proc.* **35**:2115–2118, 1976. *J. Biochem.* **247**:6039–6046, 1972. *Compar. Biochem. Physiol.* **60B**:189–193, 1978. *Compar. Biochem. Physiol.* **62A**:257–271, 1979. Comparative properties of fish Hb.

100. Roberts, J. *Biol. Bull.* **148**:85–105, 1975. Rhythmic branchial and ram gill ventilation in fishes.

101. Sacktor, B. *Adv. Insect Physiol.* **7**:267–347, 1970. *Biochem. Soc. Symp.* **41**:111–131, 1976. Biochemical adaptations for flight in insects.

102. Sassaman, C. and C. P. Mangum. *Biol. Bull.* **143**:657–678, 1972. Adaptations of sea anemones to oxygen levels.

103. Saz, H. J. *Am. Zool.* **11**:125–135, 1971. Facultative anaerobiosis in parasites.

104.   Schmidt-Nielsen, K. *Science* **177**:222–228, 1972. Energy for locomotion.

105.   Schmidt-Nielsen, K. *Animal Physiology,* 3rd ed., Cambridge University Press, New York, 1983.

106.   Schroff, G. and U. Schottler. *J. Compar. Physiol.* **116**:325–336, 1977; **138**:35–41, 1980. Anaerobic metabolism in *Arenicola.*

107.   Seymour, R. S. *Copeia* 103–115, 1973. *J. Compar. Physi.* **144**:215–227, 1984. Metabolism of toad *Scaphiopus.*

108.   Seymour, R. S. et al. *Planta* **157**:336–343, 1983. Respiration and heat production by flowers of arum lily.

109.   Singh, B. N. and G. M. Hughes. *J. Exper. Biol.* **65**:421–434, 1971. Respiration of air breathing fish.

110.   Smatresk, N. J. and J. N. Cameron. *J. Exper. Biol.* **96**:263–280, 1982. Respiration and acid-base balance in garfish.

111.   Snyder, G. K. *J. Appl. Physiol.* **42**:673–678, 679–681, 1977. *Science* **195**:412–413, 1977. Large Hb of invertebrates.

111a. Solomon, T. *Annu. Rev. Plant Physiol.* **28**:279–297, 1977. Cyanide resistant respiration in plants.

112.   Steers, E. and R. H. Davis. *Compar. Biochem. Physiol.* **62B**:393–402, 1979. Myoglobins of *Paramecium.*

113.   Storey, K. B. et al. *J. Compar. Physiol.* **131**:311–319, 1979. *Compar. Biochem. Physiol.* **73**:521–528, 1982. Octopine pathways in marine invertebrates.

113a. Storey, K. B. and J. M. Storey. Ch. 3, pp. 92–136 in *The Mollusca,* I. Academic Press, New York, 1983. Carbohydrate metabolism in cephalopods.

114.   Taylor, C. R. pp. 161–170 in *A Companion to Animal Physiology,* Ed. C. R. Taylor et al., Cambridge University Press, New York, 1982. Scaling of metabolism to body size.

115.   Taylor, E. W. and P. J. Butler. *J. Compar. Physiol.* **127**:315–323, 1978. Aquatic and aerial respiration in *Carcinus. J. Exper. Biol.* **93**:197–208, 1981.

116.   Terwilliger, R. C. et al. *Compar. Biochem. Physiol.* **55A**:51–55, 1976; **57A**:143–149, 1977; **61B**:463–469, 1978; **59A**:359–362, 1978. Respiratory pigments in polychaete worms.

117.   Ting, I. P. pp. 99–109 in *Physiological Adaptation to the Environment,* Ed. F. J. Vernberg, Intext, New York, 1975. Patterns of photosynthesis.

118.   Torre-Bueno, J. R. and J. Larochelle. *J. Exper. Biol.* **75**:223–229, 1978. Metabolic cost of flight in birds.

119.   Trayburn, P. et al. *Nature* **298**:59–60, 1982. Heat production by brown adipose tissue.

120.   Turney, L. D. and V. H. Hutchison. *Compar. Biochem. Physiol.* **49A**:583–601, 1974. Metabolism of *Rana pipiens.*

121.   Ultsch, G. R. et al. *J. Compar. Physiol.* **142**:439–443, 1981. Metabolism of toadfish.

122.   Van den Branden, C. et al. *Compar. Biochem. Physiol.* **60A**:185–187, 1978. Hb of *Artemia.*

123.   Vinogradov, S. N. et al. *Compar. Biochem. Physiol.* **67B**:1–16, 1980. Annelid Hb.

124. Vos, J. et al. *Compar. Biochem. Physiol.* **62A:**545–548, 1979. Respiration of adult *Artemia* acclimated to different O_2 concentrations.

125. Walsh, P. *J. Compar. Physiol.* **143:**213–222, 1981. Allozymes of octopine dehydrogenase.

126. Wares, W. D. and R. Igram. *Compar. Biochem. Physiol.* **62A:**351–356, 1979. O_2 consumption of minnow *Pimephales*.

127. Weber, R. E. et al. *Netherlands J. Sea Res.* **7:**316–327, 1973. *Compar. Biochem. Physiol.* **53B:**23–30, 1976; **57A:**151–155, 1977. *J. Compar. Physiol.* **123:**177–184, 1978. Hb of *Glycera. Nature* **292:**384–387, 1981.

128. Weber, R. E. *Am. Zool.* **20:**79–101, 1980. Functions of invertebrate hemoglobins in environmental hypoxia.

129. Weber, R. E. et al. *Compar. Biochem. Physiol.* **68A:**159–165, 1981. Myoglobin from a burrowing reptile.

130. Weber, R. E. and G. Lykkeboe. *J. Compar. Physiol.* **128:**127–137, 1976. Respiratory adaptations of carp blood.

131. Weibel, E. R. pp. 31–48 in *A Companion to Animal Physiology,* Ed. C. R. Taylor, Cambridge University Press, New York, 1982. Scaling of metabolism.

132. Widdows, J. et al. *Compar. Biochem. Physiol.* **62A:**301–308, 1979. O_2 consumption of molluscs in air.

133. Wilkes, P. R. H. and B. R. McMahon. *J. Exper. Biol.* **98:**119–137, 1982. Hypoxic responses in crayfish.

134. Wilps, H. and E. Zebe. *J. Compar. Physiol.* **112:**263–272, 1976. Metabolism in larvae of *Chironomus*.

135. Wittenberg, J. B. and B. A. Wittenberg. *Biol. Bull.* **146:**116–136, 1974. Oxygen secretion in eye and swimbladder of fishes. *J. Mar. Biol. Assoc. UK.* **10:**263–276, 1980. *Biol. Bull.* **161:**426–439, 1981.

136. Wittenberg, J. B. et al. *J. Biol. Chem.* **250:**9038–9043, 1975. Role of myoglobin in O_2 supply to pigeon breast muscle.

137. Wyatt, G. R. *Adv. Insect Physiol.* **4:**287–360, 1967. Enzyme pathways in insects.

138. Zammit V. A. and E. A. Newsholme. B. *Biochem. J.* **184:**313–322, 1979. Metabolites of fish and mammal hearts.

139. Zebe, E. *J. Compar. Physiol.* **101:**133–145, 1975. Glucose metabolism in *Arenicola*.

7

TEMPERATURE

ECOLOGICAL CORRELATES WITH PHYSIOLOGY

Temperature is a measure of the kinetic activity of atoms. The temperature range within which life occurs is determined by the thermal properties of water and solutions. In hot springs thermophilic bacteria and algae live at temperatures as high as $+70°C$. Living bacteria have been reported in deep oceanic thermal vents at $>250°C$ under high pressures (Ch. 8). Thermophilic ostracods from hot springs are probably the most heat-tolerant of metazoans; they go into coma at $49.5°C$ and at $50.4°C$ show a 6-minute LD_{50} (lethal dose that kills 50 percent of a group of animals) (171). In hot deserts the temperature of plants may reach $45-50°C$. Desert animals protect themselves against the full heat of midday by niche selection; some insects (cicadas), however, are active at a body temperature of $45°C$. The lowest temperatures for active aquatic life are around ice floes at temperatures near the freezing point of seawater ($-1.86°C$). Polar oceans have a relatively constant temperature a few degrees above $0°C$ at all depths below the thermocline (a layer separating surface water of variable temperature from deep water of relatively constant temperature). Some arctic terrestrial plants withstand freezing; a few invertebrates (subarctic insects and intertidal molluscs and barnacles) tolerate freezing of their tissue fluids. Antifreeze solutes in body fluids of many insects and polar fishes permit some physiological activity at several degrees below freezing. When rapidly cooled in liquid nitrogen, some cells and small animals can enter a vitreous state from which they can be revived by rapid thawing; if freeing or thawing is slow, ice crystals form and cells are damaged.

Temperature conformers are poikilotherms in which T_b, body temperature, equals T_a, ambient temperature; temperature regulators are homeotherms, in which T_b is relatively constant in a varying T_a. Some poikilotherms, such as basking insects and reptiles, utilize external heat for raising body temperature (ectothermy); some poikilotherms, such as flying or singing insects and large swimming fish, use internally produced heat (endothermy) to reach the temperature necessary for activity. Poikilothermy and homeothermy refer to variability or constancy of internal temperature; ectothermy and endothermy refer to the major heat source, external or internal. Intermediate states occur. There are some ectothermic homeotherms, such as deep-sea and Antarctic fish, which have very stable T_b because they live in constant T_a. In some homeotherms T_b declines to near T_a in nocturnal dormancy or hibernation; these animals become temporarily poikilothermic. Hibernating mammals may conform to a cooling environment but regulate T_b when T_a approaches freezing. Heterothermy refers to the state when there are body-regional differences in temperature, as between skin and liver, or temporal differences, as in dormancy and hibernation alternating with activity.

Two general ways in which the temperature of an animal is regulated are behavioral and metabolic. Most animals regulate behaviorly by aggregating at "preferred" temperatures; behavioral regulation of ciliates or vertebrates consists of a coordinated sequence of sensory, locomotor, and integrative actions. Metabolic regulation may be controlled by the same receptors and integrators as behavioral regulation; the metabolic effector system modulates heat production and controls heat loss by circulatory, insulative, and evaporative means.

The T_b of most mammals is 37–39°C; for birds T_b is slightly higher (38–40°C), and for primitive mammals somewhat lower (32–36°C). Active insects and large fishes tend to regulate the temperature of critical organs to the mid-30s. Why this temperature range has been selected so generally is not clear. One hypothesis is that it represents an economical balance between heat loss and heat production. Another hypothesis is that at this temperature range $\Delta G\ddagger$ is minimal and contributions of entropy and enthalpy to metabolic reactions are equal (98). Entropy and enthalpy compensation plots for metabolic enzymes intersect at about 40°C. Above 45°C many proteins show changes in conformation resulting in denaturation. Male gametogenesis is limited to a narrower temperature range than is tolerated by other tissues.

The rise of mammals in the late Cretaceous and early Tertiary occurred at the time of decline of dominant reptiles, the dinosaurs. Numerous hypotheses for the transition have been proposed, most of them relating to temperature tolerance during periods of climatic change in mid-latitudes. One hypothesis is that ectothermic reptiles flourished during a long period of relative warmth from late Permian to early Cretaceous. Birds and mammals, because of insulating feathers and hair, were able to maintain body temperature endothermally when climates became colder (52a). A contrary view regarding decline of reptiles was given in Ch. 4.

Naturalists have correlated distribution of plants and animals with temperature. The most apparent ecological correlations with temperature are latitudinal—tropic, temperate, and polar zones. Regional temperature gradients may delineate clines.

RESISTANCE ADAPTATION; EFFECTS OF EXTREME HEAT OR COLD

Several physiological measures of the direct effects of temperature have been correlated with ecology—critical thermal maxima (CT_{ma}), critical thermal minima (CT_{mi}), coma, death, rate of development, and behavior. These characters are to a large extent genetically determined. Plants and animals from polar regions are much more sensitive to warming than those from temperate regions, while species from the tropics are more sensitive to chilling. Antarctic fish living usually at 0°C die of heat at 5°C (probably by central nervous failure). *Pandalus* (shrimp) from Sweden at 5–7°C die if warmed to 11°C, while *Pandalus* from Plymouth from 15°C water survive at above 17°C. Jellyfish *Aurelia* from Halifax has a CT_{ma} of 29–30°C; jellyfish from Florida Tortugas 38.5°C. A tolerance polygon (Fig. 7-1) encloses the high and low thermal limits over an acclimation range, which distinguishes qualitatively between genetic and environmental determinants (Ch. 1). The tolerance polygons for a fish *Notropus* were different for populations from Ontario and from Tennessee; this indicates two genetically different populations (53a). Animals from low intertidal regions of oceans are less heat-tolerant than

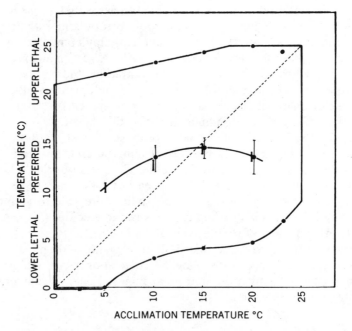

Figure 7-1. Thermal tolerance and preferred temperature of young sockeye salmon in relation to acclimation temperature. Reprinted with permission from J. R. Brett, *Am. Zool.*, Vol. 11, p. 101. Copyright 1971, American Society of Zoologists, Thousand Oaks, CA.

those from the high intertidal. Related species of snails tolerated heating in the order: high littoral > high intertidal > low intertidal > subtidal (52).

Behavior correlates with temperature tolerance. Diurnal honey ants from the desert are active over the temperature range 13–46°C and have CT_{min} of 11.8°C; nocturnal ants are active at 2–30°C and have CT_{min} of 0.4°C (83). A burrowing lizard *Sceloporus* maintains body temperature in the narrow range 32–33°C; a basking lizard is active over the range 26–36°C (51).

Adaptations for life in the desert are similar in cacti and insects: (1) cellular tolerance of high temperatures, (2) decrease of thermal load by orientation and heat reflection, (3) opening and closing of apertures for gas exchange, (4) lipid or waxy covering that reduces water loss, (5) thermolability (heterothermy), (6) absorption and utilization of moisture from air and substrate, (7) utilization of metabolic water, and (8) tolerance of cell and tissue desiccation (51).

Thermal sensitivity may change with the stage in embryonic development. Tadpoles of *Bufo woodhousii* are more sensitive to heat (CT_{ma} 37°C) during metamorphosis than in premetamorphic stages (42.5°C) or as frogs (41.1°C) (149). Development of embryos of many kinds of poikilotherms, such as crustaceans, occurs in a more limited temperature range than survival of adults (163).

Fishes show considerable metabolic acclimation, while anurans show very little. Resistance adaptation can occur without capacity adaptation. Temperate zone anurans show limited capacity acclimation, while tropical anurans and urodeles show no acclimation or inverse acclimation, presumably because their body temperature is already optimal for activity (42). A snake (*Natrix*) from a cool climate (England) has twice the standard metabolism of a *Natrix* species from a warm climate (Spain) when both are measured at an intermediate temperature (32). In 22 species of lizards in the West Indies, behavioral signs of heat intolerance were greater in cooler habitats, from sea level to mountain altitude (68). Metabolic acclimation varies with habitat and season.

SEASONAL AND HABITAT EFFECTS ON TEMPERATURE TOLERANCE

Seasonal effects on temperature tolerance and metabolism vary according to life style. Some poikilotherms that remain active over a wide temperature range show positive metabolic compensation, or acclimatization. Such eurythermal species may also compensate by continuing neural and behavioral activity at lower temperatures in winter than in summer. Examples of such compensations are freshwater fishes, such as sunfish and bass. In species that become dormant in winter no compensation or inverse acclimatization occurs, so that metabolism is lower for cold-acclimatized than for warm-acclimatized individuals. Thus energy reserves are conserved during periods of inactivity. Examples are lizards that have short breeding periods and long periods of dormancy. Some fish species show a positive acclimatization over a mid-temperature range and show negative acclimatization in a lower temperature range. Tropical lungless salamanders from very constant environmental temperature failed to thermoregulate behaviorly under natural conditions; high altitude species from a cycling environment selected low temperature (42). In a turtle, metabolism and aerobic scope are reduced by cold acclimation; this is inverse acclimation (48).

Oxygen consumption by wolf spiders is maximum at a lower temperature in January than in June (116). Heart rate in *Rana pipiens* shows a steeper dependence on temperature (Q_{10} of 4.5) in summer than in winter (Q_{10} of 1.6–2.9) (113). Temperature for heat coma in a snail *Physa* is 8.4°C higher in summer than in winter (112). The LD_{50} of a desert pupfish *Cyprinodon* in spring (temperature of water 15°C) was 39°C; in summer (water at 42°C) it was 43.5° C (16). Metabolism of medusae in winter was insensitive to low temperature, but in summer the metabolism was much reduced by cold (103). Frogs in Finland showed effects of laboratory acclimation on $\dot{V}_{O_2}$ and on LDH in winter, but not in summer (88). Active metabolism of some salmonid fish is maximum at the same temperature as the temperature for highest speed of swimming; this is the same as the preferred temperature in a gradient (Fig. 7-2) (13).

Activity of tissues correlates with habitat. From high intertidal to deep water the temperature for heat block of ciliary movement in bivalves after 10 minutes of heating correlated with the habitat: *Crassostrea* 44.7°C > *Mytilus* (intertidal) 40.5°C > *Arca* 37.1°C > *Modiolus* 36.4°C > *Pecten* (deep water) 32.2°C.

Photoperiod may modify the effects of temperature. Metabolism of sunfish shows a greater degree of compensation for cold on long days than on short days (144). The time to death of 50 percent of a population on exposure to heating at 33.7°C increased for Atlantic menhaden with long daily light period (69):

8L:16D	16L:8D	24L
52 min	79 min	96.6 min

METABOLIC CORRELATES OF GEOGRAPHIC DISTRIBUTION

Metabolism of poikilotherms compensates for temperature such that animals from cold temperate regions have higher metabolic rates than similar animals from warm regions (metabolism measured at the same intermediate temperature). Functions such as heart rate and metabolic rate correlate with geographic distribution. The heart rate of Alaska *Mytilus* is greater than that of California *Mytilus,* both measured at an intermediate temperature (135).

Metabolism of polar aquatic animals is low. Growth of antarctic marine organisms is slow, energy flow low, longevity high (25a). Measurements of

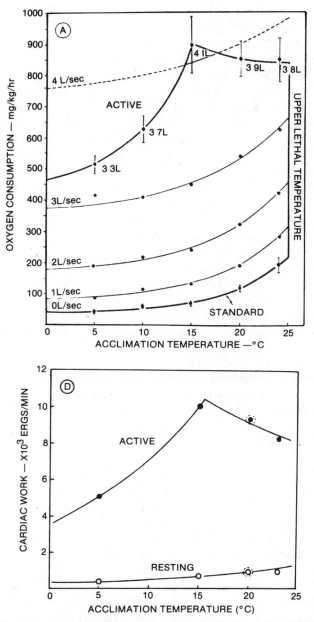

Figure 7-2. Oxygen consumption and cardiac work of salmon acclimated to different temperatures and measured at temperature of acclimation. Metabolism measured at increasing rates of swimming from resting (0 lengths/second) to maximum activity (4 lengths/second). Reprinted with permission from J. R. Brett, *Am. Zool.,* Vol. 11, p. 104. Copyright 1971, American Society of Zoologists, Thousand Oaks, CA.

O_2 consumption cannot be interpreted as standard metabolism unless allowance is made for activity. Comparison of animal species of different life styles and levels of activity are of limited validity. Complete compensation for temperature effects is rare (25a,72).

A few examples of genetically based geographic variation follow. A Cape Cod population of a polychaete *Diopatra* shows higher metabolism at all temperatures of measurement than a North Carolina *Diopatra* population; the differences indicate genetically different populations, since no change occurs during laboratory acclimation (103). Antarctic mites have standard metabolism 2–4 times greater than temperate zone mites (10). Activity metabolism correlates more closely with life habit than does rest or standard metabolism. For example, the Q_{10} for activity metabolism of tropical toad *Bufo marinus* is 1.0; Q_{10} for temperate-zone frog *Rana* is 2.4. The aerobic scope (difference between active and rest metabolism) is greater for *Bufo,* but the anaerobic scope is greater for *Rana,* which is capable of faster movements (17a).

Low arctic temperatures cause slowing of photosynthesis, decreased uptake of water and ions, lowered transpiration and growth; plants compensate for arctic and alpine conditions with short stature, low temperature optima of enzyme reactions, more growth at low temperatures, and utilization of the long day for photosynthesis. Antarctic lichens show maximum photosynthesis at 0°C (115). Sorrel plants (*Oxyria*) observed from Colorado to Point Barrow, Alaska, showed increases in flower and rhizome production farther north; northern populations photosynthesized more at low temperatures and had higher rates of respiration than southern populations at the same temperature; southern populations tolerated warm summer nights better. When both northern and southern populations were grown at 20°C, the differences persisted, hence may be genetic (118).

BEHAVIORAL AND METABOLIC REGULATION OF BODY TEMPERATURE

Neural responses may participate in capacity adaptation by bringing animals into a "preferred" temperature that is usually near the temperature for maximum energy output. In temperature gradients many aquatic crustaceans and fishes (also terrestrial lizards and insects) aggregate within narrow temperature ranges that may correspond to their prior thermal regimes. Behavioral temperature regulation is mediated by the nervous system. After lesions have been made in the anterior hypothalamus of a sunfish or goldfish, the fish no longer selects an optimum temperature. The same region of the brain (preoptic hypothalamus) that functions in ectothermic vertebrates (fish) for behavioral temperature regulation also functions in metabolic temperature regulation in endothermic vertebrates (man). When fish are left in a thermal gradient for some hours, they change location so that both cold- and warm-acclimated fish settle in the same temperature region (29); initially selected temperature is more affected by acclimation than is final preferendum.

In most animals both peripheral and central receptors for temperature initiate behavioral thermoregulation. All neurons change their activity with warming or cooling; passive changes may not be thermoregulatory. Thermoregulating centers show "set points." Some cold receptors (usually peripheral) increase discharge on cooling; warmth receptors increase discharge on warming. There is convergence from central and peripheral receptors onto integrative neurons. Outputs from a temperature regulating center go to motor centers (behavioral regulation) or to autonomic centers (metabolic and circulatory regulation) (120a).

Under natural conditions, behavioral responses bring body temperature to the level appropriate for metabolism. Sockeye salmon and zooplankton go up and down in diurnal migration (Fig. 7-3). Some ectotherms (perching insects, basking reptiles) orient in such a way as to receive solar warming, then seek and take food within a critical range of body temperature. Iguanas regulate body temperature by diurnal basking and nocturnal aggregating; their set point is 35.5°C (167). Adult tautogs along the New Jersey coast

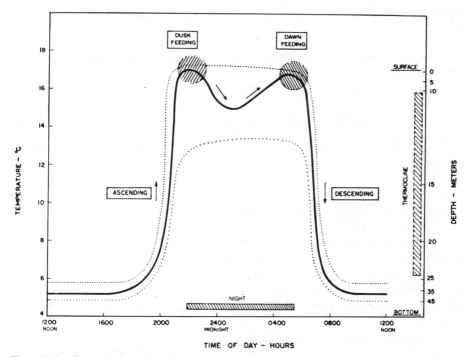

Figure 7-3. Pattern of vertical migration of sockeye salmon in midsummer. Feeding occurs near the surface at dawn and at dusk. Reprinted with permission from J. R. Brett, *Am. Zool.*, Vol. 11, p. 104. Copyright 1971, American Society of Zoologists, Thousand Oaks, CA.

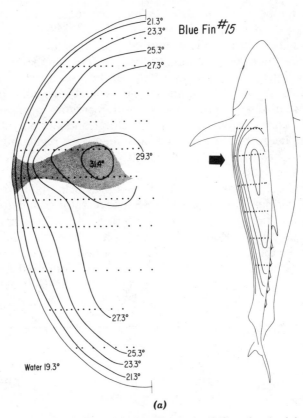

Water 19.3°

(a)

Figure 7-4. Temperature profiles of (*a*) bluefin tuna and (*b*) mako shark in water of 19° and 21°C. Isotherms show that the warmest parts of the fish are in the dark muscle on each side of the body. Reprinted with permission from F. G. Carey and J. M. Teal, *Comparative Biochemistry and Physiology,* Vol. 28, pp. 200, 206. Copyright 1969, Pergamon Press, Oxford, UK.

migrate offshore in the fall and return in the spring; migration is dependent on temperature, not on photoperiod (123).

Large marine fish, such as tuna and lamnid sharks, are fast long-distance swimmers that maintain muscle and visceral temperatures several degrees warmer than ambient water temperatures. These fish have countercurrent heat exchangers in their circulatory systems that conserve the heat produced by muscles or by heat-generating tissue at the base of the brain. (Fig. 7-4) (18).

Endothermic insects—butterflies, moths, bees, beetles—produce heat by oscillating wing muscles prior to flying. Heat production by insect muscle raises the temperature of the nervous system and muscles to the level for flight. A scarab beetle that has been cooled to 20°C increases its muscle metabolism and raises body temperature internally without locomotion

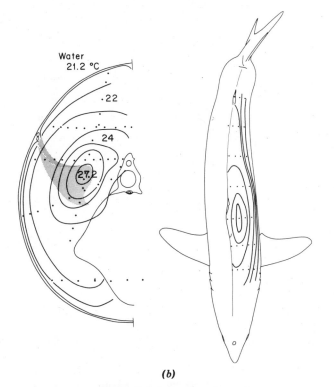

(b)

Figure 7-4. *Continued*

(119). Beetle larvae have both cold and warmth receptors on sensilla of antennae. By convergence of inputs from central and peripheral receptors onto interneurons the larva can distinguish a temperature difference of 0.7°C (96). A moth *Hyalophora* warms endothermically 4°C/min and maintains body temperature (T_b) of 32–36°C over an ambient temperature (T_a) range of 7–29°C. During warmup the moth's $\dot{V}_{O_2}$ increases by 2.3 times; body fat is the primary fuel. Flight is possible only if the nerve cord is at a critical temperature and if the thoracic ganglia are heated by a thermode, flight is initiated without warmup contractions (53,60). The moth *Manduca* flies only after its thoracic temperature reaches 38°C; during flight T_{thorax} is maintained at about 42°C (82a). Honeybees and termites maintain relatively constant nest temperature throughout the year (Fig. 7-5). In winter they generate heat by muscle activity and retain heat by whole-body clustering; in summer they cool the nest by bringing in water and evaporating by fanning. An active swarm of honeybees regulates temperature of the core of the swarm to 35 ± 1°C; when a swarm is inactive the regulated temperature falls below 35°C. The outer layer of bees of the swarm is held at above 17°C even at air temperature below 5°C; the upper set point is 36°C, the lower set point 17°C (64). In cicadas the range of body temperature is narrow for complex behavior, wide

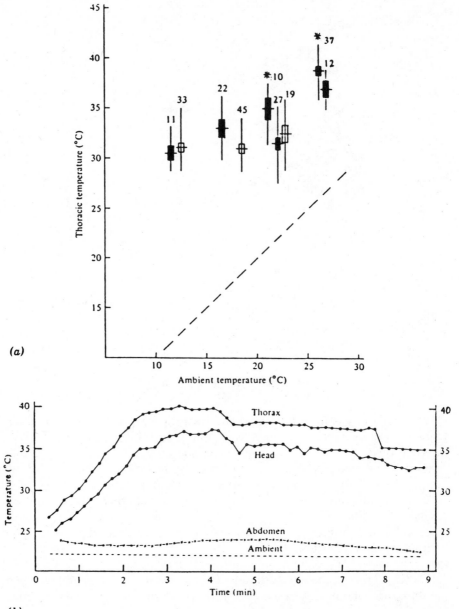

(a)

(b)

Figure 7-5. (*a*) Thoracic temperature of foraging honeybees at different ambient temperatures. (*b*) Thoracic, head, and abdominal temperatures of a bee during endothermic warmup and stabilization at constant ambient temperature. Reprinted with permission from B. Heinrich et al., *Journal of Experimental Biology,* Vol. 85, pp. 64, 224. Copyright 1979, Cambridge University Press, Cambridge, UK.

for simple behavior; 31–35°C for social behavior (chorusing); 22–36°C for courtship; 20–37°C for flight; 13–40°C for walking; and 10–43°C for feeding (60).

Endothermic vertebrates (birds and mammals) maintain body temperature within 0.1–0.3°C during standard conditions of activity. The body temperature of birds tends to be a few degrees higher than that of mammals. Primitive mammals (monotremes and some marsupials) regulate less closely and have body temperatures 10–15°C lower than eurythermic mammals. Lethargic mammals, for example, sloths, have variable T_b (body temperature). Most birds and mammals have lower T_b when asleep than when active. Some endotherms—bats, hummingbirds, poor-wills—become semidormant when at rest, and their T_b may fall to approach ambient T_a; they produce heat abruptly when they become active. A few mammals—ground squirrels, woodchucks—enter a state of hibernation in winter in which their T_b equals T_a; their thermoregulatory set point is then lower than in summer. They arouse if cooled below a critical temperature. Hibernation is a complex phenomenon to which nutritional state, endocrine activity, and ambient temperature contribute. True hibernants arouse periodically to feed, urinate, and defecate. It is calculated that the annual energy saving due to hibernation by a Richardson's ground squirrel is 88 percent (165a).

Body temperature in homeotherms is regulated by metabolism (heat production) and by insulation (peripheral blood flow, feathers, fur). As the ambient temperature declines below a critical level (T_c), metabolism increases, vasomotor reflexes shunt blood, and insulating reflexes erect feathers or fur. When T_a becomes warmer than T_b, behavioral and insulative cooling occur. A thermoneutral range between low and high critical ambient temperatures (T_c) may be relatively wide or may be a single point in body temperature. Acclimatization may alter T_c, downward in cold, upward in warmth; it may alter the amount of metabolic and vasomotor response to cold or warmth, and make structural changes in fur or feathers (54).

The relation between metabolism and temperature is given by $M = C (T_b - T_a)$ where M is metabolic rate in heat production, T_b and T_a are body and ambient temperatures, and C is bulk heat transfer coefficient, which is a measure of insulative capacity.

ADAPTATIONS TO TEMPERATURE

Neural and Membrane Adaptations

The causes of temperature-induced death are multiple and differ with the organism and with rate of heating or cooling. Disruption of selective permeability of cell membranes may be the first sign of temperature damage. As temperature extremes are approached, cell membranes become leaky and ionic gradients are not maintained. In plants, frost damage occurs well above freezing temperature; cold hardening elicits changes in membrane composition. Cold-resistant plant species and strains tolerate low temperatures; their cell membranes are less affected by cold than membranes of cold-sensitive plants. When tissues of poikilothermic animals (e.g., frog muscle) are kept for a few hours at refrigerator temperatures, their ion pumps slow down and the membranes become leaky; potassium is lost and sodium gained. The net effect is depolarization of membrane potentials. When excitable cells are warmed, there may be increased activity of ion pumps with resulting hyperpolarization. As temperature rises, membrane resistance decreases. In crayfish muscle the decrease in extrasynaptic membrane resistance short-circuits and thus diminishes amplitude of synaptic potentials (170).

As temperatures of heat coma are approached, potassium concentration in the blood of crayfish rises and central nervous function may be impaired (12). However, synaptic transmission in crayfish fails with less warming than is required for failure of postsynaptic membranes to depolarize when transmitter is applied; synaptic block appears to result from decrease in transmitter liberation at temperatures above a critical level (139a, 170). Heat or cold coma occurs at a critical thermal maximum and is usually reversible. The CT_{ma} and CT_{mi} in some poikilotherms can be extended by supplying extra O_2; thus impairment of function may be due to hypoxia rather than to temperature *per se*. When snails are subjected to warming they lose ability to adhere to a vertical surface; heat coma occurs at the temperature of cessation of spontaneous activity of neurons in central ganglia (52).

When sunfish are cooled or warmed 5–10°C beyond their acclimation temperature, the maximum velocity of induced swimming is lowered (144a). Conditioned responses to flashes of light are blocked by less cooling than are direct responses to light; still more cooling is needed to block spinal reflexes. Temperature changes have more effect on synaptic transmission than on nerve conduction (Fig. 7-6). Fish from intermediate temperature (22°C), cooled or heated by 8–10°C, become hyperactive and hypersensitive to touch. With a few more degrees of cooling or heating the fish show motor disturbances, and with still more temperature change control of equilibrium is lost. Breathing becomes shallow and the fish may enter a state of coma (45). The temperature for each of these effects is changed by acclimation. Death appears to be due not to temperature but to asphyxia. The same sequence of behavior in fish has been observed after poisoning with heavy metals or other toxic substances. Initial hyperactivity followed by motor incoordination at temperature extremes has been noted also in crayfish, a salamander, and some insects.

Recordings from fish cerebellum during brain cooling or heating show a sequence of failure of neural function: inhibitory synapses, patterned spontaneous activity, rhythmic spontaneous activity, disynaptic excitatory transmission, monosynaptic transmission, and axon conduction (e.g., antidromic spikes) (45). Inhibitory synapses have been shown to be more thermolabile than excitatory synapses in several other animal preparations.

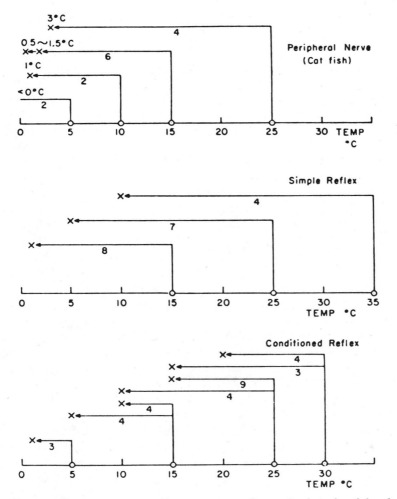

Figure 7-6. Representations of blocking temperature for conduction of peripheral nerve. For spinal reflexes and for conditioned respiratory reflexes. Reprinted with permission from C. L. Prosser and T. Nagai in *Central Nervous System and Fish Behavior,* D. Ingle, Ed., Copyright 1968, University of Chicago Press, Chicago.

In an identified neuron of the snail *Helix aspersa* spontaneous action potential, amplitude, and excitability decrease on direct cooling from 20 to 5°C; but after acclimation to low temperature, amplitude and excitability return to previous levels. The ratio of conductances for Na^+ and K^+ shows acclimation, but no acclimation was observed in duration of action potential or in Na-K pump activity (181). A burster neuron of a snail *Limax* shows temperature acclimation of electrical coupling and of frequency of bursting (138).

In sunfish skeletal muscle the resting conductance for chloride ions is six to seven times greater than for potassium; in cold acclimation, conductance for chloride is lowered more than for potassium. Membrane resistance increases upon cooling, and this may decrease the magnitude of shunting of synaptic potentials, resulting in the increased likelihood of spike electrogenesis. These membrane changes may make possible locomotion at low temperature (85a).

At synapses, a variety of events can be separated. At neuromuscular junctions and at some central nervous synapses, the postsynaptic responses are maximum in a mid-temperature range; postsynaptic potential (psp) amplitudes decrease both above and below this range (139a). One mechanism of decrease in psp at high temperature may be decreased extrasynaptic membrane resistance and increased shunting; a mechanism of reduced psp amplitude in the cold may be decreased transmitter liberation. In a locust neuromuscular junction the duration of postsynaptic response to the transmitter glutamate is longer in cold than in warmth (3). Synaptic block at both high and low temperatures is probably presynaptic in that it can be altered by extracellular Ca^{2+}; as temperature is raised, transmitter liberation (quantal content) increases up to a critical temperature at which transmission fails (170). In the stretcher muscle of a leg of the crab *Pachygrapsus,* neuromuscular response peaks at 11.2°C for 12°C acclimated crabs and at 18.6°C for 21°C acclimated crabs; facilitation is minimal at the temperature of peak response and the decline in psp on either side of maximum is compensated by increased facilitation, so that maximum tension occurs near the acclimation temperature (157). Single shocks trigger double spike responses at 23.9°C for 8°C acclimated crabs and at 27.7°C for 12°C acclimated crabs (157).

Membrane permeability may be related to molecular fluidity, which depends on both the phospholipids and proteins of the membrane. Membranes of synaptosomes from the brain of temperate zone sunfish are less fluid than those from arctic cod, and those from desert pupfish are still less fluid (28). Whether fluidity *per se* is a determinant of ion permeability is not established.

In summary, behavior is important in both resistance and capacity adaptation to temperature. In resistance to heat and cold, the nervous system is critical and fails before other systems; the order in which neural processes fail is the same for either heat or cold. Molecular adaptations include a variety of membrane changes in lipids and proteins. Metabolic responses are of secondary importance in resistance adaptation of most animals. In capacity adaptation behavioral and metabolic responses may put animals into temperatures appropriate for feeding, locomotion, and other essential functions. Regulation—homeothermy—may be by absorption of external heat in ectotherms or by metabolic production of heat and insulative conservation in endotherms. Poikilothermic animals may regulate ectothermically by body orientation or temperature selection, and may regulate endothermically by heat production by muscles. The changes in membrane properties of central nervous systems and enzyme pathways of heat production are adaptive. The interactions between behavioral, neural, and enzymatic adaptations are not well understood.

Adaptations of Proteins and RNAs to Heat and Cold

Differences in thermal limits for activity of many enzymes are correlated with the temperature or habitat of plants and animals and are genetically determined. Death or coma of whole animals and wilting of plants may occur with much less cooling or heating than will denature the proteins or alter the fluidity of lipids. It is difficult to understand how resistant proteins could have evolved when neither the nucleotides nor proteins could have been exposed *in vivo* to temperatures as extreme as are needed to inactivate them. The explanation of the correlations may lie (1) in the necessity for integrated function and the possibility that isolated proteins may be more resistant than the same proteins in an organelle or in the protoplasmic milieu of a cell, or (2) in our inability to recognize subtle properties of proteins and lipids at environmental extremes. A few examples of identified resistance adaptations of molecules will be cited.

In the laboratory of Ushakov in Leningrad, heat inactivation of enzymes, especially those of intermediary metabolism and neuromuscular function, has been correlated with habitat. Enzymes of subtidal animals are more heat-sensitive than the corresponding enzymes from high intertidal species. Two populations (races) of *Rana ridibunda,* one from the cold Baltic region and the other from the warm Transcaucasus, differ adaptively in thermostability of acetylcholine esterase and muscle ATPases. The same holds for several functions of the northern frog *R. temporaria* and southern *R. ridibunda* and for deep-water and shallow-water fishes (161), as follows:

Function Inactivated	Temperature for Inactivation in 30 Minutes in Frogs and Fishes (°C)	
	Rana temporaria	*Rana ridibunda*
reflex activity	30–32	35–36
sperm motility	39.2	41.4
muscle contraction	36.3	40.4
ciliary beat	41.7	44.5
succinate dehydrogenase (50% inactivation)	42	46
actomyosin ATPase	40.5	44.5
	Deep-Water Fishes	Shallow-Water Fishes
reflex activity	17–17.5	20
AChE (50%)	35.6–36.8	40.8

These observations and many others indicate that (1) integrated functions such as reflexes are more thermolabile than single enzymes and that (2) many proteins in a tissue are similar in thermostability and are correlated with thermosensitivity of the whole animal (161).

Myosin ATPases of many species of fishes from various habitats differ in half-time for inactivation at 37°C by more than an order of magnitude. Inactivation time at 37°C is shortest for enzymes from Antarctic species, intermediate for Mediterranean species, longest for enzymes from fishes from East African hot springs (Fig. 5-7) (81).

Fructose diphosphoaldolases of the Antarctic fish *Trematomus* and of domestic rabbit were found to be similar in amino acid composition, molecular weight, pH optima, activation energy, and apparent binding constants, but the enzyme from *Trematomus* was thermally inactivated by much less heating, apparently because of differences in the number of sulfhydryl groups (43).

Conformation of a protein is established by weak bonds, and secondary, tertiary, and quaternary structures give flexibility to protein molecules (Ch. 5). Higher-order conformation depends on primary structure that is genetically encoded. Inactivation of enzymes can result from changes in conformation such as that caused by mild heating (71). The breaking of weak bonds and consequent changes in conformation occur with less heating than changes in primary structure, and a subtle relation between primary and higher-order conformation may account for inactivation by less heating than is normally experienced by the organisms containing the enzymes.

The primary structure of collagen is correlated with temperatures of thermal transition of the molecules. When strips of collagen are warmed under slight tension they show a critical temperature for shrinkage (T_s) which reflects intermolecular and intramolecular changes. Purified or acid-treated strips show melting due to intramolecular rearrangements at transition temperatures (T_t) some 20°C below the shrinkage temperature. Comparison of the (T_t) for collagens from many animals shows correlation with the sum of proline plus hydroxypyroline content. The lower transition temperatures correlate with upper thermal limits for the animals. Selected examples follow (143):

	(Proline + Hydroxypyroline Residues)/1000	T_s (°C)	T_t (°C)
calf skin	232	59	36
Ascaris	295	59	40
beef tendon	216	60	36
earthworm cuticle	170	40	22
Trematomus (ice fish)	147	27	6
codfish	155	37	15

Some species of bacteria are adapted either obligately or facultatively to growing at low temperatures (psychrophiles) or at high temperatures (thermophiles). In thermophilic bacteria which thrive above 50°C, high heat stability of such enzymes as malate dehydrogenase, aldolase, hydrogenase, and flagellin has been demonstrated (41). A thermostable α-amylase has two cystine bridges; the enzyme from mesophilic bacteria lacks these.

A cold-requiring strain of *Pseudomonas* grows well at 0°C but fails to grow at 37°C. A mesophile *Pseudomonas aerogines* grows optimally at 37°C and shows no growth at 10°C; its enzymes for oxidation of glucose and succinate fail at high temperatures (122). Coupled phosphorylation at 7°C is three times greater in the psychrophilic than in the mesophilic strain (144b). Isocitrate lyase from cold-requiring or from mesophilic bacteria is the same in respect to thermostability and activation energy, but the optima for enzyme induction are respectively 10°C or 20–25°C and the optima for growth, 20°C or 37°C (144b). MDH from a psychrophilic *Pseudomonas* shows maximal activity at 20°C; this is also its optimum for growth (76).

Polymeric proteins may depolymerize at low temperatures. A number of mammalian enzymes that are polymeric, depolymerize into subunits at temperatures below 17°C. This has been observed for pyruvate carboxylase, glyceraldehyde 3-P-dehydrogenase and acetyl coA-carboxylase (7). Depolymerization of enzymes of cold-blooded vertebrates occurs at temperatures lower than for mammals. In the aquarium fish *Idus* the enzyme G6PDH, a tetrameric enzyme, dissociates into doublets and becomes inactive below 4°C; this effect of cooling is reversible. The fish *Idus* is unable to swim at below 4°C, possibly because of dissociation of G6PDH. Phosphofructokinase (PFK) dissociates from a tetramer to an inactive dimer at low temperature.

Thermostability of rRNA is such that the rRNA of thermophiles melts at higher temperatures (69°C) than the rRNA of mesophiles and psychrophiles (64°C). In a cold-requiring *Micrococcus cryophilus* tRNA synthetase fails when temperature rises to 25°C, and at higher temperatures its tertiary structure changes so that acylation of the tRNA no longer takes place (102).

Depolymerization of ribosomal RNA can also be induced by heating. Animal ribosomal RNAs are of two types: prostomians (ciliates, molluscs, flatworms, insects) have a 28 S rRNA which is converted to an 18 S form by heating to 45°C for 10 minutes; in deuterostomes (mainly vertebrates) the rRNA is converted by heating to units of unequal size, but not to 18 S. Some coelenterates have both types of RNA (78).

In summary, many enzymes and nucleic acids correlate in temperature of inactivation with the temperatures at which the plants and animals that contain them live. Organisms die of heat at temperatures lower than are required to denature their proteins. Psychrophiles have enzymes adapted to cold, thermophiles to heat. Usually enzyme activity continues at temperatures above or below the limits lethal for the whole organism. By some unknown mechanisms, possibly genetic linkage, the suitable enzymes have been selected.

Heat Shock Proteins

Organisms of many kinds exposed briefly to heat produce heat shock proteins (hsps). When larvae, pupae, adults, or cultured cells of *Drosophila* are warmed from 25°C to 35–37°C for 1/2 to 4 hours several heat shock proteins are synthesized in large quantity. The most abundant protein has a molecular weight of 70,000, the next most abundant is a protein at 82,000; three or four smaller proteins are formed in lesser amounts. Puffs appear on salivary chromosomes at positions corresponding to coding sites for certain mRNAs, presumably hsps. Heat treatment promotes the translation of mRNAs for hsps and represses the translation of other mRNAs (95). Different durations of exposure to temperature of 33–37°C lead to differences in amounts of the several hsps. Heat shock proteins are induced at lower temperatures in cold-acclimated catfish than in warm-acclimated fish.

The heat shock proteins are similar in size in *E. coli,* yeast, slime mold, sea urchin, salmon, embryonic cells from chickens, and some mammals. Transcription of hsps can be detected in short times (a few minutes) after heat exposure. Some mRNA already present may be degraded. Immunological cross-reactions occur between several of the hsps of yeast, *Drosophila,* and chicken cell lines. Proteins of similar size and number are synthesized by unrelated organisms. Clearly the hsps are highly conserved and must have evolved very early.

The function of the hsps is not well understood. Repression of synthesis of other proteins may have some value after heat shock. In avian and mammalian cell lines one hsp (68 kDa) is methylated, but two others are not (83 and 25 kDa) (165). A 3°C fever induces 74 kDa hsp in rabbits. Yeast cultures which are pretreated with moderate heat (36°C after culture at 23°C) show 42 percent recovery from 5 min at 52°C; cultures heated directly from 23 to 52°C show only 1.7 percent recovery (107). Mild shock at 35°C induces hsps and increases survival of Drosophila (all stages) to brief exposure to 40.1°C, which is ordinarily lethal (130). Heat shock proteins are, apparently, protective against lethal heating.

KINETIC PROPERTIES OF ENZYMES

Inactivation of enzymes by heat or cold gives some clue to the thermal ecology of plants and animals. Kinetic properties of enzymes give evidence for molecular mechanisms of adaptation. Measurement of enzyme activity as affected by thermal energy provides a connection between thermodynamics and whole-organism biology. The adaptive properties of enzyme kinetics and protein structure were outlined in Chapter 5. Examples of kinetic adaptations in relation to temperature will be given here.

The change in kinetic activity in many chemical reactions is approximately 0.3%/10°C; most enzyme-catalyzed reactions change by much more: 100–300%/10°C. A simple measure of temperature effect on a catalyzed reaction is the Q_{10}, the ratio of reaction rates at $T + 10°C$ to rates at T.

The apparent energy of activation (E_a) is given by the Arrhenius equation (Ch. 5). E_a values for purified enzymes are constant over a wide substrate concentration range. In general, enzyme reactions in poikilotherms are relatively temperature-independent—that is, they have lower E_a values or less steep slopes of the Arrhenius plot than do enzymes of homeothermic animals. Arrhenius plots of some enzyme reactions of homeotherms show breaks at critical temperatures well below T_b (Fig. 7-7a).

An apparent K_m is the concentration of substrate corresponding to half-saturation which is usually near the physiological concentration of substrates. In practice E_a is based on reaction velocity at saturating concentrations of substrate (V_{max}). In general, the position of the K_m curves of high-temperature organisms is to the right of those for low-temperature organisms.

The energy of catalyzed reactions can come from internal kinetic activity or heat content—that is, from enthalpy ($\Delta H\ddagger$)—or from environmental heat ($\Delta S\ddagger$)—that is, from entropy. The change in free energy is given by $\Delta G\ddagger = \Delta H\ddagger - T\Delta S\ddagger$. For similar changes in free energy the entropic contribution is higher for ectotherms, and the enthalpic contribution is higher for endotherms.

Kinetic properties of enzymes may be better than gross metabolism as measures of differences correlated with phylogeny and ecology. Higher activation energies for homeotherms than for poikilotherms suggest that enzymes from homeotherms are more temperature-sensitive than those from poikilotherms. The $\Delta G\ddagger$s for enzymes of species living in cold environments are lower than for the same enzymes of tropical species—that is, enthalpy differences are less in cold adapted forms—and there is compensation for the low heat content. An example is given by myofibrillar ATPases from fishes (81):

Source of Fish	V_m (mM) (mmP$_i$ mg min) at Habitat Temperature (°C)	$\Delta H\ddagger$ (cal/mol)	$\Delta S\ddagger$ (entropy units)	$\Delta G\ddagger$ (cal/mol)
Antarctic	7.7 at −1 to 2	7,400	−31	15,870
North Sea	3.5 at 3 to 12	13,530	−9.8	16,290
Indopacific	0.31 at 18 to 26	26,500	+32.5	17,620

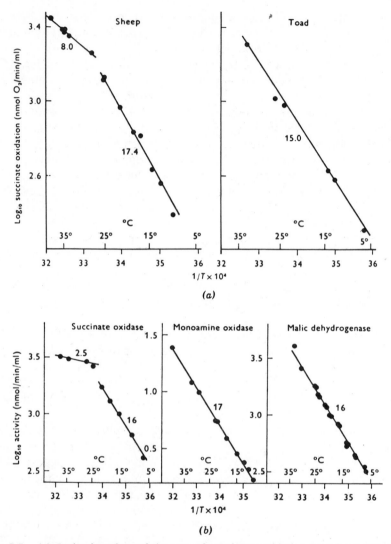

Figure 7-7. (*a*) Arrhenius plots of the rate of succinate oxidation by mitochondria from sheep and toad liver and (*b*) plots of activities of three enzymes from rat liver. (*c*) Arrhenius plots of ostate 3, ostate 4 espiration by plant mitochondria obtained from chill-sensitive and chill-resistant plants. (*d*) The effect of treatment with detergent on temperature-induced changes in E_a of succinate oxidation by mitochondria from sweet potato and rat liver. Reprinted with permission from J. K. Raisar and J. M. Lyons, *Symposium of the Society for Experimental Biology,* Vol. 27, pp. 488–491. Copyright 1973, Cambridge University Press, Cambridge, UK.

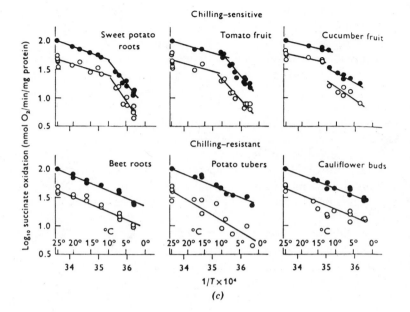

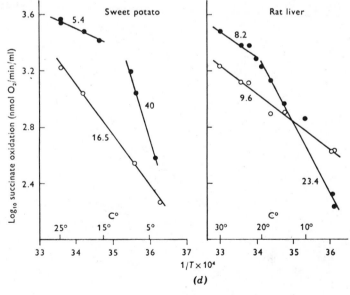

Figure 7-7. *Continued*

Ca-ATPase of muscles from Antaractic, temperate, and tropical fishes shows 7–20 times more activity in cold-water than in tropical species (Fig. 7-8). $\Delta H\ddagger$ increases in a series from cold-adapted to warm-adapted species (108). Myofibrillar Ca-ATPase of goldfish acclimated to cold shows reduced $\Delta H\ddagger$, increased k_{cat} and increased sensitivity to heat denaturation; this enzyme of trout and striped bass fails to show such changes in cold acclimation (164).

Polymerization of the muscle protein from globular (G) to fibrillar (F) state is due to exclusion of water from hydrophobic groups in the protein. $\Delta H\ddagger$ for polymerization of actin was examined in 15 species of poikilotherms; values ranged in a series from 14.5 kcal/mol in a warmth-living lizard *Dipsosaurus* to 0.67 in a cold-living fish *Coryphaeroides*. $\Delta S\ddagger$ ranged from 80 entropy units in the lizard to 39 entropy units in the deep-water fish (159). It is concluded that capacity to polymerize is achieved by differences in number of hydrophobic bonds relative to charged groups on the protein.

The position of the K_m-temperature curves shifts adaptively with the ecotype. An example is FDPase which is modulated (inhibited) by AMP. The enzyme FDPase from a fish *Genypterus* which lives at 7–13°C is modulated such that its activity at 5–15°C equals that of a rabbit at 35°C. FDPase of carp shows a decrease in AMP inhibition on cooling by 10°C so that enzyme activity (modulated) is nearly independent of temperature (104). LDH from the benthic fish *Antimora* is less sensitive than beef enzyme to temperature and pressure. In fish from temperate water the K_m for LDH is independent

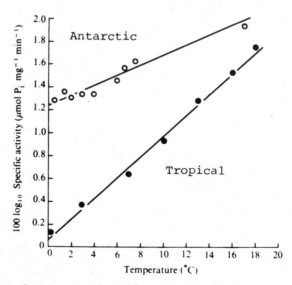

Figure 7-8. Effect of temperature on specific activity of white-muscle myofibrillar ATPase of Notothenia (Antarctic) and Amphibrion (Indian Ocean). Assays at different temperatures. Reprinted by permission from *Nature*, Vol. 254, p. 74. Copyright © 1974 Macmillan Journals Limited.

of temperature over the range 10–25°C if correction is made for temperature effect on pH. In the cold where metabolism is more aerobic, LDH has heart-like properties; in warmth where the fish are more glycolytic the LDH is musclelike (156). At the temperature normally experienced and at pHs corrected for temperature, K_m (pyruvate) of LDH of several vertebrates varied only from 0.15 to 0.35 mM; when measurements were made at a constant pH the range was 0.07 to 0.8 mM. Thus the pH-temperature effect tends to make enzyme-substrate binding relatively constant (177). Glyceraldehyde-3PDH from the Antarctic *Dissostichus* is similar in molecular weight to the enzyme from rabbit, has one fewer electrophoretic band, and is similar in K_m; but E_a is 14,500 in *Dissostichus*, 18,200 in rabbit (50).

Enzyme differences in discrete populations of a species occurring over a wide thermal range are shown in liver LDH in *Fundulus* (Ch. 4). Liver LDHB is a tetramer consisting of two forms corresponding to two alleles B^a and B^b. When the relative amounts of B^a and B^b were assayed in a population series from Halifax to Beaufort, North Carolina, the allelic frequency of B^{b}B^b was seen to increase from 0 to 100 percent and of B^{a}B^a to decrease from 87 to 2 percent (Fig. 4-1). High temperature *Fundulus* are more anaerobic, tolerate higher blood lactic acid, and show more conversion of lactate to glycogen. It appears that one allotype has an advantage in warm water, and is disadvantageous at temperatures below 12.4°C (136). In hybrids between populations at the extremes, the catalytic efficiency parallels field observations. Blood lactate and liver glycogen (but not muscle glycogen) increase as temperature rises. Similar clinal variations occur in *Fundulus* for allozymes of MDH and GPI (136). A cline in LDH for a blenny *Anoplarchus* ranges from mid-British Columbia to Tacoma, Washington. The quantity of isozyme LDHA′ is greater in the warm part of the range (80). Another example of regional differences is malic dehydrogenase in two genotypes of largemouth bass. In a northern population one allotype predominates, in a southern population another allotype (Fig. 4-2).

Clones of the plant *Lathyrus japonicus* obtained near Hudson Bay and from New Jersey were grown under the same temperature regime. The Hudson Bay clone showed higher K_m for malate dehydrogenase and was more sensitive to high temperature; both clones grown at three temperatures showed positive acclimation superposed on the genetic differences (152).

Arrhenius plots for many enzymes of mammals show a break at a mid-temperature whereas plots from most poikilotherms are linear. Rabbit mitochondrial MDH shows a break at 23°C with E_a of 13 kcal/m above and 21 kcal/m below the break; MDH from fish shows no break (Fig. 7-7a). The following enzymes from rat liver show breaks: succinic dehydrogenase, cytochrome c reductase, cytochrome c oxidase. No breaks are seen for malic dehydrogenase and monoamine oxidase. Toad liver shows no breaks for any of these enzymes (140). LDH from tropical snakes shows breaks, LDH from a temperate zone snake shows none (2).

Respiration of mitochondria from chill-sensitive plants shows breaks in

Arrhenius plots; mitochondria from chill-insensitive plants shows no breaks. For example, cucumber mitochondria show a break at 12°C with Q_{10} less than 1.5 at higher temperatures and Q_{10}'s ranging from 2.2 to 6 at lower temperatures; tomato and sweet potato showed a transition at 8°C. In contrast, potato, beet, and cauliflower mitochondrial respiration is linear over the entire temperature range (Fig. 7-7c) (140). After 10 days in a cool chamber, cucumbers were no longer chill-sensitive and respiration at 5°C became higher.

After phospholipids had been detergent-extracted from rabbit tissue mitochondria or from chill-sensitive plants, the activation energy (E_a) decreased and no breaks in the Arrhenius plot were seen for enzymes that had previously shown them (140). (Na + K)-ATPase from rabbit kidney cortex showed a transition at 16.7°C; after treatment with phospholipase A the transition temperature was raised and E_a over the lower temperature range was much higher. (Na + K)-ATPase from beef brain showed a break for ouabain-binding at 25°C; this break was abolished by treatment with phospholipase A (21). The effect of phospholipase on the transition was overcome by adding phosphatidyl serine. These observations indicate that membrane-bound enzymes are influenced by the phospholipid of the membrane to which they are bound and that they may show phase changes at critical temperatures. It appears that the phospholipid composition is adaptively different in poikilotherms and cold-insensitive plants from that in homeotherms and chill-sensitive plants. Some enzymes depolymerize into subunits below a critical temperature; depolymerization occurs at low temperatures for mammalian pyruvate carboxylase arginosuccinase, and for acetyl-CoA-carboxylase (7).

Another explanation of the transitions in Arrhenius plots was proposed for rabbit muscle aldolase (91). Kinetic quantities $\Delta G\ddagger$, $\Delta H\ddagger$, $\Delta S\ddagger$, and E_a differ above and below the transition temperatures, and these effects may be explained if two forms of an enzyme occur with different activity constants above and below the transition. However, this explanation is not applicable to enzymes which occur in one form only or to enzymes where loss of the activity on phospholipid extraction is reversed by putting back phospholipid. We conclude that kinetic measurements in relation to temperature show adaptive differences in properties of enzyme proteins.

LIPIDS

Lipids are important in both direct and acclimatory responses to temperature. Storage fats consist of neutral fats (fatty acids combined with glycerol) that become altered in fluidity according to temperature of deposition. In winter the melting point is lower than in summer for the fat of superficial regions of a bird or mammal, such as the subcutaneous fat of pig or the fat in the foot and lower leg of deer. Similar differences occur in blowfly larvae grown in cold or warmth.

Phospholipids make up 90 percent of the molecular composition of plasma membranes and organelle membranes. Phospholipids (PL) during synthesis may be altered in proportions of PL classes—phosphatidyl choline, P-ethanolamine, P-inositol, P-serine—and of cardiolipins and plasmalogens. Membranes may also show changes in proportion of cholesterol to phospholipid. The most extensive changes in PLs are in fatty-acid moieties—chain length and number of unsaturated bonds. Organisms that live at low temperatures contain more unsaturated fatty acids in neutral fats and phospholipids than organisms that live at high temperatures. Increased unsaturation makes for greater fluidity at a given temperature.

The melting point of lipids from thermophilic bacteria rises as temperature of the culture increases, whereas heat-sensitive mesophilic bacteria show no increase in melting point (39). Membrane fluidity of *Bacillus stearothermophilus* was measured by paramagnetic resonance, and lateral phase separation was correlated with temperature of culture. When the composition of membrane PL was adjusted to the same state of fluidity at different growth temperatures by fatty acids supplied in the medium, the maximal and minimal temperatures for growth changed in correspondence with membrane composition.

Freshwater crustaceans that overwinter as adults have high (50 percent) proportions of long-chain polyunsaturated fatty acids; in crustaceans that spend the winter as eggs, the proportion of polyunsaturated fatty acids is low (10 percent) (40). Mitochondrial membranes of tropical fishes have more saturated PLs than the membranes from cold-water fishes. Lipid composition of a series of Arctic animals showed greater percentages of unsaturated fatty acids than temperate-zone species as follows (93):

Fatty Acids	16:0	16:1	18:0	18:1
temperate shallow-water fish	26.4	14.3	19	25
temperate shallow-water shrimp	32.5	18.2	5.9	22
Arctic plankton	5.9	44.6	0	16.2
Arctic amphipods	9.6	39	0	30.2
deep-water crustaceans	9.8	11.1	3–6	52

Analyses of phospholipids of white muscle of several Antarctic fishes follow (127):

	Water Temperature (°C)	% Poly-unsaturated	% Mono-unsaturated	% Saturated
from surface water	23–25	33	17	49
from deep water	2–4	42	22	35

Composition of phospholipids in synaptosomal membranes was examined in brains of Arctic, temperate, and hot-springs fishes (28):

	Phosphatidyl Choline			Phosphatidyl Ethanolamine			Phosphatidyl Inositol	
	Arctic Sculpin	Desert Pupfish	Gold-fish	Arctic Sculpin	Desert Pupfish	Gold-fish	Arctic Sculpin	Pup-fish
monoun-saturated	26	34	32	22	22	41	22	31
polyun-saturated	35	15	23	57	38	55	46	29
saturated	37	49	45	21	34	34	31	37

The Arctic sculpin has relatively more of polyunsaturated fatty acids and less of saturated acids, especially in the choline phospholipid fraction than warm-water fishes. Membrane fluidity as indicated by fluorescence polarization is correlated with composition (Fig. 7-9) (28). Partial compensation occurs in that the Arctic fish and cold-acclimated temperate-zone fish have membranes that are more fluid at a given temperature than the warm-water fish, and these are more fluid than membranes of mammals. The differences in lipid composition reflect differences in desaturating enzymes that are genetically controlled.

An example of biochemical convergence is synthesis of cyclopropane fatty acids. These make up 25 percent of total fatty acids in tropical and desert millipedes (161a); similar cyclopropane fatty acids are found in thermophilic bacteria and in drought-resistant plants.

The following are some of the properties that have been correlated with phospholipid composition:

1.   Activation of membrane-bound enzymes. Succinic dehydrogenase is activated by mitochondrial phospholipid, and the PL from cold-acclimated fish activates succinic dehydrogenase to a greater extent than PL from warm-acclimated fish. The enzyme proteins are the same from both acclimated groups, but the PLs are more unsaturated from the fish acclimated in cold (58). Mitochondria from cold-acclimated carp contain more PE and less PC, hence the membranes are more negatively charged at low acclimation temperature; the fatty acids of PC are more unsaturated (175).

2.   Fluidity of liposomes prepared from membranes of cold-acclimated fish is greater than fluidity of liposomes from warm-acclimated animals. Fluidity as measured by fluorescence polarization is greater in membranes rich in unsaturated fatty acids. The temperature acclimation of membrane fluidity may not be complete; mitochondria from liver of green sunfish show about 40 percent compensation of fluidity (26,27).

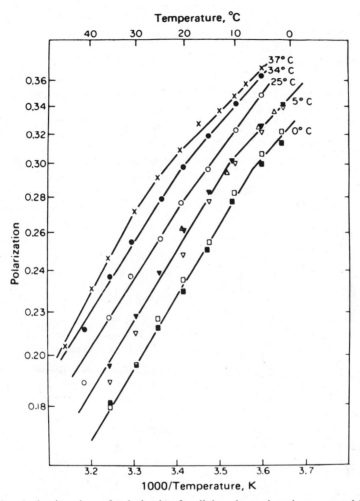

Figure 7-9. Arrhenius plots of polarization for dipheynhexatriene incorporated into synaptosome membranes from various fishes. Low values of polarization indicate high fluidity. Arctic sculpin from 0°C, goldfish acclimated at 5 and 25°C, desert pupfish from 34°C, rat 37°C. Reprinted with permission from C. L. Prosser and A. R. Cossins, *Proceedings of the National Academy of Sciences,* Vol. 75, p. 2041. Copyright 1978, National Academy of Sciences, Washington, D.C.

3.   Permeability to ions and to nonelectrolytes is enhanced by unsaturation of phospholipids, for example in *Tetrahymena* (25).

4.   Phase transitions of the lipid-protein interactions correlate in some enzymes with breaks in the Arrhenius plots; sometimes they do not so correlate. Ca-ATPase of sarcoplasmic reticulum vesicles shows a discontinuity at 15°C either with or without association with the phospholipid. Lipid composition is correlated with breaks in the Arrhenius curves but may not be

the cause of them. Arrhenius plots for (Na + K) ATPase of red blood cells show breaks, but transition temperatures for the ion pump may not be related to fluidity as reflected by bulk lipid changes (38). The phospholipid in microenvironment of ion channels is not given by bulk analyses. (Na + K) ATPase from rabbit kidney was stimulated 20-fold by addition of unsaturated phospholipids. In absence of lipids the activity was low and linear with temperature; on addition of PS, activity was high and the Arrhenius plot showed a break (85b). (Na + K) ATPase of hamster cells in culture showed a 10-fold range of activity according to membrane fluidity (153a).

5.   Growth rate of organisms is adapted to certain temperature ranges. The lowest temperature of maximum growth of *E. coli* is 34.5–35.5°C, which is where the break comes in the fluid-temperature curve (56). An auxotrophic bacterium from cattle rumen uses saturated fatty acids (palmitic and stearic), has a rigid cell membrane, and is limited to high temperature (38–39°C) for growth (56). Some bacteria which occur in ocean depths at a temperature of 5°C are psychrophiles, cold-requiring; many are psychrotrophs, cold-tolerant. When cultured at low temperatures deep-ocean bacteria have more unsaturated fatty acid, especially 16:1 and 18:1. Best growth occurs at a temperature above phase transition (67). It is concluded that appropriate lipid composition of the membrane of a bacterium is not sufficient to make it a psychrophile. A strain of *Tetrahymena pyriformis* from a hot spring in New Mexico grows well at either 39.5 or 15°C; at 15°C the cells have more unsaturated fatty acids and relatively more phosphatidyl ethanolamine than at 39.5°C (46). Spin label incorporation in this *Tetrahymena* indicates membrane composition in order of decreasing fluidity: microsomes > pellicle > cilia. Lipid composition may change with stage in growth cycle and nutrition; the changes in unsaturation/saturation (*u/s*) ratio may be determined by growth rate and not by temperature *per se* (25).

There are many instances of correlation of phospholipid composition with the temperature at which cellular membranes were synthesized; also many functional changes are correlated with membrane composition. However, exceptions have been noted. Phospholipid structure as fine-tuned by saturation/unsaturation ratio maintains a balance between liquid-crystalline and gel states. Proteins and phospholipids are in dynamic equilibrium; hydrophobic interactions due to the PLs determine the bilayer arrangement and give stability to the lateral and rotational diffusion of membrane proteins (137). The precise ways in which lipids modify the properties of membrane proteins remain to be elucidated.

Mechanisms that may account for lipid changes with temperature are: (1) alterations in digestion and absorption of specific dietary lipids; (2) changes in activity of enzymes which catalyze incorporation of fatty acids, for example, acyl transferases and phosphoglyceride transferases; and (3) enzymes of fatty acid desaturation and elongation. In mammals, alterations in proportions of phospholipid-containing fatty acids are brought about by de-acylation-reacylation reactions.

In bacteria, synthesis of desaturase is enhanced in cold. In *Bacillus megaterium* lowering of culture temperature from 30 to 20°C leads to increase in desaturase (measured with palmitic acid); the reaction is a hyperinduction, an increase much above the steady state value at 20°C. Two mRNAs are synthesized, one for the desaturase formed during the first few minutes at 20°C and one mRNA for a modulator protein formed more slowly; the modulator causes an increase in desaturasae relative to the 20°C steady state (47). In *B. licheniformis* two C_{16} polyunsaturated acids are formed at 20°C, but not at 35°C (47).

In *E. coli* high proportions of saturated fatty acids are incorporated into phospholipids at high temperature; when grown in a mixture of palmitate and oleate the ratio of palmitate/oleate incorporated is maximum at 30°C; temperature controls fatty-acid composition mainly at the transacylation of fatty acid acyl-CoA to α-glycerophosphate (153). For bacterial growth the minimum content of unsaturated fatty acids in membranes is 15–20 percent of total phospholipid. A large decrease in unsaturated acids occurs at temperatures above 34°C. Phase transition between ordered and disordered lipid state occurs at critical temperatures (30a).

In summary, temperature adaptation occurs by changes in both proteins and lipids. Cell membranes are heterogeneous systems of lipids and proteins in dynamic equilibrium. The bilayer arrangement of cell membranes is crucial for normal functioning of the membranes, and the rigidity (fluidity) of membranes is influenced by the ratio of saturated/unsaturated fatty acids in membrane phospholipids. Temperature affects the kind of lipid deposited with consequent influences on the activity of associated proteins. Permeability to specific solutes is determined by membrane proteins; how lipid composition modifies permeability has not been ascertained.

ENVIRONMENTALLY INDUCED ADAPTATIONS

The preceding sections have listed some of the properties of proteins and lipids that are adaptively correlated with temperature of habitat of organisms. These adaptations are mostly genetically or developmentally determined; however, the genetic coding for *capacity* to acclimate to temperature has received little attention. Many organisms are capable of acclimatory changes with the seasons or whenever the organisms have grown or been maintained at temperatures higher or lower than the temperature of measurement. A classification of acclimation (or acclimatization) patterns was given in Chapter 1. We consider here the extrapolations for acclimation to temperature from intact individual organisms to molecular mechanisms. The limits of acclimatory change are genetically fixed, and both capacity and resistance compensations will be considered.

The temperature during development can determine physiological properties at later stages. For example, isogenic clones from asexually reproducing sea anemones were reared at either 28°C or at 18°C, then one-half of each group was transferred to the other temperature. Animals that had developed at 18°C were larger than those that had developed at 28°C, but no differences in patterns of five enzymes were found. Irreversible physiological adaptations in developmental stages may determine population differences of adults (180). *Drosophila* reared at 25°C, then held as adults at 15°C, showed longer survival when tested at 33.5°C than *Drosophila* reared at 15°C. Acclimation during development has not been much recognized as contributing to physiological differences in adults. Positive metabolic acclimation associated with seasonal changes was detailed in Chapter 6.

Many plants show thermal compensation. Oleander bushes maintained on a cool regime showed twice the photosynthesis rate of plants on a warm schedule. The rate-limiting enzyme RuP_2-carboxylase showed a lower optimum temperature for CO_2 fixation in oleander on a 20/15°C than on a 45/32°C daily temperature cycle; leaves showed more compensation than isolated chloroplasts. Heat stability of CO_2 fixation was 6–10°C greater for oleander plants on the warm schedule (9). A desert creosote bush *Larrea* grown at 20/15°C showed twice the rate of photosynthesis of *Larrea* grown at 45/33°C when both were tested at 17°C (118). Barley and corn germinated at 10°C showed a higher rate of respiration of roots than plants germinated at 28°C; however, Rb uptake (ion absorption) by the roots was the reverse (18a).

An example of the differences in temperature effects on two related processes in one organism is photosynthesis (Ch. 6). In C-3 photosynthesis three-carbon acids are formed on the way to glucose; in C-4 photosynthesis four-carbon acids are formed. C-3 photosynthesis is the general pathway in temperate-zone plants; C-4 photosynthesis is adaptive to high temperature, dry environment, and high light intensity. In some plants C-3 is the pathway in bundle sheath cell, and C-4 in mesophyll cells. Figure 7-10 shows that the two pathways are differently affected by temperature. In C-4 plants, the quantum yield is not affected by temperature, but CO_2 fixation is more sensitive to temperature than it is in C-3 plants.

Animals from environments where there is a cycle of temperature, seasonal or diurnal, are more capable of acclimation than animals from a constant environment. An example is cytochrome oxidase activity in rockfish *Sebastes,* one species living at about 10°C throughout the year (at surface in winter, in deep water in summer) and the other near the surface where the annual temperature is 10–20°C (173). The fish from the continuous 10°C showed less acclimatory capacity than the fish from the cycling environment.

The acclimation capacity of an animal may not be the same in all parts of

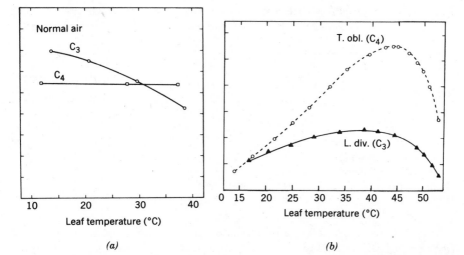

Figure 7-10. Comparison of temperature effects on C-3 and C-4 photosynthesis. (*a*) Rate of CO_2 uptake as function of temperature in C-4 and C-3 species. Reprinted with permission from O. Bjorkman, *Encyclopedia of Plant Physiology, New Series, Vol 12A, Physiological Plant Ecology, I*, O. L. Lange et al., Eds., Copyright 1981, Springer-Verlag, New York. (*b*) Quantum yield for net CO_2 uptake as function of temperature in C-4 and C-3 plants. Reprinted with permission from Osmond et al., *Physiological Processes in Plant Ecology; Toward a Synthesis with Atriplex*, Copyright 1980, Springer-Verlag, New York.

the temperature range. Metabolism of some intertidal invertebrates is relatively constant over the mid-temperature range within which they normally live. In the snail *Bullia* O_2 consumption is constant from 15 to 22.5°C (15). Myofibrillar ATPase of goldfish that had been acclimated at temperatures from 5 to 35°C showed compensation only between 10 and 30°C acclimation, not at lower or higher acclimation temperatures (128).

There is considerable specificity of enzymes and of organs in which thermal acclimation occurs. A survey of a large number of enzymes shows that, in animals, enzymes that function in energy liberation, particularly enzymes of the TCA cycle and electron transport system, show most temperature compensation; glycolytic enzymes show less compensation, sometimes none; and degradative enzymes, hydrolytic enzymes, phosphatases, and enzymes of nitrogen metabolism show no or inverse acclimation (59). Enzymes of ion transport, Na^+, K^+-ATPase, and nonspecific Mg^+-ATPases in fishes, show inverse acclimation in kidney, no acclimation in brain (26). In a cockroach brain the Na^+-K^+-ATPase showed acclimation (134). Capacity for acclimation of several enzymes and organs is illustrated in Table 7-1, (148). Comparisons of respiratory quotients, effects of enzyme inhibitors, and metabolic intermediates in striped bass acclimated to low and high temperatures indicate the following: increased activity of pentose shunt in cold, more deposition of fat in liver in cold, an increase in oxidative red muscle fibers in

Table 7-1.   Enzyme Activities in Tissues of Green Sunfish Acclimated to 5°C and to 25°C at Ambient Dissolved Oxygen Concentrations Measured at 15°C (148)

Enzyme	Muscle		Liver		Brain	
	25°C	5°C	25°C	5°C	25°C	5°C
Aldolase	1.55	0.59	0.0827	0.100		
Pyruvate kinase	2.04	1.50	0.189	1.138	1.07	1.57
Lactate dehydrogenase	2.06	1.86	0.035	0.0287	1.18	1.44
Succinic dehydrogenase	3.16	4.26	0.176	0.536	2.79	4.02
Cytochrome oxidase	0.0115	0.0223	0.0061	0.0237		
Cytochrome c	0.99	1.46				

cold (82). Liver enzymes of a blenny show acclimation to cold in the series: G6PDH > SDH > LDH (17). White muscle of brook trout shows compensatory acclimation of myofibrillar ATPase to temperature, but red muscle does not (164). How does an integrated metabolic system function at different temperatures when its components are differently affected?

Acclimation to low and high temperatures is influenced by other parameters. For example, when the photoperiod is long, metabolism of sunfish fails to compensate for cold, but it does compensate when the photoperiod is short (144). Isozymal patterns of liver LDH changed with temperature acclimation in the creek chub *Semotilus* on a short but not on a long photoperiod (84). Medusae showed temperature acclimation in summer, not in winter (103). These findings suggest that seasonal hormones may influence thermal acclimation.

In some animals capacity adaptation is separate from resistance adaptation. A frog *Rana vergatipes* shows no capacity acclimation (metabolism) but does show tolerance change (resistance adaptation) (73). Similar differences have been noted in other amphibians. Evidence for differences between field acclimatization and laboratory acclimation has frequently been found. In a lizard *Sceloporus,* metabolic compensation occurs in the laboratory but is inverse or not present in the field.

Inverse acclimation, the reduction in metabolism during cold acclimation, occurs in many organisms. This is adaptive in conserving energy reserves during periods of inactivity. In a turtle *Chelydra* metabolism and aerobic scope are reduced by cold acclimation (48). Metabolism of sunfish measured at acclimation temperatures is relatively constant in the mid-range, and thus shows positive acclimation, but at extremely low temperatures $\dot{V}_{O_2}$ remains low (144). Inverse acclimation of O_2 consumption after long periods at low T_a has been reported for lizards, snakes, hylid frogs, cricket frogs, and sea urchins. Whether the molecular mechanisms of inverse (negative) acclimation are the converse of compensatory (positive) acclimation has not been investigated. Inverse acclimation has been observed in G6PDH of frogs and in several respiratory enzymes of tropical (Puerto Rico) fiddler crabs. In any one tissue energy-yielding enzymes are positive. In general, enzymes showing no or negative acclimation include peroxidase, catalase, some hydrolytic enzymes (such as acetylcholine esterase), and enzymes of protein and urea degradation (59).

Acclimatization of mammals living outdoors in winter is not equivalent to acclimation at low temperatures in the laboratory. Photoperiod, nutrition, and social factors modify the effects of cold in nature. Vasomotor adaptations are more marked in the wild animals, tolerance of extreme cold is greater, subcutaneous fat is different, there is less shivering but more non-shivering thermogenesis. Eskimo fishermen have higher finger temperature when handling nets than non-Eskimos in comparable cold exposure.

Whether acclimation can occur in isolated tissues and cells or only in intact animals and plants has not been adequately tested. In unicellular organisms, such as yeast, there is clear evidence for acclimation of several enzyme systems (24). Ciliary beat in isolated gills of bivalve molluscs and in *Planorbis* tentacles shows compensation after several days at a new temperature (88). Cold resistance of the ciliary beat shows a 6°C difference between those acclimated to 8°C and to 25°C (136a). Tissue culture of a cell line from fish epithelium shows inverse acclimation for the enzymes that in intact fish show positive compensation. Primary cultures of catfish hepatocytes at 25, 15, and 7°C showed positive acclimatory change in the cells similar to changes *in vivo* in activity of citrate synthase, cytochrome oxidase, and glucose-6-P-dehydrogenase. However, protein synthesis in cultured cells is different from synthesis *in vivo*. Hormonal influences have not been adequately tested on cultured cells.

In summary, many poikilothermic organisms show metabolic compensation after acclimation (or acclimatization) to lower or higher temperatures. These capacity adaptations are made within genetically determined limits and are influenced by thermal range, photoperiod, and season of year, and doubtless by other factors. Animals (and probably plants) that live in thermally cycling environments are more capable of compensation than those living in constant environments. Resistance adaptations, tolerance of heat or cold, can also change with acclimation. Thermal acclimation can occur in primary cultures of hepatocytes as well as in whole fish. The nature of the feedback control in individual cells is not known.

MECHANISMS OF ACCLIMATION

Alteration in Metabolic Pathways

Energy is liberated in metabolic pathways which are parallel, convergent, or divergent. The relative importance of a pathway may vary over days or weeks according to temperature. For example, many animals that are acclimated to low temperatures synthesize fat and rely on energy from fat; but after acclimation to high temperature, carbohydrate utilization predominates. In some fishes (70,71) and in blue crabs (*Callinectes*) CO_2 production via hexose monophosphate shunt (HMP) is three times as great for cold-acclimated as for warm-acclimated animals. A limpet *Acmaea* acclimated to 8°C utilizes the HMP more than it does when it is acclimated to 18°C (105). In many aquatic poikilotherms—*Gillichthys* (156), *Acmaea* (105)—at high temperatures in low environmental O_2, lactate is formed, glycolysis increases, and activity of lactate dehydrogenase (LDH) is increased; whereas at low temperatures, where more O_2 is available, pyruvate is channeled to the TCA cycle and oxidative paths. In the cold, pyruvate inhibits LDH and total LDH activity is reduced.

Metabolic pathways may differ in cold and warm acclimation; liver of striped bass shows nearly complete compensation of $\dot{V}_{O_2}$ acclimated and measured at 15 or 25°C; carbohydrate is the predominant fuel at low temperatures, palmitate at higher temperatures. Increased use of the pentose shunt occurs in cold water. In muscle of bass the proportion of oxidative red fibers to glycolytic white fibers increases in going from 25 to 15°C, from 9 to 15 percent (82). The plant *Dianthus* grown at low temperatures shows reduced activity of amylase, LDH, acid phosphatase, and some esterases (109). Locust trees in winter lack enzymes of photosynthesis but convert carbohydrates to protein in the bark (151). These examples show changes in total pathways according to temperature.

Different Enzyme Activity in Various Organs

Metabolic pathways differ with tissue: brain and heart are highly oxidative, white muscle and liver are more glycolytic. In blue crabs *Callinectes* the midgut gland uses the HMP more than does muscle. Table 7-1 shows that in cold-acclimated green sunfish glycolytic enzymes (PK, LDH) of muscle show an inverse acclimation (Precht type V), and brain shows a positive acclimation (type III). In contrast, oxidative enzymes MDH and SDH showed positive acclimation in all organs, more in brain than in muscle (148). Differences were observed in goldfish between red muscle (oxidative) and white muscle (mainly glycolytic). In goldfish, activity of muscle aldolase and liver G-3-PDH was greater at 25°C; MDH activity was greater at 5°C (172).

Differences in Classes of Enzymes

In general, in poikilotherms that show compensatory acclimation to cold, not all enzymes follow the same pattern. Degradative enzymes, for example, are commonly noncompensatory and sometimes show inverse changes—for example, catalase and peroxidase, D-amino acid oxidase, acid phosphatase, and acetyl choline esterase. Hydrolytic enzymes (especially of digestion) may not compensate for temperature change, but nutrition may alter the compensatory response. Enzymes of nitrogen metabolism are noncompensating (59). Enzymes that show variable effects (weak compensation or no change) are glycolytic enzymes of some tissues, and various ATPases. Enzymes that show most positive acclimation (Precht types II and III) are those that liberate the most energy—those of the tricarboxylic acid (TCA) cycle and electron transport chain; cytochromes are highly compensating. There may be no correlation of capacity adaptation with enzyme stability at high temperatures; cytochrome oxidase in goldfish mitochondria shows compensation in activity but not in resistance to heat. In carp acclimated to 10 or 20°C, Mg^{2+}-ATPase and (Na^+, K^+-ATPase) showed no change in brain but in kidney (Na^+, K-ATPase) activity was greater in cold and Mg^{2+}-ATPase was more active in warm acclimation (159a). In rainbow trout, liver mitochondrial activity of NADH- and succinate-cytochrome c reductase increased in cold acclimation, and mitochondrial permeability to nonelectrolytes increased; less acclimation was found in gill mitochondria (58).

Concentrations of Specific Enzyme Proteins

One way in which changes in activity may be brought about is by change in concentration within a compartment (e.g., per mitochondrial protein). By immunological methods, cytochrome oxidase was shown to be increased in concentration in muscle of cold- as compared to warm-acclimated goldfish. By chemical analysis, the concentration of cytochrome c in sunfish liver was shown to be higher in cold as follows: $5 > 15 > 25$°C (Fig. 7-11) (150). Changes in concentration of cytochrome c can be estimated by density of electrophoretic bands. Enzymes that showed inverse acclimatory patterns had denser bands in 25°C-acclimated sunfish; muscle aldolase, P-glucomutase, liver G-3PDH (148).

Differential Synthesis of Allozymes and Isozymes

One hypothesis of compensatory acclimation is that certain isozymes and allozymes of polymorphic proteins are selectively synthesized according to temperature. Alternative forms of an enzyme may increase the capacity for acclimation.

In the liver (but not in other tissue) of the chub *Semotilus* G-PDH and LDH isozyme patterns change with temperature (84). In Alaskan king crab, PK shows a single electrophoretic band, yet two PK forms can be demonstrated kinetically; one shows apparent maximum affinity for PEP substrate at higher temperature than the other and reaches saturation (PEP as substrate) at lower temperatures of assay (156). In rainbow trout the K_m for PK is lower in winter than in summer; the proportions of isozymes present

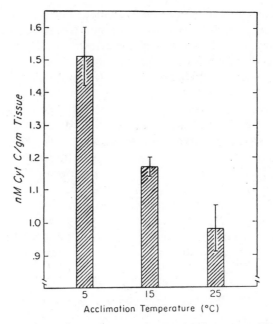

Figure 7-11. Concentrations of cytochrome c in muscle of green sunfish acclimated at 5, 15, and 25°C. Reprinted with permission from B. D. Sidell, *Journal of Experimental Zoology,* Vol. 199, p. 237. Copyright 1977, Alan R. Liss, Inc., New York.

change with the season. In trout, brain AChE occurs in two isozymal forms, one predominant in winter, the other in summer; both isozymes are present in spring and fall. The total hydrolytic activity of AChE is similar for 2°C- and 17°C-acclimated trout but the K_m minimum occurs at a lower temperature for the 2°C fish (5). The activity is the same for each isozyme in assay at 5 and 12°C (5) but the K_m of the winter form is lower at all temperatures than the K_m of summer trout (61). Changes in myosin ATPase isozymes occur in goldfish muscle according to temperature acclimation (81).

The most extensive changes in isozymes with acclimation have been described for polyploid species of salmonid and cyprinid fishes; enzymes such as LDH show many electrophoretic bands; variability in a population may be such as to mask selective synthesis of one isozyme during acclimation (172). In diploid species (*Gillichthys,* sunfish) polymorphism is virtually lacking and LDH isozymes show no correlation with temperature of acclimation. Of twenty enzymes in *Lepomis* liver, esterases 1 and 2 were present in greater quantity than esterases 3, 4, and 6 in cold acclimation, whereas in *Lepomis* eye esterases there were more of 2, 4, 6 than of 1, 3 in the cold; the functional significance of these differences is unknown (148).

In hatchery trout, isocitrate dehydrogenase (IDH) occurs in six phenotypes (bands) coded at two or more loci. The high mobility isozyme increases in summer and the low mobility isozyme increases in winter. In winter there are five anodal bands, in summer four, one of which corresponds to one of the winter isozymes. The K_m for total IDH activity in winter is minimal at lower temperature than the K_m in summer (117). A clear case of the adaptive value of isozymes is for liver G6PDH of the mullet *Mugil cephalus*. There are two isozymal forms, each a tetramer. Form I is inhibited by the cofactor NADPH, more at high than at low temperature. The relative affinity for substrate (G6P) of I decreases in the cold and affinity for cofactor increases; form II has K_m which is virtually independent of temperature, but its activity is more sensitive to pH than that of I (70). Form I increases compared to II in warm water.

Protein Synthesis and Degradation

The absolute amounts of some proteins change with acclimation due to differences in rates of protein synthesis and degradation. Incorporation of radioactive leucine into proteins of liver, muscle, and gill is greater for tissues from 5°C-acclimated goldfish than for tissue from the 25°C-acclimated (31). Enhanced protein synthesis in toadfish liver in cold acclimation and in winter has been ascribed to increase in amino acyltransferase (elongation factor I); polypeptide chain elongation was 70 percent greater in 10°C-acclimated than in 20°C-acclimated fish (55):

	Elongation Factor Activity (μM/mg RNA)	Protein Synthesis *in vivo* (mg/hr)
summer		
20°C acclimation temperature	330	0.62
10°C acclimation temperature	560	0.97
winter		
8°C acclimation temperature	640	0.93

At steady state, synthesis equals degradation, but on direct transfer from one temperature to another synthesis and degradation may be affected differently. Data for cytochrome c of green sunfish liver mitochondria follow:

Acclimation Temperature (°C)	Steady-State Concentration (nM/g)	Degradation First-Order Rate Constant [K_d] (day^{-1})	Synthesis Zero-Order Rate Constant (nM/g/day)
5	1.5	0.0505	0.076
25	0.98	0.1229	0.12

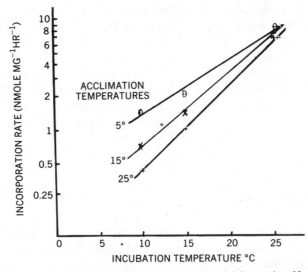

Figure 7-12. Incorporation of radioleucine into mitochondrial protein of hepatocytes from sunfish acclimated at 5, 15, and 25°C. Incorporation measured at 10, 15, and 25°C. Temperature coefficient greater for warm-acclimated than for cold-acclimated fish. Reprinted with permission from J. Kent and C. L. Prosser, *Physiological Zoology,* Vol. 53, p. 30. Copyright 1980, University of Chicago Press, Chicago.

The rate of synthesis decreased by 40 percent on going from 25 to 5°C; the degradation rate constant decreased by 60 percent, thus the concentration increased by 50 percent (150). The Q_{10} of synthesis of mitochondrial proteins measured during one hour following transfer to a different temperature was greater for warm-acclimated sunfish than for the cold-acclimated (Fig.7-12) (85). We conclude that there is compensation of enzyme concentration and that the Q_{10} of synthesis may be different from the Q_{10} of degradation. Much of the increase in amount of oxidative enzymes in cold acclimation may be due to increase in number of mitochondria per cell. In cold acclimation of eels the liver weight increased 1.8-fold; in muscle, mitochondrial protein content increased twofold (175).

Change in Size of Organs

Liver of channel catfish in acclimation to 15°C from 25°C shows 100 percent hypertrophy; there is little or no change in DNA and no hyperplasia, but protein content doubles. Because of a general increase in liver protein in catfish in the cold, enzyme activities expressed per mg protein change less than if expressed per mg DNA. Hypertrophy provides more total enzyme activity for an intact fish. Catfish hepatocytes were cultured at 25 and 7°C and protein synthesis and degradation were measured; at 7°C synthesis of protein was less depressed than degradation. Hearts also hypertrophy in cold acclimation. The number of mitochondria per cell in muscle increases in the cold, more in red than in white muscle (159b).

Membrane Phospholipids (PL)

In microorganisms, plants, and animals the phospholipids of cell membranes change during acclimation in classes of PL, chain length, and especially in percent of unsaturated fatty acids. The adaptive function seems to be maintenance of "optimal" fluidity of membranes at all temperatures.

Lipids are the only molecules that are changed in composition under temperature acclimation; changes in lipids result from compensation of enzymes of lipid metabolism. Mitochondrial enzymes are normally activated when bound to phospholipid membranes. Succinate dehydrogenase (SDH) of goldfish mitochondria is activated more by PL extracted from 5°C-acclimated fish mitochondrial membranes than by PL from 25°C fish (57). The magnitude of activation depends on the number of unsaturated fatty acids; the threshold concentration depends on the class of PL (57) (Fig. 7-13). Microsomes from goldfish intestine showed an increase in percentage of the unsaturated fatty acids 22:6 and 20:4 in PE in cold acclimation (114):

Acclimation Temperature (°C)	Percentage of Total Phospholipid		
	PE	PC	PS,PI
30	18	48	23
6	26	45	21

Synaptosomes from brain of goldfish held at 5 and 25°C showed differences in composition of fatty acids, more in PE than in other phospholipid classes (27). In *Chlorella* when culture temperature was raised from 14 to 38°C the ratio of C16/C18 fatty acids increased (125).

Changes in desaturation may be caused by four mechanisms, (1) rates of specific fatty acid desaturation, (2) selective incorporation into phospholipids by *de novo* synthesis, (3) changes in PL metabolic pathways, and (4) membrane turnover and acylation of lysophosphatides (57a). In *Bacillus megaterium* a desaturase is induced on transfer from 35 to 20°C and the activity of the desaturase decays on the reverse transfer (47). In *Tetrahymena* microsomal membrane fluidity is compensated within one hour of transfer from 39 to 15°C, apparently due to changes in deacylation reacylation (35a). Microsomal membranes of *Tetrahymena* grown at 30°C show breaks in curves of fluorescence polarization; after transfer to 15°C the breaks disappear and unsaturation index increases (35a).

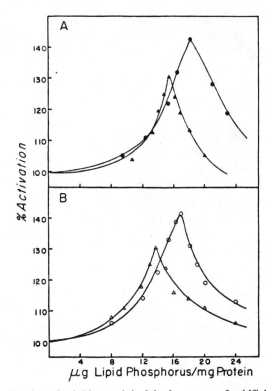

Figure 7-13.   Reactivation of soluble succinic dehydrogenase of goldfish mitochondria from which lipids had been removed. Activation by added mitochondrial phospholipid. (*a*) 5° soluble protein activated with lipid from 5° (▲) and 25° (●) goldfish. (*b*) 25° soluble protein reactivated with mitochondrial lipid from 5° (△) and 25° (○) acclimated fish. Reprinted with permission from J. R. Hazel, *Comparative Biochemistry and Physiology*, Vol. 43B, p. 860. Copyright 1972, Pergamon Press, Ltd., Oxford, UK.

In fishes, differences in phosphoglyceride transferase specificity influence phospholipid fatty acid profiles, and changes in specific desaturase and elongation enzymes occur upon acclimation to a new temperature—in trout (147), green sunfish (23), pinfish. Trout hepatocytes from 5°C-acclimated fish show complete compensation of fatty acids of the linolenic family, partial compensation of the linoleic acid family, and no compensation for the oleic acid family (147). Isolated microsomes from liver of green sunfish show compensation in Δ9 desaturases, but not in Δ6 desaturases (23). Carp liver showed increase in desaturate activity for both Δ9 and Δ6 desaturase after cold acclimation (146). Various desaturases compensate by amounts that vary with animal species, cell type, and organelle. Temperature may alter desaturase activity by affecting synthesis and degradation of specific enzymes.

Acid-Base Balance; Inorganic Ions

An effect of temperature that modifies protein function is change in pH of aqueous solutions when cooled or warmed. Neutral water declines in pH by 0.016 pH unit/°C rise in temperature; water is neutral at pH 7.0 only at 22°C. The most common buffer in biological fluids is the imidazole group of histidine in which, for different compounds, d pH/dT is 0.016–0.020; there is a change in pH with temperature, and also change in some -SH groups; the change of some -NH$_2$ groups is -0.025/°C to -0.040/°C. The second dissocation of H$_3$PO$_4$ has a pK of 6.8 and is relatively independent of temperature (174). Most biological fluids have pH-temperature curves that parallel the curve of neutral water, but the pH is higher than that of water at each temperature (Fig. 7-14). Such pH-temperature curves may be described by the ratio of hydroxyl to hydrogen ions—the relative alkalinity (168). Blood and intracellular fluids of most poikilotherms show alkalinity curves approximately parallel to the curve for water. This is true also for homeotherms over a narrow temperature range. Heterotherms are exceptions to the relation. Some heterotherms—hibernators—maintain relatively constant blood pH at different temperatures; that is, they are acidotic in the cold. The slope

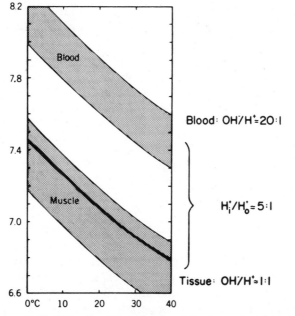

Figure 7-14. Range of pH of blood and muscle of vertebrates as function of body temperature. Solid line is muscle in relation to water. Reprinted with permission from H. Rahn et al., *American Review of Respiratory Disease,* Vol. 112, p. 167, American Lung Association, 1975 New York.

for pH as a function of temperature for blood of a dormant lizard *Scaphiopus* is −0.028, whereas for a heliothermic *Dipsosaurus* it is −0.007. Trout blood follows the imidazole pattern with d pH/dT of −0.027.

Intracellular pH can be measured by a proton-sensitive indicator and follows a relation to temperature similar to that of blood. In frog sartorius muscle the cell pH is 7.2 at 5°C, 6.8 at 30°C. Dogfish plasma and intracellular fluid (muscle) show initial pH undershoot on warming or overshoot on cooling followed by stabilization after 10–20 hours (65).

Changes with temperature tend to maintain a constant $(OH^-)/(H^+)$ ratio rather than constant pH. The ionization of dissociable groups can contribute to the enthalpy of a catalyzed reaction (174). For phosphofructokinase (PFK) the binding of the regulatory ligand ATP is determined by protonation of two ionizable groups per enzyme subunit; ATP binds best to the protonated form (131). When enzyme activities at different temperatures are measured at constant OH/H rather than at constant pH the results are more physiological and K_m values are less sensitive to temperature. At constant pH, K_m for LDH$_{pyr}$ varied according to temperature from 0.07 to 0.8 mM, but at constant OH/H K_m was 0.15–0.35 mM, in a comparison of Antarctic *Trematomus*, Pacific rockfish, Amazon fish, and laboratory mammals (177).

Respiratory and excretory control of acid-base balance becomes adjusted toward the passive pH-temperature shifts. Air-breathing poikilotherms maintain a constant blood OH/H ratio; P_{CO_2} is adjusted and HCO_3^- remains relatively constant. Water-breathing poikilotherms regulate acid-base balance mainly by varying HCO_3^-) (16a). A shark *Scylliorhinus* after a change in temperature shows complete acclimation of intracellular pH within 24 hours. The d pH/dT slope for shark blood is −0.0115 and there are compensating changes in both P_{CO_2} and HCO_3^- (16a). However, there is tissue specificity as follows:

Tissue	d pH/dT
red muscle	0.031
white muscle	0.018
heart	0.007

Enzymes with an alkaline pH optimum are less retarded by low temperature than enzymes with acid optimum or pH-independent enzymes (58). pH-sensitive binding and catalytic steps have temperature dependence altered over different temperature ranges (174). Temperature-dependent pH changes damp the effects of temperature on K_m values (166).

Besides changes in pH with temperature there are changes in concentration of some blood and tissue ions. Changes in blood and muscle K^+ and Na^+ tend to keep the ion pumps functioning with minimum energy requirement. In marine fish in winter or in the Arctic, blood NaCl is higher than in summer or in warmer water. In freshwater fish the opposite holds—blood sodium falls in winter (160). Changes in divalent cations have also been reported (75). Na, K-ATPase of toad skin shows relatively constant K_m from 8°C to 37°C, and constant enzyme activity from 20 to 37°C if OH/H is held constant (124).

In summary, metabolic acclimation (or natural acclimatization) to temperature is brought about by a number of biochemical changes. Some changes, such as adjustment of acid-base balance, take place relatively rapidly; others, such as changes in metabolic pathways and synthesis of isozymes, are determined by rates of activity of elongation factors, turnover times, synthesis and degradation of enzymes. Related to protein activities are alterations in membrane phospholipids, changes in proportions of phospholipid classes, and of ratios of unsaturated to saturated fatty acids. As acclimation goes to completion, quantitative changes in proteins are accompanied by changes in cell size, hypertrophy in the cold with little or no cell division, changes in number of mitochondria and in amount of certain critical proteins. In some organs and species, acclimation is predominantly by one method—protein synthesis—in others by lipid changes; in other organs and species compensation is by hypertrophy. The net result of capacity acclimation is constancy of energy liberation and continuance of life processes over a wide temperature range.

MOLECULAR ADAPTATIONS FOR SURVIVAL AT SUBFREEZING TEMPERATURES

Damage by freezing of plants or poikilothermic animals may be averted by several molecular mechanisms. From observations on cells, damage by freezing has been attributed to disruption of cytoplasm or plasma membrane. When tissues freeze, ice crystals form first in extracellular fluid (ECF); this leaves the remaining ECF hypertonic, and cells may lose water osmotically, with resulting changes in ionic strength and enzyme activity. Cell membranes provide some protection against freezing of cell contents, but when ice crystals form inside cells, cytoplasmic structure is usually disrupted. In plant cells the primary site of injury is the plasmalemma; mitochondria and other intracellular organelles can withstand lower temperatures than intact cells. In both plant and animal cells, chill damage may cause leakage of potassium from cytoplasm. Prior dehydration protects against freezing, especially in plants. Drought hardening of the plant *Phaseolus* by withholding water for several days diminishes susceptibility to chill injury. Susceptibility to cold shock is not the same as susceptibility to freezing. Citrus leaves accumulate free proline as a solute, more in the light than in the dark, and can tolerate temperatures as low as −6.7°C (178). In alfalfa and *Salix* stems many metabolic changes occur on cold hardening; two peroxidases are formed and amylase and catalase become more active.

Unsaturated fatty acids in cell membranes increase in amount in cold acclimation. In cold-resistant plants the percentage of unsaturated fatty acids in mitochondrial membranes is greater than in cold-sensitive plants. In cereal plants there may be a doubling in mitochondrial and microsomal content of linolenic acid during winter hardening. After 15 days of cold hardening of wheat plants the lower lethal temperature goes from -5.0 to $-17.7°C$ (34). However, strains of wheat that differ in tolerance of cold may be alike in fatty acid composition, and there may be genetic differences in fatty acid composition in plants with the same hardiness. Cold hardening can be induced by desiccation without change in lipids, and synthesis of linolenic acid can be inhibited with no resulting difference in hardiness. Evidently, for cold hardening of plants changes in fatty acid composition of membranes may not be so critical as unknown changes in membrane proteins (34).

In some plants, such as cabbage, changes in water-soluble proteins occur during cold hardening; total protein concentration may increase 2-3-fold, but sulfhydryl content of proteins may decrease by half (120). Metabolic pathways may change; treatment with cyanide shows a decrease in utilization of the cytochrome pathway (soybean cotyledons). An alternate pathway which is sensitive to salicylhydroxamic acid (SHAM) takes over in the cold (92); this is a less widely occurring path, one which, used by arums such as skunk cabbage, produces heat which volatilizes substances attractive to pollinating flies. Beans stored at $0.5°C$ show twice the O_2 consumption of those kept at $10-15°C$, both measured at an intermediate temperature.

Several types of acclimatization have evolved in animals which survive freezing temperatures. Some animals tolerate freezing of extracellular fluids; during extracellular freezing, water moves osmotically out of cells and dehydration results. Other animals are freeze-sensitive and avoid freezing by accumulating cryoprotectants both inside and outside cells.

In a few invertebrates—intertidal molluscs, barnacles, subarctic insects that tolerate freezing of extracellular fluid—probably some cells can freeze but it is unlikely that central neurons become frozen. The plasma membrane resists penetration of ice crystals. The amount of free water in cytoplasm may be reduced relative to bound water. For example, intertidal barnacles and mussels become frozen, but only 55 to 65 percent of body water is frozen; the remainder is bound to protein and other solutes. Some insects, for example, a hornet *Vespula,* have nucleator proteins in the hemolymph which may induce formation of ice extracellularly when temperature falls to $-4.6°C$. *Vespula* can survive down to $-14°C$; ice forms in its hemolymph and free water in cells is reduced.

Avoidance of freezing may be by colligative or noncolligative means. Some fishes have increased plasma NaCl concentration in winter. In other fishes plasma solutes, such as glucose, increase and lower the freezing point (160). Many subarctic insects synthesize large proportions of polyhydric alcohols which lower the freezing points of body fluids. In Vespula the extracellular glycerol concentration is 12 mg percent in summer, 3788 mg percent in winter; the melting point of hemolymph in summer is −0.72°C, in winter −2.88°C (36). Larvae of gallflies on goldenrod begin to accumulate glycerol in late August and sorbitol after the first frost; bound water in cells is high relative to free water in the frozen state (158).

Many terrestrial frogs burrow deep enough to avoid freezing. Some species produce cryoprotectants—glucose in *Rana sylvatica,* glycerol in *Hyla versicolor.* Exposure to 0–2°C stimulates liver phosphorylase, hexokinase and glucose-6-phosphatase. These anurans can survive a few days of freezing (158a).

An important molecular means of protection against freezing is synthesis of glycoproteins and polypeptides that provide for freezing point depression by a noncolligative mechanism; there is hysteresis, in that freezing temperature is lower than melting temperature (Fig. 7-15). In the supercooled state a solution is unstable, and this state cannot exist in the presence of ice. Fishes with antifreeze live in water in the presence of ice; their body fluids are stable at temperatures lower than those at which their colligative properties would result in freezing. Fish antifreezes act as protectors against freezing rather than as supercooling agents. In the Antarctic fish *Trematomus (Pagothenia)* the freezing point of the serum is −2.75°C and the melting point is −0.98°C; this contrasts with a temperate-water black perch (*Embi-*

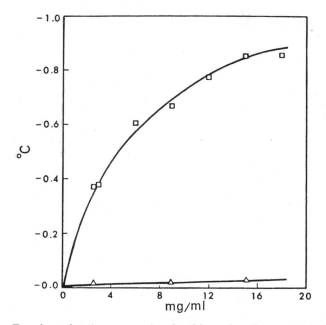

Figure 7-15. Freezing points (upper curve) and melting points (lower curve) of solutions of glycoprotein antifreezes from Antarctic *Trematomus borchgrevinki.* Reprinted with permission from A. L. DeVries, *Comparative Biochemistry and Physiology,* Vol. 73A, No. 4, p. 633. Copyright 1976, Pergamon Press, Ltd., Oxford, UK.

otoca) in which both freezing and melting points are −0.7°C (35). Electrophoresis of *Trematomus* serum shows several different glycoproteins ranging from 2600 D to 33,000 D. Each consists of a skeleton of two alanine and one threonine as repeating units with a disaccharide (a galactopyranose) attached to the threonines (Fig. 7-16a). In some of the smaller glycopeptides a proline replaces alanine. Some fish from Bering Sea have glycoprotein antifreezes that resemble the glycoproteins of Antarctic fish. Since the glycoprotein antifreezes are similar in Arctic and Antarctic fishes and these fishes are unrelated taxonomically there has either been evolutionary convergence to the same antifreeze or the two kinds of fishes have a common ancestry.

In some north temperate and subarctic fish, polypeptides serve as antifreezes. One polypeptide has a molecular weight of 3300 and consists of 24

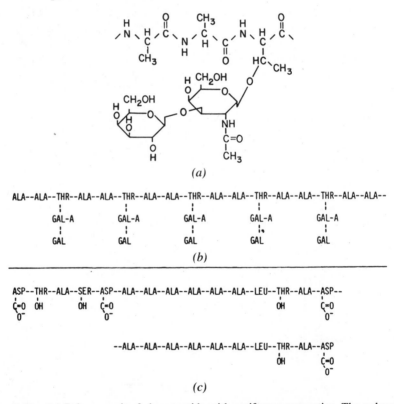

(a)

ALA--ALA--THR--ALA--ALA--THR--ALA--ALA--THR--ALA--ALA--THR--ALA--ALA--THR--ALA--ALA--
        | | | | |
      GAL-A   GAL-A   GAL-A   GAL-A   GAL-A
       | | | |
      GAL   GAL   GAL   GAL   GAL

(b)

ASP--THR--ALA--SER--ASP--ALA--ALA--ALA--ALA--ALA--ALA--LEU--THR--ALA--ASP--
C=O OH   OH   C=O   OH   C=O
| | |
O⁻   O⁻   O⁻

--ALA--ALA--ALA--ALA--ALA--ALA--LEU--THR--ALA--ASP
                                      OH   C=O
                                           |
                                           O⁻

(c)

Figure 7-16. (a) Polymer unit of glycopeptide with antifreeze properties. The polypeptide consists of a backbone of alanine and threonine (Ala-Ala-Thr) with a disaccharide joined to every threonine. (b) Structure of glycopeptide from antarctic cod (*Dissostichus mawsoni*). (c) Peptide antifreeze from blood of winter flounder *Pseudopleuronectes americanus*. Reprinted with permission from A. L. DeVries, *Comparative Biochemistry and Physiology*, Vol. 73A, No. 4, p. 630, 632. Copyright 1976, Pergamon Press, Ltd., Oxford, UK.

Ala, 5 Asp, 4 Thr, 1 Ser, 2 Leu, and 1 Lys (Fig. 7-16*b*) (35). A winter flounder *Pseudopleuronectes* from Long Island has peptide antifreeze content less than 1 mg/ml blood in summer, 25 mg/ml in winter; the melting point is essentially constant throughout the year but the freezing point goes from $-0.8°C$ (summer) to $-1.7°C$ (winter). The polypeptide is not excreted in urine, probably because its negative charges are repelled by anionic sites in the kidney (132). Antifreeze polypeptides also occur in some insects. A beetle Meracantha has an antifreeze in its tissues and hemolymph; supercooling prevents ice formation down to $-11°C$, while melting is at -7 to $-8°C$.

Several hypotheses have been proposed for the freezing point depression by the glycoproteins and polypeptides. One hypothesis is that the antifreezes hydrogen-bond to the solidifying ice and alter the growth patterns of ice formation. Some Antarctic species show little change in glycoprotein concentration with season; others, both Antarctic and Arctic species, show marked seasonal variation. Polypeptides are virtually absent from the blood of north temperate fish in the summer but are present in high concentrations (30 mg/ml serum) in winter. The poly (A)-containing mRNA which codes for one antifreeze polypeptide has been isolated from liver of winter flounder. The antifreeze mRNA may constitute as much as 1 percent of the total liver RNA in January, but is nearly absent in July, and there appear to be three different component mRNAs (94). Seasonal antifreeze biosynthesis appears to depend on synthesis and degradation of translatable mRNA, probably under hormonal control (94). Hepatocytes in culture from Antarctic *Pagothenia* can incorporate Leu and Ala into antifreeze peptides (121). The cDNA for a polypeptide antifreeze has been cloned and sequenced. The nucleotide sequences corresponding to amino acids in an antifreeze are given in Fig. 7-17.

In summary, numerous means have evolved for survival in subfreezing temperatures. Some organisms and cells are freeze-tolerant, others are damaged by ice crystals; some organisms are chill-sensitive and die at temperatures well above freezing; some freeze-sensitive species synthesize antifreeze and supercooling agents. Freezing of extracellular fluids may be tolerated by cells. Many kinds of animals die at temperatures above freezing because synaptic transmission in the central nervous system fails. Cold sensitivity of plants has been correlated with water content and with membrane phospholipids but the precise cause of chill injury remains unknown. In cold, denaturation of some critical proteins occurs. In psychrophilic bacteria, membrane fluidity is maintained by appropriate phospholipids. Proteins may be more critical than lipids for growth of cold-requiring bacteria at low temperature. Some animals accumulate in blood and cells organic molecules that lower the freezing point by their colligative properties. Some animals synthesize glycoproteins or polypeptides which lower the freezing point more than the melting point; apparently these molecules hinder formation of ice crystals. Similar antifreeze substances have evolved in unrelated animals.

TC	ACT	TTT	CAC	TGT	CGA	ACA	ATT	GAT	TTC	TTA	TTT	TGA
1												
39 / AAC		Met ATG [-21]	Arg AGA (HinfI ↑)	Ile ATC	Thr ACT	Glu GAA	Ala ·GCC	Asn AAC	Pro CCC	Asp GAC	Pro CCC	
72 /	Asp GAC	Ala GCC	Lys AAA	Ala GCC	Val GTC	Pro CCT	Ala GCC	Ala GCA	Ala GCC	Ala GCC		Pro CCA [-1]
105 /	Asp AGC [1]	Thr ACC [2]	Ala GCC [3]	Ser TCT [4]	Asp GAT [5]	Ala GCC [6]	Ala GCC [7]	Ala GCA [8]	Ala GCA [9]	Ala (GCC [10]	Ala GCC [11]	Ala GCC [12]
141 /	Thr ACC [13]	Ala GCA	Ala GCC	Thr ACC	Ala GCC	Ala GCC)	Ala GCC	Ala GCA	Ala (GCA	Ala GCC	Ala GCC	Ala GCC
174 /	Thr ACC [24]	Ala GCA	Ala GCC	Thr ACC	Ala GCC	Ala GCC)	Ala GCA	Ala GCA	Ala (GCC	Ala GCC	Ala GCC	Ala GCC
207 /	Thr ACC [35]	Ala GCA	Ala GCC	Thr ACC	Ala GCC	Ala GCC)	Lys AAA	Ala GCC	Ala GCA	Ala GCC	Ala GCC	Leu CTA
240 /	Thr ACC [46]	Ala GCC	Ala GCC	Asn AAC	Ala GCC	Ala GCC	Ala GCC	Ala GCC	Ala GCA	Ala GCA	Ala GCA	Ala GCC
273 /	Thr ACC [57]	Ala GCC	Ala GCC	Ala GCA	Ala GCC	Ala GCC	Arg AGA	GGT	TAA	GGA (Sau3AI ↓)	TCC	

B

1	2	3	4	5	6	7	8	9	10	11	12
Asp	Thr	Ala	Ser	Asp	Ala	Ala	Ala	Ala	Ala	Ala	Leu
[13]	Thr	Ala	Ala	Asn	Ala	Ala	Ala	Ala	Ala	Lys	Leu
[24]	Thr	Ala	Asp	Asn	Ala	Ala	Ala	Ala	Ala	Ala	Ala
[35]	Thr	Ala	Ala								

Figure 7-17. Nucleotide sequences of antifreeze cDNA of winter flounder. Members above amino acids refer to positions of amino acids in the peptide. Negative numbers indicate leader sequences. Three segments of repeating sequences are enclosed in parentheses. Reprinted with permission from Y. Lin and J. K. Gross, *Proceedings of the National Academy of Sciences,* Vol. 78, No. 5, p. 2828. Copyright 1981, National Academy of Sciences, Washington, D.C.

CONCLUSIONS

Of all environmental parameters which limit life, temperature is the most pervasive; hence adaptations with respect to temperature are multiple and affect every biological process. Effects of temperature are closely correlated with thermal properties of water—range of 100°C in the liquid state, high heat capacity, lower density as a solid than as a liquid, high latent heat of vaporization. The temperature range within which most life occurs is from 0°C or slightly below to between 40 and 45°C. Seawater freezes at -1.86°C. Some microorganisms, animals, and plants can tolerate the frozen state, and a very few animals are adapted by content of antifreeze or supercooling compounds for activity below the freezing point of seawater. The upper limit for life is fixed by the temperatures for denaturation of proteins and change in state of lipids. Most organisms that die of cold or heat die from other causes than protein denaturation or lipid melting. A few kinds of bacteria in deepsea vents tolerate very high temperatures at which water remains liquid under hydrostatic pressure.

The areas of the world occupied by polar, temperate, and tropical zones have frequently changed during geological time. Mean temperature in temperate zones has largely determined the history of plants and animals. Climatic changes in midlatitudes have accompanied the appearance and disappearance of major groups of plants and animals. The Permian was a warm era; frequent periods of glaciation have limited recent distribution patterns.

Cause of death in plants and animals at extremes of heat or cold is most usually failure of membrane function and increase in passive permeability, with leakage of ionic potassium out and sodium in. In animals, early signs of incipient temperature damage are failure of synaptic transmission, especially in multisynaptic pathways. Coma may lead to respiratory failure. The high or low temperature that is critical may be modified by acclimation to warmth or cold. In plants, cold hardening confers frost resistance. In some animals supercooling or antifreeze substances provide for cold resistance. Change in state of membrane lipids affects the activity of membrane-bound enzymes and permeability of cell membranes. Differences in proportion of phospholipid classes and in percentage of unsaturated fatty acids are correlated with ecology and contribute to temperature acclimation. At high temperatures heat shock proteins that are produced during brief exposures give some resistance to later excessive warming; heat shock proteins are alike in many kinds of organism.

Molecular adaptations for direct responses to changes in temperature are mainly genetic in origin. Differences in kinetic properties—activation energies, Michaelis constants (K_m), pH optima as a function of temperature, and so on—result from coded sequences of enzymes. Direct responses vary with relative hydrophobicity of critical proteins. Natural selection has favored plants and animals with kinetic properties adapted to the temperature range of their habitats.

All biochemical reactions are temperature-dependent. Enzyme-catalyzed reactions are affected much more by warming and cooling than are acceleration and deceleration of molecular motion. All living organisms have some means of homeokinesis which make them more or less independent of the physical effects of temperature. Many genetic variants of enzymes have been selected for their homeokinetic properties over certain temperature ranges. Adaptive enzyme activity can be altered by environmental induction within genetic limits. Most organisms in temperate environments are poikilothermic—microorganisms, plants, most invertebrates and "cold-blooded" vertebrates. Several general patterns of metabolic adaptation lead to temperature independence; the determinants are environmental and genetic:

1. Many proteins become altered so that the reaction rates and kinetic properties are adapted to the temperature range of the organism. An example of such adaptation for direct responses is lower temperature dependence as shown by smaller activation coefficients (E_a) for enzymes of poikilotherms than of homeotherms.

2. Compensation of enzyme functions during acclimation results in relatively constant energy production, higher at low temperatures and lower at high temperatures, than enzyme change after direct exposure to cold or heat. The capacity for compensatory changes in enzyme activity is genetically fixed and not all tissues in an organism respond similarly. Enzymes of energy liberation show positive acclimation; enzymes of degradation show negative acclimation. Compensation in some metabolic pathways may be determined by a few rate-determining steps. Disparity between tissues is not fully explained.

3. Negative metabolic adaptation is reduction or near-cessation of function at temperature extremes. When metabolism is reduced for periods of dormancy, energy reserves are conserved. Positive temperature acclimation over ranges permitting high activity, combined with inverse or negative acclimation at temperature extremes constitutes a biologically efficient set of adaptations.

Compensatory changes can occur during seasonal acclimatization. Acclimatory alterations include changes in amount of critical energy-yielding enzymes, synthesis of isozymes, shifts in metabolic pathways, changes in ionic strength and pH, and organ hypertrophy in cold with increase in total protein. A widespread adaptation is a change in percent saturation of fatty acids, and, to a lesser extent, in abundance of phospholipid classes in cell membranes. Such changes are well known in bacteria, plants, and poikilothermic animals. The net effect of phospholipid change is to maintain membrane fluidity. How gross differences in membrane phospholipids modify protein channels of permeability is not known.

Adaptive acclimatory changes, both positive and negative, are best known for intact organisms. The changes may occur in isolated cells, as was shown some years ago for yeast. Recent evidence indicates acclimatory responses in primary cell cultures that may be similar, but not identical, to those in whole animals. How temperature signals the protein synthesizing system and the enzymes of lipid metabolism to change their activity is unknown. Negative acclimation may be by reactions converse to those of positive acclimation.

4. Another type of adaptation of poikilothermic animals is behavioral—selection of temperatures appropriate for locomotion, feeding, and other vital functions. The net effect of the behavior is to maintain some constancy of body temperature. Ectothermic animals gain heat from external sources—

by basking in the sun, by absorption of warmth from the substrate. Endo-thermic animals produce heat metabolically—flying insects in thoracic mus-cles, large fish by specialized muscle bands or by brain heaters. Neural re-ceptors, peripheral and central, sense environmental temperature. In ectothermic and endothermic poikilotherms behavioral regulation of body temperature can be turned on or off according to prevailing conditions. Pre-ferred temperatures can shift with acclimation and the optimum temperature differs for different neural functions.

5.   Most free from environmental temperatures are birds and mammals—homeotherms. Their adaptations are metabolic—heat production under en-docrine regulation—and insulative—external hair, feathers, or in peripheral tissues by fat and by regulation of blood flow. Heat loss and conservation are regulated by blood flow in the appendages. Some "warm-blooded" ani-mals are heterothermic. In them, the temperature of the body may fall pe-riodically nearly to the ambient in lethargy or hibernation, or peripheral body regions may drop in temperature while internal organs are maintained at high levels. Why the 35–40°C range has been selected as the preferred temperature for mammals, birds, and endothermic insects is not clear, but it may be a compensation point between entropy and enthalpy.

Independence of living organisms from the limits imposed by tempera-ture, by whatever the means, requires energy. Cost and benefit of tempera-ture independence have been balanced in natural selection. Energy is con-served in states of dormancy and hibernation. In animals that become dormant, arousal coupled with high metabolism in the active state may re-quire much energy to make use of the periods of tolerable temperature. After a direct response to temperature change, biochemical compensation may occur by selective protein synthesis and changes in lipid composition; long-term acclimation may be morphological, especially in homeotherms with changes in insulation. Acclimatory compensations require energy for protein and lipid synthesis and cell growth. The advantages of compensatory accli-mation, positive or negative, are that organisms can have energy available suited to season, food availability, and breeding cycle. Behavioral thermo-regulation, either ectothermic or endothermic, requires energy for motor ac-tivity and has the benefit of periodic transient high temperature for optimal performance. Constant endothermy of homeotherms has very high energy requirements. Oxygen consumption by mammals is some ten times greater than by reptiles of comparable size at the same body temperature. Metabo-lism of small mammals per unit of tissue is usually many times higher than of large mammals; food requirements correspond to the elevated metabo-lism. Benefits of constant endothermy are steady state of readiness, quick reactions to predators and prey, and high levels of motor function and behavior.

Numerous ways of coping with the effects of varying environmental tem-perature have evolved. Each method has certain costs and benefits, and is suitable for a given way of life. The all-pervasiveness of temperature has been met by coordinated adaptations of all parts of organisms. Both the nature of thermal feedback to genetic mechanisms and the identity of controls—hormonal and neural—remain problem areas.

REFERENCES

1. Alahiotis, S. N. and G. Stephanou. *Compar. Biochem. Physiol.* **73B:**529–533, 1982. Heat shock protein in *Drosophila*.
2. Aleksiuk, M. *Compar. Biochem. Physiol.* **39A:**495–503, 1971. Temperature effects on metabolism of snakes.
3. Anderson, C. R. et al. *J. Physiol.* **282:**219–242, 1978. Temperature effects on glutamate current noise at synapses.
4. Argos, P. et al. *Biochemistry* **18:**5698–5703, 1979. Thermostability and protein structure.
5. Baldwin, J. *Compar. Biochem. Physiol.* **40:**181–187, 1971. *Biochem. J.* **116:**883–887, 1970. Isozyme of AchEs trout brain.
6. Berry, J. A. and J. S. Downton. Ch. 9, pp. 263–343 in *Photosynthesis,* Ed. Govindjee, Academic Press, New York, 1983. Differences in quantum yield of C4 and C3 in plants at different temperatures.
7. Beyer, R. E. pp. 17–54 in *Hibernation and Hypothermia,* Ed. F. E. South, Elsevier Pub. Co., Amsterdam, 1972. Effects of cold on mammalian enzymes.
8. Bjorkman, O. et al. *Carnegie Institute Yearbook* 1975:400–440. Heat stability of photosynthesis from heat and cold adapted plants.
9. Billings, W. D. and H. A. Mooney. *Biol. Rev.* **43:**481–529, 1968. Ecology of Arctic and alpine plants.
10. Block, W. and S. R. Young. *Compar. Biochem. Physiol.* **61A:**363–368, 1978. Metabolic adaptation of Antarctic arthropods.
11. Bourne, P. K. and A. R. Cossins. *J. Therm. Biol.* **6:**179–181, 1981. Thermal acclimation on ion flux in erythrocytes.
12. Bowler, K. et al. *Compar. Biochem. Physiol.* **45A:**441–450, 1973. Cellular heat death in crayfish.
13. Brett, J. R. *Am. Zool.* **11:**99–113, 1971. Energetic responses of salmon to temperature.
14. Brock, T. D. *Thermophilic Microorganisms and Life at High Temperatures,* Springer, New York, 1978.
15. Brown, A. C. and F. M. Da Silva. *Compar. Biochem. Physiol.* **62A:**573–576, 1979. Temperature effects on metabolism of snail.
16. Brown, J. H. and C. R. Feldmeth. *Evolution* **25:**390–398, 1971. Lethal temperatures of hot springs fishes.
16a. Cameron, J. N. *Am. J. Physiol.* **246:**R452–R459, 1984. Acid-base regulation of fish at different temperatures.
17. Campbell, S. M. and P. S. Davies. *Compar. Biochem. Physiol.* **61B:**165–167, 1978. Acclimation of fish liver enzymes.

17a. Carey, C. *Oecologia* **38**:1979. Aerobic and anaerobic metabolism in anurans.

18. Carey, F. G. pp. 216–234 in *Companion to Animal Physiology,* Ed. C. R. Taylor et al., Cambridge Press, Cambridge, U.K., 1982; *Am. Zool.* **11**:137–145, 1973. *Compar. Biochem. Physiol.* **28**:199–213, 1969. Warm fish.

18a. Carey, R. and J. Berry. *Carnegie Institute Yearbook* **75**:433–438, 1976. Temperature effects on respiration and ion uptake in barley and corn.

19. Chabot, B. F. pp. 283–304 in *Comparative Mechanisms of Cold Adaptation,* Ed. L. S. Underwood, Academic Press, New York, 1979.

20. Chabot, B. F. and W. D. Billings. *Photosynthetica* **6**:364–369, 1972. Photosynthesis in arctic and alpine plants.

21. Charnock, J. S. and L. P. Simonson. *Compar. Biochem. Physiol.* **53B**:381–387, 1977. Lipid effects on (Na + K)-ATPase.

22. Charnock, J. S. pp. 417–460 in *Biological Responses to Hibernation in Strategies in Cold,* Ed. L. Wang and J. Hudson, Academic Press, New York, 1978. Membrane lipid phase transitions.

23. Christiansen, J. *Physiol. Zool.* **57**:481–492, 1984. Desaturases in liver of sunfish acclimated to cold or warmth.

24. Christophersen, J. pp. 327–350 in *Molecular Mechanics of Temperature Adaptation,* Ed. C. L. Prosser, Am. Ass. Adv. Sci., Washington, DC, 1966. Adaptive temperature responses of microorganisms.

25. Clarke, A. pp. 54–82 in *Effects of Low Temperature on Biological Membranes,* Ed. G. J. Morris and A. Clarke, Academic Press, London, 1981. Effects of cold on *Tetrahymena* membrane.

25a. Clarke, A. *Biol. J. Linnean Soc. London* **14**:77–92, 1980. Appraisal of metabolic cold-adaptation in polar marine invertebrates.

26. Cossins, A. R. *Biochem. Biophys. Acta* **470**:395–411, 1977. pp. 83–106 in *Effects of Low Temperature on Cell Membranes,* Ed. G. J. Morris and A. Clarke, Academic Press, London, 1981. Dynamic structure of membranes in relation to temperature.

27. Cossins, A. R. et al. *J. Compar. Physiol.* **120**:109–121, 1977. Temperature acclimation of synaptic membrane.

28. Cossins, A. R. and C. L. Prosser. *Proc. Natl. Acad. Sci.* **75**:2040–2043, 1978. Evolutionary adaptation of membranes to temperature.

29. Crawshaw, L. I. *Compar. Biochem. Physiol.* **51A**:11–14, 1975; **52A**:171–173, 1975. Thermal preferendum of fish.

30. Cress, A. E. and E. W. Gerner. *Nature* **283**:677–679, 1980. Cholesterol in cells cultured at different temperatures.

30a. Cronan, J. E. and E. P. Gelmann. *Bact. Rev.* **39**:232–256, 1975. Physical properties of membrane lipids.

31. Das, A. B. and C. L. Prosser. *Compar. Biochem. Physiol.* **21**:449–467, 1967. Protein synthesis in temperature acclimation in fish.

32. Davies, P. M. C. and E. L. Bennett. *J. Compar. Physiol.* **142**:489–494, 1981. Metabolic adaptation of snakes from latitudes.

33. DeFelice, L. J. *Introduction to Membrane Noise,* Plenum, New York, 1981.

34. de la Roche, A. I. pp. 235–253 in *Comparative Mechanisms of Cold Adapta-*

tion, Ed. L. S. Underwood, Academic Press, New York, 1979. Temperature resistance and cold hardening of plants.

35. DeVries, A. L. *Oceanus* 23–31, 1976. pp. 583–607 in *Animals and Environmental Fitness,* Ed. R. Gilles, Pergamon Press, Oxford, 1980. *Compar. Biochem. Physiol.* **73A:**627–640, 1982. Antifreeze in Antarctic fish.

35a. Dickens, B. F. and G. A. Thompson. *Biochem. Biophys. Acta* **644:**211–218, 1981. *Biochemistry* **21:**3604–3611, 1982. Phospholipid changes in microsomes during acclimation of *Tetrahymena* to reduced temperature.

36. Duman, J. G. et al. *Cryobiology* **9:**469–672, 1972. *J. Exper. Zool.* **190:**89–98, 1974. *Compar. Biochem. Physiol.* **52A:**193–199, 1975. *J. Compar. Physiol.* **131:**347–352, 1979. *Compar. Biochem. Physiol.* **59A:**69–72, 1978. Freezing tolerance in insects.

37. Dye, A. H. *Compar. Biochem. Physiol.* **63A:**405–409, 1979. Seasonal effects on respiration of molluscs.

38. Ellory, J. C. and J. S. Willis. pp. 106–119 in *Effects of Low Temeprature on Biological Membranes,* Ed. G. J. Morris and A. Clarke, Academic Press, London, 1981. Passive and active ion movements in erythrocytes.

39. Esser, A. F. and K. A. Souza. *Proc. Natl. Acad. Sci.* **71:**4111–4115, 1974. Heat death and membrane fluidity in bacteria.

40. Farkas, T. *Compar. Biochem. Physiol.* **64B:**71–76, 1979. Adaptation of fatty acid to temperature in crustaceans.

41. Farrell, J. and A. H. Rose. *Annu. Rev. Microbiol.* **21:**101–120, 1967. Temperature effects on microorganisms.

42. Feder, M. E. *J. Therm. Biol.* **7:**23–28, 1982. *Ecology* **63:**1665–1674, 1982. Temperature compensation in tropical salamanders.

43. Feeney, R. E. et al. *Compar. Biochem. Physiol.* **54A:**281–286, 1976. *Adv. Protein Chem.* **32:**191–282, 1978. Adaptation of protein in antarctic fish.

44. Fitzpatrick, L. C. and A. V. Brown. *Compar. Biochem. Physiol.* **50A:**733–737, 1975. Geographic variation in salamander.

45. Friedlander, M. J. et al. *J. Compar. Physiol.* **112:**19–45, 1976. Temeprature effects on fish brain.

46. Fukushima, H. et al. *Biochem. Biophys. Acta* **431:**165–179, 1976. Membrane lipid adaptation in *Tetrahymena.*

47. Fulco, A. J. et al. *J. Biochem.* **245:**2985–2990, 1970; **247:**3503–3519, 1972; **252:**3660–3670, 1977. Temperature induction of unsaturation in bacterial fatty acids.

48. Gatten, R. E. *Compar. Biochem. Physiol.* **61A:**325–337, 1978. Thermal acclimation of metabolism of turtle.

49. Gladwell, R. T. et al. *J. Therm. Biol.* **1:**79–100, 1975. Heat death and effect on ion fluxes in tissues of crayfish.

50. Greene, F. C. and R. E. Feeney. *Biochem. Biophys. Acta* **220:**430–442, 1970. Enzymes from antarctic fish.

51. Hadley, N. F. *Am. Sci.* **60:**338–347, 1972. Desert species and adaptation.

52. Hamby, R. J. *Biol. Bull.* **149:**331–347, 1975. Heat effects on a marine snail.

52a. Hammel, H. T. *Israel J. Med. Sci.* **12:**905–915, 1976. Origin of endothermic mammals.

53. Hanegan, J. and J. Heath. *J. Exper. Biol.* **53:**629–639, 1970. Temperature dependence of neural control of moth flight.

53a. Hart, J. S. *Publ. Ontario Fish Res. Lab.* **72:**1–79, 1952. Lethal temperatures of fish from different latitudes.

54. Hart, J. S. pp. 1–149 in *Comparative Thermoregulation,* Ed. Whittow, 1971. *Canad. J. Zool.* **43:**711–716, 1965. Seasonal differences in Canadian hares.

55. Haschemeyer, A. E. V. *J. Biol. Chem.* **248:**1643–1649, 1973. *Physiol. Zool.* **50:**11–42, 1977. Protein synthesis in temperature acclimation of toadfish.

56. Hauser, H. et al. *Nature* **279:**536–538, 1979. Membrane fluidity and adaptations of bacteria.

57. Hazel, J. R. *Compar. Biochem. Physiol.* **48:**837–888, 1972. Temperature acclimation of SDH.

57a. Hazel, J. R. *Am. J. Physiol.* **246:**R460–R470, 1984. Effects of temperature on structure and metabolism of cell membranes in fish.

58. Hazel, J. R. et al. *J. Compar. Physiol.* **123:**97–104, 1978. *Am. J. Physiol.* **236:**R91–101, 1979. Temperature effects on enzymes and lipids in trout liver.

59. Hazel, J. R. and C. L. Prosser. *Z. vergleichende Physiol.* **67:**217–238, 1970. *Physiol. Rev.* **54:**620–677, 1974. Temperature adaptation of enzymes.

60. Heath, J. E. and J. Hanegan. *Am. Zool.* **11:**147–158, 1971. Temperature effects on flight in moths.

61. Hebb, J. et al. *Biochem. J.* **129:**1013–1021, 1972. Temperature effects on brain enzymes in goldfish.

62. Heikkila, J. J. et al. *J. Biol. Chem.* **257:**1200–1205, 1982. Heat shock proteins in salmon cells.

63. Heinrich, B. pp. 90–105 in *Physiology of Desert Organisms,* Ed. N. F. Hadley, 1975. *Insect Thermoregulation,* Halsted Press, New York, 1981.

64. Heinrich, B. et al. *J. Exper. Biol.* **54:**141–152, 1971; **58:**677–688, 1973; **62:**599–610, 1975; **85:**73–87, 1980; **91:**25–55, 1981. Thermal regulation in moths, flies, and bumblebees.

65. Heisler, N. et al. *Respir. Physiol.* **33:**145–160, 1978. *J. Exper. Biol.* **85:**89–110, 1980. *Am. J. Physiol.* **246:**R441–451, 1984. Acid-base adjustments to temperature in dogfish.

66. Hensel, H. *Thermoreception and Temperature Regulation,* Monographs of the Physiological Society **38,** 1981.

67. Herbert, R. A. pp. 41–54 in *Effects of Low Temperature on Biological Membranes,* Ed. G. J. Morris and A. Clarke, Academic Press, London, 1981. Cold adaptation in bacteria.

68. Hertz, P. E. *Compar. Biochem. Physiol.* **63A:**217–222, 1979. High temeprature sensitivity of lizards.

69. Hettler, W. F. and D. R. Colby. *Compar. Biochem. Physiol.* **63A:**141–143, 1979. Photoperiod effects on heat resistance of menhaden.

70. Hochachka, P. W. et al. *Marine Biol.* **18:**251–259, 1973. *Nature* **260:**648–650, 1976. Thermal acclimation of enzymes.

71. Hochachka, P. W. and G. Somero. *Strategies of Biochemical Adaptation,* Saunders, Philadelphia, 1973, Ch. 7, pp. 179–270. *Biochemical Adaptation,* Princeton University Press, Princeton, NJ, 1984, Ch. 11, pp. 359–449.

72. Holeton, G. F. *Physiol. Zool.* **47**:137–152, 1974. Metabolic adaptation of polar fish.

73. Holzman, N. and J. J. McManus. *Compar. Biochem. Physiol.* **45A**:833–842, 1973. Thermal acclimation in frogs.

74. Hoskins, M. A. H. and M. Aleksiuk. *Compar. Biochem. Physiol.* **45A**:737–756, 1973; **45B**:343–353, 1973. Seasonal effects in cold-climate snakes.

75. Houston, A. H. and K. M. Mearow. *J. Exper. Biol.* **78**:255–264, 1979. Temperature acclimation in red blood cells of goldfish.

76. Inniss, W. E. *Annu. Rev. Microbiol.* **29**:445–465, 1975. Temperature and psychrophilic microorganisms.

77. Irving, D. O. and K. Watson. *Compar. Biochem. Physiol.* **54B**:81–92, 1976. Lipids and mitochondrial enzymes in tropical fish.

78. Ishikawa, H. *Compar. Biochem. Physiol.* **46B**:217–227, 1973; **50B**:1–4, 1975. Thermal stability of ribosomal RNAs from sea anemones.

79. Jensen, D. *Compar. Biochem. Physiol.* **41A**:685–695, 1972. Cold and heat block of frog neuromuscular junction.

80. Johnson, M. S. *Marine Biol.* **41**:147–152, 1977. Allozymes and temperature gradients in blenny.

81. Johnston, I. A. et al. *Nature* **254**:74–75, 1975; **257**:620–622, 1975; *J. Comp. Physiol.* **119**:195–206, 1977; *J. Compar. Physiol.* **129**:163–177, 1979; p. 111–143 in *Cellular Acclimatisation to Environmental Change*, Ed. A. Cossins and P. Sheterline, Cambridge University Press, Cambridge, U.K., 1984. Temperature adaptation of muscle ATPase.

82. Jones, P. L. and B. D. Sidell. *J. Exp. Zool.* **219**:163–171, 1982. Metabolic response of fish muscle to temperature.

82a. Kammer, A. E. *Z. Vergl. Physiol.* **70**:45–56, 1970. Preflight warm-up in moths.

83. Kay, C. A. R. and W. G. Whitford. *Physiol. Zool.* **51**:206–213, 1978. Critical thermal limits of desert ants.

84. Kent, J. and R. G. Hart. *Comp. Biochem. Physiol.* **54B**:77–80, 1976. Effects of temperature and photoperiod on enzyme induction in fish.

85. Kent, J. and C. L. Prosser. *Physiol. Zool.* **53**:293–304, 1980. Acclimation of protein synthesis in green sunfish.

85a. Klein, M. J. *J. Exp. Biol.* *114*, 563–579, 581–598, 1985. Electrical properties of muscle of sunfish from different temperatures.

85b. Kimellerg, H. K. et al. *J. Biol. Chem.* **249**:1071–1080, 1974. Stimulation of $(Na + K)$ ATPase by phospholipids.

86. Komatsu, S. K. and R. E. Feeney. *Biochem. Biophys. Acta* **206**:305–315, 1970. Enzymes from antarctic fish.

87. Kowalski, K. et al. *J. Therm. Biol.* **3**:105–108, 1979. Species and seasonal difference in temperature tolerance of fishes.

88. Lagerspetz, K. Y. H. et al. *Compar. Biochem. Physiol.* **17**:665–671, 1966. *General Compar. Endocr.* **22**:169–176, 1974. Temperature acclimation in clams and frogs.

89. Layne, J. R. and D. L. Claussen. *J. Therm. Biol.* **7**:29–33, 1982. Acclimation of critical thermal maxima and minima in salamander.

90. Lee, R. E. and J. G. Baust. *Compar. Biochem. Physiol.* **70A**:579–582, 1981. Cold-hardiness in antarctic arthropods.

91. Lehrer, G. M. and R. Baker. *Biochemistry* **9**:1533–1541, 1970. Energetics of aldolases.

92. Leopold, A. C. and M. E. Musgrave. *Plant Physiol.* **64**:702–705, 1979. Cold injury in soybeans.

93. Lewis, R. W. *Compar. Biochem. Physiol.* **6**:75–89, 1962. Temperature effects on fatty acid of marine organisms.

94. Lin, Y. et al. *J. Biochem.* **254**:1422–1426, 1979. *Biochemistry* **19**:1111–1116, 1980. *Proc. Natl. Acad. Sci.* **78**:2825–2829, 1981. Flounder antifreeze peptide and messenger RNA.

95. Lindquist, S. *Develop. Biol.* **77**:463–479, 1980. *Nature* **293**:311–314, 1981. Synthesis of heat shock proteins in *Drosophila* and yeast.

96. Loftus, R. and G. Corbiere-Tichane. *J. Compar. Physiol.* **143**:443–452, 1981. Thermal receptors in beetles.

97. Low, P. S. and G. N. Somero. *J. Exper. Zool.* **198**:1–12, 1976. Temperature effects and kinetics of enzymes.

98. Lumry, R. Ch. 1, pp. 1–116 in *Electron and Coupled Energy Transfer in Biological Systems,* vol. 1, part A, Ed. T. E. King and M. Klingenberg, Dekker, New York, 1971. Enthalpy and entropy compensation; physical chemistry of proteins.

99. Lyman, C. P. and J. S. Hayward, pp. 346–355 in *Mammalian Hibernation,* III, Ed. K. C. Fisher et al., Oliver and Boyd, Edinburgh, 1967. Heat production by brown fat in arousal from hibernation.

100. Lyons, J. M. and J. K. Raison. *Plant Physiol.* **45**:386–389, 1970. *Compar. Biochem. Physiol.* **37**:405–411, 1970. *Cryobiology* **9**:341–350, 1972. Temperature-induced phase transitions in plant membrane.

101. Macdonald, J. A. *J. Compar. Physiol.* **142**:411–418, 1981. Temperature compensation in peripheral nerves.

102. Malcolm, N. L. *Nature* **221**:1031–1033, 1969. Molecular adaptation of psychrophilic bacteria.

103. Mangum, C. P. et al. *Marine Biol.* **15**:298–303, 1972; **17**:105–114, 1972. Temperature sensitivity of marine invertebrates.

104. Marcus, F. and J. Villanueva. *J. Biol. Chem.* **249**:745–749, 1974. Temperature adaptation of fish enzymes.

105. Markel, R. P. *Compar. Biochem. Physiol.* **53B**:81–84, 1976. Temperature acclimation of enzymes of marine molluscs.

106. Mathews, R. W. and A. E. Haschemeyer. *Compar. Biochem. Physiol.* **61B**:479–484, 1978. Temperature dependency of protein synthesis in toadfish liver.

107. McAlister, L. and D. B. Finkelstein. *Biochem. Biophys. Res. Comm.* **83**:819–824, 1980. *Develop. Biol.* **83**:173–177, 1981. Heat shock protein in yeast and sea urchins.

108. McArdle, H. J. and I. A. Johnston. *J. Therm. Biol.* **7**:63–67, 1982. Thermal adaptation of ATPases.

109. McCann, B. H. et al. *Plant Physiol.* **44:**210–216, 1969. Seasonal adaptations of enzymes in plants.

110. McCorkle, F. M. et al. *Compar. Biochem. Physiol.* **62B:**151–153, 1979. Seasonal effects on enzymes in catfish.

111. McKenzie, S. L. et al. *Proc. Natl. Acad. Sci.* **72:**1117–1121, 1975. Heat-induced messenger RNA in *Drosophila*.

112. McMahon, R. F. *Compar. Biochem. Physiol.* **55A:**23–28, 1976. Thermal tolerance of population of snail *Physa*.

113. Miller, L. C. and S. Mizell. *Compar. Biochem. Physiol.* **42A:**773–779, 1972. Seasonal effects on frog heart.

114. Miller, N. G. A. et al. *Biochem. Biophys. Acta* **455:**644–654, 1976. Phospholipid adaptation of goldfish membrane.

115. Miller, P. C. et al. pp. 181–214 in *Comparative Mechanisms in Cold Adaptation,* Ed. L. S. Underwood, Academic Press, New York, 1979. Adaptations of tundra plants.

116. Moeur, J. E. and C. H. Eriksen. *Physiol. Zool.* **45:**290–301, 1972. Temperature effects on metabolism of spider.

117. Moon, T. W. pp. 207–220 in *Isozymes,* II, Ed. C. L. Markert, Academic Press, New York, 1975. Temperature adaptation of isozymes.

118. Mooney, H. A. et al. *Ecol. Monographs* **31:**1–29, 1961. *Carnegie Yearbook* 1976–1977, 328. Temperature acclimatization of arctic, alpine and desert plants.

119. Morgan, K. R. and G. A. Bartholomew. *Science* **216:**1409–1410, 1982. Endothermy in a scarab beetle.

120. Morton, W. M. *Plant Physiol.* **44:**168–172, 1969. Freeze hardening in cabbage leaves.

120a. Nelson, D. O., J. E. Heath, and C. L. Prosser. *Am. Zool.* **24:**791–804, 1984. Evolution of temperature regulation in vertebrate nervous systems.

121. O'Grady, S. M. et al. *J. Exper. Zool.* **220:**179–189, 1982. Synthesis of glycoprotein antifreeze by hepatocytes of antarctic fish.

122. Okuyama, H. et al. *Biochemistry* **16:**2668–2673, 1977. Regulation of membrane lipid synthesis in *E. coli*.

123. Olla, B. L. et al. *Marine Biol.* **45:**369–378, 1978. Effects of high temperature on behavior of tautogs.

124. Park, Y. S. and S. K. Hong. *Am. J. Physiol.* **231:**1356–1363, 1976. Temperature effects on toad skin (Na-K)ATPase.

125. Patterson, J. L. *Lipids* **5:**597–600, 1969. Temperature effects on fatty acids of *Chlorella*.

126. Patterson, J. L. and J. G. Duman. *J. Exper. Biol.* **74:**37–45, 1978. Thermal hysteresis in *Tenebrio* larvae.

127. Patton, J. S. *Compar. Biochem. Physiol.* **52B:**105–110, 1975. Effect of temperature on phospholipids of fish muscle.

128. Penney, R. K. and G. Goldspink. *J. Therm. Biol.* **4:**269–272, 1979. Compensatory limits of fish muscle ATPase.

129. Percy, J. A. *Physiol. Zool.* **47**:163–171, 1974. Thermal acclimation in sea urchins.

130. Petersen, N. S. and H. Mitchell. *Proc. Natl. Acad. Sci.* **78**:1708–1711, 1981. Protein synthesis recovery after heat shock.

131. Pettigrew, D. W. and C. Frieden. *J. Biol. Chem.* **254**:1887–1895, 1979. Properties of PFK.

132. Petzel, D. et al. *J. Exper. Zool.* **211**:63–69, 1980. Seasonal variation in flounder antifreeze.

133. Phillip, D. P. et al. *Canad. J. Fish. Aquat. Sci.* **38**:1715–1723, 1981. Genetic variations in races of largemouth bass.

134. Piccione, W. and J. G. Baust. *Insect Biochem.* **7**:185–189, 1977. Acclimation of (Na-K) ATPase in cockroach nervous system.

135. Pickens, P. E. *Physiol. Zool.* **38**:390–405, 1965. Latitudinal and intertidal variations in *Mytilus*.

136. Powers, D. and A. Place. *Biochem. Genetics* **16**:593–607, 1978. Geographic variations in temperature effects on enzymes of *Fundulus*.

136a. Precht, H. et al. *Temperature und Leben,* Springer, Berlin, 1955.

137. Pringle, M. J. and D. Chapman. pp. 21–40 in *Effects of Low Temperature on Biological Membranes,* Ed. G. J. Morris and A. Clarke, Academic Press, London, 1981. Biomembrane structure and effects of temperature.

138. Prior, D. J. and D. S. Grega. *J. Exper. Biol.* **98**:415–428, 1982. Temperature effects on neurons in snail *Limax*.

139. Prosser, C. L. and T. Nagai. Ch. 9, pp. 171–180 in *Central Nervous System and Fish Behavior,* Ed. D. Ingle, University of Chicago Press, Chicago, 1968.

139a. Prosser, C. L. and D. O. Nelson. *Annu. Rev. Physiol.* **43**:281–300, 1981. Nervous systems in temperature adaptation of poikilotherms.

140. Raison, J. K. et al. *Arch. Biochem. Biophys.* **142**:83–90, 1971. *J. Biochem.* **246**:4036–4040, 1971. *Symp. Soc. Exper. Biol.* **27**:485–512, 1973. Temperature-induced phase change in membrane lipids of plants.

141. Raymond, J. A. and A. L. DeVries. *Proc. Natl. Acad. Sci.* **74**:2589–2593, 1977. Mechanism of freezing resistance in polar fishes.

142. Reeves, R. B. *Respir. Physiol.* **14**:219–236, 1972. Effects of temperature on acid-base regulation.

143. Rigby, B. J. *J. Therm. Biol.* **2**:89–93, 1977. Thermal effects on collagens.

144. Roberts, J. L. *Helgoland Wissenschaftliche Meeresuntersuchung* **9**:459–573, 1964. Responses of sunfish to photoperiod and temperature.

144a. Roots, B. I. and C. L. Prosser. *J. Exper. Biol.* **39**:617–629, 1962. Behavioral temperature acclimation in fish.

144b. Sasaki, S. et al. *J. General and Appl. Microbiol.* **25**:299:306, 1979; **26**:265–272, 1980. Culture of psychrophilic and mesophilic bacteria.

145. Satinoff, E. pp. 217–252 in *Physiological Mechanisms of Motivation,* Ed. D. W. Pfaff, Springer, New York, 1982. *Physiol. Behav.* **29**:537–541, 1982. Hierarchical control of temperature regulation.

146. Schünke, M. and E. Wodtke. *Biochem. Biophys. Acta* **734**:70–75, 1983. Cold-induced increase of desaturases in carp liver.

147. Sellner, P. A. and J. R. Hazel. *Arch. Biochem. Biophys.* **213**:58–66, 1982. Fatty-acid desaturation by hepatocytes from trout.

148. Shaklee, J. B. et al. *J. Exper. Zool.* **201**:1–20, 1977. Molecular temperature acclimation in sunfish.

149. Sherman, E. *Compar. Biochem. Physiol.* **65A**:227–230, 1980. Developmental changes in thermal tolerance of *Bufo*.

150. Sidell, B. D. *J. Exper. Zool.* **199**:233–250, 1977. Turnover of cytochrome c in muscle of green sunfish.

151. Siminovitch, D. et al. pp. 3–40 in *Molecular Mechanics of Temperature Adaptation,* Ed. C. L. Prosser, AAAS Publ. 84, 1966. Seasonal changes in proteins of trees.

152. Simon, J. *Plant Sci. Lett.* **1415**:113–120, 1979. Enzyme adaptations in higher plants. *Oecologia* **39**:279–287, 1979.

153. Sinensky, M. *Proc. Natl. Acad. Sci.* **71**:522–525, 1974. *J. Bacteriol.* **106**:449–455, 1971. Fatty acid changes in *E. coli* with temperature.

153a. Sinensky, M. et al. *Proc. Natl. Acad. Sci.* **76**:4893–4897, 1979. Rate limitation of (Na=K) ATPase by membrane acyl chain order (fluidity).

154. Smith, C. L. *Compar. Biochem. Physiol.* **59B**:231–237, 1978. Temperature dependence of respiration in mitochondria of rat and trout.

155. Snyder, G. K. and W. N. Weathers. *Am. Nat.* **109**:93–101, 1975. Temperature adaptation in amphibians.

156. Somero, G. N. pp. 221–234 in *Isozymes,* II, Ed. C. L. Markert, Academic Press, New York, 1975. Isozymes in temperature adaptation.

157. Stephens, P. J. and H. L. Atwood. *J. Exper. Biol.* **98**:39–47, 1982. *J. Compar. Physiol.* **142**:309–314, 1981. Temperature effects on crustacean neuromuscular junction.

158. Storey, K. *Cryobiology* **20**:365–379, 1983. Metabolism and bound water in overwintering insects.

158a. Storey, K. B. and J. M. Storey. *J. Compar. Physiol.* **155**:1–8, 1984. Adaptation for freezing in *Rana sylvatica*.

159. Swezey, R. R. and G. S. Somero. *Biochemistry* **21**:4496–4503, 1982. Stability of actins from vertebrates adapted to different temperatures.

159a. Tirri, R. et al. *J. Therm. Biol.* **3**:131–135, 1978. ATPase in carp.

159b. Tyler, S. and B. D. Sidell. *J. Exper. Zool.* **232**:1–9, 1984. Mitochondrial distribution and diffusion distances in muscle of goldfish.

160. Umminger, B. L. pp. 59–71 in *Physiological Ecology of Estuarine Organisms,* Ed. F. J. Vernberg, University of South Carolina Press, Columbia, 1975. Cold resistance in *Fundulus*.

161. Ushakov, B. P. pp. 107–130 in *Molecular Mechanisms of Temperature Adaptation,* Ed. C. L. Prosser, AAAS Publ. 84, 1966. pp. 322–334 in *The Cell and Environmental Temperature,* Ed. A. S. Troshin, Pergamon Press, London, 1967. Thermostability of cells and proteins in relation to speciation.

161a. Van der Horst, D. J. and R. C. Oudegans. *Compar. Biochem. Physiol.* **46B**:277–288, 1973. Cyclopropane fatty acids in desert millipedes.

162. Vargo, S. L. and A. M. Sastry. *Marine Biol.* **40**:165–171, 1977. Temperature tolerance of crustacean larvae.

163. Vernberg, J. and J. Costlow. *Physiol. Zool.* **39**:36–52, 1966. Geographic variation and development of crab larvae.

164. Walesby, N. J. and I. Johnston. *Experientia* **37**:716–718, 1981. Temperature acclimation of myofibrillar ATPase in trout and goldfish.

165. Wang, C. et al. *Proc. Natl. Acad. Sci.* **78**:3531–3535, 1981. Heat shock proteins in bird and mammal cells.

165a. Wang, L. C. H. *Canad. J. Zool.* **57**:149–155, 1979. Natural torpor in ground squirrels.

166. Walsh, P. J. and G. N. Somero. *Canad. J. Zool.* **60**:1293–1299, 1982. Kinetics of fish LDH in relation to temperature acclimation.

167. White, F. N. *Compar. Biochem. Physiol.* **45A**:503–513, 1973. Temperature regulation in iguana.

168. White, F. N. and G. N. Somero. *Physiol. Rev.* **62**:40–90, 1982. Effects of temperature on acid-base balance.

169. White, R. J. *Thermobiology* **1**:227–232, 1976. Junctional block by heat in frog.

170. White, R. *Physiol. Zool.* **56**:174–194, 1983. Temperature block of neuromuscular transmission in crayfish.

171. Wickstrom, C. E. and R. W. Castenholz. *Science* **181**:1063–1064, 1973. Thermophilic ostracod.

172. Wilson, F. R. et al. pp. 193–206 in *Isozymes,* II, Ed. C. L. Markert, Academic Press, New York, 1975. Isozymes in temperature acclimated fish.

173. Wilson, F. R. et al. *Compar. Biochem. Physiol.* **47B**:485–491, 1974. Temperature metabolism of two species of *Sebastes.*

174. Wilson, T. *Arch. Biochem. Biophys.* **179**:378–390, 1977; **183**:409–419, 1977. Temperature effects on enzyme kinetics.

174a. Whittow, G. C., Ed. *Comparative Physiology of Thermoregulation,* Academic Press, New York, 1970.

175. Wodtke, E. *Biochem. Biophys. Acta* **640**:698–709; 710–720; 1981. Temperature acclimation of enzymes as influenced by lipids.

176. Wood, F. E. et al. *Insect Biochem.* **7**:141–149, 1976. *J. Insect Physiol.* **22**:1665–1673, 1976. Cold induction of glycerol in arctic insects.

177. Yancey, P. H. and G. N. Somero. *J. Compar. Physiol.* **125**:129–134, 1978. Temperature dependence of intracellular pH.

178. Yelenosky, G. *Plant Physiol.* **64**:425–426, 1979. Cold hardening in leaves.

179. Zachariassen, K. E. *J. Compar. Physiol.* **140**:227–234, 1980. *Nature* **298**:865–867, 1982. Supercooling in insects.

180. Zamer, W. E. and C. P. Mangum. *Biol. Bull.* **157**:536–547, 1979. Temperature differences in populations of sea anemones.

181. Zecevic, D. and H. Levitan. *Am. J. Physiol.* **239**:C47–C57, 1980. Temperature acclimation of membranes of a snail neuron.

8

HYDROSTATIC PRESSURE

Seventy percent of the earth's surface is covered by oceans, and 63 percent of ocean water is more than 1000 meters deep. Some 75 percent of the volume of seawater is in the deep sea (Fig. 8-1) (25). Hydrostatic pressure increases 1 atm per 10 meters depth (14.7 psi or pounds per square inch, 760 mm Hg, 1.013 bar); at 1000 m depth the pressure is 101 atm or 102.3 bar. The mean depth of the Atlantic Ocean is 3332 m, and the depth of trenches in the oceans ranges from 6000 to 11,000 m (Fig. 8-2). Many deep-sea animals are luminescent. Below 1000 m there is virtually no sunlight, hence no photosynthesis, and organisms must depend for food on organic material falling from above. Exceptions are the hydrothermal vent animals and bacteria that rely on *in situ* chemolithotrophic production of nutrients (Ch. 2). Encrusting coralline algae, Rhodophyta, have been found at depths of > 250m where light intensity is less than 1 percent of that at the surface. Green algae can grow at depths of 157 m, brown algae at 88 m (12a).

The temperature of the deep ocean (benthos) is relatively uniform, 2–4°C. In the stressful environment of the deep sea live bacteria and animals which are adapted to low temperature, high pressure, and scanty food supply. Life processes are very slow and metabolism is reduced far below that of shallow-water forms. Organisms from deep thermal vents reproduce at high rates and have active metabolism; therefore pressure *per se* is evidently not responsible for reduced life processes of deep-sea forms. The age of a clam *Tindaria* from 3800 m deep was estimated by its [228]Radon content; its growth was calculated to have been 8.4 mm diameter in 100 years; this clam reaches sexual maturity at 4 mm length, at age 50–60 years (28). Pressures comparable to those in deep ocean can inactivate many enzymes of shallow-water organisms; these enzymes in deep-sea organisms are adapted to high pressure.

EFFECTS OF PRESSURE ON GROWTH OF BACTERIA AND PHYSIOLOGY OF ANIMALS

An example of retarded bacterial growth in a year at 154 atm was a meat sandwich in the submersible vessel *Alvin* which would have spoiled at 3°C

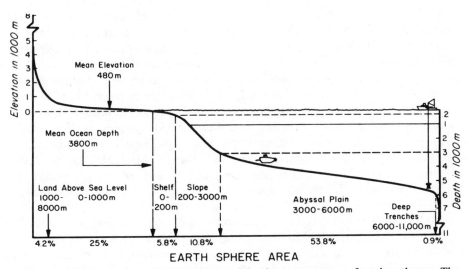

Figure 8-1. Diagram of average elevations and depths as percentage of total earth area. The deep sea (below a depth of 1000 m) covers 63 percent of the earth's surface and represents about 75 percent of the total volume of seawater. Reprinted with permission from H. W. Jannaseh, *Oceanus,* Copyright 1979, Woods Hole Oceanographic Institute, Woods Hole, MA.

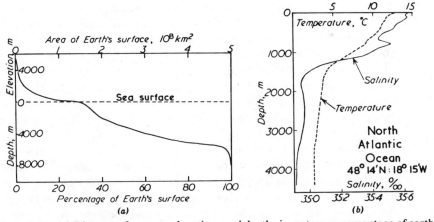

Figure 8-2. (*a*) Diagram of average elevations and depths in meters as percentage of earth's surface. H. U. Sverdrup et al., *The Oceans: Their Physics, Chemistry, and General Biology,* Copyright 1942, renewed 1970, p. 19. Reprinted by permission of Prentice-Hall, Inc., Englewood Cliffs, NJ. (*b*) Temperature and salinity profiles at 48°14′N, 18°15′W. Reprinted with permission from F. C. Fuglister, *Atlantic Ocean Atlas of Temperature and Salinity. Profiles and Data from International Geophysical Year 1952–1958,* Copyright 1960, Woods Hole Oceanographic Institute, Woods Hole, MA.

and 1 atm but did not spoil at 154 atm. Some kinds of deep-sea bacteria are barotolerant, some kinds are barophilic. Barotolerant bacteria grow well at 1 atm and grow slowly under pressure. At low temperatures and at one atm, bacteria from 300 atm pressure showed lower metabolism than surface bacteria; O_2 consumption rate was 11 times lower and carbon fixation was 64 times slower for the barotolerant bacteria. Barophilic bacteria from the deep sea grow better under pressure than at 1 atm. Bacteria cultured from a deep-water amphipod grow best at 500 bars pressure (31). A species of *Spirillum* grew only at 2–4°C; it had a generation time of 4–13 hours at 500 bars, 3–4 days at 1 bar (31). An amphipod recovered from the Mariana Trench, 10,476 m deep, yielded bacteria that grew best at 2°C and 690 bars, and did not grow at 380 bars. Generation time at 1035 bars, the pressure of its habitat, was 33 hours (31).

Bacteria have been found in hydrothermal vents at depths of 2500 m (pressure approximately 250 atm), at temperatures as high as 250°C. In the > 100° water coming out of the vent, bacterial growth is so great that the water may be milky with bacteria. No light penetrates to lower levels where bacteria occur and oxygen concentration approaches zero. The chemolithotrophic bacteria from sulfur-containing vents are of several species. The best known of them oxidize HS^- to SO_4^{-2} and have anaerobic S-metabolizing enzymes—rhodanese (anaerobic, sulfide oxidase), adenylsulfate reductase, and ATP sulfurylase. These bacteria fix CO_2 by the Calvin-Benson cycle (Ch. 2) via the enzymes ribulose-1,5, biphosphate carboxylase, and ribulose-5-phosphate kinase. The net result is ATP production by sulfide oxidation and synthesis of carbohydrate from fixed CO_2. Bacteria from the vents grow rapidly in culture (generation times 37–65 minutes) at 100°C and 1 atm (1,22). Live cells occur in water hotter than 300°C. At 265 atm pressure, water remains liquid at temperatures up to 460°C. Apparently the bacteria from deep vents have proteins which are not denatured by high temperature and high pressure, so long as the medium remains liquid. Methanogenic bacteria from 306°C and vent pressure of 265 atm produce CH_4, H_2, and CO (1). The metabolic pathways in both the anaerobic and aerobic sulfide-oxidizing bacteria are discussed in relation to the evolution of energy-yielding patterns in Chapter 2.

Pogonophoran worms and some bivalve molluscs within which HS^--oxidizing bacteria are present as symbionts occur in water emerging from a vent where it is rapidly mixed with cold seawater. The worms lack digestive tracts but trophosome tissue in the coelomic cavity of the worms and in gills of the molluscs is rich in bacteria. The worms and molluscs make use of energy and reducing carbon compounds formed by the aerobic sulfur bacteria (6). At the mouths of vents oxygen is available and bacteria synthesize carbohydrate by the reaction

$$CO_2 + O_2 + 4H_2S \rightarrow [CH_2O] + 4 S + 3 H_2O$$

In the pressure and cold of the deep sea, many kinds of animals occur—bivalve and gastropod molluscs; crinoid, holothurian, and ophiuroid echinoderms; colonial coelenterates (pennatulids); crustaceans (cumaceans, isopods); pycnogonids; fishes (elasmobranches and teleosts). Species diversity is nearly as great in the ocean depths as in tropical shallow water and is much greater than in boreal shallow waters and estuaries. The kinds of ophiuroids that shed pelagic eggs show species diversity and are eurybenthic; species diversity is less for polychaetes, cumaceans and gastropods; bivalves are intermediate in diversity (20). Amphipods collected at 5700 m (583 bars) tolerate prolonged reduction of pressure by 29 percent of habitat pressure, brief reduction of pressure by 70 percent (31). Echinoderms from ocean depths have greater enzyme polymorphism than species from shallow water.

When abyssal animals (especially invertebrates) are brought to the surface slowly and kept at low temperature they may show little change in behavior or metabolism. However, warm surface temperatures may prove lethal. Fish with swim bladders may explode or rupture the body wall as the gas expands on decompression. There is little problem with the bends unless there is a store of gas (N_2 or air) dissolved in blood at low pressure and then compressed. On rapid decompression the gas is released as bubbles. Midwater animals in diurnal migrations may be subjected to pressure changes of 30–50 atm. Shallow-water animals do not tolerate the pressure of the deep ocean; abyssal animals are adapted to high pressure and low temperature.

Relatively small increases in pressure affect organ function in animals that normally live at one atmosphere. Shallow-water vertebrates and invertebrates become hyperactive when pressurized above 50 atm; the heart rate increases and initially the O_2 consumption increases, probably because of the hyperactivity. Above 100 atm pressure activity becomes reduced. Shallow-water amphipods are hyperactive at 80–100 atm, activity becomes depressed at 120–230 atm; abyssal amphipods show convulsive behavior at 520 atm, which is 340 atm above their normal environmental pressure (7). Abyssal animals are consistently tolerant of pressure and capable of recovery (15). A deep-sea amphipod *Lancela* shows normal swimming and ventilation at pressures up to 350–400 atm; at higher pressures its activity is depressed (14). Shallow-water amphipods from Arctic and Temperate zones increase O_2 consumption as pressure is increased up to about 10 atm; at higher pressures metabolism drops. The stimulating effect of pressure is greater at 14°C than at −1°C. Deep-sea amphipods show relatively constant metabolism over a wide pressure range. A decapod crustacean from intermediate depths (sea slope) went into convulsions at 225 atm; a decapod from abyssal depths first went into convulsions at 520 atm (7). A behavioral sequence consequent upon pressure above 480 atm in oceanic amphipods is hyperactivity, impaired coordination, convulsions, paralysis, irregularity of heart and pleopod beat (14). The pressures that occasion hyperactivity, coma, and death are different for different animals. In Antarctic euphausids hyperactivity is induced at 145–180 atm in larvae, at 90–140 atm in adults (7). In fishes the sequence of effects of pressure is: increased movement, incoordination, convulsions, and coma; the nervous system is the most sensitive (14).

Metabolism is much lower in deep-sea fishes than in shallow-water fishes measured under similar conditions. The deeper-living fishes and crustaceans have significantly higher water content (27).

Applied pressure of 200 atm depolarized lobster neurons by 3 mV, and *Helix* (snail) neurons by 5–15 mV; pressure up to 408 atm increased spike amplitude in squid axons, and higher pressure decreased the action potentials. Synaptic potentials in Helix were reduced by 50 percent at 100 atm (29). In *Helix* neurons compression at 208×10^5 N/m² (approximately 200 atm) reduced inward current of an action potential and increased leak current (8a).

When a frog heart is subjected to pressure, contractions increase in amplitude to a maximum at 260–320 atm, and then at higher pressures become irregular and stop (19). Under pressure, neuromuscular junctions of lobsters show depressed synaptic potentials, probably due to decreased transmitter liberation (2). At 100–120 atm lobster heart slows. In frog neuromuscular junctions, the frequency of miniature endplate potentials decreases (14a).

There is antagonism between pressure and various anesthetics. Pressure counteracts the effects of ethanol and other anesthetics in a squid giant axon and in *Helix* nervous system. Mice and salamanders show recovery from ether anesthesia if subjected to high pressure (16). The pressure relief from anesthesia may be on synapses which use glycine as transmitter. Mitosis can be blocked by high pressure; sea urchin eggs subjected to pressure stop division and continue on decompression (14b). Amoeboid cells become spherical under pressure and their protoplasm becomes fluid (14b). In *Tetrahymena,* pressure delays division and the incorporation of leucine into protein (28a).

In *E. coli,* pressure interrupts the synthesis of galactosidase; the block is at the translational step. Pressure, like cold, increases viscosity of lipids, and thus can alter cell membrane fluidity. Arrhenius plots of membrane ATPase from *Achoplasma* and plots of nitrogenase from *Azotobacter* show discontinuities in E_a values that shift to higher temperatures under pressure; this is interpreted as due to a phase change.

It is concluded that hydrostatic pressure, like temperature, has multiple effects, and may be either stimulating or depressing on intact organisms. In the deep sea, there is considerable diversity of effects on organisms, but because of the concurrent effects of cold, pressure, and sparse nutrients, metabolism is low and growth is slow. Excitable membranes, especially synapses of shallow water animals, are reversibly damaged if subjected to high pressure, and animals under pressure show a sequence of behavior resembling the responses to heat or cold.

EFFECTS OF HYDROSTATIC PRESSURE ON ENZYME KINETICS

Enzyme-catalyzed reactions can be accompanied by volume changes of the reactants. Volume change (ΔV) between initial and final steady states in a reaction—that is, volume differences between reactants and products—are given by the relation

$$\Delta V = 2.3 \, RT \log \frac{K_{p2} - K_{p1}}{P_2 - P_1}$$

where ΔV is volume change in cm³, K_{p2} and K_{p1} are *equilibrium* constants at pressure P_2 and P_1. The change in volume is estimated from measurements of equilibrium constants at two pressures (13).

Volume of activation ($\Delta V‡$) relates to the change in free energy $\Delta G‡$ at different pressures. An energetics effect of increase in pressure-volume (PV) is an increase in energy of activation $\Delta G‡$ as described in Chapter 5. Pressure can accelerate or inhibit a reaction rate according to the relation

$$k_p = k_o \exp \frac{P\Delta V‡}{RT}$$

where k_p and k_o are *rate constants* before and after activation. $\Delta V‡$ is activation volume in cm³/mol; $\Delta V‡$ is obtained by measuring rate constants (k) at different pressures. The equation is the same as for ΔV except that rate constants are compared rather than equilibrium constants. $\Delta V‡$ is the difference in volumes of the activated enzyme-substrate complex $\Delta V_{ES}‡$ and the volume of ES at the ground state. This is summarized by the relation $\Delta V‡ = V_{ES}‡ - V_{ES}$. If the volume of ES‡ is less than V_{ES}, pressure accelerates the reaction; if the volume of ES‡ is more than V_{ES}, pressure inhibits the reaction. In pressure-activated reactions, $\Delta V‡$ is positive. For example, if the activated state is more ionized than the reactants, $\Delta V‡$ is negative and pressure speeds the reaction. If the activated state is less ionized than the reactants, $\Delta V‡$ is positive and pressure slows the reaction (12).

Figure 8-3 shows that activation volume $\Delta V‡$ for pyruvate kinase of several animals increases in proportion to the normal habitat temperature. The effect of pressure varies according to temperature. Bacterial luminescence shows an increase with pressure at 37.5°C, a decrease at 3°C at the same pressure, and little change at 26°C (Fig. 8-4) (12). A reaction in which V_{max} increases with increase in pressure is cleavage of FDP catalyzed by fructose-1,6-diphosphatase (FDPase) (9). An enzyme for which V_{max} decreases with pressure (i.e., $\Delta V‡$ is positive) is the conversion of PEP to pyruvate catalyzed by pyruvate kinase (PK) (Ch. 2) (9).

Alterations in volume of activation may be (1) structural, resulting in alteration of charge distribution during bond formation or breakage, or (2) changes in hydration of a protein, resulting in solvation or dehydration. Pres-

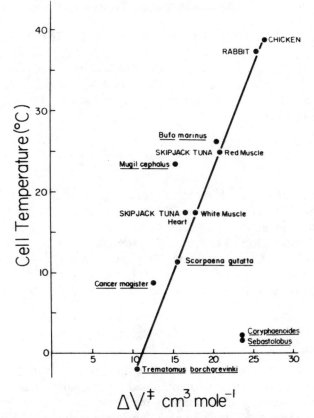

Figure 8-3. Relation between activation volume $\Delta V\ddagger$ and cell temperature for pyruvate kinase of several species. $\Delta V\ddagger$ may provide estimate of size of catalytic conformational change in the enzyme. Reprinted with permission from P. S. Low and G. N. Somero, *Journal of Experimental Zoology,* Vol. 198, p. 9. Copyright 1976, Alan R. Liss, Inc., New York.

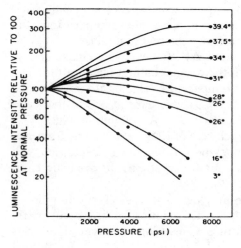

Figure 8-4. Influence of pressure on steady-state level of luminescence intensity of extracts of Achromobacter fischeri at given temperatures. Reprinted with permission from F. H. Johnson and H. Eyring. *Theory of Rate Processes in Biology and Medicine,* Copyright, 1974, John Wiley & Sons, Inc., New York.

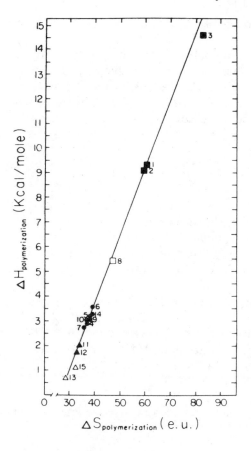

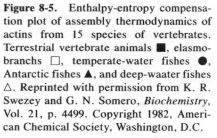

Figure 8-5. Enthalpy-entropy compensation plot of assembly thermodynamics of actins from 15 species of vertebrates. Terrestrial vertebrate animals ■, elasmobranchs □, temperate-water fishes ●, Antarctic fishes ▲, and deep-waater fishes △. Reprinted with permission from K. R. Swezey and G. N. Somero, *Biochemistry*, Vol. 21, p. 4499. Copyright 1982, American Chemical Society, Washington, D.C.

sure and temperature are interrelated such that some reactions may be inhibited by pressure at low temperatures and enhanced by pressure at high temperatures. In general, a rise in temperature increases hydrophobic and decreases ionic binding, whereas a rise in pressure increases coulombic (ionic) and decreases hydrophobic binding (10). Hydrophobic bonds are destabilized by cold and by high pressure; hydrogen bonds and ionic bonds are stabilized by cold and by high pressure. Denaturation at high pressure results from rupture of hydrophobic bonds, especially those of tertiary and quaternary structure.

In some polymer systems high pressure depolymerizes—for example, F-actin goes to G-actin. Some polymers dissociate to monomers with decrease in volume—for example, enolase, a dimer, dissociates to monomers under pressure. Enthalpy and entropy of polymerization of G- to F-actin from muscles of 15 species ranging in body temperature from $+39°C$ to $-1.9°C$ decrease, and heat stability decreases in the series from warm- to cold-bodied animals (Fig. 8-5). However, actins from deep-sea fish are exceptions to the usual trend; their protein shows high stability despite low T_b, probably because evolutionary selection for tolerance of pressure has been stronger than that for low temperature tolerance (26).

Pressure also affects lipid-protein interactions. Activity of the enzyme (Na + K)-ATPase increases with pressure and a plot of activity as a function of pressure shows a transition point which increases 36.1 atm/deg as temperature increases (5). Phospholipid bilayers and red blood cell membranes decrease in fluidity on cooling and under pressure such that 100 atm has the same effect as a 2°C decrease in temperature (14).

BIOCHEMICAL ADAPTATIONS OF DEEP-WATER ORGANISMS

Effects of pressure on enzyme kinetics may be interpreted in terms of volume changes; enzyme kinetics are correlated with adaptive differences in proteins. Two species of the fish *Sebastolobus* occur at different depths, *S. altivelis* at 550 to 1300 m, *S. alascanus* in shallower water, 180 to 440 m. The temperature range for activity is 4–12°C for both fish, and electrophoretic polymorphism is similar. Pressure brings about much less increase in K_m values of M_4 LDH for NADH and pyruvate in *altivelis* than in *alascanus* (21). In the deep-sea fish *Antimora rostrata* from depths of 1300–2500 m there is virtually no effect of pressure on K_m of LDH (19).

Differences in sensitivity to pressure of muscle LDH from deep-sea fishes and from shallow-water fishes are shown by measurements of K_m for NADH and pyruvate at 5°C (22,23):

	1 atm	68 atm	340 atm
LDH K_m for NADH			
shallow-water fishes			
Sebastolobus alascanus	23.4	44.1	68.4
Scorpaena	22.2	51.7	60.2
midwater fish			
Sebastolobus altivelis	21.7	33.7	31.4
deep-water fish			
Coryphenoides	24.6	29	29
Antimora	23.3	29.2	28.1
LDH K_m for pyruvate			
shallow-water fishes			
Scorpaena	0.373	0.523	0.511
Sebastolobus alascanus	0.399	0.508	0.550
midwater fish			
Sebastolobus altivelis	0.395	0.408	0.396
deep-water fish			
Coryphaenoides	0.353	0.368	0.331
Antimora	0.393	0.397	0.396

The K_m values of muscle LDH of deep water fishes are relatively insensitive to pressure at the low temperatures of their habitat; however, at surface temperatures the K_ms are more pressure-sensitive. The K_ms of LDH of shallow-water fishes increase more on warming at 1 atm than at 500 atm, but at 5°C K_m is similar at both high and low pressures.

The specific activity of *Antimora* M_4 LDH is 1/20 that of trout; this correlates with low metabolism of deep sea fish.

At 5°C a molecule of LDH from a deep-sea fish functions at about 60 percent the rate for shallow-water, cold-adapted fish (22). ADP competitively inhibits the binding of the coenzyme, NADH. The K_i (ADP) for LDH of beef muscle decreases as temperature rises and increases as pressure rises, whereas for the same enzyme from *Antimora* the K_i for ADP increases with rising temperature and falls slightly with increased pressure. The interpretation is that binding in beef LDH is to a hydrophobic site and shows a large positive $\Delta V\ddagger$, whereas in *Antimora* coulombic and ionic sites are dominant and $\Delta V\ddagger$ is negative (10).

In acetylcholine esterase (AChE) (purified from electric organ), two sites are identified: an anionic site of quaternary NH_4^+ to which substrate (ACh or its analog) binds; and an esteratic or catalytic site. An uncharged substrate, dimethyl acetate, and charged substrates, acetyl thiocholine and dimethyl ammonium Cl, were tested for K_m at different temperatures and pressures with results shown in Figure 8-6 (10). For the uncharged substrate (hydrophobic site) K_m decreases with rising temperature (T) and as pressure (P) increases; binding is less as P increases and T decreases. Cold and pressure weaken hydrophobic bonds. For the charged substrates K_m is little affected by T (acetylthiocholine) and decreases as P increases (dimethyl ammonium), hence its binding is coulombic.

Comparison of kinetics at 1 atm of AChEs for brain of a mammal, shallow-water fish, and deep-sea fish follows (10):

Subject	$\Delta G\ddagger$ (kJ/mol)	$\Delta H\ddagger$ (kJ/mol)	$\Delta S\ddagger$ (J/mol/K)	$\Delta V\ddagger$ at 6°C (cm³/mol)
beef	−25.2	−6.8	+59	
dolphin fish	−25.2	−1.7	+80	94
Antimora	−25.2	−26	−2.1	43

These data indicate that *Antimora* AchE has a high enthalpic and low to zero entropic energy component. The negative ΔH indicates that high temperature inhibits binding. Pressure disrupts ACh binding for shallow-water fish, not for abyssal fish (13).

Enzymes from deep-sea animals tend to be much more resistant to pressure inactivation than corresponding enzymes from shallow-water animals. Fructose diphosphatase (FDPase) from two cold-water species, abyssal rattail *Coryphaenoides* and from shallow-water Antarctic *Antimora*, was com-

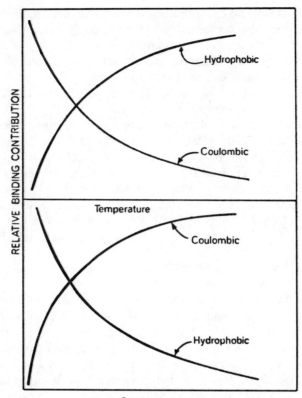

Figure 8-6. Effects of temperature and pressure on purified acetylcholine esterase. Hydrophobic contribution to binding shown by action of uncharged substrate analog 3,3-dimethylacetate; coulombic contribution shown by action at anionic site by charged inhibitor dimethylammonium ion. Coulombic contribution to binding favored by low temperature and high pressure; hydrophilic contribution disrupted by low temperature and high pressure. Reprinted with permission from P. W. Hochachka, K. B. Storey, and J. Baldwin, *Comparative Biochemistry and Physiology,* Vol. 52B, p. 14. Copyright 1974, Pergamon Press, Ltd., Oxford, UK.

pared with that from a mesopelagic fish *Stomias* and a shallow-water sheepshead *Pimelometopon.* Inactivation occurred in the enzyme from the rat-tail at 1370 atm, whereas inactivation of sheepshead enzyme was at 685 atm. The pressure sensitivity was less in the deep-water species. K_m of FDPase of deep-water fish is $0.5 \times 10^{-5}M$ and E_a is 8.5 kcal/mol at either high pressure or at 1 atm. FDPase from a shallow-water fish (sheepshead) is inhibited at 10,000 psi. A series of fish showing decreasing pressure sensitivity is *Pimelometopon* (an inshore fish) > *Stomias* (from 100 to 1000 m depth) > *Coryphaenoides* (1000 to 3000 m) (9).

Isocitrate dehydrogenase, IDH, from trout liver is sensitive to pressure and the K_m increases with pressure, whereas this enzyme from *Antimora* has a K_m which is insensitive to pressure but is temperature-sensitive (17).

Values of activation volume ($\Delta V\ddagger$) may be more closely correlated with depth of occurrence than are K_m values. The following table gives mean values of $\Delta V\ddagger$ for M_4 LDH from fishes of a series from shallow to deep water habitats (21,22):

	$\Delta V\ddagger$ (cm³/mol)
Shallow water	
Hippoglossus stenolepsis	12.7
Sebastolobus alascanus	12.0
Thunnus thynnus	10.7
Trematomus borchgrevinki	10.3
Deep water	
Sebastolobus altivelis	8.7
Lampanyctus regalis	7.4
Symbolophorus californiensis	5.8
Triphoturis mexicanus	1.8
Bajacalifornia burragei	0.3

Pyruvate kinase (PK) from muscle shows a linear relation between salt-insensitive $\Delta V\ddagger$ and cell temperature. The enzyme from rabbit or chicken has a $\Delta V\ddagger$ of 25 cm³/mol, from tuna dark and white muscle $\Delta V\ddagger$ of 18 at temperature of 17°C, from muscle of a crab *Cancer* 12 cm³/mol at a body temperature of 8–9°C. In the deep-sea fish *Trematomus* from −2°C, $\Delta V\ddagger$ is slightly negative or independent of pressure. In contrast, in another deep water fish *Coryphaenoides* $\Delta V\ddagger$ is positive by 3 ml/mol and K_m for PEP increases with pressure; its activation energy E_a is doubled when pressure is raised from 1 to 80 atm (10).

BIOCHEMICALS SENSITIVE TO HIGH PRESSURE

The preceding data indicate that animals from the deep sea have adaptations that make their enzymes less sensitive to pressure and cold than corresponding enzymes from shallow-water animals. Simpler molecules such as coenzymes (e.g., flavodoxin) are affected by pressures which are one to two orders of magnitude greater than those in deeper oceans, kilobars rather than 10–100 bars (30). Conformational changes can be observed by physical means, such as fluorescence polarization. Binding of small ligands to macromolecules may be either enhanced or destabilized and hydration of ions may be accompanied by decrease in volume. Covalent bonds are relatively incompressible but they can rotate under pressure. Larger changes occur in solvation, and decrease in molecular volume results from packing of water molecules around charged ionic groups. Some flavin complexes increase in volume, others decrease in volume under high pressure. Volume changes in flavodoxins are 5–75 ml/mol/kbar in the pressure range of 5–10 kbar. Pressurized yeast enolase dissociates to monomers, decreases in its molecular volume, and decreases fluorescence polarization. Lysozyme also decreases fluorescence and volume. Meta-myoglobin undergoes a reversible transition between native and denatured states according to pressure, temperature, and pH. Over a low temperature range the pressure required for denaturation increases to a maximum, and in a high temperature range the required pressure decreases. Structural changes in cofactors and oligomeric proteins do not occur to significant extent at the pressures at which biological effects occur at ocean depths (30).

MODIFICATION OF PRESSURE EFFECTS ON ENERGY METABOLISM

Benthic animals are subject not only to high pressures but also to cold and to scanty nutrients. The fact that chemolithotrophic bacteria can grow at high pressures indicates that pressure *per se* may not be limiting. Preceding sections show that the enzyme kinetics of benthic animals are affected by pressure differently from those of animals in shallow water.

Comparisons of enzyme activities (V_{max}) indicate the importance of factors other than pressure. Activity of muscle lactate dehydrogenase (LDH M_4) from white muscle correlates with bursts of rapid swimming. Measurements of V_{max} of several enzymes from muscle of shallow- and deep-water species of the fish *Sebastolobus* gave the following data:

	Depth of Occurrence (m)	Enzyme Activity (units/mg$_{ww}$)		
		LDH	PK	MDH
S. alascanus	100–400	57.8	7.1	4.5
S. altivelis	550–1300	24.9	4.3	3.0

The muscle of the shallow-water species has nearly twice the LDH activity of muscle from the deeper species. K_m values for M_4 LDH for the two fish are similar at 1 atm, but K_m for the shallow-dwelling species is 30 percent higher, measured at 69 atm. LDH of *S. alascanus* is pressure-sensitive and has histidine at position 115; *S. altivelus* is pressure-insensitive and has asparagine at position 115 (21).

Several species of ratfishes (macrourids) from different depths are compared for muscle enzyme activities as follows:

Species	Depth of Greatest Abundance (m)	Activity (units/g_{ww})			
		LDH	PK	MDH	CS
Nezumia bairdni	600	6.9	4.6	17.5	0.62
Coryphaenoides rupestris	1000	16.0	5.4	9.7	0.58
C. carapinus	2000	4.7	5.9	6.8	0.50
C. armatus	2900	53.1	7.2	18.5	0.79
C. leptolepis	3500	4.3	2.6	6.9	0.41

These data show little direct correlation with depth of occurrence.

Quantitative data regarding swimming activity are not available, but general field observations indicate swimming in the following series: *C. armatus* is an active swimmer for prey, *N. bairdni* and *C. carapinus* feed on benthic prey and are intermediate in activity, *C. leptolepis* is a bottom-dweller with little swimming activity (21). *C. armatus* has high glycolytic activity.

Comparison of LDH and PK activities in muscles of 16 shallow- and 11 deep-dwelling species of fish showed glycolytic activities as much as 100 times greater in active swimmers than in "sit and wait" fishes, regardless of depth (24). Activity of glycolytic enzymes in brain showed no correlation with swimming vigor or depth of habitat in any of the 27 species.

There is much evidence that energy liberation, as indicated by oxygen consumption, is lower for deep-sea animals than for shallow-water species of the same kind, when measured at 1 atm (Fig. 8-7). Values of $\dot{V}_{O_2}$ for crustaceans (3) and for fishes (4,27) show considerable diversity among species, but the mean values of shallow-water forms are ten times higher than of deep-water animals. Quantitative comparisons of nutrition are not available, but the low metabolism of benthic forms is probably the result of scanty nutrients. Acclimatization of animals to reduced food supply lowers metabolic capacity.

In summary, pressure to which animals are subjected at different depths is only one factor influencing the activity of energy-yielding enzymes. Glycolytic activity of muscle correlates more with way of life, locomotion, and nutrition than with depth of habitat. Swimming is partly determined by central nervous functioning. It would be of interest to compare synaptic functions and electrical activity of such regions of brain as cerebellum in fish from different depths.

CONCLUSIONS

Microorganisms and animals that live in deep oceans are subjected to low temperature (2°C), low nutrients, no sunlight, and high pressure (300–3000

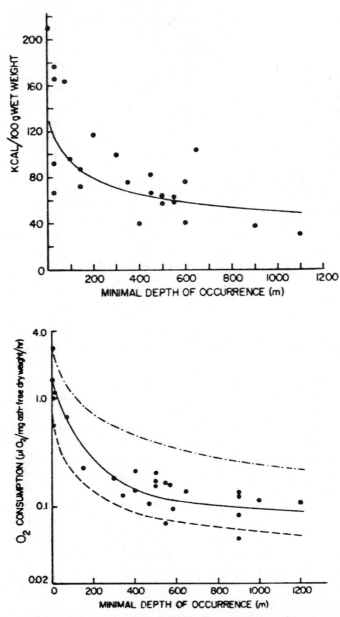

Figure 8-7. (*a*) Relation between minimal depth of occurrence and energy content (kcal/ 100 g wet w) for midwater fishes. (*b*) Relation between maximal respiration (--•--•---), respiration at 30–70 mm Hg O_2 (————), and minimal respiration (-------) and minimal depths of occurrence for midwater crustaceans. Reprinted with permission from G. N. Somero et al., in *Biochemical and Physiological Adaptations of Deep Sea Animals in the Sea,* Vol. 8, G. T. Rowe, Ed., Copyright 1983, John Wiley & Sons, Inc., New York.

atm). Oxygen is not sufficiently reduced to be limiting. Adaptations of animals in the depths include low metabolism, very slow growth, and behavioral intolerance of decompression. Some bacteria are barotolerant, others are obligate barophiles. Shallow-water animals pass through a series of behavioral responses when compressed experimentally. Some activities— heartbeat, muscle contraction—are increased by moderate pressure and are reduced by high pressure. Pressure effects on enzyme-catalyzed reactions result from changes in protein conformation, hydration, lipid-protein interaction, polymerization or depolymerization. Activation volume ($\Delta V\ddag$), the difference between energy of active enzyme-substrate (ES$\ddag$) and unactivated ES can increase or decrease rates of reaction according to whether the ΔV is negative or positive.

Hydrostatic pressure can reverse the denaturing effects of high temperature and the effects of some anesthetics; pressure can stop cell movement by decreasing cytoplasmic viscosity. Pressure effects can be enhanced by high temperature. Simple biochemical systems are not affected by pressure changes in the biological range, but can be altered structurally by very high pressure. Metabolic enzymes of deep-sea animals differ from those of shallow-water animals in having K_m, k_{cat}, E_a, and levels of pressure denaturation that are adaptive for high pressure. Much of the variability in metabolism is in the relation of V_{max} to locomotor behavior rather than to depth of occurrence. No genetic or sequence data are available to account for the observed differences in kinetics of enzymes from deep sea and from shallow water. No acclimation experiments have been made.

Some enzyme proteins of deep-sea animals have been found to be adapted in kinetic properties to high pressure. K_m values of enzymes from shallow-water forms rise steeply on compression. Activation volumes, $\Delta V\ddag$, tend to be larger for deep-sea forms; this is adaptive in that enzymes for which the energy of an active enzyme-substrate complex is much greater than the ES in its ground state, tend to be inhibited or less affected by pressure than the converse. Many nonenzyme molecules, such as cofactors and hemoglobin, are relatively unaffected by pressure unless it is extremely high. No genetic data or nucleotide sequence data are available to account for the differences in enzyme kinetics.

The interactions between pressure, temperature, and nutrition may confound attempts to identify adaptations solely to pressure. Growth and division of a few kinds of bacteria can occur under high pressure; however, mitosis of cells from one atmosphere pressure is blocked by relatively low pressure increase. Diversity of respiration values and of activity of enzymes such as those of muscle glycolysis may be more correlated with life style (feeding activity) than with pressure.

Conduction in excitable membranes and synaptic transmission (neuromuscular) of animals that live at one atmosphere can be blocked by subjection to the pressure at which deep-water animals live. No data are available regarding membrane properties of deep-water animals. Reactions of deep-water animals to decompression follow a behavioral sequence resembling that to heat or cold. Death from either high or low pressure may be due more to failure of central nervous systems than of energy supply. The kinetic adaptations of enzyme proteins may correlate with, but not be critical for, tolerance of the high pressure at ocean depths.

REFERENCES

1. Baross, J. and J. Deming. *Nature* **303**:423–426, 1983; **307**:749, 1984. Methanogenic bacteria from hot vents and high pressure.

2. Campenot, R. B. *Compar. Biochem. Physiol.* **52B**:133–140, 1975. Effects of pressure on crustacean neuromuscular junctions.

3. Childress, J. J. *Compar. Biochem. Physiol.* **50A**:787–799, 1975. Respiration of midwater crustaceans.

4. Childress, J. J. and G. N. Somero. *Marine Biol.* **52**:263–283, 1973. Enzymes of fish from different depths.

5. de Smedt, H. et al. *Biochim. Biophys. Acta* **556**:479–489, 1979. Pressure effects of lipid-protein interactions.

6. Felbeck, H. and G. N. Somero. *Trends Biochem. Sci.* **7**:201–204, 1982. Primary production in hydrothermal vent organisms.

7. George, R. L. and J. P. Maxum. *Int. Rev. gesamte Hydrobiol.* **59**:175–186, 1974. Effects of pressure on aquatic animals.

8. Hand, S. C. and G. N. Somero. *Biol. Bull.* **165**:167–181, 1983. Energy metabolism of hydrothermal vent animals.

8a. Harper, A. A. et al. *J. Physiol.* **273**:70–71P, 1977; **311**:325–339, 1981. Hydrostatic pressure effects on membrane of *Helix* neurons.

9. Hochachka, P. W. et al. *Marine Biol.* **7**:285–293, 1970. Pressure effects on FDPase.

10. Hochachka, P. W. et al. *Biochem. J.* **143**:535–539, 1974. *Compar. Biochem. Physiol.* **52B**:13–18, 1975. Pressure and temperature adaptation of acetylcholinesterase.

11. Jaenicke, R. et al. *Europ. J. Biochem.* **23**:150–159, 1971. Pressure on LDH.

11a. Jannasch, H. W. *Oceanus* **27**:73–78, 1984. *Science* **180**:641–643, 1973. *Sci. Am.* **236**:42–52, 1977. Microbial life in the deep sea.

12. Johnson, F. H. and H. Eyring. *Theory of Rate Processes in Biology and Medicine*, Wiley, New York, 1974. Ch. 4, pp. 273–369. Hydrostatic pressure.

12a. Littler, M. M. et al. *Science* **227**:57–59, 1985. Deepest known plant life.

13. Low, P. S. and G. N. Somero. *Proc. Natl. Acad. Sci.* **72**:3014–3018, 1975. *J. Exper. Zool.* **198**:1–12, 1976. Activation volume of enzymes.

13a. Macdonald, A. G. *Biochim. Biophys. Acta* **507**:26–37, 1978. Phase transitions and fluidity of membranes as affected by anesthetics and pressure.

14. MacDonald, A. G. et al. Ch. VI, pp. 385–403 in *Animals and Environmental*

Stresses, Ed. R. Gilles, Wiley, New York, 1980. Tolerance of pressure by animals.

14a. MacDonald, A. G. et al. *J. Physiol.* **292:**44P, 1979; **296:**105P, 1979. Effect of pressure on neuromuscular transmission frog and on ATPase.

14b. Marsland, D. pp. 259–312 in *High Pressure Effects on Cellular Processes,* Ed. A. M. Zimmerman, Academic Press, New York, 1970. Pressure-temperature interaction on cell division.

15. Menzies, R. J. *Int. Rev. Hydrobiol.* **59:**153–160, 199–205, 1974. Pressure tolerance by marine animals.

16. Minton, K. W., et al. *Nature* **285:**482–483, 1980. Pressure effects on anesthetized cells.

17. Moon, T. W. and K. B. Storey. *Compar. Biochem. Physiol.* **52B:**51–57, 1975. Volume activation of isocitrate dehydrogenase from *Antimora.*

18. Pequex, A. and G. Gilles. *Compar. Biochem. Physiol.* **59B:**207–212, 1978. Pressure effects on membrane ATPases.

19. Prosser, C. L. et al. *Compar. Biochem. Physiol.* **52B:**127–131, 1975. Pressure effects on heart.

20. Sanders, H. L. pp. 222–243 in *Changing Scenes in Natural Sciences,* Ed. C. S. Goueden, Philadelphia Academy of Sciences, Vol. 197. Evolutionary ecology and the deep-sea benthos.

21. Siebenaller, J. F. and G. N. Somero. *J. Compar. Physiol.* **129:**295–300, 1979. *Physiol. Zool.* **55:**171–179, 1982. *Biol. Bull.* **163:**240–249, 1982. *Bioch. Biophys. Acta* **786:**161–169, 1984. Biochemical adaptation of fishes from different depths.

22. Somero, G. N. et al. Ch. 11, pp. 257–278 in *The Environment of the Deep Sea,* II, Ed. W. G. Ernst and J. G. Morin, Prentice-Hall, Englewood Cliffs, NJ, 1982. Ch. 7, pp. 261–330 in *The Sea,* vol. 8, Ed. G. Rowe, Wiley, New York, 1983. Ch. 11, pp. 257–278 in *Environment of the Deep Sea,* II, Ed. W. G. Ernst, Prentice-Hall, Englewood Cliffs, NJ, 1982. Biochemical adaptation of deep sea crustaceans and fishes.

23. Somero, G. N. et al. *Nature* **266:**276–278, 1977; **282:**100–102, 1979. LDH of deep sea fishes.

24. Sullivan, K. M. and G. N. Somero. *Marine Biol.* **60:**91–99, 1980. Muscle and brain enzymes in relation to depth in the sea and behavior.

25. Sverdrup, H. U. et al. *The Oceans; Physics, Chemistry, and Biology,* Prentice-Hall, Englewood Cliffs, NJ, 1942.

26. Swezey, R. R. and G. N. Somero. *Biochemistry* **21:**4496–4503, 1982. Pressure effects on actins.

27. Torres, J. J. et al. *Deep-Sea Res.* **26A:**185–197, 1979. O_2 consumption of midwater fish.

28. Turekian, K. K. et al. *Proc. Natl. Acad. Sci.* **72:**2829–2832, 1975. Slow growth of deep-sea clams.

28a. Walker, E. and D. N. Wheatley. *J. Cell. Physiol.* **99:**1–14, 1979. Effects of pressure on cell division and protein synthesis in *Tetrahymena.*

29. Wann, K. T. and A. G. Macdonald. *Compar. Biochem. Physiol.* **66A:**1–12, 1980. Pressure effects on excitable cells.

30. Weber, G. and H. G. Drickamer. *Q. Rev. Biophys.* **16:**89–112, 1983. Effect of high pressure on proteins and other biochemicals. Weber, G. et al. *Proc. Natl. Acad. Sci.* **71:**1264–1266, 1974. *Biochemistry* **16:**4879–4886, 1977; **20:**2587–2593, 1981. Effects of high pressure on isoalloxazine, flavodoxin, enolase.

31. Yayanos, A. A. et al. *Science* **200:**1056–1058, 1978; **205:**808–810, 1979. *Proc. Natl. Acad. Sci.* **78:**5212–5215, 1981. Bacteria from deep-sea amphipods.

9

WATER AND IONS

Life originated in an aqueous medium, and water is the universal biological solvent. Many physical properties of water make it a unique fluid for life processes: its temperature of solidification and heat of fusion (-80 cal/g); its high temperature of vaporization, which permits water to be fluid at most temperatures on earth; its high heat of vaporization ($+836$ cal/g) and lower density as a solid than as a fluid; and water's properties as a solvent for most inorganic and many organic solutes. The high specific heat of water is important as a buffer of climate.

Aquatic organisms are different in ionic composition from the medium. In the earliest organisms a bounding membrane was essential for separating intracellular from extracellular fluids. Plasma membranes became modified when organisms occupied new ecological niches. Only after protoplasm had become separated from the medium could protein and nucleic acid synthesis, degradation, and replication have evolved. The first sign of cell death is solute leakage across a cell membrane, potassium out, sodium and calcium in. Plasma membranes regulate intracellular osmoconcentration, cell volume, and ionic composition by multiple mechanisms. Plasma membranes consist of hydrophilic proteins and hydrophobic lipids; large protein molecules extend through a bilayer of phospholipids.

The composition of living matter is different from the earth's crust; extracellular fluids of organisms resemble sea water in a general way, but some elements have been selected, others excluded. Intracellular fluids differ from extracellular fluids. Virtually all cells have higher intracellular concentrations of potassium and lower concentrations of sodium and chloride than the medium. Sea water has much more sodium (450 mM) than potassium (10 mM). One of the most constant of biological quantities is the intracellular concentration of potassium, which is usually between 100 and 150 mM. In muscles from a series of animals—marine invertebrates to mammals—the ratio of K_i/K_o (potassium inside to outside) varies only threefold, whereas the Na_i/Na_o varies more than 20-fold (15). The amounts of calcium and magnesium in cells are variable, but in general they are lower than in the medium, Mg with a steeper gradient than Ca. However, calcium is commonly bound to organic molecules and its free ionic concentration is very low in protoplasm. Hydrogen is also an ion for which a gradient exists across cell membranes. Generally the concentration of H^+ inside cells is higher than outside. The most abundant intracellular anions are organic, predominantly proteins, amino acids, and organophosphates. Osmoconcentrations in cells are more constant than extracellular osmoconcentrations. Differences in composition between intracellular and extracellular fluids indicate that selective permeability is a critical characteristic of cell membranes and that it must have evolved early.

Another property of living cells that must have evolved very early is volume regulation. As organic molecules were synthesized or retained after uptake from the medium, they added to inorganic ions and raised the intracellular osmoconcentration. Mechanisms evolved that maintained constant cell volume. Gradients of solutes and water, and constancy of volume, are maintained by the so-called Donnan equilibrium. This is a dynamic state (not a true equilibrium), in which a concentration gradient may oppose an electrical gradient. This is usually written

Inside	Outside
C_i^+	C_o^+
A_i^{-d}	A_o^{-d}
A_i^{-n}	A_o^{-n}

where C^+ represents cations inside or outside, A^{-d} represents diffusible anions, and A^{-n} nondiffusible anions. At steady state

$$C_i^+ = A_i^{-d} + A_i^{-n}$$

$$C_o^+ = A_o^{-d}$$

if $A_o^{-n} = 0$,

$$C_i^+ + (A_i^{-d} + A_i^{-n}) > C_o^+ + A_o^{-d}$$

For example, since the ratios of diffusible cations and anions are equal, $K_i/K_o = Cl_o/Cl_i$; but since $K_i = Cl_i + A_i$ it follows that $K_i > K_o$. Gradients of two cationic species may constitute a double Donnan equilibrium.

In a Donnan situation three gradients result: (1) an electrical gradient is present so that the inside is negative to the outside,

$$E = \frac{RT}{F(\ln \frac{K_i}{K_o})}$$

Where E is potential, R is universal gas constant, T is temperature Kelvin, F the Faraday constant. K_i and K_o are K^+ inside and out.

(2) an osmotic gradient consists of higher osmotic concentration (OC) inside than out $[OC_i > OC_o]$; and (3) there is an asymmetrical distribution of diffusible ions. Diffusible cations such as Na^+ tend to diffuse inward; because of the osmotic gradient water tends to diffuse inward. The net result is a tendency of the inner compartment (cell) to swell. There must be mechanisms to maintain volume and to permit retention of ions such as K^+ in preference to Na^+. Because in cells the concentrations of organic anions are generally high, there is an intracellular deficit of inorganic anions compared with their concentration in extracellular fluids.

Measurements on cells that have been put into several osmoconcentrations show that cells of many kinds—sea urchin eggs, red blood cells, muscle fibers, tumor cells—fail to swell or shrink as much as if they were simple osmometers. Several mechanisms have been proposed to account for this. One is maintenance of hydrostatic pressure by an inelastic or contractile cell wall. In bacteria, there may be steep gradients of hydrostatic pressure. Plant cells oppose the osmotic pressure with a rigid cellulose wall and thus are turgid. In animal cells the hydrostatic pressure is not sufficient to account for volume regulation; contractile cell membranes may modify cell shape. Another regulatory mechanism is that some cell water is bound to solute and thus is not available for osmotic shrinking or swelling. However, measurements indicate that differences between free and bound water are insufficient to account for disequilibrium, and that most of the cell water is available as solvent (56). A third volume-regulatory mechanism of importance is change in membrane permeability with change in cell volume. Passive fluxes of K^+ and of free amino acids decrease as osmotic concentration of the medium increases; this permits internal osmoconcentration to rise with increased external concentration.

A widespread mechanism for volume regulation of cells is active extrusion of Na^+ or H^+ ions, minimizing the electrical gradient and at the same time tending to balance the osmotic disequilibrium. Energy as ATP is provided for these pumps which are ion ATPases. A Na-K pump mechanism occurs in most animal cells, and the ratio of Na transported out to K transported in may not be unity; hence the pump may be electrogenic. Most prokaryotes and many plants have a K^+-H^+ exchange pump which maintains high intracellular potassium. Other ion exchanges and coupled movements may be one-for-one; these use carrier molecules in the membranes and derive energy from ion gradients. Details of these exchange pumps will be given later. It is concluded that since transmembrane electrical gradients are normally minimized and osmotic swelling is prevented, active ion transport must have evolved as early as cells.

SELECTION OF IONS

In all cells—bacteria, plants, and animals—the most abundant cation is potassium; impermeant organic anions serve to maintain internal electroneutrality. The mechanisms by which K is concentrated and Na, Ca, and Mg are excluded, differ in the cells of the three kinds of organisms (bacteria, plants, and animals). Why was potassium selected and the more abundant sodium excluded from primitive cells? No single answer can be given, and answers that have been proposed are in part speculative. One idea is that the hydrated sodium ion is larger (2.8 Å diameter) than hydrated potassium (2.0 Å). Nonhydrated diameters are in the opposite direction—Na 0.95 Å and K 1.33 Å (29). A related hypothesis (117) is based on the differences in free energies of ion binding. Nuclear magnetic resonance (NMR) measurements show that intracellular water is more ordered than water in free solution. When partial mol volumes of water are measured, K^+, Rb^+, and Cs^+ ions are found to perturb water structure, Na^+ and Li^+ to order water structure; ions appear to be excluded from binding to solutes as their water-ordering power increases. The free energy of ion exchange is given by

$$\Delta G_{(exchange\ equilibrium)} = \Delta G_{(binding)} - \Delta G_{(hydration)}$$

A model system for ion-selective membranes consists of ion exchangers and glass electrodes selective for five alkali cations (34). Of the 120 possible permutations of the five cations in biological systems, only 11 are found in a permeability series. Selectivity is determined by ionic size, steric effects, charge, and electrostatic effects. Entry into cells is described in terms of charged channels or of binding to proteins. Cation permeability is controlled by acidic groups on the membrane with pK_a near or above 4.5; anion permeability is controlled by basic groups with pK_a of less than 3.0 (29). A nerve membrane can shift in conductance ratios from one univalent cation sequence to another, for example, between rest and activity. Corresponding membranes of different species may not have the same cationic sequences: permeability of gallbladder of rabbit differs from that of bullfrog (29). The cation permeability sequence in bullfrog gallbladder is $NH_4^+ > Rb^+ = K^+ > Ca^+$. The cation series for depolarization of crab nerves differs from that of frog nerve.

A quantitative analysis of the interaction of cell volume, membrane potential, and ion transport indicates that in animal cells: (1) volume regulation results from Na^+ extrusion; (2) the cell interior is electrically negative by a significant amount; and (3) resting K^+ permeability (P_K) is much greater than Na^+ permeability (P_{Na}) (60). Volume regulation requires a low ratio of P_{Na}/P_K—this is observed in marine ciliates, frog oocytes, and mammalian red blood cells. Decrease in volume (osmotic shrinking) results in a slight increase in P_{Na}/P_K. Volume decrease following osmotic swelling of fish red blood cells results in a large increase in P_K relative to P_{Na} (11 times increase for potassium permeability compared with four times increase for sodium) so that P_{Na}/P_K decreases and K^+ leaks out when volume is restored at a reduced osmotic concentration (101). In crab muscle, volume regulation is not by K^+ fluxes but by gain or loss of organic solutes. Red blood cells of camels swell when the animals are hydrated and shrink when they are dehydrated.

In bacteria, as in mitochondria of eukaryotes, high internal K^+ is maintained by a proton-K exchange pump. In higher plants high K^+ in the protoplast is maintained by active K absorption, ATP dependent, but not by an Na-K exchange pump as in animal cells.

Some ions function in enzyme activities. Potassium promotes oxidation and oxidative phosphorylation in mitochondria which are in state 4 (when all P-nucleotide is present as ATP and ADP is absent). The mitochondria lose capacity to be stimulated by ADP if K_i^+ is decreased. Reduced K_i^+ decreases outward pumping of Na^+; increases in Ca_i^{2+} uncouple mitochondrial metabolism from phosphorylation. K^+ activates numerous enzymes—degradative, biosynthetic, and others (74).

Protein synthesis is promoted by intracellular K^+. A mutant of *E. coli* lacks capacity to concentrate K_i^+; its K_i^+ approaches the K_o^+ value. Such mutant bacterial cells can make RNA but not protein. Transfer of amino acids to ribosomal RNA requires K^+. Selection of potassium in relation to amino acid transfer to ribosomal RNA may have occurred early in evolution (74).

Many plants require K^+ for growth but do not require Na^+. *Neurospora* maintains K_i^+ at 180 mM when the environmental K^+ range is 0.3 to 100 mM. Growing yeast accumulates 25 times as much K^+ as Na^+.

Halophilic bacteria from salt lakes require 10–15 percent NaCl in the medium for survival, and for best growth require more than 20 percent NaCl. Salt lake bacteria concentrate K more than Na intracellularly; most need Mg for growth. One action of KCl and NaCl in saline lake bacteria is to stabilize critical proteins by molecular shielding. Preference for K^+ over Na^+ at a low level ($K^+/Na^+ > 2$) is found for LDH, MDH, and cytochrome oxidase; there is moderate preference for citrate synthetase and SDH; high preference, $K^+/Na^+ > 5$ occurs for isocitrate dehydrogenase, SDH and amino acyl transferase and RNA synthetase. More stabilization of cell membranes is accomplished by Na^+ than by K^+. At low concentrations of K^+ there may be shielding of enzyme proteins; at high concentrations of K^+ there may be stabilization of hydrophobic interactions (70,71).

In summary, numerous functions are served by the high intracellular K relative to Na.

ECOLOGY OF WATER AND IONS

Adaptations for osmotic and ionic balance enable free-living cells and multicellular organisms to survive in diverse environments—oceans, estuaries, fresh water, hosts (as parasites), moist air, and dry deserts. Most marine and parasitic unicellular organisms are isosmotic with their environment, hence need regulate only volume and ionic composition. Freshwater unicellular organisms are hyperosmotic, and most of them have some means for extruding the water that enters osmotically; extrusion of water is obligatorily coupled to loss of some ions, and a mechanism for active uptake of ions is present. A few kinds of aquatic eggs are virtually impermeable until after the embryo develops its water-excreting capacity. Aquatic plants have protoplasts and vacuoles hypertonic to the water; in land plants the cells are hyperosmotic to the extracellular fluid; cells of both aquatic and land plants have rigid cellulose walls that prevent osmotic swelling. In multicellular animals the osmoconcentration within cells is essentially the same as that of the extracellular fluid. Many aquatic animals are poikilosmotic; the body fluid osmoconcentration conforms, rises, or falls with the concentration of the medium (some marine and estuarine animals); but all aquatic animals regulate ion composition and their cells regulate volume. Some aquatic animals are homeosmotic and regulate body fluids either hyperosmotically in dilute medium or hypoosmotically in concentrated medium (Fig. 1-2). In terrestrial animals, the permeability of respiratory membranes and integument may be low and thus protect against evaporative stress. Regulation of ion composition in cells must have evolved very early. Many animals that are isosmotic to sea water regulate concentration of specific ions in body fluids. Osmoregulation at the whole animal level probably evolved relatively recently.

Salinity of aquatic environments is as follows:

Hypersaline Lakes	Sea Water	Brackish Water	Fresh Water
7–9 osmolal	1.01 osmolal		
50–250 0/00	35 0/00	5–30 0/00	<0.5 0/00
Δf.p. $-13.5°C$ to	$-1.86°C$		
$-15°C$			

Organisms that tolerate a wide range of salinity are euryhaline; those that are restricted to a narrow range are stenohaline. The highest osmotic concentrations in which aquatic organisms live are hypersaline lakes where algae such as *Dunaliella* and bacteria such as *Halobacterium* live. In some of these habitats (e.g., the Dead Sea) there may be high concentrations of un-usual salts, such as $MgSO_4$. A few algae, protozoans, and crustaceans (*Artemia*) live in saline ponds where salts are precipitating, on the edge of Great Salt Lake, or in ponds where sea salt is obtained by evaporation. Sea water is more dilute where there is inflow from rivers than in open ocean, and more concentrated where there is much evaporation. In open oceans, sea water is approximately at 1000 mosm. Geochemical evidence indicates that the con-centrations of ions in sea water have changed very little since the Precam-brian. Cations occur in sea water in the order of decreasing concentration: $Na^+ > Mg^{2+} > Ca^{2+} > K^+$. Anions occur in sea water in the order $Cl^- > SO_4^{2-} > HCO_3^- >$ phosphate (Table 9-1A and B). In estuaries bottom-dwelling organisms may be subject to tidal cycles ranging in content from nearly full-strength sea water to fresh water. During evolution, many organisms have made exit from the oceans to fresh water and some have reentered sea water. A few euryhaline animals normally alternate between sea water and fresh water, anadromous animals to breed in fresh water, catadromous to breed in the sea. Some crustacean species spend only a short time—two or three larval stages—in sea water, and the rest of their lives in fresh or brackish water. Many kinds of animals have made the transition from fresh water to land; some of them are in water for part of a life cycle only—amphibians, for example. Various terrestrial plants and animals have morphological and

Table 9-1A. Composition of Sea Water, Plasma, and Muscle of Dogfish Squalus (98)

	Sea Water	Plasma	Muscle
mOsm	965.0	993.0	1023.0
Na	453.0	296.0	42.0
K	9.6	7.2	119.0
Ca	9.9	2.9	2.1
Mg	59.6	3.5	12.9
Cl	529.0	276.0	35.9
SO_4	27.3	3.1	1.2
Total P	—	595	91.3
Lactate	—	—	24.0
Urea	—	308.0	333.0
TMAO	—	72.4	180.0
Betaine	—	9.0	100.0
Creatine	—	0.126	48.2
Creatinine	—	0.046	1.0
Amino acids	—	11.6	108.0
NH_4	—	0.4	4.7
Total nonprotein N	—	838.0	1447.0

Table 9-1B.   Ions in Plasma (P) and Muscle (M) of Eel from Sea Water and Fresh Water; Plasma of Lamprey from Fresh Water (98)

	Eel—Sea Water		Eel—Fresh Water		Lampetra—Fresh Water
	P.	M.	P.	M.	P.
Na	189.0	27.0	172.0	30.0	138.0
K	4.7	131.0	4.2	126.0	4.3
Ca	4.9	7.7	3.0	6.6	2.8
Mg	2.5	15.0	1.4	13.0	1.2
Cl	158.0	26.0	138.0	22.0	121.0

physiological adaptations for different degrees of humidity—for moist air in tropical rain forests, moisture gradients in soils, or dry air in deserts.

Organisms occur in hypersaline lakes, sea water, brackish estuaries, fresh water, moist terrestrial environments, and deserts. Specific adaptations exist for life in each habitat.

VOLUME REGULATION, OSMOREGULATION, AND IONIC REGULATION OF WHOLE ORGANISMS

Cell volume regulation is a universal necessity because of the presence of Donnan relations. When a euryhaline animal is transferred from one salinity to another, water or salts or organic solutes are lost or gained after an initial swelling or shrinking, and the volume then returns to "normal." The time required for regulation varies from hours to days. The sensing mechanisms for volume control are not well understood; how does a worm or mollusc "know" when it is at its "correct" volume and osmoconcentration? One hypothesis is that cytoskeletal proteins respond to stretching. However, volume-regulating responses in some cells (Ehrlich ascites) do not occur when osmoconcentration is increased by sucrose; but volume regulation does occur when osmoconcentration is changed by salt addition or withdrawal (57).

Several patterns of osmoregulation at the whole-animal level are recognized. Many marine conformers maintain an internal osmoconcentration higher than that of the medium by a small, constant amount (Fig. 9-1). The function of the elevated concentration in osmoconformers is unknown. Even when they are isosmotic with the medium, most conformers show ionic regulation; they reabsorb essential ions and excrete nonessential ones. Most estuarine animals show osmotic conformity when in sea water or in water of higher concentration than they normally tolerate.

Most brackish-water animals show hyperosmotic regulation when in dilute sea water. Freshwater animals regulate internal concentration when in

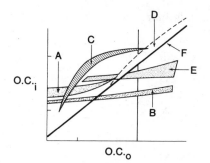

Figure 9-1. Diagram of several patterns of internal osmoticity as function of osmoticity of medium. (A) Strong hyperosmoregulators live in fresh water and are limited in capacity to live in brackish water. (B) Animals that are hyperosmotic in fresh water, hypoosmotic in sea water (e.g., euryhaline fishes). (C) Weak hyperosmotic regulators live in estuaries, do poorly in sea water (e.g., estuarine crabs and molluscs). (D) Osmoconformers are marine invertebrates which maintain slight hypertonicity above and below seawater concentration. (E) Strong hyper- and hypoosmotic regulators at high concentrations (e.g., terrestrial crabs). (F) is line of osmotic equality.

a very dilute medium. They have (1) low permeability of integument, (2) capacity to extrude water, (3) capacity to retain salts which have obtained by inward diffusion or drinking, (4) excretion of hypoosmotic urine, and (5) active absorption of ions against a concentration gradient. Maintenance of a steep gradient requires much energy—a few sluggish animals, such as freshwater bivalves, reduce the energy needed by tolerating low concentrations of body fluids; active insects, crustaceans, and vertebrates expend one-third of their metabolic energy for osmotic and ionic regulation at cell membranes. The energy used at interfaces with the environment is less.

Marine elasmobranch fishes remain hyperosmotic to the medium by retaining urea in blood and tissues. Elasmobranch tissues also have variable amino acid concentration (Table 9-1). These fish load their eggs with urea. When skates from normal sea water were acclimated to 50 percent sea water, their plasma osmotic concentration decreased from 1100 to 700 mosm; muscle cell water remained constant, muscle amino acids decreased from 214 to 144, urea went from 394 to 264, and trimethylamine oxide decreased from 64 to 36 mM. Osmoregulation in skate muscle is by alanine and sarcosine, in skate brain by taurine (123).

A few kinds of marine invertebrates, mainly those which spend much time on shore or in hypersaline lakes, are capable of hypoosmotic regulation whenever they remain in a concentrated medium (Fig. 9-1). They keep their body fluids at an osmotic concentration which is near that of sea water. Hypoosmotic regulation in normal sea water is found also in marine bony fishes, marine reptiles, and mammals. The mechanisms of hypoosmotic regulation are the reverse of those for hyperosmotic regulation; however, dif-

Table 9-2.   Composition in mM of Body Fluids of Marine Animals and of Sea Water (94a)

	Na	K	Ca	Mg	Cl	SO$_4$
Aurelia (gel)	454	10.2	9.7	51.0	554	14.6
Asterias (vascular)	428	9.5	11.7	49.2	487	26.7
Golfingia	508	11.5	11.2	38.8	561	26.8
Aplysia	492	9.7	13.3	49	543	28.2
Maja	500	12.7	13.9	45.2	569	14.3
Limulus	445	11.8	9.9	46	514	14.1
Hemilepidotus (marine teleost)	184	5.6	2.8	2.5	170	
Seawater	450–500	9.8–11	10	50–54	548–555	24–28

ferent cells are probably involved in the active transport. In terrestrial animals, the adaptations are mainly for reduction of evaporative and excretory loss of water. Water loss via the integument is lessened by waxy or horny coat in insects and reptiles. Loss of respiratory water from insects is lessened by spiracles and branched breathing tubes (tracheae). Water is conserved in plants by stomata that can be opened or closed. The urine excreted by some insects or mammals may be hyperosmotic to blood by as much as tenfold. Urine of birds and reptiles is slightly hyperosmotic.

Virtually all marine animals, even osmoconformers, regulate the ion concentrations of body fluids (Table 9-2). Sponges and some coelenterates use sea water as extracellular fluid. Jellyfish tissues may have lower SO$_4^{2-}$ than sea water. Compared with the concentrations in sea water, some invertebrates have lower Mg^{2+} and SO$_4^{2-}$, and some have higher K$^+$ and lower Na$^+$. Tissues of marine fishes, hyperosmotic elasmobranchs, and hypoosmotic teleosts contain lower and different concentrations of ions than sea water. Excretory organs evolved in osmoconformers as ion-regulating organs, and in many aquatic animals the organs used for excretion of divalent ions—Mg^{2+}, SO$_4^{2-}$, sometimes Ca^{2+}—are not the same as the organs for excretion of univalents—Na$^+$, K$^+$, Cl$^+$. A terrestrial animal has its characteristic extracellular ionic concentration; some herbivores have higher K$^+$ and lower Na$^+$ than omnivores and carnivores. Higher plants are usually lower in Na$^+$ than animals, and some plants can grow without Na$^+$. Plants such as mangroves, that grow in sea water, excrete salts, especially MgSO$_4^{2-}$, via the leaves. Halophilic bacteria that grow in 2.8–6.2 M NaCl extrude Na$^+$ and accumulate K$^+$.

Animals that are related but live in milieux of different salinities have different limits of salinity tolerance. One species of mud shrimp *Callianassa* (*C. jamaicense*) which lives in estaurine mud and tidal streams shows hyperosmotic regulation and has a low lethal salinity limit of 2 0/00; the mud shrimp *C. major* from sea beaches does not tolerate greater dilution than 15

0/00 and is an osmoconformer (37). In an anadromous population of the cyclostome *Petromyzon marinus* the ammocoete larvae hyperregulate over the range 0–15 0/00 and adults hyperregulate to 34 0/00; adults of a landlocked population regulate only to 16 0/00 (6). The crustacean genus *Gammarus* occurs in four common species in Britain, some of which live in sea water, some in estuaries or freshwater lakes; the gammarid species differ markedly in tolerance of dilution and capacity for active absorption of sodium.

Osmoregulation varies with season, state of development, and habitat. In *Callinectes* serum Cl and osmotic concentration are high in winter, low in summer (76). Of two species of sticklebacks *Gasterosteus*, *G. brachurus* is anadromous, *G. leiurus* remains in fresh water throughout life; some hybrids occur in overlap zones (51). Most elasmobranchs are hyperosmotic marine fish, but a few species occur in fresh water in Central and South American rivers. They have very low urea concentrations in their blood—2 to 3 mg urea N per 100 ml blood compared to 300 to 1300 in marine species. When freshwater rays were acclimated experimentally to salinity approaching that of sea water, their blood urea failed to increase but blood NaCl did increase (114a). It is not known whether the freshwater populations are derived from marine elasmobranchs or are relicts from fish that originated in fresh water.

Populations of *Mytilus edulis* have been examined electrophoretically for isozymes of leucine amino-dipeptidase, an enzyme that may function in osmoregulation. One isozyme (LAP-94) occurs in high concentrations in mussels in high salinity, at low concentration in low salinity. The free-swimming larvae are widely dispersed, but the attached adults occur in salinity gradients from low to high in estuaries such as Long Island Sound, and in *Mytilus* the concentrations of LAP-94 correspond to clines (68). Percentages of the isozyme LAP 94 are: at the eastern end of Long Island Sound 59, in the middle 36, in western regions of the Sound where much fresh water enters, 12–15 (68). Of two species of freshwater cladoceran, *Daphnia magna* often occurs in alkaline water of high salt content and shows maximum Na uptake of 12.5 nM/hr; the cladoceran *Acantholebris* may occur in acid water low in salt, and shows maximum uptake of 39 nM/animal/hr (94). The half-saturation concentration and maximum rate for Na^+ uptake in different animals correlate with distribution (see Table 9-7) (73,48). The contractile vacuoles of freshwater ciliates function primarily in excreting hypoosmotic fluid; in a ciliate from a saline lake the vacuolar output decreases whenever Na_o is lowered; the vacuoles then function in ionic rather than in osmotic control (52).

Summary

Many mechanisms have evolved which enable bacteria, plants, and animals to maintain steep gradients of water and ions between organisms and environment. In aquatic animals, the transport functions of cells at the interface of body fluids and external medium are specialized. The cell walls of bacteria and plants are rigid. Ion pumps and facilitated exchange are probably present in all organisms.

Among aquatic prokaryotes, plants, and animals there are some that conform osmotically to the medium, are isosmotic. Others have hypoosmotic regulation, for example, crustaceans in hypersaline lakes and bony fish in sea water; other life forms are hyperosmotic, for example, organisms living in fresh water. All terrestrial plants and animals have specialized outer layers that protect against evaporative loss, and all have specialized cells or organs for gas exchange.

All organisms, whether osmoconformers or osmoregulators, show ionic regulation. Most animals have special organs that excrete the less essential elements and/or retain essential ions. Regulations of volume, water, and ions are intimately interrelated. The adaptive mechanisms appear to be diverse, and as a consequence life forms which probably arose in the oceans now exist in hypersaline lakes, in fresh water nearly free of solutes, and in humid and dry air.

CELLULAR REGULATION: VOLUME, IONS, AND OSMOTIC CONCENTRATION

Ion fluxes maintain cell volume, concentration of ions in cells, and transmembrane potentials. Since the optimum concentrations of different ions are relatively fixed, other diffusible solutes are commonly used in osmotic and volume regulation. Volume and ion regulation seem to be precise and primitive; osmotic regulation may have evolved secondarily.

Volume regulation by ions in such cells as Ehrlich tumor cells is accomplished by (1) pump-leak movements in which pump fluxes are equal and opposite to leak fluxes; (2) loss of K^+ and Cl^- along with water from swollen cells; (3) volume-dependent cation permeability; (4) Ca^{2+}-dependent K^+ permeability; (5) volume-dependent anion flux; (6) cotransport and exchange of ions (56,57). Volume control in nucleated red blood cells (duck, Amphiuma) is by independent movements of Na^+, K^+, and Cl^-, followed by water. Chloride exit is by exchange for HCO_3^-. Fluxes of Na^+ and Cl^- can be separately blocked (69). Some cells subjected to altered extracellular osmotic concentration gain or lose ions, especially K^+, and thus recover volume. Flounder red blood cells, transferred from an isotonic to a hypotonic medium, increase in permeability to potassium (P_K) by 11 times, to P_{Na} by 4 times; upon transfer from isotonic to hypertonic medium, P_K increases 1.7 times and P_{Na} 2.5 times (101). Restoration of membrane permeability is determined by a (Ca-Mg) ATPase. Halophilic bacteria may accumulate KCl to a cellular concentration of 2 M; they have high concentrations of glutamate, aspartate, and of negatively charged proteins (11).

Gain or loss of cellular ions is tolerated by most organisms only over relatively narrow ranges. A usual method of cellular osmoregulation is the increase or decrease of the concentration of relatively neutral organic solutes (Table 9-3). This leaves the ionic composition constant although osmoconcentration has changed. Changes in intracellular concentrations of organic solutes are brought about by selective synthesis or degradation and by increases or decreases in permeability. In a clam *Rangia* the amino acid pool in gills, mantle, and muscle is proportional to the salinity of the milieu over the tolerated range (90). Heart cells of a clam *Modiolus* swell when placed in low salinity and recover normal volume by efflux of amino acids (3,91).

Nonhalophilic bacteria grown in media with increasing NaCl concentrations may make and retain amino acids. *E. coli* grown in low NaCl and transferred to normal NaCl show per increment of 1 M NaCl an increase in glutamic acid of 150 mM, increase in proline of 630 mM; *B. subtilis* increase proline by 1650 mM; *Pseudomonas* increase glutamic acid by 900 mM. In general, gram-negative bacteria accumulate glutamic acid, gram-positive bacteria proline (82). For equivalent osmotic increases, the increase in amino acids in a sucrose medium is half that in a NaCl medium. Some yeasts grow well in media of low water activity. Such yeasts increase concentrations of polyols and arabitol and glycerol to as much as 18.5 percent of dry weight (13). Algae *Dunalielli* in 1.5 M NaCl accumulate glycerol to 2 M (14). Halophytic plants such as Bermuda grass in a drought make and retain amino acids with as much as 30 percent proline (112).

Amino acid concentration changes not only *in vivo* but also in isolated tissues; the heart of the bivalve *Modiolus* in a concentrated or dilute medium maintains water and ion concentrations constant but increases or decreases concentrations of amino acids (5). If placed in a dilute medium in which Na^+, K^+, or Cl^- are decreased, there is a transient amino acid efflux; if Ca^{2+} and Mg^{2+} are also decreased, efflux of amino acids continues; permeability to amino acids is regulated by calcium. After a few hours, volume is restored. When the bivalve heart is returned to normal sea water after being in dilute sea water, or is put into hypertonic sea water, volume is restored by increase in free amino acids, initially alanine and then glycine, taurine, and proline. Recovery of *Modiolus* from hypoosmotic stress is correlated with activity of a Ca-activated ATPase (115). A similar sequence is observed over a period of days in intact clams. Neurons of a clam *Mya* exposed to reduced osmotic concentration show initial hyperpolarization, with decreased excitability, but after a few hours recovery occurs and the resting potential overshoots its normal value (9). The specificity of amino acid changes in molluscs shows that the source of amino acids is not general protein breakdown; the compensation can also occur anaerobically. The enzymatic pathways of amino acid production are not well known, but in part they consist of a metabolic shift to glycolysis, with alanine and succinate as products (91).

Permeability of bivalve tissues to amino acids increases in hypoosmotic media. The regulation of amino acid permeability is correlated with activity

Table 9-3. Osmolytes (mM/kg) in Muscles (123)

	Osmoconcentration (mosm/kg)	Inorganic Ions		Amino Acids	Betaine	TMAO	Urea
		Na	K				
Barnacle *Balanus*, S.W.	1005	45	169	503	82		
Crab *Eriocheir* S.W.	1118	144	146	341	14	75	
Echinoderm *Parastichopus* S.W.	1246	71	217	221	208		
Mollusc *Sepia* S.W.	1377	31	189	483	108	86	
Hagfish *Myxine* S.W.		96	140	291		87	2
Dogfish *Squalus* S.W.		18	130		100	180	333
Ray *Raja* S.W.		10	162	214		64	398
Fish *Pleuronectes* S.W.		15	158	71		30	
Brackish water ray		4	134	144		36	264
Eriocheir	588	55	71	158	11	47	
Fish *Pleuronectes* F.W.		10	157	44		20	

Key: S.W. = sea water; F.W. = freshwater; B.W. = brackish water.

of Ca-activated ATPase (115,91). In bivalve molluscs, mantle tissue resembles heart in amino acid changes, except that taurine does not change (5). Isolated blood cells of a bivalve mollusc *Neotia* regulate volume much as does *Modiolus* heart muscle. In hypoosmotic sea water the blood cells regulate by efflux mainly of taurine and some glycine and alanine; the efflux diminishes and volume regulation ceases in a Ca-free medium, at low temperature, or when glycolysis is blocked (3).

A molecular explanation of the use of organic osmolytes instead of inorganic ions is as follows (11). The activity of several enzymes—PK, LDH, and ribonuclease—is perturbed by increases of the concentrations of KCl, NaCl, arginine, or lysine, but not by increase of Gly, Ala, Pro, TMO, or octopine. Enzymes from echinoderm *Parastichopus*, crustaceans *Uca*, *Panulirus*, and from rabbit are all affected similarly by the same compounds. As a measure of enzyme efficiency, the apparent K_m was examined; for PK the apparent K_m (PEP) when arginine was added increased from 34 to 132 mM and K_m (ADP) from 0.18 to 0.43 mM. Similar increases in the K_m values of LDH occurred for K_m (Pyr) and K_m (NADH) when arginine was added. Urea is perturbing for enzyme proteins much as is KCl; however, the effect of urea is counteracted by solutes such as TMO and sarcosine (121). In elasmobranchs, LDH may have amino acid sequences which offset the perturbing effect of urea (121,122). In general, high concentrations of Na and low concentrations of K diminish peptide synthesis; that is, the Na/K ratio regulates ADP and ATP concentrations and K promotes oxidative phosphorylation in mitochondria (118). It is concluded that specific organic solutes, often amino acids and amines, regulate osmotic concentration and hence volume and ion balance, and enable many enzymes to function in cytoplasms with different water activities.

PERMEABILITY OF CELL MEMBRANES

Cell membranes maintain differences in composition between cell contents and extracellular fluid; they are not, however, strictly semipermeable, and to various degrees membranes permit water, O_2, CO_2, some ions, and some neutral molecules to pass. Plasma membranes consist of proteins and lipids—mostly phospholipids; in most prokaryotes, less usually in eukaryotes, there are also some membrane polysaccharides. Several models of the organization of the plasma membrane have been proposed. The simplest is a lipid bilayer bounded by protein on each side. The lipids are polar molecules with one end hydrophobic and the other hydrophilic. A more complex and more realistic model (107) provides for different protein patches, for membrane steroids as well as phospholipids, and for lateral and transverse interactions (Fig. 9-2).

The basic structure of the membrane, the genetic coding for the membrane proteins and for proteins used in lipid synthesis and organization, may

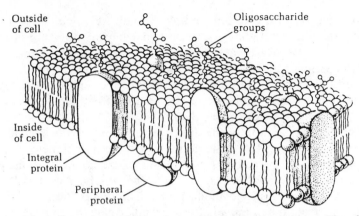

Figure 9-2.   Three dimensional picture of a model of plasma membrane. Phospholipid bilayer with hydrophobic ends of molecules together; proteins of various sorts traverse the membrane and may line channels of ion conductance. Other proteins are transport enzymes within the membrane. Reprinted with permission from A. L. Lehninger, *Principles of Biochemistry,* Worth Publishers, New York, 1982, p. 321.

have evolved in the period before enzymes of energy metabolism had evolved. The phospholipids are generally similar in prokaryotes and eukaryotes. Some of the membrane proteins are structural; others are enzymatic, for example, those used in ion transport. Bacterial membranes may contain peptidoglycans in addition to lipid and protein. Much membrane protein is neither freely mobile nor random; approximately one-third of membrane protein molecules are relatively fixed (1). Bacterial membranes may also have RNA and oxidative enzymes such as those found in the mitochondria of eukaryotes. In eukaryotes the membranes of mitochondria and chloroplasts contain phospholipid and protein, and enzymes of these organelles are relatively inactive if the phospholipids are removed. Plasma membranes contain much cholesterol. Presumably the membrane organization of mitochondria and chloroplasts was established when the precursors of the organelles were free-living prokaryotic organisms (Ch. 2).

Cell membranes separate protoplasm from the medium by mechanisms the least understood of which is selective permeability. For water there is directional selectivity, influx slower than efflux (sea urchin eggs). Diffusional permeability to water (measured by tracers) is less than osmotic permeability (measured by swelling and shrinking). For organic molecules, permeability is regulated by molecular size and lipid solubility. For inorganic ions, permeability depends on ionic size and charge. Protein channels or ionophores are used in relation with cell activity to increase passage of ions. Specific channels provide the basis for electrical events in excitable tissues. Facilitated diffusion, especially of organic molecules, depends on carrier molecules within the plasma membrane and accelerates movement, which follows a concentration gradient. Diffusive fluxes and facilitated diffusion are passive, depending on electrical and concentration gradients. Active transport is catalyzed and requires energy, glycolytic or oxidative or from the Na gradient, and implements movement of ions against concentration and electrical gradients. The best known active transport systems are proton pumps and sodium-potassium exchange pumps. In general, cell membranes are mosaics, with regions of passive permeability in parallel with ion pump regions (69). Four K^+ flux systems are Na^+/K^+ pump, Ca^{2+}-activated K^+ transport, Na^+/K^+ cotransport (exchange), and electrodiffusional leak (32). Sodium flux systems are: passive leak, active transport, and exchange with potassium or calcium. Bacterial membranes contain a number of permeases which catalyze the entry of metabolic substrates.

The word *permeability* refers to the property of the membrane barrier that regulates free diffusion. Both passive functions—fluxes and facilitated diffusion—and active functions—energy-requiring transport enzymes—participate in the selective cell barrier. Passive fluxes increase in swollen cells. At low temperatures permeability and active transport decrease; potassium fluxes diminish more than sodium transfer, and swelling results (118). The fatty acids of membrane phospholipids are more unsaturated and membranes more fluid when formed at low temperatures than at high temperatures, and changes in fluidity may affect ion transfer (Ch. 7). In a euryhaline amphipod *Gammarus duebeni* transferred from seawater to dilute (2 percent) sea water, water exchange is greater initially than after several weeks. In acclimation there is a decrease in unsaturated fatty acids in gill membranes.

Most of the cell membrane models, of which there are many, are based on indirect evidence. Electrical evidence shows a lamellar orientation of lipid and protein, thickness 75–175 mm (Fig. 9-2). Orientation of macromolecules is shown by optical, electron microscope, and X-ray studies. Permeability measurements indicate that the plasma membrane has regions of lipid solubility and regions of aqueous channels. Efforts have been made to isolate the proteins that constitute channels for specific ions (Ch. 15).

Water Permeability

Water permeability has been measured by osmotic swelling and shrinking and by diffusion of tracer water, D_2O or THO. Osmotic permeability (P_{osm}) is considerably greater than diffusive permeability (P_d). For example, the ratio P_{osm}/P_d is 5.7 for *Amoeba*, 3.1 for beef red blood cells, and 1.3 for zebra fish eggs (30) (Table 9-4). Permeability differences are also observed in intact whole animals where the water flux is mainly across respiratory epithelia. One explanation of higher P_{osm} is that an unstirred layer of water close to the plasma membrane makes the P_d appear too low. Also, osmotic flow may be high when water is carried in bulk flow due to hydrostatic pressure ex-

Table 9-4.   Osmotic Permeability (P_{osm}) and Tracer Water Permeability (P_d) for Free-Living Cells and Dissected Cells (30)

Cells	$P_{osm} \times 10^4$	$P_d \times 10^4$	P_{osm/P_d}
Frog eggs (shed)	1.3	0.75	1.7
Frog eggs (ovarian)	89	1.28	70.0
Zebra fish egg (shed)	0.45	0.36	1.3
Zebra fish egg (ovarian)	29	0.68	43
Amoeba (Chaos)	0.37	0.23	1.6
Squid axon (dissected)	11.0	1.4	7.9
Crab muscle (dissected)	96.0	1.2	80
Beef erythrocyte	156.0	51	3.1
Human erythrocyte	127	53	2.4

Table 9-5.   Osmotic Permeability (L/mosm/hr) and Sodium Permeability (L/kg/hr) (63, 64)

	P_{osm}	P_{Na}
Marine aminals		
Crab Pugettia	0.33	0.3
Crab Eupagurus	0.43	0.068
Brackish water animals		
Crab Carcinus	0.042	0.047
Crab Cancer	0.02	0.04
Freshwater animals		
Crab Eriocheir	0.003	0.009
Crayfish	0.009	0.001
Fish Carassius	0.039	0.003
Fish Salmo	0.015	0.001
Fish Anguilla	0.016	0.002

erted by contractile filaments at cell membranes. The activation energy for osmotic water permeability may be higher than for simple diffusion because water may cause rearrangement of membrane lipids. Fluxes of water are not symmetrical; usually water efflux across cell membranes is faster than water influx. In sea urchin eggs, osmotic shrinking goes faster than osmotic swelling for the same gradient. Human erythrocytes are so permeable to water than if a pressure gradient of 1 cm H_2O were maintained across the membrane, the cell would double in volume in 0.54 sec (57).

Many adaptive differences have been recognized with respect to water permeability. In euryhaline animals the permeability to water is less in a dilute medium than in seawater, where the osmotic gradient is much greater (Table 9-5). In the sea-worm *Nereis diversicolor* P_d (diffusional permeability constant in cm/s) in 100 percent S. W. is 0.131; in 20 percent S. W. the constant is 0.071 (40). The osmotic permeability (P_{osm}) of freshwater crustaceans and fishes is much less than that of marine species. Selected examples for P_{osm} in liters per mosm gradient per hour are, in marine crustaceans, *Maja* 0.10, *Eupagurus* 0.430, *Pugettia* 0.33, *Libinia* 0.185; in estuarine crabs, *Carcinus* 0.043, *Cancer* 0.02; in freshwater crustaceans, *Astacus* 0.009, *Pseudotelphusa* 0.0004, *Potamon* 0.0004 (63,64).

Water permeability is modifiable by hormones; permeability of frog skin, toad bladder, and renal tubules of most vertebrates is increased by pituitary ADH. Aqueous pores in cell membranes are more postulated than demonstrated structures. Plants regulate water loss by opening and closing stomata; plant epithelia may be virtually impermeable, especially in the desert. Insects have a waxy cuticle which is relatively impermeable to water except at high temperatures; terrestrial insects control respiratory water loss by spiracles. Horny scaled animals, such as most reptiles, are much less permeable to water loss than soft-skinned animals, such as amphibians. Explanation is not complete for molecular mechanisms of control of water movement in adaptations of water balance.

Permeability to Solutes

Membranes with selective permeability must have evolved before there could have been active cell metabolism (Ch. 2). Passive permeability to electrolytes and nonelectrolytes may be determined by charged and uncharged pathways in cell membranes.

Permeability to ions is influenced by both the potential across a membrane and by charges on micellar particles in the membrane. Measurements of permeability and of electrical properties in epithelia show that not all cells are uniformly surrounded; one face of a cell may be very different from another face, and there are patches of different permeabilities. Permeability of sheets of cells, such as epithelia, determines the distribution of solutes within most multicelllular organisms; one goal in membrane studies is to interpret epithelial properties in terms of the component cells.

Fluxes across cell membranes, J_i (influx), J_o (efflux), J_n (net flux) are measured by ion concentration analyses, by use of tracers, and indirectly by electrical measurements. Active transport of an ion can be postulated only if a flux cannot be accounted for by passive forces: diffusion down a concentration gradient, down an electrical gradient, or convection. The latter is normally disregarded although in tissue systems concentrations adjacent to a multicellular membrane may differ from bulk concentrations because of an unstirred layer near the membrane.

Measurements of fluxes by tracers combined with measurements of concentrations allow one to make the fewest assumptions about the mechanisms of ion permeation and are useful for steady state but not for brief transients.

Electrical measurements combined with stepwise ion replacements can be interpreted in terms of fluxes. The expression for passive electrodiffusive fluxes is

$$J_n = -D \left(\frac{dc}{dx} + \frac{zF}{RT} \cdot C_m \cdot \frac{dE}{dx} \right)$$

where D is diffusion coefficient in cm^2/sec, dc/dx is concentration gradient across the membrane of thickness x and in units $M/cm^3/cm$, z is valence, F is the Faraday unit, R is the gas constant, T is the absolute temperature, C_m is mean concentration within the membrane, and dE/dx is the voltage gradient in V/cm. C_m is assumed to be constant and much smaller than C_{out} or C_{in}; the partition coefficient relates C_m to C_o or C_i. Voltage gradient across the membrane is

$$V_m = \frac{RT}{zF} \ln \frac{C_o}{C_i}$$

Two equations have been commonly used to describe fluxes from measurements of membrane potential. These are discussed in detail in Chapter 15. The Goldman equation (Ch. 15) is based on the Nernst-Planck relation that describes potentials across a junction between two dissimilar solutions. It has been extended to membranes which separate two solutions, and which are of unknown thickness and contain postulated pores. The Goldman equation uses as measures of fluxes the relative permeabilities of ions across the membrane, subject to certain simplifying assumptions. Permeabilities of ions are expressed relative to each other, P_{Na}/P_K. The Goldman equation assumes a constant field within a conductive channel and a net flux of zero.

A second equation, the Hodgkin-Horowicz equation, uses parallel conductances for each ion as described by an equivalent circuit for a patch of membrane. This equation relates the current carried by an ion to the ion's relative conductance and to its driving force, which is the difference between the equilibrium potential for that ion (E_{Na+}, E_{K+}, E_{Cl-}) and membrane potential (E_m). An assumption is that conductances are proportional to ion mobilities within the membrane. Given the appropriate preceding assumptions, it is possible to obtain useful values of relative permeabilities and conductances from measurements of membrane potentials as used in the Goldman and in the Hodgkin-Horowicz equations.

Much evidence indicates that K^+ and Na^+ move through different channels that presumably bear negative charges on the bounding protein, and that the channels are of different diameters.

Ionic permeabilities and conductances vary with the state of a cell. In nerve and muscle membranes the permeability to sodium increases during the rising phase of an action potential, while the permeability to potassium rises more slowly, so that the falling phase of an action potential is caused by outward K current. In squid giant axons, for example, the resting ratio P_{Na}/P_K is 0.01, during the rising phase of an impulse P_{Na}/P_K is 12.0. In some excitable cells the inward current of the rising phase of depolarization is carried by calcium ions (Ch. 15). In general, permeability to anions is lower than to cations, presumably because most membranes bear a net negative charge. Gills of the flatfish *Platichthys* are 30 times more permeable to Na^+ than to Cl^- (92). For anions, size is often the determinant; small ions pass much more readily than large ones (Hoffmeister series). Attempts are in progress to isolate the selective proteins which determine ion permeability— that is, the ionic channels (Ch. 15). Useful information is obtained by specific blocking agents. Tetrodotoxin and saxitoxin block the active Na^+ channels in many but not in all excitable cells. Tetraethyl ammonium (TEA) delays the closing of K^+ channels; verapamil and some ions, such as Mn^{2+}, Co^{2+}, and Ni^{2+}, block opening of Ca^{2+} channels.

Not only does ion permeability change adaptively with activity, but membrane permeability changes according to the environment. The potential across the gill of a seawater-adapted eel is 25 mV (blood positive); in fresh water the potential is 40 mV (blood negative). When potential differences are taken into account the permeability ratios for eel gills are as follows (80,81):

	Sea Water	Fresh Water
P_K/P_{Na}	30	5.8
P_{Cl}/P_{Na}	0.16	0.3

When eels are transferred from sea water to fresh water the flux of K across the gills remains constant; Na^+ and Cl^- fluxes decline steeply over several days (103).

The blood of euryhaline fishes—eels, flounders, killifish—is electropositive to sea water; it is negative to fresh water if Ca^{2+} is low or absent in the water, but it may be slightly positive if Ca^{2+} is high.

The values of transmembrane potentials are not the algebraic sum of equilibrium potentials of the ions in the blood and medium but are the resultant of differences in passive fluxes and of active transport (Table 9-6).

Permeant anions tend toward electrochemical equilibrium, but some membranes are selective for chloride ion. In frog muscle, conductance for small anions $Cl > Br > NO_3 > I > TCA$ decreases as pH is lowered. Conductances to larger (mostly organic) anions—benzoate > valerate > butyrate > propionate > formate > acetate—increase as pH decreases. Apparently channels for Cl-type anions and for benzoate-type anions differ in charge distribution (121).

In crayfish muscle the invaginating membranes and T-tubules have high permeability to Cl^-. Muscle membranes of some freshwater fish are more

Table 9-6. Flux Ratio Analysis of Na⁺. Movements across Body Surface of Freshwater Animals (64)

	C_{out}/C_{in}	Transepithelial Potential (mV)	Fluxes (nM/g/hr)		J_{in}/J_{out}	
			Influx	Efflux	Calculated	Observed
Fish Salmo	0.0067	0.008	193	143	0.005	1.35
Salamander Ambystoma	0.013	0.014	130	160	0.007	0.81
Crayfish Astacus	0.0098	0.005	410	410	0.008	1.0
Earthworm Lumbricus	0.0067	−0.02	32	37	0.016	0.86
Clam Carunculina	0.067	−0.006	1114	990	0.086	1.13

permeable to Cl^- than to K^+ (65a). Inhibition at neural and inhibitory neuromuscular synapses is by an increase in Cl^- conductance (Ch. 15).

EXCHANGE AND FACILITATED DIFFUSION

The preceding section considers permeation down an electrochemical gradient. Some ions exchange for other ions of the same or a related element as shown by fluxes of radioisotopes. One-for-one exchange was discovered by use of radioactive ions in equivalent concentrations on each side of a membrane. Probably water molecules undergo similar exchange.

In intact crabs *Callinectes,* 50 percent of Na^+ efflux and 70 percent of Cl^- efflux is by exchange (16). In red blood cells of vertebrates Na^+-Na^+ exchange; also K^+-K^+ and Na^+-K^+ exchanges occur. A physiologically important exchange in red blood cells is the exchange of Cl^- for HCO_3^-; Cl^- leaves the cells in tissues and enters the cells in lungs. One component of red cell membranes, designated band 3, constitutes 30–35 percent of total cell protein. Band 3 binds anions to positively charged groups. This protein consists of (1) a 35,000 Da segment which is sensitive to pronase and is positioned on the outer face, and (2) a 65,000 Da segment which is pronase-resistant, is hydrophobic, and to which anions bind. The exact nature of the conformational change associated with the anionic exchange is not known (99).

In cultured fibroblasts two Na^+-H^+ exchange mechanisms provide for influx of Na^+, especially when cells are grown in low Na^+. One Na^+-H^- exchange is inhibited by amiloride, another is stimulated by a calcium-calmodulin complex. In human red blood cells, a Na/K cotransport mediates both uphill and downhill movement. Driving force is given by electrochemical gradients in the ratio of Na:K:Cl of 1:1:2 (24). Ascites tumor cells regulate volume by cotransport of Na^+ and Cl^- via a furosemide-sensitive channel; after volume is regulated a Na^+ is replaced by a K^+. A Ca^{2+}-activated K^+ pathway is blocked by quinine (57).

Facilitation of diffusion has been described for nonelectrolytes. Entrance of sugars into red blood cells in an isotonic medium, but down a gradient, is faster than can be accounted for by diffusion. It is postulated that a carrier molecule within the blood cell membrane, combines with the permeating substance on the high-concentration side, traverses or rotates in the membrane and releases the permeant molecule on the opposite side. Energy is provided by the concentration gradient. Separation of passive diffusion from carrier-mediated transport can be made by competition (interference) of related compounds. Aldoses penetrate faster by carrier than ketoses, less fast than ketoses by passive diffusion. In red cells the protein carriers for D-glucose are different from those for methyl glucosides. Urea enters passively into red blood cells of fishes, by facilitated transport into red blood cells of amphibians and mammals.

Ionophores are substances that facilitate ion transfer across membranes.

To serve as models for ionophores several antibiotics have been used to facilitate the net movement of ions down a transmembrane gradient. Valinomycin increases permeability to K^+, Rb^+, and Cs^+. Valinomycin facilitates movement of K^+ into mitochondria and cells. Vesicles made from plasma membranes accumulate K^+ and Rb^+; the accumulation is facilitated by valinomycin. Nigericin facilitates Na^+-K^+ and H^+-K^+ exchanges.

Natural ionophores may occur as noncatalytic binding sites on transport ATPases. $(Na^+ + K^+)$-ATPase from an electric organ of fish has a 57,000 dalton fragment which serves as a Na^+ ionophore. $(Ca^{2+} + Mg^{2+})$-ATPase from sarcoplasmic reticulum has an ionophore portion of 100,000 Da which serves transport divalent ions in the series $Ba > Ca > Sr > Mg$. Inhibitors of catalysis by these ATPases do not affect their ionophore activity (104). Antiporters are ionophores that function in natural exchange of ions of the same charge, for example, Na^+-H^+, Cl^--HCO_3^-. Symporters facilitate transport of ions of opposite charge, for example, Na^+-Cl^-, Na^+, or K^+-Cl^-. Discovery of the codons for natural ionophores may be a step toward getting at the molecular genetics of membrane selective permeability. Possibly transport by exchange diffusion preceded the evolution of ion pumps.

ACTIVE TRANSPORT

The distinction between facilitated transfer and active transport is theoretically this: if the sum of the concentration and electrical gradients is insufficient to account for the flux of a solute, then an active pump is postulated. Primary active transport requires input of metabolic energy. The energetic reaction of ATP hydrolysis may be linked directly to the solute transport. Secondary transport occurs when movement of a solute is coupled to transfer of another solute; the energy is provided by diffusion of the second solute down a gradient. Metabolic energy is expended in maintaining this gradient rather than in moving the transported solute.

An example of active transport is the permease for entry of some sugars into bacteria. The transporting protein molecules show specificity for sugars; the process can be saturated, it can go against concentration gradients, and it requires metabolic energy. Also, the carrier molecules can be induced, and the gene for one of them has been much studied—the lac gene of *E. coli,* which encodes the permease for lactose (Ch. 3).

The best identified active transport molecules are those for inorganic ions. Mechanisms for extruding ions from cells must have evolved concomitantly with codons for proteins of catabolism and anabolism. Transport mechanisms for extruding ions maintain cellular concentrations of specific ions and prevent osmotic swelling. Potential gradients are generated when the number of ions transported in one direction differs from the number moved in the opposite direction. One view is that there are numerous specific transport enzymes in different tissues and organelles; a contrary view is that all active transport can take place by the coupling of three systems—a proton pump, a Na^+-K^+ pump, and a Ca^{2+} pump. The similarity of proton pumps of bacteria, mitochondria, and chloroplasts is evidence for the prokaryote origin of eukaryotic organelles (Ch. 2).

Proton Pumps

Protons constitute a fundamental metabolic currency with numerous functions, including transfer between oxidation-reduction paths and acid-base balance and transfer between ionic and covalent bonds. Proton pumps are of two general types, those which use energy from ATP and function in a hydrolytic mode and those which are synthetases that make ATP by phosphorylation (109a). There are several ATP-requiring proton pumps which have evolved independently. It is probable that in primitive prokaryotes ATP-requiring pumps served to extrude H^+ and maintain volume. Concomitantly, Na^+/H^+ exchange excluded Na^+, and the resulting negative potential drove Cl^- out while K^+ was retained. Later, proton pumps reversed their direction and became ATP synthetases. One of these is found in halophilic bacteria, where a form of bacteriorhodopsin transfers a proton along a transmembrane channel and generates a proton motive force that drives ATP synthesis. Better known are proton pumps which are coupled with the transfer of electrons in either oxidative or photosynthetic pathways. These synthetases mediate the energetic transfer of P_i to ADP and make ATP. In bacteria, mitochondria and chloroplasts proton transport is coupled to oxidative or photosynthetic phosphorylation. The enzymes are called a f_0/f_1 complex and are large molecules—360 to 400 KDa (81a,81b). In mitochondria, according to the chemiosmotic hypothesis, the inner membrane translocates H^+ by energy from electron transport (oxidation); energy from the proton gradient drives synthesis of ATP (83). Redox-driven proton gradients result as electrons flow from negative to positive potentials. In mitochondria, the proton motive force may generate a potential of -250 mV (83).

Proton pumps that use energy from ATP have broad specificity. These ATPases function in cation exchange, such as Na^+/H^+ or K^+/H^+, possibly $Ca^{2+}/2H^+$. In both prokaryotes and eukaryotes exchange pumps are located at cell membranes; they are designated as E_1/E_2 enzymes and are of small molecular weight—100 KDa.

In most bacteria and plants, energy from hydrolysis of ATP provides for extrusion of H^+, which results in cellular electronegativity. Protons outside the cell membrane reenter by a coupled or neutral antiport in exchange for Na^+ or K^+ (Fig. 9-3). Thus, in bacteria, K^+ uptake and Na^+ efflux are generated by proton movement. In bacteria, K^+ uptake is not coupled to Na^+ extrusion, but proton extrusion is stimulated by extracellular Na^+ (83,116). Accumulation of metabolites, especially of anionic amino acids and

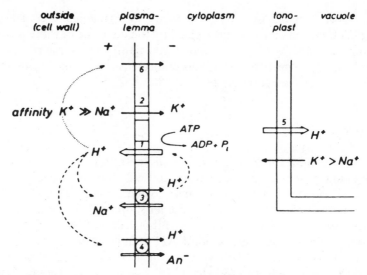

outside   plasma-   cytoplasm   tono-   vacuole
(cell wall)   lemma   plast

affinity K⁺ ≫ Na⁺

Figure 9-3. Postulated ionic movements across plasmalemma and tonoplast of typical plant cell. (1) Proton efflux by ATPase. (2) K⁺ influx. (3) Proton-sodium antiport. (4) Proton-anion symport. (5) Proton pump at tonoplast. (6) Passive reentry. Reprinted with permission from N. D. Jeschke in *Plant Membrane Transport: Current Conceptual Issues,* R. M. Spanswick, Ed., Copyright 1980, Elsevier Biomedical Press, Amsterdam, The Netherlands.

sugars, is coupled to proton gradients. In *Neurospora* and higher plants the membrane potential may be -200 mV, of which 20–60 mV may be due to a K^+ gradient, the remainder due to electrogenic extrusion of H^+. In *Neurospora* two protons are extruded for each ATP hydrolyzed (109); decrease in intracellular pH stimulates outward pumping of H^+ and hyperpolarizes (100). In a giant mutant of *E. coli* direct measurement of intracellular potential gave a slope of -22 mV/pH unit.

Extrusion of Na^+ and retention of K^+ is inhibited by proton uncouplers in *Streptococcus faecalis* (52b). In *E. coli,* protons are moved out by energy from respiration; in an anaerobic environment proton movement ceases, but resumes after O_2 is admitted. Acidification of the medium by extruded protons persists longer in the absence than in the presence of Na^+ because of antiport stimulation (116). Similar proton transport occurs in liver mitochondria (83). In photosynthetic bacteria and in photosynthesizing chloroplasts, protons are extruded and K^+ influx is by an antiporter. In plant cells, transmembrane potentials are mainly due to active transport of H^+. H^+ may be transported actively by either an anion symport or a cation antiport. A symporter facilitates movement across a membrane together with an ion of opposite sign (Fig. 9-3).

Illumination results in ATP synthesis in *Halobacterium halobium;* this provides the energy for H^+ extrusion coupled with Na^+ uptake (70,71,85a). One proton is pumped out per photon of light absorbed; K^+ is taken up and

Na^+ extruded. Another halobacterial protein is a light-driven inward Cl^- pump. A third pigment bacteriorhodopsin (Ch. 14) is sensory in function (85a).

Active transport of H^+ in plasma membranes of animals has evolved secondarily. In mammalian stomach, secretion of acid requires (H + K)-ATPase of ~100 KDa; this secretion can be inhibited by benzimidazole and uses energy from oxidation. Stimulation by histamine is mediated by cyclic AMP (30). A proton gradient of 4×10^6 exists across parietal cells. With acidification of lumen, alkalinization of the cytosolic face occurs. A Na^+-H^+ exchange is independent of and in parallel with Cl^- movement (85c).

In distal and collecting tubules of mammalian kidney and in bladder of turtle, protons are secreted into the lumen and urine is acidified. In collecting ducts, proton secretion is independent of and in other cells from Na^+ and K^+ movements; Na^+ enters down its electrochemical gradient and is pumped out to the blood. HCO_3^- is transported independently to the blood side. The nature of the H^+ transport remains unknown, but the protons are apparently derived from H_2CO_3 and the proton pump is electrogenic. Cortical collecting tubules have a transtubular potential, lumen negative; thus H^+ moves with the electrical gradient. If Na^+ and K^+ are omitted from the bathing solutions, the potential reverses to lumen positive (31,40a,65a). Collecting tubules in the deep medulla of rabbit kidney are electrically positive in lumen, and transport is exclusively of protons. In proximal tubules of kidney, exit of H^+ is by a different mechanism in which H^+ is coupled to Na^+ movement. The Na^+ is actively pumped to the blood; apparently protons come from H_2CO_3, since exit of HCO_3^- is coupled to the secretion of protons (40a).

Calcium Pumps

Calcium pumps occur in many eukaryote cells and plasma membranes. In the sarcoplasmic reticulum (SR) of vertebrate striated muscle 65 percent of the protein is a Ca-ATPase of 102,000 Da. When this protein becomes phosphorylated it binds Ca^{2+}; dephosphorylation is Mg-dependent. The SR has 75 phospholipid (PL) molecules per ATPase, and if the phospholipid is removed the activity of the Ca-ATPase is lost reversibly. Ca-ATPase vesicles or lipsomes accumulate Ca^{2+} in the presence of ATP. Sequestered Ca^{2+} is then stored by another protein of lower affinity but of higher capacity. The adaptive function of the muscle SR is to reduce ionic Ca after activation of contraction, and to store the Ca^{2+} for release in the next contraction. Vesicles containing Ca-ATPase are relevant models for muscle in that they have the same requirements for Mg and ATP, and are inhibited by ADP and by SH reagents. Calcium sequestration is inhibited by ruthenium red and enhanced by the ionophore A23187 (119). Other Ca pumps are known from mitochondria and from some cell membranes.

Sodium-Potassium Pumps

Sodium extrusion by plasma membrane of most animal cells is by $(Na^+ + K^+)$-ATPase. This enzyme requires phospholipid in the membrane in the ratio of phospholipid to protein of 120:1 to 30:1. Sodium is normally extruded and K^+ taken into the cell; $Na^+:K^+$ ratio varies with the tissue—that is, the net transfer may be electrogenic if the ratio deviates from one. In squid axon and mammalian red blood cells the ratio is 3 Na^+:2 K^+ per ATP hydrolyzed. When lipsomes are made with $(Na^+ + K^+)$-ATPase and ATP is supplied, Na^+ is transported against a gradient. The enzyme, like the cellular pumps, is inhibited by cardiac glycosides, such as ouabain. The $(Na^+ + K^+)$-ATPase probably spans the membrane; the ouabain and K^+ binding sites are on the outside, the Na and ATP attachments are on the inner side, as shown by vesicles from red blood cells made in normal and inverted orientations (Fig. 9-4). Electrogenic Na^+-K^+ pumps contribute in many cells to resting potentials.

$(Na^+ + K^+)$-ATPase-containing vesicles have been prepared; when these are applied to planar synthetic membranes containing cholesterol and alkanes, and ATP is added, there is a net transmembrane potential; ion transport occurs that can be inhibited by ouabain. Plasmalemmal vesicles with purified $(Na^+ + K^+)$-ATPase have the same pH optimum and Mg^{2+} requirement for both the enzymal activity and Na^+ uptake. The plasmalemma vesicles can accumulate Na^+ reaching a sevenfold concentration gradient from a medium of 20–38 mM Na^+ and with a ratio of Na^+/ATP of 0.3–0.4; little Na^+ is transferred in the absence of K^+ on the outside (54).

The enzyme $(Na^+ + K^+)$-ATPase has been purified from rectal gland of dogfish, electric organ of electric fish, beef brain, and salt gland of birds; from nerve, muscle, gill epithelium the enzyme has a molecular weight of 250,000 Da (58). The $(Na^+ + K^+)$-ATPase from electric eel electroplax consists of two polypeptides of 90 KDa and 53 KDa (61a): its activity is Na^+-dependent, voltage-independent and it increases the electrical conductance of lipid membranes 300-fold (104). When the $(Na^+ + K^+)$-ATPase of kidney is depleted of phospholipid it is inactive; when the enzyme is then added to a lipid membrane, activity and ouabain sensitivity are restored. Phosphorylation requires Mg^{2+} and Na^+, dephosphorylation requires Mg^{2+} and K^+. The ratio of the two subunits of the ATPase varies, 2 large:1 small in shark rectal gland, 1 large:2 small in dog kidney. The small subunit is a glycoprotein with amount of sugar varying with source; the large subunit is catalytic. The enzyme has two phosphorylation sites, and two binding sites each for ATP and ouabain (45) (Fig. 9-4) (114).

Many mammalian tissues when stored at low temperatures gain Na^+ and lose K^+; the normal gradient is restored when the Na^+-K^+ pump is reactivated by warming. In rabbit ileum, ouabain inhibits Na^+ absorption when applied on the serosal side. Binding of radioactive ouabain and histochemical tests show that the binding and ATPase activity are in the basolateral

The Sodium Pump – Na/K-ATPase

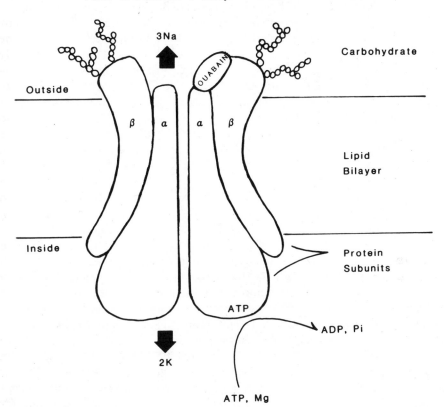

Figure 9-4.   Diagram of (Na/K)-ATPase molecule showing the α and β subunits, postulated location of ouabain binding, and of ATPase hydrolysis. Reprinted by permission of *The New England Journal of Medicine,* Vol. 302, p. 777, 1980.

membranes, not in the brush border (85). In many epithelial cells the (Na^+ + K^+)-ATPase is located on the cell face toward which Na^+ is transported (45,103).

Most red blood cells have high internal K^+ like other body cells; in certain strains of sheep there is low K and high Na^+ in blood cells due to a genetic alteration; these cells have reduced Na^+-K^+ pump activity.

Frog skin has been much used as an epithelial model for transport of Na^+ inward against a concentration gradient. When a piece of skin is mounted between two compartments, each containing Ringer solution, potentials of 50–90 mV are generated, the inner side of skin positive. When the transepithelial potential is short-circuited a current is measured which is proportional to the net Na^+ transported from *out* to *in*. The external apical face of the transporting cells shows permeability to Na^+, which may be modulated by antidiuretic hormone (ADH) from the pituitary when this is supplied from the inner side; amiloride can block the Na^+ influx. The internal basolateral face has a $(Na^+ + K^+)$-ATPase which is ouabain-sensitive and which transports Na^+ toward the inner surface. In parallel with the Na^+-K^+ pump is a furosemide-sensitive ion channel for Na^+, Cl^-, and probably K^+. In dilute medium, such as pond water, a frog actively absorbs Na^+ and Cl^- by two independent ion pumps. A frog can take up Cl^- from dilute NaCl (against a 1:1000 gradient) and Cl^- from KCl, NH_4Cl, or $CaCl_2$. It can take up Na^+ but not K^+. Frog skin is negligibly permeable to Cl^- *in vivo,* very permeable to Cl^- *in vitro* (2,14).

Other Ion Pumps

Besides the proton pumps and the Ca^{2+} and $(Na^+ + K^+)$-ATPase pumps, other pumps have been suggested but not isolated. Magnesium is transported against an electrochemical gradient in several epithelia. The kidney tubules of marine fish excrete Mg^{2+} into the urine. The antennary glands of some marine and estuarine crustaceans selectively transport Mg^{2+} outward. In larvae of a *Cecropia* moth Mg^{2+} moves from a midgut concentration of 9 mM Mg^{2+} to 35 mM Mg^{2+} in hemolymph; the transport has a half-saturation at 34 mM (53,120).

Sodium transport, not coupled to potassium and not inhibited by ouabain, has been noted in several tissues. In the gills of freshwater trout, Na^+ uptake is not K^+-dependent and uptake is not ouabain-sensitive at physiological (low) temperatures. At high temperatures the gills show a typical $(Na^+ + K^+)$-ATPase, and in seawater transport via the ouabain-sensitive path is predominant. The activation energy for the Na^+-ATPase is 7.7 kcal/M and for the $(Na^+ + K^+)$-ATPase is 19.5 kcal/m (64). Ouabain-insensitive Na transport has been noted in parallel with $(Na^+ + K^+)$-ATPase in numerous kinds of cells.

Potassium may be transported by an electrogenic system other than $(Na^+ + K^+)$-ATPase. From apical plasma membranes of cation-transporting cells of midgut, salivary glands, and Malpighian tubules of several insects, particles 7–15 μM in size were isolated as "portasomes." These particles have ATPase activity and transport K^+ electrogenically toward the lumen such that the midgut lumen is 100 mV positive to the hemolymph. Two K^+ ions are transported per ATP, and the ATPase is insensitive to ouabain (53).

Chloride is actively transported by many plant cells (Ch. 15). In the alga *Chara* Cl^- influx occurs against an electrochemical gradient; energy is from ATP (96a). Chloride pumps have been postulated for a number of animal tissues. Toad corneas transport Cl^- from endothelium to epithelium according to concentration on the endothelial side; 70 percent of the corneal short-circuit current is due to Cl^-, 30 percent to Na^+ (8). Earthworms (*Lumbricus*) absorb Cl^-, more when they are salt-depleted than at normal blood Cl^-. The uptake of Cl^- appears to be independent of cations; it is abolished by acetolamide and may be a Cl^-/HCO_3^- exchange.

In gills of several kinds of fish, Cl^- transport (by exchange with HCO_3^-) can be blocked by SCN ions; an anion-sensitive ATPase with high affinity for Cl^- has been isolated (goldfish gill) (28). Efflux of Cl^- from gills of some marine fish (*Anguilla, Pseudopleuronectes*) can be lessened by SCN^- (35). In other fishes (especially marine) the transport of Cl^- may be by a carrier, and the energy for Cl^- transport may come from movement of Na^+ down its gradient. In the rectal gland of dogfish, Cl^- is transported into the lumen by a process which moves 30 Cl^- for one O_2 consumed. Movement is against an electrochemical gradient, and Cl transport increases in parallel with increased urea concentration. Elasmobranch fishes maintain hypertonicity to sea water by retention of urea while keeping blood Na^+ and Cl^+ at much lower concentrations—260–290 meq Na^+/L—than of sea water. The rectal gland of a dogfish *Squalus* forms an NaCl-rich secretion by cotransport of Cl^- with Na^+ across basolateral membranes of epithelial cells. Potassium is required for cotransport of Na^+ and Cl^- by isolated rectal gland vesicles and the cotransport may be 1 Na : 1 K : 2 Cl. Na^+ is extruded by an ouabain-sensitive $(Na^+ + K^+)$-ATPase in basolateral folds (Fig. 9-5), and Cl^- moves down its gradient to the lumen. Secretion of Cl^- is in the ratio Cl^-/O_2 of 30. The cotransport can be inhibited by furosemide and is activated by cAMP which also stimulates the $(Na^+ + K^+)$-ATPase (35,106).

In summary, chloride may be transported by several means: (1) a Na-coupled symport, as in frog skin; (2) a $Cl^- - HCO_3^-$ antiport or countertransport, as in some fish gills and in mosquito anal papillae; (3) an anion-stimulated ATPase which requires ATP as an energy source, for example in many cells, and in locust rectum (44a).

COUPLED TRANSPORT OF SUGARS AND AMINO ACIDS

Active transport of ions provides the driving force for movement of sugars and amino acids against electrochemical concentration gradients. In the intestine of vertebrates, glucose is absorbed against a concentration gradient; aerobic energy is required and phlorizin competitively inhibits glucose transport.

Sugar accumulation in the mucosal epithelial cells and transport from lumen to the blood occurs only when sodium is present in the lumen. Across the epithelium a potential develops that is proportional to the Na^+ transported. Ouabain added on the serosal side reduces the potential and the sugar absorption. One model is: a carrier molecule on the mucosal side binds both glucose and sodium, making a ternary complex that crosses the mucosal surface down a gradient, and then dissociates. At the serosal surface a ouabain-sensitive transport of Na^+ occurs and glucose diffuses out. Absorption of glucose is inhibited if K^+ or Li^+ is substituted for Na^+. Galactose may compete with glucose for a carrier site and D-glucose but not L-glucose is moved; xylose and deoxyglucose bind to the same carrier as glucose but affinity is much less. A mucosal membrane protein added ex-

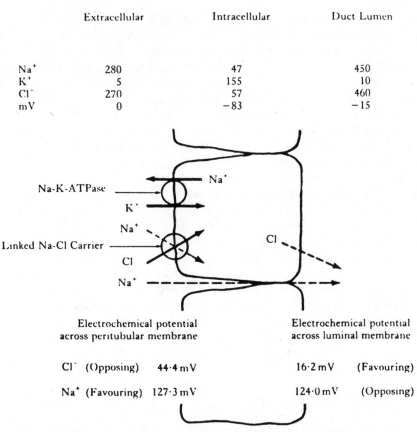

	Extracellular	Intracellular	Duct Lumen
Na⁺	280	47	450
K⁺	5	155	10
Cl⁻	270	57	460
mV	0	−83	−15

Na-K-ATPase

Linked Na-Cl Carrier

Electrochemical potential across peritubular membrane		Electrochemical potential across luminal membrane	
Cl⁻ (Opposing)	44·4 mV	16·2 mV	(Favouring)
Na⁺ (Favouring)	127·3 mV	124·0 mV	(Opposing)

Figure 9-5.   Diagram of ionic movements associated with Cl⁻ secretion by rectal gland of dogfish. Cl⁻ moves into cell via coupled NaCl carrier with energy provided by sodium gradient. Cl⁻ leaves cell across luminal border down electrochemical gradient; Na⁺ moves passively through paracellular pathway. Concentrations of Na, K, Cl given in m*M*. Intracellular and lumen potentials relative to extracellular fluid. Calculated potentials for Cl⁻ and Na⁺ across peritubular and luminal membranes. Reprinted with permission from J. H. Epstein et al., *Journal of Experimental Biology,* Vol. 106, p. 28, Fig. 2. Copyright 1983, Cambridge University Press, Cambridge, UK.

perimentally to liposomes made from mucosal membranes can facilitate entry of glucose into the liposomes if Na⁺ is present (23). The carrier protein has been only partly purified; its function is analogous to that of band 3 of red blood cells.

In plants a cotransport of protons and glucose occurs. A ternary complex of sugar, H⁺, and carrier is formed and moves into cells. The complex then dissociates and sugar is thus transported. Protonation of the carrier changes the affinity of the carrier for sugar (67). In kidney tubules, transport of glucose from lumen to extracellular fluid is by coupling with Na⁺ movement.

Water transport is coupled with active movement of Na⁺ in the intestine of eel; water may move from lumen to plasma when the osmotic concentration in the lumen is greater than in plasma. This water transport depends on active movement of Na^+ (108).

Intestinal absorption of amino acids, like absorption of glucose, occurs only when Na^+ is present on the mucosal side of the brush border. Amino acid absorption into protozoans is coupled to Na^+. Probably the uptake of amino acids against concentration gradients by marine and brackish-water invertebrates is also mediated by a carrier which requires Na^+. In nucleated red blood cells (skate) the influx of alanine decreases when Na_o^+ is diminished (47). In the ileum of rabbit or rat, uptake of anionic amino acids Asp and Glu is Na^+-dependent; cationic Lys and Arg enter independently of Na^+; neutral amino acids Ala, Leu, and Val are intermediate in that their entry has low Na^+-dependence (103). In mammalian kidney tubules both Na^+-dependent and Na^+-independent transport of amino acids from low to higher concentrations occurs. Alanine transport by kidney tubules is Na^+-dependent, transport of Leu is not. Ehrlich ascites tumor cells show Na^+-independent nonsaturable uptake of Leu, Ileu, Val, Phenyl, Met; these cells show Na^+-dependent saturable uptake of Ala, Gly, Ser, and Cys (102a).

We conclude that transport of organic molecules in many organisms is coupled with Na^+ or proton pumping.

ADAPTIVE INTEGRATION OF ION TRANSPORT SYSTEMS

The role of molecular mechanisms in osmotic and ionic balance requires balance sheets of transmembrane potential, of input and output of water and ions and, to a lesser degree, of nutrients. Usually there are multiple avenues of intake and output that are adapted to life style and habitat.

Analysis of water and ion balance in the brine shrimp *Artemia* helps to elucidate how this animal can live in salinities as high as 220 0/00 (5.8 times the concentration of sea water) or as low as 2.6 0/00 (0.07 times sea water) (Fig. 9-6) (110). *Artemia* drinks the equivalent of 2 percent of its body weight per hour when in sea water; it absorbs the water and salts, excretes NaCl by way of gills, and defecates divalent ions. The net effect is that in a medium of 58 0/00 the shrimp's hemolymph (Hl) concentration is 13 0/00. The relations between osmotic concentration of medium, gut fluid, and hemolymph are given in Figure 9-6. At milieu concentrations greater than one-fourth seawater the osmoconcentrations are in series: medium > gut fluid > hemolymph. Gills contain ATPase which in a medium at S.W. concentration is five times as active as in 1/2 S.W. Calculated equilibrium potentials and measured transmembrane potential (E_m) in seawater are

E_{Na} + 26 mV

E_K + 6 mV

E_{Cl} − 33 mV

E_m + 23 mV

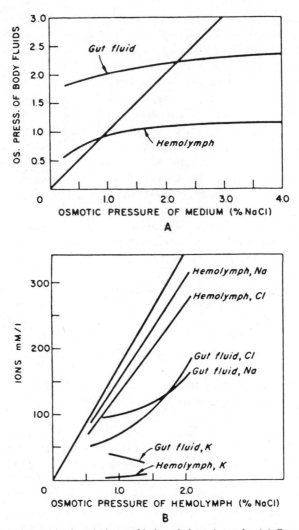

Figure 9-6.   Osmotic and ionic relations of brine shrimp Artemia. (*a*) Osmotic concentrations of gut fluid and hemolymph at different concentrations of medium in equivalent percent NaCl. (*b*) Na⁺, Cl⁻, and K⁺ in hemolymph and gut fluid at different osmotic concentrations of hemolymph. Reprinted with permission from P. C. Croghan, *Journal of Experimental Biology,* Vol. 35, pp. 223, 228. Copyright 1958, Cambridge University Press, Cambridge, UK.

Thus the measured potential is close to E_{Na}. Influx of ions continues after the mouth is ligated, hence the gills are ion-permeable.

Permeabilities calculated by the Goldman equation are

$$P_{Na} = 2.8 \times 10^{-5} \text{ cm/sec}$$

$$P_{Cl} = 3.1 \times 10^{-6} \text{ cm/sec}$$

that is, P_{Cl} is much lower than P_{Na}. However, the net effluxes (by tracers) are similar:

$$J_{Na} = 7200 \text{ pM/cm}^2\text{/sec}$$

$$J_{Cl} = 7400 \text{ pM/cm}^2\text{/sec}$$

It is concluded that Cl^- is actively extruded.

Chloride efflux is as follows:

Total Cl efflux (by tracers)	220 pM/sec
Diffusive efflux (from electrical measurement)	10 pM/sec
Active efflux (via gills and gut)	90 pM/sec
Calculated exchange diffusion	120 pM/sec

Thus 70 percent of Cl efflux is by exchange, is abolished in Cl-free medium; 30 percent is by active transport (110).

Nauplius larvae of *Artemia* have a salt gland in the neck which is rich in $(Na^+ + K^+)$-ATPase; ouabain added to the medium reduces survival of the larvae (19). In sea water of osmolality 932 mosm, the hemolymph of the larva is 161 mosm. The $(Na^+ + K^+)$-ATPase consists of one large (100 K^+) and one small (39 K^+) subunit. Cellular ATP decreases when salinity increases and both protein synthesis and ion transport increase (19). Na^+ is the ion transported predominantly by the larva, Cl by the adult *Artemia*.

ADAPTATIONS TO FRESH WATER

Freshwater fishes are hyperosmotic to the medium, hence are subject to osmotic water influx and diffusive salt efflux. Water balance has been measured for goldfish, eel, and trout. A summary diagram for goldfish is given in Figure 9-7. Sodium is actively absorbed in exchange for NH_4^+ and to a lesser extent for H^+. Na^+ uptake and NH_4^+ excretion are increased after blood sodium depletion or after blood NH_4^+ loading. Acidification of the medium decreases the influx of Na^+ and causes a transient increase in NH_4^+ efflux. Net Na^+ flux is proportional to the sum of NH_4^+ and H^+ efflux in a ratio of 4 Na^+ per 3 H^+ or 3 NH_4^+. In the absence of NH_o^+ there is excretion

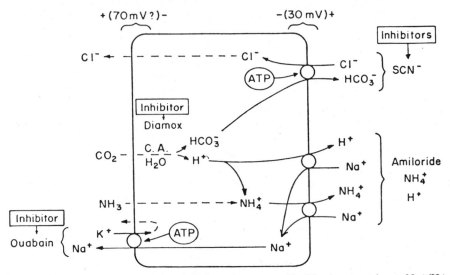

Figure 9-7. Model of active Na$^+$ and Cl$^-$ uptake across gill of freshwater teleost. Na$^+$/H$^+$ and Na$^+$/NH$_4^+$ exchanges shown as independent and in parallel. Reprinted with permission from L. B. Kirschner in *Mechanisms of Osmoregulation in Animals*, O. R. Gilles, Ed., Copyright 1979, John Wiley & Sons, Inc., New York.

of NH$_3$ by gills; if Na$^+$ is present the excretion is of NH$_4^+$. Very little water is drunk by freshwater fishes (Fig. 9-8a). Cl$^-$ is absorbed actively by Cl$^-$/ HCO$_3^-$ exchange; this anion transport can be blocked by SCN. Acetazolamide decreases Cl$^-$ influx by stopping production of HCO$_3^-$; the affinity of the anion transporter for Cl$^-$ is greater than for HCO$_3^-$ (80,81). Chloride cells with anion-sensitive ATPase occur at the distal ends of gill filaments. Efflux of Na$^+$ from freshwater fishes across the gills is 20.7, by urine 5.1 µeq/hr/g. Na$^+$ influx into a Na$^+$-depleted fish has a K_m of 260–300 µeq in fresh water (Table 9-7). A decrease in pH$_o$ by 1 pH unit decreases both influx and efflux of Na by 60 percent. Maximum uptake of Na is 125.5 µeq/hr. Blood is electronegative to the medium by 40 mV. The Q$_{10}$ for Na$^+$ influx (active) is 3.0, for Na$^+$ efflux (passive) is 1.7. The role of the kidneys in freshwater fishes is to produce urine which is hypoosmotic to extracellular fluid and thus to eliminate water. The ratio of urine/blood ions is 0.12 in goldfish. Thus extrarenal uptake of ions is essential for ionic balance.

Freshwater invertebrate animals maintain hyperosmoticity by means similar to those of fish. The excretory organs of freshwater invertebrates excrete urine that is hypoosmotic to the blood but not solute-free. Very little water is ingested by mouth and ions are actively absorbed against concentration gradients. Each transport mechanism—in gills or epidermis—may be specific for one anion and, while only Na$^+$ pumps have been isolated, it

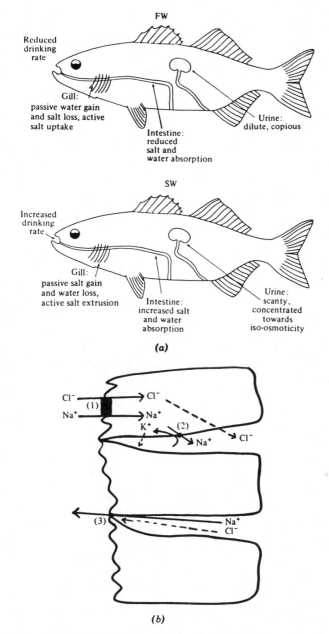

Figure 9-8. (*a*) Generalized diagrams illustrating principal route for salt and water movement in freshwater fish (upper diagram) and seawater fish (lower diagram). (*b*) Model for salt transport by intestinal cells in marine teleosts. Regulatory sites: (1) coupled mucosal Na$^+$ and Cl$^-$ entry; (2) Na-K pump in basolateral membrane; (3) junctional membrane and exit of Na through intercellular spaces. (*a*) Reprinted with permission from J. C. Ellory and J. S. Gibson in *Cellular Acclimatisation to Environmental Change*, A. R. Cossins and P. Sheterline, Eds., Cambridge University Press, Cambridge, 1983. (*b*) Reprinted with permission from M. Field, in *Membrane Transport Processes*, Vol. 1, J. F. Hoffman, Ed. Copyright 1978, Raven Press, New York.

Table 9-7.   Half-saturation Values ($K_{m,app}$) for Sodium Absorption by Selected Hyperosmotic Regulators (63, 48, 73)

	$K_{m,app}$ (mM)
Brackish water animals	
Crustaceans	
Marinogammarus	6.0–10.0
Carcinus	20.0
Gammarus tigrini	1.5
Mesidotea	12.0
Vertebrates	
Fundulus	2.0
Poecilia	8.0
Freshwater animals	
Crustaceans	
Eriocheir	0.5
Gammarus pulex	0.1–0.15
Astacus	0.15–0.25
Mollusc	
Limnea	0.25
Fish	
Carassius	0.05–0.3
Salmo gairdneri	0.5
Amphibia	
Rana	0.2–0.4
Xenopus	<0.05
Amphiuma	0.2

seems probable that there are others. The affinity of transport systems for sodium is as follows: in freshwater animals > in brackish water animals > in marine animals. The affinity for Na$^+$ is shown by the low apparent K_m values for Na$^+$ uptake by several fishes and crustaceans (Table 9-7). For example, the water flea *Daphnia magna* absorbs Na$^+$ from fresh water; its uptake is stimulated by Na depletion and can be blocked by KCN or DNP. Transport of Na can attain 6.5 mM/kg; its K_m is 0.4 mM (112). A cladoceran that lives in acid water of lower salt content than *Daphnia* shows a K_m for Na$^+$ of 0.016 mM (94).

ADAPTATIONS TO SEA WATER

A very few kinds of animals, for example, elasmobranch fishes, are hyperosmotic to sea water, hence they are subject to osmotic gradients similar to those of freshwater fishes; hyperosmoticity is maintained by retention of urea and of certain amines. Many marine animals, especially pelagic invertebrates, are isosmotic with sea water and must select appropriate ions from sea water. Other animals, especially teleost fishes, are hypoosmotic to sea water, and they not only select needed ions and extrude others, but they also reduce the osmotic leakage of water. The osmotic gradient faced by marine teleosts is opposite to that of freshwater fishes; species which migrate between fresh and salt water are able to reverse their osmotic gradients. In sea water of 1000 mosm the plasma osmolality of teleosts is 300–400 mosm and in fresh water it is only slightly less; urine osmoconcentrations are usually 150–250 mosm in fish in either medium. Urine volume is low in marine fish and many marine fish have aglomerular kidneys that can form urine only by tubular secretion. Seawater teleosts drink considerable amounts; some divalent ions are excreted in feces, and urine contains Mg^{2+}, Ca^{2+}, SO_4^{2-}, some phosphate, and potassium (Fig. 9-8a). The Na^+ and Cl^- that have been absorbed by the intestine are excreted by way of gills.

Na^+, Cl^-, and water are absorbed by isolated segments of the intestine of the marine fish plaice. Part of the Cl^- absorption is Na^+-dependent; the potential across the intestine is abolished by ouabain. Coupled entry of Na^+ and Cl^- into intestinal epithelial cells depends on the Na^+ gradient; Na^+ is pumped out and Cl^- follows. Figure 9-8b shows three sites of ion movement: (1) coupled mucosal Na^+ and Cl^- entry; (2) sodium pump in basolateral membranes drives Na^+ into the lateral intercellular spaces and Cl^- follows; (3) tight junctions between cells are cation-permeable and Na^+ diffuses back into the lumen to be recycled (34a,95a).

When euryhaline fish are transferred from sea water to fresh water the Na^+ efflux decreases, rapidly at first, then more slowly. The fast efflux is by Na^+-Na^+ exchange, since in Na-free (mannitol substitution) sea water the efflux decreases by some 47 percent (84). Gills of marine fish and crustaceans have cells which stain with silver, chloride cells. Such cells have many basal mitochondria between which are deep infoldings of the membrane. These cells may be scarce or absent in fish from fresh water, but chloride cells develop rapidly after transfer to sea water. At the same time the gills of euryhaline fishes in sea water develop an ATPase with activity ten times greater than in fresh water. Cells with high $(Na^+ + K^+)$-ATPase activity occur on basolateral membranes of the chloride cells of gills of marine fishes. The blood is electrically positive to sea water by some 20–30 mV; in dilute sea water the potential reverses. Apparently Cl^- enters cells of gills from the blood by a symporter which carries Na down its electrochemical gradient. Na^+ is then actively pumped back to the extracellular fluid (ECF) by a $(Na^+ + K^+)$-ATPase in basolateral membranes, and Cl^- diffuses down the electrical gradient through the apical surface into sea water. Extrusion of Na^+ is probably between epithelial cells down a potential gradient and up a concentration gradient (26). The overall electrochemical gradient favors passive transfer from blood and extracellular fluid to sea water.

Gill $(Na^+ + K^+)$-ATPase in crabs is of high activity in a semiterrestrial crab, intermediate in a osmoregulating species, and lower in an osmocon-former. It may be concluded that, as in other anion-transporting systems (and those in which neutral organic molecules are moved), active Cl^- extrusion in seawater involves both Na^+-anion carriers and an active Na^+-K^+ pump with Cl^- moving down the electrical gradient.

Transport of Water

An example of adaptation for water movement is the ascent of sap in green plants. Water enters the roots by diffusion and roots develop suction pressure which may be as great as 3–5 atm. Transpiration from leaves exerts evaporative force which results in negative pressure and ascent of sap in xylem vessels. In trees this may be to heights corresponding to many atmospheres. The narrow columns of water in the xylem have a tensile strength due to hydrogen bonds between water molecules; the force generated by the transpiration process may be as great as 2000 lb/in.^{2}. The rise of sap is due to the physical properties of water at interfaces in small columns in combination with evaporation from leaves and root pressure.

Another phenomenon due to the physical properties of water is the absorption of water by some terrestrial arthropods from air that is moist but not saturated. A *Tenebrio* beetle takes up atmospheric water via its rectum; a tick takes up water via its mouthparts. Water is absorbed by insects only at or above a critical humidity; absorption is independent of temperature, thus the mechanism is not by vapor pressure deficit. Several mechanisms for absorption of water from the atmosphere have evolved. The first step is condensation in a local region, such as mouth or rectum; condensation is into a secreted fluid and is followed by absorption. In some insects a condensing fluid is secreted by labial glands, in others there may be hydrophilic cuticular hairs (77).

Excretory organs can form a blood-hyperosmotic urine in insects and mammals. Birds and land crabs can produce slightly hyperosmotic urine; the formation of hyperosmotic urine appears explicable on the basis of ion pumps.

MALPIGHIAN TUBULES AND RECTAL REABSORPTION IN INSECTS

Insects differ in water and ion relations according to habitat and diet; some insects, such as flour beetles, can obtain all needed water metabolically. Most insects are capable of producing hemolymph-hyperosmotic excreta; the methods used are very different from those in mammalian kidneys. Insects have Malpighian tubules (MT) which end blindly in the abdomen and form tubular fluid by secretion, not by filtration. The principal cation which is cycled in insects is K^+, rather than Na^+, as in mammals, or Mg^{2+}, as in marine fishes. The evolutionary explanation of the use of K^+ by insects may

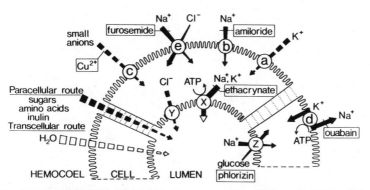

Figure 9-9. Diagram of Malpighian tubule in cross section. Pathways: transcellular routes for passive fluxes of water and for organic solutes. Lumen positive to hemocoel by 10-30 mV. Cotransport of NaCl by furosemide sensitive channels. Na$^+$ enters by amiloride sensitive channel. (Na-K) pump sensitive to ouabain. Absorption pump for Na-glucose sensitive to phlorizin. Reproduced with permission from J. Phillips, *American Journal of Physiology*, Vol. 241, p. R245. Copyright 1981, American Physiological Society, Bethesda, MD.

relate to their diet of plants rich in K and poor in Na. Figure 9-9 is a generalized diagram of the transport pathways for which evidence is available from several species. The initial secreted fluid is nearly isosmotic to hemolymph and contains a 3 to 30 times higher concentration of potassium. In some insects uric acid is secreted, in other species urate is stored in tissues rather than excreted. An active K$^+$ pump is located at the apical (lumen) face; an (Na-K)-ATPase is located on the hemocoel side, and Na$^+$ which enters by a Na-Cl or Na-K-Cl cotransporter is extruded by the Na-K pump. The lumen is electropositive and Cl$^-$ enters down its gradient. A hormone from the corpora cardiaca of Rhodnius increases cellular cAMP which stimulates the apical K-pump. Secretion by Malpighian tubules of the fly *Calliphora* is as follows (49):

	Bathing Fluid	Malpighian Tubule Lumen Fluid
Na	125 mM	20 mM
K	10	125
Mg	5	<1
Cl	120	75
Δ_{fp}	−.52°C	−.525°C

Isolated Malpighian tubules of a mosquito secrete fluid, Na$^+$, K$^+$, and Cl$^-$ into the lumen. Secretion of fluid and Na is stimulated by dibutyryl cyclic AMP; a hormone derived from the head can increase secretion (88). In a blood-sucking bug *Rhodnius* K$^+$ in hemolymph is 4 mM, in the fluid secreted into Malpighian tube K$^+$ is 70 mM, and in hindgut K$^+$ is reabsorbed and urine concentration is 3 mM.

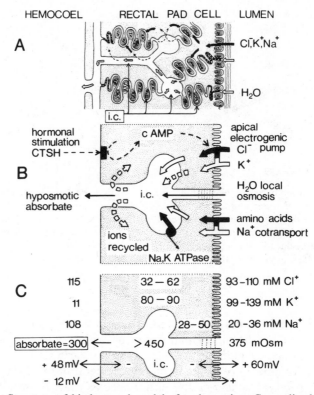

Figure 9-10. Structure of hindgut and model of reabsorption. Generalized from data for cockroach and locust. Ion transport sites: solid arrows; ion recovery sites: broken arrows. (*a*) Structure showing mitochondria-rich folds and wide collecting channels. (*b*) Diagram of two cells. Energy requiring processes: solid arrows. (*c*) Profiles of electrical and concentration gradients. Reproduced with permission from J. Phillips, *American Journal of Physiology*, Vol. 241, p. R249. Copyright 1981, American Physiological Society, Bethesda, MD.

As fluid passes down the MT and enters the upper hindgut some reabsorption of ions and water occurs, but in the lower hindgut 80 to 95 percent of Na^+, K^+, and H_2O are removed (Fig. 9-10). Electrogenic transport of Cl^- makes the hemolymph slightly (12 mV) negative. The Cl^- pump at the luminal side is under the influence of a hormone from corpora cardiaca that stimulates production of cAMP which activates the pump (111a). Transport of Cl^- is coupled with influx of K^+. Cl^- is transported at a rate approaching 10 $\mu M/hr/cm^2$. The ions move passively into large intercellular spaces, making the fluid hyperosmotic; water enters this space and P_{osm} for reabsorption is threefold greater than P_{osm} in the reverse direction. Water can be absorbed from rectum to hemocoel if the rectal fluid is a sugar solution. Some ions are reabsorbed into hindgut cells, but most enter the hemocoel and are recycled via the MT. The net effect is reabsorption to the hemocoel of a fluid

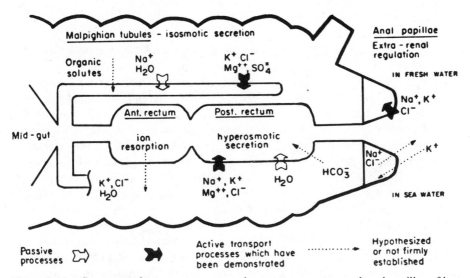

Figure 9-11. Summary of transport processes in excretory system and anal papillae of larval mosquito Aedes in fresh water or in sea water. Reproduced with permission from J. Phillips and T. J. Bradley in *Transport of Ions and Water in Animals,* B. L. Gupta, Ed., Copyright 1977, Academic Press, New York.

that is hypoosmotic to the lumen, and the final excreta may be hyperosmotic. The excreta may have a Δ_{fp} of -1.52, hindgut fluid -0.78, and hemolymph $-0.76°C$.

In flour beetle larvae *Tenebrio* the osmotic ratio of excreta to plasma is 10:1 and the osmotic concentration of urine may reach a Δ_{fp} of $-10.5°C$. In *Tenebrio* the Malpighian tubules are closely applied to the rectal wall and K^+ is absorbed from the rectum, then secreted into Malpighian tubules (MT) and passes back to the rectum. Chloride follows passively because of electropositivity of the MT fluid. Permeability of rectal membrane and MT to H_2O is high and water is absorbed osmotically. A tsetse fly makes similar use of Na^+ (89). Mosquito larvae that live in seawater or saline pools have hypoosmotic hemolymph (O.C. 348 mosm in medium of 818 mosm). Mosquitoes drink saline water and their Malpighian tubes may secrete SO_4^{2-} and Mg^{2+} against a threefold gradient (12,89) (Fig. 9-11). Marine mosquito larvae secrete hypertonic rectal fluid; the posterior rectum secretes hyperosmotic fluid (818 mosm) in which Na^+, Mg^{2+}, K^+, Cl^-, and HCO_3^{2-} are more concentrated than in hemolymph. Rectal secretion content of K^+ is much higher (185 mM) than the concentration in seawater. In marine mosquito larvae anal papillae absorb K^+.

The net result in insect excretory systems is to conserve water. This is important on a dry diet. Ions and metabolites are also conserved.

SALT GLANDS OF BIRDS AND REPTILES

Marine birds such as gulls, pelicans, and some ducks consume a salt-rich diet and conserve water by extrarenal excretion of salt; desert birds such as ostriches also have this capacity. An orbital gland forms a hyperosmotic secretion, high in NaCl content, that is excreted through the nasal passage. Lizards exude nasal secretion, sea turtles and crocodiles orbital secretion, and sea snakes oral secretion. Sea birds, sea snakes, and turtles excrete NaCl, lizards and ostriches excrete NaCl and KCl. In birds, osmoreceptors in the heart or on large blood vessels sense increase in plasma osmoconcentration. Afferent fibers are in the vagus, efferent fibers in a secretory branch of the seventh cranial nerve that activates the salt gland by acetylcholine. Some control is also by steroid hormones (87).

The secreted fluid may have four to five times higher concentrations of Na^+ and Cl^- than the plasma. Electrogenic transport of sodium from cells to lumen of the gland makes the lumen electropositive, and chloride moves passively across the cell membranes (Fig. 9-12). Sodium enters from the

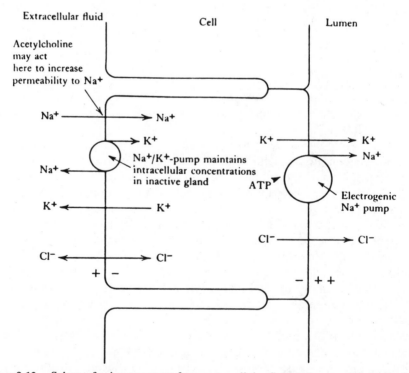

Figure 9-12. Scheme for ion transport from extracellular fluid to lumen of bird salt gland. Coupled exchange mechanisms on basolateral membrane. Reproduced with permission from M. Peaker and J. L. Linzell, *Monographs of the Physiological Society No. 32: Salt Glands in Birds and Reptiles,* Copyright 1975, Cambridge University Press, Cambridge, UK.

extracellular fluid by a proton exchange pump and Cl^- by exchange for HCO_3^-. Salt glands are necessary because reptiles and birds cannot produce such hyperosmotic urine as mammals. An extrarenal route for salt excretion is adaptive for diets high in salt and for water conservation in arid climates.

MAMMALS

Mammalian kidneys have the capacity to produce hyperosmotic urine. This process conserves water in desert mammals, and also in marine mammals in which the blood is more dilute than is sea water. Maximum osmotic concentration of urine of a few mammals in mosm is: desert mouse *Psammomys* 6000, camel 3170, laboratory rat 3000, dog 2600, rabbit 1900, human 1480, whale 1300, pig 1100, beaver 520. In all these animals the blood osmotic concentration is approximately 380–400 mosm. If the animals are on diets containing much water the urine concentrations are lower. The maximum ratio of osmoconcentration of urine to that of plasma is correlated with habitat and with need to conserve water: 14 in desert rat, 8.0 in camel, 4.0 in human, 2.4 in beaver. Osmotic concentration of urine increases not only on a diet low in water but also on a high-protein diet.

Adaptations of mammalian kidneys are largely of two kinds: countercurrent flow between adjacent tubules and selective permeabilities of tubule regions. A human kidney has 1,000,000 nephrons. A nephron in a mammal consists of a glomerulus through which plasma water, small molecules, and ions are filtered and protein retained; filtration pressure is the difference between blood pressure and colloid osmotic pressure of plasma proteins, and the amount of filtration varies with blood flow and capillary permeability. The fluid in the proximal kidney tubule is essentially isosmotic with plasma; in this tubule segment glucose, amino acids, and other small organic molecules are reabsorbed (Fig. 9-13). This segment is followed by the loop of Henle, an intermediate tubule which occurs only in mammals and birds. In general, those species of mammals with long loops which penetrate deep into the medulla to the renal papilla are able to extract more water than those with short loops. The ascending segment is thin near the loop and thicker as it approaches the distal tubule. Above the ascending arm of the loop in the renal cortex is the distal tubule; several distal tubules converge to form the collecting ducts that penetrate through the medulla; fluid from the ducts exits to the ureter (7).

Sections from various regions of the kidney show that the interstitium (extracellular fluid) of the medulla is hyperosmotic to the fluid of cortical plasma and to the fluid of the proximal tubule. As fluid in the collecting duct passes through the medulla, water moves out osmotically into the interstitium and the resulting urine may be hyperosmotic. The permeability of the collecting duct is regulated by the blood concentration of the antidiuretic

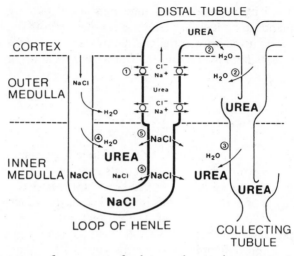

Figure 9-13. Summary of movement of solutes and water in components of mammalian nephron. Ascending limbs impermeable to water as indicated by thickened lining. (1) Active chloride reabsorption accompanied by passive Na$^+$ absorption; (2) Na$^+$ reabsorbed down its osmotic gradient, concentration of urea increases; (3) passive reabsorption of H$_2$O and urea from collecting duct—urea accumulates in medullary interstitium where it osmotically extracts water from descending limb; (4) and concentrates NaCl in descending limb; (5) ascending limb permeable to NaCl but impermeable to water, and NaCl moves down its gradient. Reprinted by permission of *The New England Journal of Medicine,* Vol. 295, pp. 1059, 1976.

hormone (ADH) from the pituitary, arginine vasopressin in most mammals, lysine vasopressin in pig and hippopotamus.

Evidence for changes in fluid within the kidney tubules and changes in the interstitial fluid of the medulla has come from analysis of slices, micropuncture of tubules *in vivo,* and perfusion of isolated tubule segments. Species differences occur in the relative importance of active transport of NaCl and of permeability of different tubule segments to water, NaCl, and urea. The general scheme is that salt is transferred, partly passively and partly actively, from the lumen of the thin ascending limb into the interstitium of the medulla; this permits water to leave the descending limb by an osmotic gradient. The loop of Henle constitutes one countercurrent multiplier. The ascending and distal tubules have low permeability to urea, and fluid in the distal tubule is nearly isosmotic or hypoosmotic to plasma. The thick ascending limb, the cortical distal tubule, and the collecting tubule constitute a second countercurrent multiplier. The collecting duct is permeable to water and urea, and urea from this duct adds to NaCl from the ascending limb to raise the osmoconcentration of the interstitium. The properties of the membrane of the tubules taking part in urine concentration are summarized in Figure 9-13 and Table 9-8.

Table 9-8.   Transport and Permeability Properties of Rabbit Nephron Segments as Defined in Isolated Perfused Tubules (66)

	Active Salt Transport	Permeability		
		H_2O	NaCl	Urea
Thin decending limb	0	+ + + +	0	+
Thin ascending limb	0	0	+ + +	+
Thick ascending limb	+ + + +	0	0	0
Cortical and outer medullary collecting duct	+	+ + +	0	0
Inner medullary collecting duct	+	+ + +	0	+ + +

CONCLUSIONS

In prebiotic evolution the organization of chemical compounds into what was to become protoplasm took place after separation of the protoplast from the medium. The first organisms probably appeared with the evolution of cell membranes. The size of cells was determined by the distance over which nutrients and gases could diffuse into the protoplast, and by the distance over which a nucleus could encode essential proteins. Water in protoplasm is structured, and some of it is bound as water shells around polar molecules. Cells and multicellular organisms live within the restrictions of properties of water and of selectively permeable membranes. There is some positive transmembrane movement of water, oxygen, ions, and metabolites.

A primitive and universal property of membranes is the regulation of cell volume; this is essential for any cell containing nonpenetrating organic molecules, which cause an osmotic gradient favoring water flow inward. Regulation of cell volume takes place via unequal distribution of water and ions. Certain ions were selected as solutes, and these participated in biochemical functions. Selection of specific ions such as K^+ rather than Na^+ as the principal intracellular cation may have been by ionic diameter, or by differences in perturbing effects on water structure and/or on enzyme function. Calcium is maintained at low intracellular concentrations by binding, but it is essential for many membrane-bound enzymes. Magnesium is variable in cell concentration. Of the abundant ions in sea water, sulfate is the most excluded by organisms; magnesium is less excluded.

Synthesis and retention of amino acids, urea, and amines increase in many cells whenever osmoconcentration increases. Increase of such organic solutes permits concentrations of ions to be relatively constant. The organic solutes that are used in osmoconformity and volume regulation are less perturbing for enzymes than would be equivalent concentrations of Na and K.

Volume and ionic regulation in cells led to osmotic regulation in organisms: isosmoticity in free-living cells in seawater or in organic media and in the cells of multicellular animals; hypoosmoticity in organisms living in hypersaline media and in a few organisms living in the sea; hyperosmoticity in cells and multicellular organisms living in dilute media (fresh water) and hyperosmoticity in cells of plants. Two general mechanisms of volume regulation are recognized: (1) Cells of most bacteria and plants have inelastic membranes, hence steep osmotic gradients between cell interior and extracellular fluid are possible. (2) Cells with elastic membranes have ion pumps which extrude protons (many prokaryotes and plants) or extrude sodium (most animals); thus volume is regulated, isosmoticity across the membrane is maintained, and transmembrane potentials result. Volume and ionic regulation probably preceded osmotic regulation. In multicellular organisms the tissues which interface with the environment have specialized regulating properties—restricted, specific permeability, active extrusion, and absorption.

As organisms invaded different environments, specific adaptations evolved and constancy of volume—and of osmotic and ionic concentrations—were maintained. The molecular adaptations to marine, brackish, hypersaline, and fresh water, moist-air and desert environments, are extensions and elaborations of mechanisms present in primitive cells. The composition of cell membranes—mainly proteins and phospholipids—was established in the earliest organisms. There are many variations in membrane composition, for example, the presence of some carbohydrate moieties, diversification of membrane proteins that provide structural and enzymatic functions, and fatty acid composition of phospholipids that maintain uniform membrane fluidity.

Movements of ions across cell membranes may be by diffusion down electrochemical gradients at rates determined by solvent properties and pore sizes and charges of membrane. More susceptible of control than passive diffusion is facilitated diffusion, in which a protein ionophore or carrier molecule binds an ion on one side of a membrane and releases it on the opposite face. Energy is given by the electrochemical gradient. Cotransport or antiport permits exchange of one ion for another of the same sign—H^+ for K^+, Na^+ for K^+, Cl^- for HCO_3^-—or exchange of two ions of the same kind. Symport exchange facilitates unidirectional movement of ions of opposite charge ($Na^+ - Cl^-$, $K^+ - Cl^-$) with maintenance of electroneutrality. The identity and binding properties of a few carrier proteins are known. Inhibitors of transporters vary among tissues, some relatively specific and others more general. Movement of water may be linked to movement of ions.

Another kind of transport requires metabolic energy and proceeds either up or down an electrochemical gradient. Ion pumps are suggested by specific ion competition, kinetics of transport, selective inhibitors and activators. The first proton transporters may have been ATPases of large size. Later in evolution proton transporters reversed function to be ATP synthetases which generate proton-motive forces and are powered from redox or photosynthetic sequences. Secondarily, ATPases of smaller size function in transport of Na$^+$, K$^+$, Ca^{2+}, and protons. Active transport probably appeared first for protons, later for Cl$^-$ in plants and for Na$^+$, K$^+$, and Ca^{2+} in animals. Facilitated transport may have preceded ion pumps which use metabolic energy. One transport ATPase, that for Na$^+$-K$^+$, has been purified and its structure ascertained. Many cells are asymmetric and by symport and antiport together with active transport—often (Na$^+$ + K$^+$)-ATPase—ions can be moved from one side of an epithelium to the other. Absorption of organic solutes—sugars and amino acids—is frequently coupled to ion transport. In multicellular animals, the regulation of facilitated transport and ion pumping is hormonal—in insects, fishes, and amphibians—or neural—in birds and mammals (85b).

REFERENCES

1. Almers, W. and C. Stirling. *J. Membrane Biol.* **77:**169–186, 1984. Distribution of transport proteins over cell membranes.

2. Alvarado, R. H. pp. 261–303 in *Comparative Physiology of Osmoregulation in Animals,* Ed. G. M. Malory, Academic Press, New York, 1979. Amphibian regulation.

3. Amende, L. M. and S. K. Pierce. *J. Compar. Physiol.* **138:**283–289, 1980. Regulation of cell volume in salinity-stressed molluscs.

4. Aomine, M. *Compar. Biochem. Physiol.* **68A:**531–540, 1981. Amino acid absorption and transport in protozoa.

5. Baginski, R. M. and S. K. Pierce. *Compar. Biochem. Physiol.* **57A:**407–412, 1977. Amino acid accumulation in cells of Modiolus in high salinity.

6. Beamish, F. W. H. et al. *Compar. Biochem. Physiol.* **60A:**435–443, 1975. Osmoregulation in lampreys.

7. Beeuwkes, R. Ch. 19, pp. 266–268 in *Companion to Animal Physiology,* Ed. C. R. Taylor et al., Cambridge University Press, New York, 1982. Renal countercurrent mechanisms.

8. Bentley, P. J. and T. Yorio. *Compar. Biochem. Physiol.* **61A:**635–641, 1978. Active Cl transport across cornea of toad.

9. Beres, L. S. and S. K. Pierce. *J. Compar. Physiol.* **144:**165–173, 1979. Salinity stress on Mya neurons.

10. Borowitzka, L. J. et al. *Science* **210:**650–651, 1980. Osmoregulation in blue-green algae.

11. Bowlus, R. D. and G. N. Somero. *J. Exper. Zool.* **208:**137–152, 1979. Solute compatibility with enzyme function.

12. Bradley, T. J. and J. E. Phillips. *J. Exper. Biol.* **66:**83–126, 1977. Excretion by mosquito larvae.

13. Brown, A. D. and J. R. Simpson. *J. General Microbiol.* **72:**589–591, 1972. Osmotic relations in yeast.

14. Brown, A. D. *Bact. Rev.* **40:**803–846, 1976. Microbial water stress; halophilic bacteria, xerotolerant yeasts, halophilic algae.

15. Burton, R. F. *Compar. Biochem. Physiol.* **27**:763–773, 1968; **44A**:781–792, 1973. Balance of cations in blood and muscle.

16. Cameron, J. N. *J. Compar. Physiol.* **123**:123–141, 1978. Acid-base balance and ion distribution in Callinectes.

17. Cirillo, V. P. *Annu. Rev. Microbiol.* **15**:197–218, 1961. Sugar transport in yeast and bacteria.

18. Clark, M. E. et al. *J. Exper. Biol.* **90**:33–41, 1981. Solutes in barnacle muscle fibers.

19. Conté, F. P. *J. Biol. Chem.* **253**:4762–4770, 1978; *Am. J. Physiol.* **238R**:269–276, 1980; *Int. Rev. Cytol.* **91**:45–106, 1984. Structure and function of larval salt gland of Artemia.

19a. Conway, E. J. *Physiol. Rev.* **37**:84–132, 1957. Na and K in skeletal muscle.

20. Cook, P. and I. D. Kuntz. *Ann. Rev. Biophys. Bioeng.* **3**:95–126, 1974. Properties of water in biological systems.

21. Cornell, J. C. *Biol. Bull.* **157**:422–433, 1979; **158**:16–25, 1980. Salt and water balance in Libinia and Pugettia.

22. Costa, C. J., S. K. Pierce, and K. Warren. *Biol. Bull.* **159**:626–638, 1980. Volume regulation by Glycera red coelomocytes.

23. Crane, R. K. *Biochem. Biophys. Acta* **553**:295–306, 1978; **554**:259–267, 1979. Sodium-dependent glucose transport.

24. Dahm, J. and B. O. Gobel. *J. Membrane Biol.* **77**:243–254, 1984. Furosemide-sensitive Na/K transport in red blood cells.

25. Dall, W. *J. Exper. Marine Biol. Ecol.* **15**:97–125, 1974. Osmotic and ionic regulation in Panulirus.

26. Degnan, K. J. and J. A. Zadunaisky. *Am. J. Physiol.* **238R**:231–239, 1980. *J. Membrane Biol.* **55**:175–185, 1980. Ion permeability and transport across opercular epithelium of Fundulus.

27. Dehnel, P. *Canad. J. Zool.* **57**:521–532, 1979. Ion and water balance in muscle of Hemigrapsus.

28. De Renzis, G. and M. Bornancin. *Biochem. Biophys. Acta* **467**:192–207, 1977. Cl/HCO_3-ATPase in gills of goldfish.

29. Diamond, J. *J. Exper. Zool.* **194**:227–240, 1975. How cells discriminate ions.

30. Dick, D. A. pp. 3–45 in *Mechanisms of Osmoregulation in Animals,* Ed. R. Gilles, Wiley, New York, 1979. pp. 211–250 in *Membranes and Ion Transport,* Ed. E. E. Bittar, III, Wiley, New York, 1970. Structure and properties of water in cells.

31. Dixon, T. E. and Q. Al-Awqati. *Proc. Natl. Acad. Sci.* **76**:3135–3138, 1979. Hydrogen ion transport in turtle bladder.

32. Dunham, P. B., G. W. Stewart, and J. C. Ellory. *Proc. Natl. Acad. Sci.* **77**:1711–1715, 1980. Cl activated potassium transport in human erythrocytes.

33. Edney, E. B. pp. 77–97 in *Physiological Adaptations to the Environment,* Ed. J. Vernberg, Intext, New York, 1977. Absorption of water from unsaturated air.

34. Eisenman, G. *Biophys. J.* **2**:259–323, 1962. Cation selective glass electrodes.

34a. Ellory, J. C. and J. S. Gibson. pp. 197–216 in *Cellular Acclimatization to En-*

vironmental Change, Ed. A. R. Cossins and P. Sheterline, Cambridge University Press, New York, 1984. Cellular aspects of salinity adaptation in fishes.

35. Epstein, J. H., J. Maetz, and G. DeRenzis. *Am. J. Physiol.* **224**:1295–1299, 1973; **238**:R246–250, 1980. *J. Exper. Biol.* **106**:25–41, 1983. Active transport of Cl by fish gill and elasmobranch rectal gland.

36. Evans, D. H. et al. *J. Exper. Biol.* **61**:277–283, 1974. *Compar. Biochem. Physiol.* **51A**:491–495, 1975. pp. 197–205 in *Epithelial Transport,* Ed. B. Lahlou, Cambridge University Press, New York, 1980.

37. Felder, D. L. *Biol. Bull.* **154**:409–429, 1978. Osmotic and ionic regulation in mud shrimps.

38. Felle, H. et al. *Biochemistry* **19**:3585–3590, 1980. Membrane potential of E. coli.

39. Fellenius, E. et al. *Nature* **290**:159–161, 1981. Gastric acid secretion.

40. Fletcher, C. R. *Compar. Biochem. Physiol.* **47A**:1199–1234, 1974. Volume and water regulation in Nereis diversicolor.

40a. Forte, J. G. et al. *Bioscience* **35**:38–42, 1985. Proton pumps in stomach, kidney, and turtle bladder.

41. Freel, R. W. et al. *Biol. Bull.* **144**:289–303, 1973. Ion regulation in euryhaline Neanthes.

42. Fugelli, K. and H. Rohrs. *Compar. Biochem. Physiol.* **67A**:545–551, 1980. Solute in osmotic regulation in flounder erythrocytes.

43. Gaines, M. S., J. Caldwell, and A. M. Vivas. *Marine Biol.* **27**:327–332, 1974. Population differences in Littorina.

44. Gennis, R. B. and A. Jonas. *Annu. Rev. Biophys. Bioeng.* **6**:195–238, 1977. Protein-lipid interaction.

44a. Gerencser, G. A. and S. H. Lee. *J. Exper. Biol.* **106**:141–161, 1983. Cl⁻-stimulated ATPase.

45. Glynn, I. M. and S. J. Karlish. *Annu. Rev. Physiol.* **37**:13–55, 1975. The sodium pump.

46. Goh, S. and J. E. Phillips. *J. Exper. Biol.* **72**:25–41, 1978. Transport properties of locust rectum.

47. Goldstein, L. and T. A. Boyd. *Compar. Biochem. Physiol.* **60A**:319–325, 1978. Amino acid transport in skate erythrocytes.

48. Greenwald, L. *Physiol. Zool.* **45**:229, 1972. Affinity of Na-transporting system in aquatic animals.

49. Gupta, B. L. et al. *Exper. Biol.* **88**:21–47, 1980. Excretion by rectal papillae of Calliphora.

50. Hadley, N., Ed. *Environmental Physiology of Desert Organisms,* Halsted Press, New York, 1975.

51. Hagen, D. W. *Journal of Fisheries Research Board of Canada* **24**:1637–1692, 1967. Species isolation in sticklebacks.

52. Hampton, J. R. and J. L. Schwartz. *Compar. Biochem. Physiol.* **55A**:1–4, 1976. Contractile vacuole function in ciliate from Great Salt Lake.

52a. Harold, F. M. and D. Papineau. *J. Membrane Biol.* **8**:45–62, 1972. Proton and Na extrusion by Streptococcus.

52b. Hanrahan, J. W. and J. E. Phillips. *J. Exper. Biol.* **106**:71–89, 1983. Control of KCl absorption in insect hindgut.

53. Harvey, W. R. et al. *Am. Zool.* **21**:775–791, 1981. Ion transport in insect midgut. *J. Exper. Biol.* **106**:91–117, 1983. K-ATPase.

54. Hilden, S. et al. *J. Biochem.* **249**:7432–7440, 1974. Na, K activated ATPase in phospholipid vesicles.

55. Ho, S. M. and D. K. Chan. *Compar. Biochem. Physiol.* **66B**:255–260, 1980. Branchial ATPase in eels.

56. Hoffman, E. K. Ch. 12, pp. 285–332 in *Transport of Ions and Water in Animals,* Ed. B. J. Gupta et al., Academic, New York, 1977.

57. Hoffman, E. K. et al. *J. Physiol.* **296**:61–84, 1979. *J. Membrane Biol.* **76**:269–280, 1983. Cl exchange in tumor cells, Na Cl cotransport activated in volume regulation.

57a. Hoffman, E. K. pp. 55–80 in *Cellular Acclimation to Environmental Change,* Ed. A. Cossins and P. Sheterline, Cambridge University Press, New York, 1984. Volume regulation by animal cells.

58. Hokin, L. E. *Ann. New York Acad. Sci.* **242**:12–13, 1974. *Arch. Biochem. Biophys.* **151**:453–463, 1972. *Exper. Zool.* **194**:197–206, 1975. Energetics of sodium pump.

59. Jaffe, L. F. *J. Theoret. Biol.* **48**:11–18, 1974. Interpretations of voltage concentration relations.

60. Jakobsson, E. *Am. J. Physiol.* **238C**:196–206, 1980. Relation of cell volume and membrane potential to ion transport.

61. Jeschke, W. D. pp. 17–28 in *Plant Membrane Transport,* Ed. R. M. Spanswick, North Holland, New York, 1980. Cation selectivity of root.

61a. Jorgensen, P. L. *Q. Rev. Biophys.* **7**:239–274, 1978. Review of isolation and properties of Na pump.

62. Karnaky, K. J. et al. *Science* **195**:203–204, 1977. Cl transport across opercular epithelium of Fundulus.

63. Kirschner, L. B. Ch. 5, pp. 157–222 in *Osmoregulation in Animals,* Ed. R. Gilles, Wiley, New York, 1979. Sodium chloride excreting cells of marine vertebrates.

64. Kirschner, L. B. *Am. J. Phys.* **238**:R219–R223, 1980; **244**:R429–R443, 1983. Ion regulation in freshwater animals.

65. Kobayashi, H. et al. *J. Biol. Chem.* **253**:2085–2092, 1978. Calcium transport in Streptococcus.

65a. Koepen, B. and S. Helman. *Am. J. Physiol.* **242**:F521–F531, 1982. Acidification by renal collecting tubules.

66. Kokko, J. P. *Fed. Proc.* **33**:25–30, 1974. Salt and water transport in loop of Henle.

67. Komor, E. and W. Tanner. pp. 247–257 in *Plant Membrane Transport,* Ed. R. M. Spanswick, Elsevier-North Holland, New York, 1980. Proton-cotransport of sugars in plants.

68. Koehn, R. K. et al. *Evolution* **30**:2–32, 1976. Population genetics in Mytilus.

69. Kregenow, F. M. *J. General Physiol.* **64:**393–412, 1974. *Annu. Rev. Physiol.* **43:**493–501, 1981. Volume regulation in red cells.

70. Lanyi, J. K. *J. Supramolec. Struct.* **13:**83–92, 1980. *J. Biochem.* **254:**10986–10994, 1979; **255:**243–250, 1980. Electrogenic pump for Na in Halobacterium.

71. Lanyi, J. K. *Bact. Rev.* **38:**272–290, 1974. Biology of halophilic bacteria.

72. Lee, K. H. and R. Blostein. *Nature* **285:**338–339, 1980. Na fluxes in red cells.

73. Lockwood, A. P. M. et al. *J. Exper. Biol.* **58:**137–148, 1973. Water fluxes in Gammarus duebeni.

74. Lubin, M. pp. 193–210 in *Cell Function and Membrane Transport,* Ed. J. F. Hoffman, Prentice-Hall, Englewood Cliffs, New Jersey, 1964. Cell K and regulation of protein synthesis.

75. Lukacovic, M. F. et al. *Fed. Proc.* **35:**3–10, 1976. Anion transport in red blood cells.

76. Lynch, M. P. et al. *Compar. Biochem. Physiol.* **44A:**719–734, 1973. Serum ions in Callinectes.

77. Machin, J. et al. *J. Exper. Zool.* **222:**309–320, 1982. Water vapor absorption in insects.

78. Maddrell, S. H. P. and J. E. Phillips. *J. Exper. Biol.* **62:**367–378, 671–683, 1975; **72:**181–202, 1978; **90:**1–15, 1981. *Perspectives Exper. Zool.* **1:**179–186, 1975. Excretion of ions in insects.

79. Maddrell, S. H. P. *Philos. Trans. R. Soc. London* **B262:**197–207, 1971. Fluid secretion by Malpighian tubules.

80. Maetz, J. *Philos. Trans. R. Soc. London* **B262:**209–249, 1971; *J. Exper. Biol.* **58:**255–275, 1973. Salt transfer across gills of freshwater fishes.

81. Maetz, J. et al. *Perspectives Exper. Biol.* **7:**77–92, 1975. Review of ion transport mechanisms.

81a. Maloney, P. C. and T. H. Wilson. *Bioscience* **35:**43–48, 1985. Evolution of ion pumps.

81b. McCarty, R. E. *Bioscience* **35:**27–41, 1985. Oxidative and photosynthetic coupling of synthesis and hydrolysis of HTP by H^+ATPase to ion transfer.

82. Measures, J. C. *Nature* **257:**398–400, 1975. Osmoregulation of bacteria.

83. Mitchell, P. and J. Moyle. *Europ. J. Biochem.* **7:**471–484, 1969. Proton transport by liver mitochondria.

84. Motais, R. et al. *J. Genetic Physiol.* **50:**391–422, 1966. Exchange diffusion across fish gills.

85. Nellans, H. N. et al. *Am. J. Physiol.* **225:**467–475, 1973. *J. Genetic Physiol.* **68:**441–463, 1976. Sodium-dependent transport in intestine.

85a. Oesterhelt, D. *Bioscience* **35:**18–21, 1985. Light-driven proton pump in Halobacteria.

85b. Pang, P. K. T. *J. Exper. Biol.* **106:**283–299, 1983. Evolution of epithelial transport in vertebrates.

85c. Paradiso, A. M. et al. *Proc. Natl. Acad. Sci.* **81:**7436–7444, 1984. Na^+-H^+ exchange in gastric glands.

86. Parker, J. C. *J. Genetic Physiol.* **62:**147–156, 1973. Salt and water regulation in red blood cells.

87. Peaker, M. and J. L. Linzell. *Salt Glands in Birds and Reptiles,* Cambridge Univ. Press, New York, 1975.

88. Phillips, J. E. *Am. Zool.* **10**:413–436, 1970; Fed. Proc. **36**:2480–2486, 1977. *Am. J. Phys.* **241**:R241–R257, 1981. Excretion in insects.

89. Phillips, J. E. and T. J. Bradley. pp. 709–734 in *Transport of Ions and Water,* Ed. B. L. Gupta, Academic Press, New York, 1977. Saline-water mosquito larvae.

90. Pierce, S. K. et al. *J. Exper. Biol.* **57**:681–692, 1972; **59**:435–440, 1973. *J. Exper. Zool.* **215**:247–257, 1981. Amino acid mediated cell volume regulation in molluscs.

91. Pierce, S. K. *Biol. Bull.* **163**:405–419, 1982. Review of invertebrate cell volume control.

92. Potts, W. T. W. et al. *J. Compar. Physiol.* **87**:21–28, 29–48, 1973. Ion regulation in flounder.

93. Potts, W. T. W. *Perspectives Exper. Biol.* **1**:65–76, 1976. Ionic and osmotic regulation in marine fishes.

94. Potts, W. T. W. and G. Fryer. *J. Compar. Physiol.* **129**:289–294, 1979. Na balance in Daphnia.

94a. Prosser, C. L. Ch. 2, pp. 79–110 in *Comparative Animal Physiology,* Ed. C. L. Prosser, Saunders, Philadelphia, 1973. Inorganic ions in tissues and body fluids.

95. Prusch, R. D. and T. Otter. *Compar. Biochem. Physiol.* **57A**:87–92, 1977. Ion transport in annelids.

95a. Ramos, M. and J. C. Ellory. *J. Exper. Biol.* **90**:123–142, 1981. Na and Cl transport across intestine of plaice.

96. Raven, J. A. and F. A. Smith. pp. 161–174 in *Plant Membrane Transport,* Ed. R. M. Spanswick et al., Elsevier-North Holland, New York, 1980. Proton pump in plant cells.

96a. Reid, R. J. and N. A. Walker. *J. Membrane Biol.* **78**:35–41, 1984. Active transport of Cl^- into Chara.

97. Rhoads, A. B. and W. Epstein. *J. Biol. Chem.* **252**:1394–1401, 1977. Bacterial uptake of K coupled to proton pump.

98. Robertson, J. D. *Biol. Bull.* **148**:309–319, 1975. *Compar. Biochem. Physiol.* **77A**:431–439, 1984. Ions in plasma and muscle of Squalus, eel, and lamprey.

99. Rothstein, A. et al. *J. Membrane Biol.* **15**:227–248, 1974. *Biol. Biophys. Res. Com.* **64**:144–150, 1975. *Fed. Proc.* **35**:3–10, 1976. Anion transport in red blood cells.

100. Sanders, D. et al. *Proc. Natl. Acad. Sci.* **78**:5903–5907, 1981. Proton pump in plant and animal cells.

101. Schmidt-Nielsen, B. *J. Exper. Zool.* **194**:207–220, 1975. Volume regulation and intracellular ions.

102. Schoffeniels, E. *Biochem. Soc. Symp.* **41**:179–204, 1976. Adaptation to salinity.

102a. Schultz, S. G. *Am. J. Physiol.* **233**:E249–E254, 1977. Sodium-coupled solute transport by small intestine.

103. Schultz, S. G. and P. F. Curran. *Physiol. Rev.* **50**:637–718, 1970. *Biochem. Biophys. Acta* **241**:857–860, 1971. Sodium-coupled amino acid transport.

104. Shamoo, A. E. and T. E. Ryan. *Ann. New York Acad. Sci.* **264**:83–96, 1975. *J. Membrane Biol.* **43**:227–242, 1973. Ionophores for calcium transport.

105. Shuttleworth, T. J. and R. F. H. Freeman. *J. Compar. Physiol.* **86**:293–314, 315–322, 323–330, 1973. Ion regulation by gills of Anguilla.

106. Silva, P. et al. *J. Exper. Zool.* **199**:419–426, 1977. *J. Membrane Biol.* **53**:215–221, 1980; **75**:105–114, 1983. *J. Compar. Physiol.* **154**:139–144, 1984. Salt secretion by gills and rectal glands of fishes.

107. Singer, S. J. and D. L. Nicolson. *Science* **175**:720–31, 1972. Mosaic model of membrane structure.

108. Skadhauge, E. *J. Exper. Biol.* **60**:535–546, 1974. Ion and water fluxes in Anguilla.

109. Slayman, C. *Biochem. Biophys. Acta* **102**:149–160, 1965. **426:732–744, 1976;** *J. Memb. Biol.* **23**:181–212, 1975; pp. 179–190 in *Plant Membrane Transport*, Ed. R. M. Spanswick, Elsevier-North Holland, New York, 1980. Ion transport in Neurospora.

109a. Slayman, C. L. *Bioscience* **35**:34–37, 1985; **35**:16–17, 1985. Proton pumps in plants and fungi.

110. Smith, P. G. *J. Exper. Biol.* **51**:727–738, 739–757, 1969. Ionic regulation in Artemia.

111. Spencer, A. M. et al. *Phys. Zool.* **52**:1–10, 1979. Gill (Na,K)-ATPase and osmoregulation in crabs.

111a. Spring, J. H. and J. E. Phillips. *J. Exper. Biol.* **86**:211–223, 1980. *Canad. J. Zool.* **56**:1879–1882, 1978. Hormonal control of Cl⁻ transport across locust rectum.

112. Stewart G. R. and J. A. Lee. *Planta* **120**:279–289, 1974. Proline accumulation in halophytes.

113. Stobbart, R. H. *J. Exper. Biol.* **60**:493–533, 1974; **69**:53–85, 1977. *Compar. Biochem. Physiol.* **58A**:299–309. Ion transport in Aedes aegypti, Daphnia.

114. Sweadner, K. J. and S. M. Goldin. *New Eng. J. Med.* **302**:377–383, 1980. Model for (Na, K)-ATPase.

114a. Thorson, T. B. *Life Sci.* **9**:893–900, 1970. Urea in freshwater elasmobranchs.

114b. Wall, B. J. *Fed. Proc.* **30**:42–48, 1971. Ultrastructure of excretory epithelium in insects.

115. Watts, J. A. and S. K. Pierce. *J. Exper. Zool.* **204**:43–56, 1978. ATPase activity and tolerance of low salinity in Modiolus.

116. West, I. and P. Mitchell. *Biochem. J.* **144**:87–90, 1974. Proton/sodium exchange in E. coli.

117. Wiggins, P. M. *J. Theoret. Biol.* **32**:131–146, 1971. Water structure in protoplasm.

118. Willis, J. S. *Cryobiology* **9**:351–366, 1972. Cellular potassium in survival of cells at low temperatures.

119. Wohlrab, H. *Biochemistry* **13**:4014–4018, 1974. Respiration-linked Ca ion uptake by muscle mitochondria.

120. Wood, J. L. et al. *J. Exper. Biol.* **62**:313–320, 1975. Transport of Mg in insect larvae.

121. Woodbury, J. W. and P. R. Miles. *J. General Physiol.* **62**:324–353, 1973. Anion conductance in muscle membrane.

122. Yancey, P. H. and G. N. Somero. *J. Compar. Physiol.* **125**:135–146, 1978. Urea-requiring LDH in elasmobranch.

123. Yancey, P. H. et al. *Science* **217**:1214–1222, 1982. Osmolytes in muscle of marine and brackish water animals.

124. Zadunaisky, J. A. and K. J. Degnan. pp. 185–196 in *Epithelial Transport in Lower Vertebrates,* Ed. B. Lahlou, Cambridge University Press, New York, 1980. Chloride transport and osmoregulation in fishes.

10

ANIMAL BEHAVIOR

The theme of this book is that the essence of evolution is adaptive diversity. Adaptations to the physical environment have been considered in the preceding chapters in terms of molecular properties that permit organisms to occupy diverse ecological niches and geographic ranges. Adaptations to the biotic environment are of critical importance in evolution but are less amenable to molecular analysis than are adaptations to the physical environment. In plants, biotic factors influence nutrition, growth, and dispersal. In animals, behavioral adaptations permit diversification. In this chapter behavioral adaptations are considered at the sensorimotor and social levels, with extrapolations to human behavior. Animal behavior is an emergent property, amenable to change within the constraint of the genome.

Social adaptations permit much diversity among animals and provide behavioral isolating mechanisms for ethospecies (species in which reproductive isolation is effected by behavior).

Although behavioral adaptations are physiologically far removed from DNA replication and protein synthesis, many behaviors are genetically determined, and many features of social behavior are transmitted genetically. Genes specify connections in nervous systems and thus modify simple behavior; genes for neurotransmitters and receptors are described in Chapter 12. Many more genes and more kinds of protein are needed for brain than for liver or kidney. The coding of complex innate patterns is not understood. Social behavior is a holistic aspect of biology for which few reductionist explanations are possible. Social behavior may be transmitted by imprinting, learning, and cultural or group conditioning. The sum of the characters which are socially transmitted are called sociotypes by analogy with phenotypes; the memory units of socially transmitted behavior are called memes by analogy with genes (37). Two species of animals that appear morphologically alike sibling species may be distinguishable by their reproductive behavior, which is presumably determined by small genetic differences. The genetic changes for effecting altered neural patterns of reproductive behavior are evolutionarily comparable to macromutations and to genetic changes in regulatory genes.

Aspects of animal social behavior are often described in anthropomorphic terms. An observer judges whether a given behavior is relevant or irrelevant on the basis of the contributions of that behavior to overall fitness or to adaptation to the biotic environment; the terminology of sociobiology will be used in this chapter despite its anthropomorphic connotation.

This chapter considers behavioral diversity in animals and relates it to "higher" functions in nervous systems. Behavior will be classified in a continuum from protozoans to humans. Succeeding chapters consider adaptive variations in "simple" nervous systems, in synaptic transmission, in sense organs of three modalities, and in the biophysics of excitable membranes. The present chapter offers examples of adaptive behavior which has been analyzed only in holistic terms.

PRINCIPLES OF ETHOLOGY

Animal Associations

Many animals live in social groups, and for such animals natural selection may act on the society rather than on individuals. Several proximal advantages of societal structure are recognized; one is division of functions among members, for feeding, protection, and reproduction. Coelenterates—coral heads, hydroid assemblages, and colonial siphonophores—share supporting skeletons and cooperate in prey capture. In siphonophores such as Portuguese men-of-war a gas-filled float is attached to a complex of feeding, defending, and reproducing polyps and sexually reproducing medusae in a loose cooperative association. Flocks of birds and herds of mammals collectively exploit food supplies that fluctuate in availability. Protection is another function of animal societies. Musk oxen keep their young in the center of a herd. Culling of marginal individuals by predators serves preserves the herd as a whole (70). Schooling of fish is a protective association. Tadpoles derived from a single clutch aggregate and may form sibling schools (170). Some protection against predators is provided by synchronous breeding, for example, in packs of wild dogs or colonies of sea birds. Flocks of passerine birds disperse for feeding but come together and make a mobbing response when threatened by a predator, such as a hawk. In social carnivores—pack hunters—cooperation increases kill success over solitary hunting. Pack hunting also enables relatively small canids, such as African hunting dogs, to kill large prey.

A simple social interaction is performed by a reproductive pair of animals. The two sexes of some kinds of animals come together only for fertilization of eggs and the association is transitory. Some male molluscs and fishes shed sperm over deposited eggs. A more complex reproductive association occurs when a nest is prepared and one or both members of a pair care for the young. Females of some fishes deposit eggs in a nest made by the male, in which eggs must be oxygenated by fanning. One or both parents of birds and of some fishes care for the young after hatching. A different sort of interaction exists between mammalian mothers and neonates. Some animals are monogamous, for example, penguins, others are polygamous—lions, seals and walruses; a few are polyandrous—some varieties of bees. Populations of wild mice consist of breeding units, demes, of approximately 10 individuals; territory is maintained by the males and the dominant male sires 90 percent of litters. Males of eighteen species of bowerbirds of New Guinea construct species-specific bowers to which they attract mates. The bowerbird is inconspicuous as an individual but decorates his bowers with colored objects; some species weave a platform on which fresh flowers are placed daily, others construct a maypole. Young males may practice bower building for two years before mating, and the bower may establish territoriality. After mating, the male drives the female out; she builds a nest and rears the young elsewhere (43). Female widowbirds may choose mates according to length of tail (3).

Elaborate societies of social insects—bees, ants, and termites—are divided into castes. A colony of bees has one reproductive individual, the queen, who lays the fertile eggs. The males are haploid drones that fertilize a queen on a nuptial flight. Workers (diploid) carry out essential functions of the hive. Some are nurses which care for the developing larvae by feeding regurgitated food. Some bee workers scout for food and recruit other bees to collect honey, pollen, and water. Some workers detect and attack intruders—predators and bees from other hives.

There are many variations in caste organization. The members of a given beehive develop from eggs laid by one queen, who may have been fertilized by several drones. Castes favor group cohesiveness and survival of the bee society. After discharging its sting a bee dies; so-called soldier bees may die in combat with intruders. Altruistic behavior favors genetic continuity of the colony after loss of some members. The evolutionary advantage of the death of some individuals to the society is the increase in reproductive success of the genes carried by the queen. Some students of Hymenoptera consider a colony to be biologically equivalent to an individual of other kinds of animals (173). Human societies use a very different strategy for cohesiveness; individuals differ in genotypes but have culturetypes in common.

Savannah baboons live in troops of 13–185 members. Trooping protects individuals from predators; males protect females and young. The range of movement of a troop is small; several troops may gather at a waterhole. Each male has several mates, and in the social hierarchy the dominant male is most frequently groomed. Hierarchical organization favors mating by strong dominant males. Communication by monkeys and apes is largely by species-specific gestures, to a lesser degree by vocalization and odors. Kin recognition is well known for primates. Macaques recognize and spend more time with closely related individuals, even after half-siblings have been reared apart and brought together at maturity (171).

In crowded rookeries sea birds—gulls, guillemots, and penguins—recognize their own chicks among hundreds in a colony, and the chicks recognize their parents, presumably by acoustic cues. In all social groups—pairs, families, or colonies—behavior toward kin is usually different from that toward non-kin. Kin recognition, chemical or visual, leads to protective behavior. Female lions live in groups consisting of kin; groups of male lions include non-kin.

The size of a social group is often regulated by available space. Survival time of individuals—average maximum age—is related to development and to strategies of reproduction. There are two fundamental ways of balancing fecundity with resources, hence establishment of population size (Chapter 4). One kind of population growth curve is geometric, the other kind of growth curve is logistic and reaches a limit or saturation density. Animal species which follow the first strategy are opportunistic, form temporary associations, show rapid development, wide dispersal, short life span, and early reproduction. Species which follow the second strategy have stable populations, are long-lived, have constant habitats, and large body size. Each strategy embodies ecological and reproductive adaptations.

Territoriality

The best-known examples of territoriality are among birds. A male establishes an area, advertises this by song, and defends it against intruders of the same species by physical combat. Even in colonies, such as in rookeries of marine birds, individual pairs defend their nest area. Territorial defense is shown also by some lizards. Some mammals, for example, wolves, mark territorial limits by depositing scented urine. Sessile sea anemones secrete a repellent chemical which prevents other anemones from settling close by (55). An adaptive consequence of territoriality is the limitation of population size and thus optimization of use of resources.

Crowding has hormonal effects. In mammals, crowding inhibits reproduction and increases ACTH, corticosterone secretion, and adrenalin levels (4). Female rats caged together tend to go into oestrus together; rats in separate cages do not synchronize, and when housed separately may not cycle at all.

White-footed bee-eater birds of Kenya roost and breed in large colonies. Each colony is divided into clans of one or more breeding pairs plus some seven helpers. The feeding site may be as much as 7 km from the roosting site, and each group site is divided into clan territories. A colony may move its roosting site without changing its feeding site (71).

Territoriality and hierarchical organization are related. Among seals and walruses a dominant male maintains a harem, and he must defend his position against challengers. In a lizard colony, individuals are aligned in a dominance hierarchy. Lesion of limbic areas of the forebrain of the dominant lizard (alpha male) results in usurpation. In an extended family group of primates a social hierarchy is established and maintained by sequential agonistic actions: threat postures, vocal noises, physical confrontation, and fighting. Limbic lesions convert aggressive primates into submissive individuals. A social system that is initially high in agonistic behavior leads to stability as aggressive acts decline with time. Spacing between individuals is fundamental to both territoriality and hierarchical organization.

Threat behavior of pairs of cichlid fish consists of the following sequence: two fish display colors, fan the tail and propel water, snap with lips, push and pull; one fish gives up and swims away. Among male iguanas the intruder approaches and struts before a defender, the defender lunges toward the intruder, one or the other is pushed back, and the loser lowers its belly in submission. Mantis shrimps show a sequence of aggressive behavior: meet, approach, spread second thoracic appendage, lunge, strike with claw, grasp, coil, release, and the loser leaves (44).

Examples of ritualized behavior are head nodding of iguanas in asserting territoriality, and in squirrel monkeys rolling onto the back signaling appeasement. Ritualized behavior consists of patterns modified over many generations to become communicatory and to reduce ambiguity in signal exchanges.

Altruism

Behavioral altruism may decrease the chance of survival of the altruist while increasing the probability of survival of others, the beneficiaries. On the approach of a hawk, one crow in a flock sounds a warning call and the flock makes evasive moves. The crow making the distress call may be more subject to attack than others in the flock. Frogs make warning calls. "Soldier" termites engage in combat, sometimes fatal, with soldiers of another colony; this is aggression of the colony, altruism of individuals. Parent mammals may risk themselves to protect their offspring. Injury feigning by a bird may expose it to attack but draws attention away from the nest. The nest of a social wasp *Polistes* is initiated by one or two foundresses; the alpha foundress (dominant) lays the most eggs. Reproductive success of a nest with two foundresses is twice that with one; the beta foundress produces few young but she cares for offspring of the alpha female and performs other duties, and her altruism favors success of the colony. If the alpha foundress dies the beta foundress takes over (111). A male praying mantis cautiously stalks the female and once he has mounted her, the female may bite off his head.

As a general rule altruism in animal societies is shown primarily toward kin—members of the same family. Occasionally altruistic behavior may extend to other social groups of the same species; in addition, warning signals may be heeded by other species. The extending of altruism to kin may provide for continuity of genes. The evolutionary value of altruism is greatest if the actor and the recipient have many genes in common. The inclusive fitness (relatedness of recipient × benefit to recipient) must exceed the individual fitness (genetic contribution to next generation) of the altruist (2,70). In humans, but rarely in other primates, altruism may extend well beyond a kin group; this has been noted as one factor in the origin of human societies. The concept of altruism in humans goes beyond that in nonhumans in that altruism is expressed in humans not only for kin and non-kin but also for other species. As a result of social conditioning altruism has different meanings in human and nonhuman evolution.

Aggression; Population Regulation

Aggression takes many forms—competition for space, food, mates, and establishment of a social hierarchy which may be manifested in fighting, predation, and antipredation. Aggression may occur between individuals of the same species, between different kin groups of the same species, and between members of a pair. In nonhumans the intensity of aggression is usually greater between different populations of a species than within a kin group. The ethologist Lorenz maintained that human intraspecies aggression is more intense than that between species in nonhumans (101). Crowding and lack of food may increase aggressiveness. Displays such as color flashes by fishes or threat postures by lizards are releasers of aggressive behavior. Seasonal changes in intraspecific aggression are correlated with concentrations of reproductive hormones. A proximate trigger of aggression may be a high concentration of stress-related hormones.

Many animals practice cannibalism and infanticide. Fish eat their fry, male lions and hyenas may eat cubs. Tribolium (flour beetle) larvae eat eggs, adults eat pupae. Lemmings and gerbils do not attack their own young but may attack other young. Cannibalism increases when population density is high; the practice maintains constancy of population levels (2).

Signal Production and Transmission

Signals have two general functions in animal behavior: (1) to test the environment, for example, echolocation to locate food; (2) to communicate with other individuals, usually conspecific, for example, recruitment of worker bees by scouts, alarm calls by birds, mating calls by anurans. Examples of each function will be presented for the following physical modes of signalling and reception: light and vision, chemical signals, electric discharge, sound and hearing, and mechanical vibration.

Communication

Communication requires (1) capacity to emit signals, (2) sensory capacity to receive signals, (3) ability to recognize meaning of sensory messages, and (4) ability to react appropriately; this may be by direct motor responses, modulation of returned signals, or stereotyped complex response signals. Physiological analysis of the capacity to emit or receive signals is straightforward, but the role the signals play in the life of the animal is difficult to ascertain. All animals use multiple channels for sending and receiving information; multiple channels of communication provide redundancy and safety factors. Stereotyped signals may be species-specific, or even specific to the individual. Stereotypy has a genetic origin. Communication is, therefore, important in natural selection.

VISION

Vision is a sensory modality much used in animal communication. The molecular basis of photoreception is discussed in Chapter 14. Many animals use visual display in aggression, courtship, and reproductive behavior. Males may attract females by conspicuous color patterns, or by transient chromatophore expansion. Many kinds of animals make ritualized displays: turkeys, love-birds, peacocks dance and expand feathers. A male fiddler crab waves its large claw in a species-specific way to attract a female (137). Some species of fishes, lizards, birds, and mammals exhibit 20–25 sequential distinguishable components of courtship display. Ducks make some 10 courtship poses and specific sequences of movement (Fig. 10-1a), bill shake, head flick, tail shake, head lift, tail lift, whistle, nod, and exposure of back of head (101). Gulls engage in some eight courtship postures, usually characteristic of one species only (Fig. 10-1b). Gulls also use certain calls and postures as threats, raising the head upright, up-down thrusting, grass-pulling (162). Species recognition may be by simple visual signs. Young hatchlings become imprinted to the markings of their own species. Laughing gulls recognize the red bill and black back of head, herring gulls the white head and yellow bill with red spot (Fig. 10-2).

   Mammals, especially primates and carnivores, make visual threats by facial grimaces, usually accompanied by vocalization. Wolves use postures to threaten, or to indicate submission. Wolf threats are graded signals: piloerection, ear, tail, and head position, facial gestures, and vocalization.

   Flashing by fireflies is an example of visual communication. In some species of *Photinus* the female is sedentary; the male flies freely and flashes periodically; a female responds to a male after a latency and the male then approaches the female. The flash and flight pattern of a male is species-specific. In some fireflies of southeast Asia the males assemble by thousands in mangrove trees and flash in synchrony. The flash rhythm is controlled by an endogenous pacemaker in the firefly brain. Flashes can be entrained over periods shorter or longer than the natural rhythm of one per second and the synchrony of many fireflies may be due to entrainment (18).

CHEMICAL COMMUNICATION

Many of the organic molecules produced by animals are used for behavioral interactions. Only a few of the compounds used in animal communication have been identified. Some chemicals used in behavior are toxins, some are repellents, some attractants. Some pheromones lead to sex attraction; other pheromones confirm the presence of conspecifics, hive mates, and individuals.

One class of active substances produced in relatively large amounts is toxins which inactivate prey and stun predators. Some toxins can block synaptic transmission or axon conduction—toxins from puffer fish, scorpions, or skin of toads. Snake venoms have multiple effects; the most common is cell lysis. Toxins produced by some jellyfish inactivate prey; toxins of Hymenopteran stings paralyze hive intruders; the sting of certain wasps im-

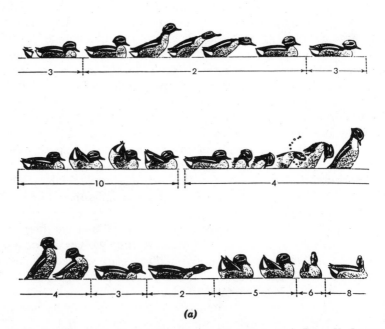

(a)

Figure 10-1. (*a*) Sequence of courtship poses of European teal. From K. Loreny, "The Evolution of Behavior." Copyright Dec. 1958 by Scientific American, Inc. All rights reserved. (*b*) Series of displays of four species of gulls. From N. Tinbergen, "The Evolution of Behavior in Gulls." Copyright Dec. 1960 by Scientific American, Inc. All rights reserved.

Figure 10-1. (*Continued*)

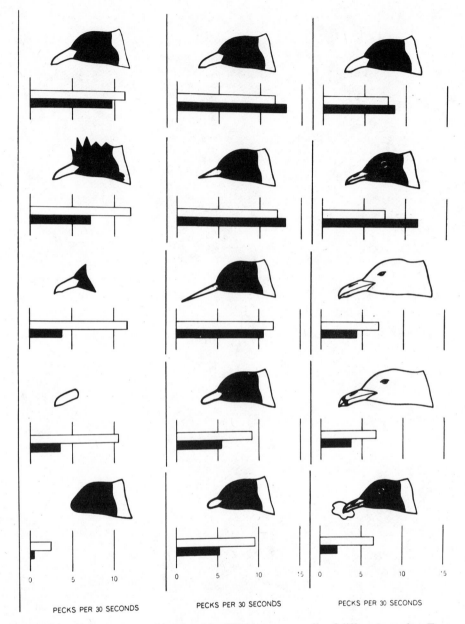

PECKS PER 30 SECONDS

PECKS PER 30 SECONDS

PECKS PER 30 SECONDS

Figure 10-2. Color patterns of head as identified by young gulls of different species. From J. Hailman, "How Instinct Is Learned." Copyright Dec. 1969 by Scientific American, Inc. All rights reserved.

406

mobilizes and preserves insect prey to be stored and used later as food by larvae.

Small amounts of some complex organic substances serve to repel attackers. Some carnivores and ungulates deposit scented urine or musk to define territory. Some insects and toads contain substances distasteful to predators. A bird learns after one or two attempts that a monarch butterfly is distasteful. A few butterflies lose their lives, but the species is protected. A bombardier beetle secretes two substances which, on combining, generate heat; the spray emerges at 100°C and toads reject the beetles, after they have tasted one (38).

Chemosensory behavior is highly developed in many insects and some mammals. Chemical communication is used for recognition of individuals in primate families. Carnivores such as dogs can distinguish single odors, as of one human, from mixtures of other odors, and may also use odors in establishing hierarchies. Glands which secrete scents used by males in marking territories and in signaling to females have been described in deer, gerbils, hamsters, and primates. Androgens increase tendency for aggression in males, and seasonal changes in hormone levels bring on responsiveness of females to male odors. In a breeding colony of mice, presence of a male can induce oestrus and accelerate puberty in females. Odor of a strange male mouse can block oestrus cycling of a female and can terminate pregnancy.

Rat pups (older than nine days) orient to and approach the home nest by odors. Olfactory orientation increases as thermotaxis decreases in young rats. Rat pups, born blind and deaf, produce ultrasonic calls in response to "stressful" stimuli—mainly odors. In the presence of the odor of a potential predator or of a strange lactating female, frequency of calls per unit time is several times greater than in the presence of odor of the home cage bedding or familiar female (103). Salmon return to streams where they were spawned by sensing the odor (11).

Garter snakes *Thamnophis radix* winter in underground dens and the males emerge prior to females. A pheromone in the skin of a female attracts males, and after one male succeeds in copulation and deposits a pheromone plug the other males leave. The female pheromone is vitellogenin which, after mating, is transported from skin to the ovarian follicles where it initiates making of yolk (34).

Organic molecules, as pheromones, serve many functions in insect communication—species recognition, attraction of the opposite sex, aggregation of both sexes, sexual stimulation, establishment of trails. The first sex pheromone to be identified chemically was bombykol, by which female silkmoths attract males from distances of several kilometers. The antenna of a male moth has 17,000 odor receptor hairs, each with a pair of sense cells (141). Bombykol is trans-10-cis-12 hexadecadieniol (Fig. 10-3). The trans-cis compound is more effective than the cis-trans, and each is more effective than either the cis-cis or trans-trans form alone. The odor is detected in an airstream containing only a few hundred molecules per milliliter from a

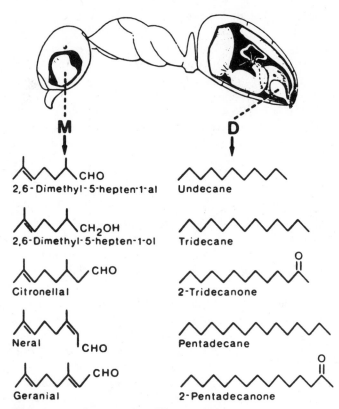

Figure 10-3. Pheromone substances found in mandibular and Dufour's glands of an ant. Reprinted with permission from B. Holldobler in *How Animals Communicate*, R. A. Sebeok, Ed. Copyright 1977, Indiana University Press, Bloomington.

source (female or treated paper) of 3×10^{-12} g. The concentration for eliciting flight by a male *Bombyx* is one-hundredth the concentration that evokes an electrical response from his single antennal nerve fiber (141). A male moth can respond when only 40 of its 40,000 receptors receive one molecule of bombykol per second (141).

Most moth pheromones are long-chain unsaturated acetates, alcohols, or aldehydes. A pheromone may be a mixture of isomers, and a precise ratio of geometric isomers may be prerequisite to a behavioral response. The pheromones of some species contain epoxides, ketones, or hydrocarbons (148).

Many butterflies of the genus *Danaus* (monarch and queen butterflies) use two pheromones in reproductive behavior. The male synthesizes a ketone (2,3,-dihydro-7-me-pyrrolizin). This accumulates in a pocket in the wing; the male has hair pencils at the tip of the abdomen which dip into the wing pocket and take up the ketone. When a male approaches a female in nuptial flight he dusts the hair pencils onto the antennae of the female and this induces her to mate. In addition, another compound, a terpenoid (141), is needed to cause the hair pencils to stick to the female. Synthesis of the terpenoid requires a plant source, heliotrope; the African monarch butterfly fails to mate unless the male can feed on heliotrope in addition to the milkweed on which all danaids feed. American monarchs do not require the second plant source. Sensory receptors for these pheromones are present on both male and female butterflies (48,141). The scent glands of an Asian arctiid moth synthesize a complex pheromone only if the larvae have ingested certain plant alkaloids (141).

Bees produce pheromones in several glands (mandibular, Dufours) according to caste. A queen bee's mandibular gland produces some 30 compounds which are potential pheromones; only a few of these have been characterized. One queen substance (trans-9-keto-2-decenoic acid) is taken from the queen by workers, about 0.1 µg/day (173). This substance prevents the workers from rearing larvae to become queens, and it prevents workers' ovaries from developing. The same or a related compound from a queen serves as a sex attractant for drones in nuptial flight. Another queen pheromone, α-hydroxy-2-decenoic acid, causes clustering of workers in a swarm (174).

Ants, particularly army ant scouts, lay down odor trails which guide other workers of the colony to a food source.

Many insects produce pheromones that evoke aggregation (75). The aggregation pheromone of the western pine beetle consists of a compound made by the male, a compound made by the female, and a component from the host tree. European elm bark beetles are attracted by a pheromone of two compounds secreted by the female and one component from the host tree. A male cotton boll weevil produces four synergistic compounds which cause aggregation of both sexes. These and other aggregation pheromones have been synthesized and used to control pest insects (148).

Chemoreceptors of insects are located on appendages—antennae, mouthparts, and tarsi. Thresholds of chemoreceptors vary with physiological state, and are higher after feeding. Proboscis extension by a fly is initiated by taste stimulation of tarsi and by olfactory stimulation of antennae.

Electrical recordings from chemoreceptors give sensory potentials which may initiate trains of spikes. The amplitude of chemoreceptor potentials and frequency, latency, and persistence of firing of spikes are indicators of specificity of receptors. Each taste hair of the labella or tarsi of a blowfly has five sensory cells—two for salt, one for sugar, one for water, and one for movement. Relative sensitivities of sugar receptors are: sucrose 1, glucose 0.71, fructose 0.46; salt receptors are stimulated mainly by cations (42). Maxillary sensilla of lepidopteran larvae have four chemosensory neurons per peg. Some cells respond to both salt and certain amino acids. Chemo-

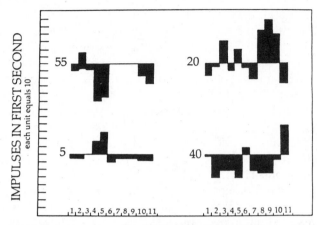

IMPULSES IN FIRST SECOND
each unit equals 10

Figure 10-4. Activity spectra in spike frequency of four olfactory receptors of tobacco hornworm, Manduca sexta, in response to eleven different odors. Reprinted with permission from V. G. Dethier in *Handbook of Physiology, Sec. B, Alimentary Canal*, C. F. Code, Ed., Vol. 1, p. 90. Copyright 1967, American Physiological Society, Washington, D.C.

sensitive sensilla of a carrion beetle respond to fatty acids; those with chain lengths longer than six carbons are depolarizing, while shorter chain acids are hyperpolarizing (12).

Many chemoreceptors of insects are not specific but are susceptible in different degrees to stimulation by several substances. Fifty olfactory cells on antennae of saturnid moths were tested for response to 13 odors and no two cells gave the same pattern of sensitivities. Figure 10-4 shows the relative sensitivity of four receptor cells of a tobacco hornworm to 11 odors, and no two receptors have the same pattern (42). Twenty-seven receptors of a moth *Antherea* showed different patterns of response to 10 compounds (12). Cross-fiber patterning is an adaptive response to complex odors and taste, for example, odor of carrion for certain beetles and flies. Responses of a number of receptors establish a neural pattern; the behavioral response depends on integration of a complex of many odors or tastes, not on separate signals initiated by single compounds (42).

Information transfer by specific chemicals has been important in the co-evolution of plants and insects. Many plants produce compounds repellent to herbivorous insects. A few insects have mutated in such a way as to be attracted by these secondary plant products (54). Extensive reciprocal co-evolution of plants and insects occurred during the Cretaceous period. Insects are attracted to flowers by odors.

Most phytophagous insects feed on only a few kinds of plants. Potato beetles and tobacco hornworms are restricted to Solanaceae. Potato beetles feed on the solanacean potato, but the content of the alkaloid nicotine makes the solanacean tobacco unacceptable and the glycoside tomatin makes the solanacean tomato unacceptable to them. Potato plants treated with nicotine or tomatin become unacceptable to the potato beetle. Hornworms feed on tomato or tobacco but normally not on potato plants. Of two species of the butterfly genus *Papilio,* one feeds only on plants of the family Rutaceae, the other mostly on Umbelliferae. Caterpillars conditioned to carrot are attracted to methyl chavicol; those conditioned to rue are attracted to methylnonyl ketone (42). The active agents attracting cabbage butterflies (Pieris) are the glucosides sinagrin and sinalbin; a *Pieris* caterpillar will feed on noncrucifer plants that have been painted with mustard oil glucosides. Larval conditioning can alter chemobehavior of adult insects; *Drosophila* larvae fed on a medium containing peppermint develop into flies with positive chemotaxis to peppermint.

Secondary plant products, kairomones, serve as attractants for some herbivorous insects and as repellents for other insects. One cucurbitacin is repellent to larvae of many kinds of beetles, but members of one family, cucumber beetles, have evolved a positive response (110).

In summary, there is great chemical diversity of pheromones and of plant secondary compounds. One compound may serve to induce two actions, for example, aggregation and reproductive behavior; a secondary plant product can be repellent for some insects, attractant for others. Some insects produce substances which are distasteful to predators. Pheromones in mammals trigger chemotaxis of young, recognition of adults for sex and species, and warning against aggressors. To be functional in behavior, both the synthesis of pheromones and chemoreceptors must be present. Their simultaneous evolution seems unlikely. One suggestion is that the substances or related compounds may have evolved having other functions; receptors and neural connections for interpreting the behavioral significance may have evolved later. Transition of a secondary plant repellent to attractiveness may well have resulted from mutations which altered the animal feeding on the plant in respect to the receptor and central neuronal responses. Highly specific membranes of receptors detect pheromones and plant secondary products; food selection is characterized by broad spectra of sensitivity and crossneuronal patterns. Plants and herbivorous insects have coevolved. There has been little investigation of central nervous processing for either high selectivity or multiple-channel inputs from chemoreceptors.

ELECTRIC AND MAGNETIC SENSE

Many organisms can detect static and dynamic magnetic fields. Organisms which orient by magnetic detection contain magnetite, but so do some nonorienting organisms. The relation of magnetite to the nervous system has yet to be established. Some mud-dwelling bacteria contain magnetite and orient in a magnetic field of 0.5 G or higher (10,88). Behavioral evidence indicates that foraging bees can make use of magnetic fields for orientation. Homing pigeons can find their way by visual landmarks (131), and may become disoriented if small magnets have been attached to them or whenever they fly in a region where the earth's magnetic field is distorted (64). Bees use the earth's magnetic field as one of several kinds of sensory input in communication of direction and distance of food sources to hivemates. On a horizontal surface, bees communicate by dances and use the angle of the sun or of the polarized light of the sky for orientation. On a vertical surface in darkness bees normally orient by gravity, but if they can see the sky they bisect the angles indicated by sun and gravity. On a horizontal surface in the shade they use gravity but with modification according to the earth's magnetic field. Deposits of iron oxides occur inside cells in the abdomen of bees, and iron deposits have been observed in lacrimal glands of birds. However, the necessity of magnetite for the detection of a magnetic field has been questioned (99).

Many fishes which have specialized electroreceptors in the skin—lampreys, elasmobranchs, lungfish, sturgeons, catfish, and electric fish—can detect small electric fields, such as the potentials generated by respiratory movements of nearby fish. Sharks and rays are able to orient to electrical currents induced when they swim or drift across the earth's magnetic field (88). A marine fish swimming at velocity of 10 cm/sec generates a field of 0.05 μV/cm. Stingrays have been conditioned for changes in heart rate with an electric field as the conditioned stimulus; responses could be obtained at 0.005–0.01 μV/cm. These fish can sense direction, polarity, and magnitude of a low-frequency (0.1–16 Hz) field (88). Cellular mechanisms of electric discharge and of electroreception are given in Chapter 13.

Electrocommunication

Fish that have electroreceptors sense electric fields in two modes: (1) passive detection of electric currents emitted by another electric fish or from muscle potentials of a nearby fish; (2) active detection of the distortion of its own electric field by objects of different electric conductances. Electric discharge is used by some fishes for communication, for object detection, and for paralysis of prey.

Electric communication is based on electromotor organs for sending electric signals, electroreceptor organs for receiving the signals, and a central processing system. Electric organs have evolved in at least seven different families of fishes; similarities indicate evolutionary convergence (19).

Two types of electric organ discharge are (1) tonic, continuous, sinusoidal discharge, and (2) spike-like, pulsed discharge (Ch. 13). Two morphological types of electroreceptors are known, tuberous and nontuberous. The nontuberous or ampullar receptors detect low-frequency fields and may be sensitive to currents induced by motion through magnetic fields. Tuberous receptors are sensitive to high-frequency waves (50–15,000 Hz in different species). Some tuberous receptors give tonic responses, others give phasic responses. Some electric fish, for example, African *Gymnarchus,* detect very small electric fields. Behaviorally it has been shown that these fishes respond to 2×10^{-5} μA/cm^2 in the water outside the fish. This corresponds to 300 electrons/cell/sec or to a sensitivity of 0.003×10^{-12} amp/receptor; this is some 5 to 10 orders of magnitude less than the minimum current that stimulates a squid giant axon (19,100).

Electric sensing by fish is used to locate objects of electrical conductivity different from the surrounding water. Sensitivity to distortions in the fish's electric field serves both the detection of objects and the recognition of prey. Electric organ discharge is also employed in social behavior. Many species briefly interrupt the steady discharge of the electric organ as an agonistic or submissive signal. Changes in frequency accompany changes in posture, motor attacks, and dominance relations. Another function of electric discharge patterning is for sexual interaction; *Sternopygous* males and females have nonoverlapping discharge frequency ranges. The discharges of mature males of *Eigenmannia* average about 403 Hz, of females 452 Hz. In the presence of a female, the discharge of a male (77) becomes frequently interrupted. *Eigenmannia* and others with wave-type discharges have a species range in excess of 100 Hz. Individual fishes have a preferred frequency range within which discharge can be maintained with a coefficient of variation of 0.01 over tens of thousands of cycles of neuronally paced discharge (19). Electric rays (*Torpedo*), electric eels (*Electrophorus*), and catfish (*Malopterus*) appear to stun prey with a brief burst of intense discharge.

When a wave-type electric fish detects the discharge of another electric fish swimming nearby and discharging close to its own frequency (within 6–8 Hz) the two fish avoid jamming by changing discharge rate up or down (jamming avoidance response—JAR) until the difference is some 10–20 Hz (Figs. 10-5, 13-3). By jamming avoidance a fish in a group can maintain its electrolocation (7,72). Methods of producing JARs are described in Chapter 13. Electrocommunication signals are species-specific in that a species has not only a characteristic rate but shape of electric organ discharge (EOD) (77). The sensory systems can discriminate when the EOD is less than 0.3 msec duration.

Pulse-type species have less regular discharges than wave-type species but both have some degree of JAR. Electrocommunication is unique in that the response (electric discharge) is in the same physical mode as the sensory input.

COMMUNICATION BY SOUND

Sound emission (vocalization) and sound reception (hearing) are used by many animals for communication and for localization of objects. The pe-

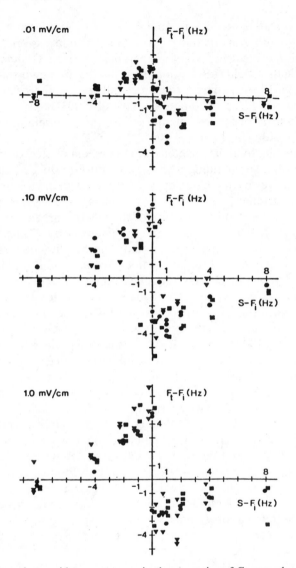

Figure 10-5. Jamming avoidance response in three species of Gymnarchus. S − F_i is frequency stimulus difference (+ or −) from normal electric organ discharge; F_t − F_i is response (EOD) difference for three different intensities of stimulus, S. The response is positive for a reduction in stimulus frequency and negative for an increase in stimulus frequency. Reprinted with permission from T. H. Bullock et al., *Journal of Comparative Physiology*, Vol. 103, p. 106. Copyright 1975, Springer-Verlag, New York.

ripheral receptors and the central sensory pathway are often closely connected in the nervous system to the sound-producing system. The sound
pattern is "meaningful" for a behavioral response. Sound patterns are established by inheritance and modified by experience.

Insects

A few kinds of insects (crickets, etc.) make characteristic calls by moving
one wing (the scraper) over the other (the file) to vibrate cuticular teeth.
Some kinds of grasshoppers call by vibrating tymbal structures on the hind
legs.

Male crickets sing three types of songs: calling, courtship, and aggression. A female may respond to a calling song by moving toward a calling
male; after light antennal or cervical contact, the male sings a courtship
song, which may elicit a postural response by the female; the male then
transfers a spermatophore. If the male antennae are vigorously contacted by
the female, the male may sing the aggressive song and the female may then
either fight or flee. The aggression song can also be produced by the male
after vigorous antennal and foreleg contact with another male (82). The
male's calling follows a daily rhythm, with highest frequency in the evening;
timing is controlled by the supraesophageal ganglion (brain) (79,130). A female grasshopper recognizes the calling song of a conspecific male and sings
a response song as she approaches the male; this is usually followed by the
male's courtship song to which the female may or may not be responsive
(89).

The calling song of a male cricket *Gryllus bimaculatus* consists of syllables, each a series of chirps (Fig. 10-6); chirps are of 3–5 pulses with the
primary carrier frequency within a pulse of 4.5–5 kHz. The courtship call
has major pulses separated by small pulses, primary frequency 16 kHz and
intervals 300–400 msec between major pulses; the aggression song resembles
the calling song in primary frequency (4.5 kHz), but has more pulses per
chirp (134). There is species variation in duration and frequency of the syllables, in intervals between chirps and number of vibrations per chirp (Fig.
10-6) (82). If the corpora pedunculata and central bodies of the supraesophageal ganglion are lesioned the pattern is disrupted and singing may be continuous. Electrical stimulation of the supraesophgeal ganglion by implanted
electrodes can elicit typical songs (Fig. 13-4). Stimulation at 50–70 cps elicits
the calling song; stimulation at 100 cps elicits the aggression song (8). Patterns are generated in a central neural network. The song rhythm can be
recorded in discharges of motorneurons in an isolated nervous system; thus
sensory feedback is not necessary. In a grasshopper, the song pattern is
generated in a thoracic ganglion, probably by an oscillator network of interneurons, and is modulated by afferent signals (8). The song patterns pro-

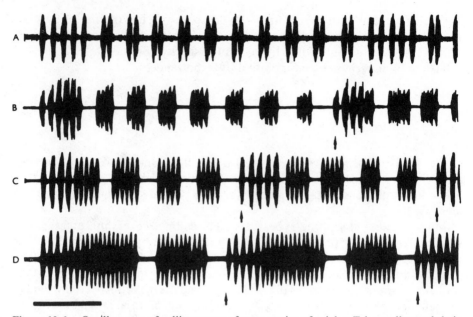

Figure 10-6.   Oscillograms of calling songs of two species of cricket Teleogryllus and their hybrids. A, T. oceanicus; D, T. commodus; B, cross between female T. oceanicus and male T. commodus; C, the converse cross. Reprinted with permission from R. R. Hoy, *American Zoologist,* Vol. 14, p. 1074. Copyright 1974, American Society of Zoologists, Thousand Oaks, California.

duced by vibrations of left and right hindlegs may be different and can vary from time to time.

Cricket species have characteristic songs; recognition is based on syllable period and chirp duration. Genes for song pattern are carried on X-chromosomes, and a hybrid female usually prefers the song of a nonsibling male to the song of her hybrid brother. One cricket mutant sings three rather than two vibrations per trill (49). In hybrids of two grasshopper species, the calling and rivalry songs resemble those of one parent; the courtship song resembles that of the other parent (Fig. 10-6). It is postulated that the network pattern of an interneuron oscillator is genetically controlled (79).

Drosophila males produce species-specific songs by characteristic wing movements. Two sibling species, *Drosophila pseudoobscura* and *D. persimilis,* are distinguished behaviorally by the sounds they produce. *D. pseudoobscura* produces two songs: one with 523 Hz vibrations at 6/sec, the other with 250 Hz at 24/sec. *D. persimilis* has only one song, at 525 Hz repeated at 15/sec. No hybrids are found in the field but crosses can occur in the laboratory; the frequency and pulse repetition rates are controlled by a gene on the X-chromosome. When a *D. persimilis* male is crossed with a *D. pseudoobscura* female the song of the F_1 is of a *pseudoobscura* type, the reverse cross results in songs of the *persimilis*—the female type. The genetic control of the song may be by tension in one or several flight muscles; the difference in calls results in reproductive isolation (51).

Some arthropods communicate by vibrating the substratum. Male fiddler crabs (Uca) inhabit and defend burrows in the sand in the intertidal zone. They attract females into the burrow for mating by waving their large claws during the day and by vibrating the substratum with their legs at night. The female crabs detect the vibration by mechanoreceptors on their legs; one set of these receptors responds best at about 600 Hz, another at 1500 Hz. Species of fiddler crabs differ in patterns of vibration (137).

Male wolf spiders vibrate the substratum (leaves and rubble). Females respond only to the kind of vibration made by males of their own species. Laboratory crosses between two species (by artificial mating) yield normal offspring (158); the natural reproductive isolation is by the vibration patterns.

Anurans

Frogs use patterns of sound for communication. The bullfrog *Rana catesbeiana* gives at least five distinct calls: mating, release, distress, warning, and rain. Mating calls of anurans are species-specific and permit species isolation. A receptive female frog moves toward a mating call made by a male of her species but not toward the call made by a male of any other species (50). Mating calls or modifications of them may function also in aggregation and chorusing of male frogs and in spacing between calling males. Release calls are emitted by a male or by a female when clasped by an undesired male. When a predator is detected, a frog makes the warning call. Distress calls are given when a frog is seized by a predator, such as a bird or turtle, and may alert other frogs.

Large males of *R. catesbeiana* emit mating calls of lower mean frequency than small frogs, and this may give them some reproductive advantage. The component in a male call essential for recognition by a female is the frequency of maximum energy (159). The mating call of *Rana pipiens* is a series of 3–5 trills, each with pulses 6–10 msec duration at 15–30 pulses per second. The trills have maximum energy in two regions—100–600 Hz and 1100–1500 Hz. The release call consists of 5–15 trills, some in pulses 175–400 msec duration, others of pulse durations 20–30 msec; the energy spectrum of the release call has a broad peak at 500–1000 Hz (26).

Geographic variation in the anuran call may indicate subspecies. Data for populations of the tree toad *Pseudacris nigrita* are:

	Texas	Oklahoma	Colorado
Pulse/second	18.8	28.6	31.2
Pulse duration (sec)	0.68	0.58	0.43

Hyla versicolor from Oklahoma and Arkansas have a pulse rate of 28.6/sec; from west Louisiana 50.9/sec (9). *Hyla cinerea* and *H. gratiosa* have a wide zone of sympatry and are interfertile; a female exhibits preference for the call of a conspecific male (58).

In some species of *Hyla,* the temporal pattern of calls is critical. *Hyla ewings* is distinguished from *H. verreauxi* by pulse repetition rate (159). When *Hyla regilla* and *H. cadaverina* were tested with synthetic calls the pulse repetition rate determined whether or not the female responded. Cricket frogs *Acris gryllus* and *A. crepitans* are sympatric in a zone from Virginia to Louisiana. A call of *A. crepitans* has a low-frequency component (550 Hz) which is similar in each geographic race; however, high-frequency peaks differ: in New Jersey (3350 Hz), South Dakota (2900 Hz), Kansas (3750 Hz). In the medullary auditory nucleus the high-frequency units give optimum response to the local peak in the call. Female frogs respond best to the local dialects (26). *Hyla chrysoscellis* and *H. versicolor* increase the number of pulses per second with warming and at any given temperature *H. chrysoscellis* has more pulses than *H. versicolor;* the female of each responds to a synthetic call corresponding to the call of the conspecific male (58).

A neotropical tree frog *Eleutherodactylus coqui* has a two-note call, a "Co" (1 kHz) and a "Qui" (2 kHz). Males and females respond differently to the two components. In the auditory nerve three populations of fibers were found, tuned respectively to low, mediuim, and high frequency. The sex difference was greatest in the high-frequency units (115). Males answer to a playback of the "Co" alone or with the "Qui"; they do not respond to the "Qui" alone. Probably "Co" functions in male-male encounters, and "Qui" attracts females (115).

In *Rana pipiens,* chorusing can be induced by playback of the conspecific call if the listening frog is hormonally primed. Implantation of testosterone in the anterior preoptic nucleus (APON) of the frog favors an acoustically evoked call. Pituitary injections increase the number of responding neurons in the APON or in the pretrigeminal nucleus. In some pituitary-injected male frogs a few neural units that were responsive to conspecific calls also responded to heterospecific calls (165).

Birds

Communication by song is highly developed in birds. Some birds, such as mockingbirds and mynas, are imitators and have extensive repertoires. Other birds, for example, eastern robins, sing only a few songs, and these are species-specific. Songs of many birds become modified by experience. The relative importance of songs and feather color patterns for territoriality and for courtship varies with species and habitat. Singing may be more important in woodlands; color display may be more important in open fields.

Most songbirds have several calling patterns, each serving a different function. The best known are the calls which announce territorial occupancy. In chickens the territorial and dominance crowing of a rooster is unlike the cackling of a hen and her clucking to chicks, or the alarm cry at sighting a hawk. Chickens recognize some 25 different calls.

A songbird derives its song pattern from its innate template, from hearing other birds, and from hearing its own song. The relative importance of the three components varies with species and can be assessed by (1) rearing birds in isolation, (2) permitting them to hear but not to see other birds, and (3) rearing birds after deafening them as hatchlings. A male song sparrow's innate template is sufficient for development of the specific song if the bird can hear itself; a white-crowned sparrow can sing only if it has heard other sparrows. A zebra finch memorizes its song components during the first 14–45 days after hatching, well before it begins to sing. It does not need to practice (107). Indigo buntings in isolation develop abnormal songs; reared in groups of ten, the songs of individuals are normal for the species but lack detailed adult pattern. Young buntings copy adult tutors better when able to both see and hear than if the bird can only hear the tutor (123).

Local dialects have been recognized in several bird species. White-crowned sparrows (*Zonotrichia leucophrys*) from separate valleys in the San Francisco Bay area have calls characteristic of the locale (Fig. 10-7*a*). Birds kept in isolation fail to develop the local dialect; a deafened bird gives a call which is unlike that of an isolated but hearing bird; apparently an internal model is used and compared with the bird's own performance, with the song of neighbors, and especially with the song of its parents (95) (Fig. 10-7*b*). There is little dispersal of *Zonotrichia* outside their breeding territory, and migration of individuals from two populations into areas where dialects were different was only 1/15 that predicted if movement were random. A few hybrid songs were found in the zone of contact (6).

Song dialects of white-crowned sparrows showed little correlation with genetic constitution as measured by zymograms of alleles of several enzymes. In Colorado some altitudinal differences were found for enzymes, but not for dialects; in California three of six genetic loci differed along with dialects (6).

On an island near New Zealand, a flock of some 20 birds of one species was found to sing with its group dialect. Within a flock each individual had its own territory. When a young male moved from one flock to another it began to sing the song of the new group. These findings are interpreted as an example of cultural transmission of a behavior (86). A European nightingale may use 61 different phrases in regular succession; day and night songs differ in duration, but not in order of phrases (163).

Male canaries sing only during the breeding season and may sing 25 patterns of syllables grouped in phrases. The repertoire is redeveloped each spring and new songs are added. If a singing canary is deafened, its old patterns remain but new ones do not develop. If the left hypoglossal nerve

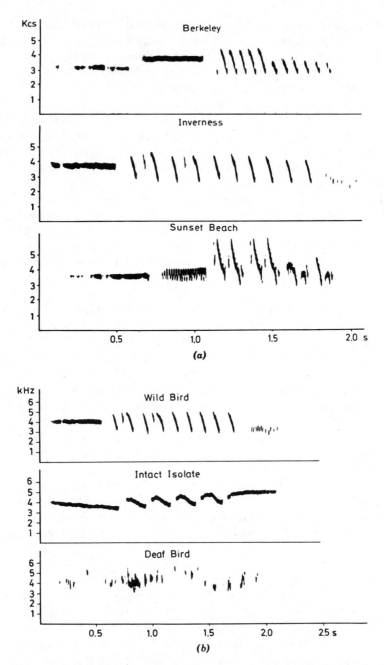

Figure 10-7. (*a*) Sound spectrograms of white-crowned sparrows from three localities in Northern California. (*b*) Sound spectrograms from wild, intact isolate and deafened white-crowned sparrows. Reprinted with permission from M. Konishi in *Handbook of Sensory Physiology,* R. Held et al., Eds., Vol. VIII, pp. 295, 296. Copyright 1978, Springer-Verlag, New York.

is transected, the song becomes disturbed, but cutting the right motor nerve has no such effect. However, if the left side is denervated in a young canary, singing can be learned with the right side (119). Singing by female canaries can be induced by giving them testosterone. Ring doves' cooing is activated by presence of testosterone in the nuclei that control vocalization (119).

In canaries three forebrain nuclei must be present for normal singing—the hyperstriatum ventrale, pars centrale (HVc), the nucleus robustus archistriatalis (RA), and a poorly defined (x). These regions are larger in male canaries than in females, and lesioning results in deficits of singing in proportion to the amount of tissue removed (109). The two nuclei (HVc and RA) are normally 99% and 76% larger in the spring, when new songs are learned, than in the fall. Neurons show increased dendritic branching in the spring, possibly some hyperplasia (119). HVc is also involved in hearing, and some units respond to tape-recorded playback but not to the same song emitted voluntarily.

Bats

Many bats use acoustic sonar for locating food and for orienting. A few bats use vision and some use both visual and audio cues. For a bat to emit, perceive, and respond to sound signals it must have audio reception, central modulation of responses to echo, neural coupling between hearing and emission of calls, and central control of vocalization (66). Fossil bats from the Eocene resemble modern bats; coordinated adaptations of the central nervous and peripheral mechanisms of vocalization and hearing appear to have evolved in synchrony.

Many bats call through the open mouth, a few emit calls through nostrils. Microchiropterans produce sounds by the larynx, megachiropterans by tongue clicking. The larynx of bats is structured for high-frequency resonance, and the cricothyroid muscles, activated by the vagus nerve, tense the largyngeal membranes and are among the fastest muscles known in vertebrates. Other fast muscles in bats are the middle ear muscles. Two kinds of vocalization are (1) constant frequency (CF) pulses, and (2) frequency-modulated (FM) calls. A megachiropteran *Rousettus* emits clicks with maximum energy at 12 kHz in pulses at 7/sec. These bats are not so efficient at dodging wires as microchiropterans. The FM pulses of *Myotis* (microchiropteran) start at 78 kHz, drop to 39 kHz, and have maximum energy at 50 kHz; pulse repetition rate increases from 8 to 30/sec during cruising, to 150 to 200/sec on nearing a flying prey insect. *Myotis* can dodge barriers of wires of 3 mm diameter from 2 m distance and wires of 0.18 mm diameter from 1 m (139). The CF pulses are longer than FM pulses and are at frequency of 40 kHz in large bats, 120 kHz in small microchiropterans.

During flight the bat *Myotis* emits pulsed cries, each pulse of 2.3 msec duration starting at 70–80 kHz. Pulses by *Eptesicus* start at 30–77 kHz, are of 50–100 msec duration, and terminate at 19–23 kHz, 5 msec (66). Pulses are repeated by the bat about 30/sec, and as it approaches an object the repetition rate rises to 50/s or even to 150/sec. From the echoes a bat gauges the size, distance, velocity, and position of the object ahead. In addition, *Myotis* emits low-frequency sounds (7 kHz) which are audible to man (145).

The horseshoe bat *Rhinolophus* and the mustache bat *Pteronotus* emit calls with two components (Fig. 13-11). A constant-frequency component (CF) in *Rhinolophus* is at about 81–83 kHz; on approaching a target the echo frequency rises in a Doppler shift and the bat compensates by lowering the CF frequency so that the echo frequency remains constant. The initial frequency on take-off is 82.4–83.3 kHz; in flight the echo frequency is 150 Hz above the emitted frequency because of Doppler shift; this difference depends on flight speed, headwind, and sound speed (142). The frequency of CF pulses sweeps downward during the last few ms of approach to a target. The second component is frequency-modulated (FM); the frequency decreases as the target is approached. Sound pressure of the echo is maximum from the direction ahead of the bat and decreases by about 6 dB at 24° above or 23° to one side of the bat. Target localization depends on hearing by both ears.

Vocal communication from one bat to another has not been interpreted, but colonies of bats are noisy. The horseshoe bat *Pteropterus policephalus* of Australia emits some 20 distinct vocalizations. Vocalization by a California bat *Antrozous pallidus* under laboratory conditions showed four communicative sounds: (1) directional for orientation between bats, (2) squabble notes in a colony, (3) irritation buzzes, in the colony and as alarm cries, and (4) frequency-modulated orientation pulses. Bats are born in an altricial state, and when mothers return from a feeding flight, they recognize their own infants by auditory and/or olfactory cues (17). Neonate bats emit infrasonic calls that are species-specific (62).

Some moths which are prey for bats attempt to escape by altering flight patterns, diving and looping. Noctuid moths have an ear on each side at the rear of the thorax; hawk moths have ears on labial palps. Each moth ear has one mechanosensitive cell and two sound receptors which detect the pulsed high-frequency cries of bats. The sensitivity of a hawk moth ear is 0.01–0.03 dyne/cm² for sounds in the range of 15–170 kHz (135). A moth can detect a bat at about 30 meters distance. When the moth wings are in the up position the ipsilateral stimulation by sound is 20–48 dB more effective than the contralateral stimulation; when the wings are in the down position sound from below is more effective than sound from above. By comparing intensities with its wings up or down the moth may gauge direction and distance of the sound source (Fig. 10-8) (125). The bat can detect the echo from a moth at some 3 m; the echo is maximum when the moth wings are in the up position. A bat can fly faster than a moth, but the moth detects from greater distances and makes evasive movements.

Bat and moth are well matched; sometimes a moth is captured, sometimes

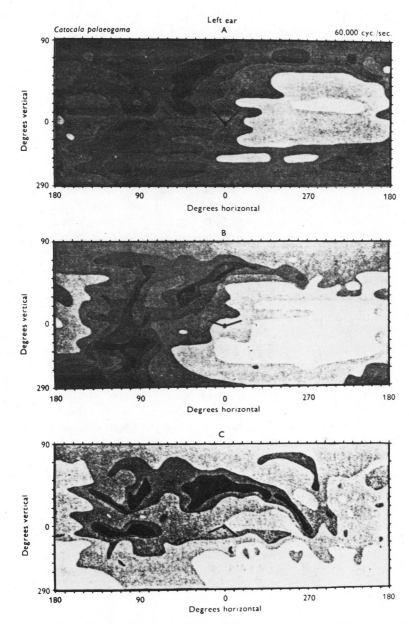

Figure 10-8. Mercator projections of sound intensity originating from different positions as judged from sensory response in left ear of moth when wings were (A) raised, (B) horizontal, and (C) lowered. Reprinted with permission from R. S. Payne, D. K. Roeder and J. Williams, *Journal of Experimental Biology,* Vol. 44, plate 3. Copyright 1966, Biochemical Society, Colchester, Essex, England.

it escapes. Other flying insects such as dipterans, do not have integrated hearing and motor mechanisms.

Cetaceans

Whales, dolphins, and porpoises emit several kinds of sound and respond to sounds over a wide frequency range, infrasonic and ultrasonic. Cetaceans may respond over as wide a range as 10 octaves; they show pitch discrimination, sound localization, memory of sounds, and ability to echolocate. Cetaceans have large dorsal parietal lobes and well-developed internal auditory organs. How they produce sounds is not well understood; whales may use the blowhole, porpoises apparently focus their sound into a narrow beam forward of the head by the fatty "melon." Dolphins make complex sounds, whistles, and clicks, which they increase in frequency as they approach a target. Humpback whales emit patterned calls. Records of their songs show individual repertoires; the songs last many minutes and are sung repeatedly from beginning to end (124). Calls have been recorded at 160 km distance from the presumed singing whale and it is calculated that some components of the sound can travel farther if sender and receiver are at the same depth. Whistlers (dolphins, porpoises) live part of the time in social groups and their songs may function in aggregation and spacing of individuals. Songs of lone males are identified (Fig. 10-9) by a pattern, but some phrases may be dropped and others added every few months. The toothed whales (Odontoceti), porpoises, dolphins, and sperm whales produce ultrasonic clicking. Some whales, whistlers, and nonwhistlers show individually distinguishable sequences of vocalization. The baleen whales (Mysticetes) emit calls of narrow bandwidth, and low-frequency and long-duration tones, grunts, and clicks.

Audiograms (behavioral) from porpoises show lowest thresholds at 40–70 kHz, with slightly less sensitivity at 15–40 and 70–100 kHz. Neural units in the inferior colliculus of an anesthetized porpoise (Tursiops) showed no responses below 5 kHz but cortical responses up to 10–15 kHz (21). Porpoises emit pulses at 600/s in echolocation. Blindfolded porpoises can avoid wire barriers, distinguish between spheres 6.3 and 5 cm in diameter, distinguish objects of Cu, Al, and brass (92). For cetacean behavior, audition is more important than vision (73,92).

BEES—RECRUITMENT BEHAVIOR

Communication for recruitment of workers is essential for colony life of many hymenopterans—ants and bees. Bee scouts returning from a food source to a hive transmit to hive mates direction signals derived from polarized light, angle of the sun from the zenith, landmarks such as trees, hills,

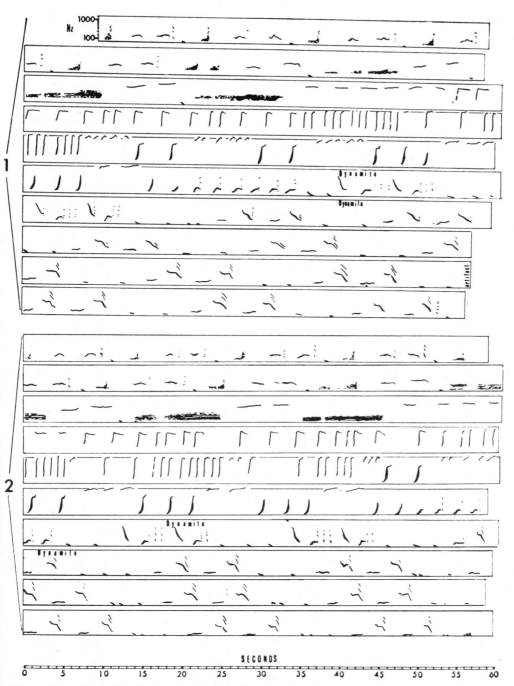

Figure 10-9. Song of single humpback whale. Tracings 1 and 2 represent two repetitions of same song. Each song approximately 30 minutes in duration. Reprinted with permission from R. S. Payne and S. McVay, *Science*, Vol. 173, p. 587. Copyright 1973 by the AAAS.

and ponds, odor and color of flowers, endogenous time sense, gravity, and magnetic field. The scouts signal to recruits by visual patterns of dances when on a lighted comb; in a dark hive, by tactile stimuli, sounds and vibrations, transfer of honey and scents. The exploring bee has left its pheromone at the food site. The scout bee in the hive dance indicates distance, direction, kind, and relative amount of food. Recruitment communication makes use of several modalities according to environmental conditions; the behavior appears to be genetically programmed.

A simple type of bee language is the round dance; the bee reverses the direction of circling every 15 sec and gives traces of nectar to recruit workers at hand (Fig. 10-10) (99a,169). The round dance signals a nearby food source not farther than 25–100 m for domestic bees. Recruits alerted by a round dance cruise around the hive in all directions. For food at greater distances the scout bees perform a waggle dance, a straight run with the tip of the abdomen waggling from side to side; the scout bee then returns via a hemicircle; the side of the circling return reverses from dance to dance (Fig. 10-10). The length of the waggle (straight run) and the number of dances are proportional to the distance of the food from the hive, faster dances for nearer food. If a bee faced headwinds or went over a hill or building in her outward flight, she gives the signal as greater distance than a level flight. On a horizontal platform in front of a hive, or an open horizontal comb, the dance is toward the sun if the food is in line with this direction; the waggle run deviates to left or right according to the angle from a path toward the sun. The direction of a horizontal run points toward the goal and the angle is the same as the one the scout had taken on her flight from hive to feeding place. She can orient equally well by the polarized light of blue sky or by the sun itself. If neither the sun nor blue sky is visible the dance is disoriented. On a vertical comb in darkness, inside the hive, the dancer transposes the solar angle to a gravitational angle, as if the direction toward the sun were upward; the angle of the bee's run is then to the left or right of vertical. Gravity receptors at the base of the legs govern the transposition. The bee transposes a visual angle to a gravity angle during the dance if the inclination of the comb is greater than 15° to the horizontal. The bee is uncertain at an angle of 10°, which is the threshold for detecting gravity. The sun moves 15°/hour and bees change the angle of their dance correspondingly by reference to their internal clock. Experienced worker bees dance more precisely than inexperienced ones. In bee language, there is no "word" for upward or downard; bees are unable to indicate food directly above or below the hive.

Domesticated honeybees do not signal sugar water; the odor of flowers or scent is necessary. During the waggle dance, sounds are emitted according to quality of food, in pulses of 12 msec duration and at 12 msec intervals; in each pulse vibrations are at 287 Hz (93).

Scout domestic bees communicate by using sunlight when on a horizontal surface, by gravity on a vertical surface in a dark hive. If on a vertical sur-

The round dance. The dancer
is followed by three bees who trip along
after her and receive the information.

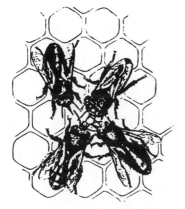

A forager (lower left) who
has returned home and is giving nectar to
three other bees.

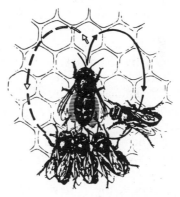

The tail-wagging dance.
Four followers are receiving the message.

Figure 10-10. Representations of round dance, tail-wagging dance, and sickle-shaped dance performed by returning scout honeybees in signalling to recruits. Reprinted, by permission from Karl von Frisch, *The Dance Language and Orientation of Bees*. Copyright 1967, Harvard University Press, Cambridge, Massachusetts.

face the comb is exposed so that a dancer can see the sun, she is confused and the run bisects the angle between orientation to light and to gravity. Recruits translate the new angle back to a correct angle for flight. However, the direction of the dance is not strictly an average of gravity and light. The deviation is called misdirection (*Missweisung*). The misdirection is less than 10° from a bisect and it shows a regular variation with time of day. The deviation is greater when the magnetic field is stronger; misdirection is cancelled when the magnetic field is neutralized. No misdirection occurs on a horizontal platform, but if bees have been restricted for two weeks to a horizontal field under shade, they develop runs which use magnetic fields as cues. The use of magnetic field by bees varies with normal diurnal variations in the earth's field. Such variation results from solar winds in the upper atmosphere and amounts to about 0.05% of the total field. Increasing the magnetic field from 1.0 to 1.2–3.0 oersted leads to greater response of the bees and neutralizing the field by applying an opposite magnetic field eliminates the response (93). Demagnetized bees become disoriented under conditions where they would normally rely on magnetic field (63). The diurnal aspect of the sound pattern also becomes altered with change in magnetic field. At about midday vibration frequency is minimal and pulse duration maximal (93).

The compound eyes of bees detect the plane of polarization (E vector) of polarized light; this sensory cue can be abolished by rotating a polarizing glass over bees which are using blue sky for orientation. They respond to polarized light mainly in ultraviolet (350 μm), but bees can detect polarization of longer wavelengths (14). Also, a bee trained to a route for food follows the same route after a day without practice. Conditioning to odor combined with sugar can occur in one trial, to a color in 5 trials, and to the shape of an object visible under a food dish in 20 trials. Conditioning can be accomplished by exposure to the sugar and odor reward for as little as 100 msec. Both short-term and long-term learning take place. Bees remember landmarks and use these as cues on cloudy days (63). Strains of honeybees, for example, the Italian strain of *Apis mellifera,* show regional "dialects" in their dances.

Other types of communal activity of workers are tactile begging for food (trophallaxis), water collecting, fanning for air cooling of the hive, and clustering for warmth.

In summary, the patterns of communication between worker bees are highly stereotyped and largely genetically determined, but they are modified by experience.

GENETIC BASIS OF BEHAVIOR

Many behavior patterns are innate and are modifiable by experience only within genetic limits. A few examples of behavior which is genetically coded follow. Genetic specification of neuronal development has been shown by location and connections of neurons in a central nervous system, by dendritic patterns, and by axonal branching. Genetic coding of the enzymes associated with neurotransmitters, and the coding of membrane proteins which specify ion conductances, indicate that molecular mechanisms underlie behavior (Ch. 12,15).

There are genetic strains of Paramecium in which behavior has mutated. Taxes and kinases have become altered. Abnormal kinesis is shown by the "spinner" mutant. A mutant paramecium "pawn" has disrupted conductance channels for Ca^{2+}; another mutant has abnormal K^+ channels (98).

Numerous behavioral mutants are recognized in *Drosophila melanogaster*. Motor patterns of flight are affected by some 50 genes on the X-chromosome. Courtship and egg laying are controlled by the brain, copulation by the thoracic ganglia. One mutant of Drosophila sings an altered courtship song in which pulses of wing vibration and intervals between pulses are longer and the amplitude of pulses greater than in normal males (69,166). Several genes in Drosophila, at least 50 on the X-chromosome and 6 on autosomes, code for temperature-sensitive behaviors; when warmed to approximately 30°C reversible paralysis occurs. One X-linked mutant shows defective receptor potentials in the eye when warmed to 35°C. A mutant for a gene on chromosome III has reduced amounts of rhodopsin in the visual receptors (122). One Drosophila mutant has atypical connections between two neurons of the giant-fiber system (161).

In mice several mutations carry changes in central nervous function. One mutant has defective myelin-forming cells. Another mutant shows degeneration or failure of development of deep cerebellar nuclei. Mutants called rollers or leaners show degeneration or failure of migration of Purkinje cells or of granule cells in the cerebellum. Staggerers and weavers show degeneration of granule cells restricted to some cerebellar layers (13,164).

Strains of domestic dogs show characteristic social roles in competition and cooperation (143). Breeds of dogs are selected in part by differences in behavior. In Siamese cats the axons from retinal ganglion cells make connections in the lateral geniculate nucleus that are different from connections in other kinds of cats.

Evidence for inherited human psychological disorders and for exceptional skills corroborates the concept of genetic control of development. Calculations based on mRNAs needed to specify proteins of various organs indicate that four to six times more genes are needed for encoding in the brain than in organs such as liver or kidney. Rat brain has DNA sequences sufficient for 240,000 different proteins, liver for 57,000, and kidney for 37,000 proteins (28,29). Brain mRNAs are both polyadenylated and nonpolyadenylated. Some of the poly-A$^+$ mRNAs are similar to those in liver and kidney and are transcribed prior to birth. The poly-A$^-$ mRNAs are mostly unique to brain and are formed during postnatal development (28,29).

Most of the genetic variants that have been described are deleterious. Many neural defects and abnormalities in development are known. Benefiting mutations are less common. An example of a positive genetic effect is tolerance of insecticides by mutant insects, which extends their ecological range.

INTEGRATION OF BEHAVIOR; EMBRYOS, NEONATES, ADULTS

Some behaviors are genetically determined by the action of genes on membranes and neural connections. Some behaviors are ontogenetically determined; adult behavior depends on neural connections that are formed at specific stages in development. Many behavior patterns in adults result from environmental (social) influences; they are conditioned or learned. The interaction of genes, development, and environment is integrated in the central nervous system (CNS). Fractionating a behavior pattern into components is a first step from a holistic description toward a molecular analysis; each component can then be examined separately. A more difficult task is to put the components together. This is systems analysis, which aims to define relationships between the parts. The integrated whole, when considered in its physical and social milieu, has properties which are different from those of the separate components (Ch. 1). Each component of behavior can be interpreted in the context of exogenous and endogenous factors, of environment and internal state.

The components of behavior are organized at discrete levels in nervous systems, and these levels may be arranged in hierarchies with feedback between levels and from motor effectors to CNS. As behavioral components become more closely organized, their interactions with the environment are diminished. The system shifts from an interactive to a self-organized state. With increased independence, the range for facilitation decreases and behavioral outputs become more tightly organized. Relationships between input and output include (1) proportionate transfer, (2) transformation of input signals with quantitative changes in output, (3) trigger of output patterns, and (4) spontaneous output in absence of input (52,53). Isolated nervous systems (crayfish, insect) or slices of brain (mammalian hippocampus, cerebellum) show patterned endogenous activity which can be modified by stimulation of afferent neurons.

Whether a pattern of behavior is initiated by an environmental stimulus or arises endogenously within the nervous system, it is subject to tonic modifying influences, which may be internal (hormonal state) or external (environmental temperature), as well as to neural and sensory inputs. Tonic modifications may affect (1) probability of occurrence of a given behavior, (2) latency of performance, (3) number of elements taking part in an activity, and (4) amplitude, speed, and structure of performance (52,53). Tonic inputs may modulate behavior patterns in nonspecific as well as in specific ways. In an integrated system, multiple tonic influence may converge to affect a single output and multiple outputs may diverge from a single input. External tonic controls are day-length, temperature, and humidity. Examples of internal tonic influence are: in the breeding season, a male wolf or deer changes in approach to a female from indifference to courting. Male songbirds start to sing and to establish territories when testosterone in the brain reaches a critical level. Grooming by mice becomes slowed and less precise after dorsal roots are cut and mechanosensory inputs reduced. Stimulation of the putamen in the mouse may evoke grooming and the threshold for grooming is altered with illumination, red or white. Rats made aphagic by hypothalamic lesion can be made to eat if given a slight tail pinch (52,160).

A behavior pattern of interaction between animals may reverse according to internal state from negative to positive, from aggressive to cooperative, from hungry or thirsty to satiated.

Many behavior patterns are programmed in sequences of acts. Mouse grooming follows a certain order—face, belly, back—and has been analyzed as eight discrete foreleg movements which are not random and which differ in various genetic strains. If a limb-bud is amputated in a newborn mouse, the central pattern of grooming develops even though the limb does not make the movement. As mice recover from anesthesia, reflexes recover in the sequence: scratching, gnawing, face wiping. If a mild tail pinch is given early during the recovery from anesthesia, the sequence is facilitated; if the pinch is strong, the sequence is blocked (53). Wolf postures range from self-confidence in facial expression with tail high to submission with tail between the hind legs. Many programmed sequences are released by specific features of a complex stimulus. A hatchling herring gull, only a few hours out of its shell, pecks at the red dot on the tip of the parent's bill when the beak is swung back and forth. By use of models, Tinbergen (162) showed the releaser to be a pointed object (bill), a color contrast (dot), or a movement (bill swinging). As the chicks develop they become more discriminating (2,162).

Much behavior leads to selection of "preferred" body states. Humans report sensory "pleasure" on cooling when the body has been hyperthermic or on warming after chilling. In a temperature gradient paramecia, fish, and lizards spend most time at a "preferred" temperature. Selected temperature depends on prior temperature experience. The degree of sweetness as judged by feeding in rats depends on recency of eating and/or blood sugar level.

Patterns of behavior may be determined by interactive properties of regions in the central nervous system. The lateral hypothalamus of mammals controls or influences numerous endogenously triggered behaviors. When electrodes are implanted in the hypothalamus, cats and rats stimulate themselves persistently as if for pleasure; stimulation of hypothalamus can reinforce learned bar pressing, and can elicit eating, drinking, mating, or attack. If the lateral hypothalamus is damaged bilaterally, aphagia and adipsia result, which can lead to death. If, however, the animals are force-fed after the lesion, recovery occurs after some weeks in a definite behavior sequence from aphagia and adipsia, to eating wet food, then dry food. The sequence of recovery resembles the sequence in weaning (160). Animals such as cats in which large lesions of the lateral hypothalamus have been made, may lose motivated behavior, become akinetic and even cataleptic. Such a cat crouches in a corner and permits its leg to be raised and placed in bizarre postures. However, if the cat is dropped, it orients and lands normally on its feet (160). These observations indicate redundancy of central pathways and hierarchical control.

In amphibian and earthworm embryos, myogenic movements generally occur before neurally controlled movement. Anterior parts of the body make neurogenic movements before posterior parts. In chickens, gross movements of limbs develop before fractional movements, and spontaneous movements usually occur before reflex responses. Spontaneous electrical activity occurs in motor centers when the first movements are made. In chick embryos and sheep fetuses the myogenic stage may be omitted or attenuated.

Sleep pattern illustrates a sequence of encephalization. Sleep by an infant is largely REM sleep (rapid eye movement), whereas in adults it is 60 percent slow-wave sleep with slow electroencephalographic rhythms.

Experience and learning exert tonic influences on behavior patterns. Association of a food with previous illness (as when sugar is given with an emetic) conditions a rat against taking the food (sugar alone). Habituation, reduction in synaptic transmission on repetition, leads to behavioral decrement by central modification. Habituation may be short-term or long-term; it may develop rapidly or slowly and may last variable lengths of time. Examples of behavioral habituation brought on by repeated stimulation are: disappearance of tail flip in fish, cessation of fleeing by ducklings, cessation of prey catching by spiders. Cellular mechanisms of neural habituation are described in Chapter 11. After habituation, recovery may be spontaneous or dishabituation may occur on vigorous nonspecific stimulation. In spinal rats, reflex lifting of a hindleg habituates but is disinhibited by a pinch to the foot (131a).

One effect of experience is to establish a central engram such that behavior may be anticipatory. Another persisting result of experience is establishment of correlation with a visual or auditory model. A chimpanzee responds with excitement to a model of a chimpanzee head; birds recognize features of models of birds of their own species.

The capacity of neurons in the visual cortex of cats to distinguish visual patterns is impaired if one eye is occluded at a critical stage in development (81).

Early experience affects general behavior in later life. Children were reared for months or years in orphanages in drab cribs without stimulation of a spoken language; when removed at several years of age to a normal stimulating environment they were retarded not only in language development but also in general intelligence (83).

Development of rat brain and behavior differs according to whether the animals have been kept for weeks in (1) an enriched environment (EC) with lights, play objects, patterned surround, and other rats, (2) in a social environment (SC), caged with other rats but lacking visual variety, or (3) caged in isolation (IC). The EC rats showed superior performance when tested for maze learning. The dendritic field size of pyramidal neurons of visual cortex and the number of synapses per neuron were 20 percent larger in the EC than in the IC rats. Under EC conditions increases in dendritic branching can occur in adults but to a lesser degree than in young rats. Some stimulation of synapse formation occurs in regions of the brain other than visual cortex, particularly in the cerebellum; brain metabolism as indicated by uptake of ^{14}C-deoxyglucose is increased more in EC rats than in IC rats (65).

LEVELS OF ADAPTIVE BEHAVIOR

It is difficult, if not impossible, for human observers to describe animal behavior objectively. Descriptions of postures, movements, and vocalizations can be quantitative and objective, but the interpretations of these behaviors are in words with connotations of human experience. At one extreme are persons who ascribe self-awareness (consciousness), pain, and pleasure to all animals, and to plants. They describe animal behavior in terms of human experience and motivation, sensation, and intelligence. At the opposite extreme are those who consider all behavior as mechanistic or forced, predictable if all endogenous and environmental conditions are known. In this view the brain is regarded as a computer. Not only does every animal have its own sensory world, but the description of an animal's world varies with the viewpoint of the human observer. The problem is acute because of lack of communication between humans and other animals; the world is interpreted from a human viewpoint, not as viewed by the animal. Among humans with different languages and traditions, equivalent descriptions of behavior are not easily made. It is not appropriate to speak of lower and higher animals; each species that has survived is adapted behaviorally to its way of life.

Although completely objective analysis of behavior is impossible, a classification can be attempted and a hierarchy of behavior complexity arranged. The following categories (often given in subjective terms) are not to be considered as distinct but rather as parts of a continuum. Each type of behavior has survival value for the organisms performing it.

1. The simplest behavior is direct response to environmental changes or stimuli. Responses may be positive or negative, often toward a preferred or neutral environment, satisfying some need, such as food, or avoiding some harmful condition. Protozoa (especially ciliates), as well as animals with nervous systems, respond directly to differences in temperature, light, gravity, and chemicals (food or toxins). Responses show thresholds, gradation with intensity of stimulus, and some sensory adaptation on repetition. Direct responses are genetically programmed and stereotyped. There may be convergent effects of multiple stimuli, differing in direction and strength, and the responses may be hierarchical—for example, the responses of a snail to food, touch, sex. The direct response may serve to maintain the position of an animal (e.g., echolocation in bats). The direct response may also be modified according to internal state (e.g., the movement of a lizard into or out of sunlight according to body temperature). Direct responses of male moths to female pheromones depend on reproductive state.

2.   Direct responses, while genetically programmed, may be modified by experience. Modification may be short-term or long-term. Habituation occurs in ciliates as well as in animals with nervous systems. Central nervous habituation is distinct from sensory adaptation or motor fatigue. Conditioning is the association of paired stimuli, either sequentially or spatially. Classical (Pavlovian) conditioning pairs a conditioning stimulus with either a noxious stimulus or a reward (unconditioned stimulus). In operant conditioning, the performance of an act constitutes a reward. Reversal learning occurs when preference for one stimulus S^+ is established and the experimenter then reverses the reward so that a reverse stimulus S^- is to be selected. Probability conditioning requires either learning of delays between conditioning stimuli or learning the number of presentations per reward. Neural mechanisms of simple behavior modification are discussed in Chapter 11.

3.   A level of behavior that is more complex than conditioning of direct responses requires reaction to a complex environment and selection of specific elements as relevant. Honeybees can be conditioned to a particular food and learn to come at a given time of day for feeding. This is conditioning of an innate behavior pattern. Homing pigeons learn the environs surrounding the home loft, a modified behavior that requires considerable experience. Dolphins are conditioned to sounds or visual stimuli, behavior which is interpreted anthropomorphically by some observers as purposive. This level is Gestalt behavior. Two levels of complexity are: intrinsic complexity (genetic and prenatal environment) and complexity of ability to modify recognition and response to the total environment.

4.   Behaviors involving communication may be complex. Many of these are innate and stereotyped. Communicatory behaviors are most frequently between animals of the same species, rarely with other species. Male frogs or songbirds have vocal repertoires of signals. Much communication is related to reproductive behavior and is modified according to internal state, for example, in mammals when the female is in oestrous. Viewed objectively, the dance of a scout honeybee on return to the hive is an innately programmed communication; the bee has found food of a given kind at a certain distance and direction, and it stereotypically communicates this information to workers in the hive. Another example of complex stereotyped communication is the jamming avoidance response of electric fish.

5.   Much communication, and other motor response, is not programmed genetically but is acquired by imprinting or by learning processes at certain stages in development. Not only are appropriate genetic and environmental conditions requisite but also developmental patterns. In some species of birds a dialect song pattern develops endogenously; in others, dialect songs must be heard at a critical age as sung by others of the same kind (95). Young carnivores in a den play and learn varied behaviors. Hunting is learned by the young following and imitating the parent. A great deal of the behavior of birds and mammals is acquired by social learning within genetic limits; it is

culturally transmitted from generation to generation. Behavior that is learned from family members or associates emphasizes the importance of natural selection of groups as opposed to individuals.

6. The language of communication acquires symbolic meaning in that a sign elicits a sequence of behavior, not a direct stereotyped response. Symbols are substitutes for objects or recognized actions, and can initiate behavior similarly. Some uses of symbols are genetically or developmentally programmed; other uses of symbols are acquired; these include the use of symbols unrelated to the objects they symbolize. Some ethologists consider that the dances of bees are symbols of routes to be followed. Many animals recognize other individuals of the same family group by odors and color patterns which are symbolic signs. Facial expressions symbolize moods in chimpanzees. A tool can take on symbolic meaning as a trigger for a given action: desert birds use cactus spines for digging in cacti for grubs, gulls drop shells onto rocks and eat the snails, sea otters crack sea urchins on rocks held on their bellies, a wasp closes the entrance to its nest with a pebble, archer fish aim water streams at insects, chimpanzees use sticks as whips and clubs, and to poke into holes or as levers.

The stuffed head of a female turkey is a symbol which evokes sexual responses in a male. A male gull reponds to the bill and face markings of another gull. A dog can learn to select a particular odor from a complex mixture of smells. Chimpanzees have been trained to use sign language, to match plastic cards and to punch computer keys in order to designate objects, other chimpanzees, or the trainer, and to combine pairs of signs. A few chimpanzees have learned more than 100 such signs. This is associative symbolism which permits communication (57).

7. The most complex level of behavior is the capacity for abstraction as shown by humans. Whether abstract thinking occurs in nonhuman animals has been much debated and is not agreed upon (145). Cognition allows initiation of behavior and development of abstract concepts in the absence of symbols. Communication is not needed for cognition. Thought is considered by some psychologists as covert verbal behavior. Language has both phonetic (speech) and psychic (idea) components. The neural mechanisms for language are presented later. Every human has a private consciousness and each of us assumes that others experience the same interpretations, the same concepts, the same abstractions. Philosophy, logic, and the manipulation of abstract concepts represent extremes of the most complex level of behavior.

Consciousness, Intelligence, Sensation

The question of whether nonhuman animals—octopus, bees, pigeons, porpoises—have awareness cannot be answered simply "yes" or "no." The concept of awareness in animals is generally couched in anthropomorphic terms; and awareness itself is a term with several meanings, none of which is strictly objective. Human infants gradually develop self-awareness, and patients with neurological defects have varying degrees of awareness. The evolutionary view is that awareness and use of symbolic language have been brought about in degrees by natural selection. There appear to be many levels that differ in degree to the extent that awareness and language in'adult humans are qualitatively different from these functions in nonhumans.

Criteria of self-awareness or consciousness are:

1. Decisions or choices are made; a bee selects flowers of one color or odor, ciliates select a preferred temperature. Such decisions are made in neurally preprogrammed behavior.
2. Choices are made on the basis of experience. Most mammals and birds when first born or hatched are unable to make critical choices. Their decisions are based on later experience. A laboratory rat would not do well in an environment where wild rats flourish.
3. Behavior patterns, some conflicting with others, occur in hierarchies. In a mollusc Pleurobranchaea escape swimming takes precedence over feeding, gill and siphon withdrawal, and copulation (Ch. 11). This behavior indicates choices made according to strength of sensory input, state of responsiveness, feedback from one neural level to another.
4. Anticipatory awareness. A type of awareness shared by all adult humans is expectation of death; whether this is genetically programmed or not is unknown. Whether nonhuman animals are aware that they will die cannot be ascertained. Preparation by hibernating mammals for winter, and provisioning by some hymenopteran insects for offspring could be interpreted as anticipatory awareness, but are more likely neurally programmed behaviors selected over evolutionary history.

We conclude that animal awareness is a property deduced by anthropomorphic interpretations and that it can be known only in humans.

Intelligence is a behavioral concept for which there are numerous definitions:

1. One measure of intelligence is conditionability or simple learning (behavior level 2 above). Vertebrates of all classes display classical conditioning, instrumental conditioning, reversal and probability learning. It has been suggested (105) that on the basis of the capacity to be conditioned, fishes are as intelligent as rats. One difficulty with equating conditionability with intelligence is that the tests which are used are devised by experimenters to suit experimental protocols.
2. Intelligence tests of humans provide intelligence quotients (IQ scores). These tests were devised to assess progress of school children but despite efforts to make the tests culture free, experience and social back-

ground affect performance. (Clearly, intelligence tests cannot be used with nonhuman animals.)

3. A vague but useful measure of intelligence is demonstrable knowledge of and ability to deal with a total environment. The subject constructs his own cognitive map by which behavior is programmed. Although difficult to measure objectively, the ability to cope with a changing environment is an indication of intelligence, whether genetically, developmentally, or experientially determined.

4. A related definition of intelligence is an ethological evaluation of behavior under natural conditions. This definition, while anthropomorphic, has intuitive value. Field observations of feeding, reproducing, territoriality, and other actions are more relevant than laboratory tests and may, once quantified, come close to a true adaptational meaning of intelligence. Laboratory-reared animals are less intelligent by such criteria than wild animals of the same kind.

Sensation, like awareness and intelligence, is a concept which for nonhumans can only be inferred from reactive behavior. Transduction of environmental stimuli at sense organs is the first step. This is followed by central processing and finally by recognition or sensation, which may or may not lead to a motor act. One neurophysiological view is that the complexities of sensation reside in single neurons, called "pontifical" neurons, at the apex of a hierarchical organization. An alternative neurophysiological view regards sensation as a property of neural networks, that is, of interneuron interactions. This implies interaction among sensory, memory storage, and motor components, with feedback among these elements. However, as more detailed analyses are made of single-unit activity, cells with very complex properties are being discovered. Examples of cells which respond to and are modified by multiple inputs are found in animals with limited numbers of neurons—molluscs and arthropods (Ch. 11). No single neuron of the auditory system of a cricket or frog has yet been found that is responsive only to the species-specific mating call. Toads respond behaviorally to wormlike objects moving in the long axis. Some neurons in visual areas of toad optic tectum and thalamic pretectum respond most vigorously to visual fields that expand in the direction of movement; other neurons respond to elongate objects moved parallel to long axis, not to perpendicular movement (50). Examples from awake monkeys of precisely specified sensory associations are given later in this chapter. Sensations may be coded in specialized neurons and in networks.

The preceding classification of the continuum of behavioral levels and summary of subjectively defined properties have heuristic value; this classification is clearly holistic and not presently amenable to molecular explanations. As noted by numerous philosophers, brain as presently understood is not equivalent to mind; yet mind is dependent on, is emergent from, brain. In no area of adaptational physiology is it more clearly shown that the whole is greater than the sum of its parts (Ch. 1). Esthetic experience results from multiple components of an event. Auditory experience combines rhythm, harmony, melody, and volume, and one listener may focus on sounds which another fails to note. Brain waves from musicians differ from those of non-musicians during similar musical stimulation. Appreciation of a summer morning depends on combined inputs to eyes, ears, and nose to different degrees. How much of such appreciation is innate and how much due to cultural conditioning is uncertain. An understanding of consciousness is not likely to come from detailed analysis of neuronal function (152). Neuronal analysis provides element detection, not a Gestalt picture.

NEURAL MECHANISMS OF COMPLEX BEHAVIOR IN VERTEBRATES

Integrative properties that can be studied in simple nervous systems include neural relays, transfer functions of sensory input and motor output, patterned output, controls by command neurons and networks, plasticity of synaptic function, and simple conditioning (Ch. 11). Complex functions—language, symbolism, and abstraction—are properties of human brains, and to some degree of brains of other primates. No single method provides a complete picture of neural function. Evidence is derived from developmental and experimental anatomy, from maps of central connections, from electrical measures of unit activity and of summed or evoked potentials. The neural basis for levels 5 to 7 of the preceding classification is indicated by comparative behavioral observations of several species, and by assessing effects of lesions and of localized stimulation. In a vertebrate embryo, the nervous system begins as a neural tube; early in development, cephalization, the enlargement of what will be the head end, occurs. Next, bilaterality is established and evaginations and contacts with ectoderm lead to the formation of sense organs (eyes and ears), and myotomes become organized as muscles.

Cephalization or expansion of anterior (rostral) brain occurs during embryonic development. Cephalization is well shown in segmentally organized invertebrates—annelids and arthropods. Cephalization, head dominance in nervous systems has two functions: (1) the enlargement of sensory regions correlates with concentration in the head of some organs, especially eyes, taste and olfactory organs, and ears (of vertebrates). (2) Integrative association is concentrated in anterior regions of a brain. Fishes and amphibians use midbrain—tectum and tegmentum—for integrating some complex behavior; they have also sensory projections to diencephalon (thalamus) and telencephalon (forebrain). In birds and mammals integration of behavior is centered more in forebrain than in midbrain.

The following organizational characteristics of the nervous system of vertebrates are best known for mammals (16):

1. Hierarchical (series) organization is indicated by relays of neurons from one level to another. At each level, sensory messages change during

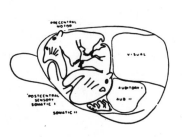
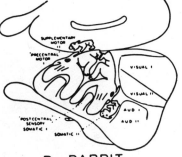

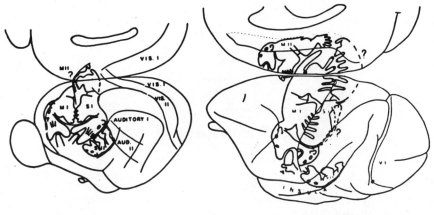

A. RAT

B. RABBIT

C. CAT

D. MONKEY

Figure 10-11. Diagrams of cerebral cortex of rat, rabbit, cat, and monkey, showing primary and secondary motor (MI and MII) and somatic sensory areas (SI and SII), also primary and secondary visual and auditory areas. Orientation of body represented as animalculus. Each brain outline corresponds to lateral surface of hemisphere; upper portion of each brain corresponds to mesial surface. From H. Harlow and C. Woolsey, Eds., *Biological and Biochemical Bases of Behavior.* © 1958 by the Regents of the University of Wisconsin, used by permission of The University of Wisconsin Press.

synaptic interactions. Motor functions are arranged in hierarchical levels.

2. Redundancy and parallel pathways are indicated by the partial persistence of a function after a few fibers or a brain region are damaged. Behavioral deficits in maze learning by rats are greater after a large amount of "association" cortex has been removed than when a small region is removed. Several neural routes transmit signals from retina or cochlea to cortex, each carrying slightly different information. There are usually several cortical projection areas for a given sense, a primary area and two or more secondary areas. These regions are not strictly equivalent, but one can serve in the absence of others.

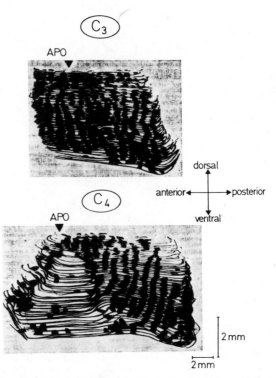

Figure 10-12. Three-dimensional reconstructions of orientation column system in area 17 (visual) of four cats. Columns visualized by activity as shown by incorporation of deoxyglucose. Reprinted with permission from W. Singer, *Experimental Brain Research,* Vol. 44, p. 435. Copyright 1981, Springer-Verlag, New York.

3. Divergence and convergence are frequent consequences of parallel pathways. The parameters of a sensory message are altered as a result of convergence in integrative cortex.

4. Feedback is common to all nervous systems. Neurons from a higher level in a hierarchical chain project back to a lower level. Loops and circles are established in sensory pathways. For some modalities (e.g., hearing), there are efferents to the receptor organ. An association center may thus modulate its own input.

5. Localization of function holds in general for the lobes of the cortex— vision in occipital lobes, hearing in temporal lobes, somatosensory in postcentral gyrus of parietal lobes, motor in precentral gyrus of frontal lobe facing the somatosensory area (Fig. 10-11). Electrical stimulation of motor cortex elicits movements of regions of the body. Large regions of the cortex, mainly parietal and frontal lobes, serve associative functions.

6. Organization of the cortex differs in different areas, but in general it is

organized into six layers. The outer molecular layer consists of small axons and neurons. Layer two consists of small pyramidal cells, layer three of medium and large pyramidal cells. Layer four contains small stellate cells; layers five and six are deep layers of large pyramidal cells. Within each sensory area, neurons are organized in columns which traverse the six layers (Fig. 10-12). All of the neurons in one column receive input from a small region of the relevant sense organ. Columns provide for mapping or projection and for vertical integration of cells serving a similar function. A column may have inhibitory action on adjacent columns. Stratified organization also occurs in cerebellum, and in optic tectum in amphibians and fishes.

7. Field effects or averaged evoked potentials combine with spikes in single units to indicate regional activity. Field effects correlate with behavior in terms of latency correlated with attention. Single units show diversity of coding. Whether field potentials have physiological meaning for single neurons is debatable. However, the wide occurrence of field potentials and their correlations with behavior favor some function other than as byproducts of unit activity.

Localization of Function in Sensory Areas of Mammalian Cortex
Somatosensory System

Neural units in somatosensory cortex I respond to contralateral tactile stimulation. In both the thalamus and the cortex the body surface is projected as an animalculus with representation approximately according to the importance of the appendage for the animal—mouth parts in rabbit and sheep; claws and forelimbs in cat; hand and face areas and vibrissae in rat and rabbit; forepaws in racoon and monkey. Somatosensory cortex II is in the parietal lobe and has both contra- and ipsi-lateral projections. A column in the somatosensory cortex receives projections from tactile receptors in a small region of skin; each minicolumn contains only 110 cells (33,114). A column provides a two-dimensional matrix within which vertical integration occurs. In rodents columns in parts of primary somatosensory cortex for the snout take the form of barrels, each receiving from receptors of individual vibrissae. The barrels are in five rows, each in an arc of eight barrels (178).

Auditory System

The mammalian auditory system has a hierarchical organization with parallel projections and with extensive feedback between levels as well as to the sensory organ of Corti. A generalized diagram is given in Figure 10-13.

The primary auditory area of the cortex in mammals is in the temporal

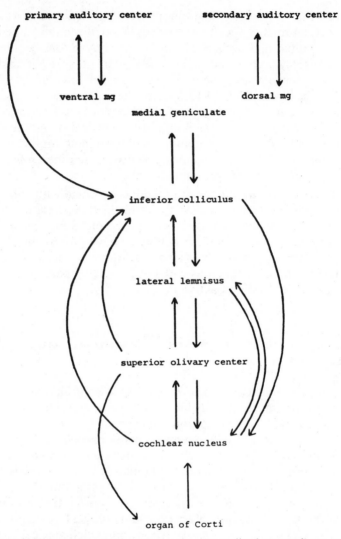

primary auditory center   secondary auditory center

ventral mg   dorsal mg

medial geniculate

inferior colliculus

lateral lemnisus

superior olivary center

cochlear nucleus

organ of Corti

Figure 10-13.   Schema of auditory pathways in a generalized mammal; mg, medial geniculate. By personal communication from Albert Feng.

lobe. Neurons are arranged in tonotopic representation, corresponding to
the frequency map in the cochlea. Each column contains cells with the same
best frequency from each ear. Many columns have projections from both
ears, others from only one ear. After removal of auditory cortical areas, cats
and monkeys can relearn responses to sounds, but in monkeys removal of
cortical auditory areas results in loss of discrimination of tone patterns or
sequences. Monkeys hear in the same general range of frequencies as man
but monkeys also hear one octave higher to about 45 kHz. In some echolo-

cating bats, different cortical neurons respond to the constant-frequency component of a call or to a harmonic of that frequency, other neurons to frequency-modulated components of the emitted calls (Ch. 13).

Central areas for hearing in frogs and the regions for hearing and singing by birds are discussed in Chapter 13.

Visual Systems

From the retina, axons of ganglion cells go to sensory relay stations in mesencephalon and diencephalon. In mammals the mesencephalic nuclei are in the superior colliculus (SC); in frogs and fishes the large optic tectum constitutes the principal integrative region for visual processing. The diencephalic nuclei in mammals are in the lateral geniculate (LGN), a laminated structure, and in the pulvinar. Ganglion cells of the retina are characterized as x-cells which give brisk tonic responses, linear with light intensity and on-off responses; y-cells give phasic nonlinear responses: w-cells give sluggish responses and have slow axon conduction. Ganglion cell fields have predominantly center-surround organization which is preserved in units of the LGN. If area 17 of the cortex is removed from a cat, visual discrimination of patterns persists but acuity is reduced. Considerable processing occurs between LGN and area 17 in that some cortical neurons respond to oriented fields and to movement in a "preferred" direction. In each cortical column all cells respond to a particular orientation of the stimulus and there is interaction between the six layers within a column.

Cortical cells are simple, complex, or hypercomplex (61,81,147). In simple cells the excitatory and inhibitory areas are not concentric, as in the LGN, but these areas lie along a boundary at a definite angle; some cells respond to on, some to off, and some to on-off light stimuli. Most neurons respond best to a moving rather than a stationary boundary. Complex cells give high-frequency responses to movements in a given direction and at a certain angle; responses are persistent or tonic. Hypercomplex cells respond to moving oriented boundaries; excited lengths are flanked by inhibitory zones that limit length of the most effective stimulus. In hypercomplex cells, extension of a stimulating line diminishes the response; a bar or line must terminate within the borders of the receptive field. The LGN is organized into four lobes and each of these has distinct laminae. These different regions of the LGN receive x, y, and w fibers in different proportions as shown in Figure 10-14. In cat, one lamina of the A region of LGN receives contralateral, the other lamina ipsilateral input.

In the C-lobe two laminae receive contralateral, one lamina receives ipsilateral, and a fourth lamina receives from the superior colliculus (SC). Input from LGN to cortex was formerly thought to be organized in orientation-selective patterns corresponding to LGN cells. However, blocking the upper

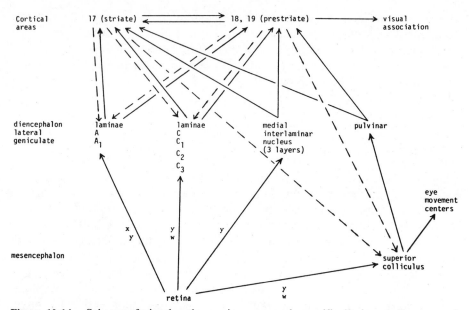

Figure 10-14. Schema of visual pathways in a mammal, specifically in cat. By personal communication from Joseph Malpeli.

laminae of LGN by cooling shuts off the deep-layer responses of the striate cortex, leaving the upper layers responsive; this indicates that properties of cortical cells do not depend on column organization in LGN (106). As shown in Figure 10-14, the superior colliculus sends fibers to eye movement nuclei, and to the pulvinar which in turn sends fibers to both striate and prestriate cortex. The different parts of the LGN project to both areas 17 and 18–19. In addition to the parallel pathways to the visual cortex there is much divergence. It is estimated that approximately 10^6 optic nerve fibers terminate in one LGN, that each LGN has 2×10^6 cells, that visual cortex (monkey) has 150,000 neurons per square millimeter, amounting to a total of some 200×10^6 cells in the striate cortex on one side. Thus there are more than 200 cortical cells per each LGN cell and twice that number for each retinal ganglion cell (129). In addition to the parallel ascending pathways and the divergence there is considerable feedback from both primary and secondary visual cortex to the geniculate (Fig. 10-14). In the absence of primary visual cortex the superior colliculus can mediate gross localization of a visually perceived object.

Inferotemporal, posterior parietal, and prefrontal regions have long been considered as "association" areas. Their functions in vision are integrative—object recognition, egocentric orientation in space and time; individual units in the inferotemporal cortex differ in sensitivity to contrast, color, size, shape, orientation and direction of movement of visual stimuli (68). Lesions in the inferotemporal lobe lead to psychic blindness. The perception and processing of visual stimuli are shaped by attention. Units in the parietal lobe have been recorded by chronically implanted electrodes while the animals performed tasks which required them to fixate on visual targets, to track targets, and to make saccadic movements. Unit firing frequency declined after a reward was given, even without a shift of gaze. Other units in the same area fired when the eyes were in slow pursuit of a target, and still others decreased in frequency during a saccade to a new target (114).

Recordings from single units of dorsolateral frontal cortex of awake monkeys show highly specific responses to problem situations. An individual unit responds in its own period of delayed response and spatial presentation; some units are active only during the required delay between stimulus (patterned light) and motor response (pushing a lever for reward). Some units respond 200 msec before muscle activity of the response; some are differentially active for right or left presentations, some units depend on stimulus location, others during the delay period correlate with direction of impending response (97,117). In the superior temporal sulcus of awake monkeys, some individual neurons respond to visual stimulation by a face—human or monkey; some cells are sensitive to orientation, full face or profile, many respond to gaze, some to rotation of the face (126). It is concluded that single neurons in the visual association cortex have properties of complex integration, properties expected of short-term memory and decision making. The visual system illustrates several properties of complex central integration.

Two regions of the brain concerned with autonomic functions are present in various configurations in all classes of vertebrates. In mammals, the limbic system of the forebrain lies beneath the cortex and consists of amygdala in the depth of the temporal lobe, the cortical hippocampus, the septum and olfactory projection fields. These regions are concerned with emotions, motivation, and some types of learning (hippocampus). Amnesia results from lesions in hippocampus of humans. Both amygdala and hippocampus are needed for associative memory. In the diencephalon beneath the thalamus is located the hypothalamus, which has regions concerned with emotional behavior, temperature regulation, water balance, food intake, and circadian rhythms. Self-stimulation is most actively pursued when electrodes are in lateral hypothalamus. Certain cells in the hypothalamus synthesize peptide hormones that are transported to the pituitary for secretion (Ch. 12). The functions of the limbic system and hypothalamus have changed during vertebrate evolution. In fishes, the precursors of limbic structures are primarily concerned with olfaction.

In mammals, aggressive behavior is under the control of the limbic system and hypothalamus. In a cat, stimulation of hippocampus, limbic cortex, and regions of hypothalamus elicits defensive behavior—growling, hissing, lowering of head, hunching of back, dilation of pupil. Recordings show electrical activity in the hypothalamus whenever threat behavior is made. Ablation of amygdala leads to compulsive oral responses, loss of fear, hypersexuality, drop in social dominance. Stimulation of a lateral region of hypothalamus evokes predatory stalking; stimulation of the medial region, defensive reaction; and of a dorsal region, retreat. In pigeons a lesion to archistriatum reduces agonistic behavior. In fishes and rodents local stimulation, especially in thalamic centers, can evoke sex behavior. As shown by immuno-cytochemical staining, the thalamic regions contain neurons rich in reproductive hormones (40). In mammals and birds behavior evoked by brain stimulation is influenced by objects seen, such as predators.

LATERALIZATION OF BRAIN FUNCTION

Lateralization of brain function is most evident in humans. The two cerebral hemispheres are connected by three commissures; the largest, the corpus callosum, contains some 200×10^6 fibers. These commissures have been transected in some intractable epileptics, leaving the two hemispheres disconnected. Responses to visual, auditory, and tactile stimuli to each side reveal considerable independence of the hemispheres, each functioning separately.

In 92 percent of right-handed persons the left hemisphere is dominant and controls speech; in 69 percent of left-handed or ambidextrous persons the left hemisphere is dominant. The right hemisphere is dominant in 4 percent of right-handed and in 15 percent of non–right handed persons (152). In general, the left hemisphere controls speech, language use, aggressiveness, mathematical ability. The right hemisphere is nonlinguistic in that it cannot respond by speech but recognizes visual stimuli, remembers, and can direct responses by matching objects; it is concerned with spatial orientation and nonverbal memory. The right hemisphere makes emotional responses and does some reasoning. After the hemispheres are separated, old memories are retained in each but recent memories are lost. Lesions to the right hemisphere impair tone discrimination and recognition of music; melodies are perceived by the left. Each hemisphere receives input from the whole visual field of both eyes but processes the inputs differently; when a chimeric face picture is presented the right hemisphere responds to the whole face as a perceptual unit, the left notes salient features and labels them verbally (15).

Injection of the anesthetic sodium amytal into either left or right carotid artery of persons with brain intact confirms the results of split brain studies (152); after one side of the brain is anesthetized the other side controls motor behavior and performs some cognitive functions. In normal brains the two halves operate together and the whole is more capable of function than the sum of the two halves.

Some lateralization is present in monkeys and rats. Rhesus monkeys have a hand preference (41). The left hemisphere is more effective for processing conspecific relevant calls, the right for visuospatial perception and emotional reactions. In rats there is a preferred side in a T-maze and a preferred direction of turning in an open field. Dopamine concentration in the striatum is higher on the preferred side. Handling young rats during development results in enlargement of the left cortex. Possibly lateralization takes place during development.

In songbirds such as chaffinch and canary, the vocalization centers in the left side of the striatum are dominant and if the left centers have been lesioned a song is sung but it is not characteristic for the species. Australian cockatoos and parrots show a high percentage of left footedness, 100 percent in several species (119).

Pigeons trained for a visual discrimination with one eye (opposite eye blindfolded) can transfer the learned response capability to the other eye; also if pigeons are trained using one eye they can later more readily be trained using the other eye. Pigeons have interhemispheric transfer of perception of color, pattern, and movement of visual objects. When the two hemispheres are separated by cutting the decussating tract, no transfer occurs and there is no improved learning with the second eye after training with the first (35).

Asymmetry in the autonomic system is well recognized. The left and right vagi have different actions on the heart in mammals, turtles, and frogs.

The evolutionary origin of lateralization in vertebrates is indicated by differences in structure of the habenula, a forebrain nucleus which receives multiple sensory inputs, especially olfactory. The habenular nuclei are asymmetric in fishes and cyclostomes. Goldfish show a preference for turns to the right. In Rana the left habenula has more complex organization than the right. In petromyzonts the right habenula is larger than the left, in elasmobranchs the left is larger than the right. No functional tests to compare the two sides of the habenula in fishes or amphibians have been described.

In summary, cephalization and bilaterality permit maximization of sensory and motor function and enhance integration. Attention and experience can modify sensory and motor processing in cortical units. Functions such as abstract reasoning and emotional responses require neural interactions in areas which have been experimentally localized by lesion, by stimulation, and by separating the two hemispheres. In mammalian cortex different regions have gross localization of function; there is extensive interaction between regions and between cortical layers of a region. Recording from highly specific neurons may lead to localization of the sites of discrimination and decision making.

EVOLUTION OF LANGUAGE

Communication underlies many aspects of behavior and is amenable to neural and, ultimately, to molecular analysis. Many animals communicate by sounds that have meaning for their conspecifics—sounds connoting aggression, alarm, and courtship. These are stereotyped and largely genetically controlled. In the preceding classification of levels of complexity of behavior, the use of verbal language for communication and thought is considered as distinct from simple use of symbols. Human speech shows cognition and lability in contrast to the stereotyped calls of animals. One opinion concerning the origin of human language is that hand gestures, facial expressions, posture, and vocal calls were precursors of words. Another opinion is that the use of words came by coevolution of association cortex with neural and laryngeal vocal mechanisms, and that language is a part of general intelligence. Human language is more than communicative sounds; language has a cognitive component by which sounds and words take on meaning.

One view is that thought is covert speech, that the same areas of the cortex are active whether a phrase is spoken or is thought without being spoken. A contrary view is that thought is not implicit speech. This view is supported by the fact that, in child development, imitation and make-believe precede language. Animals show symbolic associations without verbalization.

A deaf (and mute) person can communicate by gestures, finger movements, and facial expression, and can reason and think if visually experienced. Deaf, mute, and blind persons, such as Helen Keller, also learn to use language cognitively; tactile input supplies a training cue.

Chomsky has argued that humans have a genetically coded innate capacity for language in the sense of syntax. This view deemphasizes the role of imitation and learning and argues for some similarity in origin between human speech and inherited but modifiable patterns of vocalization in birds.

Human language is considered by linguists to have the following components, in ascending order of complexity: (1) phonology, sound making, communicating by gestures, imitative vocalization; (2) semantics, symbolization, recognition of objects' color, shape, number, and assignment of words to them; (3) syntax, grammar, arrangement of words as subject and predicate, sentence structure; (4) abstraction, reasoning, forming conclusions based on prior experience.

In observations of children, Piaget identified several stages in language and psychological development; these have been summarized in detail by Hunt (83). The first or sensorimotor stage, which covers the period from birth to about 24 months, can be sequenced as follows: (1) The initial period is exercise and modification of ready-made systems, such as those for sucking, head turning, vocalization, moving arms and legs, looking, and listening. When something heard becomes something to look at, the sight of an object is assimilated into listening. (2) Circular reactions appear when a child comes to anticipate the outcome of an action based on results which are contingent in time on performance of the act, coordination between sensations and motor acts. (3) Elementary constructs of reality, concepts of sequence or time and of spatial relations of objects. (4) Period of imitation, play, symbolic imagery, curiosity. (5) Invention, imitation, internalization of experience.

The second major stage in child development is preconceptual. This extends from a few months of age until puberty and the time for given developments varies considerably with the subject's experience in infancy. This is the period of symbolic play, imaginative actions, development of language. It includes intuitive actions and play. The next period is adolescence, in which actions become less egocentric, more reversible; form may be more important than content and thought directs observations and action.

Development of vocal language occurs much sooner in children who as infants associate sounds with objects than in children lacking such experience. The development of language is basic for the maturation of intelligence.

Pathology of human speech may occur in increasingly severe forms as follows: loss of primary perception (deafness), agnosis or loss of recognition, loss of ability for conceptualization, aphasia, apraxia, loss of precise vocalization, and loss of articulate speech due to paralysis. Speech pathology may result from lesions in speech areas of the brain (90).

Lesions anterior to the central sulcus reduce verbal output; posterior lesions cause incorrect word usage. One type of lesion leads to deficit in auditory comprehension, another type to motor aphasia. Local stimulation during brain surgery may either elicit sensations of specific words or may cause the patient to speak (36).

Whether chimpanzees are capable of language in the human sense has been questioned. Apes can use symbols for objects and sensations, and can form cross-modal associations—that is, they associate specific sounds with visual stimuli and come to recognize one when the other is presented—but so can many animals (for example, honeybees). Many animals (including molluscs) remember signs for hours or days and make decisions. Apes use symbols to express internal state—attitudes, emotions—but this is also true of most vertebrates, especially for reproductive behavior signs.

In their natural environment primates communicate by calls—gorillas 22 calls, chimpanzees 23, and macaques 20–30. Macaques deafened shortly after birth make abnormal calls; deafened squirrel monkeys show little impairment. A rhesus monkey reared in isolation fails to develop the normal repertoire of calls (107). Field observations indicate that "higher" apes can make two-word sentences based on symbolic use of sounds.

Apes lack the vocal apparatus for word enunciation and extensive vocalization but have been trained to recognize many gestures and objects, to use a sign language, to match plastic chips, and to press the keys of a computer. Communication has been established between an ape and a trainer, and to a lesser extent between two apes. The responses of individual chimpanzees can be conditioned and indicate symbolic associations. One trained ape recognized 130 separate signs and could put these together in "sentences" (57). However, there is no convincing evidence that a chimpanzee uses language in the sense of abstract concepts. In the human cerebral cortex the speech area (hearing and vocalization) is several times larger than in other primates and lateralization is much greater.

A conservative view of the origin of language takes into account the fact that humans have not been derived from any existing nonhuman primate. Apes can learn the use of symbols and can use them in different contexts; the ape cortex has the rudiments of the speech association areas. However, the differences in brain and vocal apparatus between ape and man are so great as to represent a quantum jump in complexity and to result in the uniqueness of humans (155). Such major changes as the development of a new area of cortex are unlikely to have occurred without precursor structures. Quantitative changes in cortex were probably of adaptive value in primate evolution. Fossil evidence indicates divergence of the stocks of ape and man at about 20 million years ago. There were hominid species which disappeared when *Homo sapiens* became dominant. Possibly the capacity for language in its dual aspect of speech and abstraction was selected in early hominids. My conclusion is that a chimpanzee may achieve stage 6 but not stage 7 as classified earlier in this chapter.

THE ORIGIN AND NATURE OF HUMANS; ANTHROPOLOGICAL EVIDENCE

The evolution of complex human behavior and the relative contributions of genetic and cultural inheritance are partially brought out by study of primate evolution (Fig. 10-15). Behavior capabilities can be inferred from skeletons (fossil and modern) and, in recent hominids, from tools and diet residues. The oldest fossils of primates are dated from 60–65 myr ago. The divergence of old-world monkeys from anthropoids occurred 30–35 myr ago. Molecular dating puts the anthropoid-monkey divergence as somewhat more recent than do fossil data (127). The biochemical difference of one kind of primate from another is far less than the differences in the skeleton and nervous system. In humans, gorillas, and chimpanzees there is now 98–99 percent identity of many proteins, serum and mitochondrial (94) (Ch. 4).

Several species of apes (pongids) and hominids apparently lived at the same time and died out after a few million years, for example, *Australopithecus robustus*. Several species of *Ramapithecus* (formerly *Sivapithecus*) lived in the Miocene 8 to 12 million years ago (Fig. 10-15). Divergence of chimpanzees and gorillas from the human line probably occurred about 6–7 myr ago. Australopithecus sites have been dated from the Pliocene (3.5 myr) to the Pleistocene (less than 1 myr). *Homo habilis* coexisted with *Australopithecus* in Africa. *Homo erectus* is dated from 1.2 to 0.25 myr, the early Pleistocene (59).

Homo habilis had a brain size of approximately 800 ml. *Homo erectus* brain capacity was 1100–1200 ml. *Homo erectus* dispersed from Africa into what is now Asia Minor, Europe, and Eastern Asia (59).

Fossil footprints dated about 3.5 myr ago show that Australopithecus stood upright. In pongids (chimpanzees and gorillas) the structure of elbows,

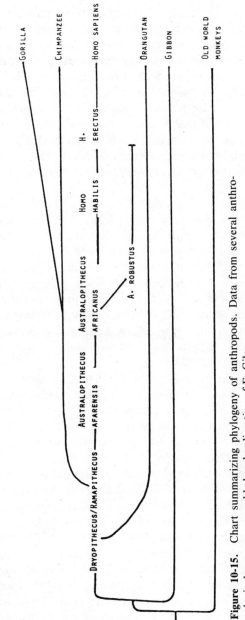

Figure 10-15. Chart summarizing phylogeny of anthropods. Data from several anthropological sources, assembled under direction of E. Giles.

shoulders, and pelvis with muscle attachments is adapted for climbing and swinging from tree branches. Present-day pongids are essentially quadripedal; a gorilla walks on its knuckles. Bipedalism is characteristic of the Australopithecus-Homo line; this freed the arms and hands for other functions.

Tooth structure indicates diet and behavior. Sexual dimorphism of teeth is high in gorilla, low in Homo; high dimorphism suggests considerable male-male competition and sex differences in predation (91). The large canines in pongids are adapted for stripping branches, for male-male aggressive behavior, and for defense. In the hominid line canines became reduced, molars flattened, and it is inferred that primitive hominids were omnivorous. Fossil evidence indicates that with upright posture and diversified diet the hominids left the forest and became savannah dwellers. Other characteristics of primates are reduction in sense of smell, and improved forward vision and color vision.

Evidence of cooperative hunting by hominids has been uncovered in anthropological and archaeological excavations from Pliocene sites; *Homo erectus* (probably also *H. habilis*) appears to have lived in communities with social organization. Stone implements were probably used by *Australopithecus* 2 myr ago. Fabricated implements are found along with skeletons of *H. habilis*. Stone tools were chipped to form sharp edges. Chimpanzees and gorillas throw stones and use sticks to obtain food, but their throwing accuracy is poor. Some stone tools made by early hominids were from types of rock from distant regions; it is inferred that early Homo, possibly Australopithecus, transported stones, and probably also food, over many kilometers.

The making of stone tools, communal living, and associated behavior could hardly have been accomplished without communication; language must have developed very early among hominids. The first use of fire, as indicated by ashes in sites in China, was by *H. erectus*. *Homo erectus* spread widely, from Africa to Europe and into southeastern Asia. Artifacts are much the same in excavations in all of these regions and little change in stone tools occurred during at least one million years. The oldest deposits of Olduvai man date to 1.9 myr, Peking man from 400 kyr (thousand years). Pleistocene populations in different regions differed in cranial capacity and in facial structure. *Homo erectus* was gradually supplanted by *H. sapiens,* whose cranial capacity was some 11 percent larger (177). Neanderthal man was present during the upper Pleistocene, 35 to 150 myr ago. The cranial capacity in Neanderthal man, as of modern Australian aborigines, was variable and was large compared to body size. During the early Pleistocene, stone tools were improved in design and variety, and in Africa, Europe, and eastern Asia cave paintings from 30 myr show a high degree of representational art. Burial sites from this period indicate concern for the deceased.

One school of anthropology has postulated the sudden emergence of human culture; a contrary view is that human cultural activities developed slowly and by small increments. Other primates (especially chimpanzees and gorillas) exhibit complex behavior—use of tools, use of symbols for communication and to identify objects, memory of events and processes, care of young during prolonged maturation. Differences between nonhuman and human are more quantitative than qualitative. However, quantitative changes may be so great as to provide for social behavior several orders of magnitude more complex in hominids than in pongids.

CULTURAL TRANSMISSION OF BEHAVIOR

The behavior of a mammal is the result of genetic programming, developmental experience, and cultural transmission. Genetic coding for ionic conductances and for synthesis of neurotransmitters is well recognized, and genetically defective mutants show abnormal behavior. The genotype specifies reflex patterns of locomotion, feeding, reproduction, and cross-modality sensory interactions. Genetic coding of neural circuits specifies developmental sequences, and resulting behavior may be altered by conditioning. The neural nature of genetically programmed or instinctive behavior is yet to be understood but is probably established via development of brain structures.

The genotype furnishes a template on which social (cultural) forces act. Imprinting of birds and social mammals by experience at certain stages in development was mentioned previously in this chapter. In humans the family and educational system indoctrinate children with local mores. Is cultural transmission like the imprinting of young birds and social mammals, or are there general attitudes and behaviors which are transmitted culturally in all societies and specific patterns for individual societies? Cultural evolution proceeds faster than biological evolution; some culturally transmitted characters probably have evolved since the Pleistocene, and were not present in primitive humans.

Homo sapiens is a single species throughout the world capable of interbreeding. Archaeological evidence shows that human populations expanded in wide geographic areas during the early Pleistocene. It is probable that some categories of socially transmitted characters exist in all cultures and that local societies refine and specify patterns. A diagrammatic representation is

Biological Heritage   Cultural Heritage
genotype   general culture
    ↓   ↓
development   specific culture
    ↓
conditioning
                    → Homo ←

Coadaptation of the two types of inheritance is required for a complete human being. For some general characters there is overlap of genetic and cultural determination. General social characteristics are common to all cultures. Specific or imprinted characteristics are local, and a person can go from one culture to another and adopt the new local characteristics.

The following is a list of general kinds of culturally transmitted characters:

1. *Societies.* Social organization is indicated for primitive man and occurs in all cultures. Interactions in human societies are more varied and more labile than those of other social mammals, fishes, or hymenopterans. Animal societies function mainly to promote feeding, breeding, and protection from predators. Altruistic behavior in human societies extends to non-kin (also to nonhuman species), whereas in animal societies altruism is generally restricted to kin. There are dominance hierarchies in animal societies, but in humans rules of interactive behavior become formulated which differ in detail in different cultures. Such rules form the basis of moral codes.

2. *Language.* In behavior and in brain structure the most distinctive characteristic of humans is language. This serves two functions, communication and formulating concepts and abstract ideas (i.e., thought without speech). Human language differs from communication in animals, chimpanzees, and so on in that humans use language less symbolically and more cognitively. Human language is also highly flexible as to meanings. Monkeys make use of different sounds according to moods and social status; frogs and crickets have repertoires of several calls that are stereotyped and species-specific. Apes learn a limited number of signs that have symbolic meaning and can be used for communication. The genetically coded speech area of the human brain and laryngeal structures provide the capacity for language. It seems likely from archaeological evidence that the use of words and sentences for abstract as well as symbolic language characterized early human societies, and that language is a general property which is culturally transmitted. Attempts by linguists to find the most primitive language (*Ursprache*) have not revealed a common precursor of known languages. It is postulated that human language evolved from multiple sounds when specific sounds came to reduce ambiguities of meanings. Thus there is much local modification but the use of language appears to be a general cultural property based on biological capacity. The range of language may be restricted by brain structure.

3. *Technology.* An early technology was fabrication of stone tools for cutting, a culturally transmitted trait. Many animals use sticks and stones as tools for obtaining food, sticks and grass for making nests, but primitive man was unique in making stone tools by sharpening them. Man is properly called the tool-making animal. Invention of wheels, pulleys, levers, occurred relatively recently, and their use spread rapidly in prehistoric cultures.

The earliest use of fire for cooking was probably by *Homo erectus*. In the early Pleistocene, caves were much used for habitation; fire and fur clothing probably helped survival. Fire was also probably useful in protecting Pleistocene families from predator animals. Cooking may have had survival value in destroying parasites.

Another general aspect of technology is the use of nonhuman sources of energy for performing work. Early humans harnessed the energy of wind and water. Use of some sources of energy, such as fossil fuels, evolved recently.

A type of technology characteristic of recent human cultures is agriculture. Analogies have been made to ant-fungus cultures and ant-aphid symbioses, but these are stereotyped and genetically coded; it is not appropriate to compare them to human technology. Early humans harvested wild grain, and later humans planted crops. The kind of grain which is cultivated varies from society to society, but the general activity of agriculture is common to most societies.

4. *Arts.* Drawing and painting appeared in early Pleistocene and spread to all regions inhabited by humans. The paintings which have persisted to the present were mostly in caves. Body decoration with paints, beads, and jewels is virtually universal. Rhythmic sounds made by instruments, tuneful or not, are used in all present-day cultures in entertainment and rituals. Dances often accompany sounds and serve numerous functions. The specific forms of art, music, and dance vary locally and are acquired by children. The general practices of art, painting, music, and dance have been selected in all cultures. This general property of culture is the basis for esthetics and clearly separates human from nonhuman primates.

5. *Religion and animism.* Religion is used in the restricted sense that whenever man was unable to explain natural phenomena by what he could sense he attributed them to gods. There was also belief in magic, for instance, in the curative power of objects. A second aspect of religion is the nearly universal provision for the dead. Animism may have evolved in an effort to explain death. Burial rites vary in different cultures, but all have in common the idea of continuity after death, and this belief may assuage fear of death. The earliest burial sites date from 20–30 kyr. A broader definition of religion includes all of the mores and rules of a society which are socially transmitted.

6. *Taboos.* A taboo found in all cultures is that against incest, sex with close relatives. This probably evolved in early human societies and is transmitted culturally from generation to generation. A psychological explanation of this taboo is that it is derived from attitudes toward male and female roles in society. An evolutionary explanation is that outbreeding increases genetic diversity. Many societies have also a taboo against cannibalism.

7. *Division of labor.* In all cultures there is division of labor between the sexes. Hunting has been primarily a male activity; harvesting grain and

processing food a female activity. This division of labor is based on the biological function of women to produce and care for offspring. Societies of hunters-gatherers preceded the practice of agriculture. In human societies division of labor is labile and dependent on the social organization.

8. *Population dispersal.*   Human societies are characterized by a tendency for movement from one locality to another. Beginning with *Homo erectus,* human populations spread throughout the habitable world. This general cultural trait led to continuity of gene flow and to early occurrence of common cultural characteristics. Cultural changes encouraged dispersal. In some recent cultures the tendency for population spread has not continued. Some animals, especially birds, go on long migrations annually, but these are seasonal and the migrants return to the original regions for breeding.

The preceding list includes cultural characters common to most human societies. Some biologists have suggested that certain of these characters are genetically transmitted—grammatical language, religion in the sense of belief in magic. The lability of all of the listed characters is such that they seem more likely to have been culturally transmitted. Genetic inheritance provides the *capacity* to process culturally transmitted general characters. It seems likely that the general characters are part of the culturetype, and that specific societal characters, such as specific languages, are transmitted culturally.

In summary, the genome specifies the neural connections that are made during development. Genetically inherited behavior is relatively rigid; the cultural inheritance provides the plasticity of human nature. General cultural characters are found in all societies and probably appeared early in the evolution of Homo. The details of culture—specifics of instrument making, kind of language, artistic forms—vary from society to society, and an individual human can change from one specific form to another.

NEURAL BASIS OF HUMAN BEHAVIOR

Properties of the brain allow coadaptations of genetic and cultural inheritance. The genotype codes the templates within which developmental processes make neuronal connections. The genetically and developmentally determined neural patterns provide the substrate on which culture is transmitted. What structural properties in the brain allow for the great functional differences between man and pongids and between anthropoids and other mammals? Several structural differences are as follows (60):

1.   Brain size in many kinds of vertebrates is a function of body weight (cephalization coefficient) and follows an exponential relation, brain volume increasing as the 2/3 power of body mass. The slope for present-day animals

is similar to that for fossil forms as estimated from casts of cranial capacity. However, the position of the curve for mammals is higher than for birds and both curves are above those for reptiles and fishes of comparable body size. When cephalization quotient is plotted against evolutionary time, dolphins and humans show the largest increase of all mammals (87). Large relative size of brain may correlate with properties other than intelligence.

Data for brain volume of primates are (5):

	Brain Volume (ml)	Body Weight (g)	Cephalic Index ($\times 10^2$)
Human	1,330	65,000	2
Chimpanzee	420	46,000	0.9
Gorilla	465	165,000	0.2

The convoluted cerebral cortex provides a large surface area for synaptic interaction; the surface area of the cortex of several species is: mouse 4 cm^2, Homo 2000 cm^2, chimpanzee 800 cm^2, dolphin 3000 cm^2. An increase in gyri and depth of sulci results in the enlargement of surface area. The volume of neocortex in proportion to the volume in an insectivore taken as unity is: Homo 156$\times$, chimpanzee 58$\times$, gorilla 32$\times$, Miopithecus 60$\times$, lemur 20$\times$. The size of striatum is second only to neocortex in the series (157). The ratio of primary projection areas to association areas decreases through the series of primates. The thalamus of man contains more neurons than thalamus of ape, especially in ventrolateral and ventrobasal nuclei; the lateral geniculate correlates more with cortical size than does medial geniculate (157). In all primates, but especially in man, the asymmetry of the two sides of the cortex appears to be more pronounced than in nonprimates.

2.   The most important differences of the human brain from pongids and monkeys are in the speech area, which consists of both motor laryngeal (phonation) and associative auditory areas (Fig. 10-16). The speech area extends from part of the primary auditory area of temporal lobe (areas 41 and 42, Wernicke's area in Brodman classification) forward to a large motor region (areas 44 and 45, Broca's areas). Broca's areas constitute the posterior third of the inferior frontal gyrus anterior to the head and neck motor regions. A third language area integrates sounds with speech. Lesion of the speech area of the left hemisphere causes aphasia. Broca's area of the left hemisphere connects with both auditory and visual cortex. Speech requires interaction between temporal and frontal cortex. Chimpanzees make little use of vocal communication but can use manual gestures, and it is probable that primitive hominids used both gestures and vocalization for communication.

3.   Small neurons in cerebrum and in some other brain regions, especially cerebellum, are significantly more numerous in man and in those primates which are capable of complex symbolic behavior. The human dentate gyrus has many more type II granule cells than this region in chimpanzee

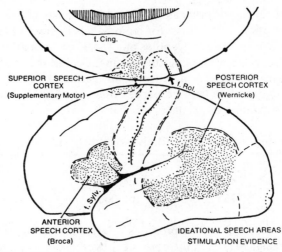

Figure 10-16. Map of human cerebral cortex showing areas in dominant hemisphere from which speech has been obtained on electrical stimulation. From W. Penfield and L. Roberts, *Speech and Brain Mechanism.* Copyright © 1959 by Princeton University Press. Figure reprinted with permission of the literary executors of the Penfield Papers and Princeton University Press.

brain. The percentage of pyramidal cells and fusiform cells in visual and motor cortex of monkey and cat is similar, but stellate cells in monkey are twice as abundant as in cat (113). In cerebellum the ratios of granule cells to Purkinje cells are 100 in frog, 950 in monkeys, and 2500 in human (85). Some electric fishes have high ratios of small to large neurons (19). Relative abundance of small neurons varies so much that generalizations are not possible.

In addition to the greater numbers of small neurons, there are more cortical layers in primates than in nonprimate mammals. It is of interest that in the octopus (the most intelligent of invertebrates), the highest integrative center of the brain, the verticalis complex, has predominantly small neurons.

4.   The crudely estimated information capacity of the brain increases with allometric increase in brain size, cortical surface area, and abundance of small neurons. Species differences reside more in networks than in qualitative differences in brain regions. The exponential increases in capacity for information processing with number of elements is well known for computers, in which the storage capacity increases nonlinearly with increase in number of elements (Ch. 1). As information capacity increases, properties of the neural network emerge that are far more complex than the properties in those networks with few elements. The coding properties of the cortex depend on the number of synapses that can be increased by enriched experience.

5.   Variation in embryonic nervous systems is greater than in adult ner-

vous systems. In most nervous systems many more neuroblasts are formed than mature as neurons; some 50 to 70 percent of embryonic nerve cells degenerate. Neural circuits develop at critical times and the determination of which cells are retained may depend on extrinsic factors or on appropriate connections for reflex activity. In a monkey, all visual neurons are established two months before birth; columns in the cortex are established by three weeks before birth (132). Thus neuronal networks are formed before they are used. The long period of maturation in humans permits some synapses to be added and others subtracted. Synaptic density in cortical area 17 increases to peak at 8–11 months of age (84). Columnar organization in visual cortex becomes modified in kittens if the eyelids are sewn shut at birth and opened a few weeks later. Transplantation of limbs in amphibian embryos can induce formation of spinal motorneurons (hyperplasia) in the recipient region and reduction in number of motorneurons in the region from which the limb bud was removed (hypoplasia) (121). During development, changes occur in sensitivity to nerve growth factor and to neurotransmitters.

In addition to cell degeneration during development, selective elimination of neuronal collaterals occurs. Staining by retrogradely transported dyes shows that in newborn rats, axons of corpus callosum are widely dispersed in parietal cortex. Similarly, in neonates collaterals of pyramidal neurons are distributed throughout layer V of neocortex. By 15 days after birth the collaterals of corpus callosum and by 48 days collaterals of pyramidals have become restricted to narrow areas (120,154).

It is postulated that there may be more selection in nervous systems of embryos than of adults (32,118). Ontogenetic selection may be for delayed properties, functions which appear after maturation. Unless a primate has developed a certain network as an embryo, it may not survive in behavioral competition.

6. Sexual dimorphism in brains of anthropoids differs markedly from that in brains of other mammals. Differences in the role of the hypothalamus in triggering the synthesis and liberation of reproductive hormones alter the ages of maturity and of senility of males and females. Brain weight is 100–200 g greater in human males than in females of the same body size. This relation also holds for greater apes and rats but not for hamsters. Also, the shape of the corpus callosum differs in the two sexes. Human and ape females show no oestrus period and can be sexually receptive at all stages of an ovarian cycle. Differences in the neuroendocrine system result in an extended maturation time, providing an opportunity for the social conditioning important in human societies.

7. Size of the olfactory bulb is relatively small in higher primates compared to its size in primitive primates, carnivores, and rodents. The olfactory bulb constitutes 17 percent of forebrain in a primitive insectivore, 4 percent in a monkey (157). Use of olfaction has become diminished in apes and humans.

8.   Humans are capable of more diverse facial expressions than other primates. Facial muscles and their controlling motor nuclei are more extensive in humans.

In summary, there are general correlations between cortical structure and behavioral complexity. However, brains of different vertebrate classes and species show so much variation that correlations of size and structure with general intelligence have been denied (105). Unfortunately, capacity for conditioning is a gross measure of intelligence. Much more specific behavioral and neurohistological measurements are likely to show adaptive correlations.

HUMAN NATURE

The preceding sections summarize some of the principles of ethology and give examples of neural mechanisms of complex behavior. Simple behavior and its underlying molecular mechanisms will be presented in the following three chapters. An important, all-inclusive viewpoint is that of sociobiology, which attempts to bridge the gap between the biological and social sciences. To what extent can ethological principles be applied to human societies?

The basic premises of sociobiology are that behavior, like all physiological characters, has genetic, developmental, and environmental bases, and that behavior patterns are subject to natural selection according to their adaptive value. Selection may be of social units rather than of individuals. Interactions of humans and of other social animals are, therefore, determined by their biological history; survival of behavior patterns, such as mate selection and reactions to kin and non-kin, optimizes success of the species. Human social phenomena are expressions of behavioral dispositions and emotions. The preceding sections give evidence that behavior is highly adaptive for the evolution of animals, including humans.

How the *genotype* determines behavior and sets limits on behavioral plasticity is far from understood; neuronal connections and motor patterns are dependent upon specific genes. In parallel with genetic determination via neural structures, there occurs cultural or social transmission of behavior patterns. *Culturetypes* are constellations of behavioral characters that are not transmitted genetically but are transmitted culturally. Evidence was presented previously for both the genetic and social determination for such characteristics as bird songs. In humans the contribution of culturetypes is much greater than in any other species. How much of human behavior is genetically programmed, how much is developmentally determined, and how much is culturally transmitted is difficult to estimate. General traits which are culturally transmitted in virtually all societies have been compared with specific traits of local cultures earlier in this chapter.

The units of culturally transmitted characters are called *culturegenes* or *memes* (memory units) by analogy with genes (37,174,175). The term *meme* has not been favorably received by social scientists, who contend that each cultural character has meaning only in a particular context; the term *meme* is used here to refer to culturally transmitted characters of a general nature. One can speak of multiple memes for components of language, of art, of religion and animism, of technologies, by analogy with genes for enzyme pathways. No mechanistic similarities between memes and genes are implied. Dawkins (37) notes that after death both genes and memes continue on, genes in the reproductive line, memes in cultural transmission. Genes are active agents working for their own survival in a host organism. Memes are self-perpetuating because of their cultural and sociological impact. Genes survive and reproduce within a host animal; memes are carried and expanded within a society. Memes are transmitted by social communication; as culturetypes, customs, and beliefs they are transmitted from brain to brain, often by nonverbal means. The theory of biological evolution has been applied to cultural evolution, but the mechanisms of transmission and selection are so different that similarities are more formal than real. Some modes of cultural transmission are relatively rapid, for example, formal education or television; others are slow, from one segment of society to another; still others are intermediate, from parent to child (27).

It must be emphasized that *general* cultural traits constitute heritages common to all human societies. *Specific* behavior patterns are restricted to individual societies. Specialized patterns are very labile, general cultural patterns less labile, and genetically coded behavior patterns are stereotyped. Selection of genetic variants of behavior is speeded up or slowed down by their survival value in a social environment. In this way cultural inheritance influences the rate of biological evolution. A summary diagram of adaptive limits at different levels of biological organization is given in Figure 10-17. The genetically determined limits for a function (resistance adaptation) are wide for molecules, intermediate for cells and tissues, and narrow for integrated whole organisms (Ch. 1). Cultural inheritance reverses the genetic trend and widens survival limits; the adaptiveness of societies is greater than that of individuals.

A number of biological properties of humans correlate with their social inheritance, and provide the basis for applying the principles of sociobiology to human societies (2):

1. The long maturation time of humans has been selected because of the biological value of social determination. Prolonged development permits extensive information storage and neural modification.
2. Homo is now one species, and gene exchange can take place among all populations. Both gene and meme exchange are restricted as long as societies maintain some separation.
3. Cultural transmission is more rapid than genetic; evolution of culture

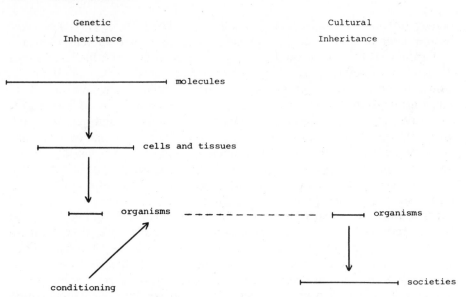

Figure 10-17.   Diagram of ranges of genetically transmitted functional limits of molecules, cells and tissues, and organisms, showing decreasing ranges with integration, and of culturally transmitted functional ranges of organisms and societies showing increasing ranges in more complex systems.

    patterns occurred in tens of thousands of years rather than in the millions of years required for genetic separation of hominids.

4.   Nonhuman animals tend to show altruism only within a kin group; humans extend altruistic behavior to non-kin (175).

5.   The most important biologically developed property of man is the presence of language centers in the association areas of the cerebral cortex which serve both speech and abstraction aspects of language. The association areas in human brain represent a quantitative (probably not qualitative) major step in primate evolution. Language has both symbolic and cognitive functions, and is associated with the capacity for abstract thought.

6.   Pair bonding is a characteristic of humans and of some other mammals and of some birds. Sex activity without reproduction functions for bonding. Whether or not there is genetic determination for aggressive behavior, grammar, and religion (174) is not agreed upon.

    Many social scientists object to applying the principles of sociobiology (136) to man. They contend that in human societies (1) reproductive units may be more socially than biologically determined; (2) mates are not selected for strength or weakness in reproductive capacity (Mates must not be blood relations; taboos ban incest.); (3) behavior is more affected by societal values than by biological values (Culture frees humanity from some of the restrictions of biological emotions and motivations.); (4) thoughts, beliefs, and culture patterns are epiphenomena, properties emergent from neural systems (Humans have "escaped" from biological constraints.); (5) the concepts of social science and philosophy are qualitatively different from the concepts of biology and must be described by a different language (Inquiry into thought and culture are separate from inquiry into the physical properties of life.); and (6) language-culture patterns limit social evolution and may eventually lead to the decline and demise of Homo; thus the species may be an evolutionary failure because of cultural, not biological weaknesses.

An alternative to the positions of the social and biological sciences has been provided by writers such as the psychologist Campbell (259) and the scientific theologian Burhoe (22). They maintain that man represents a symbiosis of genotype with culturetype. Interaction of the two heritages provides positive reinforcement resulting in social stability. The neural basis for the coadaptation is genetically coded and was probably selected under social pressures. Under societal influences the brain becomes somewhat independent of its genes.

The dual nature of man is reflected in the two types of inheritance and the two types of information—genetic and cultural. Memes persist through generations over the time spans of cultural evolution. The genes code for neural structures—the association and language areas, the synaptic capacities for memory and endogenous behavior; thus the culturetype requires a genetically determined brain. From the unified but symbiotic nature of man properties emerge which require the combination of genotype and sociotype. Natural selection acts on both adaptive genes and on societies. Such a view of humans is a dualism with a sound biological basis. It resolves many philosophical conflicts and avoids vitalistic implications.

GENERAL SUMMARY

Behavior is of primary importance in animal evolution. Behavior is not subject to molecular analysis, consequently it must be considered in holistic terms. Animals exhibit behavior that is adaptive to both the physical and biotic components of the environment. The determinants of behavior are (1) genetic (innate), (2) developmental (maturational change at critical stages), and (3) conditioning (short-term and long-term learning). The genome of an animal encodes its neural connections; development and experience act within the limits of the neural patterns. The relative importance of genotype, development, and experience differs according to an animal's behavior and life style.

The simplest behaviors are kineses and taxes, direct responses to stimuli, demonstrated as genetically programmed by comparing normal individuals with mutants. Simple reactions can be modified by experience—by habituation, facilitation, and classical conditioning. More complicated behavior involves communication, interactions between individuals of the same species, and defense of territory—often learned from a conspecific at a critical stage of development. Communication takes place by exchange of information between individuals, usually of the same species, and is based on a common language.

Direct responses, conditioned behavior, and communication underlie social behavior within families, colonies, demes. Individual actions establish territory and perform altruistic acts, defense, and aggression. An altruist may lose its own life and thereby favor reproduction and continuity of the society or species. Aggression toward predators or prey is usually different from aggression toward conspecifics.

Different kinds and patterns of behavior are adaptive for evolution and speciation. Natural selection favors adaptive behaviors which promote survival of certain genotypes, reproductive units, or societies. Adaptive behavior furthers exploitation of physical and biotic resources; genetic variants which extend ecological ranges eventually become fixed in the genomes. There are numerous species which differ little in morphology but which are isolated reproductively by behavior differences.

Language in its broadest sense is the medium of communication, usually to conspecifics, rarely, as in threats, to other kinds of animal. Methods of communication are visual signs, sounds, electric discharges, chemical pheromones. In some insects, amphibians, birds, and mammals, vocalization and hearing provide the principal language of communication. Vocal calls by many kinds of animals (crickets, cicadas, frogs, songbirds) are stereotyped and genetically based. Vocalization, hearing, and locomotion are integrated within a nervous system; how all these precisely timed and patterned components coevolved is not understood. Bird songs are patterned partly by the genetic template, but also by hearing conspecific songs and by hearing the bird's own vocalization. Bats use stereotyped calls for echolocation and for communication between neonates and adults; both vocalization and hearing mechanisms are tuned to certain frequencies and sound patterns. Cetaceans have repertoires of many calls.

In many mammals specific sounds come to be associated with actions and objects, moods and social status. Symbolic language is well developed in primates; the symbolic, syntactic, and cognitive functions of spoken language are uniquely characteristic of humans.

Electric fishes used patterned electric discharges to stun prey, to locate objects in turbid water, and to signal to other electric fish. The coordination between electric organs, receptors, and motor centers is precise.

Chemical signals, pheromones, are used by many insects as sex attractants and are species-specific. Some insects discharge repellents or toxins against predators or prey. In the coevolution of plants and insects some plant products came to be repellent to many insects, attractant to a few. Stereotyped behavior, as in Hymenoptera, may be very complex, with a minimum of learning, and may have functioned unchanged during long periods of evolution. The interaction between genetically coded neural patterns and modulation of these by conditioning of behavior such as the sequence of digging, ovipositing, and provisioning by digger wasps, are laid down in nervous systems; how they evolved remains a question in developmental and evolutionary biology.

It has not been possible to find single neurons, "pontifical" cells, which respond only to specific signals (for example, the species calls by crickets or frogs), and an organized neural network seems to serve the function. However, in the association cortex of awake monkeys there are highly specific neurons which respond to visual patterns, such as faces, or to orientation or movement of objects. Similarly, specific interneurons may well be found in other animals when techniques of recording from awake reacting animals are extended. Many more neurons are formed early in development than are retained in adult animals. The cells that persist may be those selected over many generations for use by adults. The number of synapses per neuron changes with maturation, and with experience. To translate and encode the proteins of a brain requires action of more kinds of mRNA than to code the proteins of liver or kidney. Some mRNAs are specific to brain, some are common to all tissues.

All nervous systems have features in common: convergence of inputs, parallel pathways, divergence of inputs and of sensory and integrative processing at several levels. Plasticity is a property of some synapses and neural networks (Ch. 11). Experience plays a role in much behavior; short-term memory probably differs from long-term memory in mechanism and occurrence in various animals. Human brains are capable of longer memory than brains of nonhumans. Information processing increases nonlinearly with the number of neurons, especially of small cells. Quantitative differences in behavioral capacities and brain structure between human and nonhuman primates are so great as to make humans qualitatively unique. Chimpanzees and humans are very similar in sequences of blood proteins but are very different in their behavior, especially in speech and interpretation of sounds and in cognitive functions. Different levels of complexity build on preexisting structures; the precursors of structures in the human brain occur in other mammals.

The most complex of genetic and environmental interactions occur in humans; cultural conditioning is especially important. Cultural evolution proceeds more rapidly than genetic evolution. Some general classes of cultural characters are common to all human societies and are modified in detail by individual societies. The genotype specifies the structure of a brain and the culturetype modulates the function of the brain. Human behavior depends on both genetic and cultural inheritance.

Some ethological concepts are difficult to separate from anthropomorphic interpretations: animal self-awareness, intelligence, cognition, and sensation. To understand the adaptive significance of each aspect of behavior requires knowledge of life style in relation to an animal's total natural environment and of the animal's ability to sense and respond to environmental change.

REFERENCES

1. Alcock, J. *Am. Sci.* **68**:146–153, 1980. Mating systems of solitary bees.
2. Alcock, J. *Animal Behavior,* 3rd ed., Sinauer, Sunderland, MA, 1984.
3. Andersson, M. *Nature* **299**:818–820, 1982. Choice of males by female widowbirds by length of tail.
4. Andrews, R. V. *Compar. Biochem. Physiol.* **63A**:1–6, 1979. Physiology of crowding.
5. Armstrong, E. and D. Falk, Eds. *Primate Brain Evolution,* Plenum, New York, 1982.
6. Baker, M. C. and L. R. Mewaldt. *Ecology* **32**:712–722, 1978. *Evolution* **29**:226–241, 1975. Song dialects as barriers to dispersal in white-crowned sparrows.
7. Bastian, J. and W. Heiligenberg. *J. Compar. Physiol.* **136**:135–152, 1980. Neural correlates of JAR in electric fish Eigenmannia.
8. Bentley, D. *J. Compar. Physiol.* **116**:30–38, 1977. Song patterns, crickets.
9. Blair, W. F. *Am. Nat.* **92**:27–51, 1958. Mating calls in speciation of anurans.
10. Blakemore, R. P. and R. B. Frankel. *Science* **190**:377–379, 1975; **203**:1355–1356, 1979; **212**:1269–1270, 1981. *Nature* **280**:384–385, 1980. Magnetic orientation in bacteria.
11. Bodznick, D. *Compar. Physiol.* **127**:139–166, 1978. Olfactory stimuli in migration of salmon.
12. Boeck, J. *J. Compar. Physiol.* **90**:183–205, 1974. pp. 239–245 in *Olfaction and Taste,* vol. 5, Ed. D. Denton, Academic, New York, 1975. Coding, odor quality, olfactory pathways, insects.
13. Breakenfield, X. O., Ed. *Neurogenetics.* Elsevier, New York, 1979.
14. Brines, M. L. and J. L. Gould. *J. Exper. Biol.* **96**:69–91, 1982. Sky light polarization patterns and orientation of insects.
15. Broadbent, D. E. pp. 31–41 in *Neurosciences 3rd Study Program,* MIT Press, Cambridge, MA, 1974. Laterality in man.
16. Brodal, A. *Neurological Anatomy,* 3rd ed., Oxford University Press, New York, 1981.
17. Brown, P. E. et al. *J. Compar. Physiol.* **126**:167–182, 1978. Development of hearing, infant bats.
18. Buck, J. et al. *J. Compar. Physiol.* **144**:277–286, 1981. Flashing rhythms in fireflies.
19. Bullock, T. H. et al. *Brain Res. Rev.* **6**:25–46, 1983. Phylogeny of electroreception.
20. Bullock, T. H. et al. *Proc. Natl. Acad. Sci.* **54**:422–429, 1965. *J. Compar. Physiol.* **77**:1–48, 1972. Sensory coding in electric behavior in fishes.

21. Bullock, T. H. and S. H. Ridgway. *J. Neurobiol.* **3**:79–99, 1972. Central nervous responses of porpoises to sound.

22. Burhoe, R. W. *Toward a Scientific Theology,* Christian Journals, Ottawa, 1981. *Zygon* **11**:156–162, 1976.

23. Bygott, J. D. et al. *Nature* **282**:837–846, 1979. Behavior of male lions.

24. Cabanac, M. *Q. Rev. Biol.* **54**:1–29, 1979. Sensory pleasure.

25. Camhi, J. N. *Neuroethology.* Sinauer, Sunderland, MA, 1984.

25a. Campbell, D. T. *Am. Psychol.* Dec. 1975:1103–1126; *Zygon* **11**:167–208, 1976. Biological and social evolution.

26. Capranica, R. *Vocalization and Hearing of Frogs,* M. I. T. Press, Cambridge, MA, 1965.

27. Cavalli-Spose, L. et al. *Science* **218**:19–27, 1982. Rate of transmission of culture.

28. Chaudhari, N. and W. Hahn. *Science* **200**:924–928, 1983; **220**:824–828, 1983. Genetic expression of brain.

29. Chikaraishi, D. M. *Biochemistry* **18**:3249–3256, 1979. Cytoplasmic poly-A$^+$ and poly-A$^-$ RNA in rat brain.

30. Cohen, J. *J. Compar. Physiol. Psychol.* **95**:512–528, 1981. Hormones and midbrain control of behavior in male ring dove.

31. Comer, C. and P. Grobstein. *J. Compar. Physiol.* **142**:151–160, 1981. Midbrain function in tactually and visually elicited behavior in Rana pipiens.

32. Cowan, W. M. pp. 59–79 in *Neurosciences 4th Study Program* M.I.T. Press, Cambridge, MA, 1979. Neuronal selection in developing nervous systems.

33. Creutzfeldt, O. *Prog. Brain Res.* **45**:451–462, 1976. Columnar organization of cerebral cortex.

34. Crews, D. and W. R. Garstka. *Sci. Am.* **247**:159–168, 1982. Physiology of reproductive behavior in garter snake.

35. Cuenod, M. pp. 21–29 in *Neurosciences 3rd Study Program,* M. I. T. Press, Cambridge, MA, 1984. Interhemispheric transfer in pigeon.

36. Damasi, A. R. and N. Geschwind. *Annu. Rev. Neurosci.* **7**:127–147, 1984. Neural basis of language.

37. Dawkins, R. *The Selfish Gene,* Oxford University Press, New York, 1976.

38. Dean, J. *J. Compar. Physiol.* **135**:41–50, 1980. Encounters between bombardier beetles and toads.

39. Dehenberg, P. H. *Behavior and Brain Science* **4**:1–50, 1981. Hemispheric laterality and effects of early experience.

40. Demski, L. S., *Am. Zool.* **24**:809–830, 1984. Evolution of brain regions, fish reproduction.

41. Denel and Dunlap. *Arch. Neurol.* **37**:217–221, 1980. Hand preference in rhesus monkey.

42. Dethier, V. *The Hungry Fly.* Harvard University Press, Cambridge, MA, 1976.

43. Diamond, J. *Nature* **297**:99–102, 1982. Behavior of bowerbirds. *Nature* **301**:288–289, 1983. Mound building birds.

44. Dingle, H. *Animal Behav.* **17**:561–575, 1969. Aggressive behavior in mantis shrimp.

45. Distel, H. and H. Hallander. *J. Compar. Neurol.* **192:**505–518, 1980. Subcortical projections in fetal rabbits.

46. Doty, R. L. *Mammalian Olfaction in Reproductive Behavior,* Academic Press, New York, 1976.

47. Ebbesson, S. O. E. *Cell Tissue Res.* **213:**179–212, 1980. Parcellation theory, brain organization, evolutionary and ontogenetic development.

48. Eisner, T. et al. *Science* **164:**1170–1171, 1174–1175, 1969. Queen butterfly sex pheromones.

49. Esch, H. et al. *J. Compar. Physiol.* **137:**27–38, 1980. Primary auditory neurons in crickets.

50. Ewert, J. P. *J. Compar. Physiol.* **126:**43–47, 1978. Selectivity of tectal neurons in toad Bufo.

51. Ewing, A. W. *Animal Behav.* **17:**555–560, 1969. Genetic basis of sound production in Drosophila.

52. Fentress, J. C. *Ann. New York Acad. Sci.* **209:**370–395, 1977. Tonic hypotheses and the patterning of behavior.

53. Fentress, J. C., Ed. *Simpler Networks and Behavior,* Sinauer, Sunderland, MA, 1976.

54. Fraenkel, G. *Science* **129:**1466–1470, 1959. *Entomol. Exper. Appl.* **12:**473–486, 1969. Chemical interactions between plants and insects.

55. Francis, L. *Biol. Bull.* **144:**64–72, 73–92, 1973; **150:**361–376, 1976. Self-protection in sea anemones.

56. Fuzessery, Z. M. and A. S. Feng. *J. Compar. Physiol.* **150:**107–119, 1983. Frequency selectivity in midbrain of Rana pipiens as basis for species mating.

57. Gardner, B. T. and R. A. Gardner. *J. Exper. Psychol.* **104:**244–267, 1975. Sentence constituents in child and chimpanzee.

58. Gerhardt, C. *J. Exper. Biol.* **74:**59–73, 1978. *Science* **199:**992–994, 1978. Mating calls of Hyla.

59. Gertz, C. Ch. 3, pp. 37–48 in *Horizons of Anthropology,* Ed. S. Tax, Aldine, Chicago, 1964. Homo erectus.

60. Geschwind, N. *Sci. Am.* **241:**180–199, 1979. Specialization of human brain.

61. Gilbert, C. D. *Annu. Rev. Neurosci.* **6:**217–247, 1983. Microcircuitry of visual cortex.

62. Gould, E. *Am. Zool.* **19:**481–491, 1979. Bat neonatal calls.

63. Gould, J. L. *Nature* **252:**300–301, 1974. *Q. Rev. Biol.* **51:**211–244, 1976. *Science* **189:**685–693, 1975; **214:**1041–1042, 1981. Communication and orientation of bees.

64. Gould, J. L. et al. *J. Exper. Biol.* **86:**1–8, 1980. *Nature* **296:**205–211, 1982. Orientation and magnetic detection by birds.

65. Greenough, W. T. Ch. 16 in *Neural Mechanisms of Learning and Memory,* Ed. Rosenzweig and Bennett, M. I. T. Press, Cambridge, MA, 1976. *Trends Neurosci.* **7:**229–233, 1984. Ch. 4, pp. 69–91 in *Continuities and Discontinuities in Development,* Ed. R. Emde and R. Harman, Plenum, New York, 1984. Enduring effects on brain histology of experience and training.

66. Griffin, D. R. *Listening in the Dark,* Yale University Press, New Haven, CT, 1958.

67. Griffin, D. R. *The Question of Animal Awareness,* Rockefeller University Press, New York, 1976. *Am. Sci.* **72:**456–464, 1985. Animal thinking.

68. Gross, C. G. et al. *J. Neurophysiol.* **35:**96–111, 1972. Visual properties of units in inferotemporal cortex, macaque.

69. Hall, J. C. *Bioscience* **31:**125–130, 1981. Sex behavior mutants in Drosophila.

70. Hamilton, W. D. *J. Theoret. Biol.* **7:**17–52, 1964; **31:**295–311, 1971. *Annu. Rev. Evol. Syst.* **3:**193–232, 1971. Evolutionary function of herding and altruism.

71. Hegner, R. E. et al. *Nature* **298:**264–266, 1982. Behavior of African bee-eaters.

72. Heiligenberg, W. *Naturwissenschaften* **67:**499–507, 1980. *J. Compar. Physiol.* **136:**115–152, 1980. Jamming avoidance responses, electric fishes.

73. Herman, L. M., Ed. *Cetacean Behavior,* Wiley, New York, 1980.

74. Hinde, R. A. *Animal Behavior. A Synthesis of Ethology and Comparative Physiology,* McGraw-Hill, New York, 1970.

75. Holldobler, B. pp. 418–471 in *How Animals Communicate,* Ed. T. A. Sebeok, Indiana University Press, Bloomington, 1977. Pheromones and communication in insects.

76. Holloway, R. L. and D. Post, pp. 57–76 in *Primate Brain Evolution,* Eds. E. Armstrong and D. Falk, Plenum, New York, 1982. Relative brain sizes in species of primates.

77. Hopkins, C. *Behavior* **50:**270–305, 1974. *Am. Sci.* **62:**426–437, 1974. Ch. 13, pp. 263–289 in *How Animals Communicate,* Ed. T. A. Sebeok, Indiana University Press, Bloomington, 1977. *Science* **212:**85–87, 1981. Electrical communication between electric fish.

78. Horn, G. and R. Hinde. *Short-Term Changes in Neural Activity and Behavior,* Cambridge University Press, New York, 1970.

79. Hoy, R. P. et al. *Science* **180:**82–83, 1973; **195:**82–84, 1977. *Am. Zool.* **14:**1067–1080, 1974. Genetic control of song in crickets.

80. Hubel, D. G. *Nature* **299:**515–524, 1982. The primary visual cortex.

81. Hubel, D. and T. Wiesel. *J. Physiol.* **195:**215–243, 1968. *Proc. R. Soc. London* **B198:**1–59, 1977. Organization of monkey striate cortex. *J. Compar. Neurol.* **158:**267–293, 1974; **177:**361–380, 1978.

82. Huber, F. pp. 15–66 in *Rheinisch-Westfalisch Akad. Wissenschaften* **265:**1977. Song generation in crickets.

83. Hunt, J. McV. *Intelligence and Experience,* Ronald Press, New York, 1961. Ch. 8, pp. 169–202 in *Advances in Intrinsic Motivation and Aesthetics,* Ed. H. I. Day, Plenum, New York, 1981.

84. Huttenlocher, P. R. et al. *Brain Res.* **163:**195–205, 1979. *Neurosci. Lett.* **33:**247–252, 1982. Elimination of synapses during human development.

85. Jacobson, M. pp. 147–151 in *Golgi Centennial Volume,* Ed. M. Santini, Raven Press, New York, 1975. Cell types in mammalian brain.

86. Jenkins, B. F. *Animal Behav.* **25:**50–78, 1977. Dialects of songs in passerine birds in islands near New Zealand.

87. Jerison, N. J. *Evolution of Brain,* Academic Press, New York, 1973. *Sci. Am.* **234:**90–101, 1976. Paleoneurology and evolution of mind.

88. Kalmijn, A. *IEEE Trans. Magnetics* **17:**1113–1124, 1981. *Science* **218:**916–918, 1982. Detection of magnetic and electric fields.

89. Kalmring, K. and R. Kuhne. *J. Compar. Physiol.* **139:**267–275, 1980. Auditory and vibration senses in a grasshopper.

90. Karlin I. D. et al. *Disorders of Speech,* C. C. Thomas, Springfield, IL, 1977.

91. Kay, R. F. *Proc. Natl. Acad. Sci.* **79:**209–212, 1982. Sexual dimorphism in Miocene hominids.

92. Kellogg, W. N. *Porpoises and Sonar,* University of Chicago Press, Chicago, 1961.

93. Kilbert, K. *J. Compar. Physiol.* **132:**11–25, 1979. Sound analysis in dancing bees under different magnetic fields.

94. King, M. and A. C. Wilson. *Science* **188:**107–116, 1975. Biochemical evolution in humans and chimpanzees.

95. Konishi, M. Ch. 9, pp. 289–309 in *Handbook of Sensory Physiology,* vol. VIII, Ed. R. Held et al., Springer, Heidelberg, 1978. Ch. 4, pp. 105–118 in *Perception and Experience,* Ed. R. D. Walk and H. L. Pick, Plenum, New York, 1978. *Annu. Rev. Neurosci.* **8:**125–170, 1985. Ethological aspects of auditory pattern recognition mostly, in birds.

96. Konishi, M. and E. I. Knudsen. *Science* **204:**425–427, 1979. Hearing in oilbirds.

97. Kubota, K. and H. Niki. *J. Neurophysiol.* **34:**337–347, 1971. Prefrontal cortical unit activity in monkeys.

98. Kung, A. et al. *Science* **188:**898–904, 1975. Behavioral mutants, Paramecium.

99. Kuterbach, D. B. et al. *Science* **218:**696–697, 1982. Iron-containing cells in bees.

99a. Lindauer, M. *Naturwissenschaften* **57:**463–467, 1970. Learning and memory in honeybees.

100. Lissman, H. W. *J. Exper. Biol.* **35:**451–486, 1958; **37:**801–811, 1960. Electrolocation by African electric fish.

101. Lorenz, K. *Studies in Animal and Human Behavior,* Harvard University Press, Cambridge, MA, 1971.

102. Lumsden, C. J. and E. O. Wilson. *Behav. Brain Sci.* **5:**1–37, 1982. Precis of genes, mind and culture. *Proc. Natl. Acad. Sci.* **77:**6248–6250, 1980.

103. Lyons, D. N. and E. Banks. *Develop. Psychobiol.* **15:**455–460, 1982. Ultrasounds in neonatal rats.

104. MacDonald, D. W. *Nature* **301:**379–384, 1983. Ecology of carnivore social behavior.

105. MacPhail, E. M. *Brain and Intelligence in Vertebrates,* Clarendon Press, Oxford, 1982.

106. Malpeli, J. *J. Neurophysiol.* **46:**1102–1119, 1981; **49:**595–610, 1983. Responses of area 17 cells in cat in absence of input from layer A of LGN.

107. Marler, P. R. *Behaviour* **11:**13–39, 1957. pp. 348–367 in *Animal Sound and Communication,* Ed. W. E. Lanyon and W. N. Tavolga, American Institute of Biological Science, Washington, DC, 1960; *Science* **157:**769–774, 1967. pp. 389–450 in *Ontogeny of Vertebrate Behavior,* Ed. H. Moltz, American Institute of Biological Science, Washington, DC, 1971. pp. 314–329 in *Acquisition of Bird Songs and Bird Communication,* Ed. J. C. Fentress, Sinauer, Sunderland, MA, 1976. Sensory templates in species specific behavior.

108. Martin, H. and M. Lindauer. *J. Compar. Physiol.* **122:**145–187, 1977. Effects of earth magnetic field on gravity orientation of honeybee.

109. McCasland, J. S. and M. Konishi. *Proc. Natl. Acad. Sci.* **78:**7815–7819, 1981. Avian song control nucleus in brain.

110. Metcalf, R. L. et al. *Proc. Natl. Acad. Sci.* **77:**3769–3772; **78:**4007–4010, 1981. Molecular parameters and olfaction in a fruit fly; secondary plant substances and food selection by beetles. *Bull Entomol. Soc. Am.* **25:**30–35, 1979.

111. Metcalf, R. W. *Annu. Rev. Ecol. Syst.* **3:**193–232, 1972. Social behavior of wasp Polistes.

112. Milner, M. pp. 75–89 in *Neurosciences 3rd Study Program,* M.I.T. Press, Cambridge, MA, 1974. Hemispheric specialization in brain.

113. Mitra, N. L. *J. Anat.* **89:**467–483, 1955. Cell types in mammalian cortex.

114. Mountcastle, V. B. et al. *J. Neurophysiol.* **40:**362–389, 1977. Ch. 2, pp. 21–42 in *Neurosciences 4th Study Program,* M. I. T. Press, Cambridge, MA, 1979. *J. Neurol. Sci.* **1:**1218–1235, 1981. Functions of parietal lobe in visual attention.

115. Narins, P. M. and R. R. Capranica. *Brain Behav. Evol.* **17:**48–66, 1980. Neural adaptation for two-note call of treefrog. *Science* **192:**378–380, 1976.

116. Neuweiler, G. et al. *J. Acoust. Soc. Am.* **68:**641–653, 1980. *J. Compar. Physiol.* **152:**421–432, 1983. Species differences in sonar behavior of bats.

117. Niki, H. *Brain Res.* **68:**185–301, 1974; **70:**346–349, 1974. Prefrontal unit responses during right and left delayed responses in monkeys.

118. Northcutt, R. G. *Annu. Rev. Neurosci.* **4:**301–350, 1981. Evolution of telencephalon in nonmammals.

119. Nottebohn, F. *J. Compar. Neurol.* **165:**457–486, 1976. *J. Exp. Zool.* **177:**229–262, 1971. *Science* **214:**1369–1370, 1980. *J. Comp. Physiol.* **108:**171–192, 1976. Neural templates of song patterns in hemispheric dominance in birds.

120. O'Leary, D. D., B. Stanfield and W. M. Cowan. *Develop. Brain. Res.* **1:**607–617, 1981. Changes in corpus callosum during development.

121. Oppenheim, R. W. Ch. 5, pp. 73–133 in *Studies in Developmental Neurobiology,* Ed. W. M. Cowan, Oxford Univ. Press, New York, 1981. Neuron death in development.

122. Pak, W. *Handbook of Genetics* **3:**703–733, 1975. Vision mutations, Drosophila.

123. Payne, P. *Animal Behav.* **29:**688–697, 1981. Song learning and social interaction in indigo buntings.

124. Payne, R. S. *Ann. New York Acad. Sci.* **88:**110–141, 1971. *Science* **173:**585–597, 1971. Song of whales.

125. Payne, R. S. and K. Roeder. *J. Exper. Biol.* **44:**17–31, 1966. Sound fields around moths with wings in different positions.

126. Perrett, D. S. et al. *Neurosci. Newslett.* Suppl 3, p. 5358, 1979. *Exper. Brain Res.* **47:**329–342, 1982. *Proc. R. Soc. London* **B223:**293–317, 1985. Temporal lobe neurons responding to faces.

127. Pilbeam, D. *Nature* **295:**232–234, 1982. Evolution of hominids.

128. Ploog, D. *Brain Res. Rev.* **3:**35–61, 1981. *Neurosciences* **2:**349–361, 1970. Neurobiology of primate audiovocal behavior.

129. Poggio, G. F. Ch. 16, pp. 497–535 in *Medical Physiology,* Ed. V. Mountcastle, Mosby, St. Louis, 1974. Divergence in visual pathways in brain.

130. Pollack, F. S. and R. R. Hoy. *Science* **204**:429–432, 1979. Species specific song recognition in crickets. *J. Compar. Physiol.* **144**:367–373, 1981.

131. Presti, D. and J. D. Pettigrew. *Nature* **285**:99–100, 1980. Magnetic sense in migrating birds.

131a. Prosser, C. L. and W. S. Hunter. *Am. J. Phys.* **117**:609–618, 1936. Extinction of spinal reflexes.

132. Rakic, P. *Trans. R. Soc. Lond.* **B287**:245–260, 1977. Prenatal development of visual system in Rhesus monkey.

133. Reissland, A. and P. Gorner. *J. Compar. Physiol.* **123**:59–69, 1978. Mechano-receptors in spiders.

134. Rheinlander, J. et al. *J. Compar. Physiol.* **110**:251–269, 1976. Central processing of sound signals in crickets.

135. Roeder, K. D. *Nerve Cells in Insect Behavior,* Harvard University Press, Cambridge, MA, 1967. *J. Insect Physiol.* **13**:873–888, 1967; **16**:1069–1086, 1970. *Science* **159**:331–333, 1978. Auditory sense and orientation of moths to bat sounds.

136. Sahlins, M. D. *Use and Abuse of Biology,* University of Michigan Press, Ann Arbor, 1976.

137. Salmon, M. et al. *Evolution* **33**:182–191, 1979. Behavioral species of fiddler crabs.

138. Sarnat, H. B. and M. G. Netsky. *Evolution of Nervous Systems,* Oxford University Press, New York, 1981.

139. Sayles, G. and D. Pye, Eds. *Ultrasonic Communication by Animals,* Chapman and Hall, London, 1974.

140. Scheich, H. *J. Compar. Physiol.* **113**:207–227, 1977. Electrobehavior in fish Eigenmannia.

141. Schneider, D. *Symp. Soc. Exp. Biol.* **20**:273–297, 1965. *Science* **165**:1031–1037, 1969. pp. 173–193 in *Sensory Physiology and Behavior,* Ed. R. Galum et al., Plenum, New York, 1975. *J. Compar. Physiol.* **97**:245–256, 1975. Pheromone communication in butterflies.

142. Schnitzler, H. J. *Compar. Physiol.* **82**:79–92, 1973. Compensation for Doppler shift in bats.

143. Scott, J. P. *Behav. Genet.* **7**:327–346, 1977. Social genetics of dogs.

144. Searcy, W. A. and P. Marler. *Science* **213**:926–928, 1981. Responsiveness to song, female sparrows.

145. Sebeok, T. A., Ed. *How Animals Communicate,* Indiana University Press, Bloomington, 1977.

146. Sefarth, A. M. and D. L. Cheney *Trends in Neuroscience* **7**:66–73, 1983. Vocalizations of nonhuman primates.

147. Sherman, M. and P. Spear *Physiol. Rev.* **62**:738–855, 1982. Neuronal properties, visual cortex.

148. Silverstein, R. M. *Science* **213**:1326–1332, 1981. Pheromones in insect pest control.

149. Simmons, J. A. et al. *Am. Sci.* **63**:204–215, 1976. Species differences in bat calls.

150. Simmons, P. and D. Young. *J. Exper. Biol.* **76**:27–45, 1978. Song production in a cicada.

151. Singer, W. *Exper. Brain Res.* **44**:431–436, 1981. Orientation or columns, visual cortex of cat.

152. Sperry, R. W. pp. 5–19 in *Neurosciences 3rd Study Program,* M.I.T. Press, Cambridge, MA, 1974. Harvey Lectures **62**:293–323, Academic Press, New York, 1968. Lateral specialization in cortical hemispheres.

153. Spitzer, N. C. and J. E. Laborghini. Ch. 12, pp. 261–287 in *Developmental Neurobiology,* Ed. W. M. Cowan, Oxford University Press, New York, 1981. Programs of connection in early neuronal development.

154. Stanfield, B. B. et al. *Nature* **298**:371–373, 1982. Selective elimination of collaterals of pyramidal neurons in young rats.

155. Steklis, H. D. and M. J. Raleigh. *Neurobiology of Social Communication in Primates,* Academic Press, New York, 1979.

156. Stent, G. S. *Annu. Rev. Neurosci.* **4**:163–194, 1981. Genetics of development of nervous system.

157. Stephan, H. and O. J. Andy. *Am. Zool.* **4**:59–74, 1964. Quantitative studies on primate brain structure; *Ann. New York Acad. Sci.* **167**:370–387, 1969.

158. Stratton, G. E. and G. W. Uetz. *Science* **234**:575–577, 1981. Species-specific sounds in wolf spiders.

159. Straughan, I. R. *Copeia* 415–424, 1975. Mating call discrimination in species of Hyla. Ch. 10, pp. 321–327 in *Evolutionary Biology of Anurans,* Ed. L. J. Vial, University of Missouri Press, Columbia, 1973.

160. Teitelbaum, P. et al. p. 127–143 in *Mechanisms of Goal-Directed Behavior,* Ed. R. Thompson et al., Academic Press, New York, 1980, pp. 357–385 in *Behavioral Models and Analysis of Drug Action,* Eds. M. Spiegelstein and A. Levi, Elsevier Amsterdam, 1982. Behavioral functions of lateral hypothalamus. *Exper. Neurol.* **77**:286–294, 1982.

161. Thomas, J. B. and R. J. Wyman. *Nature* **298**:650–651, 1982. Mutation which alters behavior in Drosophila.

162. Tinbergen, N. *The Animal in Its World.* Harvard Univ. Press, Cambridge, MA, 1972. *Social Behaviour in Animals,* Methuen, London, 1965.

163. Todt, D. *Z. vergl. Physiol.* **71**:262–285, 1971. Vocal behavior of nightingales.

164. Trenkaer, E. *Nature* **277**:566–567, 1971. Cerebellar development in mutant mice.

165. Urano, H. and A. Gorbman. *J. Compar. Phys.* **141**:163–171, 1981. Hormone effects on frog cells.

166. vanSchilcher, F. *Behav. Genet.* **7**:251–259, 1977. Mutant of courtship in Drosophila.

167. Vogler, B. and G. Neuweiler, *J. Compar. Physiol.* **152**:421–432, 1983. Sound patterns, bats.

168. Vogt, R. G. and L. Riddiford in *Peptides,* Ed. F. Bloom, North Holland, Amsterdam, 1980. Pheromones in moths.

169. Von Frisch, K. *Dance Language and Orientation of Bees,* Harvard Univ. Press, Cambridge, MA, 1967.

170. Waldman, B. and K. Adler. *Nature* **282:**611–613, 1979. Sibling identification by toad tadpoles.

171. Washburn, S. *Bull. Am. Acad. Arts Sci.* **35:**25–39, 1982. Behavior in primates. *Neurosciences 2nd Study Program,* M. I. T. Press, Cambridge, MA, 1970.

172. Wiesel, T. N. *Nature* **299:**583–591, 1982. Postnatal development of visual cortex.

173. Wilson, E. O. *Insect Societies,* Harvard University Press, Cambridge, MA, 1971.

174. Wilson, E. O. *Sociobiology,* Harvard University Press, Cambridge, MA, 1975.

175. Wilson, E. O. *On Human Nature,* Harvard University Press, Cambridge, MA, 1978.

176. Winter, P. et al. *Exper. Brain Res.* **1:**359–384, 1966. Vocal repertoire of squirrel monkey.

177. Wolpoff, M. H. *Paleo-Anthropology,* Knopf, New York, 1980.

178. Woolsey, T. A. et al. *J. Compar. Neurol.* **164:**79–94, 1975. *Somatosensory Res.* **1:**207–245, 1984. *J. Compar. Neurol.* **171:**54–560, 1977. *Brain Res.* **17:**205–242, 1970. Barrel organization in somato-sensory cortex of mouse and rat.

11

SIMPLE NERVOUS SYSTEMS; FIXED ACTION PATTERNS

MORPHOLOGY OF NERVOUS SYSTEMS

Neurobiology analyzes the neuronal basis of animal behavior in terms of (1) social interactions, (2) cognitive activity and learning, (3) integrative actions of brain function, (4) properties of so-called simple nervous systems and of reflex centers in complex nervous systems, (5) synaptic transmission and modulation, and (6) ionic mechanisms of membrane activity. Each level of neural function is based on genetic and developmental determinants modulated by environmental factors. Physiological measurements include sensory responses, central nervous interactions and motor output, synaptic transmission, and changes in ionic conductances of membranes.

One view of behavior has been that it can be described in stimulus-response terms. Forced movements are direct responses to environmental stimuli—tropisms, taxes, kineses. Reflexes are mediated by local nerve centers and may modify direct responses by repetitive discharges, facilitation, inhibition, habituation, and post-tetanic potentiation. One opinion postulates control by the environment with no independent action of the central nervous system (CNS). A contrary opinion is that nervous systems can initiate and modify behavior irrespective of sensory input. It is now clear that both intrinsic and extrinsic inputs contribute to behavior.

Patterned behavior is the outcome of interactions between neurons, for example, in a network circuit. Motor acts of animals have been shaped by natural selection, which works on action patterns, not on single neurons per se. Reflex chains are not sufficient; rather, centrally patterned chains of responses are triggered by unpatterned sensory inputs or by command neurons in regions of a nervous system. *Fixed action patterns* are motor reactions which are partly programmed (innate) and are subject to modulation by feedback. Many fixed action patterns are rhythmic. Some nonrhythmic fixed action patterns are episodic, occurring once or seldom in a lifetime, some action patterns are circadian (daily) or circannual (yearly), some are triggered by sensory input, others by endogenous rhythms (2,3). A distinction can be made between a reflex and a fixed action pattern; in the regions of nervous systems that mediate reflexes there are input-output connections, direct or via interneurons, from sense cells to motorneurons serving one or a few muscles. For fixed action patterns a ganglion or neural network co-ordinates a number of muscles to act synergistically, antagonistically, or sequentially. A relatively small number of neurons economically pattern a co-ordinated movement of one or more appendages or segments.

There are many kinds of central nervous systems generating rhythmic activity which is not paced by, but may be modulated by, extrinsic signals. Two modes of frequency control of rhythmic activity are (1) endogenous rhythms in single or electrically coupled neurons and (2) recurrent interactions between neurons in a reverberating circuit or network. Endogenous cellular rhythms are based on time- and voltage-dependent ionic currents, usually of Ca^{2+} or K^+ (Ch. 15). Network loops involve a minimum of three neurons. Three cells may constitute a self-exciting network in which accelerating impulse production in cells A and B is terminated by inhibition from C (Fig. 11-1a). In a reciprocal inhibition network, cell A is tonically (endogenously) active and drives B and C, which are interconnected and which adapt during continued firing (Fig. 11-1b). Figure 11-1c, d shows three-cell and four-cell networks with recurrent cyclic inhibition. Cells A, B, and C make reciprocal inhibitory contact; if C is in a depolarized impulse-generating state, postsynaptic cell B is hyperpolarized and inactive—its presynaptic cell A is recovering from prior inhibition. After A recovers, C becomes inhibited and this disinhibits B. When B is active it inhibits A and C; when C recovers, B is made inactive and A is recovering. Thus sequential firing of the three cells yields a continuing rhythm (53).

There are two ways of interpreting electrical measurements of central nervous activity. One view is that all nerve signals are unitary, by all-or-none impulses or by graded local potentials; these are measured as unit spikes or synaptic potentials. A second view is that variable spontaneous and evoked potentials (near field and far field potentials) of many neurons together provide significant neural information. Field potentials result from synchronized activity, which often is graded, is produced by large numbers of interacting neurons; is distinguished by latencies, wave-forms, and electrotonic spread. Evoked responses may induce behavior. Neurons within an electric field become modified in excitability.

Nervous systems can be arranged in an evolutionary series by structure. The most generalized type is a diffuse neural network within which interconnecting neurons receive sensory signals; the network controls muscles and glands of a body region. The nervous systems of cnidarians (coelenterates and ctenophores) consist of one or more nerve nets, in tentacles, oral structures, and/or mesenteries. Nerve nets connect one polyp to another in coral. In addition, in coelenterates there are aggregations of neurons into rings, marginal or circumoral, each with some autonomy. Peripheral neural networks exert local motor control (flatworms, molluscs, echinoderms) or regulate visceral functions (in arthropods, vertebrates). Nerve nets and

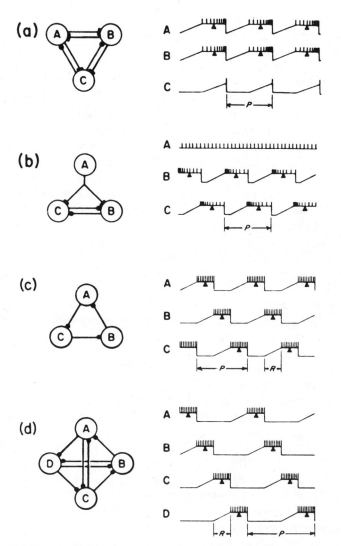

Figure 11-1. Models of rhythmic neural networks. T-bar junctions indicate excitatory, filled circles, inhibitory synaptic connections. Traces at right of each diagram represent membrane potential and impulse burst activity in individual cells. (*a*) Self-excitatory network in which accelerating impulse production in cells A and B is terminated by activation of inhibitory cell C. (*b*) Reciprocal inhibition network in which B and C are driven to produce alternating bursts of impulses by the tonically active cell A. (*c*) Three-cell and (*d*) four-cell networks with recurrent cyclic inhibition. Reprinted with permission from G. S. Stent and W. B. Kristan in *Neurobiology of the Leech*, K. J. Muller, J. G. Nicholls, and G. S. Stent, Eds. Copyright 1981, Cold Spring Harbor Laboratory, Cold Spring Harbor, New York.

nerve rings coordinate activity in radially symmetrical animals—echinoderms and cnidarians. A nerve net with diffuse conduction is especially suitable for radially symmetrical animals; tubular or ladder-type nervous systems are adapted to bilaterally symmetrical animals.

In bilaterally symmetrical animals, a more complex neural organization
occurs with ganglia condensed into chains, such as the paired segmental
ganglia of annelids or arthropods. Ganglia are joined longitudinally by connectives, transversely by commissures; innervation of each abdominal segment is by three pairs of nerves, each containing sensory and motor fibers.
The thoracic chain in crustaceans and insects results from fusion of several
of the ganglia which innervate appendages. Molluscs have several fused
paired ganglia, each serving a region of the body. Most interneurons of annelids and arthropods are unipolar, with somata (cell bodies) in a cortex
around a central neuropil of the ganglion. In the neural chains of annelids
and arthropods there are also giant neurons, some with somata in anterior
ganglia, some with somata in segments.

A further morphological specialization of the nervous system in flatworms, annelids, arthropods, and molluscs is cephalization. In these animals a cerebral ganglion (brain) is formed in the embryo by fusion of ganglia,
and lies at the anterior end and dorsal to the digestive tract. Cephalization
probably evolved very early, along with bilateral symmetry, and is associated with (1) concentration of sense organs—eyes, statocysts, chemoreceptors—in the anterior end of the animal, and (2) integration of sensory information in anterior central nervous centers.

Vertebrate animals are bilaterally symmetrical and have highly developed
cephalization. The nervous system originates as a neural tube; each segment
has a dorsal and a ventral root on the right side and also on the left side.
There are ten or twelve paired cranial nerves. Much vertebrate behavior
consists of (1) reflexes (locomotor, respiratory, and autonomic responses),
and (2) fixed action patterns (coordinated locomotion, saccadic movements
of eyes). Fixed action patterns have been described from many kinds of
animals, but are most completely analyzed for arthropods, annelids, and
molluscs. A few examples follow.

FIXED ACTION PATTERNS

Nerve Nets

In coelenterates and ctenophores two conducting systems are in parallel:
conduction over a sheet of epithelial cells (ectoderm and endoderm) and
conduction in nerve nets. In the hydrozoan jellyfish *Polyorchis* three discrete nerve networks have been described. They consist of anastomosing
cells. One net is of motorneurons which mediate swimming; the other two
nets integrate sensory information. When neurons are filled with the dye

lucifer-yellow, it spreads from cell to cell; the cells are electrically coupled. There are also chemical synapses between neurons of the different nets, and transmission from a motor nerve net to muscle is chemical. Anthozoa and Scyphozoa differ from Hydrozoa in having chemical synapses, but probably not electrical transmission (74). It has been suggested that neurons may have evolved as modified epithelial cells, but this is less evident than the modification of epithelium into muscle and sense cells. The fact that some interneuronal conduction is electrical, as it also is in hydrozoan epithelia, suggests that electrical neurotransmission preceded chemical neurotransmission.

In an effort to find a primitive "simple" nervous system, comparative neurophysiologists have examined the behavior of cnidarians—free-swimming and sessile, isolated and colonial. The axons of the nerve nets are very small, and thus only a few single-unit recordings have been possible; also, the nature of chemical transmitters in coelenterate nervous systems is unknown. However, as judged by behavior, the coelenterate nervous system has the following capabilities: specialized conduction in nerve tracts, spatial and temporal facilitation, inhibition, endogenous rhythmicity, and some plasticity. In a typical nerve net, conduction proceeds in all directions, impulses pass around corners, cross tissue bridges, and diverge beyond a narrow neck. This was demonstrated many years ago with medusae cut into bizarre patterns. Most coelenterates have two nets in parallel, one fast and the other slow; these show reciprocal innervation of muscle (47). In a hydrozoan jellyfish *Aglantha* giant neurons in the subumbrella and a single giant fiber in each tentacle mediate rapid escape swimming (21).

Coelenterate nerve nets show facilitation; for example, in some corals a wave of excitation spreads throughout a colony of polyps in response to a single localized stimulus; in some other kinds of corals many facilitating stimuli are needed for maximal spread. By blocking with agents such as magnesium, it is concluded that much of the facilitation is synaptic, and some is neuromuscular (5a).

Some nerve nets initiate spontaneous rhythmic behavior. Marginal bodies of jellyfish initiate rhythmic contractions of the bell, and if all marginal bodies but one are removed a contraction wave passes in both directions from the one remaining marginal body. A scyphozoan cubomedusa *Carybdea* has four pacemaker organs, rhopalia, of which the one with shortest recovery intervals is the pacemaker (71). Conduction systems in tentacles for swimming are in different neurons from those for feeding. Jellyfish effector systems are contractile cells (myoepithelia) and luminescent organs.

Nerve nets mediate complex behaviors which are fixed action patterns: rejection of some foods, acceptance of others, pointing of the manubrium (hydromedusae); slow body expansion and contraction, elongation, and swaying (anemones). Chemical stimulus from a starfish initiates the complex movements of swimming away by a motile anemone *Stomphia*. The anemone *Calliactis* normally lives on snail shells inhabited by hermit crabs and whenever the crab transfers to a larger shell the anemone, by a series of maneuvers, moves with the crab to the new shell (69a). Sea anemones have been conditioned by classical methods to attach to specific textures of substrate (49).

Nerve nets, the most primitive known nervous systems, show many of the complex properties of so-called "higher" nervous systems. Coelenterate behavior, highly stereotyped, shows some plasticity.

Swimmeret Motor System

In such crustaceans as crayfish, lobsters, and shrimp, abdominal ganglia set the pace of the rhythmic beat of the five segmental pairs of swimmerets. The swimmeret beat is bilaterally symmetrical in a crayfish; a metachronal wave passes from the fifth segment forward. The time for transmission forward in each ganglion is 150 msec; sequences consist of repeated swimmeret beats, each of 1.4 sec duration, for 10–20 sec, followed by a pause. Motor impulses leave the nerve cord in the first nerve of each segment; rhythmic motor discharges can still be recorded after deafferentation, and thus sensory input is not essential. The patterned discharge starts in ganglion VI, but if the connective between V and VI is cut the impulses start in nerves of segment V; or if the cut is anterior to V the discharge starts in IV. Five paired large interneurons in the thorax are command neurons; stimulation at some 30/sec increases the frequency of the neural bursts to swimmerets and can either initiate or terminate the patterned activity. No single motorneuron is an endogenous pacemaker; the oscillatory discharge is a property of the ganglionic network and some 50 motorneurons coordinate sequential contraction in 11 muscles to move the swimmeret (33).

Stomatogastric System in Crustaceans

Stomachs of crustaceans are muscular structures, driven to contract rhythmically by oscillatory discharges from the stomatogastric ganglion. The number and arrangement of neurons differ with the crab species. There are two regions of the stomach. In crayfish the anterior stomach has a gastric mill containing teeth moved by 25 muscles driven by 9 motorneurons. The pyloric stomach is driven by 14 motorneurons. The stomatogastric ganglion of the lobster *Panulirus* contains 33 neurons, 12 neurons in the gastric mill and 14 in the pyloric region, activating a total of 30 muscles (Fig. 11-2). The rhythmic activity of the gastric mill consists of (1) opening and closing of two lateral teeth and (2) power and return strokes of a single medial tooth (53a,72). Movement of the pylorus consists of alternating dilation and constriction. The neurons for each region are coupled by electrical and chemical synapses and generate separate interconnected rhythms. Pyloric bursts oc-

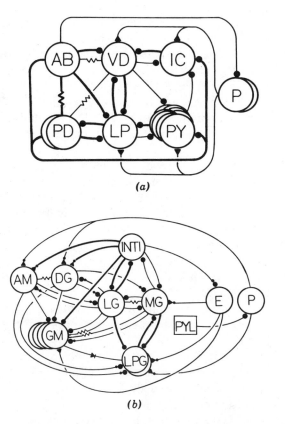

(a)

(b)

Figure 11-2. Abbreviated circuit diagrams of stomatogastric ganglia of lobster. (*a*) Pyloric and (*b*) gastric pattern generators. Inhibitory synapses indicated by solid circles, excitatory synapses by triangles, electrical junctions by resistors. Only strong synapses shown; synergistic neurons grouped. (*a*) Pyloric generator: AB and P endogenous bursters, inhibitory to other pyloric neurons, PD pyloric dilators, VD ventricular dilator, IC inferior cardiac, LP and PY pyloric neurons. P in commissural ganglion (of CNS) is general phasic excitor of network. (*b*) Gastric pattern generator: E and P in commissural ganglion are phasic excitors, INTI inhibits E, LG lateral gastric and MG median gastric close medial tooth, LPG lateral posterior gastric opens lateral teeth, DG dorsal gastric and AM anterior median, GM and AM are triggered bursters tonically active when E cells are active. P indicated as tonically acting on both circuits. Reprinted with permission from A. I. Selversten, P. Miller, and M. Wadepuhl, *Society for Experimental Biology Symposium,* Vol. 37, p. 35. Copyright 1983, Cambridge University Press, Cambridge, England.

cur at 0.5-sec intervals, gastric bursts at 3-sec intervals. The functions of neurons of the stomatogastric ganglion have been ascertained by (1) simultaneous recordings from each of several neurons, (2) photodestruction of single cells by spot illumination after staining with lucifer yellow, and (3) reversible blocking of cells by hyperpolarizing them.

Two different mechanisms generate stomatogastric rhythms. The pyloric rhythm is generated by cellular bursting and synaptic interaction (Fig. 11-

2*a*). The AB (anterior burster) cell is an endogenous burster but the other cells can be made to burst when supplied with input from extrinsic ganglia. Reciprocal inhibition, especially between the PD (pyloric dilator) and LP (lateral pyloric) cells also contributes to alternating bursting. The synapses are arranged so that when PD and AB fire, all of the other cells are inhibited. On release from inhibition, LP and IC (inferior cardiac) fire first, then cells PY (pyloric) and VD (ventricular dilator) (72). Thus all connections between the cells are inhibitory; discharge normally occurs after stepwise removal of inhibition.

Muscles of the stomach are activated by neurons in succession. The basic gastric mill rhythm appears to be due to reciprocal inhibition between the LG-MG (lateral gastric-medial gastric) pair (which close the lateral teeth) and Int-1 (interneuron 1) (Fig. 11-2*b*). This network produces alternate bursting between the two components. When the LG and MG fire, they inhibit the LPGs (lateral posterior gastrics) which open the lateral teeth. When interneuron 1 fires, it excites the DG-AM (dorsal gastric-anterior median) which resets the medial tooth and inhibits the GM (gastric mill) cells which pull the medial tooth forward. The GMs are also inhibited by the DG and AM so that these two groups fire out of phase with each other. The gastric mill network has been simulated by a computer program that fires in a pattern like the ganglion.

The mechanism of endogenous bursting in pyloric pacemakers in crustaceans consists of voltage-sensitive slow waves, much like those in bursting molluscan neurons. If the cells (PD) are hyperpolarized, spikes are stopped and sinusoidal membrane oscillations occur. Depolarizing current is carried by Na^+ and Ca^{2+} and outward currents are carried by K^+.

The gastric and pyloric rhythms are endogenous but are modulated by input from other neural centers. For example, Int-1 cells inhibit E (excitatory) cells of the commissural ganglion, which in turn excite gastric mill neurons. Coupling between pyloric and gastric mill systems results in synchrony of bursts. Command neurons in the central nervous system excite PD neurons and probably food in the stomach reflexly activates PD (Fig. 11-2B, cells E, P). No single mechanism is adequate, but one can substitute for another; there is redundancy in the total stomach system.

Swimming by Leeches

A leech is an annelid with 21 segments, each containing a double ganglion with some 175 neurons per hemiganglion; successive ganglia are connected by two lateral and one medial connective. A swimming wave is a phase-linked series of segmental oscillations, with the motorneurons activating dorsal muscles out of phase with those activating ventral muscles. Each hemiganglion receives sensory processes from three touch (*T*), two pressure

(*P*), and two noxious-substance (*N*) sense endings. The sensory neurons connect with motorneurons, groups of which are electrically coupled by contralateral processes. Touch afferents activate longitudinal motorneurons (*L*) via rectifying electrical synapses, *N* afferents activate *L* neurons by chemical synapses, *P* afferents act by combined electrical and chemical synapses. *N* and *P* receptors connect by chemical synapses to two motorneurons, one of which raises annuli, while the other shortens segments. The axons in longitudinal connectives are not symmetrical; anterior-to-posterior connections are excitatory only, posterior-to-anterior neurons are both inhibitory and excitatory (52,53).

The metachronal rhythm of swimming can be generated by several ganglia in series. The neural network in a single ganglion can generate the swimming sequence, but two adjacent ganglia are required to initiate and maintain the swim motor pattern (central pattern generator, CPG). Swim-initiating axons are in the median connective, pattern-generating axons run in the lateral interganglionic connectives. Stimulation of one of two unpaired segmentally-interconnected neurons initiates and maintains the swim motor pattern in as few as two ganglia (82). Cell 108 is an unpaired intersegmental interneuron, stimulation of which can reset the swim pattern. This cell links swim-initiating neurons and motorneurons; it is inhibited by N input (82).

Motorneurons of an isolated nerve cord can produce motor discharges corresponding to the swimming activity pattern (53a). A network of interneurons coupled with motorneurons generates the normal locomotor rhythm. Each segmental ganglion has interneurons which show polarization rhythms phaselocked with impulse bursts of motorneurons and stimulation of these interneurons shifts the phase of the burst rhythm of motorneurons. Each segmental ganglion has four pairs of oscillatory interneurons; the paired interneurons are reciprocally connected by inhibitory links. Figure 11-3 is a summary circuit diagram of identified synaptic connections between interneurons (shaded circles), motorneurons (open circles) and longitudinal muscles of two segments. Each interneuron of the oscillator makes inhibitory connections with a cell that leads it by a phase of 90° in the swim cycle. Between ganglia the three interneurons (cells 28, 33, and 27) make inhibitory connections in anterior ganglia with corresponding cells to which they connect in their own ganglion. One interneuron (cell 123) makes inhibitory connections in a posterior ganglion with the serial homolog (cell 28) that follows it (cell 123) by a phase angle of 90° (76).

A computer model of the leech network corresponds to the antiphasic oscillations and reproduces many features of the swimming pattern. No one cell is endogenously rhythmic, but when the network is tonically excited, the total system oscillates. In addition to the intraganglionic network there are forward and backward links, excitatory and inhibitory. Patterned rhythmicity is determined by several neurons, most of which are connected by inhibitory chemical synapses. Once a swim-initiating neuron is in the activated state, the nerve cord can implement rhythmic swimming (25).

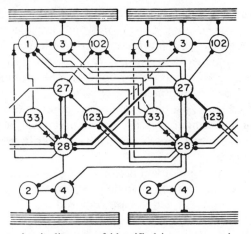

Figure 11-3. Summary circuit diagram of identified interneurons in swimming by leech. Interneurons (27, 28, 33, 123) and motorneurons (1, 3, 2, 4, 102) and longitudinal muscles used in swimming rhythms. Bars designate excitatory junctions, solid circles inhibition, and diodes rectifying electrical junctions. Reprinted with permission from G. S. Stent and W. B. Kristan in *Neurobiology of the Leech,* K. J. Muller, J. G. Nicholls, and G. S. Stent, Eds. Copyright 1981, Cold Spring Harbor Laboratory, Cold Spring Harbor, New York.

Walking Rhythms in Insects

Hexapods have patterned sequences of movements of the three pairs of legs, modified with gait. The leg sequence in a mantis is as follows (68a):

Slow walking	L 3, R 3, R 2
Running	L 2, R 2, L 3 with R 3
Climbing	L 1, R 1, L 2 with R 3; L 1 with R 2

A cockroach's six legs form two alternating tripods; these persist at all speeds and rates of leg protraction and retraction. Leg movement pattern is centrally controlled but the timing is influenced by feedback from mechanoreceptors. After deafferentation, bursts in mesothoracic efferent neurons begin near the end of bursts in metathoracic efferents. In the eight-leg walking sequences of scorpions and spiders, changes in patterns of leg movement occur with gait and number of legs present (5b).

The pattern of motorneuron discharge in sequential leg movements of cockroach or locust is determined by the neural network of mesothoracic and metathoracic ganglia. During very slow walking, flexor and extensor bursts synchronize with movements of a leg; as speed increases, the movements lag behind the bursts from motorneurons. Extensor and flexor discharges are 180° out of phase with each other. Input to a motor network is from sense organs and from command neurons in the supraesophageal ganglion; frequency of firing of a command neuron sets the period of bursts in the motorneurons. In the ganglionic neuropil are small interneurons that do not spike and modulate motorneurons via chemical synapses. When a sinusoidal depolarization is applied to an interneuron, a motorneuron fires one burst per wave. Three types of nonspiking interneurons have been recognized in the cockroach metathoracic ganglion. One type strongly excites flexor motorneurons; during spontaneous leg movement the membrane potential of the interneuron oscillates with a depolarizing phase which occurs during flexion. Another kind of interneuron, when depolarized, inhibits flexor activity. Two other types of interneurons specifically affect extensor motorneurons. Command neurons are activated by sense organs—for example, anal cerci of cockroach stimulated by wind, eyes of locust stimulated by moving light (6,19,59).

Each abdominal cercus of a cockroach has nine columns of mechanosensory hairs; all the hairs in one column are excited by wind in one compass direction. Sensory cells of all nine ipsilateral columns make synaptic contact with two giant interneurons that differ in response with wind direction (16). In a locust, more types of premotor interneurons occur than in cockroach. As many as six nonspiking interneurons may affect one motorneuron; some interneurons may affect both flexor and extensor motorneurons. Further, some spiking interneurons synaptically excite or inhibit motorneurons. The extensor and flexor muscles of a given segment of a leg are activated reciprocally, and one interneuron can affect movement of one or two joints.

Insect Flying and Jumping

The contraction of many muscles of flying insects coordinates liftoff, change in speed, yawing, turning, and other maneuvers. Flight can be initiated by one of several stimuli—lifting tarsi from the substrate or blowing wind on the head. Once started, the patterned sequence of flying continues after the sensory input has ceased. In *Drosophila* a pair of giant neurons with somas in the supraesophageal ganglion are activated by an air puff on the head; in a mutant these neurons are activated by a flash of light (44).

The neurons serving flight muscles in locusts have been identified. A neuronal network in the metathoracic ganglion generates an oscillating pattern of muscle contractions. The sequential contractions of the muscles follow a fixed pattern, but sensory stimuli, such as wind on the head, may modify the number of motor impulses per oscillation and the synergism and antagonism between motorneurons (11). Recording by electrodes implanted in flight muscles of tethered insects permits measurement of patterns and intervals between bursts from motorneurons supplying single muscles. It has also been possible to record directly from identified motorneurons. The pattern of activity of motorneurons in the oscillator is set by reciprocal connections and delays without modulation.

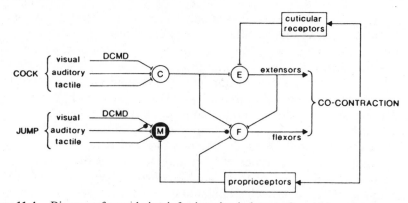

Figure 11-4. Diagram of neural circuit for jumping in locust. C and M are interneurons, E fast extensor tibiae motorneurons, F flexor tibiae motorneurons, DCMD descending contralateral movement detector neuron. Reprinted with permission from K. G. Pearson, *Journal de Physiologie* (Paris), Vol. 78, p. 770. Copyright 1982–1983, Masson S. A., Paris.

Recordings from metathoracic flight motorneurons show one neuron that drives a depressor muscle, and one that drives an elevator muscle (59). Patterns recorded from the motorneurons correspond to recordings from the muscles during flight. The motorneurons serving depressors on the two sides are synchronous; those serving the elevators are in antiphase to the depressors. The delay pathway contains interneurons capable of reverberation. Flight is modulated by input from stretch receptors in the wings; EPSPs (excitatory postsynaptic potentials) in depressor motorneurons synchronize with IPSPs (inhibitory postsynaptic potentials) in elevator motorneurons. Visual signals from contralateral eye fields cause EPSPs with 9.9 msec delay. If sensory nerves are cut the flight pattern persists at reduced frequency.

In large insects—butterflies, locusts, cicadas—muscle action potentials synchronize with contractions. In flies and bees the muscles are asynchronous, and a wing may beat several times per burst of impulses (Ch. 16). However, there is some phasing of motorneuron discharge with wing movement. In Drosophila about 26 "giant" muscle fibers are programmed in the flight pattern.

Many insects jump as an escape reaction or to move from one food site to another. A click beetle can jump 30 cm upward with initial acceleration of 400 g. The jump of a locust consists of three phases: (1) cocking by flexion of hindleg tibia in full flexion; (2) cocontraction of hindleg extensors and flexors and storage of energy in elastic elements of the leg; (3) triggering by sudden inhibition of flexors, shortening of contracted extensors and release of the stored energy. The neural circuit for jumping (Fig. 11-4) consists of the following pairs of identified neurons (59): pairs of interneurons that receive inputs from auditory and tactile receptors and that evoke cocking by neurons (C) and triggering of a jump by neurons (M), two large descending neurons that detect visual movement (DCMDs), excitatory interneurons (E and F) that activate extensor and flexor motorneurons simultaneously, positive feedback from cuticle to extensor neurons (E) during co-contraction, and feedback from proprioceptors to jump neurons (M) and flexor neurons (F) (58a).

Mutants of Drosophila show disturbances of neural circuits, such as altered balance between excitation and inhibition for certain motorneurons. A hyperkinetic mutant Drosophila after anesthesia vibrates its legs; another mutant becomes paralyzed if warmed. Three major factors in neural pattern generation are (1) resetting of a motorneuron after it has spiked, (2) sharing of inputs between motorneurons, and (3) reciprocal inhibition. The structure of the neural circuit is genetically preset (2).

Escape Behavior of Crayfish; Giant Neuron Function

A tap on antenna or rostrum of a crayfish or stimulation of the eye by an oncoming object triggers a tail flip. This is mediated by the median giant fibers, which extend from cell bodies in the supraoesophageal ganglion. A tap or puff on sensory hairs on the abdomen initiates escape mediated by the lateral giant fibers. The lateral giant fibers (LGs) conduct forward. Median giant fibers (MGs) conduct rearward. Nonspecific stimuli to many parts of the body trigger escape mediated by nongiant fibers, latency 220 msec, compared to 8–20 msec for a giant fiber response. The escape tail flip starts with contraction of abdominal flexor muscles, then extensor muscles of many segments contract synchronously and propel the crayfish forward. Many of the neurons in the circuit for escape behavior have been identified. There is redundancy in parallel pathways; some synapses are electrical (rectifying or nonrectifying), some are chemical. A balance between excitation and inhibition makes for precise control of action in the large muscle mass.

Diagrams summarizing the circuit for escape behavior are given in Figure 11-5 (84). The most direct path to flexors (Fig. 11-5a) is from a giant command fiber (LG or MG) via a segmental giant motorneuron (MoG) which excites the muscles. A parallel path runs from nongiant interneurons to fast flexor motorneurons (FF). A third excitatory path is from either the lateral or median giant to interneurons (CD 2/3 or SG) which activate the fast flexor motorneurons (FF). A branch from this latter path is inhibitory to the giant motor neurons. Another inhibitory neuron (FI) is activated by an interneuron (CDI) or from (FF); FI inhibits the muscle directly. Each hemisegment of the first five abdominal ganglia has one MoG, one FI, and five to nine motorneurons.

The extensor muscles are excited by extensor motorneurons which are activated within each segment by sensory interneurons. Stretch receptors, stimulated by abdominal flexion, are excitatory to extensor motorneurons.

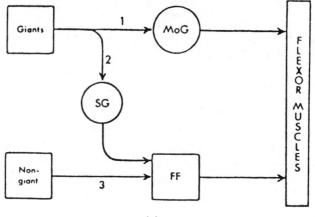

(a)

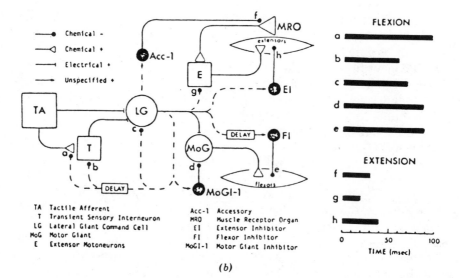

(b)

Figure 11-5. Circuit diagrams for escape behavior of crayfish. (*a*) Schema of excitatory flexion circuit for one side of an anterior abdominal segment. Pathway 1 is disynaptic from giant fibers via motor giants (MoG). Pathway 3 is from nongiants to fast flexor motorneurons. (*b*) Pathways for command-driven inhibition to show eight inhibitory actions. (*c*) Summary of known connections to one ganglion. (*a, b*) Reprinted with permission from J. J. Wine and F. B. Krasne in *Biology of Crustacea,* Vol. 4, D. Bliss, Ed. Copyright 1982, Academic Press, New York. (*c*) Reprinted with permission from J. J. Wine et al., *Journal of Neurophysiology,* Vol. 47, No. 5, p. 777. Copyright 1982, American Physiological Society, Bethesda, Md.

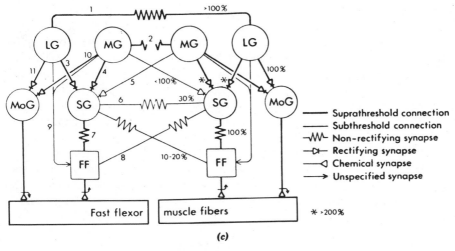

Figure 11-5. *Continued*

The giant fibers (LG and MG) cause a short-latency inhibition of extensor motorneurons. Figure 11-5*b* shows inhibition from LG via delay paths of small interneurons. Reciprocal inhibitory paths cause inhibition of extensor motorneurons when flexion occurs, and inhibition of flexor motorneurons upon extension.

Between each pair of abdominal segments are two pairs of stretch receptors, the muscle receptor organs MRO_1 (slow-adapting) and MRO_2 (fast-adapting). Each consists of a thin muscle which is embraced by finger-like dendrites of a sensory neuron of which the axon passes centrally. The muscle fibers of MRO_1 are innervated by the same motor axons that innervate superficial extensors, MRO_2s by axons to deep extensors. Each receptor also receives inhibitory innervation. During flexion of the abdomen the segmental stretch receptors are stimulated and MRO_1 inhibits extensor motorneurons. Other functions of the RMOs are as error detectors in a load-sensing system (84).

Multiplicity of modes of transmission within the escape reflex circuit is shown in Figure 11-5*c*, which indicates the distribution of nonrectifying and rectifying electrical, and inhibitory and excitatory chemical synapses. The uropods and telson are moved by seven muscles activated by the lateral and median giant neurons. Motorneurons to muscles of telson and uropods are inhibited so that these appendages do not flare out during a tail flip.

Giant nerve fibers have evolved in several kinds of animals. In crayfish normal swimming and exploratory movements are carried out by nongiant neurons in the nerve cord; the giant fibers participate only in rapid stereotyped reactions. In some invertebrate animals the giant fibers arise from very large single neurons, as in the median giants of crayfish. In some animals giant fibers are formed by large segmental neurons connected by gap junctions; electrotonic coupling provides continuous conduction. This is the anatomy of earthworm giant fibers and of the lateral giant fibers of crayfish. In some cephalopod molluscs (squid), processes from a cluster of cell bodies fuse to form giant axons. In crustaceans such as crayfish and many annelids, conduction in giant axons is an order of magnitude faster than in the network of small neurons. The primitive escape response was probably mediated by a network of small neurons; later, giant axons enabled crustaceans to move as fast as vertebrate predators with myelinated axons.

Stimulation of certain nongiant fibers of crustaceans in the circumpharyngeal connectives can elicit discrete postures, bilateral or ipsilateral appendage pronation, cheliped lifting, abdominal slow flexion and extension, turning, and walking backward or forward.

Neural Patterns of Behavior in Gastropod Moluscs

The ganglia of gastropod molluscs have fixed numbers of large neurons and clusters of small neurons. Each neuron is identified by (1) location, form, and color, (2) input and output recorded intracellularly, and (3) responses to putative transmitters iontophoretically applied. *Aplysia* has about 15,000 neurons in 9 ganglia (4 pairs plus one unpaired); about 100–150 large neurons have been identified so far, and about 10–20 clusters of smaller neurons. Stimulating electrodes were implanted into certain identified interneurons in Aplysia; the mollusc was then suspended in water; stimulation of a few interneurons evoked the complex movements of swimming (83). Besides central ganglia, a peripheral nerve network takes part in Aplysia coordination (45).

Reflexive Withdrawal; Inking

Tactile stimulation of the *Aplysia* siphon or mantle shelf elicits withdrawal of mantle organs. First the siphon is drawn in, then the gill spreads and closes, then pinnules bunch and rotate. Stimulation of the siphon skin activates mechanoreceptor cells that excite gill and siphon motorneurons, excitatory interneurons (L22 and L23), and an inhibitory interneuron (L16) (Fig. 11-6). The same stimulus can elicit withdrawal of gill (by motorneuron L7) and withdrawal of siphon. More than half of the input to gill motorneurons is monosynaptic. The peripheral nervous system includes neurons in the gill. They can initiate and mediate flexion of gill pinnae without the CNS, but the CNS provides variability and suppresses some peripheral network reflexes (59a). Siphon withdrawal is mediated by 8 central and 25–30 peripheral motorneurons.

Respiratory pumping and rotation of the gills of Aplysia circulates water in the mantle cavity and aerates the gills. Some of the same motorneurons serve both respiratory pumping and the gill withdrawal reflex. The same

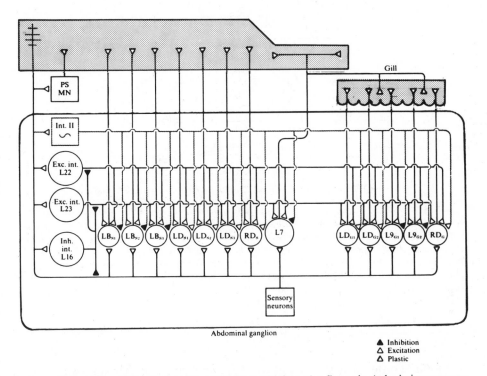

Figure 11-6. Neural circuit for gill- and siphon-withdrawal reflexes in Aplysia in response to tactile stimulation of siphon skin. A weak stimulus activates mechanoreceptors that excite gill and siphon motorneurons as well as two excitatory interneurons (L22 and L23) and one inhibitory interneuron (L16). A stronger stimulus also excites command elements for respiratory pumping (Interneuron II). Reproduced with permission from E. R. Kandel, *Behavioral Biology of Aplysia*. Copyright 1979, W. H. Freeman and Co., San Francisco.

sensory neurons can activate gill, withdrawal, and ink gland motorneurons; the threshold for gill reflex is low, stronger stimuli trigger inking (Fig. 11-17) (7,8). The ink gland motorneurons receive several inputs and can give two types of EPSPs and one type of IPSP. A fast EPSP is 0.1–0.2 sec duration, has reversal potential near 0 mV, and is due to an increase in membrane conductance; a slow EPSP is 10 sec to 3 min duration, reverses at −70 mV, and is due to a decrease in resting potassium conductance. Two clusters of mechanoreceptors make monosynaptic contact with gill and siphon motorneurons (L7) and with ink-discharge motorneurons (L14). Gill motorneurons are not coupled, have low resting potentials and low threshold, and may be spontaneously active; ink gland motorneurons are electrotonically coupled, have high resting potentials and high threshold, are not spontaneously active, and discharge a burst of impulses only when activated. Some interneurons go to both sets of motorneurons, some to only one.

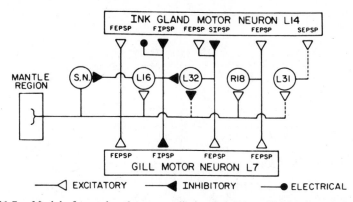

Figure 11-7.   Model of neural pathways mediating defensive gill-withdrawal and inking behavior in Aplysia. Same sensory neurons in mantle activate gill motorneurons L7 and inking motorneurons L14. Activation of sensory neuron R18 produces fast excitation in L7 and L14. The sensory neurons also connect with R18 which elicits a fast EPSP in L7 and L14. The L16 is an inhibitory neuron which also receives connections from sensory neurons; L16 produces fast inhibitory PSPs in L7, L14, and L32. Reprinted with permission from J. H. Byrne, *Journal of Neurophysiology,* Vol. 45, No. 1, p. 100. Copyright 1981, American Physiological Society, Bethesda, MD.

Escape Swimming

If the initial reflex withdrawal fails to move an Aplysia or Tritonia away from a noxious stimulus, escape swimming occurs. Chemical stimulation from tube feet of starfish elicits escape swimming. The behavioral sequence is: local withdrawal of stimulated region, general withdrawal, then the alternating ventral and dorsal flexions of swimming. Approximately 60–80 sensory neurons and 8–10 small interneurons converge on swim interneurons. Withdrawal is initiated by low-intensity stimulation of sensory neurons, which activate interneurons that converge on dorsal and central flexor motorneurons. Strong sensory stimulation activates command interneurons in the cerebral ganglion which inhibit the reflex withdrawal interneurons and trigger the swimming pattern generators. In Tritonia dorsal swim interneurons (DSI) burst in phase with dorsal flexor motorneurons (DFN); ventral swim interneurons (VSI) activate ventral flexor motorneurons (VFN) in an alternating cycle (27). Dorsal flexor motorneurons are of two sorts: DFN-A have prolonged bursts, DFN-B short bursts. Reflexive withdrawal requires activity in both DFN and VFN cells (27).

Swimming consists of alternating ventral and dorsal flexions. The circuit for rhythmic swimming is schematized in Figure 11-8. This consists of two halves; on one side are DSI (dorsal swim interneuron) and C2; on the other side are VSI-A and VSI-B (ventral swim interneurons). In initiation of swimming, DSI fires first, then 0.5 to 4 seconds later a burst appears in C2. DSI inhibits VSI-B and C2 excites it. The A current (Ch. 15) in VSI-B together with inhibition from DSI reduce or delay excitation by C2. When the A cur-

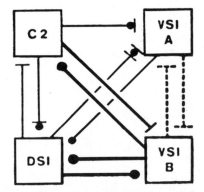

Figure 11-8. Circuit diagram of synaptic connections in neural network for swimming in Tritonia. VSI-A and VSI-B are ventral swim interneurons. Other interneurons are cerebral cell 2 (C 2) and dorsal swim interneuron (DSI). Monosynaptic pathways are indicated as solid lines, functional interactions by broken lines. Excitatory synapses shown as T-bars, inhibitory synapses as solid circles, multicomponent synapses (excitatory-inhibitory) as mixed symbols. The temporal order of multicomponent synapses is given by order of symbols from presynaptic to postsynaptic cell. Reproduced with permission from P. A. Getting et al., *Journal of Neurophysiology*, Vol. 49, p. 1048. Copyright 1983, American Physiological Society, Bethesda, MD.

rent inactivates, VSI-B starts to fire and this reciprocally excites VSI-A. Since DSI excites dorsal flexor motorneurons (DFN) and VSI-A and B excite ventral flexor motorneurons (VFN), the two sets of flexors are out of phase and swimming results. Termination of the DSI-C2 positive feedback results from adaptation of spike frequency connection from C2 to DSI by multiaction synapse (excitation plus inhibition) and delayed excitation of C2 by the two VSIs. A computer model of the postulated circuit oscillates like the swimming pattern of Tritonia (37).

Feeding

Neural mechanisms of feeding have been studied in opisthobranchs *Aplysia*, *Pleurobranchaea*, and *Navanax*, the pulmonate snail *Helisoma*, and the slug *Limax*. Feeding consists of two series of actions. The first is appetitive— orientation and seeking in response to chemical signals, and assumption of posture for oral contact and intake of food. Next is a series of stereotyped actions leading to ingestion; these are movements of the odontophore cartilage and radula by alternate protraction and retraction of muscular buccal mass and contraction of radula muscles. *Helisoma* has seven pairs of radula protractor muscles and four pairs of retractors. The initial orientation is mediated by pedal ganglia. The movements of odontophores are controlled by cerebral and buccal ganglia; in *Pleurobranchaea* at least two oscillators are present, one in buccal and one in cerebral ganglia. These molluscs feed via a central program after extensive sensory input. The central pattern of

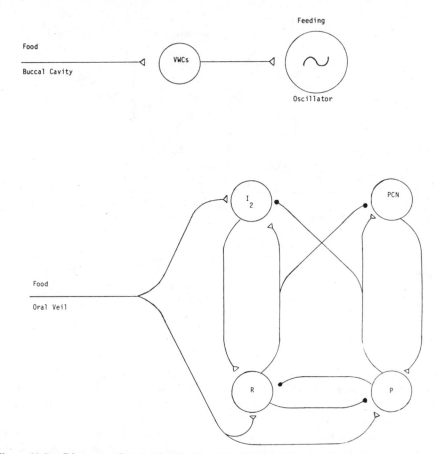

Figure 11-9. Diagrams of appetitive feeding circuit (lower figure) and consummatory (feeding and swallowing) circuit (upper diagram) in Pleurobranchaea. Input from the veil is distributed to a neural oscillator for intake of food. Input from buccal cavity activates ventral white cells which drive the same oscillator as oral veil but more vigorously. Personal communication from R. Gillette.

movements is modulated by sensory input from receptors in the buccal mass.

In *Helisoma* feeding, five pairs of bilateral protractor motorneurons are electrically coupled and fire together, 180° out of phase with eight bilateral retractor motorneurons that are also electrically coupled. The motorneurons show EPSPs and IPSPs driven by metacerebral command neurons (40).

A model of the network of 20 coupled neurons in *Helisoma* replicates the behavior (49a). In *Limax* a semiisolated preparation of ganglia connected to muscles shows the feeding motor program (26). In *Lymnaea* sensory signals from the lips go to command neurons in the brain which activate motorneurons in buccal ganglia. Sensory signals modify the feeding pattern; bite cycle frequency is higher on soft than on hard food; inflation of the crop decreases cycle frequency and may terminate feeding (1).

Feeding in the marine carnivore mollusc *Pleurobranchaea californica* can be divided into appetitive and consummatory phases. The appetitive phase is initiated by stimulation of chemoreceptors on the oral veil which send excitatory synaptic inputs to several populations of interneurons, the most dominant of which are the I_2s and motorneurons that drive retraction (R) and protraction (P) of the proboscis (Fig. 11-9). The I_2 and retractor motorneurons are connected by reciprocal inhibition to paracerebral (PCN) neurons and protractor motorneurons. This network is located within the feeding oscillator, which includes components in both buccal and cerebral ganglion; this oscillator controls the decision to feed as well as rhythmic movements of the feeding apparatus. The consummatory phase is stimulated by food acting on chemoreceptors in the buccal (mouth) cavity. These receptors activate ventral white cells (VWCs) of buccal ganglia, which are command neurons for the fixed-action pattern of intense activity. Chemical stimulation of the oral veil and of buccal chemoreceptors activates the same interneuron motorneuron oscillator, weakly for biting food and vigorously (via the ventral white cells) for swallowing it. When a food mass enters the esophagus, inhibition of PCNs occurs and feeding ceases. Stimulation of oral veil by distasteful stimuli inhibits PCNs, but activity of the VWCs can override the responses of PCNs (28).

Hierarchy of Behavior

In *Pleurobranchaea* six different behaviors are controlled by the same set of neural centers. These are related as follows (28):

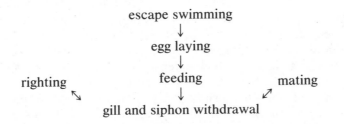

Feeding is dominant over righting, withdrawal, and mating, but feeding is suppressed during egg laying. Only escape swimming is dominant over all other behaviors. The neural hierarchy is expressed also in the semiisolated nervous system; neurally induced feeding suppresses gill and siphon withdrawal. Two neurons in the buccal ganglion fire during output of the feeding network, and if these neurons are hyperpolarized feeding no longer suppresses withdrawal. The usual hierarchy can be reversed by avoidance conditioning—presentation of food together with shock (17,18,28,29).

Egg laying is evoked by discharge of 250–400 neurosecretory cells (bag cells) that release a peptide hormone into the nerve sheath; the discharge is not synchronous on the two sides. Egg laying requires a strong stimulus and is episodic (76a).

VERTEBRATES; FIXED ACTION PATTERNS AND LOCOMOTOR RHYTHMS

A pair of very large neurons, the Mauthner cells, occur in the medulla of many teleost fishes and receive input, mainly auditory and vestibular, also from eyes and lateral lines. Some of the converging afferent fibers make chemical synapses and others make electrical synapses on the Mauthner cells. Recurrent inhibition by a branch from the Mauthner axon is by an unusual extra-axonal hyperpolarizing electric field. Stimulation of a Mauthner neuron triggers a startle response—tail flip and body flexion. A Mauthner cell can fire within 6 msec from the start of a vibration stimulus; muscles contract 2 msec later (zebrafish). In goldfish the Mauthner spike has a latency of 2 msec after an acoustic stimulus (88). The startle response upon vibration stimulus showed latencies of 5–10 msec in 11 species of teleosts. The patterned response consists of body bend of some 20 msec duration, displacement of head and tail to one side and return flip of tail to the opposite side. The effect is to propel a fish forward about 0.5–1.5 body lengths within 100 msec (23). Responses similar but less spectacular are mediated by Müller cells in lampreys.

Locomotion by vertebrates of many kinds is patterned in the spinal cord. Dogfish after a high spinal transection makes undulatory swimming movements if tactile stimuli are applied briefly to one side of the body.

Locomotor movements of the four legs of a turtle are coordinated by a spinal pattern triggered by command cells in the brain. Each turtle limb has its control center in the spinal cord; the diagonal pairs of legs move in phase. Stimulation of the dorsal lateral funiculus of the cervical cord of the turtle starts patterned swimming—limb extension, paddling, withdrawal (75).

A cat with spinal cord transected in the upper thoracic region and with sensory roots cut can generate a normal locomotor pattern. Recordings from motor nerves or muscles show alternating activity of flexor and extensor muscles. Reflex modulation begins with muscle spindles and tendon organs. In acute and chronically prepared cats, once a locomotor rhythm is started, the spinal cord can continue it (31). In walking ungulates, one hind leg synchronizes with the opposite front leg; in galloping the two forelegs move together, then the two hind legs.

Motorneuron pools in the vertebrate spinal cord and brainstem are analogous to single or clustered motorneurons of invertebrates in mediating genetically fixed patterns; after the circuit is triggered by command signals, motor activity follows sequentially.

RHYTHMS OF LONG PERIOD

Long-lasting central nervous activities are found in rhythms of hourly, diurnal, and episodic events. In some jellyfish, rhythmic contractions of swimming are triggered by bursts of impulses, about one per minute, originating in sense organs at the margins. Irrigation movements are made by some tube-dwelling polychaete worms at 20-minute intervals. In limulus the respiratory book gills move 10–30 times per minute, depending on the oxygen supply; every few (2+) minutes the respiratory movements are interrupted and gill plate appendages sweep over the gills and clean them. An isolated abdominal nerve cord of limulus discharges in bursts corresponding to the periods of ventilation and gill cleaning (81,86). In many kinds of animals, respiratory rhythms are intrinsic to the nervous system. In some—turtles—breathing is rhythmic, with interruptions of 0.5 to 3.5 minute periods of apnea.

Circadian rhythmic activity is endogenous in some nervous systems. Many animals—cockroaches, rodents—show a free-running daily rhythm, usually entrained by the light-dark cycle. In darkness the eye and optic ganglion of *Aplysia* in culture discharges a burst at the time corresponding to dawn; in continuous darkness this recurs each morning for days. Circadian discharges have been recorded in the genital nerve attached to the isolated abdominal ganglion of *Aplysia*. There is circadian discharge in crayfish in the second root of the sixth abdominal ganglion. The circadian rhythm of locomotor activity in cockroach is abolished by ablation of optic lobes or by transection of the optic tract. The locus of the circadian clock in rats is mainly in the supraoptic nucleus of hypothalamus (69a).

In Lepidoptera an episodic programmed motor behavior is eclosion from pupa to moth. Patterned behaviors are molting and pupation, initiated by a hormone acting on the nervous system. If the eclosion hormone is experimentally injected into preemergent pupae of cecropia silkmoths, the following behavior occurs after a latency of 15 minutes: (1) abdominal rotation repeated for 30-minute periods, (2) quiescence for 30-minute periods between abdominal rotations, (3) rhythmic contractions of the abdomen at 3 to 5 per minute, (4) emergence. If the hormone is washed out after the rhythm starts, the eclosion behavior continues. If the eclosion hormone is applied directly to an isolated nerve cord, bursts of spikes are elicited in a pattern at lower frequency than in intact pupae. Sensory feedback influences the frequency of the rotatory bursts but does not affect the pattern of eclosion movements. In the last larval instar of cecropia the fourth abdominal ganglion has 70 motorneurons in 10 cell groups; on transition from larva to pupa, 12 larval neurons disappear and 8 new neurons appear. No new motorneurons are added in the adult. Eclosion is genetically programmed and is triggered hormonally (88).

The moth *Hyalophora* emerges shortly after dawn; the moth *Antherea* emerges near sunset. Emergence is controlled by a hormone from the brain, and if the brain of *Hyalophora* is transplanted to the abdomen of a pupa of *Antherea,* the adult emerges at the time for *Hyalophora* (76b,78).

PLASTICITY OF NEURAL FUNCTION

Nerve impulses are transient brief events, lasting milliseconds to tenths of seconds. Some afterpotentials may persist for a second or more, and synpatic and pacemaker potentials are in the range of tens to hundreds of milliseconds. Paired stimuli show that after synaptic transmission, periods of facilitation or inhibition may outlast spikes and synaptic potentials. Facilitatory and inhibitory changes in excitability following synaptic input are relatively fixed in time.

Nervous systems mediating animal behavior show plasticity as well as persistence. One form of plasticity is substitution of one neural pattern for another. This makes use of neural redundancy, hierarchical levels, and parallel pathways. A simple example is shown by arthropods with six or eight legs. When one or several legs have been removed other legs immediately change sequence; preconnected patterns of postural support come into use. In intact arthropods the sequence of leg stepping changes with speed of locomotion and with angle and kind of substrate.

Neural plasticity is shown by changes in synaptic excitability. One of these is posttetanic potentiation (PTP), during which a burst of repetitive firing is followed by enhanced excitability, demonstrated by responses to single pulses. A converse effect is habituation, decline in excitability following repeated activation. Habituation is to be distinguished from sensory adaptation. One criterion of habituation is that it can be reversed (dishabituated) by vigorous input via a converging pathway. A spinal reflex such as leg flexion habituates over a period of many seconds or a few minutes, and the response returns immediately (is dishabituated) after generalized arousal of the animal (63a).

Conditioned Behavior

Conditioned behavior is a change in a response as a result of experience, usually of two or more sensory inputs. A brief classification and analysis of conditioned behavior follows. A conditioned stimulus (CS) does not initially evoke the response which the unconditioned stimulus (UCS) initially elicits (UCR). In classical conditioning, CS is followed by or overlaps UCS and on repetition a state is reached where CS alone elicits a response (CR) similar to that evoked by UCS. Conditioning can occur in one, or in few or many trials, which may be spaced over several days. Temporal relations between the two stimuli and the responses are critical. Usually CR, once established, has a longer latency than UCR. Duration of persistence of the conditioned state varies from brief (hours) to long (days); CS may be a nonspecific stimulation. UCS may be positive, by reward, or it may be negative, by punishment, such as a shock. Positive conditioning may lead to approach behavior, negative conditioning to avoidance.

Operant or instrumental conditioning differs from classical conditioning in that a motor response or action provides the positive or negative reinforcement; this is the basis for maze learning, problem solving, and similar performances. Trial-and-error learning is a kind of operant conditioning.

Controls are necessary to ascertain that a given behavior is conditioning and not sensitization of sensory pathways. Presentation of UCS prior to rather than after CS should not result in CR. Associative learning is distinguished from exploratory (orthokinetic) behavior. The relation of conditioning to learning and memory is complex and varies with type of behavior and animal. From the viewpont of neurophysiology, there may be a gradation of temporal events from synaptic facilitation and inhibition, through such changes as habituation and heterosynaptic facilitation, to short-term and long-term memory.

The probable events in conditioning are (1) sensory input (CS), (2) classification or coding of converging inputs at sensory nuclei, (3) consolidation of the classified information, (4) closure, persistent storage, (5) retrieval, readout. The molecular mechanisms for each of these are probably different, and likely differ with memory duration. There is no single mechanism of learning.

Conditioned response capacity, or at least some plasticity of behavior, is present in all animals with nervous systems. Ciliate protozoa can be "trained" to avoidance, but whether this is truly associative is questionable. Modification in a given neural center may be ascertained by observing effects of lesions, by use of synaptic blocking drugs, or by electrical recording. An example is the nervous system of an octopus. Octopus has several million small neurons located in a series of ganglia: buccal ganglia receive input from taste receptors of mouth; inferior frontal lobes pain and touch impulses from the arms; optic lobes visual impulses from the eyes (Fig. 11-10). In the laboratory, octopuses have been conditioned to visual and tactile stimuli (87). The standard technique for positive visual conditioning is to reward with food an attack by the octopus on a given object, such as a patterned disc; and, for negative visual conditioning, to punish with a shock. Learning by octopus is fast. By reward and punishment the animal learns to distinguish between two disc patterns in a few trials. Octopus's visual discrimination is better for horizontal than for vertical figures. It has been proposed that each neuron in the deep plexiform layer of optic lobes responds to a certain ratio of horizontal to vertical signals. The optic lobes send efferent fibers direct to the motor centers and elicit attacks. Tactile discrimination by octopus of plastic cylinders with different degrees of roughness or grooving can be demonstrated by conditioning. The sensory neurons from the oc-

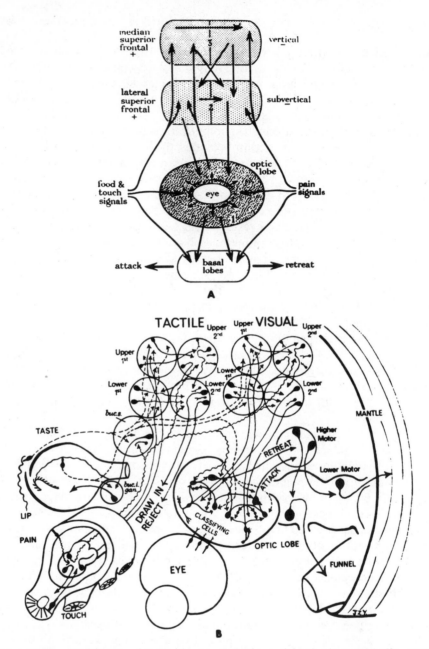

Figure 11-10. (A) Diagram of centers in octopus brain participating in control of locomotor activity. 1, 2, and 3 are the three levels at which attack and retreat are controlled in response to visual stimulus. Reproduced with permission from J. Z. Young, *Proceedings of the Royal Society of London (Biological Sciences)*, Vol. 159, p. 565. Copyright 1964, Royal Society of London, London, England. (B) Diagram of two sets of paired centers in visual and tactile memory systems of Octopus. Reproduced with permission from J. Z. Young, *Proceedings of the Royal Society of London (Biological Sciences)*, Vol. 163, p. 285. Copyright 1965, Royal Society of London, London, England.

topus arms connect to inferior frontal and subfrontal lobes (Fig. 11-10), removal of which abolishes tactile discrimination of objects. By chemical sense an octopus can distinguish between a clam shell containing a live clam and a shell containing wax.

The superior frontal and vertical lobes of the octopus nervous system apparently preserve representations received from either optic (eyes) or subfrontal (arms) lobes. Electrical stimulation of superior or vertical lobes does not elicit direct motor reactions. Axons from the superior frontal lobe pass to the vertical lobe, which is divided into five lobules. The output of the vertical lobes is via the subvertical to the optic and superior and inferior frontal lobes, that is, back to the regions from which the input came. For both visual and chemotactile systems there are two parallel loops, one via lateral superior frontal and subvertical and the other via median superior frontal and vertical lobes. After removal of the vertical lobes, learning is possible but is slower and more mistakes are made than normally; learning capacity is reduced by 75 percent. More important, the pattern learned without the vertical lobes persists only a few hours; normal memory may last for days or weeks. The learning defect is proportional to the mass of the tissue removed from the vertical lobes.

The hypothesis developed by J. Z. Young (87) for learning of a visual discrimination by an octopus is as follows: (1) extraction of information and reduction of redundancy in the two plexiform layers (optic lobes) according to a two-letter code (vertical and horizontal); (2) redistribution and mixing of channels in the superior frontal lobe; (3) increase in number of parallel channels in the vertical lobes; and (4) transmission to the optic (or subfrontal) over a reduced number of channels ending in a two-letter code—to attack or not to attack. It is proposed that the initial codification occurs within the optic or subfrontal lobes and that short-term memory can occur there, but that long-term memory occurs in the vertical lobe. Codification occurs in an oriented elaborate dendritic system of cells arranged for certain input/output ratios. Long term storage occurs in the rich neuropil where the number of channels is very great. Electrical correlates of the postulated patterning in octopus nervous system have not been obtained.

In many vertebrate animals, heart rate and breathing rhythms can be conditioned and altered by light or sound as CS. When heart rate in pigeons is conditioned to light or sound, the rate is accelerated; in mammals it is slowed; in fishes, breathing is interrupted.

In the optic tectum of goldfish a conditioned evoked potential appeared when visual stimuli were appropriately paired with shocks to the flank. When the brain of a conditioned goldfish was cooled, the conditioned response disappeared reversibly. The conditioned response proved more labile than the direct response to light (63a). Modifications of evoked potentials have been found in several regions of mammalian brain. Habituation of click-evoked acoustic responses has been described for cochlear nucleus, superior olive, inferior colliculus; partial habituation has been observed in medial geniculate and auditory cortex. The separate components of evoked potentials with specific time relations are differently affected (80). Behavioral habituation parallels the electrical effects. After auditory conditioning of rats, single units in auditory nuclei of pontine, ventral tegmentum, posterior thalamus, frontal cortex, and hippocampus show changes in latency and spike rate.

Rats continue to learn mazes after all but a small region of association cortex has been removed. Contraction of the nictitating membrane of a rabbit, or leg retraction, can be conditioned to a tone or to white noise. Conditioning of the nictitating membrane was possible after all of the forebrain and thalamus had been removed. The latency of the conditioned movements was longer than that of responses to the UCS (shock); by comparison of latencies of electrical reponses it is concluded that the site of conditioning is not in sensory or motor centers. However, basal nuclei of the cerebellum must be present, and their unit responses indicate that these nuclei are sites of the memory engram. Refinement of the CR and associations of more complex stimuli take place in the hippocampus (77). Some sorts of learning may be localized in one region of the brain, some learning occurs in several regions, and much learning has not been localized (46).

Conditioning of a fixed action behavior can occur in a decapitated cockroach or locust. A foot of the insect was shocked whenever it was extended to a given position, and after a series of shocks the foot was kept flexed. A second animal yoked to the first so as to receive shocks unrelated to where its foot might be showed no such learning of foot flexion. Tonic impulses recorded from motorneurons in a thoracic ganglion corresponded to the behavioral changes. Shocks delivered to the foot whenever the frequency of tonic neuronal firing decreased led to 140 percent increase in basal frequency. Shocks delivered whenever the frequency was increasing led to a decrease. The conditioned responses were maintained for as many as eight hours without reinforcement and, with reinforcement, for the life of the preparation. It was possible to get transfer of conditioning from one leg to another, from right prothoracic to left metathoracic (35b,35c,85,79).

Mechanisms of Neural Plasticity

Many possible biochemical correlates with plasticity of nervous systems have been suggested based on effects of blocking drugs and on chemical analyses before and after neural modification. However, the secondary effects of these treatments and the possibility of multiple interpretations have led to the rejection of many claims of biochemical mechanisms of neural plasticity. It appears probable that short-term memory may be based on biochemical and ionic changes. Long-term memory appears to be based on morphological changes—dendritic growth, branching, and changes in neuronal membranes.

In rats and goldfish, inhibitors of protein synthesis (actinomycin-D, puromycin, cycloheximide) impair retention of conditioned responses. Unidentified proteins have appeared in brain regions after behavioral modification. A phosphorylated protein (20,000 Da) increases in concentration in the eye of a snail after conditioning (54). In rat hippocampus a protein fraction is phosphorylated over a time-course similar to longer-term potentiation. In brains of animals exposed to enriched environments, the content of RNA has been found to increase (73). Injection of anticholinergic drugs has been claimed to decrease retention of conditioned responses.

Post-tetanic potentiation (PTP) is observed in some Aplysia synapses as homosynaptic facilitation. PTP has been shown to result from presynaptic accumulation of Ca^{2+} in neuromuscular junctions; increase in intracellular Ca^{2+} also occurs in Aplysia presynaptic synapses. Habituation is measured as homosynaptic depression, and in Aplysia this results from inactivation of a transient inward Ca^{2+} current.

A mechanism of heterosynaptic facilitation (short-term memory) may be similar in gastropod ganglia and rat hippocampus. In the molluscs Aplysia and Pleurobranchaea aversive conditioning occurs when an attractant (shrimp or squid juice) is presented as a CS paired with a shock to the head as UCS. After several paired presentations head and siphon withdrawal and escape locomotion occur when the food is presented without a shock. Subthreshold tactile stimulation can also be the CS and can be made suprathreshold by conditioning. In parallel with the withdrawal response, synaptic potentials in motorneurons are enhanced. Electrical stimulation of an interneuron as UCS facilitates the motorneuron synaptic potential by action on the presynaptic ending of the sensory neuron (41). The facilitating neuron probably liberates serotonin as transmitter, which modifies the calcium inward current in the presynaptic terminal (Fig. 11-11). Serotonin enhances and the 5HT blocker cinanserin blocks the facilitation. Bathing the ganglion with Ca^{2+} at reduced concentration stops the facilitation. Injection of the sensory neuron with cyclic AMP (cAMP) enhances the synaptic response of the presynaptic neuron. Current measurements made under voltage clamp of presynaptic neurons show a voltage-sensitive inward Ca^{2+} current which is a prolonged plateau when outward K^+ current is reduced by TEA. When 5HT is applied or cAMP injected, the Ca^{2+} current increases and the K^+ current decreases, hence the plateau and increased Ca^{2+} influx result from decreased presynaptic current. Applied 5HT activates adenylate cyclase. Injection of a catalytic amount of a mammalian protein kinase has the same effects as the transmitter. Apparently a protein kinase, activated by cAMP in the sensory neuron, phosphorylates a membrane protein, possibly in the K^+ channel, and K^+ efflux is reduced, allowing an increase in Ca^{2+} influx. This enhances liberation of an unidentified transmitter from the sensory axon and both EPSP and the withdrawal reflex are enhanced (Fig. 11-11) (41).

Drosophila learn to avoid an odor when its presentation is paired with a

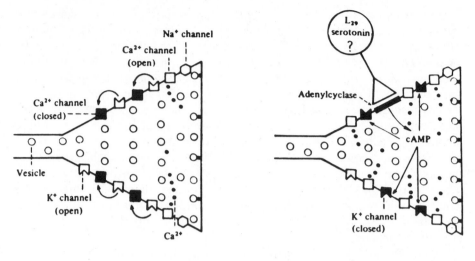

Figure 11-11. Schematic model of presynaptic facilitation underlying sensitization in an Aplysia synapse. Control represents response when presynaptic impulse reaches ending. Calcium channels are represented by squares, unfilled for open channels, filled squares for closed. Potassium channels are represented by notched squares. Arrows from K^+ channels and Ca^{2+} channels represent flow of hyperpolarizing current (through K^+ channels) which closes the open Ca^{2+} channels. Small solid dots represent calcium ions entering ending. Open circles represent vesicles containing transmitter. On sensitization the facilitating transmitter (probably serotonin) is released by the facilitating neuron (L^{29}) and acts on terminals of sensory neurons to increase the level of cAMP. The cAMP acts on K^+ channels to decrease K^+ current, which in turn prolongs the Ca^{2+} influx into the terminal and increases transmitter release. Reproduced with permission from M. Klein, E. R. Kandel, and E. Shapiro, *Journal of Experimental BIology*, Vol. 89, p. 152. Copyright 1980, Biochemical Society, Colchester, Essex, England.

shock. Mutants have been selected which learn poorly, or learn and forget within an hour. Two genes for poor learning are located near genes for cAMP metabolism. Two learning mutants have altered cAMP metabolism: "dunce" has decreased phosphodiesterase, and the mutant "rutabaga" shows decreased adenylate cyclase (63b).

In rat hippocampus the cortex shows plasticity of responses in isolated slices (3a,5). Long-term potentiation (LTP) follows activation of pyramidal neurons by trains of afferent impulses at physiological frequencies (10–400 Hz). Responses to a single volley can remain enhanced for hours, days, or weeks depending on the stimulus paradigm. Nonstimulated pathways do not show long-term changes. Both presynaptic and postsynaptic sites have been implicated, and all evidence indicates that Ca^{2+} is critically involved; hippocampal slices incubated in solution low in calcium do not show LTP. Evidence for presynaptic change is: uptake of Ca^{2+} is enhanced during LTP, and release of prelabeled transmitter glutamate is increased in parallel with increased field potentials (20a). Evidence for postsynaptic change is: intracellular injection of EGTA which binds Ca^{2+}, blocks LTP, binding sites for glutamate increase with LTP or increased Ca^{2+} (46). Increased Ca^{2+} in postsynaptic neurons may activate a proteinase which degrades a membrane protein, and this uncovers glutamate receptors on the dendritic spines (46a). It is agreed that both short-term and long-term potentiation involve increase of Ca^{2+} uptake, but that long-term change involves further steps, presynaptic or postsynaptic or both. Perfusion of hippocampal slices with norepinephrine during high-frequency stimulation of mossy fibers increases the probability of induction of long-term potentiation (35a).

Another hypothesis for learning has been based on measurements in tonically active thoracic motorneurons in cockroaches, which show learned positioning of legs. Changes in frequency of tonic firing of single motorneurons are associated with changes in membrane resistance and resting potentials as measured in motorneuron somata. It is postulated that the changes in tonic firing result from feedback of information from the conditioned act. Learning is postulated to occur in the cellular events of tonic discharge, not in synaptic alterations (85).

Morphological Changes with Experience

Central nervous connections are formed during embryonic development, but how neuron processes find their targets is not known. In insects, sensory axons grow inward guided by rows of landmark cells. In early development of most animals neurons are plastic as to how they become specified; neurons may develop according to their cellular environment. In late stages, neurons are relatively fixed and connections have become specified.

By transplanting patches of skin of amphibian embryos it has been shown that spatial specificity of sensory neurons becomes established in early tadpoles. Sensory neurons specify properties of interneurons; the interneurons may fail to develop dendrites if they receive no sensory input. When the optic nerve is cut and the eye is rotated 180° in adult frog or goldfish, axons from the retina make new connections to the optic tectum and the projection of the retina is inverted, i.e., nerve fibers go to the same target areas on the tectum to which they have previously gone. If the change in eye-position is made on early tadpoles the projection is such that the field is normal even though the retina is inverted. It is postulated that during development some type of affinity is established between axons and receptor regions in the CNS. In kittens, closure of the eyelids maintained for a few days during the age of four to six weeks prevents development of pattern discrimination in the grown cats (Ch. 10). In chicks, if dorsal roots are cut some motorneurons fail to develop. Motorneurons may degenerate if afferent signals are absent; the development of neuronal processes is much influenced by use.

In some invertebrate nervous systems certain neurons have characteristic positions and forms of processes. In cockroach thoracic ganglia, interneurons have been identified by their shape, branching of dendrites, and course of axons. After axons are cut, the dendrites may retract and cellular reorganization occur, with increase in RNA synthesis. The dendritic processes become reformed in essentially the same pattern as previously (14a).

The pattern of dendrites in cortical pyramidal cells of rats differs according to whether the animals were reared in isolation, in a drab social environment, or in an enriched environment. Ability to learn a maze is better in rats from the enriched environment; dry weight of brain is greater, acetylcholine content is less, ratio of RNA/DNA greater. In rats from enriched environment, the pattern of dendritic trees shows more branching, especially of third-, fourth-, and fifth-order branches (30). Synaptic areas are also larger. The effect of the enriched environment can be noted after three days or after 30 days of two hours per day. The effect of experience on neuronal structure is greater in young than in old rats (30). In honeybees and jewelfish dendritic growth and number of synaptic boutons have been shown to increase in correlation with experience (4).

Neuronal structure, particularly of dendrite processes, is a labile character, related to experience. Morphological changes reflect increased RNA turnover and protein synthesis.

CONCLUSIONS

Attempts to construct phylogenetic trees for nervous systems have failed because "primitive" nervous systems have many of the same general properties as "higher" nervous systems. One-to-one synaptic transmission, monosynaptic reflexes, and both nonpolarized and polarized synapses occur in animals of most phyletic levels. Electrical junctions may have preceded chemical synapses (Ch. 12); both are present in coelenterate nerve nets. Electrical synapses of more recently evolved animals (e.g., vertebrates) may have appeared secondarily (46a). The chemical transmission process has many components, any one of which might be subject to change on conditioning: current invasion of afferent action potentials into a presynaptic ending, calcium entry into the terminal, synthesis and packaging of a transmitter, activation of transmitter release, binding of transmitter with postsynaptic receptor, conductance change resulting in postsynaptic potentials, initiation of postsynaptic spikes. Functional properties of synapses in reflex behavior are synaptic delay, liberation of transmitter in quanta, temporal facilitation (fast and slow), early and late inhibition, and spatial facilitation (by convergence). Synaptic plasticity that may permit some behavioral modification can be by way of posttetanic potentiation or depression, habituation, short-term and long-term facilitation (heterosynaptic), enhancement of inhibition, and rhythmic postsynaptic discharges. Monosynaptic reflexes which are direct sensorimotor connections limit plasticity of response; more usual are disynaptic and trisynaptic reflexes with one or more interneurons interposed. Chains of neurons, linear, branched, or circular, provide the structures for variable responses, for convergence of several sensory modalities, and for divergence to more than one motor action.

Very early in the evolution of nervous systems endogenous activity appeared, which led to fixed action patterns of behavior. Endogenous activity is expressed in (1) rhythmic neurons that may discharge continuously or in bursts, and (2) repetitive output from a network. The network is often a loop with inhibitory feedback so that a neuron is interrupted periodically in its firing (Fig. 11-1). The output from endogenously active centers may trigger a sequence of muscle contractions making for coordinated movements. Most fixed action patterns are triggered by command neurons which diverge to more than one network and are activated either by sensory inputs or by endogenously active neural complexes of higher order. Some command neurons are known to be single cells in "simple" nervous systems; groups of neurons in "complex" nervous systems may issue commands.

In most animals, including humans, much behavior is carried out in direct reflexes and fixed action patterns. The evolutionary and developmental transitions from synaptic transmission to reflex reactions and from these to fixed action patterns are direct, but each level of organization has properties not fully predictable from analysis at the simpler levels. A programmed pattern in the CNS is not static and its output depends on dynamic properties; it contains redundant elements with parallel networks. D. Kennedy has used the analogy of a baseball game, where noting the movements of individual players does not provide an explanation of the game as a whole. Clearly, different properties are emergent from reflex connections, from the central pattern generators, and from complex integrating centers. Central nervous systems are programmable, and many of them show much plasticity.

Granted that much behavior—all behavior in many animals—consists of fixed action patterns, how do these develop in an individual and how did they evolve? Examples of adaptive programmed behaviors, several of which were described in this and the preceding chapter, are: (1) escape reactions— gill withdrawal in *Aplysia,* tail flip in crayfish, swimming by sea anemones; (2) rhythms of metabolic function—respiratory rhythms of swimmerets in crayfish, of book gills in limulus, neurogenic heartbeat of crustaceans, pyloric and gastric mill rhythms in crustaceans; (3) locomotor rhythms—undulating swimming by leech, walking sequence in cockroach and locust, flight of locust or fly, walking by spinal cat, swimming by spinal turtle; (4) orientation—selection of favored temperature, light intensity, position with respect to gravity by paramecia, fishes; basking of butterflies and lizards; (5) reproductive sequences of behavior; species-specific display—mate attraction, signaling, mating—by Lepidoptera, crickets, fishes, reptiles, birds; singing patterns; (6) episodic patterns—emergence of moths from pupae; (7) entrained rhythms—circadian rhythms of activity in rat, *Aplysia* and cockroach, emergence of *Drosophila.*

It can be assumed that central nervous pattern generators are genetically determined and that each pattern is modifiable within limits by experience. It is also probable that behavior patterns have been selected for adaptive value and that each pattern has a neural basis. The possibilities that modification can take place are greater in central pattern generators than in reflex relays. The molecular connection between inheritance and the neural basis of behavior is little understood. It is assumed that many neural connections in embryos could be random and that genetic programs provide for development and late expression of connections of adaptive action patterns. Mutations or genic rearrangements lead to new patterns. A few examples from behavioral genetics are: in *Paramecium* the mutant "pawn" has an alteration in protein of its Ca^{2+} channels. A hyperkinetic mutant of *Drosophila* vibrates its legs when etherized; another mutant becomes reversibly paralyzed at high temperature. Several mutants of a nematode show abnormal posture. A cricket's song pattern is controlled by a single gene, and in one mutant cricket, the trill components have changed order in the song. In humans several mutations affect the biochemistry of neurotransmitters. Small genetic changes may alter proteins which are important in neural development. Some genetic changes alter nerve membrane conductance and synaptic transmission, other changes alter the connections in a neural network.

Modifiability of function is a general property of nervous systems. Neural plasticity was first inferred from behavior; recently several different long-term alterations in synaptic transmission have been described. Several hypotheses of mechanisms, presynaptic and postsynaptic, have been proposed. A conservative view is that there are multiple kinds of plasticity, and that short-term modification is different from long-term. It is unlikely that there is a single mechanism of conditioning or that persisting synaptic modification can account for learning.

To what extent is it possible to extrapolate to highly integrated brains from the pattern generators of nerve nets, ganglia, and preprogrammed neural centers of jellyfish, snails and insects? Fixed-action behavior is stereotyped, based on genetically determined structures, yet it is ontogenetically variable and environmentally modifiable; fixed behavior exerts sensory control and reafferent feedback on endogenous patterns. Can the complex behavior of amniote animals, humans in particular, result from interacting central pattern generators? Can the ganglionic and interconnected nuclei within "simple" nervous systems be used as models for "higher" order integration? One approach is that complex behavior can be analyzed by analogy to the neural mechanisms of simple behavior. Another approach is that the complexities of nervous integration are so great that extrapolation from one level of CNS organization to another is impossible. Description of one patterned system of interactions between a few neurons (10–100) is not an explanation of how that system functions as a whole, even if it can be modeled. A third view is that each level of neural organization has properties that are unique and emergent from the system as a whole. Some neural circuits have been well described, others are incompletely known. Neural networks mediate adaptive behavior. Natural selection acts on patterns of behavior and thus indirectly on neural organization.

REFERENCES

1. Benjamin, P. R. and R. M. Rose. *J. Exper. Biol.* **80**:93–118, 137–163, 1979. Central pattern in feeding cycle in Lymnaea.

2. Bentley, D. R. *J. Insect Physiol.* **15**:677–699, 1969. *Science* **174**:1139–1141, 1971; **187**:760–764, 1975. Genetic control of song generating network in cricket.

3. Bentley, D. and M. Konishi. *Annu. Rev. Neurosci.* **1**:35–59, 1978. Neural control of behavior.

4. Brandon, J. G. and R. G. Coss. *Brain Res.* **252**:51–61, 1982. Dendritic changes in one trial learning in honeybee.

5. Browning, M. et al. *Science* **203**:60–62, 1979. *Brain Res.* **198**:478–484, 1980. Long-term learning in hippocampal slices.

5a. Bullock, T. H. and A. Horridge. *Structure and Function of the Nervous Systems of Invertebrates,* 2 vols., Freeman, San Francisco, 1965.

5b. Burns, M. D. and P. N. Usherwood. *J. Exper. Biol.* **79**:69–98, 1979. Control of walking in orthopterans.

6. Burrows, M. *J. Compar. Physiol.* **83**:135–164, 1973; **85**:221–234, 1973. *J. Exper. Biol.* **62**:189–200, 1973; **74**:175–186, 1978. *J. Physiol.* **298**:213–233, 1980. Neuronal connections for behavior patterns in locust.

7. Byrne, J. H. et al. *J. Neurophysiol.* **43**:581–594, 630–668, 896–911, 1980; **48**:1347–1361, 1982. Neural control of secretions from opaline gland in Aplysia.

8. Byrne, J. H. et al. *J. Neurophysiol.* **42**:1233–1250, 1979; **45**:98–106, 1981. Neural circuits for inking in Aplysia.

9. Byrne, J. H. et al. *J. Neurophysiol.* **37**:1041–1064, 1974; **41**:402–430, 1978; **49**:491–508, 1983. Neuron circuits for withdrawal reflex and respiratory pumping in Aplysia.

10. Calabrese, R. L. *J. Exper. Biol.* **82**:163–176, 1979. Neural control of heartbeat rhythm in leech.

11. Camhi, J. M. et al. *J. Exper. Biol.* **60**:477–492, 1974. *J. Compar. Physiol.* **121**:307–324, 1977; **128**:193–212, 1978. *Neuroethology,* Sinauer, Sunderland, MA, 1984. Mechanoreceptors in control of flight in locust and of running in cockroach.

12. Carew, T. J. and E. Kandel. *Science* **211**:501–504, 1981. Conditioned fear in Aplysia.

13. Castellucci, V. F. et al. *Proc. Natl. Acad. Sci.* **77**:1185–1189, 7492–7496, 1980. Chemical basis for synaptic modification in Aplysia.

14. Chang, J. J. *Proc. Natl. Acad. Sci.* **77**:6204–6206, 1980. Taste-aversion learning by isolated molluscan nervous system.

14a. Cohen, M. J. Ch. 70, pp. 798–812 in *Neurosciences 2nd Study Program,* M. I. T. Press, Cambridge, MA, 1970. RNA changes in neurons after axotomy.

15. Cotman, C. W., Ed. *Neuronal Plasticity,* Raven Press, New York, 1978.

16. Daley, D. L. *Brain Res.* **238**:211–216, 1982. Neural basis of wind-receptive fields of cockroach giant interneurons.

17. Davis, W. J. et al. *J. Exper. Biol.* **51**:547–563, 1969. *J. Neurophysiol.* **35**:1–12, 1972. Reflex organization of swimmeret system in lobster.

18. Davis, W. J. et al. *Science* **199**:801–804, 1978. *J. Compar. Physiol.* **138**:157–165, 1980. Neural correlates of plasticity in feeding and withdrawal of Pleurobranchaea.

19. Delcomyn, F. *J. Exper. Biol.* **59**:643–654, 1973. Neural control of walking pattern in cockroach.

20. Deutsch, J. A. *Physiological Basis of Memory,* Academic Press, New York, 1973.

20a. Dolphin, A. C. et al. *Nature* **297**:496–497, 1982. LTP is due to more transmitter release.

21. Donaldson, S. et al. *Canad. J. Zool.* **58**:549–552, 1980. Escape swimming in giant neurons in Aglantha.

22. Drews, C. D. et al. *Compar. Biochem. Physiol.* **66A**:315–321, 1980; **67A**:659–667, 1980. Conduction in giant nerve fibers of earthworms.

22a. Evans, M. E. G. *J. Zool.* (London) **169**:181–194, 1973. Jump of click beetles.

23. Faber, D. S. and H. Korn. *Science* **208**:612–614, 1980. Inhibitory postsynaptic potentials in Mauthner neurons.

24. Fourtner, C. R. pp. 401–418 in *Neural Control of Locomotion,* Ed. R. M. Herman, Plenum, New York, 1976. Central nervous control of cockroach walking.

25. Friesen, W. O. et al. *Proc. Natl. Acad. Sci.* **73**:3734–3738, 1976. *J. Exper. Biol.* **75**:25–43, 1978. Neuronal circuits for locomotion rhythms in leech.

26. Gelperin, A. et al. *J. Neurobiol.* **9**:285–300, 1978. Ch. 16, pp. 239–246 in *Central Nervous System,* Ed. J. Fentress, Sinauer, Sunderland, MA, 1976. Feeding motor program in molluscs.

27. Getting, P. A. et al. *J. Compar. Physiol.* **110**:271–286, 1976; **121**:325–342, 1977. *J. Neurophysiol.* **46**:65–79, 1981; **47**:60–102, 1982; **49**:1017–1035, 1036–1050, 1983. pp. 89–128 in *Soc. Exp. Biol. Sympos.* Vol. 37, Cambridge University Press, Cambridge, U.K., 1983. Pattern generation in neural control of swimming in Tritonia.

28. Gillette, M. W. and Gillette, R. *J. Neurosci.* **3**:1791–1806, 1983. Bursting neurons command feeding behavior in *Pleurobranchaea.*

29. Gillette, R. et al. *Science* **199**:798–801, 1978. Command neurons in motor network of Pleurobranchaea.

30. Greenough, W. T. et al. *Behav. Neurol. Biol.* **26**:287–297, 1979. *Brain Res.* **264**:233–240, 1983. *Trends in Neurosciences* **7**:229–233, 1984. Morphological changes in pyramidal neurons of rats according to experience.

31. Grillner, S. *Exper. Brain Res.* **34**:241–261, 1979. *J. Exper. Biol.* **98**:1–22, 1982. Central generation of locomotion in spinal cord of cat and dogfish.

32. Hartline, D. K. and D. M. Maynard. *J. Exper. Biol.* **62**:405–420, 1975. Motor patterns in stomato-gastric ganglion of Panulirus.

33. Heitler, W. J. *Nature* **275**:231–234, 1978. Coupled motorneurons as part of cray-

fish swimmeret central oscillator.

34. Hermann, A. *J. Compar. Physiol.* **130:**221–239, 1979. Fixed motor patterns in stomatogastric ganglion of cancer.

35. Hermann, R. A., Ed. *Neural Control of Locomotion,* Plenum, New York, 1976.

35a. Hopkins, W. and G. Johnston. *Science* **226:**350–352, 1984. Noradrenergic modulation of LTP in hippocampus.

35b. Horridge, G. A. *Interneurons: Origin, Specificity, Growth, Plasticity,* Freeman, San Francisco, 1968. *Proc. R. Soc. London* **157:**35–52, 1962.

35c. Hoyle, G. *Adv. Insect. Physiol.* **7:**349–444, 1970. *J. Exper. Zool.* **194:**51–74, 1975. Single neuron analysis of behavior in invertebrates. *Identified Neurons and Behavior of Arthropods,* Plenum, New York, 1977.

36. Ikeda, K. and W. D. Kaplan. *Am. Zool.* **14:**1055–1066, 1974. Neurophysiological genetics of Drosophila.

36a. Jacobson, M. Ch. 12, pp. 116–129 in *Neurosciences 2nd Study Program,* M. I. T. Press, Cambridge, MA, 1970. Development of specificity of neuronal connections.

37. Kandel, E. R. *Cellular Basis of Behavior,* Freeman, San Francisco, 1976. *Behavioral Biology of Aplysia,* Freeman, San Francisco, 1979.

38. Kandel, E. R. et al. *J. Neurophysiol.* **37:**996–1019, 1974. *Brain Res.* **121:**1–20, 1977. *J. Neurophysiol.* **44:**555–569, 1980. Neuronal circuits for behavior in Aplysia.

39. Kandel, E. R. et al. *Proc. Natl. Acad. Sci.* **77:**629–633, 1980. *Science* **211:**501–506, 1981. *Nature* **293:**697–700, 1981. Effects of calcium on neural modification in Aplysia ganglia.

40. Kater, S. B. et al. *J. Neurophysiol.* **36:**142–155, 1973. *Brain Res.* **146:**1–21, 1978; **159:**331–349, 1980. *J. Neurobiol.* **11:**73–102, 1980. Central program for feeding in snail Helisoma.

41. Klein, M. and E. R. Kandel. *Proc. Natl. Acad. Sci.* **75:**3512–3516, 1978; **77:**6912–6916, 1980. *J. Exper. Biol.* **89:**117–157, 1980. Analysis of presynaptic facilitation and potentiation of synaptic responses in Aplysia ganglia.

42. Kramer, A. P. et al. *Science* **214:**810–814, 1981. Command neurons in crayfish tailflip circuitry.

43. Kristan, W. B. et al. *J. Exper. Biol.* **65:**643–668, 1976; **96:**143–160, 1982. Neuronal patterns for swimming of leeches; generation of rhythmic motor patterns. Ch. 14, pp. 241–261 in *Information Processing in Nervous Systems,* Ed. H. Pinsker and D. Willis, Raven Press, New York, 1980.

44. Levin, J. *J. Compar. Physiol.* **87:**213–235, 1973; **93:**265–285, 1974. Giant neurons which initiate flight in Drosophila.

45. Lukowiak, K. and B. Peretz. *J. Compar. Physiol.* **117:**219–244, 1977. Interaction between central and peripheral nervous system in Aplysia.

46. Lynch, G. et al. *Nature* **266:**737–739, 1977; **278:**273–275, 1979; **305:**717–721, 1983. *J. Physiol.* **276:**353–367, 1978; pp. 21–31 in *Changing Concepts of the Nervous System,* Ed. A. R. Morrison and P. L. Strick, Academic Press, New York, 1982. *Brain Res.* **169:**103–110, 1979. Long-term potentiation of responses in hippocampus.

46a. Lynch G. and M. Baudry. *Science* **224**:1057–1063, 1984. Hypothesis for bio-chemistry of memory.

46b. Mackie, G. O., Ed. *Coelenterate Ecology and Behavior,* Plenum, New York, 1976.

47.   Mackie, G. O. and O. Singla. *J. Neurobiol.* **6**:339–378, 1975. Neurobiology of the hydrozoan Aglantha. *Am. Zool.* **5**:439–453, 1965. Epithelial conduction.

47a. Mackie, G. O. et al. *Biol. Bull.* **167**:120–123, 1984. Gap junctions in hydrozoans, not in anthrozoans and scyphozoans.

48.   McCormick, D. A. et al. *Proc. Natl. Acad. Sci.* **79**:2731–2735, 1982. Localization of memory traces in cerebellum.

49.   McFarlane, J. D. *J. Exper. Biol.* **61**:129–143, 1974; **64**:431–445, 1976; **65**:539–552, 1976. pp. 599–607 in *Coelenterate Ecology and Behavior,* Ed. G. O. Mackie, Plenum, New York, 1976. Electrical properties parallel conduction systems of anthozoan coelenterates.

49a. Merickel, M. et al. *Brain Res.* **146**:1–21, 1978; **159**:331–349, 1978. *J. Neurobiol.* **11**:73–102, 1980. Cycles of activity in snail neurons system.

50.   Miller, J. P. and A. I. Selverston. *J. Neurol. Phys.* **44**:1102–1121, 1980; **48**:1416–1432, 1982. Network properties of pyloric system in stomatogastric ganglion.

51.   Moore, M. J. *J. Exper. Biol.* **83**:231–238, 1979. Sensory fields activating giant fibers in earthworms.

52.   Muller, K. and S. H. Scott. *J. Physiol.* **311**:565–583, 1981. Sensory connections of interneuron of leech.

53.   Mulloney, B. *J. Compar. Physiol.* **122**:227–240, 1977. Synaptic connections of stomatogastric ganglion.

53a. Muller, K. J., J. G. Nicholls, and G. S. Stent, Eds. *Neurobiology of the Leech,* Cold Spring Harbor Lab, 1981.

54.   Neary, J. T. et al. *Nature* **293**:658–660, 1981. *Science* **244**:1254–1257, 1984. Ca-stimulated phosphorylation of channel proteins in visual conditioning in snail Hermissenla.

55.   Nicholls, J. G. and D. Purves. *J. Physiol.* **225**:637–656, 1972. Physiology of the nervous system of the leech.

56.   Ort, C. A. et al. *J. Compar. Physiol.* **94**:121–154, 1974. Motor connections in leech.

57.   Page, C. H. *J. Compar. Physiol.* **102**:65–76, 1975. Command fiber control of the abdominal movement of crayfish.

58.   Pearson, K. G. Ch. 7, pp. 99–110 in *Simpler Networks,* Ed. J. Fentress, Sinauer, Sunderland, MA, 1976. Nerve cells without action potentials.

58a. Pearson, K. G. *J. Physiol.* (Paris) **78**:765–771, 1983. Neural circuits for jumping by locusts.

59.   Pearson, K. G. and C. R. Fourtner. *J. Neurol. Physiol.* **38**:33–52, 1975. Nonspiking interneurons in walking system of cockroach.

59a. Peretz, B. *Science* **169**:379–381, 1970. *J. Compar. Physiol.* **84**:1–18, 1973; **103**:1–17, 1975. Habituation of gill withdrawal in Alpysia mediated by peripheral nervous system.

60.   Pinsker, H. M. *J. Neurophysiol.* **40**:527–552, 1977. Bursting neurons as endog-

enous oscillators in Aplysia.

61. Prior, D. J. and A. Gelperin. *J. Compar. Physiol.* **114**:217–232, 1977. Feeding responses in Limax.

62. Prior, D. J. et al. *J. Exper. Biol.* **78**:59–75, 1979. Evasive behavior of clam Spisula.

63. Poon, M. et al. *J. Exper. Biol.* **75**:45–63, 1978. Oscillator network in leech ganglia.

63a. Prosser, C. L. and T. Nagai. pp. 171–180 in *Central Nervous Systems and Fish Behavior,* Ed. D. Ingle, University of Chicago Press, Chicago, 1968. Conditioning of brain in goldfish.

63b. Quinn, W. G. and R. Greenspan. *Annu. Rev. Neurosci.* **7**:67–93, 1984. Learning in Drosophila.

63c. Rautenberg, A. *Prog. Brain Res.* **56**:349–374, 1982. Phosphorylation of a brain protein during long-term potentiation of brain responses.

64. Reingold, S. C. and A. Gelperin. *J. Exper. Biol.* **85**:1–19, 1980. Feeding motor program in Limax.

65. Ritzmann, R. E. et al. *J. Compar. Physiol.* **125**:305–316, 1978; **143**:61–70, 1981. Motor responses of giant interneurons in cockroach.

66. Roberts, A. and G. O. Mackie. *J. Exper. Biol.* **84**:303–318, 1980. Giant axon escape system of medusa Aglantha.

67. Roberts, A. et al. *J. Neurophysiol.* **47**:761–781, 1982. Function of segmental giant fibers in crayfish nervous system.

68. Roberts, A. and G. D. Block. *Science* **221**:87–189, 1983. Mutual coupling between circadian pacemaker neurons.

68a. Roeder, K. D. *Nerve Cells and Insect Behavior,* Harvard University Press, Cambridge, MA, 1967.

69. Rose, B. M. and P. R. Benjamin. *J. Exper. Biol.* **92**:187–201, 1981. Control of feeding in Lymnaea.

69a. Ross, D. M. *Proc. R. Soc. London* **B155**:266–281, 1961. Ch. 7, pp 281–312 in *Coelenterate Biology,* L. Muscatine and H. Lenhoff, Eds., Academic Press, New York, 1974. Behavior of sea anemones.

69b. Rovainen, C. M. *Physiol. Rev.* **59**:1007–1077, 1979. Neurobiology of lampreys.

69c. Rusak B. and I. Zucker. *Physiol. Rev.* **59**:449–526, 1979. Suprachiasmatic nucleus is essential for circadian rhythms in rodents.

70. Russell, D. F. and D. K. Hartline. *J. Neurophysiol.* **48**:914–937, 1982. Slow potentials and bursting motor patterns in pyloric network of lobster.

71. Satterlie, R. A. *J. Compar. Physiol.* **133**:357–367, 1969. Nerve net control of swimming in scyphozoan jellyfish.

72. Selverston, A. I. et al. *J. Compar. Physiol.* **91**:33–51, 1974. *Prog. Neurobiol.* **7**:215–290, 1976. Ch. 12, pp. 209–225, in *Identified Neurons of Arthropods,* Ed. G. Hoyle, Plenum, New York, 1977. Ch. 6, pp. 82–98 in *Simpler Networks,* Ed. J. Fentress, Sinauer, Sunderland, MA, 1979. *J. Compar. Physiol.* **129**:5–17, 1979. *Behav. Brain Sci.* **3**:535–575, 1980. *J. Neurophysiol.* **44**:1102–1121, 1980; **48**:1416–1432, 1982. *J. Compar. Physiol.* **145**:191–207, 1981. Network analysis of patterned discharge in stomatogastric ganglion of Panulirus.

73. Shashoua, V. E. *Proc. Natl. Acad. Sci.* **65**:160–167, 1970; **68**:2835–2838, 1971; **74**:1743–1747, 1979. *Brain Res.* **148**:441–449, 1978; **166**:349–358, 1979. Changes in protein and RNA in brain during learning.

74. Spencer, A. N. *J. Exper. Biol.* **93**:33–50, 1981. *J. Compar. Physiol.* **148**:353–363, 1982. Physiology of nerve net and neuromuscular junction in jellyfish.

75. Stein, P. S. et al. *J. Neurophysiol.* **40**:768–778, 1977. *Annu. Rev. Neurosci.* **1**:61–81, 1978. *J. Compar. Physiol.* **124**:203–210, 1978; **140**:287–294, 1980; **146**:402–409, 1982. Neural control of swimming and spinal reflexes in turtle.

76. Stent, G. S. et al. *J. Compar. Physiol.* **94**:121–154, 1974. *Proc. Natl. Acad. Sci.* **73**:3734–3738, 1976. *Science* **200**;1348–1357, 1978. Ch. 7, pp. 113–146 in *Neurobiology of Leech,* Ed. K. Muller et al., Cold Spr. Hbr. Lab, Cold Spring Harbor, NY, 1981. Models of circuitry of leech nervous system.

76a. Taylor, H. and J. W. Truman. *J. Compar. Phys.* **90**:367–388, 1974. Metamorphosis of abdominal ganglia of Manduca.

77. Thompson, R. F. et al. *Science* **216**:434–436, 1982. *Proc. Natl. Acad. Sci.* **79**:2731–2735, 1982. Localization of memory trace in hippocampus and cerebellum.

78. Truman, J. W. et al. *Science* **175**:1491–1493, 1972. *J. Exper. Biol.* **74**:151–173, 1978; **83**:239–253, 1979. Silk moth eclosion.

79. Tosney, T. and G. Hoyle. *Proc. R. Soc. London* **B195**:365–393, 1977. Computer-controlled learning in a simple nervous system.

80. Vrensen, G. and J. N. Cardoza. *Brain Res.* **218**:79–91, 1981. Synaptic changes in hippocampus after training.

81. Watson, W. H. and G. A. Wyse. *J. Compar. Phys.* **124**:267–275, 1978. Rhythms of ventilation, swimming and gill cleaning limulus.

82. Weeks, J. C. *J. Exper. Biol.* **77**:71–88, 1978. *J. Neurophysiol.* **45**:698–723, 1981. *J. Compar. Physiol.* **142**:253–263, 1982; **148**:265–279, 1982. Initiation and maintenance of swimming in leech.

83. Willows, A. O. D. et al. *J. Neurobiol.* **4**:207–237, 255–285, 287–300, 1973. *J. Compar. Physiol.* **100**:117–133, 1975. Neuronal mechanisms of fixed action patterns in Tritonia.

84. Wine, J. J. et al. *J. Exper. Biol.* **56**:1–18, 1972. *J. Compar. Physiol.* **121**:145–203, 1977. *J. Neurophysiol.* **40**:904–925, 1078–1097, 1977. *J. Compar. Physiol.* **142**:281–294, 1981. *J. Compar. Physiol.* **149**:155–162, 1982; **148**:143–157, 1982. *J. Neurophysiol.* **45**:550–573, 1981; **47**:761–786, 1982. pp. 242–292 in *Biology of Crustacea,* vol. 4, Ed. D. Bliss, Academic Press, New York, 1982. Function of giant fibers in escape behavior of crayfish.

85. Woolacott, M. and G. Hoyle. *Proc. R. Soc. London* **B195**:395–415, 1977. Neuronal events in learning by locust Schistocerca.

86. Wyse, G. et al. *J. Compar. Physiol.* **141**:85–92, 1982. Respiratory rhythm of Limulus.

87. Young, J. Z. *The Anatomy of the Nervous System of Octopus vulgaris,* Oxford Univ. Press, New York, 1972.

88. Zottoli, S. J. and D. Faber. *Neurosci.* **5**:1278–1302, 1980. Functions of Mauthner neurons in fishes.

12

TRANSMISSION
BETWEEN NEURONS

Animal behavior is determined by neural circuits which are controlled by transmission at synapses. Neurons are diverse in structure, electrical properties, and biochemistry. How nervous systems originated cannot be deduced from knowledge of synaptic function in present-day animals. The nerve nets of primitive diploblastic animals—the coelenterates—have many of the functional properties of complex nervous systems (Ch. 11).

The organization of living material into cells with resulting constraints was discussed in Chapter 2. As organisms became larger and cells differentiated into tissues, communication between cells became essential for the integration of whole plants or animals. Early in evolution, transmission from cell to cell was probably by modes similar to those now used in early embryos. The following general types of communication are recognized: (1) *electrotonic coupling* between cells by junctions across which charged ions and small molecules can pass; (2) secretion of chemical agents, *neurotransmitters*, which act on adjacent cells; (3) secretion of agents which diffuse for some distance and *modulate* the action of transmitters; (4) secretion of agents (*hormones*) which are transported in extracellular fluid and act on remote tissues; and (5) release into air or water of *pheromones*—attractants or repellents.

Coelenterates have in their epithelia two slowly conducting systems—ectodermal and endodermal—in which conduction is by flow of electrical currents between cells. Epithelial electrical conduction is in parallel with one or two nerve nets, in which transmission is usually chemical between neurons and between nerve nets and myoepithelial cells (137). The advantages of nervous over epithelial conduction are (1) greater transmission distances per cell, (2) faster conduction, and (3) modifiability of signals. Short neurons may use graded electrical signals without spikes but with the liberation of transmitter, for example, neurons in the nerve net of Hydra; neurons of the vertebrate retina; short neurons connecting mechanoreceptors to nearby central ganglia in crustaceans; sensory cells, such as hair cells of cochlea. Most neurons have long processes and transmit signals encoded as spikes which release chemical agents at their terminals. In some nervous systems, rapid transmission occurs at nexuses (gap junctions) by electrical coupling. Chemical synapses make possible more diversity and lability of behavior than electrical junctions and conducting axons.

ELECTRICAL TRANSMISSION IN EPITHELIA AND AT NEURON JUNCTIONS

Electrical coupling probably occurred before there were neurons and makes possible rapid conduction of signals from cell to cell. The degree of coupling is high whenever intercellular electrical resistance is low; coupling can be calculated by injecting current into one cell and measuring the voltage in adjacent cells. The junctional membranes between well-coupled cells may be permeable to ions and to small molecules, such as some dyes. The junctional regions are seen as gap junctions or nexuses in which extracellular space is locally reduced or absent. Nexuses are regions of membrane with many patches containing particles which cross the intervening space and bridge between two double membranes. The particles can be seen by scanning electron microscopy of membranes which have been freeze-cleaved (Fig. 12-1). Cleavage is along the hydrophobic layer within one member of the opposing double membranes. One face (P) is on the plasma membrane side, the other face (E) is on the extracellular space side. Particles on one face fit into pits in the opposite half of the same membrane. The fracture may be within either membrane, never in the gap between the double membranes. In membranes of electrotonic junctions between giant lateral axons and giant motor neurons of crayfish, center-to-center spacings are 20 nm and particles are 12.5 nm in diameter. In the septa between segments of earthworm giant axons the nexal membrane occupies 4.5 percent of the surface area, particles are 10.0 nm apart, and particle diameters are 11–15 nm (27). Junctional particles also occur in epithelia; in insects they are larger than in vertebrate embryos.

Many kinds of early embryos are functional syncytia; the blastomeres are electrically coupled. In development the duration of the coupled condition varies; it lasts through early cleavage stages of sea urchins, during blastula stages of *Limulus* and fish (*Fundulus*). Injected dyes can pass from cell to cell, but dyes from the medium do not enter embryonic cells (19,138). In most animal epithelia, especially during early development, signals are conducted electrically for some distance at a rate that is slow compared to the rate in nerve axons. Intercellular conduction may occur in developing embryos before nerves become differentiated. Epithelial conduction has been described in amphibian embryos and adult salamanders. Epidermal cells of salamander skin are electrically coupled; if a wound is made, cells become decoupled and after the two sides of the wound heal together the cells again become electrically coupled (93). Conduction between amphibian blastomeres decreases as transjunctional voltage input is increased in either polarity (138).

Sponges lack nervous systems but many sponges have contractile cells and most sponges have flagellated cells lining water channels. In most sponges, there is no conducting system, but in some (*Tethya*) very slow conduction (10–20 mm/sec) may result in generalized spread of excitation. In one group, the siliceous sponges, Hexactinellida, basal membranes form trabeculae which are syncytia with intercellular "plugs" analogous to nexuses. Conduction spreads at 0.6 cm/sec throughout an individual and from one sponge to another in a colony. Feeding currents generated by flagellated cells are stopped by the conducted waves of activity (98a).

Epithelial conduction is one normal mechanism of coordination in coelenterates (hydroids and medusae) (137). In a tunicate, epithelial conduction activates underlying muscles (25) and in jellyfish epithelia may activate the nerve net. The large epithelial cells of dipteran salivary glands are well coupled and appear to synchronize secretion; molecules of 1200 daltons but not of 1900 may pass between cells. Cells of sponges dissociated in culture make electrically coupled connections when they reassociate. Cancerous epithelia appear to have lost electrical connectivity (93). The cells of some kinds of plants (*Nitella*) are electrically coupled.

Electrotonic coupling in epithelia (especially in insects) is dependent on calcium concentration in the cells, and decoupling occurs when intracellular calcium increases above normal levels. In embryos (fish, amphibians) electrotonic coupling depends on pH; acidification by CO_2 reduces the coupling with a pK of 7.3; a small change in pH_i causes a large change in coupling (19,138). Junctions in Purkinje cells of heart and in smooth muscle are more sensitive to pH_i than to pCa_i.

Electrotonic coupling between neurons occurs wherever the electrical resistance of adjacent membranes is low. Coupled neurons usually pass low frequencies of current better than high frequencies. Electrical junctions implement the synchronization of coupled neurons. In electric fishes, medullary centers which trigger the electric organ discharge contain motorneurons that are electrotonically coupled. In leech ganglia 14 motorneurons on a side are electrically coupled; in *Aplysia* 30 neurons in the pleural ganglion are electrically coupled. In the giant axons of earthworms the septate junctions are of low resistance and current flows from segment to segment; small molecules, such as fluorescein, can pass across the septa. The bag cells of *Aplysia* abdominal ganglion are coupled and synchronize in release of egg-laying hormone and in egg-laying behavior (23). Electrotonic junctions in molluscs and earthworms provide for rapid transmission, near synchrony of coupled neurons, and conduction in one direction. In *Tritonia* swimming and withdrawal are controlled by the pedal ganglion with 30 coupled nerve cells; coupling is lessened if input frequencies are above 0.1 Hz.

Some electrotonic junctions between neurons are polarized, passing current in one direction only. An example is the junction between lateral giant axons and giant motorneurons in crayfish. A few junctions are doubly rectifying, with depolarizing potentials passing in each direction and hyperpo-

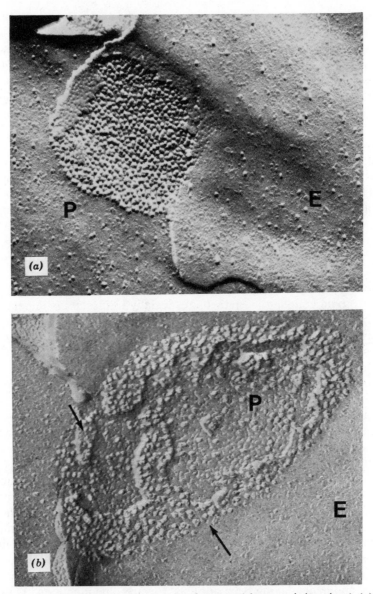

Figure 12-1. Scanning electron micrographs of nexuses (electrotonic junctions). (*a*) Freeze-fracture of anterior byssus retractor muscle of Mytilus. P face has 9–11 nm particles, pits in E face. (*b*) Freeze-fracture of lobster nerve cord particle size 10–13 nm in E face. (*c*) Freeze-fracture of earthworm nerve cord, particle in upper face, pits in lower face. (*d*) Nexuses on septate membrane of earthworm giant fibers, pits and particles arranged in strands. Personal communication, Peter Brink.

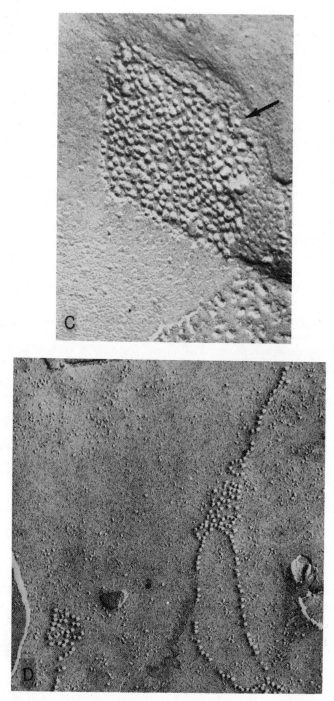

Figure 12-1. *Continued*

larizing pulses passing in neither direction. An example is the junction of sensory and motorneurons of leech ganglia (107). The ultrastructure of a one-way rectifying junction is not different from that of a nonpolarized junction, but the molecular organization must be different.

One limitation of electrical transmission between neurons is that it is usually excitatory, taking place only in response to depolarizing impulses. A few electrotonic junctions are inhibitory. One kind of electrical inhibition occurs in a goldfish Mauthner neuron where hyperpolarization is induced by an action potential in a coil of axons around the axon hillock. Another mechanism of inhibition found in *Aplysia* ganglia is by a long-lasting or late hyperpolarizing postsynaptic potential (PSP), such as occurs after a burst of impulses. Purkinje cells of rat cerebellum show fast electrical inhibition which precedes chemical excitation.

Many synapses transmit both electrically and chemically, for example, the chick ciliary ganglion. In the sea slug *Navanax* 10 neurons in the buccal ganglion are electrically coupled in a network. A chemical synapse onto one or more of these network cells reduces nonjunctional resistance, thereby shunting the electric current and decreasing the electrotonic coupling; the chemical synapse modulates electrical transmission (137a).

In lamprey spinal cord the most common type of excitatory synapse between giant axons and nerve cells uses both electrotonic and chemical transmission. A fast component of EPSP is electrotonic, a second component is chemical. Electrotonic coupling is rectifying (123a).

Both chemical and electrical transmission between cells probably occurred early in evolution. Chemical agents may first have functioned as pheromones. For example, the separate cells of slime molds release cAMP that induces cells to aggregate. Chemical transmission occurs in all nervous systems. In coelenterates, electrical conduction in epithelia takes place along with chemical conduction in nerve nets. Nervous systems use electrical coupling where speed and synchrony are important; in some neural systems electrical synapses may have evolved secondarily, as in giant fiber complexes.

CHEMICAL NEUROTRANSMISSION

During biochemical evolution, compounds may take on new functions, sometimes after they have become modified. Completely new compounds are rare. Intercellular signalling in nonneuronal cells may be chemical, for example, by cyclic aggregating factors in dissociated sponges, or by growth factors and hormones in complex plants and animals. All chemical neurotransmitters are preexisting compounds or modifications of them. Transmission at a chemical synapse requires the following sequence of events and agents: a transmitter substance, a series of enzyme reactions that synthesize the transmitter, reactions for packaging, presynaptic currents (usually inward Ca^{2+}) that trigger transmitter release, receptor molecules and associated mediators that initiate conductance or other changes in postsynaptic membranes, agents that remove or inactivate the transmitter, and modulators that control the liberation and postsynaptic action of a transmitter. The distinction between a transmitter, a modulator, and a neurohormone is not sharp. A *neurotransmitter* may be considered as an agent that causes an ion conductance change in the postsynaptic element receiving the innervation. A *modulator* is an agent that either modifies the liberation of a transmitter, or alters excitability of receptor cells at some distance. A *neurohormone* is an agent that is transported in the blood to a target site at a distance from the site of liberation. Some neuroactive agents serve more than one of these functions. Some transmitters are substances whose functions have changed during evolution; a given transmitter may function differently in several kinds of animal or in one animal at different synapses. Many transmitters are known, but few postsynaptic receptor molecules have been isolated and characterized.

It was formerly believed that a single neuron makes use of only one kind of transmitter, though the neuron may have several processes. Recent evidence demonstrates the presence of at least two transmitters in many neurons (53,17), but whatever the number of substances produced by a neuron, all the terminals of that neuron release the same set of products. Similarly, a neuron may have several kinds of receptor molecules, but there is no evidence for segregation of these. A neuron is adapted in its entirety—transmitters and receptors (132).

A survey of neurotransmitters indicates that the most primitive are amino acids, and that simple modifications of amino acids result in a variety of small-molecule transmitters. Polypeptides, many with specific neural actions, are modulators or transmitters.

AMINO ACIDS

Most neurotransmitters are either amino acids, their direct derivatives, or peptide polymers. Some twenty amino acids prevailed prebiotically as the essential building blocks for proteins. The selective forces for fixing these particular amino acids are not known. Amino acids served as feeding stimulators and as pheromones before they became neurotransmitters.

Behavioral attractants and repellants may have originated by selection of a few amino acids that had excitatory actions on cell membranes. These amino acids or their derivatives became transmitters after neurons evolved. Many cells are sensitive to amino acids at low concentrations (Fig. 12-2). *E. coli* shows positive chemotaxis to some amino acids in concentrations of 10^{-6} to 10^{-8} M, and shows negative taxis away from some other amino acids at concentrations somewhat higher: 10^{-3} to 10^{-4} M (3). Some attractants elicit chemotaxis and are transported into the cells. Receptor proteins in cell

E. coli Positive Chemotaxis	*E. coli* Negative Chemotaxis
Asp 6 $\times$ 10^{-8}M	Leu 1 $\times$ 10^{-4}M
Glut 5 $\times$ 10^{-6}M	Ileu 1.5 $\times$ 10^{-4}M
Thr 1 $\times$ 10^{-6}M	Val 2.4 $\times$ 10^{-4}M
Ala 7 $\times$ 10^{-5}M	Norleu 3 $\times$ 10^{-4}M
Ser 3 $\times$ 10^{-7}M	Others 10^{-3}M

Figure 12-2. Amino thresholds for positive and negative chemotaxis of *E. coli*. Reprinted with permission from J. Adler et al., *Journal of Bacteriology,* Vol. 118, No. 2, p. 566–567. Copyright 1974, American Society for Microbiology, Washington, DC.

Negatively charged amino acids (acidic)
   aspartic (2 carboxyls)
   glutamic (2 carboxyls)

Positively charged amino acids (basic)
   lysine has + amino group
   arginine has = guanidinium
   histidine has + imidazolium

Uncharged (hydrophilic) polar, neutral amino acids
   ser, thr, tyr polar by OH group
   asp, glyn polar by amide
   cysteine polar by SH
   glycine

Nonpolar (hydrophobic) amino acids
   ala, leu, ileu, val, pro (aliphatic)
   phenylala, ala, tryp (aromatic)
   methionine (S)

Figure 12-3. Table of amino acids according to structure.

membranes are specific for certain classes of amino acids. After binding to a receptor in bacteria, methylation occurs and locomotor or other response is then initiated (98b). Two receptor proteins have been chemically isolated, one for the negatively charged amino acids Asp and Glu, and one for the uncharged but polar amino acids Ala, Gly, Ser (Fig. 12-3). Each receptor is approximately 60,000 daltons in size (65,153). The gene for the aspartate receptor of *Salmonella* consists of 1656 nucleotides corresponding to a 59,416-Da protein (124a). Leukocytes are attracted by amino acids, eosinophils best by a tetrapeptide. When exposed to a chemoattractant, the membranes become depolarized after several minutes, from −15 to −50 mV (135a).

    Feeding reactions in coelenterates are initiated by amino acids; some species respond to proline, others to tyrosine, some to proline and the tripeptide

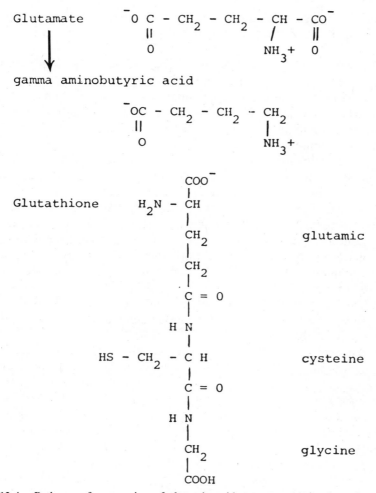

Glutamate

gamma aminobutyric acid

Glutathione

glutamic

cysteine

glycine

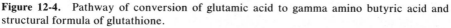

Figure 12-4. Pathway of conversion of glutamic acid to gamma amino butyric acid and structural formula of glutathione.

glutathione, a few (Hydra and others) to glutathione only (glutamyl-cysteine-glycine) (Fig. 12-4) (89). In amoebas and leucocytes, pinocytosis is stimulated by glutamic acid or by aspartic acid—both negatively charged at physiological pH.

In fishes (salmon) olfactory receptors are sensitive to some amino acids (alanine at $10^{-7}M$). In flies there are no specific amino acid taste receptors, but "salt" cells are known to be stimulated by either of two amino acids, "sugar" receptors by any of six amino acids. Each type of fly chemoreceptor is inhibited by certain amino acids (134). Chemoreceptors of a crab Cancer respond to a low concentration of glutamate, but not to glutamine. The steric specificity of action of amino acids for receptor molecules has not been much studied; acidic amino acids with negative charge are generally excitatory.

Several nutritionally essential amino acids function unmodified as neurotransmitters in complex animals. Glutamic acid is excitatory at many postsynaptic membranes; glycine is inhibitory. Aspartic acid may be an excitatory transmitter at some junctions. A simple derivative of glutamic acid—gamma amino butyric acid (GABA)—is an important inhibitory transmitter in many animals.

Glutamic and aspartic acids are excitatory when applied iontophoretically in low concentrations to frog retina; glycine, alanine, and GABA are depressant. In bipolar cells of carp retina, aspartate depolarizes and decreases the response to light in H (hyperpolarizing) cells; aspartate hyperpolarizes D (depolarizing) cells (159). Glutamate, aspartate, and GABA are present in high concentration in retinal neurons of frog, rat, and guinea pig (6). GABA content is especially high in retinal horizontal cells, taurine in photoreceptors. In goldfish retina, GABA is concentrated in the horizontal cells connected to cones and also in amacrine cells of one class; illumination increases uptake of GABA by cone horizontal cells. Cones are hyperpolarizing to specific horizontal cells and amacrine cells, an effect apparently mediated by GABA (85).

In Retzius cells (large neurons) of leech ganglia, L-glutamate is excitatory, increasing g_{Na} and g_K, inhibitory (prior to depolarization) increasing g_{Cl}. Neurons of *Aplysia* ganglia differ two- to fourfold in content of Asp and Glu. Evidence for glutamate as an excitatory transmitter was obtained for crustacean and locust neuromuscular junctions where $4.5 \times 10^{-6}M$ glutamate causes depolarization of some 3.5 mV (148). Glutamate is also an excitatory transmitter at many insect neuromuscular junctions. At crustacean neuromuscular junctions, GABA increases chloride conductance, decreases excitatory junction potentials, and is probably the agent for both presynaptic and postsynaptic inhibition (12). The inhibitory action of GABA at crustacean and insect neuromuscular junctions can be prevented by picrotoxin or bicuculline. Cockroach hindgut muscles are excited to contract by L-glutamate and by L-asparate. Body wall muscles of fly larvae are excited by L-glutamate with a reversal potential of -1.3 mV, or by asparate with reversal at -4.5 mV.

In vertebrate spinal cord, glycine appears to be an inhibitory transmitter; the action of glycine is blocked by strychnine. In other regions of the vertebrate nervous system GABA is the inhibitory transmitter. Synaptosomes from rat brain have high affinity for GABA. In cat visual cortex GABA is inhibitory and bicuculline blocks the physiological inhibition (115). Glycine and GABA are inhibitory transmitters in lamprey brain (99). In mammalian spinal cord aspartate is excitatory for polysynaptic synapses. Glutamate may be one excitatory transmitter in dorsal root afferents (160). In cerebral cortex, glutamate receptors occur, some of them Na-dependent and some

Na-independent (16). Cerebellar granule cells in culture release Glu upon stimulation.

Questions have been raised as to how amino acids can be neurotransmitters since they are present in quantity in blood and interstitial fluids. Calculations show that the effective concentration reached in the synaptic cleft is significantly higher than in the blood.

Each of the excitatory amino acids (Asp, Glu) has a second COOH which provides a negative charge at physiological pH. The excitatory amino acids are lipid soluble. The inhibitory amino acids glycine, alanine, and GABA have one COOH (Fig. 12-3).

AMINO ACID DERIVATIVES WITH METHYLATED QUATERNARY NITROGENS

Several classes of compounds of importance for membrane functions are formed by relatively direct changes in amino acids. Amines of one class have quaternary nitrogens that can become methylated, often by three methyl (Me) groups. Methylated quaternary amines occur in bacteria, plants, and animals (Fig. 12-5). The most common methylating agent is the amino acid methionine (Met), which in animal tissues is known to donate methyl groups to at least 40 different compounds (79a).

Betaine is formed from glycine by methylation of its nitrogen. Betaine is a relatively inert anion in many cells, and functions in volume regulation of numerous euryhaline molluscs and crustaceans (Ch. 9).

Taurine is formed from cysteine by way of cysteinesulfenic acid (70). Taurine constitutes 50 percent of the free amino acids in mammalian heart; there is more in auricles than in ventricles. Taurine occurs in high concentrations in the outer segments of retinal photoreceptors; stimulation of frog or chick retina by light causes release of taurine. Enzymes of taurine metabolism are abundant in rat brain; application of taurine depresses brain activity. Taurine may regulate calcium influx into presynaptic terminals (113). Taurine occurs in tissues of invertebrates, especially molluscs. Chemoreceptors of crustaceans respond to concentrations of taurine 10^{-12} M; crabs and shrimp respond to extract of molluscs at low concentrations of 10^{-18} to 10^{-13} g tissue/li (10).

Homarine is formed by reaction of glycine with succinyl co-A-N-succinyl glycine; homarine is a methylated ring (Fig. 12-5). Homarine can serve as a methylating agent to form methylamines choline and betaine (106). Homarine is abundant in the nervous system of crustaceans; its function is unknown.

Choline is formed from serine in the sequence: serine → ethanolamine → choline (Fig. 12-6). All cell membranes contain phospholipids; the principal ones are phosphatidyl-ethanolamine, phosphatidyl-choline, phosphatidyl-

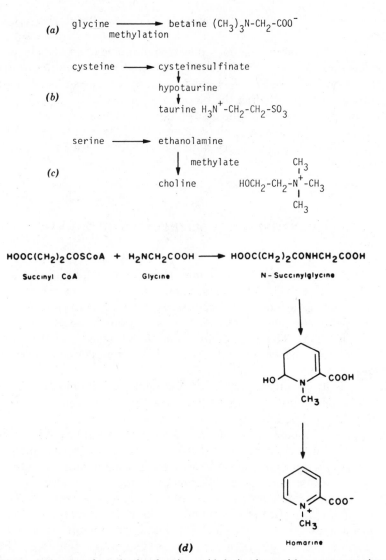

(a) glycine $\xrightarrow[\text{methylation}]{}$ betaine $(CH_3)_3N-CH_2-COO^-$

(b)

cysteine $\longrightarrow$ cysteinesulfinate

$\downarrow$

hypotaurine

$\downarrow$

taurine $H_3N^+-CH_2-CH_2-SO_3$

(c)

serine $\longrightarrow$ ethanolamine

$\downarrow$ methylate

choline $HOCH_2-CH_2-\overset{\overset{CH_3}{|}}{\underset{\underset{CH_3}{|}}{N^+}}-CH_3$

$HOOC(CH_2)_2COSCoA$ + H_2NCH_2COOH $\longrightarrow$ $HOOC(CH_2)_2CONHCH_2COOH$

Succinyl CoA   **Glycine**   **N-Succinylglycine**

(d)

Homarine

Figure 12-5. Pathways of synthesis of amino acid derivatives with quaternary nitrogens. (*a*) Glycine to betaine. (*b*) Cysteine to taurine. (*c*) Serine to choline. (*d*) Glycine to homarine. Reprinted with permission from J. C. Netherton, III and S. Gurin, *Journal of Biological Chemistry,* Vol. 257, No. 20, p. 11974. Copyright 1982, American Society of Biological Chemists, Bethesda, MD.

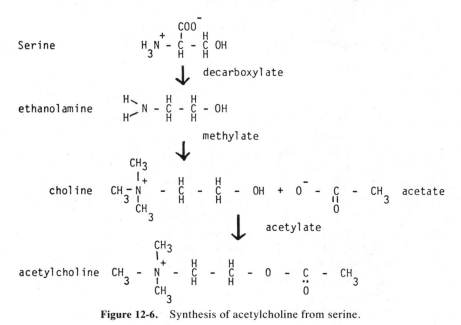

Figure 12-6. Synthesis of acetylcholine from serine.

serine, and phosphatidyl-inositol. Many animals require choline as a vitamin in the diet.

ACETYLCHOLINE; TYPES OF CHOLINERGIC NEURONS

Of all the methylated amines the most widespread in neural transmission is acetylated choline (ACh). Why acetylated choline rather than one of the other methylated quaternary amines became a universal neurotransmitter is an unanswered question. Hypotheses are:

1. Choline is readily acetylated and the enzymes of acetylation are widely distributed.
2. An enzyme (acetylcholine esterase, AChE) hydrolyzes ACh with a high turnover rate.
3. Acetylcholine is a relatively long, flexible molecule with charged groups at each end (43); other quaternary compounds may be more rigid and hence less reactive (58).
4. Selection of acetylcholine may have occurred because of the presence of several potential receptor proteins. How receptors evolved is unknown. Could they have evolved independently of a transmitter and with some other function? How could coevolution of transmitter and

receptor have occurred? A cholinergic receptor protein has been isolated and sequenced, its gene cloned, and its mRNA sequenced.

Acetylcholine (Fig. 12-6) is formed by the reaction

$$\text{acetyl CoA} + \text{choline} \xrightarrow{\text{choline acetyl transferase}} \text{ACh} + \text{CoA}$$

Acetylcholine is inactivated by hydrolysis (Fig. 12-7a); some hydrolytic enzymes (acetylcholine esterases, AChE) are relatively specific for the acetic acid ester, others are specific for other choline esters. Acetylcholine esterases occur in postsynaptic membranes and also in nonnervous tissues, red blood cells, submaxillary gland, glia, placenta, and plants.

The presence of acetylcholine has been reported in some bacteria and in plants—the alga *Nitella*. Acetylcholine is present in tips of secondary roots and in buds of mung beans; mung bean also contains AChE, which, like animal AChE, is inhibited by neostigmine (74a). When root tips of mung bean are illuminated with red light, ACh is released; local treatment with ACh induces an efflux of H^+ from the growing tips. ACh may function in plants as in animals to mediate ion fluxes. Application of ACh to the fungus *Trichoderma* brings about sporulation; ACh mimics red light in photogenic processes in which phytochrome is the photoreceptor. Honey may contain as much as 40 μg Ach/g. ACh affects ion permeability in many nonneural cells.

Acetylcholine esterase is present in trypanosomes, ciliates, flatworms, and with high activity in squid and insect ganglia. The activity of the enzyme in trout brain is very high; in mammals the level of activity varies with the region of the nervous system; ventral roots of spinal cord have 300–400 times as much AChE as dorsal roots. AChE is formed at neuromuscular junctions in cultured innervated muscle, but it is not synthesized in noninnervated muscle. Acetylcholine concentrations parallel AChE activity in many animals; the concentration of ACh in a honeybee brain is estimated at 2.7 mM, in Octopus optic ganglion 2.1 mM, in guinea pig cerebral cortex 0.013 mM. In cultured sympathetic neurons synthesis of ACh is increased within 48 hours by 100–1000 fold whenever nonneural glia cells are added. At central and peripheral synapses ACh is usually contained in clear-core vesicles of 500 Å diameter.

Several functional types of ACh receptors are recognized. Nicotinic receptors occur in vertebrate striated muscles, electric organs, and sympathetic ganglia. Nicotinic receptors can be selectively blocked by D-tubocurarine. These receptors respond to ACh within milliseconds. In vertebrate autonomic effectors, such as gastrointestinal muscle and heart, there are muscarinic receptors; these are blocked by atropine or hexamethonium; muscarinic responses may last for seconds. A third type of receptor is indicated in the heart of some bivalve molluscs and in neurons of *Aplysia* and

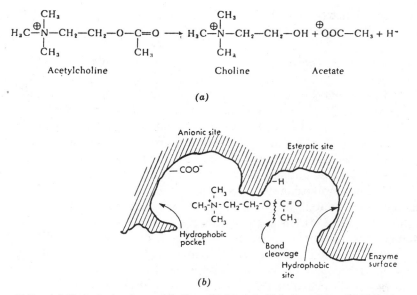

Figure 12-7. (a) Hydrolysis of acetylcholine. (b) Model of binding sites of AChEs for acetylcholine. Modified from P. H. Hochachka and G. N. Somero, *Strategies of Biochemical Adaptation*. Copyright 1973.

Helix; cholinergic stimulation of these molluscan receptors is not sensitive to nicotinic or muscarinic blockers, but is blocked by benzoquinonium, tetraethylammonium, or methylxylocholine (49).

In mammalian cerebral cortex 30 percent of neurons are responsive to ACh; some are excited, some inhibited. In pons and medulla 33 percent of cells are excited, 22 percent inhibited; in cerebellum the synapses between parallel fibers and Purkinje cells are probably cholinergic. An example of a cholinergic synapse in mammals occurs in the spinal cord, where the feedback collateral from a motorneuron onto an interneuron blocks further discharge of the motorneuron; this junction is nicotinic. In vertebrate brain, most of the ACh receptors are muscarinic; in spinal cord the receptors are nicotinic.

Neurons of sympathetic ganglia show several types of responses to orthodromic input and to applied transmitter substances. The sequence is similar in frog and rabbit but there may be differences in transmitters—see Figure 12-13. The first response to an incoming volley is a fast depolarizing synaptic potential (EPSP) of nicotinic receptors due to an increase in sodium conductance (g_{Na}). This is followed by slow excitatory potential generated by muscarinic receptors and due to decreasing g_K (91). A slow hyperpolarizing IPSP results from increase in Ca-dependent K-current elicited by acetylcholine (frog) or by dopamine (rabbit). A fourth response is a late slow depolarization which may last for a minute or more. This is due to reduction in steady outward current of K^+, the so-called M current, and may be induced by a muscarinic action of ACh or by a neuropeptide, LHRH (76). The function of the late slow depolarization may be as a brake on rapid depolarization (2).

In neurons of *Aplysia* three types of response to ACh are recognized, a fast depolarization due to increase in g_{Na}, a fast hyperpolarization by increase in G_{Cl}, and slow hyperpolarization by increase in g_K. Some molluscan neurons (with nicotinic receptors) are depolarized when ACh is applied and show an increase in g_{Na}; in these cells (D cells) the resting potential (E_m) is low (-33 to -37 mV) and E_{ACh} is similar to E_m (-33 mV); other neurons in the same ganglion are hyperpolarized by an increase in g_{Cl} and in them (H cells) the resting potential is more negative (-45 to -51 mV) as is E_{ACh} (-51 mV). In other molluscan neurons which have neither nicotinic nor muscarinic receptors there is slow hyperpolarization due to increase in g_K and E_{ACh} is -80 mV (78,133). In some molluscan hearts (*Mytilus, Modiolus*) ACh excitation is accompanied by depolarization, decrease in membrane resistance and increase g_{Na}; in oyster (*Crassostrea*) ACh is inhibitory by increase in g_{Cl}; in *Mercenaria* cardio-inhibition is due to an increase in g_K (49). *Mercenaria* heart can be inhibited by ACh concentrations as low as 10^{-12} M (71).

Acetylcholine is probably an excitatory transmitter in insects at some central sensory synapses and at some neuromuscular junctions. Receptors on crustacean muscle resemble but are not identical to nicotinic receptors of vertebrates. Echinoderm muscles are very sensitive to ACh and have nicotinic receptors. Acetylcholine receptors are localized at motor endplates of vertebrate muscles; after the motor nerve is cut and has degenerated, receptor molecules develop over the entire muscle fiber membrane.

The choline end of acetylcholine has a quaternary ammonium ion and the acetyl end forms esteratic linkages (Fig. 12-7*b*). The membrane receptor molecules and the hydrolytic esterase must accommodate both end groups of ACh. Properties of the receptor site on AChE are shown by measurement of binding of analogs of ACh. An anionic site with negative charge interacts with the quaternary ammonium nitrogen; hydrophobic amino acids near the anionic site are compatible with the three methyl residues. If the methyl groups of ACh are replaced by hydrogens, affinity between enzyme and substrate is much reduced (seven-fold decrease after one H^+ is substituted for a Me^+). The catalytic (esteratic) site is hydrogen-bonded to AChE and cleavage occurs upon the addition of water. At low temperatures the electrostatic interactions between the anionic site of AChE and the quaternary ammonium ion are stabilized and hydrophobic interactions are weakened. Opposite effects occur when hydrostatic pressure is applied, coulombic binding increases, and hydrophobic binding decreases (67a). Adaptive differences in AChE molecules have been ascertained. The enzyme of a deep-sea fish *Antimora* has smaller hydrophobic pocket, more electrostatic binding, and a higher enthalpic contribution that the enzyme from either warm-water fish or mammals (Ch. 5) (67a).

The nicotinic receptor (AChR) of the electroplax of *Torpedo* has been characterized. The receptor protein has been separated from AChE by differential salt extraction. AChR is a protein of molecular weight 250,000 daltons and consists of five subunits, the smallest two of which are identical. Designations and approximate molecular weights of the nicotinic receptor in daltons are as follows: α 40K, β 50K, γ 60K, and δ 65K, in the ratio 2:1:1:1. Each of the subunits is a single chain, and some regions show 35 to 50 percent homology. Genes for each subunit have been cloned; the α gene consists of 13,500 base pairs which correspond to 937 amino acid sequences; the α precursor has 545 amino acids (Fig. 12-8) (47). The β subunit is a glycopeptide of 469 amino acid sequences, γ of 489, and δ of 501 sequences (108a, 120). All four subunits of AChR are needed for the nicotinic response to ACh; only the α subunit is required for the binding of the toxin α-bungarotoxin (102a).

The α-receptor has three major domains: the upper hydrophilic one faces the synaptic cleft where the NH_2 terminal fits; an intermediate domain is hydrophobic and occurs in the lipid bilayer; the lower domain is hydrophilic and faces the cytoplasm (47). Use of monoclonal antibodies for the muscarinic receptor indicates extreme conservatism, with the structure similar in *Drosophila* head, rat brain, dog heart, and monkey ciliary muscle. The homologies suggest that the subunits evolved from a single ancestor, that an initial gene duplication separated the genes for two subunits, and that each of these underwent subsequent gene duplication (44,120). The α subunit has seven cysteines and linkage occurs near a disulfide bridge (108). Acetylcholine binds to the α subunits. Binding of ACh to its nicotinic receptor is very fast, with binding rate constants of approximately 2.4×15^{-9}/M/sec (44).

AChR subunits have been prepared from denervated mammalian muscles; these nicotinic receptors differ in details from those from *Torpedo* electric organ. Rat muscle AChR shows two major and three minor subunits. Cat muscle AChR appears to consist of one polypeptide subunit (97). Receptors of the denervated muscle may be different from those of innervated muscle. The AChR of *Torpedo* electric organ is different from that of *Torpedo* brain (11). The nicotinic receptor from electric organ has been inserted into a lipid bilayer and ionic conductance of single channels measured. The conductance on addition of ACh is 25 pS (picosiemens); channel conductance can be blocked by curare. A single AChR molecule spans the membrane and has a diameter of 8.5 nm (142) (Fig. 12-9).

Both the abundance of nicotinic receptors and the effectiveness of ACh as a transmitter have been estimated for muscle endplates and parasympathetic cardiac ganglion cells of frog. Terminal boutons cover 3 percent of the surface of parasympathetic ganglion cells. In these ganglion cells and in frog muscle sensitivity to ACh is found only at bouton or endplate postsynaptic membranes; after denervation the entire membrane becomes sensitive to ACh. It is calculated that the concentration of ACh reaches 50 nM over an

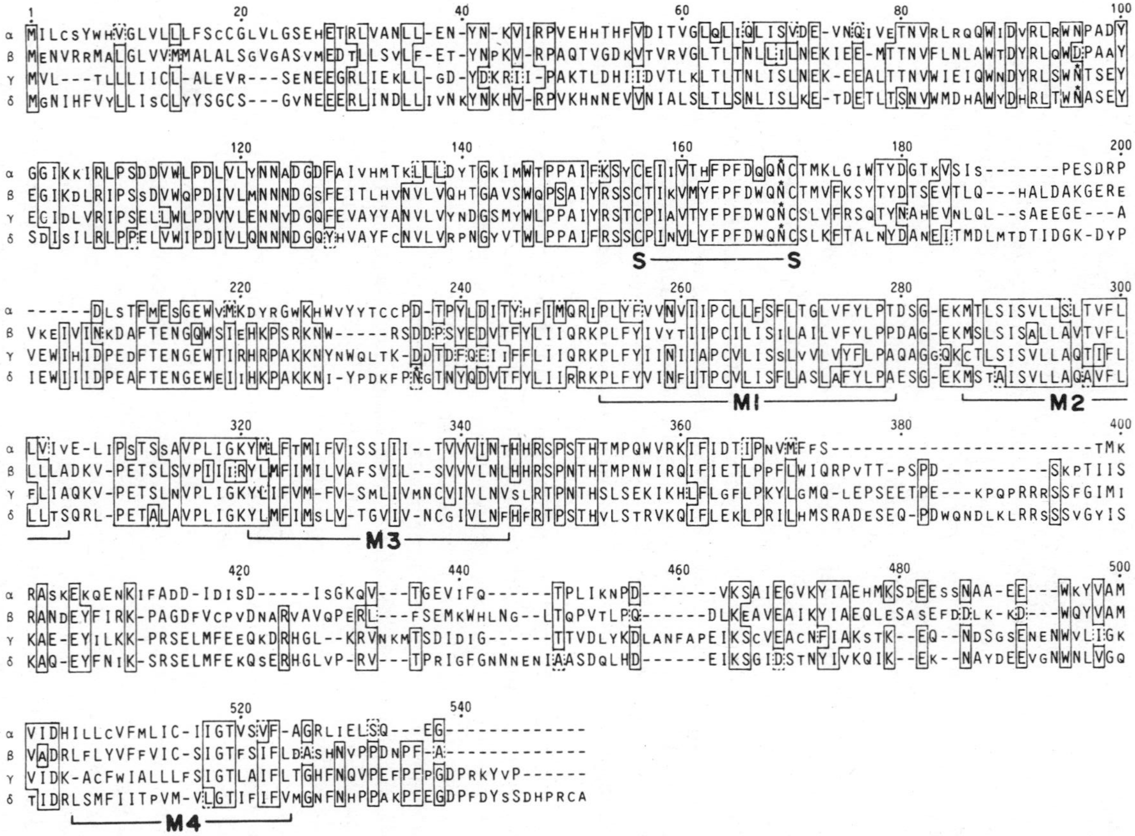

Figure 12-8. Comparison of amino acid sequences of the four AChR subunit precursors. One-letter notation of amino acids. Sequences of α, β, γ, and δ subunit precursors, positions are aligned. Amino acids deduced from hybridization analysis of electroplax RNAs by cDNA probes. Reprinted by permission from M. Noda, et al., *Nature*, Vol. 302, No. 3908, pp. 528–532. Copyright 1983, Macmillan Journals Ltd.

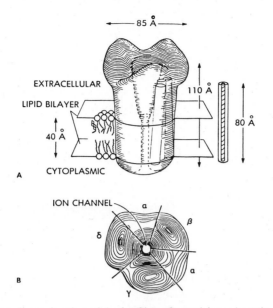

Figure 12-9. Three-dimensional model of AChR. Central ionophoretic channel traverses 110 Å length of the molecule. Funnel shape visualized by EM. Receptor extends 15 Å on cytoplasmic side and 55 Å on synaptic side. Reproduced from Kistler, J. et al., *Biophysical Journal*, 1982, Vol. 37, pp. 371–383 by permission of the Biophysical Society.

area of 1 mm², giving a concentration of 3. × 10^{-4} M; one quantum or packet of ACh liberated from a bouton contains 10,000 molecules (83). In cultured cells of embryonic muscle, the receptors occur in clusters of hot spots. The ion conductance for one miniature endplate potential corresponds to an ionic current of 2.5 pA at a single channel; 1 or 2 ACh receptors occur per channel (142). The abundance of receptors in mouse and bat muscle (8500 sites/μm²) was estimated from binding of tritiated bungarotoxin, electron micrographs, and radioautography.

In summary, the presence of acetylcholine in neural and nonneural tissues of animals, and in plants, and the wide occurrence of choline esterases of different specificities indicate that general cell permeability may be influenced by ACh. Acetylcholine is an ancient modulator of ion fluxes, and it has prevailed over other quaternary amines, probably because of its reactivity and structural flexibility. The extreme sensitivity to ACh of many postsynaptic membranes suggest that they contain high concentrations of receptor molecules. The diversity of effects of agonists and antagonists and of resulting ionic responses shows receptor molecular specialization in both invertebrate and vertebrate animals, and that there are several different ACh receptor types. For the first time, a synaptic receptor molecule of behavioral significance—the nicotinic acetylcholine receptor—has been characterized, its component amino acids have been sequenced, its gene cloned, and the nucleotides of its messenger RNA sequenced.

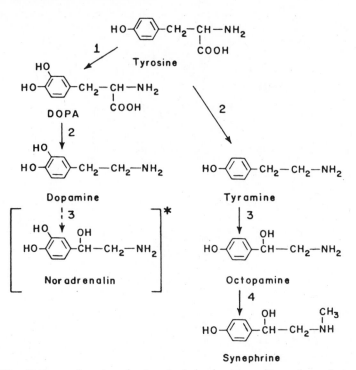

Figure 12-10. Pathways from tyrosine to catecholamine transmitters; dopa, dopamine, norepinephrine, and octopamine. Reprinted by permission from D. L. Barker and E. A. Kravitz, *Nature New Biology*, Vol. 236, No. 63, p. 62. Copyright 1973, Macmillan Journals Ltd.

MONOAMINES; AMINERGIC NEURONS

Monoamines function widely as transmitters and are derived from essential amino acids. Tyrosine can be converted to catecholamines (CA) in either of two directions: (1) via DOPA, dopamine, to norepinephrine and epinephrine, or (2) via tyramine to octopamine (Fig. 12-10). Tryptophan can be converted via 5-hydroxytryptophan to 5-hydroxytryptamine (5HT, serotonin) (Fig. 12-11).

Catecholamines

Neurons which use catecholamines as transmitters are designated adrenergic. Catecholamines act on many tissues other than neurons and muscles. Vertebrate receptors for norepinephrine (NE) and epinephrine (E) are identified by pharmacological methods as of two types, α and β; the α receptors are α_1 and α_2. Commonly used α blockers are phenoxybenzamine and phen-

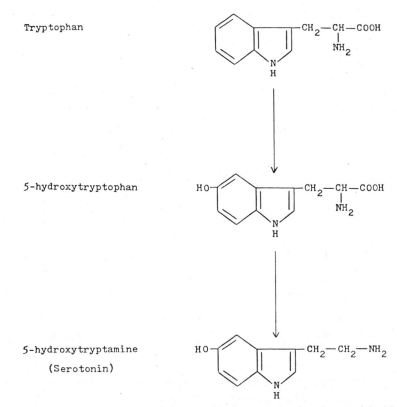

Tryptophan

5-hydroxytryptophan

5-hydroxytryptamine

(Serotonin)

Figure 12-11. Pathway from tryptophan to 5-hydroxytryptamine (serotonin). Modified from A. L. Lehninger, *Principles of Biochemistry,* Worth Publishers, New York, 1982, p. 539.

tolamine; β blockers are propranolol and dichlorophentolamine. Alpha receptors are predominant in some tissues, beta receptors in others. There is interaction between α and β receptors in some tissues (e.g., rat kidney). The β receptor is coupled to adenylate cyclase, which catalyzes synthesis of c-AMP (cyclic adenine monophosphate) (59,67); intracellular c-AMP serves as a second cellular messenger (Fig. 12-12), but the active receptor is not adenylate cyclase per se. Intracellular c-AMP activates kinases for protein synthesis.

Red blood cells (especially reticulocytes) are stimulated by catechol-amines to make membrane phospholipid; the initial step is methylation, which may be a response to a catecholamine acting on a β receptor. Turkey red blood cell ghosts are calculated to have 500–1000 receptors per cell; binding of the agonist propranolol shows a K_d (half saturation) of 2.5 mM; adenylate cyclase is lost as reticulocytes mature but catecholamines continue to bind to the cells. The β receptor is stereospecific for the L-isomer of agonists (89a). Blood platelet aggregation is stimulated by E or NE action

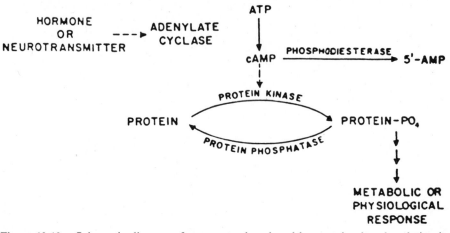

Figure 12-12. Schematic diagram of apparent roles played by protein phosphorylation in mediating effects of hormones and neurotransmitters that act via cyclic AMP. Reprinted with permission from P. Greengard, *Science,* Vol. 199, p. 147. Copyright 1978 by the AAAS.

on α_2 receptors. α-Adrenergic receptors from frog red blood cells have a ligand binding site plus a site for activating adenylate cylase; the α receptor has a molecular weight of 58–64 KDa (39). Frog erythrocyte B_2 receptor is a single polypeptide of 58 KDa.

Catecholamines act on mammalian smooth muscles via α_1, α_2, and β receptors. Activation of β receptors is relaxant; α action is usually excitatory, but in some visceral muscles it is relaxant. Relative amounts of the three types of adrenergic receptor vary in different smooth muscles and have been evaluated by actions of agonists and antagonists. The α_1 receptors are predominant in aorta, pulmonary artery, and ileocecal sphincter (guinea pig); α_2 action is seen on cholinergic neurons of ileum and on rat vas deferens (158). The iris dilator of rabbit shows β inhibition and α excitation. Inhibitory α action occurs in spontaneously active visceral smooth muscle; cell membranes of intestinal muscle become hyperpolarized and spiking ceases. Excitatory α action is seen especially in muscles of reproductive and urinary tracts, where cell membranes become depolarized. Inhibition via β receptors is by lowered spontaneous electrical activity, and by decreased contractions after intracellular free calcium concentration has been lowered.

Regions of the brain differ in the proportions of α_1 and α_2 receptors. Locus coeruleus contains neurons that are stimulated by acetylcholine to produce norepinephrine, which modulates activity in midbrain (lateral geniculate). When applied to cerebral cortex, NE increases amount of c-AMP by action on β receptors.

Both dopamine and DOPA have synaptic function in mammalian brain. Dopamine (DA) increases c-AMP in postsynaptic cells of sympathetic ganglia and hyperpolarizes the cells. DA is responsible for a late inhibition in

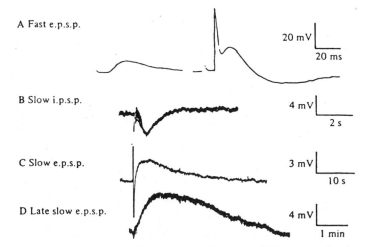

A Fast e.p.s.p.

20 mV

20 ms

B Slow i.p.s.p.

4 mV

2 s

C Slow e.p.s.p.

3 mV

10 s

D Late slow e.p.s.p.

4 mV

1 min

Figure 12-13. Four types of intracellularly recorded synaptic responses in neurons of the tenth sympathetic ganglion of mudpuppy Necturus. (*a*) Single preganglionic stimulus produces fast subthreshold EPSP (left); stronger stimulus brings in a second axon and produces a larger EPSP and an impulse (right). Fast EPSPs nicotinic cholinergic. (*b*) Slow IPSP (lasting 2 sec) results from stimulation by 13 impulses at 20/sec, muscarinic cholinergic. (*c*) Slow EPSP lasting 30–60 sec results from muscarinic cholinergic site. (*d*) Late slow EPSP lasting about 300 sec caused by stimulating specific spinal nerves at 10/sec, peptidergic, probably LHRH transmitter. Reprinted with permission from S. W. Kuffler, *Journal of Experimental Biology*, Vol. 89, p. 270. Copyright 1980, Biochemical Society, Colchester, Essex, England.

sympathetic ganglia in rabbit; activity of adenylate cyclase is increased and c-AMP accumulates (59). One class of amacrine cells in the turtle retina is dopaminergic. In retina of carp, dopamine decreases lateral spread of responses in horizontal cells, probably by uncoupling nexuses (151). DA receptors occur in basal ganglia; deficiency of DA in humans results in Parkinson's disease. DA is formed from L-DOPA in centers such as caudate nucleus. Interaction of DA, GABA, and acetylcholine is indicated by a circuit in caudate nucleus. Neural crest cells which are normally cholinergic can produce norepinephrine if cultured with nonneural tissue (103). Presumptive sympathetic neurons in culture are initially adrenergic, but can be induced to produce ACh if treated with a factor from heart or serum.

Stimulation of preganglionic nerves in bullfrog or mudpuppy initiates four different types of synaptic potentials in the cells of sympathetic ganglia (Fig. 12-13). The observed responses include fast EPSPs which are nicotinic cholinergic responses; slow IPSPs, muscarinic cholinergic but of long (2 sec) duration; a slow EPSP (30 sec) muscarinic cholinergic; and a very slow EPSP (5–10 min) for which the transmitter is probably the polypeptide LHRH (83). Cells of rabbit sympathetic ganglia have both nicotinic and muscarinic excitatory regions; also there are interneurons activated by ACh but liberating dopamine which acts on an adrenergic receptor region (91).

Catecholamines are widely distributed in invertebrate animals. In many bivalve molluscs dopamine probably functions as a transmitter. DA is the only CA in the pedal ganglion of some bivalves. After afferent nerve stimulation DA is present in the perfusate from the pedal ganglion (125). Dopamine-containing neurons occur in ganglia of a pulmonate snail *Lymnaea* (158a). DA applied to the visceral ganglion of *Mytilus* activates those neurons that stop activity of cilia. DA has been found in nerve terminals to the salivary gland of the moth *Manduca* and in salivary glands of cockroaches. The light-producing organs of fireflies have α receptors; E is more effective than NE in triggering light emission.

Octopamine (OCT) is a neurotransmitter or neuromodulator in many invertebrates; it is formed from tyrosine, with tyramine as an intermediate (Fig. 12-10). The nervous system of lobster *Homarus* contains large amounts, most in thoracic connectives and ganglia (81). Lobster supraesophageal ganglion contains 225 μg/g of octopamine, and lacks the enzyme for converting DA to NE (13). Octopamine applied to abdominal ganglia of lobster is excitatory to extensor motorneurons, inhibitory to flexor neurons. Octopamine is excitatory at neuromuscular junctions of crayfish and locust. In insects, octopamine has presynaptic and postsynaptic effects. In locust leg extensor muscle PSPs are enhanced by octopamine at 10^{-10} to 10^{-9} M. In a locust extensor-tibiae neuromuscular preparation, three classes of octopamine receptors have been identified according to sensitivity to blocking drugs (50). Octopamine stimulates feeding behavior by blowfly. When treated with octopamine at 5×10^{-9} M, heart beat of crayfish or of a crab (*Eriphia*) increases in amplitude and frequency. An octopus salivary gland contains tyramine, octopamine, serotonin, NE, and dopamine. Octopamine is present in quantity in efferent fibers to photoreceptors of *Limulus* eye (15) and in many neurons of leech ganglia. Octopamine and dopamine also occur in neurons of a nematode.

Serotonin

Serotonin, 5-hydroxy tryptamine (5HT), was discovered simultaneously as a probable transmitter in molluscs and in the autonomic nervous system of mammals. 5HT is formed from tryptophan via the intermediate 5-OH-tryptophan (Fig. 12-11). 5HT, like catecholamines, is detected in neurons by its fluorescence, which is of different color from CAs. 5HT is widely distributed in mammalian brain—in thalamus, hypothalamus, reticular formation (especially raphe), limbic system—but not in cerebral cortex or cerebellum. Inhibition of synthesis of 5HT or lesions in the reticular system lead to insomnia, after which injection of the intermediate 5-OH-tryptophan restores capacity for sleep. Some neurons of the myenteric plexus of the intestine synthesize and accumulate 5HT; this inhibits motility (48). The most dopamine-binding sites, shown by binding of radioactive analogs, are in striatum, with fewer in olfactory tubercle, hypothalamus, pons, and medulla. The most serotonin sites are in frontal cortex and posterior cortex; there are few in striatum, hypothalamus, and thalamus.

Serotonin (5HT) is a transmitter in molluscs. Many molluscan neurons contain 5HT, the enzymes for making it, and an amine oxidase which inactivates it. 5HT is released from clam hearts upon stimulation of cardioexcitatory nerves. In a catch muscle of *Mytilus* contraction is mediated by the excitatory transmitter ACh (Ch. 16). 5HT relaxes "catch" and increases the amplitude of phasic contractions (104). In a freshwater bivalve the cerebral ganglion contains 72.5 μg of 5HT/g of tissue, the pedal ganglion 66.5 μg, and the visceral ganglion 43.5 μg (124b). Excitatory serotonergic innervation occurs in the rectum of the bivalve *Tapes*. Serotonin may be the transmitter of a snail neuron which initiates feeding.

Two giant neurons in the cerebral ganglion of *Aplysia* synthesize 5HT, and similar giant serotonin-containing cells are found in *Helix, Limax,* and *Tritonia*. Acting on one group of small neurons in the cerebral ganglion of *Aplysia*, 5HT excites by increasing sodium conductance (g_{Na}); on other cells 5HT inhibits by increasing g_K, and on others the transmitter inhibits by decreasing both g_{Na} and g_K (54). The ciliated epithelium of the gill of Mytilus has dual innervation: stimulation of cerebral connectives or perfusion of the visceral ganglion with 5HT increases activity of cilia; methysergide antagonizes this. Stimulation of other neurons or perfusion with dopamine or epinephrine decreases activity of cilia (38).

The nervous system of a leech has neurons containing 5HT; in the earthworm *Lumbricus* the presence of 5HT in many neurons has been demonstrated by fluorescence. 5HT is additive with nerve activation in a crustacean muscle; threshold for 5HT effect on muscle or hearts is $10^{-10}M$ (51).

It has been proposed (59) that in mammalian serotonergic junctions 5HT acts by stimulating adenylate cyclase to catalyze synthesis of adenosine 3', 5'-cyclic phosphoric acid (cAMP) (Fig. 12-12). This may be a very general action of 5HT and of catecholamines. Either 5HT or epinephrine activates adenylate cyclase; c-AMP activates kinases and enzymes of glycogenolysis and stabilizes assembly of neurotubules. In a sea urchin embryo the concentration of 5HT correlates with morphogenetic movements; 5HT stimulates synthesis of c-AMP in *Tetrahymena* and *Euglena*. Two kinds of receptors for 5HT have been identified (114) in mammalian brain. Class 1 receptors are regulated by guanine nucleotides and are linked to adenylate cyclase; class 2 receptors are not influenced by the nucleotides. Serotonin is probably a facilitatory transmitter which may triple the c-AMP concentration in interneurons of *Aplysia* (19a).

PURINES

Transmitters that are not derived from amino acids are the purine nucleotides adenosine, adenosine monophosphate (AMP), and adenosine triphosphate (ATP). ATP is a metabolic intermediate in all prokaryotes and eukaryotes. Nerves which liberate purine nucleotides are called purinergic. Pyrimidines are not effective as neurotransmitters. Purines were discovered as noncholinergic, nonadrenergic inhibitors to several autonomic effectors. Purinergic nerve endings contain large opaque vesicles which concentrate adenosine. Intestinal muscle is relaxed by ATP at about $10^{-7}M$ (35,36).

ATP appears to be the transmitter in nonadrenergic, noncholinergic inhibitory nerves to smooth muscle of esophagus, stomach, intestine, and some sphincters. Purinergic nerves are inhibitory to the lung smooth muscle in amphibians and reptiles. ATP and adenosine are dilators of coronary arteries of mammals. Purinergic nerves relax the gastro-esophageal sphincter and inhibit motility of stomach in fishes, amphibians, and mammals. Repetitive stimulation of purinergic nerves in the gut wall elicits facilitating inhibitory junction potentials (IPSPs) that may be due to increase in K conductance; the inhibition is followed by rebound excitation. The purinergic neurons in Auerbach's plexus are activated by cholinergic neurons in the vagal and pelvic parasympathetic. Purinergic receptors are abundant in caudate nucleus and hippocampus; adenosine and ATP decrease neural activity and ^{3}H-adenosine binds at two binding sites (35).

Two classes of purinergic receptors are recognized: P_1 receptors are sensitive to adenosine and are blocked by methylxanthine; P_2 receptors are more sensitive to ATP and are blocked by quinidine imidoazolines (35). Adenosine derivatives are excitatory for many organisms. Cyclic AMP is an attractant for slime molds. Crustaceans such as shrimp *Palaemonetes* are attracted to the nucleotide AMP of dead prey; AMP is 170 times more effective than ADP; ATP, adenosine, and adenine are ineffective (37). Tests of nucleotide attractants for shrimp show that both adenine and ribose phosphate must be present; C-2 and C-3 hydroxyls on the ribose ring are also required for activity, but change in C2 and C6 positions of the purine makes little difference in attractant properties (37). Purines are much more active than pyrimidines. External chemoreceptors in crustaceans are sensitive to AMP, ATP, and to GABA, Glu, and taurine (37,1).

In summary, ATP is stored in specific vesicles and is released on nerve stimulation. Applied ATP mimics stimulation of purinergic nerves; action of ATP or adenosine is blocked by specific antagonists. Two kinds of receptors have been identified, and breakdown of ATP by ATPase occurs. Purinergic nerves are found in vertebrates of all classes.

POLYPEPTIDES

Many neuroactive compounds are polypeptides—chains of amino acids. More than 30 neuroactive peptides have been found in mammalian brain, a few of which are given in Figure 12-14 (74). Known neuropeptides range in size from 3 to 41 amino acids, in genetically specified order. Neuropeptides are so diverse in structure, function, and mode of synthesis that they cannot

Cholecystokinin
  -Asp-Arg-Asp-Tyr-Get-Gly-Trp-Met-Asp-Phe-NH$_2$

Neurotensin
  Glu-Leu-Tyr-Glu-Asn-Lys-Pro-Arg-Arg-Pro-Tyr-Ileu-Leu-OH

β endorphin H-Try-Gly-Gly-Phe-Met-Thr-Ser-Glu-Lys-Ser-Gln-Thr-Pro-Leu-Val-
  Thr-OH

met enkephalin H-Try-Gly-Gly-Phe-Met-OH
Leu enkephalin H-Try-Gly-Gly-Phe-Leu-OH

Mollusc cardiac accelerator H-Phe-Met-Arg-Phe-NH$_2$

Substance P H-Arg-Pro-Lys-Pro-Gln-Gln-Phe-Phe-Gly-Leu-Met-NH$_2$

Proctolin Arg-Tyr-Leu-Pro-Thr

LHRF Glu-His-Try-Ser-Gly-Leu-Arg-Pro-Gly-NH$_2$

Egg-releasing hormone (Aplysia) N-Ileu-Ser-Ileu-Asn-Gln-Asp-Leu-Lys-Ala-Ileu-
  Thr-Asp-Met-Leu-Leu-Thr-Glu-Glu-Ileu-Arg-Glu-Arg-Gln-Arg-Tyr-Leu-Ala-Asp-
  Leu-Arg-Gln-Arg-Leu-Leu-Glu-Lys-OH (36 amino acids)

Figure 12-14. Amino acid sequences of ten neuropeptides.

be arranged in any evolutionary pattern. Neuropeptides can, however, be grouped in families according to structure and pathway of synthesis. Peptides serve as neurotransmitters, synaptic modulators—positive or negative—and as bloodborne hormones. The same polypeptides have been found to be synthesized in several regions of the body—in endocrine cells and neurons, gut epithelium and brain; primitively the peptides may have been products of neurosecretory cells. Secretion into extracellular fluid may have preceded present-day membrane transmission via liberation of transmitter at nerve terminals. In the coelenterate *Hydra,* epitheliosecretory cells make compounds that function for growth and reproduction. Ganglion cells of jellyfish produce peptides that affect locomotion. The nervous systems of flatworms and annelids have peptide-secreting neurons. Biopeptide functions differ according to their target cells; some act on excitable membranes, some act intracellularly on metabolic pathways, particularly for protein synthesis; these regulatory peptides often use an intracellular second messenger, such as c-AMP. The principal function of a neuropeptide may differ with the kind of animal; some vertebrate polypeptides occur also in invertebrates, but with other functions.

The messenger RNAs for a few bioactive peptides have been sequenced and their genes cloned. Each peptide chain is synthesized as part of a long precursor which is cleaved posttranslationally to yield the secreted polypeptide; chains are not formed by simple addition of amino acids in sequence. Several polypeptides may be coded by a single large gene. By synthesis of long precursors the neuropeptides differ from small neurotransmitters, which are synthesized directly from amino acids. Neuropeptides can be identified and localized in cells histochemically by immunofluorescence or radioautography. Some neuropeptides may be of recent origin, some may have once served very different functions from their present ones. Examples of the better-known biopolypeptides follow.

Polypeptides of Brain and Gut

Many bioactive neuropeptides were first discovered in mammalian small intestine and later in brain; in the intestine most neuropeptides are synthesized in secretory cells of the mucosa and in neurons of the myenteric plexus. In brain many neuropeptides are synthesized in neurons, especially in hypothalamus.

One brain-gut polypeptide which is a neurotransmitter and a synaptic modulator is *substance P* (SP) (Fig. 12-14). SP is an undecapeptide: Arg-Pro-Lys-Pro-Gln-Gln-Phe-Phe-Gly-Leu-Met-NH_2. Substance P has been isolated from chromaffin cells of adrenal gland, neurons of myenteric plexus, amacrine cells of retina, dorsal root ganglia, and several brain regions. Two types of receptors for SP are distinguished by use of analogs. SP can have either excitatory or negative modulatory actions, according to the effector. SP elicits contraction of intestinal muscle when applied at $10^{-10}M$ or higher concentrations; it is liberated from axons of the myenteric plexus in the intestine. SP may elicit noncholinergic slow EPSPs in mesenteric ganglion and slow EPSPs in spinal cord (111). SP is present in rat dorsal roots at 24–130 pM/g_{ww}. SP depolarizes sensory interneurons in toad spinal cord 100 times as effectively as glutamate. SP can be synthesized in cultured chick sensory neurons. Its action at primary afferent synapses is antagonized by baclofen. SP is excitatory to neurons in sensory pathways of the brain, such as cuneate nucleus. The sensory neurons whose action is mediated by SP are small fibers that carry pain sensations (135); SP can initiate release of the opiate endorphin. SP diminishes nicotinic excitation of spinal interneurons by acetylcholine and decreases response of adrenal chromaffin cells to ACh. SP also attenuates responses of central interneurons to glutamate.

Vasoactive intestinal polypeptide (VIP) is an octosapeptide with N terminal histidine, C terminal threonine in chick, asparagine in pig. VIP is found in neurons of myenteric plexus and of ventromedial hypothalamus. VIP is a vasodilator and is probably a noncatecholamine inhibitor in the intestine. VIP has been identified in intestine of mammals, birds, teleosts, and elasmobranch fishes.

Neurotensin (NT) is a neuropeptide found in hypothalamus and gastrointestinal gland cells, not in plexus neurons. It contains 13 amino acids; it decreases esophageal sphincter contractions and inhibits gastric acid secretion (88).

Cholecystokinin (CCK) occurs in two forms, one with 33 amino acids, the other with residues. The 8-amino acid form constitutes 95 percent of the CCK in brain and myenteric neurons; the 33-amino acid form is a hormone in intestinal secretory cells. Apparently the 8-amino acid form is a neuromodulator, the 33-amino acid form a hormone. Two forms of *gastrin* (G) are known, one with 34 and the other 17 amino acids.

Somatostatin (SO) is a quadradecapeptide: Ala-Gly-Cys-Lys-Asn-Phe-Phe-Trp-Lys-Thr-Phe-Thr-Ser-Cys. Somatostatin, formed in the hypothalamus, inhibits and regulates release of growth hormone from pituitary. Somatostatin occurs in neurons of myenteric plexus of intestine; it inhibits the secretion of glucagon and insulin from pancreas, gastrin from gastrointestinal glands, and inhibits liberation of thyrotropic hormones. SO occurs in cortical bipolar neurons, and increases in Alzheimer's disease in humans. Sensory neurons which normally synthesize SP can produce SO if cultured together with nonneural cells (103).

Two biopeptides were discovered as toxins from amphibian skin; these occur also in hypothalamus and intestine. *Caerulein* is composed of 10 amino acids, of which the C-terminal is the same as CCK-8. *Bombesin*, found in frogskin, has 14 amino acids; it occurs in mammalian brain and intestine, where it has been found in myenteric plexus neurons.

Hypothalamic-Anterior Pituitary Factors

Bioactive polypeptides of another class are produced in the hypothalamus and have been identified as releasing factors for hormones of the anterior pituitary. *Corticotropin* (ACTH) occurs in two molecular sizes, 99 and 39 amino acids. ACTH stimulates the adrenal cortex. *Melanocyte-stimulating hormone* (MSH) occurs in two forms, one of which is identical with the first 13 amino acids of ACTH.

Several releasing factors for anterior pituitary hormones have recently been suggested as neurotransmitters or modulators. One is *luteinizing hormone-releasing factor* (LHRF), a multifunctional decapeptide: Glu-His-Trp-Ser-Tyr-Gly-Leu-Arg-Pro-Gly-NH$_2$. LHRH or a LHRH-like peptide has been found to be an excitatory transmitter in sympathetic ganglia of amphibians; it is present in the ganglia, becomes reduced in amount on degeneration of presynaptic fibers, causes postsynaptic depolarization lasting for minutes (like that elicited by some presynaptic nerve fibers); its action is blocked by LHRH analogs (83). LHRH-like peptide depolarizes by inactivating a slow outward K current, the M current (Ch. 15). Probably ACh and LHRH occur in the same small preganglionic axons in frog. Other releasing hormones made in hypothalamus and acting on anterior pituitary are *thyroxin-releasing hormone* (TRH) and *corticotropin-releasing hormone* (CRH).

Hypophyseal Polypeptides Which Are Not Neuropeptides

A polypeptide of multiple functions, synthesized in and released from anterior pituitary, is *prolactin* or luteotropin, a large molecule of 32,000 daltons.

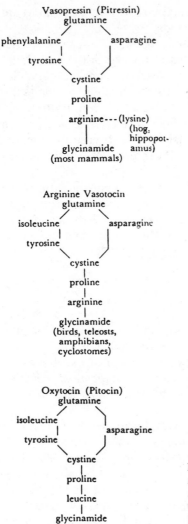

Figure 12-15. Diagrams of structure of three posterior pituitary circular polypeptides. From *Comparative Animal Physiology,* ed. 2, C. L. Prosser, Ed. Copyright 1961 by W. B. Saunders, Company, Philadelphia.

Prolactin was named from its property of initiating milk secretion by mammary glands; prolactin also stimulates growth of the crop sac or milk gland of pigeons. It stimulates corpus luteum to secrete progesterone. Prolactin is widely functional also in poikilothermic vertebrates in relation to water balance. In fishes it functions in controlling ion transport across gills. Prolactin evolved in ionic regulation in fishes and acquired very different functions in mammals.

The posterior pituitary of all vertebrates elaborates nonapeptides in which the amino acids form a ring (Fig. 12-15). Substitutions occur at positions 4 and 8, rarely at 2 and 3. One compound is oxytocin, which functions in parturition to activate uterine contraction (mammals). The posterior pituitary elaborates also vasopressin (mammals) or vasotocin (amphibians and fishes). Vasopressin and vasotocin function in water and ion balance (Ch. 9). Cyclostomes have only one nonapeptide (arginine vasotocin); all other vertebrates have two.

Bio-opiates

Some polypeptides found in brain have an opiate action. The same opiates occur in some invertebrates and may have evolved as general analgesics in response to pain sensations. At least 18 opioid peptides are known (156).

Endorphins are large molecules. β-*endorphin* is a hexadecapeptide: H_2N-Tyr-Gly-Gly-Phe-Met-Thr-Ser-Glu-Lys-Ser-Gln-Thr-Pro-Leu-Val-Thr-Leu-Phe-Lys-Asn-Ala-Ile-Lys-Asn-Ala-Tyr-Lys-Glu-Glu. α-*endorphin* is H_2N-Tyr-Gly-Gly-Phy-Met-Thr-Ser-Glu-Lys-Ser-Gln-Thr-Pro-Leu-Val-Thr. In rabbits high doses of endorphin cause hypothermia, low doses cause hyperthermia. Both endorphins modify excitability of central neurons and depress activity in cortex and thalamus; endorphin action is antagonized by naloxone. Endorphins decrease secretion of insulin and glucagon from the pancreas; thus they are hypoglycemic. They diminish responses to substance P, a transmitter of pain afferents.

Enkephalins are smaller molecules than endorphins. *Leu-enkephalin* is H_2N-Tyr-Gly-Gly-Leu-OH; *met-enkephalin* is H_2N-Tyr-Gly-Gly-Phe-Met-OH, which is the same as the sequence at the NH_2 end of β-endorphin (90). However, endorphins are derived from a different precursor from enkephalins, hence they are encoded by different genes. The ratio of the amounts of the two enkephalins is specific to species and organ. Enkephalins have a morphine-like action which, like that of endorphins, is blocked by naloxone. Enkephalins attenuate GABA-ergic inhibition of hippocampal pyramidal neurons and diminish glutamate-induced activity in the cortex. Concentration is high in toad and turtle brain. Enkephalins decrease the number of quanta of transmitter per endplate potential at frog neuromuscular junctions; met-enkephalin decreases the Ca-dependent presynaptic events (22a). Enkephalins at $10^{-8}M$ relax the intestine; they are present in some neurons of the myenteric plexus. Enkephalins inhibit the action of dopaminergic neurons of the striatum and reduce the excitatory action of SP in sensory fibers of spinal cord. Enkephalins are inactivated by proteases, especially aminopeptidases (69). Compared to ACh and to monoamines, actions of enkephalins are slow; they are primarily presynaptic modulators, not transmitters.

Dynorphin is an opioid tridecapeptide with a sequence at its NH_2 terminus resembling Leu-enkephalin. Pituitary dynorphin is (H) Tyr, Gly-Gly, Phe, Leu, Arg-Arg-Ile-Arg, Pro-Lys-Leu-Lys-Trp-Asp-Glu(OH) (56). Dynorphin (Dyn) is found in the posterior lobe of the pituitary, in several regions of brain, including hypothalamus, and in spinal cord; Dyn is 700 times more potent than leu-enkephalin on neurons of submucous plexus of guinea pig ileum, and is 1/13 as sensitive to naloxone (21). Dynorphin and vasopressin occur in the same cells in the hypothalamus in the rat. Dynorphin occurs in two forms, A and B (156).

Two principal receptor types for endorphins and enkephalins are recognized and are named μ and δ; two others, α and K, are less well known (135). Receptors of the μ type are localized in sensory areas of cerebral cortex and in plexus neurons of ileum; they have been identified in amphibians and mammals (90). α receptors are more uniformly distributed throughout the cortex; they bind strongly to enkephalins, less to naloxone and endorphins. Endorphins and enkephalins tend to bind to μ receptors, dynorphins to K receptors.

All three classes of opiates hyperpolarize gastric plexus neurons and inhibit ACh release from the plexus (21). All three classes of opiates reduce the effect of sensory input transmitted by SP. Enkephalins occur in the retina of goldfish, especially in amacrine cells, and reduce K-induced release of GABA.

Other Biopolypeptides

Several important polypeptides do not fall into any of the categories mentioned. *Angiotensin II* is formed in the blood by the action of the enzyme renin from kidney on *angiotensin I*. Angiotensin reduces blood pressure and is an important vascular regulator. The renin-angiotensin system is found in most, perhaps all, vertebrates.

Nerve growth factor (NGF) contains 118 amino acids; it stimulates neural growth and proliferation. Its action has been examined on sympathetic ganglion cells and other neuron clusters in culture. A source much used in experiments is the submaxillary gland.

Carnitine occurs widely in striated muscle, and is abundant in the olfactory tract of mammals. Carnitine has been suggested as a possible transmitter of olfaction. It is an essential nutrient for some beetles.

Synthesis of Polypeptides

The chemical sequence in the synthesis of each bioactive polypeptide is complex and the origin and adaptive function of the synthetic process are not understood. For each of the polypeptides, a long messenger RNA is transcribed and a large protein is formed which is then cleaved posttranslationally into polypeptides (Fig. 12-16). The synthesis of precursor molecules was described in Chapter 3 for insulin, relaxin, and parathormone. In all eukaryotes, proteins or polypeptides that are to be secreted are formed as

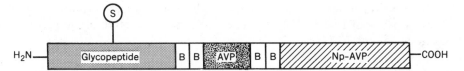

Figure 12-16. Proposed structure of vasopressin precursor. S is sugar moiety of the glycopeptide, B are basic amino acid residue fillers, AVP arginine vasopressin and Np-AVP vasopressin-associated neurophysin. Reprinted with permission from M. J. Brownstein et al., *Science*, Vol. 207, p. 378. Copyright 1980 by the AAAS.

parts of precursors that are then cleaved by proteolytic enzymes. Proteins that are retained in cells (e.g., enzymes) are not usually synthesized via long precursors. Most peptide chains are formed with pro- and pre-pro-segments. One precursor may yield several final products. Diverse precursors permit related proteins to be made in different cell types—brain, adrenal. A neuropeptide precursor may be ten times as long as the secreted polypeptide. All peptides which are secreted from or stored in cells have an N-terminal amino acid signal sequence which is hydrophobic, and as a result the molecule can enter and pass through a lipid membrane.

To sequence the mRNA experimentally for a precursor of a specific polypeptide, the mRNA for the peptide is prepared from a tissue in which synthesis of the polypeptide is active. By treatment of mRNA as a template with a reverse transcriptase, cDNA is made and this is cloned in an appropriate plasmid (usually in *E. coli*), where quantities of the DNA are made. The plasmid is treated with a restriction enzyme which removes the newly synthesized DNA. The 5' end of each DNA strand is labelled with p^{32} and the strands of DNA are separated either by denaturation or by treatment with a second restriction enzyme. Treatment of the labelled single strands of DNA with various reagents (Maxam and Gilbert technique, Chapter 3) cleaves at specific bases (G, C, A, etc.). A series of fragments result; these are electrophoresed on a gel. The radioactivity of bands is measured, and from the order and number of the bands the nucleotide sequence is obtained. From the DNA sequence, the corresponding mRNA sequence is deduced, and from this the amino acid sequence of the precursor. The deduced sequence is compared to that obtained by amino acid sequence analysis of the peptide. The mRNAs code for long precursor proteins within which are the polypeptides of interest. In addition to polypeptides, the mRNAs have long leader and termination sequences. A few examples of precursors follow.

Vasopressin (AVP) and oxytocin (OT) are formed in magnocellular neurons in the hypothalamus and are transported down the axons to the neurohypophysis. The precursors which are synthesized in the neurons are cleaved during axonal transport. The precursor for vasopressin has molecular weight 20,000 daltons and includes sequences for a large glycoprotein, arginine vasopressin, and an associated neurophysin (32).

The mRNA for nerve growth factor (NGF) was sequenced from its cDNA which consists of 1068 base pairs. The nucleotides for NGF are at base pair positions 657 to 1010 (Fig. 12-17) (131).

Pro-CCK consists of 150 residues; the principal form of active CCK is made of 8 amino acids. Pro-gastrin consists of 110 amino acids; the active polypeptide has either 34 or 17.

A precursor for several bioactive polypeptides is POMC, pro-opiomelano-cortin. The cDNA for the mRNAs for this large precursor protein has been cloned from pituitary and has 1091 base pairs (66,82,105a) (Fig. 12-18). The precursor protein consists of 132 amino acids; it has a terminal fragment or signal protein which contains within it the sequence for γ-MSH (melanocyte-stimulating hormone). Next is the ACTH region (adrenal-corticotropic hormone) of 39 amino acid sequences; this includes α-MSH (13 amino acids at positions 82–99) and CLIP (corticotropin-like intermediate lobe peptide) of 21 amino acids. The next long region of the precursor is β-LPT (β-lipotropin) of 91 amino acids (sequences 42–132) which contains sequences for γ-LPT (58 amino acids, sequences 42 to 101) and β-endorphin (30 amino acids, sequences 104 to 134). Thus, the POMC precursor contains sequences for MSH and β-endorphin, LPT and ACTH, plus a leader. Between component peptides are pairs of amino acids Lys-Arg. POMC occurs in anterior pituitary, intermediate lobe of pituitary, brain (hypothalamus) and reproductive tract; in each the precursor may be processed slightly differently. For example, in anterior pituitary ACTH and β-LPT are predominant products, while in intermediate lobe and brain α-MSH and β-endophin predominate. Immunoreactive POMC-related peptides have been found in earthworm, two molluscs, insects, a tunicate, and Tetrahymena (57).

When enkephalins were discovered it was thought that they were derived from β-endorphin, but present evidence is against this. Met-enkephalin is bound, as shown by antibody labeling, in one kind of cell in the brain, endorphin in another kind of cell; β-endorphin, but not Met-enkephalin, is made in high concentration by the pituitary. Both Met- and Leu-enkephalin but not endorphin are present in chromaffin cells of adrenal medulla. The nucleotides of cDNA for bovine adrenal pre-pro-enkephalin have been sequenced (61,69,108). The gene consists of 1222 nucleotides for seven enkephalin molecules. Figure 12-19 gives the nucleotide sequence and the deduced enkephalins: four copies of Met-enkephalin, one copy each of Leu-enkephalin and of Met-enk-Arg-Gly-Leu and Met-enk-Arg-Phe. Between the enkephalin coding sequences are signals of basic amino acids for release of

---

Figure 12-17. Complete nucleotide sequence of nerve growth factor as obtained from cloned cDNA, terminal nucleotides at beginning and end of sequence. Amino acids deduced from the nucleotide triplets, nerve growth factor underlined. Diagram of the precursor given below. Reprinted by permission from J. Scott, *Nature,* Vol. 302, p. 539. Copyright 1983 Macmillan Journals Limited.

GAGCGCCUGGAGCCGGAGGGGAGCGCAUCGAGUGACUUUGGAGCUGGCCUUAUAUUUGG   59

AUCUCCCGGGCAGCUUUUUUGGAAACUCCUAGUGAAC
                                  1
                                met leu cys leu lys
                                AUG CUG UGC CUC AAG   110

                10   20
pro val lys leu gly ser leu glu val gly his gly gln his gly
CCA GUG AAA UUA GGC UCC CUG GAG GUG GGA CAC GGG CAG CAU GGU   155

gly val leu ala cys gly arg ala val gln 30 gly ala gly trp his
GGA GUU UUG GCC UGU GGU CGU GCA GUC CAG GGG GCU GGA UGG CAU   200

            40   50
ala gly pro lys leu thr ser val ser gly pro asn lys gly phe
GCU GGA CCC AAG CUC ACC UCA GUG UCU GGG CCC AAU AAA GGU UUU   245

ala lys asp ala ala phe tyr thr gly 60 arg ser glu val his ser
GCC AAG GAC GCA GCU UUC UAU ACU GGC CGC AGU GAG GUG CAU AGC   290

          70   80
val met ser met leu phe tyr thr leu ile thr ala phe leu ile
GUA AUG UCC AUG UUG UUC UAC ACU CUG AUC ACU GCG UUU UUG AUC   335

gly val gln ala glu pro tyr 90 thr asp ser asn val pro glu gly
GGC GUA CAG GCA GAA CCG UAC ACA GAU AGC AAU GUC CCA GAA GGA   380

            100
asp ser val pro glu ala his trp thr lys leu gln his ser 110 leu
GAC UCU GUC CCU GAA GCC CAC UGG ACU AAA CUU CAG CAU UCC CUU   425

asp thr ala leu arg arg ala arg ser 120 ala pro thr ala pro ile
GAC ACA GCC CUC CGC AGA GCC CGC AGU GCC CCU ACU GCA CCA AUA   470

          130
ala ala arg val thr gly gln thr arg asn ile thr val asp 140 pro
GCU GCC CGA GUG ACA GGG CAG ACC CGC AAC AUC ACU GUA GAC CCC   515

arg leu phe lys lys arg arg leu his 150 ser pro arg val leu phe
AGA CUG UUU AAG AAA CGG AGA CUC CAC UCA CCC CGU GUG CUG UUC   560

ser thr gln pro 160 pro pro thr ser ser asp thr leu asp leu asp
AGC ACC CAG CCU CCA CCC ACC UCU UCA GAC ACU CUG GAU CUA GAC   605

phe gln ala his gly thr ile pro phe 180 asn arg thr his arg ser
UUC CAG GCC CAU GGU ACA AUC CCU UUC AAC AGG ACU CAC CGG AGC   650

        Nerve growth factor
lys arg ser ser thr his pro val phe his met gly glu phe 200 ser
AAG CGC UCA UCC ACC CAC CCA GUC UUC CAC AUG GGG GAG UUC UCA   695

val cys asp ser val ser val trp val 210 gly asp lys thr thr ala
GUG UGU GAC AGU GUC AGU GUG UGG GUU GGA GAU AAG ACC ACA GCC   740

          220
thr asp ile lys gly lys glu val thr val leu ala glu val 230 asn
ACA GAC AUC AAG GGC AAG GAG GUG ACA GUG CUG GCC GAG GUG AAC   785

ile asn asn ser val phe arg gln tyr 240 phe glu thr lys cys
AUU AAC AAC AGU GUA UUC AGA CAG UAC UUU UUU GAG ACC AAG UGC   830

arg ala ser asn 250 pro val glu ser gly cys arg gly ile asp 260 ser
CGA GCC UCC AAU CCU GUU GAG AGU GGG UGC CGG GGC AUC GAC UCC   875

lys his trp asn ser tyr cys thr thr 270 his thr phe val lys
AAA CAC UGG AAC UCA UAC UGC ACC ACG ACU CAC ACC UUC GUC AAG   920

          280
ala leu thr thr asp glu lys gln ala ala trp arg phe ile 290 arg
GCG UUG ACA ACA GAU GAG AAG CAG GCU GCC UGG AGG UUC AUC CGG   965

ile asp thr ala cys val cys val leu ser 300 arg lys ala thr arg
AUA GAC ACA GCC UGU GUG UGU GUG CUC AGC AGG AAG GCU ACA AGA   1,010

      307
arg gly OP
AGA GGC UGA CUUGCCUGCAGCCCCCUUCCCCACCUGCCCCCUCCACACUCUCUUGGG   1,067

CCCCUCCCUACCUCAGCCUGUAAAUUAUUUUUAAAUUAUAAGGACUGCAUGAUAAUUUAUC   1,127

GUUUAUACAAUUUUAAAGACAUUAUUUAUUUAAAAUUUUCAAAGCAUCCUGAAAAAAAA

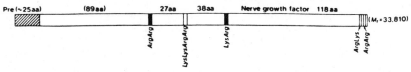

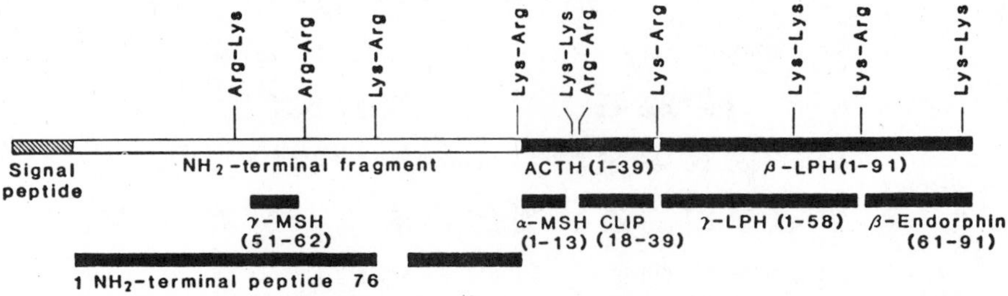

Figure 12-18. Structure of the ACTH/LPH precursor indicating formation of two neuropeptides, MSH and β-endorphin. Reprinted with permission from P. A. Rosa and E. Herbert, *Journal of Experimental Biology,* Vol. 89, p. 217. Copyright 1980, Biochemical Society, Colchester, Essex, England.

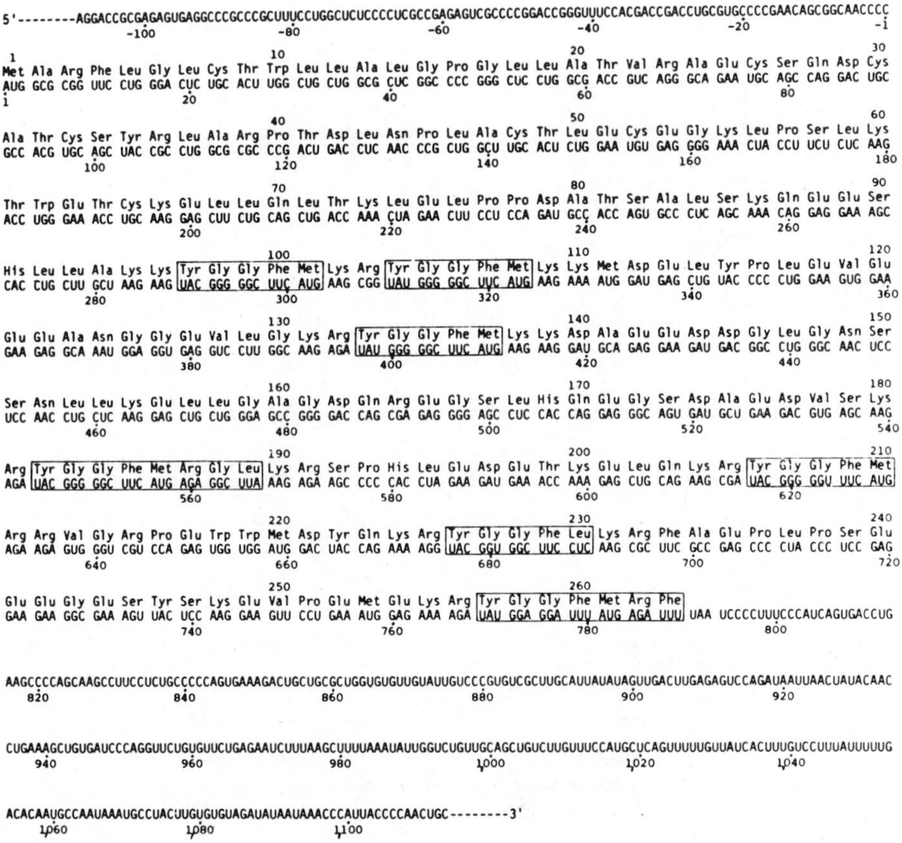

```
5'--------AGGACCGCGAGAGUGAGGCCCGCCCGCGUUUCCUGGCUCUCCCCUCGCCGAGAGUCGCCCCGGACCGGGUUUCCACGACCGACCUGCGUGCCCCGAACAGCGGCAACCCC
           -100           -80           -60           -40           -20        -1

  1                        10                       20                30
Met Ala Arg Phe Leu Gly Leu Cys Thr Trp Leu Leu Ala Leu Gly Pro Gly Leu Leu Ala Thr Val Arg Ala Glu Cys Ser Gln Asp Cys
AUG GCG CGG UUC CUG GGA CUC UGC ACU UGG CUG CUG GCG CUC GGC CCC GGG CUC CUG GCG ACC GUC AGG GCA GAA UGC AGC CAG GAC UGC
  1             20            40            60            80

                           40                       50                60
Ala Thr Cys Ser Tyr Arg Leu Ala Arg Pro Thr Asp Leu Asn Pro Leu Ala Cys Thr Leu Glu Cys Glu Gly Lys Leu Pro Ser Leu Lys
GCC ACG UGC AGC UAC CGC CUG GCG CGC CCG ACU GAC CUC AAC CCG CUG GCU UGC ACU CUG GAA UGU GAG GGG AAA CUA CCU UCU CUC AAG
              100           120           140           160           180

                           70                       80                90
Thr Trp Glu Thr Cys Lys Glu Leu Leu Gln Leu Thr Lys Leu Glu Leu Pro Pro Asp Ala Thr Ser Ala Leu Ser Lys Gln Glu Glu Ser
ACC UGG GAA ACC UGC AAG GAG CUU CUG CAG CUG ACC AAA CUA GAA CUU CCU CCA GAU GCC ACC AGU GCC CUC AGC AAA CAG GAG GAA AGC
              200           220           240           260

                           100                      110               120
His Leu Leu Ala Lys Lys [Tyr Gly Gly Phe Met] Lys Arg [Tyr Gly Gly Phe Met] Lys Lys Met Asp Glu Leu Tyr Pro Leu Glu Val Glu
CAC CUG CUU GCU AAG AAG  UAC GGG GGC UUC AUG  AAG CGG  UAU GGG GGC UUC AUG  AAG AAA AUG GAU GAG CUG UAC CCC CUG GAA GUG GAA
              280           300           320           340           360

                           130                      140               150
Glu Glu Ala Asn Gly Gly Glu Val Leu Gly Lys Arg [Tyr Gly Gly Phe Met] Lys Lys Asp Ala Glu Glu Asp Asp Gly Leu Gly Asn Ser
GAA GAG GCA AAU GGA GGU GAG GUC CUU GGC AAG AGA  UAU GGG GGC UUC AUG  AAG AAG GAU GCA GAG GAA GAU GAC GGC CUG GGC AAC UCC
              380           400           420           440

                           160                      170               180
Ser Asn Leu Leu Lys Glu Leu Leu Gly Ala Gly Asp Gln Arg Glu Gly Ser Leu His Gln Glu Gly Ser Asp Ala Glu Asp Val Ser Lys
UCC AAC CUG CUC AAG GAG CUG CUG GGA GCC GGG GAC CAG CGA GAG GGG AGC CUC CAC CAG GAG GGC AGU GAU GCU GAA GAC GUG AGC AAG
              460           480           500           520           540

                           190                      200               210
Arg [Tyr Gly Gly Phe Met Arg Gly Leu] Lys Arg Ser Pro His Leu Glu Asp Glu Thr Lys Glu Leu Gln Lys Arg [Tyr Gly Gly Phe Met]
AGA  UAC GGG GGC UUC AUG AGA GGC UUA  AAG AGA AGC CCC CAC CUA GAA GAU GAA ACC AAA GAG CUG CAG AAG CGA  UAC GGG GGU UUC AUG
              560           580           600           620

                           220                      230               240
Arg Arg Val Gly Arg Pro Glu Trp Trp Met Asp Tyr Gln Lys Arg [Tyr Gly Gly Phe Leu] Lys Arg Phe Ala Glu Pro Leu Pro Ser Glu
AGA AGA GUC GGU CGU CCA GAG UGG UGG AUG GAC UAC CAG AAA AGG  UAC GGG GGC UUC CUC  AAG CGC UUC GCC GAG CCC CUA CCC UCC GAG
              640           660           680           700           720

                           250                      260
Glu Glu Gly Glu Ser Tyr Ser Lys Glu Val Pro Glu Met Glu Lys Arg [Tyr Gly Gly Phe Met Arg Phe]
GAA GAA GGC GAA AGU UAC UCC AAG GAA GUU CCU GAA AUG GAG AAA AGA  UAU GGA GGA UUU AUG AGA UUU  UAA UCCCCUUUCCCAUCAGUGACCUG
              740           760           780           800


AAGCCCCAGCAAGCCUUCCUCUGCCCCCAGUGAAAGACUGCUGCGCUGGUGUGUUGUAUUGUCCCGUGUCGCUUGCAUUAUAUAGUUGACUUGAGAGUCCAGAUAAUUAACUAUACAAC
   820          840          860          880          900          920


CUGAAAGCUGUGAUCCCAGGUUCUGUGUUCUGAGAAUCUUUAAGCUUUUAAAUAUUGGUCUGUUGCAGCUGUCUUGUUUCCAUGCUCAGUUUUUGUUAUCACUUUGUCUUUAUUUUUG
   940          960          980          1000         1020         1040


ACACAAUGCCAUAAAUGCCUACUUGUGUGUAGAUAUAAAUAAACCCAUUACCCCAACUGC--------3'
   1060         1080         1100
```

Figure 12-19. Primary structure of bovine preproenkephalin mRNA. Nucleotide residues numbered in 5′ to 3′ direction; nucleotides on lead side of coded amino acids are indicated by negative numbers. Predicted amino acid sequence is displayed above the nucleotide sequence, beginning with initiating methionenes. Sequence of Met-enk, Leu-enk, and Met-enk with carboxyl extensions are bounded by paired basic amino acids. Reprinted with permission from M. Noda et al., *Nature,* Vol. 295, p. 204. Copyright 1982, Macmillan Journals Ltd.

the components; these releasing pairs are Lys-Arg or Lys-Lys. The pre-pro-enkephalin is a multi-peptide precursor analogous to corticotropin-β-lipotropin. A precursor gives rise on cleavage to neo-endorphin and two dynorphins (156).

A considerable amount of selective cleavage by proteolytic enzymes occurs before the products of precursor proteins are liberated. The nature of the cleavage has some tissue specificity.

Neuropeptides of Invertebrate Animals

Bioactive polypeptides of invertebrates have been investigated by means of (1) extracting active substances from tissues, (2) sequencing the active chains, (3) histochemical identification by labeled antibodies for vertebrate polypeptides, and (4) applying polypeptides from vertebrates or from other invertebrates to muscles, neurons, and secretory cells and observing responses. Active polypeptides occur in the unicellular eukaryotes Tetrahymena, Neurospora, and in all animals with nervous systems—Hydra, annelids, molluscs, arthropods. Responses to application of vertebrate neuropeptides and staining by labelled antibodies to vertebrate peptides often occur in invertebrates. No one of these procedures is adequate in itself to establish a polypeptide as a natural active molecule.

Neuropeptides of invertebrates are of two types: (1) substances that are endogenous or specific to certain invertebrate animals, and (2) polypeptides which occur in both invertebrates and vertebrates (57). A few examples of each follow.

Crustaceans have numerous peptides in the nervous system, particularly in neurosecretory cells of sinus gland and some ganglia; they are *cardioactive principles, hyperglycemic peptide,* and several peptides which activate chromatophores. One that acts on red chromatophores is Glu-Leu-Asn-Phe-Ser-Pro-Gly-Trp-NH_2 (50a).

Insects (locusts, bees) make at least two metabolically active polypeptides in secretory cells of brain and corpora cardiaca. These polypeptides mobilize fat utilization in some insects and in others cause hyperglycemia. The two active polypeptides contain 10 and 8 amino acids respectively.

In cockroaches a neuropeptide from the ventral nerve cord stimulates contraction of hindgut muscles; this is *proctolin,* molecular weight 648, composition Arg-Tyr-Leu-Pro-Thr. Proctolin is localized in neurons of abdominal ganglia; at concentration of $10^{-9}M$ it contracts cockroach gut muscle. Proctolin has been found in insects of six different orders. The pericardial organ of crabs makes two peptides which are excitatory to the neurogenic heart. The smaller peptide increases burst frequency, burst duration, and number of spikes; it increases concentration of c-AMP (88a). One of these peptides is probably proctolin. Proctolin at about 10^{-9} M is a modulator of excitation in the neurogenic heart of Limulus.

Several active peptides have been isolated from the nervous systems of molluscs; one from the nervous system of a clam is a tetrapeptide: Phe-Met-Arg-Phe-NH$_2$ or FMRFamide. It is excitatory to hearts of some clams (*Macrocallista*) and inhibitory to hearts of others (*Lampsilis*). FMRFamide at 10^{-9} to $10^{-8}M$ elicits contraction of the radula protractor of *Busycon* (57,112). FMRFamide sometimes, and serotonin always, stimulate adenylate cyclase and increase c-AMP content in excited target tissues. A closely related compound in the snail *Helix* is Glu-Asp-Pro-Phe-Leu-Arg-PheNH$_2$ (118). The nervous system and gut of *Aplysia* yield several cardioactive peptides, one which is active at $10^{-10}M$; it is Met-Ans-Tyr-Leu-Ala-Phe-Pro-Arg-Met-NH$_2$ (92).

In *Aplysia* a cluster of 400 neurons, the bag cells, make a neuropeptide, the egg-laying hormone (ELH), which induces complex behavior leading to egg laying. ELH contains 36 amino acids and has molecular weight of 4400 daltons. Another group of neurosecretory cells in the Aplysia atrial gland synthesizes two peptides (A and B) of 34 amino acids each. Bag cells are excited by peptides A and B or by afferent nerve impulses. The bag cells are electrotonically coupled; stimulation causes their axons to release ELH into interstitial space of the abdominal ganglion. ELH causes persistent discharge in neuron R15 (100,144,145,146).

By use of mRNAs from *Aplysia* sperm the cDNAs for ELH and proteins A and B have been cloned. There is 90 percent homology between the genes for the A and B proteins (Fig. 12-20). A sequence of 108 nucleotides codes for the 26 amino acids of ELH (129).

Immunocytochemical tests have identified enkephalin/endorphin in several invertebrates. β-endorphin and an enkephalin have been indicated in cerebral ganglion of earthworm (7) and α-endorphin in the subpharyngeal ganglion. One of 400 neurons in a ganglion of the posterior midbody of a leech reacts with antiserum to leu-enkephalin. The sequence of the tetrapeptide FMRFamide occurs in pre-pro-enkephalin where it is followed by the termination codon. It is possible that FMRFamide and the enkephalins diverged from an ancestral sequence Tyr-Gly-Gly-Phe-Met-Arg-Phe-NH$_2$. Enkephalins occur in ganglia of the bivalve Mytilus and in the nervous system of Octopus. Applied opiates modulate the action of monoamines in molluscan nervous systems (140). Enkephalins reduce the stimulation of adenylate cyclase by dopamine (DA) and this inhibition is blocked by naloxone. In the clam Anodonta, applied met-enkephalin and leu-enkephalin increase DA synthesis. Mammalian enkephalins stimulate DA turnover. A snail subjected to repeated noxious stimuli stops reacting after a time. It is postulated that met-enkephalin in the nervous system has induced insensitivity; such acquired tolerance is reversed by naloxone. In snails the enkephalins may have a function comparable to opiates in mammals (140).

A decapeptide from Hydra stimulates cell differentiation; it induces interstitial cells to become nerve cells (127). A substance similar to insulin has been found by radioimmunochemical tests in *Tetrahymena, Neurospora, As-*

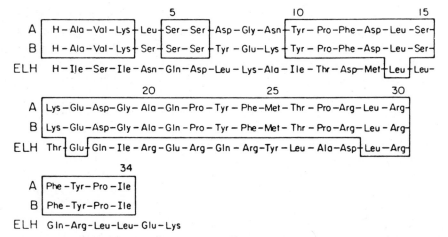

Figure 12-20. Comparison of amino acid sequences of egg-laying hormone (ELH) of bag cells and peptides A and B from atrial gland of Aplysia. Boxes identify homologous amino acid sequence. Reprinted with permission from F. Strumwasser in *Brain Peptides,* D. Krieger et al., Eds. Copyright 1983, John Wiley & Sons, Inc., New York.

pergillus, and in annelids, molluscs, echinoderms, and insects. In Hydra, cells reactive to antibodies for substance P occur in tentacles and basal stalk, cells reactive to antibodies for gastrin and CCK in the oral region, and cells reactive to antibodies for neurotensin in the mouth (60). Whether the reacting substances are the same as the mammalian peptide antigens is open to question.

Evolutionary Considerations

No single procedure can establish the presence of a neuropeptide. Staining by labeled antibodies indicates reactivity. However, specificity may be for different parts of the antibody molecule, and, while indicative, it is only a general clue to the presence of a certain neuropeptide. Another method of identification of a neuropeptide is the application of biogenic agents from an unrelated animal. Many mammalian peptides are found to be effective on nervous systems of other vertebrates and invertebrates. The reaction obtained in nonmammals with applied peptides from mammals is a first step toward identification of a neuropeptide; however, dissimilar compounds may have similar effects. Sequencing of the amino acids in a polypeptide gives a chemical description of structure. Frequently short sequences in a long polypeptide correspond to the sequences in known short polypeptides. Such homologies are not indicative of evolutionary relations or of common ancestry (57). The probability of occurrence of a particular sequence by chance depends on the length of the sequence. Calculations for a tetrapeptide such as FMRFamide indicate that the probability that a sequence will occur in any two related species approaches 100 percent if the entire genome is considered as a nucleotide source (118). To establish a polypeptide as a neurotransmitter, modulator, or hormone, purification and sequencing to determine structure, imunochemistry to determine concentration and distribution, and physiological and pharmacological studies to determine function are all necessary.

Families of neuropeptides may indicate some ancestral relations; diagrams based on numbers of sequence differences provide the basis for tentative phylogenetic trees. Bioactive polypeptides occur in great numbers, the functions of some compounds appear to have changed during the course of evolution, and the evolution of polypeptides is probably continuing. New ones are being discovered each year. Similarity of structure and function confers redundancy.

An evolutionary question that has not been answered is the significance of complex patterns of synthesis of polypeptides. In all eukaryotes for which data are available, bioactive peptides are derived from long precursor proteins which are cleaved by peptidases into several components, one of which is the active polypeptide. How the genotype could have evolved for precursors as much as 10 times as long as the product, precursors of which most of the molecule will be discarded posttranslationally, is a major evolutionary question. There may be several copies of a gene and several mRNAs may combine at the ribosomes in one synthesis. Different proteolytic cleavages may occur in different tissues, for example, some enkephalins in pituitary and others in adrenal. Synthesis of precursors and subsequent cleavage occur for all polypeptides which are synthesized at one cellular site and secreted for use elsewhere. The long precursor chains may protect a polypeptide from hydrolysis before it is secreted. Some precursor proteins may have had different functions from those of the present active polypeptides; the precursors may have had long phylogenetic histories. Natural selection can occur at three levels: at the DNA for different genes; at the mRNAs which code for long proteins, including leader and termination sequences and active peptides; and at the precursor proteins and their processing.

Many polypeptides now have several functions. Substance P appears to be a transmitter for pain sensory fibers, and it has relaxing actions on intestine and on vascular smooth muscle as a vasodilator. LHRH was first recognized as a releaser for luteinizing hormone; now it appears also to be a neurotransmitter in sympathetic ganglia. Prolactin functions in ionic regulation in fishes and in milk secretion in mammals. The same substances, synthesized in gut epithelial cells, in peripheral neurons, and in brain (especially hypothalamus) do not have the same function in each organ.

Examples were given for synthesis of several related polypeptides in a given tissue. Recent evidence shows that two neuroactive substances may be formed in a single neuron, and probably stored in separate vesicles. For example, a sympathetic neuron may produce NE and SO, or NE and Enk; some neurons produce NE early in development and ACh later. Some cells of adrenal medulla have NE and enkephalin, others have NE and SO, others E and NT. In mammalian hindbrain there are cells with both serotonin and SP, other cells with serotonin and TRH, others with SP and TRH (68). The most frequent dual transmitters are a peptide and amine together, which may be antagonistic in action. The presence in one nerve ending of two transmitters which have different synaptic functions provides for some modulation of transmission at a single synapse.

Neuropeptides which act at specific synapses may have distinct behavioral effects. Substance P mediates pain inputs, CCK leads to the sensation of food satiety, endorphin-enkephalins alleviate pain, LHRH stimulates sex behavior, angiotensin increases blood pressure and stimulates drinking. Functional analysis of the action of bioactive polypeptides is made by genetic, biochemical, physiological, and behavioral observations.

Conceivably, neuroactive agents could have evolved from toxins used as repellants against predators or as substances for immobilizing prey. However, it is more likely that plant and animal toxins evolved secondarily, some of them relatively recently, from preexisting precursors. Some plant toxins are amines, such as histamine; several venoms contain much histamine. Many animal toxins are small polypeptides, others are large molecules. The venom of a snake *Bothrops* contains seven peptides, ranging from two to ten amino acids. One snake venom contains nerve growth factors of molecular weight 30,000; the same factor occurs with no known function in mouse submaxillary gland and is a trophic agent in growing cultures of neurons. Scorpion venoms are small proteins, 60–70 amino acids each, in coiled structure and with relatively high content of cysteine.

Many toxins are hydrolytic enzymes: phospholipases are found in bee venom, scorpion venom, and poisons of annelids; other toxins are phosphatases or esterases and peptidases or hyaluronidases. Jellyfish contain many kinds of toxic neuropeptides. Like sequences are found in four snake toxins, three scorpion toxins, and two bee toxins. Many plant and animal toxins have specific neurotoxic action: d-tubocurarine blocks nicotinic cholinergic synapses; tetrodotoxin and bufotoxin block sodium channels in excitable membranes. The development of toxic agents has had survival value; the molecular evolution of the agents remains obscure. It is unlikely that neurotoxins were precursors of polypeptide modulators or transmitters.

CONCLUSIONS

In living organisms transmission from cell to cell can be either electrical or chemical. Electrical coupling is probably primitive; it occurs between epithelial cells and between cells of embryos. Primitive nervous systems (nerve nets) have chemical transmission; electrical transmission between some neurons probably evolved secondarily. In those central nervous systems where electrical transmission occurs it is rapid, allows synchronization of membrane responses, and usually goes in either direction between cells, rarely in only one direction. Chemical transmission provides for polarized conduction, excitation, and inhibition, and for degrees of integration which are not possible with electrical transmission. At some synapses both chemical and electrical transmission occur.

Neuroactive agents are biochemically conservative. Neurons liberate transmitters, modulators, and hormones that are synthesized from a small number of widely occurring substances. Some amino acids have specific membrane effects—excitatory attractant or inhibitory repellent on cell surfaces of bacteria, on hybroids, and on sense organs of more complex animals. In plants, some amino acids regulate ion permeability.

Starting with a few neuroactive amino acids, most transmitters are made as follows: monoamines, quaternary amines (including acetylcholine), and the many neuropolypeptides. Monoamine derivatives of several amino acids—dopamine, norepinephrine, serotonin—function in transmission at synapses. Quaternary amines become methylated and several of them—betaine, taurine, homarine—occur in neural tissues. One methylated quaternary amine, choline, is probably a universal constituent of cell membranes, and is acetylated as a neuroactive agent in most animals. Acetylcholine may regulate ion permeability in nonneural cells, and it is present in plants. Purines—adenosine and adenosine triphosphate—are autonomic transmitters. Other purines—c-AMP and c-GMP—are intracellular links in neural activation. A large number of polypeptides, some as short as three amino acids and some as long as 35 residues, are neuromodulators or neurohormones, each with multiple functions. Polypeptides are probably very ancient and may have had other functions before they became modified into neural agents.

Redundancy is characteristic of neurotransmitters; several transmitters may have similar or complementary excitatory or inhibitory actions. Various neurons of the mammalian myenteric plexus show immunochemical reactions for some 10 to 12 different agents, mostly polypeptides, some of which are excitatory, others inhibitory. These agents are: acetylcholine, norepinephrine, GABA, 5HT, substance P, cholecystokinin, somatostatin, bradykinin, ATP, enkephalin, bombesin, and VIP.

Two transmitter agents may occur in the same neuron. Cultured sympathetic neurons show vesicles, dense core catecholamine-liberating and small clear acetylcholine-liberating (110). Some central neurons contain both amines and polypeptides; other neurons have substance P and leu-enkephalin, and others 5HT and a polypeptide. VIP has been found in cholinergic neurons of lumbosacral sympathetics, CCK in dopamine neurons. Some autonomic neurons are cholinergic, some are adrenergic, some are both; action can be blocked by propranolol or atropine (53). Neural crest cells are at first adrenergic and become cholinergic later in development (34).

Many polypeptides are made in the hypothalamus and also in other organs—pituitary, adrenal medulla, myenteric plexus, secretory epithelium of intestine, and elsewhere. The same hormones are synthesized in gastrointestinal tract and in brain. It is not known in which organ specific polypeptides were first made.

The development of nerve endings and their contained agents can be influenced by external factors. Sympathetic neurons in culture develop more cholinergic endings in the presence of nonneural cells than in their absence. After denervation the distribution of receptors on postsynaptic membranes becomes altered. In muscle, receptors appear over the surface of the fiber membrane, but the properties of these receptors may be qualitatively different from those of the endplate receptors, which disappear after a motor nerve degenerates. A polypeptide, nerve growth factor, stimulates axon sprouting and differentiation. The nature of the feedback from a postsynaptic element to a motor neuron is unknown.

There is close resemblance of neurons to endocrine glands. Many central neurons, especially those in the hypothalamus, have secretory functions. Neurosecretory cells have been identified in nervous systems of earthworms, crustaceans, insects, and molluscs. Glands such as corpora cardiaca in insects are derived from neural structures; the sinus gland and associated structures in crustaceans are modified neural tissue. Neurons which synthesize transmitter in the cell body and secrete it from distal endings can be considered as extended gland cells.

Coevolution of transmitters and receptors is not understood. The steric properties of acetylcholinesterase are such that an appropriate fit of the quaternary ammonium and the esteratic portions of ACh occurs. The receptor for ACh which mediates ion conductance changes is a large tetrameric protein. How the ionic channels are altered whenever a transmitter impinges on a postsynaptic membrane is under active investigation. Different but related receptors for catecholamines and for 5HT activate adenylate cyclase and thus increase intracellular cAMP.

A number of transmitters act via intracellular c-AMP. In frog retina, illumination increases concentration of c-AMP in the cones, of c-GMP in the rods. L-DOPA increases c-AMP in rat caudate nucleus by way of dopamine. Catecholamine receptors which lead to activation of adenylate cyclase are stereospecific (59). Both VIP and NE activate the synthesis of c-AMP in receptor cells (24a).

The modes of action of neuromodulators are little known. Substance P and VIP are liberated from neurons of the myenteric plexus, diffuse over many muscle cell lengths, and modulate myogenic rhythms. The egg-laying hormone of *Aplysia* is liberated from axons of bag cell neurons into the interstitium of the abdominal ganglion, where it affects motorneurons which implement egg laying. The action of a transmitter may be conditioned by the state of the receptor; action of axons from locus coeruleus on hippocampus depends on behavioral state (24a).

Neurohormones stored in nerve endings, such as the neurohypophysis, may be released into the blood. Hormones from hypothalamic neurons release or restrict secretion from distant glands; for example, a hypothalamic factor inhibits release of insulin from the pancreas.

A substance may be a transmitter, a neuromodulator, and/or a hormone. Substance P is a transmitter at dorsal root endings and a modulator in the intestinal plexus. LHRH appears to be a transmitter of slow excitation in frog sympathetic ganglion; in mammals LHRH is a neurohormone causing release of luteinizing hormone, and it is also a neuromodulator affecting sex behavior.

A single neuron may synthesize several transmitters. A postsynaptic neuron may have several kinds of receptor molecules; those of one kind may occur together on the membrane (132).

Chains of neurons may use different transmitters at successive synapses. An example is dopaminergic interneurons interposed between cholinergic synapses in rabbit superior cervical ganglion. Another example is parasympathetic cholinergic neurons acting via substance P on intestine.

A single agent may have multiple actions in neurotransmission. As identified by blocking agents, there are at least three kinds of ACh receptors and three kinds of norepinephrine receptors. Analysis of ionic conductance effects shows one or two transmitters depolarizing by increasing conductance for sodium or calcium, and hyperpolarizing by increasing conductance for Cl and for K, and by stimulating a sodium pump. Some transmitter actions are diphasic for ion effects (depolarizing and hyperpolarizing). In sympathetic ganglia of some amphibians there is a fast nicotinic EPSP due to increase in Na conductance, a muscarinic slow IPSP due to increase in K conductance (Necturus) or to decrease in resting Na conductance (frog), and a very late slow EPSP mediated by LHRH. In *Aplysia*, 5HT may have any of three excitatory and three inhibitory actions; ACh and dopamine also have several effects.

Neurohormones have numerous and overlapping actions, many of which have changed function during vertebrate evolution. Prolactin is an important hormone for water balance in fishes and amphibians, and it is a milk letdown factor in mammals. Bio-opiates function in modulation in nervous systems in general, and as analgesics in some animals. The functions of bio-opiates in earthworms and molluscs are unknown. The recognized number of neuronal polypeptides is increasing as more fractions from more species are purified. Most polypeptides of molluscs and arthropods differ in sequences from mammalian polypeptides; homologies of sequences occur, but no evolutionary pattern is apparent.

The evolution of polypeptide synthetic pathways remains largely unknown. In no instances do the polypeptides seem to be built by adding one amino acid to another to make a chain. Rather, polypeptides are coded by large mRNAs with long leader nucleotide sequences; these RNAs are transcribed from complex genes. The few polypeptides for which the synthesis is known, have pro- and pre- segments which are cleaved off posttranslationally. In a few cases one large protein gives rise to several independently active polypeptides—as in pro-opiomelanocortin (POMC). How the DNA code for the large proteins evolved, especially since much of the product protein is ultimately to be cleaved away, is difficult to understand. Possibly the proteins are very ancient and originally had different functions. The discarded segments are usually lipid-soluble and may function in transport of the precursor across intracellular membranes before the active polypeptide is secreted.

Pharmacology of neurotransmitters, modulators, and hormones is an aspect of behavioral analysis. Examples can be cited from the clinical literature. Amphetamine promotes catecholamine release and blocks its reuptake. Phenothiazine has similar action, especially for dopamine. Serotonin decreases avoidance behavior in rats, while acetylcholine increases avoidance behavior (8). Elevation of 5HT content in telencephalon and diencephalon of rats depresses approach behavior. Sedation results from depletion of norepinephrine and dopamine in forebrain. Angiotensin stimulates drinking; increase in the octapeptide CCK in dopaminergic neurons elicits signs of satiety. Secretion of LHRH stimulates sex behavior. Analgesia is caused by the natural opiates enkephalins and endorphin; the behavioral effects are important for survival.

The preceding account points out the molecular basis for some of the neural components of behavior. Chemical transmission of nerve signals allows more versatility of response, more plasticity, and less stereotypy than electrotonic transmission. A few amino acids were probably primitive transmitters and have persisted as such; from amino acids there evolved monoamines, quaternary amines, and polypeptides. Redundancy and biochemical conservatism are characteristic of neuroactive agents.

REFERENCES

1. Ache, B. W. pp. 369–398 in *Biology of Crustaceans,* Academic Press, New York, 1982. Chemoreception in crustaceans.
2. Adams, P. R. et al. *J. Physiol.* **330:**537–572, 1982. M current and other K currents in bullfrog sympathetic neurons.
3. Adler, J. and A. Mesclov. *J. Bacteriol.* **112:**315–326, 1972; **118:**560–576, 1974. *Annu. Rev. Biochem.* **44:**341–356, 1975. *Nature* **280:**279–284, 1979. Amino acid attractants and repellants for bacteria.
4. Albuquerque, E. X. et al. *Proc. Natl. Acad. Sci.* **71:**2818–2822, 1974. ACh receptors in muscle endplates.
5. Altmann, H. et al. *Brain Res.* **111:**337–345, 1976. Inhibitory transmitters in brain of cat.
6. Altschuler, R. A. et al. *Nature* **298:**657–659, 1982. Aspartate/glutamate in guinea pig photoreceptors.
7. Alumets, J. et al. *Nature* **279:**805–806, 1979. Enkephalin and endorphin in earthworms.

8. Aprison, M. H. et al. *Fed. Proc.* **34**:1813–1822, 1975. Behavioral effects of serotonin and acetylcholine.

9. Arch, S. *J. Gen. Physiol.* **60**:102–119, 1972. Biochemical isolation, physiological identification and biosynthesis of ELHG of Aplysia.

10. Archer, R. *Proc. R. Soc. London* **B210**:23–43, 1980. *Trends Neurosci.* **4**:225–229, 1981. Evolution of neuropeptides.

11. Ballivet, M. et al. *Proc. Natl. Acad. Sci.* **79**:4466–4470, 1982. Sequencing of AChR.

12. Balashov, N. et al. *British J. Pharmacol.* **54**:383–388, 1975. Inhibitory transmitters in crab neuromuscular junction.

13. Barker, D. L. and E. A. Kravitz. *Nature* **63**:61–63, 1972. Octopamine in lobster nervous system.

14. Barker, J. L. and B. R. Ransom. *J. Physiol.* **280**:331–354, 1978. Effects of amino acids on neurons in tissue culture.

15. Battelle, B. A. et al. *Science* **216**:1250–1252, 1982. Octopamine in Limulus eye.

16. Baudry, M. and G. Lynch. *Nature* **282**:748–751, 1979. *Proc. Natl. Acad. Sci.* **77**:2298–2304, 1980. Glutamate receptors in hippocampus.

17. Bauer, B. et al. *Vision Res.* **21**:1665–1672, 1981. Neurotransmitters in retina.

18. Bennett, M. V. L. *Brain Res.* **143**:43–60, 1978. Mixed synapses in spinal cord of fish.

19. Bennett, M. V. L. *Develop. Biol.* **65**:114–125, 1978. Gap junctions in Fundulus embryos.

19a. Bernier, L. et al. *J. Neurosci.* **2**:1682–1691, 1982. Serotonin increases c-AMP in sensory interneurons of Aplysia.

20. Bishop, C. A. et al. *Proc. Natl. Acad. Sci.* **78**:5899–5002, 1981. Proctolin in insect nervous system.

21. Bitar, K. N. et al. *Nature* **297**:72–74, 1982. Opiate receptors on gastric smooth muscle cells.

22. Bittiger, H. et al. *Nature* **287**:645–647, 1980. Neurotransmitters in retina.

22a. Bixby, J. L. et al. *Nature* **301**:431–432, 1983. Enkephalin reduces quantal content.

23. Blankenship, J. E. et al. *Fed. Proc.* **42**:96–100, 1983. Peptide control of egg laying in Aplysia.

24. Bloom, F. E. *Peptides: Integrators of Cell and Tissue Function,* Raven, New York, 1980.

24a. Bloom, F. E. *Am. J. Physiol.* **246**:C184–C194, 1984. Functional significance of neurotransmitter diversity.

25. Bone, Q. and G. O. Mackie. *Biol. Bull.* **149**:267–286, 1975. Skin impulses in a tunicate.

26. Breen, C. A. and H. L. Atwood. *Nature* **303**:716–718, 1983. Octopamine in a neurohormone with presynaptic activity on crayfish.

27. Brink, P. et al. *J. Gen. Physiol.* **69**:517–526, 1977; **72**:67–86, 1978. Electrical properties of septum in giant axons of earthworms.

28. Brown, B. E. et al. *J. Insect Physiol.* **21:**1879–1881, 1975; **23:**861–864, 1977. Proctolin: a transmitter in insects.

29. Brown, D. A. et al. *Fed. Proc.* **40:**2625–2630, 1981. Slow cholinergic and peptidergic transmission in sympathetic ganglia.

30. Brown, D. A. *TINS* **6:**302–307, 1983. Slow cholinergic excitation in sympathetic ganglia.

31. Brown, M. et al. *Life Sci.* **21:**1729–1734, 1977. *Trends Neurosci.* **2:**95–97, 1979. Bombesin.

32. Brownstein, M. J. et al. *Science* **207:**373–378, 1980. Synthesis and transport of neurohypophysial peptides.

33. Brownstein, M. J. et al. *Brain Res.* **116:**299–305, 1976. Distribution of substance P in brain of rat.

34. Bunge, R. et al. *Science* **199:**1409–1416, 1978. Synthesis of transmitters by autonomic neurons in culture.

35. Burnstock, G. *Purinergic Receptors,* Chapman and Hall, New York, 1981.

36. Burnstock, G. *J. Physiol.* **313:**1–35, 1981. Neurotransmitters in autonomic nervous system. *Pharmacol. Rev.* **24:**509–581, 1972. Purinergic neurotransmission.

37. Carr, W. E. Personal Communication 1982. Adenosine monophosphate as chemoattractant in shrimp.

38. Catapane, E. J. et al. *J. Exper. Biol.* **74:**101–113, 1978; **83:**315–323, 1979. Pharmacology of ciliated epithelium in Mytilus.

39. Cerione, R. A. et al. *Nature* **306;**562–566, 1983. Adrenergic receptors.

40. Chan-Palay, V. et al. *Proc. Natl. Acad. Sci.* **79:**3355–3359, 1982. Effects of neuropeptides on rabbit brain.

41. Chavkin, C. and A. Goldstein. *Proc. Natl. Acad. Sci.* **78:**6543–6547, 1981. Receptor for opioid peptides.

42. Chium, A. Y. et al. *Proc. Natl. Acad. Sci.* **76:**6656–6600, 1979. Purification and primary structure of neuropeptide egg-laying hormone of Aplysia.

43. Clothia, C. and P. Pauling. *Nature* **219:**1156–1157, 1968. Conformations of ACh.

44. Conti-Tronconi, B. M. *Annu. Rev. Biochem.* **51:**491–530, 1982. Nicotinic cholinergic receptors.

45. David, J. C. *Compar. Biochem. Physiol.* **64C:**161–164, 1979. Octopamine in locust nervous system.

46. DeFeudis, F. V. and P. Mandel. *Amino Acids as Neurotransmitters,* Academic Press, New York, 1981.

47. Devillers-Thiery, A. et al. *Proc. Natl. Acad. Sci.* **80:**2067–2071, 1983. RNA coding sequence of ACh binding subunit of Torpedo ACh receptor.

48. Dreyfus, C. F. and M. D. Gershon. *Brain Res.* **128:**125–139, 1977. Synthesis of 5-HT by neurons in Auerbach plexus.

49. Elliott, E. J. *J. Compar. Physiol.* **129:**61–66, 1979. Cholinergic responses in heart of clam.

50. Evans, P. D. *J. Physiol.* **318:**99–122, 1981. Octopamine in insects.

50a. Fernlund, P. *Biochem. Biophys. Acta* **371**:304–341, 1974. Structure of red-pigment concentrating hormone of shrimp Pandalus.

51. Florey, E. and M. Rathmayer. *Compar. Biochem. Physiol.* **61C**:229–237, 1978. Effects of amines on muscles of crustaceans.

52. Froehner, S. C. et al. *Proc. Natl. Acad. Sci.* **78**:5230–5234, 1981. Localization of AChR in rat muscle.

53. Furshpan, E. J. et al. *Proc. Natl. Acad. Sci.* **73**:4225–4229, 1976. *Harvey Lectures* **76**:149–191, 1981. Chemical transmitters in sympathetic ganglion.

54. Gerschenfeld, H. M. et al. *J. Physiol.* **243**:427–456, 1974; **274**:265–278, 1978. Neurotransmitters in Aplysia.

55. Gold, G. H. and J. Dowling. *J. Neurophysiol.* **42**:292–310, 311–328, 1979. Coupling in toad retina.

56. Goldstein, A. et al. *Proc. Natl. Acad. Sci.* **78**:7219–7223, 1981. Pituitary dynorphin hepatodecapeptide.

57. Greenberg, M. J. and D. A. Price. pp. 107–126 in *Peptides,* Ed. F. E. Bloom, Raven, New York, 1980. *Annu. Rev. Physiol.* **45**:271–288, 1983. Invertebrate neuropeptides.

58. Greenberg, M. J. *Compar. Biochem. Physiol.* **33**:259–294, 1970. ACh structure-activity relation.

59. Greengard, P. et al. *Fed. Proc.* **33**:1059–1067, 1974. *Nature* **260**;101–108, 1976. *Harvey Lectures* **75**:277–331, 1981. Role of cyclic AMP in neurotransmission.

60. Grimmelikhuijzen, C. J. P. *Histochemistry* **69**:61–68, 1980; **71**:225–333, 1981. *Trends Biochem. Sci.* **4**:265–267, 1979. Immunoreactivity of *Hydra* for polypeptides.

61. Gubler, U. et al. *Nature* **295**:206–208, 1982. Proenkephalin precursor of enkephalin-containing peptides.

62. Harmar, A. et al. *Nature* **284**:267–269, 1980. Synthesis of substance P by dorsal root ganglia.

63. Hayashi, S. et al. *Compar. Biochem. Physiol.* **58C**:183–191, 1977. Octopamine metabolism in invertebrates.

64. Haynes, L. W. *Prog. Neurobiol.* **15**:205–245, 1980. Peptide neuroregulators in invertebrates.

65. Hedblom, M. L. and J. Adler. *J. Bact.* **144**:1048–1060, 1980. Receptor molecules for amino acids in *E. coli.*

66. Herbert, E. et al. *Neurosci. Newslett.* **5**:16–27, 1981. Review of neuropeptides in pituitary and brain.

67. Hirata, F. et al. *Proc. Natl. Acad. Sci.* **76**:368–372, 1979. Beta adrenergic receptors.

67a. Hochachka, P. Ch. 7., pp. 179–270 in *Strategies of Biochemical Adaptation,* ed. P. Hochachka and G. Somero, Saunders, Philadelphia, 1973.

68. Hokfelt, T. et al. *Nature* **284**:515–521, 1980. Peptidergic neurons.

69. Hughes, J. T. et al. *Brain Res.* **88**:295–308, 1975. *Nature* **258**:577–579, 1975. *J. Exper. Biol.* **89**:239–355, 1980. Brain opioids.

70. Huxtable, R. J., Ed. *Taurine in Nutrition and Neurology* (Advances in Exper-

imental Medicine and Biology, vol. 139). Plenum, New York, 1982.

71. Irisawa, H. et al. *Compar. Biochem. Physiol.* **45A:**653–666, 1973. Transmitter effects on membrane conductance in molluscan heart.

72. Ishida, A. T. et al. *Proc. Natl. Acad. Sci.* **78:**5890–5894, 1981. Amino acid transmitters in retina.

73. Inturrisi, C. E. et al. *Proc. Natl. Acad. Sci.* **77:**5512–5514, 1980. Enkephalins from adrenal and brain.

74. Iversen, L. L. *Science* **188:**1084–1089, 1975. *Sci. Am.* **241:**118–149, 1979. *TINS* **6:**293–295, 1983. Neuropeptides in brain.

74a. Jaffe, M. J. et al. *Plant Physiology* **51:**520–524, 1973; **54:**797–798, 1984. Acetylcholine esterase in plants.

75. James, V. A. and R. J. Walker. *Compar. Biochem. Physiol.* **64C:**53–59, 1979. Transmitters in limulus central nervous system.

76. Jan, Y. N. and S. W. Kuffler. *Proc. Natl. Acad. Sci.* **77:**5008–5012, 1980. Transmitters in bullfrog sympathetic ganglion.

77. Jessen, K. R. et al. *Nature* **281:**71–74, 1979. Transmitters in cultured myenteric pleux.

78. Kehoe, J. S. *J. Physiol.* **225:**85–172, 1972. Properties of synapses in Aplysia.

79. Kistler, J. et al. *Biophys. J.* **37:**371–383, 1982. Structure of an ACh receptor.

79a. Kocsis, J., S. I. Bashim, and S. W. Schaffer. *Effects of Taurine on Excitable Tissues.* Spectrum, New York, 1981.

80. Konishi, S. et al. *Nature* **282:**515–516, 1979; **294:**80–82, 1981. Enkephalin as transmitter in sympathetic ganglion.

81. Kravitz, E. A. et al. *J. Exper. Biol.* **89:**159–175, 1980. Amines and peptides as neurohormones in lobster.

82. Krieger, D. *Science* **22:**975–985, 1983. Brain peptides.

83. Kuffler, S. W. *J. Exper. Biol.* **89:**257–286, 1980. Synaptic responses in autonomic ganglia of amphibians.

84. Lafon-Cazal, M. et al. *Compar. Biochem. Physiol.* **73C:**293–296, 1982. Octopamine in ganglia of Locusta.

85. Lam, D. M. K. et al. *Proc. Natl. Acad. Sci.* **75:**6310–6313, 1973. *J. Compar. Neurol.* **182:**221–246, 1978. *Nature* **278:**565–567, 1979. Transmitters in fish retina.

86. Langer, S. F. *Biochem. Pharmacol.* **23:**1793–1800, 1974. α_1 and α_2 catecholamine receptors.

87. Lawn, I. D. *J. Exper. Biol.* **83:**45–52, 1980. Conduction in sea anemone.

88. Leeman, S. E. *J. Exper. Biol.* **89:**193–200, 1980. Substance P and neurotensin.

88a. Lemos, J. R. and A. Berlind. *J. Exper. Biol.* **80:**307–326, 1981. Transmitters in lobster cardiac ganglion.

89. Lenhoff, H. M. et al. pp. 571–579 in *Coelenterate Ecology and Behavior.* Ed. G. Mackie. Plenum, New York, 1976. Chemoreception in hydra.

89a. Levitzki, A. et al. *Proc. Natl. Acad. Sci.* **71:**2773–2276; 4247–4248, 1974. B-receptors of erythrocytes.

90. Lewis, M. E. et al. *Science* **211:**1166–1169, 1981. Opiate receptors in brain.

91. Libet, B. et al. *J. Neurophysiol.* **37**:805–814, 1974. *J. Physiol.* **237**:635–662, 1974. Transmitters in sympathetic ganglia of frog and rabbit.

92. Lloyd, P. E. et al. *J. Compar. Physiol.* **138**:265–270, **139**:333–339, 1980. *Proc. Natl. Acad. Sci.* **81**:2934–2937, 1984. Neuropeptides in molluscs.

93. Loewenstein, W. R. *Fed. Proc.* **32**:60–64, 1973. Intracellular conduction in liver and epithelium.

94. Loh, Y. P. *Proc. Natl. Acad. Sci.* **76**:796–800, 1979. Precursors of endorphin in pituitary.

95. Londos, C. et al. *Proc. Natl. Acad. Sci.* **77**:2551–2554, 1980. Adenosine receptors.

96. Lundberg, J. M. and T. Hokfelt. *TINS* **6**:325–333, 1983. Coexistence of peptides and classical transmitters.

97. Lyddiatt, A. et al. *FEBS Lett.* **108**:20–24, 1979. ACh receptors.

98. Mackie, G. O. pp. 647–657 in *Coelenterate Ecology,* Ed. G. O. Mackie, Plenum, New York, 1976. Nerve net conduction.

98a. Mackie, G. O. et al. *Phil. Trans. R. Soc. London* **B301**:365–428, 1983. Conduction in hexactinellid sponges.

98b. Maeda, K. and Y. Imae. *Proc. Natl. Acad. Sci.* **76**:91–95, 1979. Chemosensory transduction in E. coli.

99. Martin, R. J. *Compar. Biochem. Physiol.* **61C**:37–40, 1978. Glycine and GABA receptors in lamprey brain.

100. Mayeri, E. *J. Neurophysiol.* **42**:1165–1197, 1979. Neurohormones in Aplysia.

101. McGinty, J. F. et al. *Proc. Natl. Acad. Sci.* **80**:589–593, 1983. Dynorphin in hippocampus.

102. Miller, R. J. et al. *Science* **198**:748–750, 1977. pp. 127–146 in *Peptides,* Ed. F. Bloom, Raven, New York, 1980. Transmitters in retina and in sympathetic neurons.

102a. Mishima, M. et al. *Nature* **307**:604–608, 1984. Functional ACh receptor from cloned cDNAs.

103. Mudge, A. W. et al. *Nature* **292**:764–766, 1981. *Proc. Natl. Acad. Sci.* **76**:526–530, 1979. Neurotransmitters in sensory ganglia in culture.

104. Muneokam, Y. et al. *Compar. Biochem. Physiol.* **73C**:149–156, 1982. Neurotransmitter in muscle of Mytilus.

105. Myers, P. R. and D. C. Sweeney. *Compar. General Pharmacol.* **3**:277–282, 1972; **4**:321–325, 1973. *Tissue and Cell* **6**:49–54, 1974. Catecholamines in molluscan ganglia.

105a. Nakanishi, S. et al. *Nature* **278**:423–427, 1979. Sequenced nucleotides of cloned cDNA of bovine corticotropin-βlipotropin precursor.

106. Netherton, J. C. and S. Gurin. *J. Biol. Chem.* **257**:11971–11975, 1982. Biosynthesis of homarine.

107. Nicholls, J. and B. G. Wallace. *J. Physiol.* **281**:157–185, 1978. Synaptic function in nervous system in leech.

108. Noda, M. et al. *Nature* **295**:202–206, 1982. Sequence of cDNA for adrenal preproenkephalin.

108a. Noda, M. et al. *Nature* **302**:251–255, 528–532, 1983. Acetylcholine receptor structure.

109. Ochoa, E. *Compar. Biochem. Physiol.* **76:**99–103, 1980. Multiple forms of ACh receptor.

110. O'Lague, P. H. et al. *Proc. Natl. Acad. Sci.* **71:**3602–3606, 1974. Synapses in cultured sympathetic neurons.

111. Otsuka, M. and S. Konishi. *TINS* **6:**317–320, 1983. Substance P.

112. Painter, S. D. et al. *Biol. Bull.* **162:**311–332, 1982. *Life Sci.* **31:**2471–2478, 1982. Molluscan neuropeptide FMRF amide.

113. Pasantes-Morales, H. et al. pp. 273–292 in *Taurine in Nutrition and Neurology,* Ed. R. J. Huxtable, Plenum, New York, 1982. Effects of taurine on a heart.

114. Peroutka, S. J. et al. *Science* **212:**827–829, 1981. Central serotonin reception.

115. Pettigrew, J. D. et al. *Science* **182:**81–83, 1973. Inhibitory transmitters in cortex.

116. Piggott, S. M. et al. *Compar. Biochem. Physiol.* **51C:**91–100, 1975. Glutamate receptor in Helix.

117. Potter, D. D. et al. *J. Exper. Biol.* **89:**57–71, 1980. Properties of neurons in cultured sympathetic ganglion.

118. Price, D. A. et al. *Science* **197:**670–671, 1979. *Compar. Biochem. Physiol.* **72C:**325–328, 1982. Neuropeptides in molluscs.

118a. Prosser, C. L. *Biol. Bull.* **78:**92–102, 1940. ACh effect and nervous inhibition of heart of clam.

119. Quirion, R. et al. *Nature* **303:**714–716, 1983. Substance P receptors in rat nervous system.

120. Raftery, M. A. et al. *Science* **208:**1454–1456, 1980. ACh receptors.

121. Ritchie, A. K. and D. M. Fambrough. *J. Gen. Physiol.* **65:**751–767, 1975; **66:**327–355, 1975. ACh receptors in cultured myotubes.

122. Robertson, H. A. *Int. Rev. Neurobiol.* **19:**173–224, 1976. Octopamine in invertebrates.

123. Rosa, P. A. and E. Herbert. *J. Exper. Biol.* **89:**215–237, 1980. Precursors of peptides in pituitary.

123a. Rovainen, C. M. *Physiol. Rev.* **59:**1008–1077, 1979. Neurobiology of lampreys.

124. Rubin, L. L. et al. *Nature* **283:**264–267, 1980. AChEs at neuromuscular junctions.

124a. Russo, A. F. and D. E. Koshland. *Science* **220:**1016–1020, 1983. Aspartate receptor in bacteria.

124b. Salanki, J. Recent Developments in Neurobiology, *Acad. Sci. Hungary* **3:**67–89, 1972. Serotonin in ganglia of bivalves.

125. Satchell, D. G. et al. *Compar. Biochem. Physiol.* **64C:**231–235, 1979. Dopamine in Mytilus muscle.

126. Schaffer, S. W. et al., eds. Effects of Taurine on Excitable Tissues, Medical Science Books, 1981.

127. Schaller, H. C. et al. *Proc. Natl. Acad. Sci.* **78:**7000–7004, 1981. Morphogenetic peptides from *Hydra.*

128. Schally, A. V. et al. *Proc. Natl. Acad. Sci.* **77:**4489–4493, 1980. Precursor of somatostatin.

129. Scheller, P. H. et al. *Cell* **28:**707–719, 1982; **32:**7–22, 1983. Sequences of genes that encode egg-laying hormone of Aplysia.

130. Schulman, J. A. and F. F. Weight. *Science* **194**:1437–1439, 1976. Synaptic transmission in frog sympathetic ganglia.

131. Scott, J. et al. *Nature* **302**:538–540, 1983. Precursors for nerve growth factors.

132. Segal, M. M. *TINS* **6**:118–121, 1983. Specification of synaptic action.

133. Shain, W. et al. *Brain Res.* **72**:225–240, 1974. ACh receptors in Aplysia.

134. Shimoda, I. and T. Tanimura. *J. Gen. Physiol.* **77**:23–39, 1981. Amino acid receptor.

135. Snyder, S. H. et al. *Fed. Proc.* **34**:1915–1921, 1975. *Brain Res.* **111**:204–211, 1976. *Science* **209**:976–983, 1980. Neuropeptides as transmitters and their receptors.

135a. Snyderman, R. and E. Goetzl. *Science* **213**:830–837, 1981. Chemotaxis of leukocytes.

136. Solceda, R. et al. *Brain Res.* **135**:186–191, 1977. Taurine in frog retinal rod outer segments.

137. Spencer, A. N. *J. Exper. Biol.* **93**:33–50, 1981. Electrical coupling between neurons in Hydrozoan.

137a. Spira, M. E. and M. V. L. Bennett. *Brain Research* **37**:294–300, 1972. *Science* **194**:1065–1067, 1976. Synaptic control of electrotonic coupling between neurons.

138. Spray, D. C. et al. *J. Gen. Physiol.* **77**:77–93, 1981. *Proc. Natl. Acad. Sci.* **79**:441–445, 1982. Electrical coupling between embryonic cells.

139. Srikant, C. B. et al. *Nature* **294**:259–261, 1981. Somatostatin.

140. Stefano, G. et al. *Life Sci.* **24**:1617–1622, 1979. *Cell. and Molec. Neurobiol.* **1**:57–68, 1981. Opioids in molluscs.

141. Stene-Larsen, G. *Compar. Biochem. Physiol.* **70C**:1–12, 1981. Adrenoreceptors in vertebrate heart.

142. Stevens, C. F. *Nature* **287**:13–14, 1980. ACh receptors.

143. Stewart, J. M. et al. *Fed. Proc.* **38**:2302–2308, 1979. Evolution of neuropeptides.

144. Strumwasser, F. et al. pp. 197–218 in *Peptides,* Ed. F. E. Bloom, Raven, New York, 1980. Peptides in Aplysia.

145. Strumwasser, F. Ch. 7, pp. 183–215 in *Brain Peptides,* Ed. D. Krieger, Wiley, New York, 1983. Peptidergic neurons and neuroactive peptides in molluscs.

146. Stuart, D. and F. Strumwasser. *J. Neurophysiol.* **43**:488–498, 1980. Egg-laying hormones in Aplysia.

147. Sugiyama, H. et al. *Proc. Natl. Acad. Sci.* **74**:5524–5528, 1977. ACh receptors in retina.

148. Takeuchi, A. et al. *Nature New Biology* **242**:124–126, 1973. *Compar. Biochem. Physiol.* **72C**:237–239, 1982. Amino acid transmitters in crayfish.

149. Tam, S. W. *Proc. Natl. Acad. Sci.* **80**:6703–6707, 1983. Naloxone blocking, μ, α, K, but not β receptors in rat brain.

150. Tayhert, P. H. and J. W. Truman. *J. Exper. Biol.* **98**:373–383, 1982. Neurohormone in insects.

151. Teranishi, T. et al. *Nature* **301**:243–246, 1983. Dopamine effects on carp retina.

151a. Tokimasa, T. and A. North. *Nature* **294**:162–163, 1981. Opiates in plexus neurons.

152. Valente, D. et al. *Compar. Biochem. Physiol.* **69C**:161–164, 1981. ACh in honey.

153. Wang, E. A. and D. Kochland. *Proc. Natl. Acad. Sci.* **77**:7157–7161, 1980. Receptor structure in bacterial sensing systems.

154. Watling, K. et al. *Nature* **281**:578–580, 1979. Dopamine receptors in retina.

155. Webb, R. G. and I. Orchard. *Compar. Biochem. Physiol.* **70C**:201–207, 1981. Octopamine in leech.

156. Weber, E. et al. *TINS* **6**:333–336, 1983. Multiple ligands for opioid receptors.

157. Weber, I. T. et al. *Proc. Natl. Acad. Sci.* **79**:7679–7683, 1982. Cyclic AMP binding sites.

158. Wickberg, J. E. S. *Nature* **273**:164–166, 1978. β-receptors.

158a. Winlow, W. et al. *J. Exper. Biol.* **94**:137–148, 1981. Dopamine in Lymnaea.

159. Wu, S. M. and J. E. Dowling. *Proc. Natl. Acad. Sci.* **75**:5205–5209, 1978. Aspartate as transmitter in retina.

160. Young, A. B. and S. H. Snyder. *Proc. Natl. Acad. Sci.* **71**:4002–4005, 1974. Glycine receptors.

161. Zakarian, S. et al. *Nature* **296**:250–252, 1982. Endorphin in regions of brain.

13

ELECTRORECEPTION, AUDIORECEPTION, AND BEHAVIOR

Electroreception and acoustic communicative behaviors were described in Chapter 10. The diversity of receptors and behavior correlates with life habits. Electric organs and electroreceptors have evolved independently several times in several kinds of animals with convergence of structure. Sound communication and reception have evolved differently in insects from vertebrates. Audioreceptors of insects are modified neurons, and they function mainly as pulse detectors for temporal discrimination; the audioreceptors of vertebrates are epithelial sense cells which activate primary neurons and are capable of frequency discrimination. Some mammals (bats and rodents) emit and perceive very high frequencies; many birds and amphibians perceive sound only at relatively low frequencies. Insects and a few crustaceans produce sounds by rubbing skeletal elements together; fish vibrate muscles under a resonant swimbladder; frogs, birds, and mammals have sound-producing organs in respiratory tubes. A major question is how receptors, coordinating interneurons, and organs for emission of electric discharge or of sounds have evolved in coordinated fashion. This chapter briefly summarizes the production, reception, and central processing of electronical and sound signals.

ELECTRORECEPTORS AND ELECTRIC DISCHARGE ORGANS

Behavior resulting from the sensing of magnetic and electric fields was discussed in Chapter 10. Magnetic sense has been demonstrated in bacteria, bees, snails, and birds.

The best known electric organs of fishes that emit electric discharges are derived from muscle cells that do not develop contractile elements. Most electric organs consist of rows of electrocytes, each of which is a cellular unit equivalent to a modified motor endplate. In some electrocytes, depolarization is graded; in some, spikes are elicited. Depolarization can be lim-

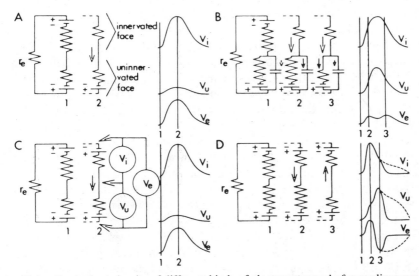

Figure 13-1. Equivalent circuits of different kinds of electrocytes and of recordings across innervated surface (V_i), across uninnervated surface (V_u), and across the entire cell (V_e). Changes from rest to activity indicated by numbered equivalent circuits; time changes as numbered on responses. (*a*) A strongly electric marine fish, innervated surface generates an EPSP. (*b*) Electrocyte of rajid fish, innervated surface generates an EPSP and uninnervated surface shows delayed rectification. (*c*) Electrocyte of electric eel, innervated surface generates overshooting spike. (*d*) Electrocyte of some mormyrids and gymnotids, both faces generate spikes and V_e is diphasic. Reprinted with permission from M. Bennett in *Fish Physiology*, Vol. 5, W. S. Hoar and D. J. Randall, Eds. Copyright 1971, Academic Press, New York.

ited to the innervated face or can spread over the entire surface of some kinds of electrocytes; others show capacitative changes (2) (Fig. 13-1).

In one genus (*Astroscopus*) extraocular muscles have developed into electric organs. In *Apteronotus* the electric organ consists of a backfolded spinal nerve which lies ventrally to vertebrae and ends blindly; the resulting organ produces an anteroposterior electric field. In *Odontosternarchus* an accessory electric organ has been derived from sensory neurons.

In the strongly electric fish *Torpedo* (a marine elasmobranch), each electric organ consists of some 45 dorsoventral columns of 700 electrocytes each. The electrocytes in a column are in series electrically; the columns are in parallel. *Torpedo* can discharge a potential of 30–50 V; in sea water this produces a current of several amperes. All of the electrocytes in a column are thickly innervated on the ventral surface and the neuro-electrocyte junctions generate graded nonpropagating postsynaptic potentials (PSPs). The innervated surface is cholinoceptive. Responses are blocked by tubocurarine and are prolonged by physostigmine. The noninnervated face of each electric plate is of lower resistance than the innervated face, and the fish skin is of low resistance; hence current can flow from dorsal to ventral surface of the fish. A *Torpedo*'s discharge starts 80 msec after its first movement toward a prey; just before "leveling in" the frequency is 140–290/sec. The two organs of *Torpedo* discharge within a few tenths of a millisecond of each other. In some other kinds of elasmobranch, for example, the Rajiidae (skates), disc electrocytes have noninnervated faces which show a decrease in resistance during the falling phase of the EPSP, thus increasing the external current.

The strongly electric fish *Electrophorus* from the Amazon has several longitudinal electric organs, the main one with some 1000 plates per column. A pulse can reach some 400 V, which amounts to 200–300 mA in the high-resistance fresh water. Each high-voltage discharge is a rapid train of 3 to 30 pulses. A motor nerve impulse elicits a graded EPSP which produces a spike that may overshoot zero potential by as much as 60 mV; across the noninnervated face there is a resting potential of 90 mV; this face does not depolarize, hence the two surfaces give a total cell potential of 150 mV. How much of this can be felt or recorded extracellularly depends on skin resistance and short-circuiting within the organ and outside. The spike is due to increased Na^+ conductance and can be blocked by TTX. During repolarization there is increased resistance (anomalous rectification). The electric organ is rich in ACh receptors and AChE. Central nervous control of the many electrocytes of an *Electrophorus* permits near synchrony of thousands of electrocytes, within 1.5 ms. Besides the main electric organ, used for stunning prey, *Electrophorus* continuously fires low-voltage electric discharges in lateral organs which constantly probe the muddy water. Many other species of weakly electric fishes use repetitive low-voltage discharge to probe for altered electrical conditions in the environment, thus for detecting other fish (Ch. 10). Mormyrids have a basal rate of one to ten discharges per second; when actively hunting prey the rate may increase to 130/sec. In all electric fish the electric organ discharge (EOD) is commanded from a nucleus in the medulla, even when the EOD is of high frequency (up to 2000 Hz), very regular, and quasi-sinusoidal. Waveforms of electric discharge of different species of gymnotids are shown in Fig. 13-2.

Electroreceptors evolved in ancestors of nonteleost fishes. Ampullar receptors are present in lampreys, elasmobranchs, holocephalans, and Osteichthyes such as dipnoans and polypterans. Of 30 orders of Teleostei (bony fish), electroreceptors are clearly established in three: gymnotiforms, mormyrids, and siluriforms (catfish).

The ampullar receptors of nonteleosts are called Lorenzini ampullae; each has a large kinocilium. The ampullae are excited by a Ca^{2+} current entering at the apical end; a large octavolateral region of the medulla is the sensory center. The ampullae of teleosts, especially of catfish, have microvillae and no kinocilium; they are excited by inward basal current. The most extensively studied of electrosensitive teleosts, mormyrids and gymnotids, have in addition to ampullae a different type of receptor, the tuberous recep-

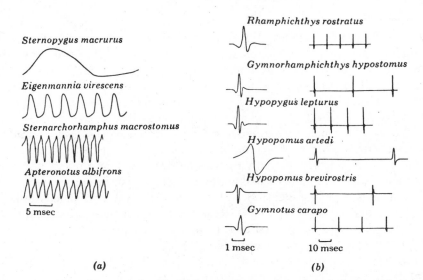

Rhamphichthys rostratus

Sternopygus macrurus

Gymnorhamphichthys hypostomus

Eigenmannia virescens

Hypopygus lepturus

Sternarchorhamphus macrostomus

Hypopomus artedi

Apteronotus albifrons

Hypopomus brevirostris

5 msec

Gymnotus carapo

1 msec   10 msec

(a)   (b)

Figure 13-2. Waveforms of electric discharges of several species of electric fish (gymnotids) from rivers of Guyana. (*a*) Tonic-type discharges with each pulse of long duration relative to intervals. (*b*) Pulse-type discharges with pulses brief relative to intervals. Reprinted from "Electric Communication in Fish," by C. D. Hopkins, *American Scientist* **62**:430 (1974).

tor. Gymnotid and mormyrid fishes detect small differences in distortion of a surrounding electric field and frequency modulations of the field. Tuberous receptors are sensitive to high-frequency pulses, 50–2000 Hz; eight functional types of tuberous receptors have been identified (6). In pulse marker units a single spike follows the fish's own electric discharge (EOD). Phase coders follow EODs with latency varying according to intensity. Other tuberous receptors are burst duration coders. Probability coders are tightly phase-locked to a sinusoidal electrical stimulus. Tuberous receptors are used in two ways: (1) to detect objects in the water that distort the fish's own EOD (electrolocation); and (2) to sense the EODs of other electric fish. In mormyriforms two types of tuberous receptors are specialized for these two functions. Tuberous receptors of similar form have evolved independently in South American gymnotiforms and in African mormyrids and gymnarchids (8). Ampullar receptors are spontaneously active, respond at low frequencies, and are not in synchrony with an electrical discharge; anodal stimulation increases the frequency of endogenous firing while cathodal stimulation decreases it. Responses in nonteleost fish are opposite to those of teleosts in respect to field polarity (6,7).

In a South American gymnotiform *Eigenmannia* the weak electrical discharge is between 250 and 500 Hz and at characteristic frequencies for individual fishes; in each discharge the head is positive to the tail. From distortions of its field *Eigenmannia* detects plants, other objects in the water, the bottom and the surface. When another electric fish emitting nearly the same frequency is nearby, a perceiving fish adjusts its own firing rate either up or down and increases the difference in frequency (ΔF) between its own rate (F_1) and the rate of the neighbor (F_2); ΔF may be 10–20 Hz. *Eigenmannia* increases its firing rate for a negative ΔF and decreases its firing rate for a positive ΔF (23,24). This response is a jamming avoidance response (JAR), and can be experimentally simulated in curarized fishes (24). Discharge becomes silenced after an *Eigenmannia* is curarized, and the discharge may then be replaced experimentally by passing sinusoidal currents (S_1) between head and tail. A stimulus (S_2) simulating a nearby fish discharge can be applied through a pair of electrodes across the fish. The two signals differ in amplitude and phase (Fig. 13-3a). Responses are recorded from neurons in the torus (Fig. 13-3b); the neuronal response shifts in position according to whether ΔF is positive or negative (Fig. 13-3b).

P receptors measure amplitude as it changes through the ΔF cycle; T receptors follow the zero-time crossing of the two sine waves (S_1 and S_2). Motor discharge is driven by the difference between P and T sensory input. P receptors increase or decrease frequency according to amplitude of field stimulation. T receptors mark the time of each cycle; they fire later according to the phase of the reflected stimulus (S_2) in relation to the EOD (S_1); hence T receptors reflect the phase of two sinusoidal stimuli in a given part of the body. The JAR is a shift in pacemaker frequency in the direction to increase ΔF, optimally when a neighbor or emitted stimulus is within 10–15 Hz of the EOD (24). When the two stimuli (S_1) and (S_2) are applied together to different body regions the fish responds by a jamming avoidance response; the nervous system integrates place as well as phase (23,24). Receptive field areas are 8.5–9 mm in diameter (1).

In *Eigenmannia* and other electric fish, sensory impulses travel in the lateral line nerve to the electroreceptive nucleus in the medulla where P and T impulses are projected. The next central relay is in the torus semicircularis, in which neurons are arranged topographically in columns corresponding to areas of skin in the parts of the body. Certain neurons of the torus respond to the sign of ΔF, to beats between S_1 and S_2 waves without phase modulation; input from different patches of skin are additive. Impulses from the torus go directly or indirectly to a pacemaker nucleus which drives motorneurons whose axons activate the electric organs. The jamming avoidance response (JAR) is precise and stereotyped according to the ΔF and the intensity of S_2, nearness of other fish.

Recordings from the *Eigenmannia* lateral line lobe and the torus semicircularis show the presence of central neurons that make responses corresponding to the characteristics of P and T receptors. The combination of P and T signals yields a neural representation of the amplitude-phase relations of the external electrical field (24).

In weakly electric fish, the amplitude of the electric organ discharge is changed when a metal object is placed in the electric field. Amplitude in-

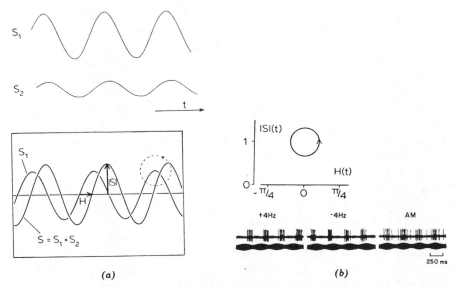

(a)   *(b)*

Figure 13-3. *(a)* Model of experiment on curarized Eigenmania. Stimulus S_1 represents normal electric discharge, S_2 represents a smaller signal of slightly different frequency coming from another fish. In the lower figure a simultaneous display of mixed signals $S_1 + S_2$ yields a sinusoidal signal with momentary amplitude (*ISI*) and phase angle (*H*) modulated according to frequency (ΔF), the difference between frequencies of S_1 and S_2. As result of the modulation of S and H the peak of $S_1 + S_2$ rotates. *(b)* is Lissajous display of (S) versus H in two-dimensional plane. Below are spikes in torus, lower record shows waxing and waning of summed stimuli corresponding to ΔF. Spikes on rising or falling phase according to whether difference between S_1 and S_2 is positive or negative. Reprinted with permission from W. Heiligenberg, *Naturwissenschaften*, Vol. 67, p. 502. Copyright 1980, Springer-Verlag, Berlin.

creases in proportion to the 1.6 power of the diameter of the object, and decreases with distance; the posterior lateral line lobe has some interneurons which increase firing and others which decrease firing rate in response.

Central control of electric discharge differs according to species of fish. In the electric catfish *Malapterus* a motor neuron on each side of the first spinal segment fires the electric organ on that side; two medullary command (pacemaker) neurons are electrotonically coupled. In gymnotids, a cluster of pacemaker cells in the dorsal medulla activates a ventral medullary relay nucleus which in turn activates spinal motorneurons; electrotonic coupling occurs at both the medullary and spinal levels. A mormyrid has a bilateral pacemaker nucleus in the midbrain which activates a second-order nucleus in the medulla which fires spinal neurons.

MECHANORECEPTORS AND AUDIORECEPTORS

Cell membranes of many organisms become depolarized when deformed. Action potentials can be initiated in algal cells by a touch, plants such as

Mimosa (sensitive plant) and *Impatiens* (touch-me-not) give spectacular responses to touch. Mechanosensitivity is high in animals with specialized receptors in skin, joints, and muscles. Receptors for sound (hair cells of cochlea) respond to minuscule deformation of nanometer amplitude. A myelinated nerve fiber can be stimulated by a 2 to 5 μm displacement of its membrane; a Pacinian corpuscle from mammalian mesentery or tendon is stimulated by displacement of 0.2–0.5 μm. Many mechanoreceptors have sensory hairs; some receptors have extensive branching of dendritic processes spreading over a muscle or tendon. Some mechanoreceptors adapt rapidly, others give tonic responses. The molecular mechanisms of stimulation are not well understood; depolarization takes place in many receptors by an increase in Na^+ conductance; the response is not sensitive to tetrodotoxin. Some sodium channels are highly sensitive to deformation and others are not, perhaps because of the geometry or the nature of the channel proteins. Some mechanoreceptors—Pacinian corpuscles of mammals, stretch receptors of crustaceans—are specialized neurons. Graded sensory potentials are initiated by deformation and, at a threshold depolarization, spikes are triggered. A neuronal mechanoreceptor may give a single spike or burst at the "on" of deformation; other receptors continue to discharge so long as deformation is maintained. The sound receptors of insects in tympanal organs are neurons which respond to frequencies of vibration that are also in the audible range of vertebrates. Another type of mechanoreceptor is a hair cell which activates nerve endings at its base—for example, in lateral line organs of fishes, and in equilibrium and auditory receptors of vertebrates. Transmission from a hair cell to nerve ending is probably chemical.

A functional series of receptors can be arranged from those sensitive to vibrations at low frequency to those sensitive to high-frequency sound. Vibration receptors of insects may respond to displacements as small as 10^{-6} cm; cockroach cercal receptors are stimulated at 30–50 Hz to vibrations of amplitude 10^{-6} cm, at 1000–5000 Hz of amplitude of 10^{-7} cm or less. Antennal organs of some beetles and flies control orientation in wind. In honeybees, antennal receptors discharge repetitively; in wind of 7.4 m/sec, the sensory flagellum vibrates at 37 Hz. A backswimmer *Notonecta* locates prey by surface waves in the water; it is most sensitive to 100–150 Hz, and receptors in the distal part of the first and second legs can detect water movements of 1 μm amplitude. Web-spinning spiders can discriminate live from dead prey by vibration on the web; the period for maximum vibration of the web is 50 Hz.

AUDIORECEPTION AND CENTRAL PROCESSING

The following is a brief survey of the physiology of sound reception and neural integration of auditory responses. Examples of behavioral communication by sound are presented in Chapter 10.

Insects

Phonoreceptors of many insects—crickets, cicadas, some moths—are membranous structures, tympanic organs consisting of an external membrane, air sac, and inner thin ligament bearing neural receptor cells. Many orthopteran insects produce sound by stridulating organs consisting of toothed scrapers and files; frequencies vary from a few hundred Hz to 110 kHz. Maximum energy of some insects' songs (katydids) is at 25–50 kHz, of others (crickets) at 10–25 kHz (3). Species-specific songs can be evoked by stimulation of central interneurons at the proper frequencies (Fig. 13-4).

Responses of receptor cells of an insect are most sensitive to the frequencies emitted by their own kind or by predators. Arctuid moths do not make sounds but have tympanic receptors mainly tuned to the cries of bat predators (41a). In a noctuid moth the receptor has three cells—a low frequency-sensitive mechanoreceptor and two phonoreceptors. The phonoreceptor cells in moth sensory nerves discharge a spike for each pulse of sound; if stimulated by a pure tone the spike frequency declines with continuous stimulation. The moth ear is a pulse detector, not a frequency discriminator; these receptors signal changes in sound intensity by varying frequency of spikes, by shifting latency, and by after-discharge following cessation of sound (41a).

Inputs from the tympanal receptors have been traced into the thoracic ganglion of a noctuid moth (41a). One interneuron on each side is a spike marker which follows each spike in the train from a phonoreceptor cell; this interneuron does not synchronize with sound waves but discharges a burst of impulses of decreasing frequency during stimulation by a tone. Another interneuron is a pulse marker, firing once (or twice) per afferent train, thus marking the train interval. A third interneuron responds to each train with a series of spikes independent of input frequency, thus marking train duration. Neurons near the midline sum the input from the two ears; one type of neuron responds to the contralateral ear but is inhibited by the ipsilateral one. A moth normally flies irregularly up and down but when a bat's pips at a distance are perceived, the moth dives steeply, probably on activation of a phonoreceptor cell (Fig. 13-5). A moth's accuracy in detecting direction of sound is shown by mercator projections of sound reception with the wings shielding or exposing the ear (Fig. 10-8).

Cricket ears have five primary fiber types, each with relatively specific carrier frequency. The primary responses reflect syllable and verse structure of songs (17). The auditory system of crickets matches the motor system of cricket sound production. Acoustically sensitive interneurons in the prothoracic ganglia of crickets have been identified by intracellular recording followed by staining (Fig. 13-6). One type of cell is spontaneously active, is inhibited by 5 kHz tones, has ipsilateral sensory connections, and connects to the brain. Another type of auditory interneuron in crickets is excited tonically at 5 kHz, follows the structure of a calling song, and gives off pro-

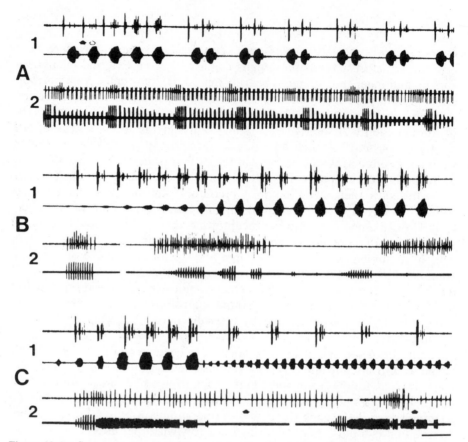

Figure 13-4. Song patterns elicited in a cricket by stimulation of neural filaments between sub-esophageal ganglion and prothoracic ganglion, compared with normal songs. In each pair of traces the upper trace shows the corresponding sound pulses in a singing animal. Stimulation frequencies (*a*) 60 Hz, (*b*) 110 Hz, (*c*) 40 Hz. (1) Low-speed records showing pattern details; (2) High-speed records showing sustained output patterns. (*a*) Calling song, each phrase with chirp of 5–7 pulses followed by a series of 2-pulse trills. (*b*) Aggressive song of 10–20 pulses at chirp rhythm starting at low intensity and building to high intensity. (*c*) Courtship song similar to normal song, begins with calling chirp, continues with trains interrupted by pauses. Reprinted with permission from D. Bentley, *Journal of Comparative Physiology*, Vol. 116, p. 30. Copyright 1977, Springer-Verlag, Berlin.

cesses to each side of the thoracic ganglion (4,11). Another kind of interneuron follows the temporal structure of a call, and is tuned sharply at 5 kHz. A fourth is a broad-band neuron that gives a burst of impulses for each pulse and is most sensitive at 15 kHz, the carrier frequency of the courtship song; it projects to the brain (39). In the brain, auditory interneurons of three types are found. All three are most responsive to sounds of 16 kHz, the primary frequency of the courtship call. These neurons differ in morphology; one is confined to the protocerebrum, two are plurisegmental within the supra-

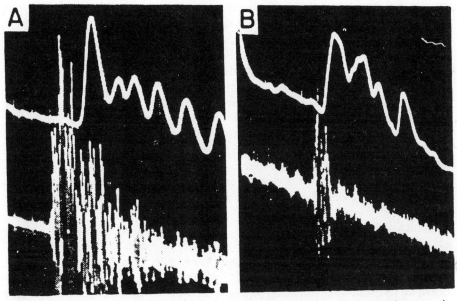

Figure 13-5. Simultaneous recording of cry of bat (lower traces) and responses in tympanic nerve of moth Prodenia. (*a*) Bat cry at range of 10–15 kHz; (*b*) Short cry at above 20 kHz. Reprinted with permission from K. Roeder and A. E. Treat, *Journal of Experimental Zoology,* Vol. 134, p. 147. Copyright 1957, Alan R. Liss, Inc., New York.

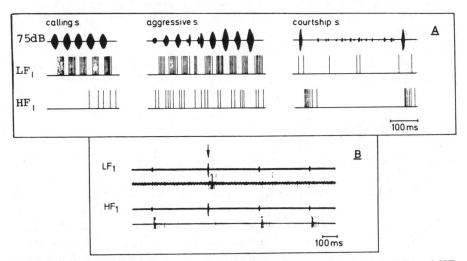

Figure 13-6. Response patterns of cricket recorded from two interneurons LF_1 and HF_1 obtained by stimulation with calling, aggressive, and courtship songs. Upper trace is sound pattern, lower two traces are responses of the two interneurons. Reprinted with permission from J. Rheinlaender et al., *Journal of Comparative Physiology,* Vol. 110, p. 259. Copyright 1976, Springer-Verlag, Berlin.

esophageal ganglion (4). There are also neurons which respond to sounds other than the courtship call; no single neuron is activated exclusvely by a species-specific song. Tettigonids (katydids) emit calls ranging from a few hundred cycles to 110 kHz, some calls with a frequency sweep, others at a constant frequency. Each tympanic organ has 100 to 300 sense cells. Large T-neurons of the ventral cord of a cricket have the lowest threshold at 10–20 kHz. Interneurons of the prothoracic ganglion of tettigonid are organized tonotopically (41). The coupling in crickets between auditory neurons and the motor neurons to sound-producing tymbal organs has not been traced.

In grasshoppers the same thoracic neurons receive the sensory signals of airborne sounds and vibrations of substrate. Male tettigonids simultaneously produce sounds and vibrations; the impulses from the two modalities converge in a female on a single neuron which facilitates the location of a male (29).

Tympanic organs of locusts show maximum frequency sensitivity at 12 kHz. Measurements of amplitude of vibration of a cricket tympanum show maximum vibration at 4–6 kHz and at 14–15 kHz. The tympanum is tuned to a narrow band and threshold is estimated at 0.5×10^{-4} μm (28a). In *Locusta* 15 ascending interneurons on each side of the nerve cord transmit responses to sound. Neurons differ in response patterns; some make ipsilateral responses, others respond bilaterally; interneurons code frequency and intensity, verse duration, and syllable rate. Responses of interneurons are more complex than those of receptors (29).

In summary, insect phonoreceptors are modified neurons on a thin membrane stretched between exoskeletal structures. These sense cells are broadband receptors; they may be most sensitive in some low-frequency range; they function mainly as detectors of sound pulses and pattern. Interneurons respond selectively to different components of the input. The central neurons activate motor responses in moths, or stereotyped motor patterns of songs in crickets.

Audition in Vertebrates

In the ears of many vertebrates the auditory mechanism lies close to the equilibrium receptors. The labyrinth has chambers which detect position, and others that detect acceleration and deceleration. Auditory portions of ears are adapted to detecting sound that is conducted in water, air, or bone. The external ear is reduced or absent in aquatic vertebrates. Internal ears detect sound that is transmitted in water or bone in fish, or in oil-filled channels of bone in whales. In snakes, high-intensity, low-frequency vibration is detected by cutaneous receptors, in fishes by lateral line organs. Many kinds of bony fishes have gas-filled swim bladders. If the swim bladder connects by ossicles to the ear, it increases sensitivity to sounds, especially of high frequencies. Elasmobranch fishes lack swim bladders; they continue to be able to respond to low-frequency sound after the lateral line nerve has been cut, but not after the auditory (eighth) nerve is cut. The external ears of mammals aid in localization of sound. The physical properties of eardrum and middle ear correlate with the limits of hearing; in a bat the speed of eardrum vibration is constant over the range from 10 to 70 kHz; in some reptiles and amphibians the eardrum does not vibrate at frequency above several kHz; in cat and guinea pig, velocity and amplitude of movement of tympanic membrane and ossicles are constant from a few hundred Hz to above 10 kHz. The middle ear ossicles couple sound waves in air to the fluid (endolymph) of the inner ear and compensate for the transmission loss at the interface.

In fishes three ampullae (chambers)—utriculus, sacculus, lagena—have hair cells functioning for sound reception. The auditory system of fishes shows considerable species diversity. The hair cells respond to change in pressure and velocity (38). The utriculus of fishes has some receptors that respond to compression, others to decompression (15).

Amphibians have two sensory papillae—the basilar papilla, a broad-band receiver, and the amphibian papilla, in which there is tonotopic organization (Fig. 13-7). In both papillae the hair cells are covered by a tectorial membrane. The hair cells are attached to the bony cochlear wall, as in fishes, and there is no underlying basilar membrane. Reptiles and birds have elongated papillae which bear hair cells, over some of which lies a tectorial membrane. In mammals the basilar membrane runs the length of the coiled cochlea. Amphibians, birds, and reptiles have two fluid chambers in the auditory organ. In mammals there are three chambers: the scala media con-

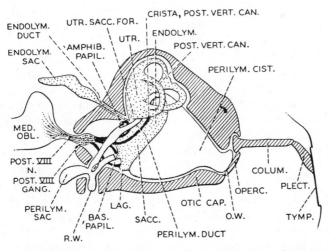

Figure 13-7. Schematic diagram of anuran ear illustrating relation between eardrum and middle ear to perilymphatic system and between perilymphatic system and amphibian and basilar papillae. Reprinted with permission from R. R. Capranica et al. *Proceedings of the IEEE*, Vol. 56, No. 6, p. 970. Copyright 1968, IEEE, New York.

tains endolymph which bathes the hair cells, the scala tympani lies on the basal side of the basilar membrane, and the scala vestibuli is separated by a membrane from the scala media.

Hair cells in vertebrate ears have two types of projecting process: (1) kinocilia, true cilia with a pattern of nine peripheral plus two central doublets of tubulin filaments and a basal body; (2) stereocilia, microvillous structures with filaments of actin (28). Many phono- and equilibrium-receptors of invertebrates have true cilia at the active endings (58). In fishes, the hair cells of the lateral line have one kinocilium plus 20–40 stereocilia. A kinocilium is approximately twice the diameter of a stereocilium (19b). Both types of cilia occur on hair cells in the phono-sensitive saccular macula of the fish ear, one kinocilium to 56–60 stereocilia (22a). Two sets of hair cells have opposite orientation on the dorsal and ventral sides of a macula. In amphibians, each hair cell has one kinocilium and up to 300 stereocilia. Around each cell are supporting cells whose surfaces are studded with microvillae (28) (Fig. 13-8). The hair cell bundle is polarized with the kinocilium at one side. Reptiles, like amphibians, have both kinocilia and stereocilia (32b). In reptiles and birds, there are two sizes of hair cells, long and short. The short hair cells of birds have many stereocilia, and few or no kinocilia (56a). In mammals all hair cells have stereocilia only. Why there are not mammalian kinocilia is not known.

Receptor potentials recorded intracellularly from hair cells of frogs are depolarizing when the hair is deflected toward the side of the kinocilium, hyperpolarizing when the hair is deflected in the opposite direction. If the kinocilium has been detached, the normal response to a tone remains; evidently the active sensory response is caused by movement of the stereocilia. Kinocilia may function for mechanical coupling; the stereocilia serve to transduce a mechanical deflection to an electrical response (28). Kinocilia seem to have diminished in importance during evolution.

Hair cells may respond to movement as small as 200 pm (2×10^{-10} m). An acceleration which is perceived as sound may be less than an attowatt (10^{-18}) of power (28). Intracellular recordings from mechanically stimulated hair cell of guinea-pig cochlea give potential deflections in either direction about the resting potential, with depolarization about three times greater than the hyperpolarization. Besides the alternating response there is a direct current component that persists for the duration of a tone (43).

In mammals, auditory nerve fibers arising from the basal turn of the cochlea carry responses to tones of high frequency; auditory nerve fibers from the upper turn carry responses to low frequencies. The width of the basilar membrane increases gradually from base to apex of the cochlea. The normal human ear is sensitive to frequencies in the range from 15–20 Hz to 16,000–20,000 Hz; human auditory threshold is lowest at 1500–2000 Hz. Tuning for frequency sensitivity is primarily determined by the basilar membrane in mammals and presumably in birds, in turtles and lizards by electrical prop-

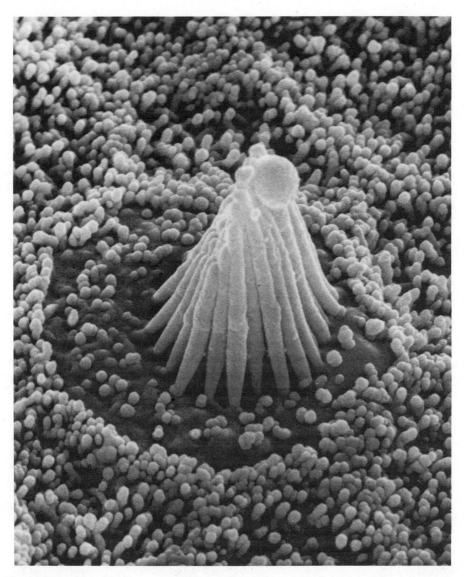

Figure 13-8. Scanning electron micrograph of apical surface of a hair cell from frog saccularis. Kinocilium shown by bulbous tip, surrounded by stereocilia. Reprinted with permission from R. Jacobs and A. J. Hudspeth, *Trends in Neuroscience,* Vol. 6, p. 366. Copyright 1983, Elsevier Science Publishers, Cambridge, England.

erties of hair cells (19a), and in frogs probably by the mass of the overlying tectorial membrane (12,13).

The fluid in the endolymph in the scala media has higher K^+ concentration than in either of the two other chambers; the K^+ is higher in endolymph than in plasma. The endolymph is electropositive to surrounding tissue fluid. A steady potential, the endocochlear potential (EP), of 160–180 mV exists between the fluid above the hair cells and the fluid surrounding them.

When the ear is stimulated by sound an alternating microphonic potential (MP) can be recorded inside the cochlea; the MP corresponds to the AC response of the hair cells. An additional extracellular response to a tone is the summating potential (SP), which corresponds to the direct-current response of hair cells (48). The hair cells contain many vesicles which resemble synaptic vesicles. Probably the response potential in the hair cells causes the release of transmitter which excites nerve endings. Single afferent fibers show relatively narrow response areas, with frequencies of maximum sensitivity according to the region of the basilar membrane where they arise. At low frequencies one fiber may give one or more spikes per sinusoidal cycle; at middle frequencies, spikes may occur as some multiple of the wave cycle; at high frequencies spikes may be random. Fibers fire in characteristic temporal patterns. During excitation of a sensory neuron by tone bursts of optimum frequency, the response can be inhibited by a second tone at a frequency on either side of the excitation range.

The selective pressures which have resulted in marked changes in auditory mechanisms at organ and cellular levels remain for further investigation.

Frogs

The auditory part of the ear of a bullfrog consists of two papillae the amphibian and the basilar; the amphibian papilla has two adjoining sensory epithelia, one of which senses low frequencies (150–500 Hz) while the other senses intermediate frequencies (500–1800 Hz) (18). Frequency is coded by place on the epithelium (31b). The basilar papilla is a tuned resonator in which all receptors respond only to high frequencies (1000–2000 Hz). An important feature of the low frequency-sensitive fibers of the amphibian papilla is two-tone suppression; low-frequency responses can be suppressed by intermediate frequencies; mid-frequency fibers are noninhibitable. The suppression may well be mechanical and not neural, since it occurs in the microphonic potential as well as in the auditory nerve responses, and is still observed after section of the auditory nerve (10).

Since sound signals are essential for mating behavior and species isolation in frogs, their auditory pathways have received much attention. In the central auditory pathway of frogs the first synaptic station is in a dorsal medullary nucleus in which single neurons have V-shaped tuning curves (20).

Low- and medium-frequency fibers from the amphibian papilla are widely distributed in the dorsal medullary nucleus, with more low-frequency fibers ventrally, and high-frequency fibers from the basilar papilla terminating dorsally. Low-frequency units are especially subject to two-tone inhibition (20). In the dorsal medullary nucleus one-half of the units are binaural (18).

The next center in the auditory tract is the superior olivary nucleus, in which single neurons are mostly unimodal (i.e., have a single frequency of maximum sensitivity); half of the neurons are binaural (18). All neurons are susceptible to two-tone inhibition; inhibitory tones may be of higher or lower frequency than the excitatory tones (Fig. 13-9). The frequency of spikes is maximum at a particular intensity-frequency combination.

The third region in the auditory tract is the midbrain torus semicircularis, where units show a variety of frequency tuning curves: V-shaped, narrow, broad, oval, bimodal (sensitive in two frequency ranges), with midrange eliciting no response (Fig. 13-9). Some 86 percent of neurons respond to single tones; 14 percent respond only to two or more simultaneous tones; a few units respond only when both low- and high-frequency tones are present. Some neurons requiring two or more tones for maximal excitation show narrowly separated excitatory frequencies, others require widely separated tones. Two-thirds of the responding cells showed inhibition by a second tone. In general, the most effective inhibitory frequency is lower than the most effective excitatory frequency (20). Thresholds for two-tone inhibition by frequencies lower than the best excitatory frequency may be independent of the excitatory tone. Sound localization (behavioral) requires that both ears be functional. Some units in the torus respond maximally to sound from either side, minimally to sound in front of the animal; other units give maximal response to contralateral stimuli (18).

Beyond the midbrain are diencephalic auditory nuclei. In the posterior auditory thalamus, evoked potentials (10) as well as unit responses (20) to tone combinations are larger than linear summation of responses to a single tone. Thalamic units are excited by two or more tones together, one tone of high and the other of low frequency. Two tones are also more effective in inhibition than a single tone. Two telencephalic regions give evoked potentials in response to tone combinations (10).

There may be direct descending fibers from preoptic nucleus to motor centers. Vocalization can be elicited by stimulation in the pretectum, anterior preoptic area, the torus, the pretrigeminal nucleus, or the laryngeal motor nucleus (45).

The population and species differences in call patterns and behavior responses in anurans indicate genetic control of sound communication; there is no evidence for a learned component, as there is in many birds. Peripheral sensitivity and central recognition and processing are matched. Input from both amphibian and basilar papillae is essential for eliciting behavior. Inhibition of low-frequency receptor responses by intermediate frequencies may sharpen the selectivity of mating calls. In some species the dominant spec-

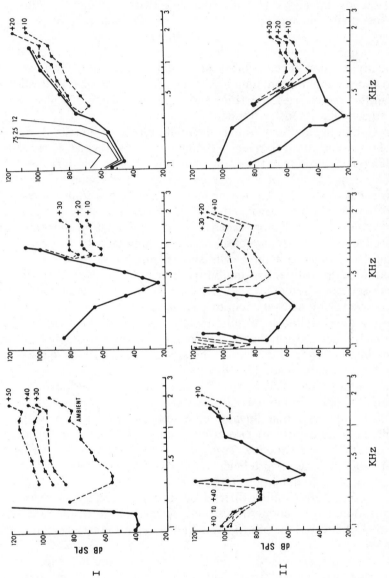

Figure 13-9. Audiograms of different single units in regions of auditory tract of frog Rana pipiens. (I) Dorsal medullary nucleus, excitation in solid curves, inhibition in broken lines. (II) Superior olivary nucleus, symbols as in I. (III) Torus semicircularis, numbers on excitatory curves give numbers of spikes for 0.5 sec of sound stimulation, inhibition indicated by thin or broken lines. (IV) Posterior thalamus, stimuli described on figure. Reprinted with permission from Z. M. Fuzessery and A. S. Feng, *Journal of Comparative Physiology*, Vol. 150, pp. 111, 115, 339; Vol. 146, p. 475. Copyright 1983, 1982, Springer-Verlag, Berlin.

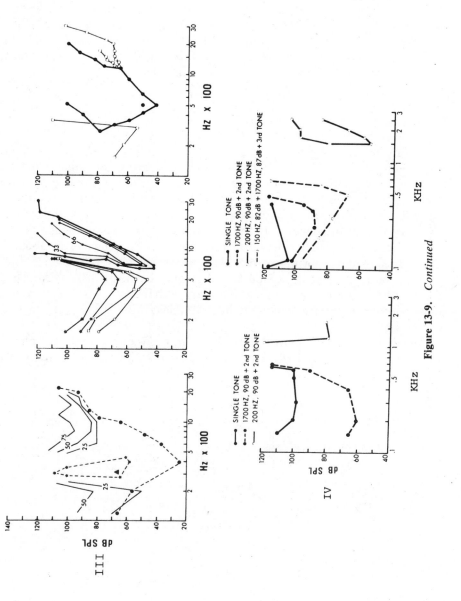

Figure 13-9. *Continued*

tral peak of the mating call need not match the best frequency for behavioral response, and in call patterning pulse duration may be more important than tone frequency. No single neuron has been found in any center which responds only to the mating call of that species. A selective response may depend not on a single species-specific neuron but on an organized network. However, the frequency selectivities of multiple tone responses match the frequencies of the species mating call (20). It is probable that the same genes code for sound emitting and receiving systems.

Birds

Properties of bird songs and their dependence on imprinting were described in Chapter 10. The cochlea of birds is straight or curved, not coiled. Cochlea microphonic potentials have electropositive and electronegative components in polarities relative to the scala tympani. The cochlear nucleus has a tonotopic distribution of units, high frequencies represented in the upper neuronal layer, low frequencies in the lower layer. In several songbirds the neural thresholds match behavioral thresholds, most sensitive at about 3 kHz with a range from 0.4 to 6.5 kHz in canary. Low frequencies may not be produced as song components but may be heard by a bird. Unit responses in the neostriatum of starlings vary with frequency and bilaterality. Some neurons detect frequency-modulated tones; these show phase-locked responses.

Connections occur within the neostriatum between sensory and motor (singing) regions (Fig. 13-10). Activity associated with singing is recorded

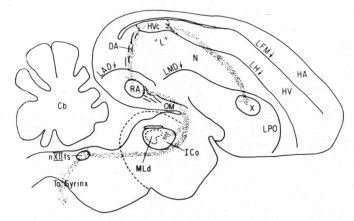

Figure 13-10. Diagram ot sagittal section of canary brain in which brain nuclei and pathways of song control are outlined. Hyperstriatum ventrale, pars caudale (HVc). Nucleus robustus archistrialis (RA). Reprinted with permission from F. Nottebohm et al., *Journal of Comparative Neurology*, Vol. 165, p. 459. Copyright 1976, Alan R. Liss, Inc., New York.

from the telencephalic nuclei of vocalization (HVc or hyperstriatum ventral pars caudale) of white-crowned sparrows and zebra finches. Cells of the HVc respond also to auditory stimulation and may convey auditory signals to the vocalization system (Chapter 10). Playback of a song while the bird is singing inhibits auditory responses in HVc. The HVc may function as a locus of convergence of sensory and motor action (31a, 32a). Some units in the HVc are selective for specific calls of species and races; these units require a temporal and spectral pattern and some units show two-tone inhibition (32a). Midbrain units in an owl give information regarding azimuth and elevation by time differences of responses to high-frequency sound between the two ears (31a).

Bats

Sound is used by many bats in orientation, social communication, and echolocation of environmental objects and insect prey. Cruising bats catch small flying insects. In laboratory experiments bats dodge suspended wires. Properties of emitted bat calls are detailed in Chapter 10. The Panama mustached bat *Pteronotus* emits orientation calls of 20–37 msec duration and repetition rate 5–10/sec; on approaching a target, such as an insect, the duration of a call shortens to 6 msec and repetition rate increases. Orientation calls are composed of four harmonics, the second of which is predominant. Each harmonic consists of a constant frequency (CF) and a frequency-modulated (FM) component (Fig. 13-11). The four frequencies are approximately 30, 60, 90, and 120 kHz. Echoes overlap the emitted sounds and the brain must separate the returning signal from the emitted call. The long CF is suitable for Doppler measurements, and gives information regarding target movement and velocity. The short FM component is used for ranging, localizing, and characterizing a target, since the sound energy is distributed over many frequencies (50,52). Different bat species are adapted in sound signaling and echo processing to their types of food prey.

The inner ear of bats is relatively large, and the unusually large basal turn of the cochlea is the detector of high frequencies. Some bats move the pinnae through a wide angle, at times alternating the two ears; in some bats the pinnae are fixed in position. Input from both ears is essential for target localization. The horseshoe bat, by alternating back-and-forth movement of the two ears, enhances the directionality of response to echoes by some 80 percent. This bat can avoid wires of 0.08 mm diameter (46,34).

In microchiropterans the stapedius muscle to ossicles of the middle ear contracts just before emission of each pulse; this reduces auditory sensitivity to the emitted call; immediately after the call the muscle relaxes and cochlear sensitivity is restored in time for detecting the echo. Cochlear microphonics, electrical response of the cochlear membrane, recorded at round

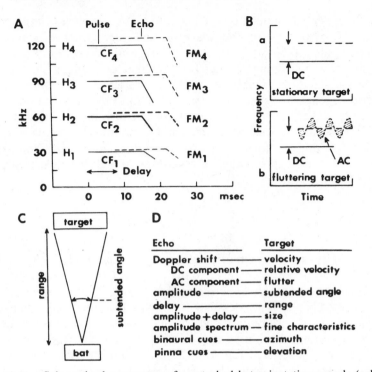

Figure 13-11. Schematized sonograms of mustached bat orientation sounds (solid lines) and echoes (dashed lines) during a pulse. The four harmonics H_1–H_4 each contains a constant frequency component CF_1–CF_4 and a frequency-modulated component FM_1–FM_4. Amplitudes of harmonics $H_2 > H_1$, $H_4 > H_3$. Echo delay given by interval between onset of corresponding components of orientation sound and echo. Reprinted with permission from N. Suga et al., in *Advances in Neuroethology*, J. P. Ewert, R. R. Capranica, D. J. Ingle, Eds. Copyright 1983, Plenum Publishing Corp., New York.

window, match the energy maximum of an echo (Fig. 13-12). In *Pteronotus* cochlear microphonics are sharply tuned at 61 kHz.

The tuning curves of peripheral neurons of bats are very sharp; "on" responses can be suppressed by "off" responses to a different frequency (21). Responses of the auditory fibers in *Myotis* (an FM bat) are very sensitive to frequency modulation and can alter their firing with changes in carrier frequency as small as 0.01 percent. The audiograms of *Myotis* nerve fibers show a broad range of sensitivity, with best frequency at 40 kHz. Auditory nerve fibers give "on" and "off" responses to brief tone pips. An "off" response can suppress a succeeding "on" response, mainly by mechanical effects on the hair cells (21). *Rhinolophus* (a CF bat) shows a sharp peak of sensitivity of auditory fiber responses in the range of its CF at 81–83 kHz. Both "on" and "off" units are found in the auditory nerve. *Pteronotus* depends on harmonic, two-time facilitation and inhibition.

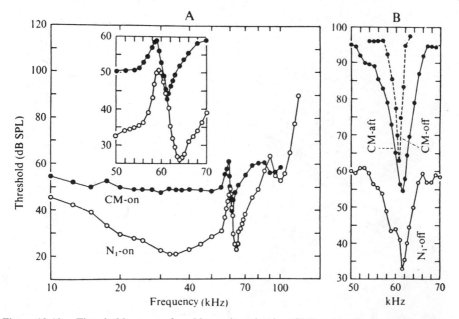

Figure 13-12.   Threshold curves of cochlear microphonics (CM) and auditory neuron spikes (N_1) in CF-FM bats. CM shown by solid circles, N_1 by open circles. Reprinted with permission from N. Suga and P. H. S. Jen, *Journal of Experimental Biology,* Vol. 69, p. 209. Copyright 1977, Biochemical Society, Colchester, Essex, England.

The auditory tract of a bat is large compared to other parts of its brain and consists of cochlear nucleus, superior olivary complex, lateral lemniscus nuclei, inferior colliculus, medial geniculate body, and auditory cortex. Responses to emitted sounds recorded in neurons of the lateral lemniscus are smaller than responses to playback sounds of the same pulses, even when the auditory nerve responses are similar. Responses to self-vocalization are attenuated by 25 dB between the cochlea and inferior colliculus, apparently by neural inhibition. The net effect of the inhibition is to decrease responses to the bat's own vocalization and to increase responses to echoes. At the level of the intercollicular nucleus of *Eptesicus,* some neurons do not respond to a single echolocating pulse alone, but respond to paired pulses which are separated by a few milliseconds. These neurons are presumably important for target ranging, since pulse interval is the most important cue.

Single neurons of the inferior colliculus are sharply tuned, and responses are lower by as much as 35 dB/kHz on either side of the best frequency. Most units show maximum sensitivity at the dominant frequency of echo tones for the species. For example, in *Rhinolophus* evoked potential responses in colliculus show a very sharp increase (lowest threshold) at the frequency of the CF, 83.3 kHz (34,46). There are large areas of inhibition in collicular neurons and inhibitory effects occur both below and above the

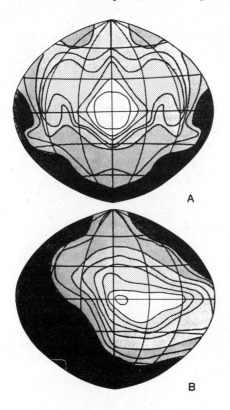

A

B

Figure 13-13. Mercator directional projections of (a) binaural hearing and (b) monaural hearing of the bat Rhinolphus. Reprinted with permission from A. D. Grinnell and H. U. Schnitzler, *Journal of Comparative Physiology*, Vol. 116, p. 71. Copyright 1977, Springer-Verlag, Berlin.

range of excitation (33). Collicular responses show directional localization, binaural interaction, and maximum responses to a source 15–45° contralateral and a minimum response to a source 60–90° ipsilateral (Fig. 13-13). Lesions to the ventral half of the inferior colliculus diminish capacity to avoid obstacles.

The primary auditory cortex occupies a large part of the bat cerebrum. In *Pteronotus* there are separate areas for CF and FM responses (Fig. 13-14); the largest central area is sensitive to 61–61.5 kHz tones, the second harmonic of the CF. Some cortical neurons respond only to paired sounds, not to single tones. The cortex is organized in columns, each column with cells of similar optimum delays between emitted sound and echo (37). Single units are tuned to both frequency and sound intensity, and as intensity increases above a maximum the spike response decreases—that is, the intensity-frequency contour is circular. Tuning is much sharper in the central area of the cortex than in the periphery.

In the absence of Doppler-shifted echoes, orientation sounds with CF_2 at about 61 kHz are emitted. When a Doppler-shifted echo is received (e.g., at 63 kHz), the bat reduces the call frequency by 1–2 kHz so that the echo is

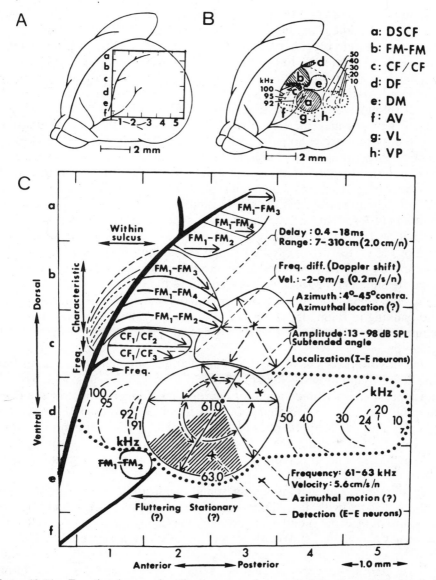

Figure 13-14. Functional map of auditory cortex of mustached bat Pteronotus. (*a*) Left cerebral hemisphere. (*b*) Subdivisions of left auditory cortex; regions a–h represent different kinds of biosonar information, as detailed in C. (*c*) Summary of functional organization of auditory cortex. CF areas represent constant echo frequency (61–63 kHz). FM areas of three types of facilitatory neurons, each representing a target range according to echo delay. FM-FM and CF/CF neurons show facilitation to paired FM and CF tones. DSCF, Doppler-shifted constant-frequency processing area. Reprinted with permission from N. Suga, *Trends in Neuroscience,* Vol. 7, No. 1, p. 22. Copyright 1984, Elsevier Science Publishers, Cambridge, England.

just above 61 kHz (54). The central CF area of the primary auditory cortex occupies 30 percent of the cortex and best frequency is 61–63 kHz; unit responses decline on either side of the best frequency. Units in the CF/CF area respond maximally to combinations of two CF tones at a critical separation; they show facilitation as measured by least latency and maximum response. Units in the FM/FM areas respond to optimal combinations of paired frequency-modulated tones (55) (Figs. 13-14, 13-15). Facilitation in CF/CF neurons may increase sensitivity by 6300 times; FM/FM units respond only to pulse-echo pairs with specific delays, and these are used for tracking target range. A bat colony is a noisy place; jamming is reduced by (1) directionality of orientation calls, (2) directional sensitivity of the ear, (3) binaural comparison, (4) sequential processing of echoes, and (5) individual bat call signatures (52,53). Synaptic inhibition delays the response to an emitted pulse so that the response arrives synchronously with the echo; this adaptation intensifies response to the echo (56). Some neurons in the CF area of the cortex are excited by sound stimulation from either side, other neurons are inhibited by sound from one side and excited by sound from the opposite site. The FM area is anterolateral to the CF area (50).

Several adaptations in the motor system of feeding bats permit rapid turns and banking in flight. If a food insect is not directly in line with the bat's mouth, the bat may scoop it with its wings and grasp it with the claws at the wing tips.

Insectivorous bats have a series of adaptations which permit them to capture small flying insects. Suga (55) has calculated that a mustache bat, because of its sharply tuned auditory neurons and compensation for the Doppler shift, can detect a modulation of echo of as little as 0.01 percent. A Doppler shift of 0.02 percent at 61 kHz is 12 Hz; a 1 cm long moth beating its wings at 50/s and moving at 3.4 cm/s would cause a 350 Hz shift (52). A bat has been found to capture as many as 500 small flying insects per hour (20a).

In summary, the adaptations for sound localization by bats are: (1) specialized larynx with fast, large cricothyroid muscle; (2) sound-focusing structure of face and mouth; (3) vocalization of constant frequency (CF) and frequency modulated (FM) pulses, and compensation of the CF pulses for Doppler effect; (4) ears for collecting sound; (5) contraction of stapedius muscle before the pulse is emitted, reducing cochlear sensitivity during sound emission; (6) presence of acoustic fovea; (7) large area of cochlea devoted to frequency of echo to CF pulse—auditory nerve fibers and central units very sharply tuned; (8) neurons in lateral lemniscus nuclei which are tuned to echoes and which inhibit responses to emitted pulses; (9) auditory cortex with different areas for processing FM and CF sounds, neurons selective for interval between two inputs, delay of response to emitted pulse, with maximum facilitation at synchrony of pulse and echo; and (10) motor coordination of flight system for capture of prey. The integration of these components provides for highly adaptive behavior of the echo-locating bat.

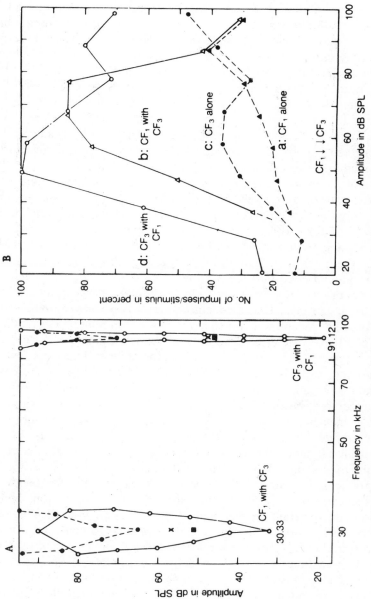

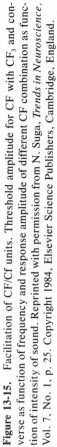

Figure 13-15. Facilitation of CF/Cf units. Threshold amplitude for CF with CF_3 and converse as function of frequency and response amplitude of different CF combination as function of intensity of sound. Reprinted with permission from N. Suga, *Trends in Neuroscience*, Vol. 7, No. 1, p. 25. Copyright 1984, Elsevier Science Publishers, Cambridge, England.

CONCLUSIONS

The emission and reception of sound and of electric discharge are highly specialized behavior. Each is used in conspecific communication in dominance and in reproductive behavior. Both songs and electric discharges are species-specific and contribute to behavioral isolation of species. Electrosonar is used by electric fish to locate prey, phonosonar by bats to locate and track flying insects. Electric discharge by strongly electric fish stuns prey. Song or electric discharge is stereotyped, but most animals which use one or the other mode of communication are capable of several patterns: cricket chirps or bird song variations, electric fish jamming avoidance reactions, bat frequency modulation. Many species that do not produce electric discharge can perceive electric signals. Phonobehavior and electrobehavior are of adaptive value for survival and reproduction.

Sending and receiving structures for electro- and phonocommunication have evolved, with modifications, in several kinds of animals. Electric organs are usually derived embryologically from myoblasts; the best known are specialized motor endplates. The electrocytes are flattened structures, and the membrane responses of the two faces vary with species—graded potentials on the innervated face only, conducted action potentials from innervated to noninnervated face, delayed rectification.

A few kinds of electric organs are formed by folded nerves; one derives from extraocular muscles. Electroreceptors are of two sorts, ampullar and tuberous. Ampullar receptors evolved at least twice: in nonteleosts and in a few orders of teleosts. Tuberous receptors have microvilli and occur only in teleosts. Both types of receptor are connected to the central nervous system by the lateral line, and possibly both cell types are derived from neuromasts of the lateral line organ. Not only are there several types of electric organs and electroreceptors, but the neural circuits for linking reception with discharge show species variation. Many electroreceptors have membranes which are extremely sensitive to electric currents, with thresholds several orders of magnitude lower than those of nerves; some "marker" electroreceptors signal pulses and do not display high sensitivity.

Sound-producing organs are vibrating structures, and the frequency of vibration is usually higher than that of nerve fibers of activation, hence sound production depends on mechanical properties of vibrating organs. Some singing insects use a file and scraper (crickets), others use stiff vibrating tympani (cicadas), some produce sound by wingbeat (*Drosophila*, bees). Sound production by some fishes is by vibration of a muscle attached to the swim bladder; a bird has a syrinx, a mammal a larynx, and a frog has species-specific out-pouchings of respiratory tubes. Sensory cells of phonoreceptors are sensitive to membrane deformation. The sensing cells for sound and vibration have small protruding processes which transduce deflection into membrane response. Two types of processes are cilia and microvilli; both occur in many hair cells. Ciliary processes are most common in invertebrate receptors (molluscs); single cilia (kinocilia) are surrounded by many microvillous stereocilia in all vertebrates except mammals, which have only stereocilia in the cochlea. It is of interest that light receptors also contain microvillous and ciliary endings in different proportions (Ch. 14).

A phonoreceptor can be stimulated (depolarized) by a deformation as small as a few nanometers. Receptor potentials can follow high frequency vibrations which depend on resonant properties of the supporting membranes or shearing of cilia on the receptor cells. The cellular events are similar in tympanal organs of insects, the papillae of inner ear of frogs, and cochleas of mammals. Responses in different receptors range from constant deformation (direct current) to 100 kHz. The central nervous connections between phonoreceptors and sound-emitting organs depend on innate stereotyped circuits, for example, those producing species-specific patterns of song (crickets, frogs, bat), or songs in which central patterns depend on what has been heard at critical stages of development or during conditioning of adults (birds). Genetic variants occur, as observed in *Drosophila* and in some frogs. Within genetic limits, modifiability of sound emission and central processing provide individual and species adaptability.

Electroreception and phonoreception provide many examples of convergent evolution. For each modality there has been separate development of receptors and central processes that are functionally similar but not anatomically homologous. Examples are the several types of electric organs and electroreceptors which occur in unrelated fishes, and sound-producing and detecting organs in insects and vertebrates. The connections of interneurons to motor output are less well known than sensory output, but resulting behavior is similar in many groups of animals that make use of electric and sound signals.

Electrobehavior and phonobehavior use complex and precise coordination of receptors and integrative central nervous networks, together with patterned motor outputs. How did so many precisely-timed components evolve? It is improbable that many genetic changes occurred simultaneously in all components of one modality. More likely, many small changes occurred over long periods and the coordination of these had selective advantages. Mutants for different central patterns support this hypothesis. Once a coincidence of the components of behavior sequences that provided for species isolation became established, for successful feeding by individuals, or for reproductive behavior, consolidation of entire sequences occurred. Whatever many have been the manner of its origin, the speed and precision of response of animals that use electro- or phonocommunication and echolocation provide a high degree of adaptive behavior.

REFERENCES

1. Bastian, J. et al. *J. Compar. Physiol.* **144**:465–479; **136**:135–152, 1981. Electroreception in a weakly electric fish.
2. Bennett, M. V. L. pp. 147–169 in *Central Nervous System and Fish Behavior,*

Ed. D. Ingle, University of Chicago Press, Chicago, 1968. pp. 347–491 in *Fish Physiology,* vol. 5., Ed. W. S. Hoar and D. J. Randall, Academic Press, New York, 1971. Electric organ, electroreceptors, and neural control of electrobehavior in fishes.

3. Bentley, D. *J. Compar. Physiol.* **116**:19–38, 1977. *Annu. Rev. Neurosci.* **1**:35–59, 1978. Neural control of song patterns in crickets.

4. Boyan, G. S. *J. Compar. Physiol.* **130**:137–150, 151–159, 1979; **140**:81–93, 1980. Auditory neurons in brain of a cricket.

5. Brown, P. *Z. Tierpsychol.* **41**:34–54, 1976. Individual recognition by bats.

6. Bullock, T. H. *Annu. Rev. Neurosci.* **5**:121–170, 1982. Electroreception.

7. Bullock, T. H. et al. *Trends Neurosci.* **5**:50–53, 1982. Evolution of electroreception.

8. Bullock, T. H. et al. *Brain Res. Rev.* **6**:25–46, 1983. Phylogeny of electroreceptors.

9. Bullock, T. H. and S. H. Ridgway. *J. Neurobiol.* **3**:79–99, 1972. Central auditory system of porpoises.

10. Capranica, R. R. *Evoked Vocal Responses of Bullfrog,* pp. 1–110, Res. Monog. M. I. T. Press, Cambridge, MA, 1965. pp. 551–575 in *Frog Neurobiology,* Ed. R. Llinas and W. Precht, Springer, New York, 1976. Auditory system of frog.

11. Cassaday, G. B. and R. Hoy. *J. Compar. Physiol.* **121**:1–43, 1977. Acoustic interneurons of a cricket.

12. Corey, D. P. and A. J. Hudspeth. *Nature* **281**:675–677, 1979. *J. Neurosci.* **3**:943–961, 1983. Receptor potentials in hair cells.

13. Dallos, P. et al. *Science* **218**:583–585, 1982. Electrical responses of cochlear hair cells.

14. Dallos, P. and D. Harris. *J. Neurophysiol.* **41**:365–383, 1978. Auditory nerve responses in absence of outer hair cells.

15. Denton, E. J. and J. A. B. Gray. *Nature* **282**:406–407, 1979. Analysis of sound by the ear of fishes.

16. Ehret, G. et al. *J. Compar. Physiol.* **148**:237–244, 1982. Frequency resolution in ear of a cricket.

17. Esch, H. et al. *J. Compar. Physiol.* **137**:27–38, 1980. Primary auditory neurons in crickets.

18. Feng, A. et al. *J. Compar. Physiol.* **100**:221–229, 1975; **107**:241–252, 1976; **144**:419–428, 1981. Responses of auditory neuron in brain of bullfrog.

19. Feng, A. and T. H. Bullock. *J. Neurobiol.* **9**:255–266, 1978. Synaptic potentials in command neurons of Eigenmannia.

19a. Fettiplace, R. and A. C. Crawford. *Proc. R. Soc. London* **B203**:209–219, 1978. ·*J. Physiol.* **315**:317–338, 1981. Responses from single hair cells of turtle retina.

19b. Flock, A. and J. Wersall. *J. Cell Biol.* **15**:19–29, 1962. Structure of hair cells of lateral line organ in fishes.

20. Fuzessary, Z. M. and A. Feng. *J. Compar. Physiol.* **143**:339–347, 1981; **146**:471–484, 1982; **150**:107–119, 333–344, 1983. Auditory responses in dorsal medullary nucleus, torus semicircularis and thalamus of frogs.

20a. Griffin, D. R. *Listening in the Dark,* Yale University Press, New Haven, 1959.

21. Grinnell, A. D. *J. Compar. Physiol.* **82:**179–194, 1973. Lateral suppression in bat cochlea.

22. Hagiwara, S. *Physiol. Compar. Oecol.* **4:**142–153, 1955. Sound production in cicada.

22a. Hama, K. *Z. Zellforsch.* **94:**155–171, 1969. Fine structure of saccular macula hair cells of goldfish.

23. Heiligenberg, W. and B. L. Partridge. *Naturwissenschaften* **67:**499–507, 1980. *J. Compar. Physiol.* **145:**153–168, 1981. Review of electric responses in Eigenmannia.

24. Heiligenberg, W. *J. Compar. Physiol.* **87:**137–164, 1973; **103:**97–121, 1975; **124:**211–214, 1978; **136:**135–164, 1980; **142:**295–308, 1981. Electrolocation and activity of pacemaker neuron in Eigenmannia.

25. Hill, K. G. *J. Compar. Physiol.* **152:**475–482, 1983. Responses of locust auditory receptors.

26. Hopkins, C. D. *Trends Neurosci.* **4:**4–6, 1981. Neuroethology of electric communication.

27. Horner, K. et al. *J. Exper. Biol.* **76:**1506–1509, 1979; **85:**323–331, 1980. *J. Neurosci.* **3:**942–976, 1983. *Proc. Natl. Acad. Sci.* **74:**2407–2411, 1977. Binaural interaction in cod.

28. Hudspeth, A. J. *Trends Neurosci.* **6:**366–369, 1983. Transduction and tuning by vertebrate hair cells; ultrastructure, hair cells.

28a. Johnstone, B. M. *Nature* **227:**625–626, 1970. Tuning of tympanic organ of cricket.

29. Kalmring, K. et al. *J. Compar. Physiol.* **128:**213–226, 1978. Coding of sound signals in auditory system in tettigonid grasshoppers.

30. Katz, L. C. and M. E. Gurney. *Brain Res.* **211:**192–197, 1981. Zebra finch motor system for song.

31. Knudsen, E. I. and M. Konishi. *J. Neurophysiol.* **41:**870–884, 1978. Moiseff, A. and M. Konishi. *J. Neurosci.* **1:**40–48, 1981. Responses in auditory midbrain of owl.

31a. Konishi, M. *Handbook of Sensory Physiol.*, vol. VIII, Ch. 9, pp. 289–309, 1978. *J. Neurosci.* **2:**1177–1194, 1982. Hearing in birds.

31b. Lewis, T. et al. *Science* **215:**1641–1643, 1982. Sensitivity of regions of frog ear.

32. Manley, J. A. *Z. vergl. Physiol.* **71:**255–261, 1971. Auditory responses of caiman lizards.

32a. Margoliash, D. *J. Neurosci.* **3:**1039–1057, 1983. Responses of song-specific neurons in brain, white-crowned sparrow.

32b. Miller, M. R. *Am. J. Anat.* **151:**409–436, 1978. Ultrastructure of hair cells of papilla basilaris, turtles and snakes.

33. Moller, J. et al. *J. Compar. Physiol.* **125:**217–229, 1978. Midbrain responses of bat to high frequency sound.

34. Neuweiler, G. *Z. vergl. Physiol.* **67:**273–306, 1970. Echolocation in bat Rhinolophus.

35. Nocke, H. *J. Compar. Phys.* **80:**141–172, 1972. Responses of cricket auditory nerve.

36. Nottebohm, F. et al. *J. Compar. Neurol.* **207:**344–356, 1982. Vocal centers in forebrain of canary.

37. O'Neill, W. E. and N. Suga. *J. Neurosci.* **2:**17–31, 1982. Echo-encoding of target range and its representation in cortex of Pteronotus.

38. Popper, A. N. and S. Coombs. *Am. Sci.* **68:**429–440, 1980. Auditory mechanisms in teleost fishes.

39. Rheinlaender, J. et al. *J. Compar. Physiol.* **110:**251–269, 1976. Brain projections of auditory system in crickets.

40. Roberts, A. *J. Compar. Physiol.* **135:**341–348, 1980. Responses of mechanoreceptive nerve endings in amphibian embryos.

41. Romer, H. *Nature* **306:**60–62, 1983. Tonotopic organization of auditory neuropile in bushcricket.

41a. Roeder, K. D. *Nerve Cells and Insect Behavior,* Harvard University Press, Cambridge, MA, 1967.

42. Russell, I. J. *J. Compar. Physiol.* **111:**335–358, 1976. Responses of lateral line in cochlea of goldfish.

43. Russell, I. J. and P. M. Sellick. *J. Physiol.* **284:**261–290, 1978; **338:**179–206, 1983. Intracellular recording from hair cells of guinea pig cochlea.

44. Sales, G. and D. Pye. *Ultrasonic Communication by Animals,* Chapman and Hall, London, 1974.

45. Schmidt, R. S. *J. Compar. Physiol.* **108:**99–113, 1976. Neural correlates of frog calling.

46. Schnitzler, H.-U. *J. Compar. Physiol.* **82:**79–92, 1978. with A. D. Goinnell, *J. Compar. Physiol.* **116:**51–61, 1977. Control of Doppler shift in Rhinolophus.

47. Schuller, G. *J. Compar. Physiol.* **139:**349–356, 1980. CF-FM bats of India.

48. Sellick, P. M . *Trends in Neurosci.* **2:**114–116, 1979. Receptor potentials in cochlea.

49. Simmons, J. A. and R. A. Stein. *J. Compar. Physiol.* **135:**61–84, 1980. Acoustic imaging in bat sonar.

50. Suga, N. et al. *J. Compar. Physiol.* **106:**111–125, 1976. *J. Exper. Biol.* **69:**207–232, 1977. *Science* **200:**778–781, 1978. pp. 157–218 in *Cortical Sensory Organization,* vol. 3, Ed. C. N. Woolsey, Hermann Press, Clifton, NJ, 1982. *J. Neurosci.* **2:**17–31, 1982. *J. Neurophysiol.* **47:**225–255, 1982. Functional organization of auditory cortex of mustached bat Rhinolophus.

51. Suga, N. and P. H. S. Jen. *J. Exper. Biol.* **69:**207–232, 1977. CM-FM bats specialized for fine frequency analysis of Doppler shift.

52. Suga, N. et al. pp. 1970–219 in *Neuronal Mechanisms of Hearing,* Ed. J. Syka and L. Aitken, Plenum, New York, 1981. Prevention of jamming by bats.

53. Suga, N. et al. pp. 829–867 in *Advances in Vertebrate Neuroethology,* Ed. J. P. Ewert and R. Capranica, Plenum, New York, 1983. Biosonar representation in cortex of mustached bat.

54. Suga, N. et al. *J. Neurophysiol.* **49:**1573–1626, 1983. Biosonar signals in auditory cortex of bats.

55. Suga, N. *Trends Neurosci.* **7:**20–27, 1984. Neural mechanisms for echolocation.

56.   Sullivan, W. E. *J. Neurophysiol.* **48**:1011–1032, 1982. Neural representation of target distance in auditory cortex of Myotis.

56a. Tanaka, K. and C. A. Smith. *Am. J. Anat.* **153**:251–272, 1978. Structure of inner ear of chicken.

57.   Varanka, I. et al. *Compar. Biochem. Physiol.* **48A**:411–426, 1974. Interneurons of wind-sensitive hair-receptors on head of Locusta.

58.   Wiederhold, M. L. *Annu. Rev. Biophys. Bioeng.* **5**:39–62, 1976. Mechanosensory transduction.

14

PHOTORECEPTION AND VISION

Behavioral aspects of vision were considered in Chapter 10. This chapter deals with the activation of photoreceptor molecules, transduction in the receptor membranes to produce nerve signals, and the neural integration in eyes of visual signals. Life originated under the influence of sunlight, and ultraviolet (UV) light provided some of the energy for early synthesis of organic substances (Ch. 2). Responses of plants and animals to light vary with wavelength. To be effective, light must be absorbed by receptor molecules.

Light of wavelengths longer than UV is absorbed by chlorophyll in photosynthesis (Ch. 2). Shoots of growing plants are phototropic and bend toward a light source. Phytochromes and flavin-containing blue pigments initiate the bending responses. Most animals respond to light in the spectral range from 385 to 650 nm, a few detect red light (650–750 nm), and some animals respond to ultraviolet light.

BEHAVIOR; ECOLOGICAL CORRELATIONS OF PHOTORECEPTION

Photoreception and visual behavior can be described at levels of (1) the behavior of whole animals, (2) sensory signals and their neural integration, (3) molecular events of photochemistry, and ·(4) initiation of membrane responses. Adaptative responses to visual stimuli provide considerable lability of behavior. Differences in light intensity and color influence behavior; in water and under the canopy of tropical vegetation light is graded in intensity and color.

One direct response of animals to light is withdrawal; an amoeba pulls back a pseudopod which is exposed to bright light. Withdrawal responses occur also in animals with photoreceptors and nervous systems; in a few animals free nerve ending are sensitive to direct light, for example, axon endings in clam siphons.

Orienting responses to light have been classified by Fraenkel and Gunn (34). Photokineses are changes in locomotor activity in proportion to light intensity. Orthokinesis refers to changes in speed or distances of straight locomotion; klinokinesis refers to increases or decreases in turnings per unit of time; each kind of kinesis may result in aggregation in moderate light or in darkness. Paramecia show positive kinesis to light. Taxes are directional locomotion, positive toward a light source, negative away from light. Klinotaxis is effected by successive lateral deviations (from side to side) of the body or of a body part. Orientation is accomplished by sequential or temporal comparison of light intensity, sometimes by an animal's single receptor, as in *Euglena,* or by paired receptors, as in fly larvae.

Tropotaxis is direct turning toward a light after simultaneous comparison of illumination on both sides. When two lights are presented to a positively phototactic animal, it goes between them, veering toward the brighter one until it comes near the lights, then going directly to the brighter light. Tropotaxis is shown by isopods and by larvae of the moth *Ephestia*. Telotaxis is direct orientation without comparison; if one eye is blacked over, orientation is direct; if two unequal lights are presented, a positively phototactic animal with paired functional eyes goes directly toward the brighter—for example, adult insects (bees) and crabs (*Eupagurus*). Light compass reactions refer to locomotion at a fixed angle to the direction of light rays, for example, the flight of bees or the crawling of ants in a determined direction with respect to the sun; the angle is remembered according to signals received from colony mates (Ch. 10). Dorsal (or ventral) light reaction describes orientation which maintains illumination perpendicular to one axis of the body so that the dorsal (or ventral) surface is lighted. Illumination from beneath an animal may cause it to turn over; such dorsal light response is exhibited by small crustaceans. Although direct behavioral reactions—kineses, tropisms, taxes—are stereotyped and genetically programmed, they are modifiable by endogenous diurnal rhythms, nutrition, and internal state. These responses and orientations have the adaptive function of putting an animal into "favorable" illumination.

Responses to patterns, to oriented bars of light, and to moving objects are more complex than responses to single light sources. The Weber-Fechner ratio $\Delta I/I$ is the least difference discriminable between two lights or between a spot of light and background. This ratio or least discriminable difference decreases with increase in brightness; it is linear in midranges and nonlinear at low and high intensities. In a given vertebrate eye, receptors of one kind (rods) are more sensitive (i.e., have lower $\Delta I/I$) than other receptors (cones). Behavioral tests show that, in midrange, intensity discrimination is 20 times greater for human eyes than for honeybees (87). The product of intensity (I) and time of presentation (T), IT, is linear only for midintensities.

Visual acuity is the least angle discriminable between two light stimuli. Visual acuity can be measured in animals by presenting a slowly moving field of alternating light and dark stripes by means of a rotating drum. Visual acuity is given by the reciprocal of the minimal angle of discrimination between two points or between stripes of different widths. Discrimination is indicated by head or eye movement. The limit to resolution of eyes of different sizes varies with the construction of the eye. Large animals with single lens eyes, such as vertebrates, have resolution superior to faceted compound eyes of the same size (48a). Acuity improves in proportion to increase in log *I*. Under moderate light intensities the visual acuity of a human is 100 times as great as a honeybee's. In a bee, horizontal acuity is greater than vertical acuity. In *Drosophila*, each receptor unit (ommatidium) subtends 4°, maximum visual acuity is 9.3°. In a fly *Musca* the interommatidial angle is approximately 2° (48a).

Visual acuity is a measure of spatial resolution. Flicker fusion frequency measures temporal resolution, a function which increases sigmoidally with increase in log *I*. Animals with image-forming eyes (compound eyes or camera eyes) show optokinetic responses to moving stripes; an eye fixates on a point or stripe and then moves to maintain the fixed image. Nystagmus is found generally in mammals; the shifts of gaze have both fast and slow phases. Insects' and crabs' optokinetic movements of eyestalk or head indicate acuity and flicker fusion. If after a crab has fixated on a striped pattern the light is turned off and the stripes are moved, then reilluminated, the eyestalk moves as if to follow the stripes. The crab has a memory of 0.5° movement after 2 minutes in the dark or of 1° after 15 minutes in the dark. The crab *Carcinus* responds to movements as small as 0.0017°/sec and its eye moves some 0.001°/sec. The average movement of the stimulus across the visual field is a slip speed of 0.0007°/sec, which corresponds to an angular movement of 0.17 revolution/24 hours. The crab sees the edges of stripes; tremor of the object improves the perception (46). Horridge calculated from a crab's optomotor behavior that it could detect the movement of sun or moon; optokinetic memory could account for light compass reactions. Visual acuity of man is about 0.0166°, which is 50 times as sharp as that of a rat, and the rat's acuity is 10 times as sharp as that of most insects (46).

All responses to light depend on absorption of photic energy by sensitive pigments which have optima of wavelength sensitivity (λ_{max}). Action spectra measure response—behavioral, electrical, photochemical—as a function of wavelength. Differences in sensitivity to different wavelengths do not imply color vision; color vision per se is apparent from behavioral matching of colors. Wavelength sensitivity in animals is measured by orientation or by conditioned reflexes; color filters or colored plaques need to be matched for luminosity. Monochromat humans vary in sensitivity to different wavelengths. Animals with color vision have several types of receptors, each with a characteristic range of wavelength sensitivity. Colors are discriminated by insects, fishes, turtles, birds, and some mammals. Color discrimination is determined not only by receptor pigments but also by filter pigments, for example, by oil droplets in the retina of some birds which filter colors.

Moving objects are more effective as visual stimuli than stationary objects. Image formation and color discrimination widen the range of visual behavior; for example, color patches on wings or bills enhance species and sex recognition by birds. The importance of color is indicated by the variety of patterns in animals. In general, males are more colorful than females. Many animals change color by fast or slow expansion and contraction of chromatophores. Chromatophore responses may be mediated by reflexes, by hormones, or by direct responses of the chromatophores to light. Some animals—reptiles, birds—display patterns of movement in courtship and in aggression. The dorsal surface of most animals is dark in color; this absorbs solar radiation which may warm an animal, and it provides protective camouflage. Some kinds of animals have fixed color patterns that resemble the habitat background; some animals, for example, flatfish and fiddler crabs, change color to match the background. Mimicry of color patterns is often found in insects; some butterflies resemble other species that are distasteful to birds. Some small fishes have a pattern like that of larger fish with repellent taste and thus escape predation. Squids have chromatophores of several colors that change rapidly in color with activity, disturbance, and illumination.

The importance of color is also indicated by the variety of pigments in integuments (33). Some pigments are contained in chromatophores which expand and contract under neural or hormonal control. Chromatophores of cephalopod molluscs are sacs of pigment—red, yellow, brown—which expand or contract by muscular strands in the integument. In crustaceans—shrimps, prawns, and crabs—pigment granules migrate slowly by amoeboid movement; the white chromatophores are fixed. In fishes some chromatophores show streaming; the blue are fixed and iridescent. In many animals the red, yellow, and brown pigments are structural, for example in lepidopteran scales and bird feathers.

Animal pigments (zoochromes) are variants of several classes of compounds (33). Melanins are synthesized from tyrosine and may be red, orange, or brown. Carotenoids are of various colors, are derived from the plant diet and transformed in situ, for example, by oxidation of carotene to xanthophylls. Flavins are yellow, purines and guanines white, pterins yellow, orange, red, or white.

The colors and shapes of flowers have coevolved with the visual senses of insects and birds (Ch. 10). The spectral sensitivity of visual pigments of insects and birds correlates with the wavelength of floral color; an example is sensitivity of honeybee visual pigment to ultraviolet, a color reflected from some flowers.

Visual sensitivity increases or decreases with light or dark adaptation. Vertebrate eyes adapt by dilatation or constriction (change in size of the pupil), by migration of shielding pigments in the pigmented layer of epithelium, by rate of regeneration of a photoreceptive pigment, and by changes in neural sensitivity. Retinal rod receptors may extend or retract by means of contractile myoids. Cyclic migration of shielding pigments provides circadian rhythms of sensitivity. In rhabdomeric eyes (e.g., crayfish), pigment migrates in and out along each ommatidium. Light intensity in full sunlight is 6 log units brighter than under a full moon, and this is 2 log units brighter than a starlit night. Eyes of animals that are visually active in bright or dim light differ structurally. Diurnally active birds have colored oil droplets that absorb some light in the retina; some birds, especially birds of prey, have a shielding tissue, the pecten, which projects in front of the retina. Diurnal species of mammals have relatively more cones than nocturnal animals. Some nocturnal mammals have a reflecting tapetum derived from the choroid which increases the light reaching the retina. Reflecting devices, for example the tapetal layer in nocturnal moths, occur in the eyes of many invertebrates. Animals, such as cave crayfish and fishes, which spend their lives in virtual darkness have regressed eyes that have become genetically fixed.

Light is absorbed by water, so that at depth of about 1000 meters the intensity is too faint for rod vision. Fish and invertebrates living permanently below 1000 meters have regressed eyes or have enlarged eyes that detect bioluminescent organisms. The decrease in light intensity with depth varies with turbidity and, in general, less light penetrates to a given depth in fresh water than in seawater. Fresh water tends to transmit more red, seawater more blue. Adaptations in the spectral properties of photoreceptor pigments correlate with depth and with freshwater-seawater absorption differences. The absorption peak of rhodopsin in the deep-water beaked-nose whale is 481 nm; in humpback whale which dives between shallow and deep water the peak is 491 nm; and in the gray whale, primarily a shallow water animal, the λ_{max} is 497 nm (56). Eyes of crustaceans from midwater oceans have λ_{max} values ranging from 462 to 480 nm, and those from shallow water 500 to 525 nm (37). Fishes from deep ocean where the maximum light is toward the blue have a pigment with λ_{max} of 470–480 nm. Marine fish from shallow water have λ_{max} at 500 nm.

PHOTORECEPTOR ORGANS

Unicellular organisms (amoebae and some bacteria) may have protoplasmic sensitivity to light, evidenced by predominantly negative photokinesis and phototaxis. Photosynthetic bacteria use the same pigment for photoklinokinesis as for photosynthesis. Flagellates show positive klinotaxis and have a photosensitive region below a pigmented "eyespot." The flagellate swims in a spiral; consequently the photoreceptor is alternately shaded by the pigment and exposed to light. Spines of sea urchins *Diadema* show negative pointing of spines to light detected by nerve axons in the spines (65). In earthworms, photosensitive cells are distributed over the body surface, with highest density in the prostomium; a lenslike structure in each cell appears to focus light on an area of the cytoplasm.

The simplest "eyes" that presumably form images are in molluscs, where an epithelial cup has a narrow aperture through which light projects to sensitive cells inside the cup. The flatworm *Planaria* has two cup-eyes with pigment cells and receptor endings that are parts of neurons.

There are two main eye types, ciliary and microvillar (rhabdomeric), for which several evolutionary schemes have been proposed. In flagellates photoreceptors often occur at the base of flagellae. In multicellular animals ciliated photoreceptors are derived from epithelial cells and occur in cnidarians, bryozoans, some echinoderms, and all vertebrates. Ciliary photoreceptors have the filament pattern and centriole characteristic of all cilia. Microvillar photoreceptors have many outfolded membranes which form stacks of plates, the rhabdomes (84). Microvillar eyes occur in all arthropods, most annelids, cephalopod molluscs, and in some transitional invertebrates, such as sipunculids. A functional correlation based on a few species is that responses of ciliary photoreceptors (vertebrate rods and cones) are hyperpolarizing and that rhabdomeric photoreceptors (insects, squid) are depolarizing.

It has been noted (25) that most of the animals with microvillar eyes are proterostomes, and those with ciliary eyes are deuterostomes (the two major divisions of animal phyla). However, there are many exceptions: some animals have both ciliary and microvillar photoreceptors, and transitions may occur between types during development (75,82). The bivalve *Pecten* has rows of eyes along the mantle edge; each eye is composed of distal and proximal layers behind the lens. The distal retina contains photoreceptors derived from cilia; the proximal retina has rhabdomeric receptors. Microvilli occur in visual receptor cells of some hemichordates. Dual eyes (ciliary and microvillous) occur in some polychaetes and coelenterates (84). Among flatworms the tubularians have ciliary receptors, the trematode have rhabdomeric receptors. Photoreceptors of asteroids and holothurians are microvillous, those of other echinoderms are ciliary (25). The evolutionary pattern of photoreceptors is not simple, and it is probable that other factors than receptor cell type are important in determining adaptive visual mechanisms.

The photoreceptors of vertebrates are of two kinds, rods and cones; each consists of an outer segment joined by a ciliary structure to an inner segment. The outer segment of a rod contains a stack of 500 to 1000 discs, the membranes of which have been pinched off from the plasma membrane (Fig. 14-1). The tips of the outer segments are shed continuously and new discs are added daily. In embryonic development the outer segment is formed before the synaptic apparatus of the inner segment. The cones also have discs, but their outer membranes are continuous with the plasma membrane (Fig. 14-2). Rods mediate vision in dim light, and have a lower light threshold than cones, which function in bright light. In many vertebrates the cones provide for color vision, and different kinds of cones usually have different pigments. Double cones (one large and one small) occur in birds, amphibians, and reptiles; twin cones (equal in size) occur in teleost fishes. Cones containing differently colored pigments are arranged in geometric patterns in fish reti-

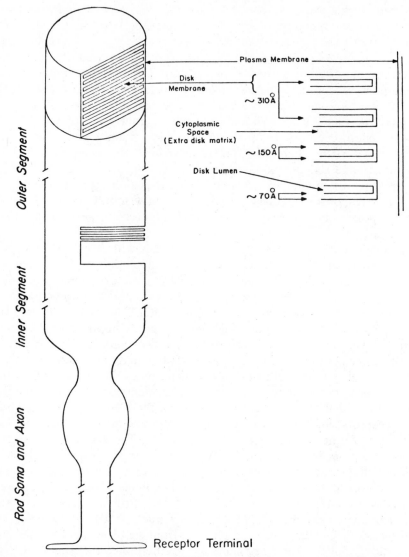

Figure 14-1. Diagram of vertebrate rod. Narrow region between outer and inner segment is ciliated neck. Reprinted with permission from T. G. Ebrey and B. Honig, *Quarterly Reviews of Biophysics,* Vol. 8, No. 2, p. 150. Copyright 1975, Cambridge University Press, London.

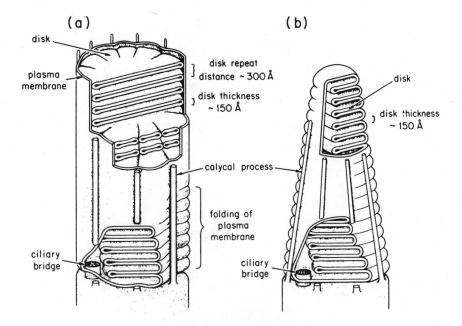

Figure 14-2. Diagram of outer segments of (*a*) rod and (*b*) cone. Discs inside plasma membrane in rod, saccules shown as invaginations in cone. From *Visual Cells* by R. W. Young. Copyright © October 1970 by Scientific American, Inc. All rights reserved.

nas. Mesopelagic fish have very few red cones, shallow water fish a few red cones, and freshwater fish many red cones (56). The proportion of rods to cones varies with species. Nocturnal mammals have relatively more rods than diurnal mammals; a diurnal ground squirrel has a retina entirely of cones. In a cat retina, the area centralis has 16,500 cones and 280,000 rods/ mm^2; at 15° outside the area centralis there are 6000 cones and 450,000 rods/ mm^2 (82b).

A dark-adapted human eye sees only about 10 percent of impinging light, the remainder being reflected by the cornea, absorbed by the lens, or transmitted without being absorbed through the retina; threshold is 5–14 quanta. A 10′ visual field has 500 rods, and if 5–10 quanta are absorbed by 500 rods there is very low probability that 2 quanta will impinge on a single rod; hence it is inferred that a single quantum can excite a rod (though more than one photon capture within a pool of 500 rods is necessary for a person to "see" light) (42).

Rhabdomeric eyes have microvilli at one end of a receptor cell and a nerve process at the other end. In arthropods the band of microvilli projects to the center of a rosette of retinular cells to form an axial rhabdom (Figs. 14-3, 14-4). Each cluster of retinulae and its sleeve of filter pigment cells constitutes an ommatidium. Ommatidia of most species have a complement of eight retinular cells; frequently these are arranged with bilateral symme-

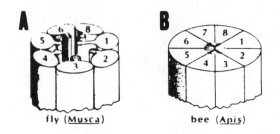

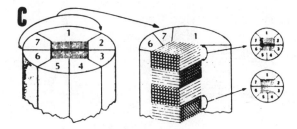

Figure 14-3. Diagrams of representative rhabdoms of ommatidia. (*a*) Fly (*Musca*) in which rhabdomeres project into central cavity and remain separate. (*b*) Honeybee (*Apis*) in which rhabdomeres fuse in center of rhabdom. (*c*) Crayfish in which microvilli project as tongues in alternate directions. (*d*) Octopus in which each receptor cell has an elongate outer segment that consists of central cytoplasm and two lateral borders of microvilli. From T. Goldsmith, *Comparative Animal Physiology,* Third Edition, edited by C. Ladd Prosser. Copyright © 1973 by Saunders College Publishing. Reprinted by permission of CBS College Publishing.

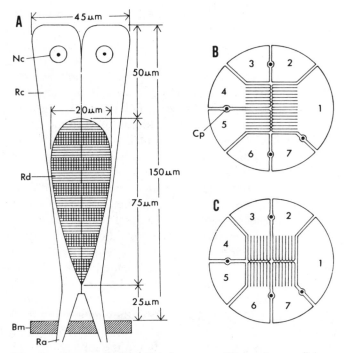

Figure 14-4. Diagram of two retinular cells of an ommatidium of a crayfish eye, showing two sets of orthogonal microvilli originating in different retinular cells. R_1, R_4, and R_5 maximally sensitive to horizontal plane of linear polarization of light, R_2, R_3, R_6, and R_7 sensitive to vertical plane. (*a*) Longitudinal section; (*b, c*) cross sections at two levels. Reprinted with permission from T. H. Waterman, *Journal of Experimental Zoology,* Vol. 294, p. 326. Copyright 1975, Alan R. Liss, Inc., New York.

try, but always with radial asymmetry. In insects and decapod crustaceans asymmetry is often due to a pair of cells with long visual fibers that project to the medulla, bypassing the lamina (see Fig. 14-14 below). Some insects have a rudimentary basal cell. *Limulus* has one eccentric retinal cell which differs in sensitivity from the others.

POLARIZED LIGHT DETECTION

The plane of polarization of light is readily distinguished by many crustaceans, insects, cephalopods, and probably by some other animals (e.g., fishes). Light from blue sky is polarized; maximum absorption of polarized light by sensitive receptors occurs when the plane of the *e* vector (direction of polarization) is parallel to the axis of photosensitive pigment molecules aligned on a membrane in the eye.

Honeybees (*Apis mellifera*) make use of the plane of polarization of light from blue sky when they are signalling to hive mates the direction of a food source. Many insects show polarotaxis and orient relative to the *e* vector (84). In water the *e* vector is perpendicular to the plane given by the position of the sun relative to an observation point. Refraction at a water surface decreases the amount of polarization; in water the polarization can be further altered by light scattering from particulates. Polarotaxis is exhibited by all decapod crustaceans, many insects, and some cephalopods (octopus). Polarotaxis may be used in diurnal migration of planktonic crustaceans. The water flea *Daphnia* orients to a horizontal beam of polarized light, most strongly to light perpendicular to the *e* vector, less to a light beam oriented at $+45°$ or $-45°$ (37).

Recordings from retinular cells and electroretinograms from intact eyes, as functions of light intensity, show lowest thresholds at rotational angles of $90°$ and $270°$ to stimulating light, and maximal responses at $+45$ and $-45°$ (37). The rhabdomeric structure of an ommatidium serves as a polarization analyzer. Details of rhabdomeric structure vary greatly among species, but usually the microvilli of a rhabdom are oriented either in layers with the microvilli oriented perpendicular to one another and to the optic axis, or in sectors, paired or threefold (Fig. 14-5). The microvilli of one retinular cell extend halfway across the rhabdom, where they meet corresponding microvilli of the opposite retinular cell. Major dichroic absorption occurs in rhabdomeric eyes when the *e* vector is parallel to the long axis of microvilli, but in vertebrate eyes maximum absorption by rods occurs when the *e* vector is perpendicular to the disc axis. In honeybee eyes eight retinular cells of one ommatidium are electrically coupled and show no dichroism; a ninth cell not in the central rosette is responsive to polarized light (61). Crayfish retinular cells are more sensitive to polarized light aligned parallel to their microvillous axis than to light orthogonal to it. In each crayfish ommatidium there are two types of cells with axes orthogonal to each other. These types of cells differ in spectral sensitivity; one is more sensitive to orange light, the other, R8, to violet. Cell pairs having the same angle of maximum sensitivity to polarized light are electrically coupled (Fig. 14-5).

Whether retinas of the ciliary type are sufficiently sensitive to *e* vector for polarization to be important in behavior is uncertain. Polarotaxis has been described for several fishes, for example, the halfbeak (*Zenarchopterus*), in which preferred orientation is at $90°$ to the *e* vector (84). In goldfish, neurons of the optic tectum show greater visual excitation to one plane of polarized light. Behavioral evidence indicates polarization selectivity in pigeons. Humans see a brush-like image when a small spot of polarized light is rotated, with minimum effect parallel to the *e* vector and maximum effect perpendicular to it. The greater sensitivity of vertebrate eyes may be due not to detection by visual pigment cells but to intensity changes resulting from scattering by filter cells. In summary, in arthropods there is a clear correlation of sensitivity to polarization of light with morphology of the retinular cells, and probably with orientation of receptor pigment molecules.

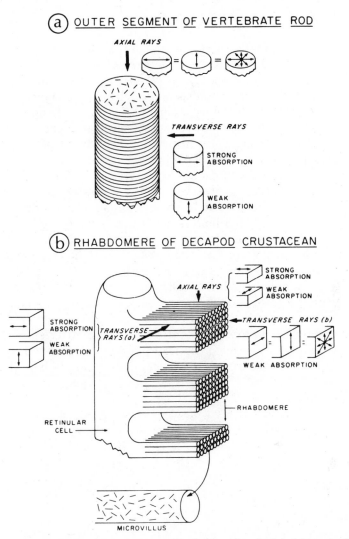

Figure 14-5. Diagrams of outer segment of vertebrate rod and of rhabdomere of crayfish. Heavy arrows indicate orientation of planes of polarization of light. In vertebrate rod light polarized parallel to planes of discs is strongly absorbed, when polarized perpendicular to planes of discs is weakly absorbed. In rhabdomeres (shown as single retinula cell) absorption is maximal when light is polarized parallel to microvillar axes. From T. Goldsmith in *Comparative Animal Physiology,* Third Edition, edited by C. Ladd Prosser. Copyright © 1973 by Saunders College Publishing. Reprinted by permission of CBS College Publishing.

PHOTOPIGMENTS AND COLOR VISION

To be an effective stimulus, light must be of the wavelength to be absorbed by a receptor molecule (17). Receptor molecules have been identified by (1) chemical extraction of receptor layers, (2) microspectrophotometry of single cells in vivo, (3) measured action spectra of electrical responses of receptor cells, and (4) psychophysical measurements. The photosensitive pigments in both ciliary and rhabdomeric types of eyes, as well as in photosensitive halobacteria, are called rhodopsins. All rhodopsins consist of an aldehyde of vitamin A (retinene) combined with a protein, opsin. Retinal has one double bond between C_5 and C_6 of the ionone ring; retinal can be reduced to retinol (vitamin A). Vitamin A is synthesized from β carotene, derived from plant sources. Opsin is a glycoprotein of some 303 amino acids with a molecular weight of 38,000 (68). Some lipid is needed for regeneration of rhodopsin; rods (bovine) have 60–90 lipid molecules per rhodopsin.

Rhodopsin molecules are oriented in bilayers on the membranes of discs of rod (or cone) outer segments or on microvilli of rhabdomes. A rat rod has some 3×10^4 rhodopsin molecules aligned in each disc.

The chromophore of rhodopsin, retinal, can occur in several stereoisomers. In the unbleached state one of these, the stereoisomer 11-cis retinal, is attached through a protonated Schiff's base linkage to a lysine on the opsin molecule (Fig. 14-6). The maximal absorption of 11-cis rhodopsin (bovine) is at 498 nm. Light isomerizes the chromophore from 11-cis to the all-trans

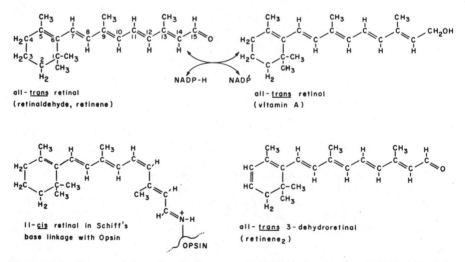

Figure 14-6. Structural formulae of all-trans retinal$_1$ of 11-cis retinal$_1$ in Schiff's base linkage with opsin and all-trans 3-dehydroretinal$_2$. From T. Goldsmith in *Comparative Animal Physiology,* Third Edition, edited by C. Ladd Prosser. Copyright © 1973 by Saunders College Publishing, Holt, Rinehart and Winston. Reprinted by permission of CBS College Publishing.

configuration (Fig. 14-7a). Isomerization is followed by a series of conformational changes in the molecule. The intermediates in the sequence can be stabilized by low temperatures. The primary photo-product is bathorhodopsin; it is formed in less than 6 psec at room temperature (72). This primary step consists of a twisting about the 11–12 double bond, to form a distorted all-trans chromophore. Bathorhodopsin, previously known as prelumirhodopsin, has its maximum absorption (λ_{max}) at 543 nm (27).

The energy of the photon is used in the initial isomerization to move a low-energy species rhodopsin to a high-energy species bathorhodopsin (45); succeeding steps are thermal (Fig. 14-7a).

Bathorhodopsin decays to lumirhodopsin (λ_{max} 497 nm); this is converted

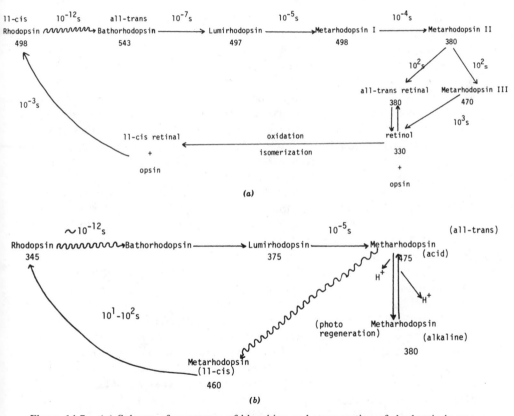

Figure 14-7. (a) Schema of sequences of bleaching and regeneration of rhodopsin in vertebrate eye. Times for fast reactions estimated from measurements at extremely low temperature. Modified and reprinted with permission from T. G. Ebrey and B. Honig, *Quarterly Reviews of Biophysics*, Vol. 8, No. 2. Copyright 1975, Cambridge University Press, London. (b) Visual cycle of UV-absorbing pigment of a fly Ascalaphus. Wavy lines indicate light-sensitive reactions. Modified and reprinted with permission from A. Fein and E. Z. Szuts, *Photoreceptors: Their Role in Vision.* Copyright 1982, Cambridge University Press, London.

to metarhodopsin, of which two (or three) forms are recognized; metarhodopsin I absorbs at 478 nm, meta-II at 380 nm (32). Meta II decays by two parallel paths to all-*trans* retinal and all-*trans* meta III; these converge to all-*trans* retinol plus opsin. The proportions of the two paths differ with species—most human meta II going via metarhodopsin III, carp meta II going via retinal to retinol. Retinol undergoes isomerization and oxidation in pigment epithelium to form 11-*cis* retinal which combines with opsin to form 11-*cis* rhodopsin. Also in vertebrates all-trans retinol (vitamin A_1) may go via the blood to liver where, with other molecules of vitamin A_1, it is isomerized to the cis form, which is transported to the retina. In many arthropods and in cephalopods (octopus) the photoreaction sequence stops at a stable intermediate analogous to meta I, and regeneration occurs directly (32,82a) (Fig. 14-7*b*).

Instead of rhodopsin, the rods of many fresh-water vertebrates contain a purple pigment, porphyropsin. The retinal porphyropsin is 11-*cis*, 3-dehydroretinal; its λ_{max} is 400 nm. The aldehyde retinal$_2$ and the corresponding alcohol (vitamin A_2) have an additional double bond, between carbons 3 and 4, of the ionone ring (Fig. 14-6). By its longer λ_{max} porphyropsin is adapted to environments where the light shifts from maximum energy in the blue-green to maximum energy at longer wavelengths. Porphyropsin is found in fresh-water fishes, larval amphibians, fresh-water turtles, and anadromous fish in their fresh-water phase. Most amphibian tadpoles have porphyropsin with λ_{max} of 522 nm; adult frogs have rhodopsin with λ_{max} 501–505 nm. Amphibians which are permanently aquatic, for example, *Xenopus* and *Necturus,* retain some porphyropsin after metamorphosis. Two species of toad *Bufo* have rhodopsin in both tadpole and adult stages. Although most fresh-water fish have porphyropsin predominantly, a few have both rhodopsin and porphyropsin in outer segments of the same rod. On migration from ocean to fresh water, most salmon eye pigments shift from porphyropsin (λ_{max} 482 nm) to rhodopsin (λ_{max} 501 nm). However, in sockeye salmon this change is only in relative proportion of the two pigments (56). Exposure of juvenile salmon to higher intensities of light can accelerate the shift to synthesis of rhodopsin. Eels which live in fresh water have porphyropsin of λ_{max} 523 nm; during migration to the sea where they spawn their pigment changes to two rhodopsins of λ_{max} 482 and 501 nm (56). Genetic control of absorption properties is shown by two species of trout: *Salvelinus fontinalis* (brook trout) has a pigment with λ_{max} 503 nm, lake trout *S. namaycush* λ_{max} 512; in hybrids both proteins are present (57a).

Pigments based on a given chromophore (retinal$_1$ or retinal$_2$) differ in their absorption properties because of differences in opsins. Species differences in rhodopsins and porphyropsins, listed above, represent different opsins. The opsins have been only partly characterized, and there are presumably many gene differences for coding these proteins. Color vision in vertebrates is performed by cones that differ from rods in having different opsins. It has long been known that the spectral optimum shifts toward shorter wavelengths with lower light intensity (Purkinje shift), and that this is due to the greater sensitivity of rods than of cones. The same chromophore is used with different proteins to provide the cone pigments for color vision. Some fish have red cones with λ_{max} of 620 nm. In double cones of some fishes (goldfish) the pigment of one cone of the pair may differ from that of the other cone. Cones of goldfish absorb maximally at 455, 535, and 570 nm; those of monkey at 440, 535, and 575 nm (16). The chicken retina has ten times more cones than rods. The pigment of chicken cones is iodopsin, and its λ_{max} is 562 nm.

Most of the receptors in the fundus of the human retina are cones. The wavelengths of maximum absorption were measured on pieces of an isolated retina and three cone types were found: blue 420 nm, green 534 nm, red 564 nm; rods had λ_{max} at 498 nm (10) (Fig. 14-8). Rushton (74) identified different cones by reflection densitometry on regions of the fundus in humans with various defects of color vision. He found a red cone 570 nm, and a green 536; blue pigment was too dilute for measurement. Of color-blind human dichromats, protanopes lack the red pigment, deuteranopes lack the green. The eyes of anomalous trichomats have three cone pigments, any one of which may have a λ_{max} different from dichromats, who lack one pigment.

The spectral absorption curves for different pigments overlap (Fig. 14-8) and identification of color is made by the ratio of absorption by cones of overlapping ranges. The long wavelength peak of humans (λ_{max} 564 nm) is shorter than the color red (λ 600–625 nm), and the color red is apparently seen by comparison of the responses of the 534 and 564 cones with stimulation by 600–625 nm light. Illumination of one type of cone with different colors gives a different subjective response according to the overlap with the absorption curve for another cone type. It is concluded that vertebrate visual pigments are adapted in both the retinaldehyde and opsin for the max-

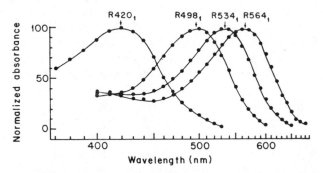

Figure 14-8. Absorption curves of the four visual pigments in man. R498 rods; cones: blue R420, green R534, red R564. Reprinted with permission from J. K. Bowmaker and H. J. A. Dartnell, *Journal of Physiology,* Vol. 298, p. 505. Copyright 1980, Cambridge University Press, Cambridge, England.

imum wavelength of light usual in the animal's environment and for selective sensitivity to colors.

Signals from two types of cones converge on a cortical neuron and one color response may cancel another. However, the sensation of color is not well accounted for by the color opponent theory but is better described by the retinex theory, according to which brightness, wavelength, and comparison between test spot and surround interact. When rods and red cones only were stimulated by adjustment of intensities of white and red light, colors other than red were seen. Experimentally three wavelengths—long, medium, and short—are presented at varying intensities for test spots and background. The brain compares the two fields of presentation and computes ratios of intensities for each of the three wavelengths. Color sensation depends on context and intensity as well as on wavelength (51a). Conditioning experiments with goldfish show that fish perceive color in the same way as humans.

PIGMENTS OF RHABDOMERIC EYES

Behavioral observations show that many arthropods—insects and crustaceans—can distinguish colors. Microspectrophotometric measurements and electrical measurements show the presence of receptors that are maximally sensitive in different regions of the spectrum. Insects have good color vision, especially in the ultraviolet, but generally do not respond to red. A single ommatidium usually has retinular cells with different pigments. In flies *Calliphora* and *Musca,* the six peripheral retinular cells have a visual pigment with λ_{max} 490 nm, while retinular cell R8 has λ_{max} 540 nm and R7, a central retinular cell, falls into one of two spectral subtypes with λ_{max} 340–360 nm, but with different sideband sensitivities in the green (41).

In bees the electroretinogram is maximal when stimulated by ultraviolet light (335–340 nm). A bee ocellus shows two peaks of sensitivity, at 335–340 and at 490 nm. Bees can distinguish ultraviolet 300–400 nm, blue 400–480 nm, green 480–500 nm, and yellow 500–605 nm. Retinal extracted from dark-adapted bee eyes has λ_{max} of 440 nm, bleached 370 nm. Worker bees have λ_{max} of receptor responses at 340, 440, and 540 nm (61). The dorsal part of the eye of a drone bee is more sensitive to blue, the ventral part more sensitive to green; this correlates with hovering at mating sites. *Bombus* retinula cells have λ_{max} values in nm of UV 353, blue 430, green 549, ocellus 353 with a secondary peak 519 (62). Single retinula cells give maximum electrical responses at peaks corresponding to the pigments; green-sensitive cells respond also to UV. In general, the eyes of invertebrates are insensitive in the red, and many invertebrates, especially insects, are more sensitive to ultraviolet than to longer wavelengths. Presumably bees see the colors of flowers as they appear in photographs made in ultraviolet light. Microspectrophotometry of ocelli on the head of larvae of mosquitoes shows a rhodopsin with λ_{max} of 515 nm; on bleaching it is in equilibrium with metarhodopsin of λ_{max} 480 nm, which can then be reduced to retinal (vitamin A) of λ_{max} 340 nm. The frontal eye of a neuropteran *Ascalaphus* is sensitive to UV of 345 nm, and after bleaching, the UV-sensitive pigment is regenerated in blue light; the blue sky maintains UV-sensitivity (Fig. 14-7*b*) (40).

In a dragonfly eye the retinula pairs R5/8 and R2/3 respond to green, R1/4 to ultraviolet and orange, R7 to violet, and R6 to green. Receptors in the ventral half of the eye are not sensitive to polarization (except for orange light), whereas in the dorsal eye the ultraviolet receptors are very sensitive to polarized light. The ventral eye is therefore, adapted to light reflected from the water surface over which a dragonfly flies (60). Voltage-intensity curves indicate that gain in transduction of light to electrical response is higher for ultraviolet receptors than for those of other spectral sensitivities (54). This greater amplification by UV cells may result from their smaller size, hence greater specific transmembrane current. In a butterfly *Papilio*, receptors are identified for ultraviolet (390), blue (450), green (540), and red-orange (610 nm). Many retinular cells show double spectral peaks, an indication of interactions between cells (57).

A moth *Manduca* has four pigments with λ_{max} values of 350, 450, 490, and 530 nm (14). In a moth *Deilephila* electrical responses peak in the ultraviolet 350 nm, blue 450 nm, and yellow 525 nm; apparently there are three pigments, probably in separate retinula cells (44). Lobster eyes have rhodopsin of λ_{max} 515 nm, which bleaches to stable metarhodopsin of λ_{max} 490 nm (82a).

Arthropod metarhodopsin is more stable than vertebrate metarhodopsin and in light metarhodopsin it is converted to rhodopsin. On illumination of rhodopsin the chromophore does not dissociate from opsin in arthropods and cephalopods as it does in vertebrates (Fig. 14-7).

TRANSDUCTION OF PHOTOCHEMICAL RESPONSES

Rods and Cones

In a dark-adapted human eye a single quantum of light absorbed by each of 5 to 14 rods in a spatially restricted area can be seen as light. The initial photoisomerization initiates a chain of reactions, and amplification of the initial transduction occurs in the sequence from photon capture to membrane activation. Intracellular recordings have been made from cones and rods of fish, salamander, frog, toad, lizard and turtle; in all of these the receptor cells are larger than in mammals (5). Measurements of transmembrane current have also been made from an outer or inner segment of a rod (toad) that had been sucked into a small capillary while the opposing segment was bathed with various solutions (47).

In the dark there is an inward sodium current through the membrane of the outer segment and an outward positive current due to an electrogenic

Na-K pump. On illumination the photocurrent flows in opposite direction to that at rest and is measured as a net decrease in dark current (Fig. 14-9). Direct measurements show that a single photon can lead to a decrease in the dark current equivalent to 10^6 univalent (Na^+) ions (11,39). A single photo-isomerization generates a response of 10^{-10} V.

The active receptor potential consists of two parts, the early receptor potential and a late receptor potential (Fig. 14-10). The early potential is the electrical sign of the charge movements associated with the photochemical event and subsequent rhodopsin changes. The late potential is a hyperpolarization graded according to light intensity. Electrical pulses applied during the receptor hyperpolarization show that the resistance increases by several megohms during illumination. Currents calculated from extracellular recordings at different positions show that current is maximal just distal to the junction between outer and inner segments.

The transmembrane potential (resting potential) of the outer segment of a receptor is about -30 mV—that is, it is positive to E_K. The dark current is due to inward flux of Na ions down their electrochemical gradient into the outer segment. Sodium which enters the outer segment in the dark is pumped outward by a Na-K exchange pump in the inner segment. Illumination results in a decrease in permeability of the membrane of the outer segment to Na^+, hence this segment hyperpolarizes toward E_K. The response to light is thus the closing of Na^+ channels, which are open in the dark (44).

According to Hagins (39) an outer segment of a rod in rat retina has 1000 discs and contains 3×10^7 molecules of rhodopsin. The plasma membrane surface is 2 to 3×10^{-5} cm^2 and the resistance (R_m) 10^4 Ω cm^2. The membrane potential change on illumination is sufficient to give a movement of 2000 ions per photon. The dark current is 25 pA, and the area of an outer segment is 1200 μm^2, hence the dark current is 2 μA/cm^2, which is a thousandfold less than the 2 mA/cm^2 in a squid axon spike (30,31). The dark current in a rat rod is large enough to turn over all of the univalent cations in the cytoplasm in one minute. The dark current increases when Na^+_o is increased. Light suppresses Na permeability and hyperpolarizes the plasma membrane; the light response decreases when Na_o is reduced.

Light has two actions, (1) initiation of primary response and (2) decrease of sensitivity to illumination. A steady background light of low intensity decreases the sensitivity to flashes.

Two hypotheses have been proposed as possible mechanisms of transduction of the rhodopsin light response closing the Na^+ channels (11). One is that Ca^{2+} is released from the discs and that Ca^{2+} then acts to close the Na^+ channels of the plasma membrane. Measurements with Ca^{2+}-sensitive electrodes show that on illumination, Ca^{2+} increases outside the outer segment.

When external Ca^{2+} is reduced to near zero the dark current increases and the light response increases by 20-fold (88). Increase in intracellular

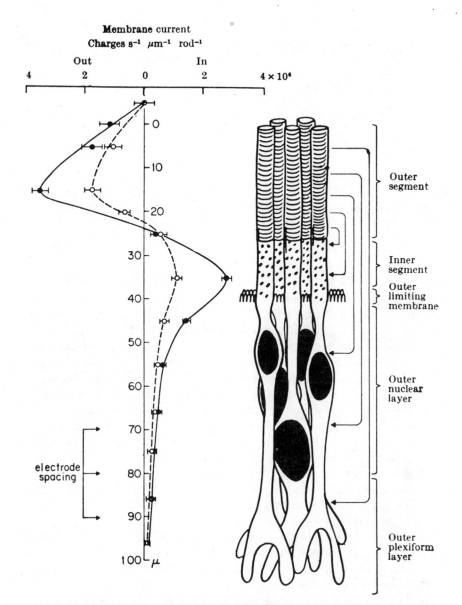

Figure 14-9. Diagrams relating photocurrent of receptors to structure of rods in rat retina. Current responses to two intensities of light. Outward current in outer segment enters the inner segment (arrows to right of diagram). Reversal is at junction between center and inner segments. Reprinted by permission from R. D. Penn and W. A. Hagins, *Nature,* Vol. 223, p. 204. Copyright 1969, Macmillan Journals, Ltd.

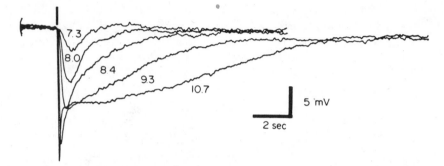

Figure 14-10. Intracellular voltage responses of toad rod to 100 msec, 500 nm flashes at different light intensities. Early and late responses shown at high intensities. Reprinted with permission from G. L. Fain and J. E. Lisman, *Progress in Biophysics and Molecular Biology,* Vol. 37, p. 18. Copyright 1981, Pergamon Press, Oxford.

Ca^{2+} decreases the light-sensitive membrane conductance. The hyperpolarizing response is accompanied by a decrease in membrane conductance and specifically by a decrease in g_{Na} (31). Reduction in Ca_o depolarizes in the dark and increases the light response. Increase of Ca^{2+} inside the outer segment blocks inward Na current—that is, it mimics the effect of illumination. Reduction of Na_o prolongs the response to light much as does increase in Ca_o, and release of Ca^{2+} from inside to outside is Na-sensitive, hence there may be Na^+-Ca^{2+} exchange across the outer membrane. Reduction of internal Ca^{2+} by injection of EGTA has two effects, a depolarization and a decrease in the response to light (12,67,69). A steady background illumination also desensitizes, but the time course is faster than for the same amount of desensitization by low Ca_i. Opposing the idea of calcium as a coupling agent are measurements of intracellular Ca^{2+}, which decreases on illumination, rather than increasing as it would if Ca^{2+} were released. Rods in Ca^{2+}-free medium give normal responses. Intracellular Ca^{2+} remains low during illumination (1,88).

The second hypothesis for coupling between light-activated rhodopsin and the turning-off of Na^+ dark current in outer segments is that phosphodiesterase is stimulated. A decrease in intracellular cGMP results in Na^+ channels becoming closed. Measurements by patch clamping of rod membrane show conductance increase when cGMP is added from the inner side. Ion channels close when cGMP is reduced. Calcium ions do not alter conductance. It is concluded that cGMP is a coupling agent, that Ca^{2+} movement occurs, but that the exact function of Ca^{2+} is not understood (1,32a).

Electroretinograms

All eyes produce an electroretinogram (ERG) that can be recorded between the front and back of the eye as a sequence of potentials in response to illumination. In the vertebrate eye the cornea becomes positive to the back of the eye; in arthropod eyes the cornea becomes negative. The electroretinogram of most vertebrates consists of a small initial negative-going a-wave followed by a large b-wave, then a maintained c-wave, and finally a positive d-wave at cessation of illumination. The principal response, the positive c-wave, is lost after poisoning the pigmented epithelium, and intracellular recordings from pigmented epithelial cells give large deflections of the same time course but opposite polarity to the ERG (79). Measurements by a potassium-sensitive electrode of extracellular potassium show a decrease in K_o and hyperpolarization of pigment epithelium during illumination. The c-wave corresponds to the hyperpolarization of pigmented epithelium as K_o decreases due to influx into receptors during illumination. The other components of the ERG are generated in cells other than the pigmented epithelium.

The ERG of insects (*Drosophila*) consists of an initial rapid positivity of the cornea followed by negativity maintained during illumination and a final negative "off" deflection. The maintained negative response reflects current generated in the receptors; the "on" and "off" probably originate in the lamina or first neural relay station (71).

Transduction in Rhabdomeric Eyes

Transduction of the photochemical responses in microvillous, rhabdomeric eyes differs from that in ciliary eyes. This transduction has been examined in squid retina, limulus ventral eye, and compound eyes of some insects. In most invertebrate eyes the Na^+ channels are closed in the dark and open in the light. An active retinular cell becomes depolarized on illumination, the opposite of the process in vertebrate receptors.

In receptor cells of limulus, crayfish, and barnacle, intracellular measurements show an increase in Na^+ conductance on illumination. In limulus an increase in intracellular Ca^{2+} was shown by a luminescent flash from aequorin injected into a receptor. Contrary to the situation in vertebrates, a depolarizing response can occur in a limulus receptor when it is in a Ca-free medium (plus EGTA), hence Ca_o is not essential for the response. In limulus, Ca^{2+} modulates photosensitivity, that is, light adaptation.

It is estimated that in the limulus ventral eye, each receptor cell has approximately 10^9 molecules of rhodopsin in a density of 8000 molecules/μm^2 (55). In dim light there is a background of small depolarizations or bumps which spread by electrotonic coupling between receptors. The electrical responses to prolonged illumination have two components, fast (150 msec) and slow (minutes). Most measurements concern only the fast response (Fig. 14-11) (56a). If Na_o is replaced by sucrose, the reversal potential is reduced from $+18$ to -50 mV and amplitude as a function of Na_o shows a 55 mV slope per 10-fold change in Na. Li^+ can substitute for Na^+ in the limulus receptor but not in vertebrate eyes. The outward current is carried by potassium.

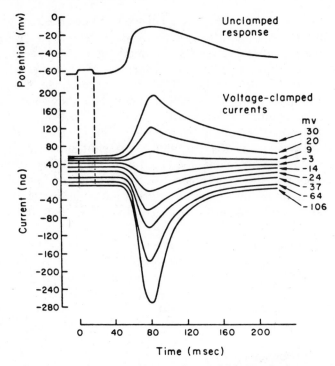

Figure 14-11. Responses of receptor in ventral eye of limulus. Receptor potential (upper trace) and light induced responses under voltage clamp at indicated voltages. Reproduced from Millechia and Mauro, from *The Journal of General Physiology*, 1969, Vol. 54, p. 338. Reprinted with permission from Rockefeller Univ. Press.

Desensitization—a reduction in the amount of depolarization per photon—appears to be due to release of Ca^{2+} and to its effect on the plasma membrane. Hyperpolarization following a depolarizing response is caused by a Na-K pump (31). A further difference from the vertebrate is the response after bleaching; retinal does not dissociate from opsin; also metarhodopsin can go directly back to rhodopsin in light. It is concluded that Ca^{2+} is released from photopigment membrane in both types of eye, but in the rhabdomeric eye the liberated Ca^{2+} decreases photosensitivity but does not change membrane conductance (55).

Retinular cells of honeybees have a high concentration of Ca^{2+} bound to intracellular membranes. In bee and limulus on illumination the extracellular K^+ increases, but in vertebrates it decreases.

Several mutants of *Drosophila melanogaster* are behaviorally blind. In one mutant the normal increase in g_{Na} on illumination of receptors fails. Another mutant gives a normal electrical response at 17°C, a prolonged response at 25°C, and no response at 34°C. Another mutant has a defect in the response of lamina neurons. Genetic control of specific membrane properties is indicated (71).

The rhabdomeric ocellus of a barnacle adapted to blue light responds to a red light with depolarization that persists up to 30 min after the red stimulus is removed. However, presentation of blue light during the after-depolarization eliminates the persistent response to red. The λ_{max} for red stimulation is 532 nm, for blue 495 nm, corresponding to a two-state interaction between rhodopsin and metarhodopsin (44a).

In scallop eyes the proximal retina is rhabdomeric and the distal retina ciliary. The proximal receptors depolarize to a light stimulus after a 30–80 msec latency. The depolarization is an increase in g_{Na}, an "on" response with reversal potential of $+10$ mV. The distal receptors give a hyperpolarizing "off" response with reversal potential of -70 to -80 mV after a 15–25 msec latency. The hyperpolarizing response is sensitive to K_o with a slope of 52 mV. It is concluded that the response of the distal retinal cells is due to increase in potassium conductance. The depolarizing cells are more sensitive to light than the hyperpolarizing ones by 2–3 log units (58).

Neural Processing in Retinas

Vertebrate Eyes

The retina develops as part of the brain, and both electrical and chemical interactions occur between neural elements in the retina (Figs. 14-12, 14-13). The outer plexiform layer is the region where rods and cones interact with horizontal and bipolar cells (21). The inner plexiform layer is the region of interactions of bipolar amacrines and ganglion cells. There is considerable species variation in synaptic structure. Electrical coupling occurs between receptor cells and between horizontal cells. Coupling of receptors is shown by an electrical response in a receptor cell to illumination of receptors at some distance. Coupling is also indicated by spread of current injected into a receptor cell and recorded at several cells distant. Electrotonic coupling between receptors has been demonstrated in fish, toad, and turtle. Receptors, bipolars, and horizontal cells do not spike but activate other cells by transmitters and modulators. Amacrines function by both graded responses and by all-or-none spikes. Ganglion cells have spikes which conduct to the brain.

Receptors activate bipolar and horizontal cells by chemical synapses. Transmitter is released continuously by the slightly depolarized receptor cells in darkness; during the hyperpolarizing response to light the liberation of transmitter is reduced. Receptors may make synaptic contact at ribbon synapses that are abundant between horizontal and bipolar cells where the central element is a bipolar cell, and the two flanking elements are endings of horizontal cells (Fig. 14-12) (53).

The primary function of horizontal cells is to transmit visual signals laterally. Some horizontal cells transmit to bipolar cells and others feed back onto cones. Horizontal cells which are activated from cones are of two

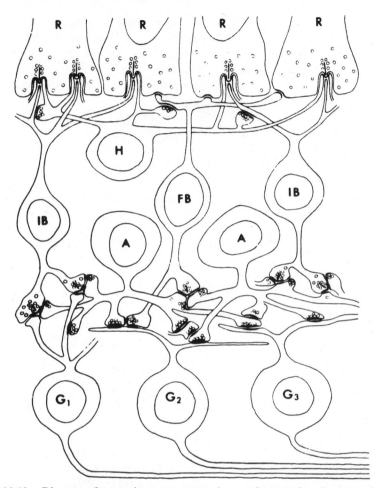

Figure 14-12.   Diagram of synaptic arrangements in vertebrate retina. In outer plexiform layer receptor terminals (R) form ribbon synapses with invaginating bipolar (FB and IB) and flanking horizontal (H) cells. In inner plexiform layer bipolar cells contact ganglion cells (G) and amacrine cells (A). Reprinted with permission from J. E. Dowling, *Proceedings Royal Society of London,* Vol. B170, p. 222. Copyright 1968, Royal Society of London, London.

types. Chromaticity (C) cells depolarize in response to light stimulation in one region of the spectrum and hyperpolarize to light in other spectral regions. Luminosity (L) cells hyperpolarize in response to light of all wavelengths. Depolarization of horizontal cells is due to an increase in g_{Na}, hyperpolarization to increase in g_K or g_{Cl} (36).

Bipolar cells are of two classes: those which hyperpolarize by an increase in g_{Na} when rods are stimulated, and those activated by cones, which hyperpolarize due to increase in g_K or g_{Cl}. Some bipolar cells show depolariz-

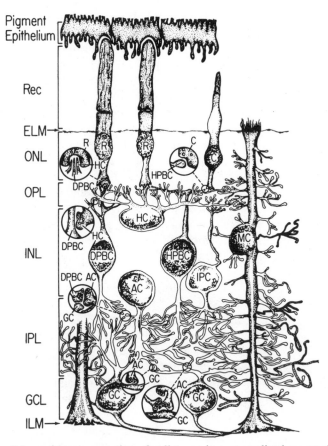

Figure 14-13. Schematic representation of cell types in a generalized mammalian retina. Receptive layer (Rec) has rods (R) and cones (C). ELM external limiting membrane, ONL outer nuclear layer, OPL outer plexiform layer, INL inner nuclear layer, IPL inner plexiform layer, ILM internal limiting membrane, BC bipolar cell, HC horizontal cell, AC and IPC, amacrine cell, GC ganglion cell of several types, MC Muller cell. Personal communication from R. F. Miller.

ing "on" responses and hyperpolarizing "off" responses; they may depolarize in response to stimulation in the center of a visual field and hyperpolarize to stimulation in an annulus. There may be either antagonism or synergism between central spot and annulus. Amacrine cells show both depolarizing and hyperpolarizing responses; they receive activation mainly from bipolar cells and other amacrine cells. Some amacrines give responses that are linear with respect to intensity, while others are nonlinear (66). These cells are axonless and receive ribbon synapses from bipolars and other amacrines (Fig. 14-12).

The axons of retinal ganglion cells are the fibers of the optic nerve; the responses of ganglion cells are conducted spikes. Recordings from 80 percent of ganglion cells of cat and monkey show responses to static visual fields, many of which are on and off responses with antagonism between center and surround. Ganglion cells receive inhibition from amacrines (73). In frog retina some 10 classes of ganglion cells have been found. In frog and pigeon, some ganglion cells respond to bars, spots, edges, others only to moving spots; some cells respond to movement in one direction only. The visual fields of ganglion cells of leopard frogs, that feed on small insects, average 5.2°; the visual fields of bullfrogs, which feed on large insects, are 7.7° (56). In ground squirrel, cat, rabbit, and monkey about half of the ganglion cells respond to static fields, half to complex stimuli, as in frogs. Some ganglion cells of rabbit are directionally sensitive, responding mainly to stimuli moving in one direction, and are weakly stimulated by movement in reverse direction. The weak response results from inhibitory interactions between stimulated areas (4).

Interplexiform cells are centripetal neurons which provide feedback from amacrines and inner plexiform layer to bipolars and horizontals of outer plexiform layer.

Evidence of probable transmitters between retinal cell types includes measurements of effects of putative transmitters, their agonists and antagonists, uptake of agents by cell types, and staining with antibodies against probable transmitters. The acidic amino acids glutamate (Glu) and aspartate (Asp), each with two carboxyl and one amino group, are the likely major excitatory transmitters liberated by rods and cones. These amino acids are also putative excitatory transmitters from some bipolar and amacrine cells. In frog retina, glutamate mimics the effects of rods and cones on horizontal cells and bipolar cells (63). In fish retinas, Asp and Glu excite some bipolars by depolarization and increase in membrane conductance; in other bipolars these agents excite by a decrease in conductance; separate receptors are postulated for the two effects. In goldfish retina GABA mimics the inhibitory feedback from horizontal cells to cones (48).

Necturus cones probably use Glu, not Asp; in rabbit retina Glu excites some bipolar synapses, and Asp acts at ribbon synapses. Comparisons with different agonists indicate that the extended conformation of Glu is effective on "on" bipolars, the folded conformation on "on-off" bipolars and on horizontals (63).

Some amacrine cells are excitatory to ganglion cells by Glu and Asp, other amacrines use ACh (56b). Dopamine on some amacrine cells changes a center-surround field to pure center (18). One class of amacrine cells uses dopamine to activate other amacrines, but not ganglion cells (22). Some amacrine cells are inhibitory to ganglion cells, and GABA and Gly are probable transmitters. In rabbit, picrotoxin (antagonist of GABA) blocks the directional responses of some ganglion cells; strychnine (antagonist of Gly) abolishes field size specification but not directionality of ganglion cell responses. Both GABA and Gly are taken up by amacrine cells, and these substances are probable transmitters of lateral inhibition on ganglion cells (13). Interplexiform cells may use dopamine as a transmitter to horizontal cells (23).

Taurine occurs in retinas, particularly in the outer segments of receptors, and illumination of chicken retinas brings about taurine release. Possibly taurine is a synaptic modulator. It is concluded that the retina, a complex part of the brain, uses several neurotransmitters and that there may be more than one receptor for some transmitters (28).

Rhabdomeric Eyes

Ommatidia are differently organized in various arthropods and molluscs. In lateral eyes of limulus each ommatidium contains a variable number (6–12) of retinular cells, which are electrically coupled and have no axons but give depolarizing responses to flashes of light. In addition, each ommatidium has one eccentric cell with an axon; this cell shows both a generator potential and a spike. Presumably the potentials of the nonaxonal retinular cells reinforce the response of the coupled eccentric cell.

Three components of the visual system in insects, each with very short axons, are (1) the photosensitive retinula cells, (2) the neurons of the lamina, and (3) the neurons of the medulla. Processes of receptors together with dendrites of lamina cells form cartridges which have precise connections between ommatidia and lamina. In many large insects, such as dragonflies, all receptors of a single ommatidium receive from the same field of view, and all receptor axons of an ommatidium project to a single underlying cartridge in the lamina. The projection is retinotopic in that an array of cartridges underlies the array of ommatidia (59,60). The second type of organization occurs in some beetles and flies. The receptors of a single ommatidium have divergent fields of view because they are at different positions behind the lenslets and light enters them at different angles. The divergence angle between adjacent receptors is the same as the divergence angle between adjacent ommatidia caused by the curvature of the eye. Consequently each receptor has the same view as seven others in adjacent ommatidia, and the projection patterns are such that all those receptors with a common view have axons that interweave so as to terminate at one cartridge in the lamina. One cartridge receives axons from all six receptors which have the same field of view. Axons of receptors 7 and 8 pass through the lamina without synapsing and terminate in the medulla (Fig. 14-14). In a fly, each receptor terminal R1–6 forms about 200 tetrad synapses on a lamina neuron. Other postsynaptic elements of the tetrad are contributed by glia and amacrine cells, the function of which is not understood. Centrifugal fibers from the medulla to lamina are known, but there are no efferent fibers from the supraesophageal ganglion. The development of the distribution of receptor axons is a complex sequence of extreme precision (60a).

Retinula cells are electrically coupled. Responses of receptors are graded

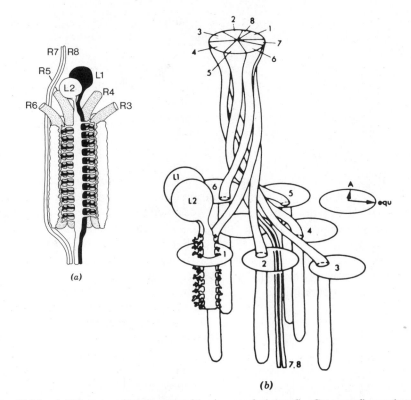

Figure 14-14. (*a*) Diagram of laminar cartridge in eye of a housefly. Cutaway figure showing R1 to R6, four of the six photoreceptor terminals, and L1 and L2, two of the monopolar laminar cells. Receptor axons R7 and R8 bypass the lamina and go directly to the medulla. Reprinted with permission from A. Frohlich and I. A. Meinertzhager, *The Journal of Neuroscience,* Vol. 3, No. 11, p. 2338. Copyright 1983, The Society of Neuroscience, Bethesda, MD. (*b*) Projection of receptor axons from a single ommatidium upon underlying lamina. Six receptors, R1–6 distribute in an asymmetrical divergent pattern upon 6 underlying cartridges. R7 and R8 bypass cartridge and end in medulla. Only L1 and L2 of cartridge receiving R1 are shown. Each cartridge also contains other interneurons. Reprinted with permission from I. A. Meinertzhager, *Symposium Society of Neuroscience,* Vol. 2, p. 96. Copyright 1977, The Society of Neuroscience, Bethseda, MD.

and are depolarizing. Transmission at the first synapse in the lamina occurs for small presynaptic depolarizations; here there is considerable amplification.

Ocelli are the simple eyes in insects which contain a small fixed number of retinular cells and are usually located on the head between the compound eyes. Ocelli are focused behind the retina and function for detection of movement and light intensity, not for image formation. Ocelli detect changes in light intensity during flight. In locusts the ocelli may function as horizon detectors during flight. An ocellus is more sensitive to light and has a larger field of view than a compound eye. In a dragonfly the primary peak of sensitivity of an ocellus is in the UV at 366 nm, a secondary peak is in the green at 516 nm; the compound eye of dragonfly has pigments for these wavelengths, and also for blue (440 nm) (60). Ocellar receptors give graded depolarizing responses which are presynaptic to dendrites of ocellar neurons (15).

In locust, second-order neurons of ocelli are 2.7 log units more sensitive than retinular cells of the compound eyes; in dragonfly and fly (*Calliphora*) the ocellar neurons are 5 times more sensitive to a point light source, 5000 times more sensitive to an extended light source than the retinular cells (86). The visual angle of detection is larger and spectral sensitivity broader for an ocellar unit (86). Responses of ocellar neurons are graded and nonregenerative, but may have fast (spikelike) components on slow potential responses.

BACTERIORHODOPSIN

Halobacterium halobium is a salt-requiring bacterium of the Archaebacteria (Ch. 2). It has a cell wall of glycoprotein plus lipid. The plasma membrane is red due to carotenoids. This color may have protective effect against intense sunlight in the saline pools where the organisms live. Under conditions of decreased oxygen tension the protein of a purple membrane is inserted into the plasma membrane. The purple membrane contains only one protein, the pigment bacteriorhodopsin, which resembles in some ways the visual pigments of animals. The purple membrane is 25 percent lipid, 75 percent protein, and has a retinal bound to a lysine of a protein, opsin, by a protonated Schiff base linkage. Bacteriorhodopsin has a molecular weight of 26,000 and consists of 248 amino acids (24). The polypeptide chains of each protein molecule are arranged in seven 40 Å long helical segments which extend through the membrane perpendicular to its surface. The width of the lipid layer is about the same (45 Å) as the thickness of each protein molecule (Fig. 14-15).

The retinal in dark-adapted cells of *Halobacterium* is a mixture of all-trans and 13-cis isomers; on illumination the 13-cis chromophores are isomerized to the all-trans form to give the light-adapted form of the pigment, λ_{max} of 570 nm; the λ_{max} is 558 nm when dark-adapted. The photochemical cycle of bleaching is similar to that of visual rhodopsin. Its absorption properties resemble those of red cones. Rates of isomerization have been estimated from measurements at very low temperatures (70 K). The bacteriorhodopsin cycle is given in Fig. 14-16. Regeneration occurs spontaneously in the dark (26).

The function of bacteriorhodopsin in purple bacteria is to convert energy from sunlight for ATP synthesis. When halobacteria are grown in light the rate of oxygen consumption decreases and the content of ATP increases. Protons are ejected and the medium becomes acidic (80). The energy for the

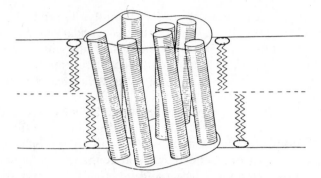

Figure 14-15. Representation of bacteriorhodopsin molecule in membrane of Halobacterium. Protein molecule in seven 40A long helical segments which traverse in membrane. Reprinted with permission from P. N. T. Unwin and R. Henderson, *Biophysics of Structure and Mechanism,* Vol. 3, p. 121, Springer-Verlag, New York.

proton ejection comes from absorbed light, $h\nu$, not from carbohydrate metabolism, and the proton gradient provides energy for Na^+ influx and for amino acid uptake. A gradient of 100–110 mV is generated. Halobacteria are phototactic; red light is attractant, blue light repellent. The pigment of color discrimination is different from bacteriorhodopsin (77a). The presence of similar rhodopsins in halobacteria and in photoreceptors of most animals is an example of convergent evolution.

EVOLUTIONARY CONSIDERATIONS; CONCLUSIONS

Life evolved on earth under solar radiation, and sunlight is now the ultimate source of energy for all vital processes. The present-day spectrum of sunlight at the earth's surface extends from the ultraviolet at 300 nm to the far infrared at 970 nm. The ultraviolet in the prebiotic period was much more intense than it is now; little UV reaches the biosphere at present because it is absorbed by the ozone of the upper atmosphere (Ch. 2). Far infrared heat rays are partially absorbed by the water in the atmosphere. The energy per photon of shorter wavelengths is higher than of longer wavelengths, 71.5 kcal/einstein for violet at 400 nm, 47.5 for orange (600 nm), and 40.7 kcal/einstein for far red (700 nm). The spectrum which is visible to the human eye extends from 400 to 650 nm. Many kinds of filters have evolved—shielding pigment cells, movable irises, and shadowing oil droplets. Many kinds of eyes detect ultraviolet—insects', and probably small birds', for example, hummingbird (38). One advantage of the narrowing of the visible spectrum by means of filters is that chromatic aberration is reduced. This is more advantageous for the camera eyes of land vertebrates than it is for appositional rhabdomeric eyes.

The energy of a photon is given in quanta by the product of frequency at

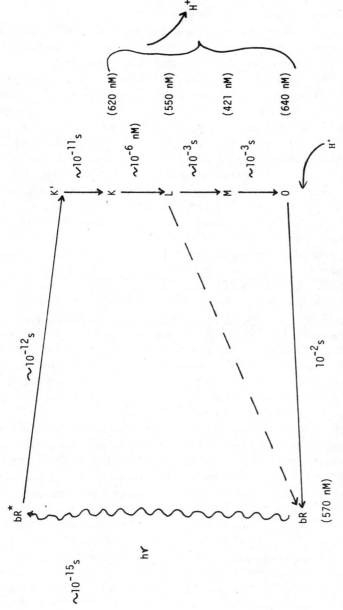

Figure 14-16. Schema of photochemical and thermal reactions of bacteriorrodopsin. Rates of initial isomerization from measurements at very low temperature (70°). Photoconversion step indicated by absorption of light energy converting BR to high energy state. Modified and reprinted with permission from T. G. Ebrey, *Biochemical Events Probed by Ultrafast Laser Spectroscopy.* Copyright 1982, Academic Press, New York.

a wavelength (v) times Planck's constant: E (joules) = hv. The quantal efficiency of a photochemical reaction is given by the number of quanta needed to provide energy for the reaction. In photosynthesis this is measured as the number of quanta absorbed per molecule CO_2 fixed or O_2 liberated. The theoretical minimum is 2 quanta per molecule of CO_2 fixed. The measured quantum efficiency for photosynthesis is 8 quanta. For the retinal photoreaction one photon is absorbed by a molecule of rhodopsin. However, a dark-adapted human reports seeing light if 8–11 photons are absorbed in an area of some 500 rods. It is calculated that one rod (which has several million molecules of rhodopsin) may be excited by a single photon. In general, in both photosynthesis and vision, the efficiency of photochemical reactions is high and a threshold may be less than 10 quanta. However, one quantum contains considerable energy—40 to 70 kcal/mol—and this is sufficient to drive a series of reactions (Fig. 14-7a). An electron of an absorbing molecule (rhodopsin, chlorophyll) is raised to an excited state by one quantum and the energy is transmitted in a series of reaction steps to a transducing mechanism. The primary photochemical steps, such as photoisomerization of retinal, are extremely fast—they occur in picoseconds. Photopigments, when light impinges, activate chains of events in photosynthesis and vision.

Highly energetic wavelengths of light are disruptive to living cells. Ultraviolet light may damage DNA molecules by causing pyrimidines to attach to each other. Repair of UV-damaged DNA can occur by action of a photosensitive photoreactivating enzyme which combines reversibly with the pyrimidine dimer in damaged DNA. Upon absorption of light in the blue and near-UV this enzyme-substrate complex dissociates as enzyme plus repaired DNA. Photoreactivation occurs in prokaryotes and in eukaryotes; the repair enzyme may have functioned before oxygen reached its present levels and before the ozone layer was formed (19).

The evolution of specific photosensitive pigments exemplifies biochemical conservatism. In the evolution of two major photochemical systems—photosynthesis and vision—use was made of preexisting molecules. Chlorophyll is built from a porphyrin; metalloporphyrins probably evolved in relation to electron transfer before chlorophyll appeared. Photosynthesis was mediated by other pigments before chlorophyll evolved (Ch. 2). Rhodopsins are built from aldehydes of carotenoids, pigments which probably were present as shielding colors long before visual systems evolved.

Why were porphyrin-containing molecules used in chlorophyll and a carotenoid-derived pigment used in rhodopsin? These molecules provide two different mechanisms for energy transfer. Chlorophyll is a rigid molecule which functions to transfer electrons to acceptor molecules in a chain of reactions. Hemes are also built from porphyrins and function in electron transport in the respiratory chain of oxidative phosphorylation, in mitochondrial respiration, and in O_2 transport in hemoglobin. When a quantum of light impinges on a molecule of chlorophyll, an electron is removed from a chlorophyll pair (electron donor). Delbruck (19) noted that the rigid frame of the porphyrin of chlorophyll is ideally suited for the maneuver of charge separation and electron transfer.

Rhodopsin consists of a chromophore, retinal, which is bound by a protonated Schiff base to a protein. Retinal is a long aldehyde which can bend at a particular carbon bond, that is, change from *cis* to all-*trans* configuration. The twisting of the chain alters the position of the proton relative to the protein, charges are transferred, and energy is then redistributed along the protein (Fig. 14-6). In summary, energy transfer in two kinds of light-absorbing molecules is accomplished by electron transfer in a rigid molecule (chlorophyll) or by rotation of a carbon chain (rhodopsin).

After the photochemical reaction of rhodopsin, amplification occurs. The energy of a proton initiates a chain of reactions. The final step is release of Ca^{2+} or a decrease in cGMP which then brings about a conductance change—closing Na^+ channels in rods and cones, opening Na^+ channels in rhabdomeric eyes. The sequence of amplification from the photochemical reaction to the ion movement is undergoing active investigation.

Virtually all photoreceptors have the chromophore retinal$_1$ complexed to a protein opsin; eyes of a few freshwater vertebrates have retinal$_2$, which is suited for vision in reddish water. Selection has occurred of pigments suited to properties of sunlight and its absorption by air, fresh water, or ocean water at different depths. The greatest diversity is not in the retinals but in the proteins or opsins, of which there are many. Visual pigments have broad spectral absorption curves, and in most eyes which contain several pigments, the absorption curves overlap. Optimal vision is not at the wavelength of maximum absorption of a pigment, but is determined in the nervous system by the ratio between activation of inputs from different curves with two pigments with overlapping absorption curves. A retina with rods containing one pigment and different cones with three (four in some fishes) pigments contains four or five opsins, and it is the proteins which determine the spectral sensitivity. The λ_{max} of rods or of retinular cells is adapted to the ecology of the animals in which they occur. Aquatic species from ocean depths at which the penetrating light is mainly blue have rods or retinular cells more sensitive to short wavelengths than surface fishes.

Color vision in the human eye is made possible by three different pigments—blue, green, red—based on three opsins with retinal$_1$. Measurements on persons with various degrees and forms of color blindness show that their retinas differ in the relative abundance of red, green, and blue pigment-sensitive cones. There may be some color-blind eye mutants of opsins among anomalous trichromats.

In most insects a red-sensitive pigment is lacking but an ultraviolet pigment is present. Absorption properties of retinular cells of bees and butterflies differ according to the colors of flowers they visit. Dragonflies distinguish between their own kind and other species by color. Apparently many forms of opsins occur in insects, and one retina may contain several different opsins.

The basic biochemistry of rhodopsins is highly conserved. The structure of bacteriorhodopsin is remarkably like that in invertebrate and vertebrate eyes. Several smooth muscles which contract in response to light, such as frog iris, have a rhodopsin like that in the rods. Many variants of opsins occur; relatively little is known of their structure. New cytogenetic techniques of cloning and monoclonal antibody production may make identification of different opsins possible.

In addition to chlorophylls and rhodopsin, other photopigments are known. In mold Phycomyces growth of sporangiophores is stimulated by blue light (λ_{max} 450 nm); growth is also stimulated by orange light at very high intensities (λ_{max} 595 nm). The action spectrum in Phycomyces is characteristic of riboflavin. The pigment cryptochrome photoactivates growth in fungi, some plants, and bacteria. In higher plants a similar photopigment, phytochrome, photoactives growth. Phytochrome is a tetrapyrrole linked to a protein; it occurs in two isomeric states, one absorbing in the red (665 nm) the other in the far red (730 nm) (19).

Another adaptation to light is the diversity of pigments which are not photoreactive but absorb or reflect certain wavelengths. Biochromes differ widely in chemical composition; they occur in chromatophores, epithelia, and pigment cells. Biochromes are behavioral signals for identifying mates, individual animals, and for camouflage and mimicry.

An important biological effect of sunlight is due to its periodicity. Most plants and animals have circadian rhythms entrained by light-dark cycles. Reproduction in most animals, and maturation in plants, are controlled by relative length of dark and light periods.

Sunlight is an important environmental parameter in evolution. Many biochemical compounds have been selected for reactions to sunlight. The genes for photosensitive compounds and the feedback control of expression of such genes constitute a large area of adaptational biology.

REFERENCES

1. Altman, J. *Nature* **313**:264–265, 1985. Calcium versus cGMP in coupling between receptors and membranes in rods.

1a. Ariel, M. and N. W. Daw. *J. Physiol.* **324**:135–160, 161–185, 1982. Acetylcholine as transmitter from amacrine to ganglion cells, rabbit retina.

2. Avery, J. A. et al. *Nature* **298**:62–63, 1982. Visual pigments in freshwater fish.

3. Barlow, H. B. and P. Fatt., Eds. *Vertebrate Photoreception,* Academic Press, New York, 1977.

4. Barlow, H. B. and W. R. Levick. *J. Physiol.* **178**:477–507, 1965. Directionally selective responses of ganglion in rabbit retina.

5. Baylor, D. A. et al. *J. Physiol.* **242:**1–26, 685–727, 729–758, 1974. Responses of turtle cones. Electrical models of receptors.

6. Baylor, D. A. et al. *J. Physiol.* **214:**265–294, 1971. Negative feedback from horizontal cells to cones in turtle.

7. Baylor, D. A. et al. *J. Physiol.* **288:**589–611, 1979. Membrane current of rod outer segment.

8. Blest, A. D. et al. *J. Compar. Physiol.* **145:**227–239, 1981. Spectral sensitivities of eyes of jumping spider.

9. Bogomolni, R. A. and J. L. Spudnich. *Proc. Natl. Acad. Sci.* **79:**6250–6254, 1982. Photopigment in Halobacterium.

10. Bowmaker, J. K. and H. J. A. Dartnall. *J. Physiol.* **298:**501–511, 1980. Microspectrophotometry of human retina.

11. Bownds, M. D. *TINS* **4:**214–217, 1981. Molecular mechanisms of visual transduction.

12. Brown, J. E. and L. H. Pinto. *J. Physiol.* **236:**575–591, 1974. Photoreceptor potential in Bufo rods.

13. Caldwell, J. H. and N. Daw. *J. Physiol.* **276:**257–276, 277–298, 299–310, 1978. Transmitters in rabbit retina.

14. Carlson, S. D. and B. Philipson. *J. Insect Physiol.* **18:**1721–1731, 1972. Photopigment in moth Manduca.

15. Chappell, R. L. and J. E. Dowling. *J. Gen. Physiol.* **60:**121–147, 1972. Electrophysiology of dragonfly ocellus.

16. Crescitelli, F. pp. 301–350 in *Biochemical Evolution,* Ed. H. Gutfreund, Cambridge University Press, New York, 1981. Review of vertebrate visual pigments.

17. Dartnall, H. J. and J. N. Lythgoe. *Vision Res.* **5:**81–100, 1965. Absorbance by visual pigments.

18. Daw, N. W. et al. *Retina* **2:**322–331, 1983. *J. Physiol.* **324:**135–160, 167–185, 1982. Neurotransmitters in retina, especially from amacrines to ganglion cells.

19. Delbruck, M. et al. *Proc. Natl. Acad. Sci.* **73:**1969–1973, 1976. Photo responses of fungi and higher plants.

20. Detwiler, P. B. et al. *Nature* **300:**59–61, 1982. Patch clamp recordings from outer segments of retinal rods.

21. Dowling, J. Ch. 10, pp. 163–181 in *Neuroscience 4th Study Program,* M. I. T. Press, Cambridge, MA, 1979. Vertebrate retina as model of the brain.

22. Dowling, J. E. et al. *Vision Res.* **23:**421–432, 1983. Pharmacology of isolated fish horizontal cells.

23. Dowling, J. et al. *J. Compar. Neurol.* **180:**203–220, 1978. Dopaminergic neurons in rabbit retina.

24. Dunn, R. et al. *Proc. Natl. Acad. Sci.* **78:**6744–6748, 1981. Bacteriorhodopsin genes.

25. Eakin, R. M. pp. 91–105 in *Visual Cells in Evolution,* Ed. J. Westfall, Raven Press, New York, 1982. Phylogeny of ciliary and microvillous photoreceptors.

26. Ebrey, T. Ch. 11, pp. 271–280 in *Biological Events Probed by Ultrafast Laser Spectroscopy,* R. R. Alfano, ed., Academic Press, New York, 1982. Primary events in bacteriorhodopsin function.

27. Ebrey, T. G. and B. Honig. *Q. Rev. Biophys.* **8**:129–184, 1975. Molecular aspects of photoreceptor function.

28. Ehinger, B. *Retina* **2**:305–321, 1983. Neurotransmitter systems in retina.

29. Ewert, J. P. Ch. 5, pp. 142–202 in *Amphibian Visual System,* Ed. K. Fite, Academic Press, New York, 1976. Visual behavior in amphibian.

30. Fain, G. L. et al. *J. Physiol.* **303**:495–513, 1980; **297**:493–520, 1979. pp. 29–60 in *Neurons without Impulses,* Ed. A. Roberts and B. Bush, Cambridge, University Press, New York, 1981. Electrical events in rods of toad retina.

31. Fain, G. L. and G. E. Lisman. *Prog. Biophys. Molec. Biol.* **37**:91–147, 1981. Membrane conductance changes in photoreceptors.

32. Fein, A. and E. Z. Szuts. *Photoreceptors and Their Role in Vision,* Cambridge University Press, New York, 1982.

32a. Fesenko, E. E. et al. *Nature* **313**:310–313, 1985. Induction by cGMP of cationic conductance in rod outer segments.

33. Fox, D. L. *Animal Biochromes and Structural Colors,* University California Press, Berkeley, 1976. *Biochromy: Natural Coloration of Living Things,* University of California Press, Berkeley, 1979.

34. Fraenkel, G. and D. L. Gunn. *The Orientation of Animals,* Clarendon Press, Oxford, 1940. Reprinted Dover, New York, 1961.

35. George, J. S. and W. A. Hagins. *Nature* **303**:344–348, 1983. Control of Ca^{2+} in rod outer segments by light and cGMP.

36. Gerschenfeld, H. M. et al. Ch. 13, pp. 213–226 in *Neuroscience 4th Study Program,* M. I. T. Press, Cambridge, MA, 1979. Pharmacology of cones and horizontal cells.

37. Goldsmith, T. Ch. 14, pp. 577–632 in *Comparative Animal Physiology,* 3rd ed., Ed. C. L. Prosser, Saunders, Philadelphia, 1973. Vision.

38. Goldsmith, T. *Science* **207**:786–788, 1980. Color vision in hummingbirds.

39. Hagins, W. A. Ch. 11, pp. 183–191 in *Neuroscience 4th Study Program,* M. I. T. Press, Cambridge, MA, 1979. Excitation in vertebrate photoreceptors.

40. Hamdorf, K. et al. *J. Compar. Physiol.* **86**:231–245, 1973. Photopigment regeneration in ultraviolet receptors.

41. Hardie, R. C. et al. *J. Compar. Physiol.* **145**:139–152, 1982. Photopigment in Musca.

42. Hecht, S. et al. *J. Gen. Physiol.* **25**:819–840, 1942. Sensitivity of human eye.

43. Hochstein, S. *J. Gen. Physiol.* **62**:105–128, 1973. Spectral sensitivity of eye of barnacle.

44. Hodgkin, A. L. et al. *J. Physiol.* **267**:737–766, 1977. Early receptor potentials in turtle cones.

44a. Hoglund, G. et al. *J. Compar. Physiol.* **86**:265–279, 1973. Trichromatic visual system in a moth.

45. Honig, B. and T. G. Ebrey. *Annu. Rev. Biophys. Bioeng.* **3**:151–177, 1974. *Proc. Natl. Acad. Sci.* **76**:2503–2507, 1979. *J. Am. Chem Soc.* **101**:7084–7086, 1979. *Biochemistry* **15**:4593–4599, 1976. Photoisomerization and energy transfer in chromophores of visual pigments.

46.   Horridge, G. A. *J. Exper. Biol.* **44**:233–296, 1966. Optokinetic memory in crab *Carcinus*.

47.   Hubbell, W. P. et al. *Annu. Rev. Neurosci.* **2**:17–34, 1979. Visual responses of frog rods.

48.   Kaneko, A. et al. *Cold Spring Harbor Symposium* **40**:537–546, 1975. *Annu. Rev. Neurosci.* **2**:169–191, 1979. *Proc. Natl. Acad. Sci.* **81**:7961–7964, 1984. Retinal responses and transmitters in goldfish frog and turtles.

48a. Kirschfeld, K. *Exper. Brain Res.* **3**:248:270, 1967. pp. 354–372 in *Neural Principles in Vision,* Ed. F. Zettler and R. Merler, Springer, Berlin, 1976. Resolution by compound eyes.

49.   Kropp, A. Ch. 2., pp. 15–28 in *Vertebrate Photoreception,* Eds. H. B. Barlow and P. Fatt, Academic Press, New York, 1977. Energetics of photochemical reactions.

50.   Kreithen, M. L. and W. T. Keeton. *J. Compar. Physiol.* **89**:83–92, 1974. Detection of polarized light in homing pigeons.

51.   Lam, D. *Cold Spring Harbor Symposium* **40**:571–579, 1975. Synaptic chemistry of identified retinal cells.

51a. Land, E. H. *Scientific American* **237**:108–128, 1977. Land E. H. et al. *Nature* **303**:616–618, 1983. Retinex theory of color vision.

52.   Lasater, E. M. and J. E. Dowling, *Proc. Natl. Acad. Sci.* **79**:963–940, 1982. Responses of carp horizontal cells to neurotransmitters.

53.   Lasansky, A. *Investigational Ophthalmology* **11**:245–275, 1972. Cell junctions in outer synaptic layer of retina.

54.   Laughlin, S. B. *J. Compar. Physiol.* **84**:335–355, 1973; **92**:377–396, 1974; **111**:221–247, 1976; **112**:199–211, 1976; **145**:169–177, 1981. Photoreception and light adaptation in insect eyes.

55.   Lisman, J. E. et al. *J. Gen. Physiol.* **70**:621–633, 1977; **73**:219–243, 1979. Limulus photoreceptors.

56.   Lythgoe, J. N. *The Ecology of Vision,* Clarendon Press, Oxford, 1979.

56a. Mallechia, R. and A. Mauro. *J. Gen. Physiol.* **54**:289–351, 1969. Intracellular responses of visual cells in limulus.

56b. Masland, R. H. et al. *J. Cell Biol.* **83**:159–178, 1979. *J. Neurophysiol.* **39**:1220–1235, 1976. Acetyecholine as transmitter to retinal ganglion cells.

57.   Matic, T. *J. Compar. Physiol.* **145**:169–177, 1981; **152**:169–182, 1983. Electrical interactions between receptors in eye of butterfly Papilio.

57a. McFarland, W. W. and F. W. Mung. *Science* **150**:1055–1056, 1965. Visual pigments in hybrid fishes.

58.   McReynolds, J. S. and A. Gorman. *J. Gen. Physiol.* **56**:376–402, 1970. *Science* **183**:658–659; **185**:620–621, 1974. Ionic mechanisms of receptor potential in eye of scallop.

59.   Meinertzhagen, I. A. *Phil. Trans. R. Soc. London* **B297**:27–49, 1982. *Soc. Neurosci. Symp.* **2**:92–119, 1977. Organization of lamina cells of optic lobe of dragonfly and of flies.

60. Meinertzhagen, I. A. et al. *J. Compar. Physiol.* **151**:295–310, 1983. Receptor types in retina and lamina of dragonfly.

60a. Meinertzhagen, I. A. and A. Frolich. *TINS* **6**:223–228, 1983. Synapse formation in fly's visual system.

61. Menzel, R. and A. W. Snyder. *J. Compar. Physiol.* **88**:247–270, 1974. Polarized light detection in honeybee.

62. Meyer-Rochow, V. B. *J. Compar. Physiol.* **139**:261–266, 1980. Spectral efficiency of compound eye and ocellus of bumblebee.

63. Miller, R. J. and M. M. Slaughter. *Nature* **303**:537–538, 1983. Excitatory amino acid receptors in vertebrate retina.

64. Miller, R. J. and D. F. Dacheux. *Vision Res.* **23**:399–411, 1983. Intracellular chloride in retinal neurons.

65. Millott, N. *J. Exper. Biol.* **37**:363–375, 376–396, 1960. Shadow reactions of sea urchins.

66. Naka, K. *J. Neurophysiol.* **38**:53–71, 72–91, 1975. Catfish retinal neurons.

67. Oakley, B. *J. Gen. Physiol.* **74**:713–737, 1979. *J. Physiol.* **339**:273–298, 1983. *Vision Res.* **22**:767–773, 1983. Light-evoked changes in extracellular K in retina.

68. O'Brien, D. F. *Science* **218**:961–966, 1982. Chemistry of vision receptors.

69. Owen, W. G. and V. Torre. Ch. 3, pp. 33–57 in *Current Topics in Membranes and Transport,* Vol. 15, Academic Press, New York, 1981. Ion movements in isolated receptors.

70. Ozawa, S. et al. *Cold Spring Harbor Symposium* **40**:563–570, 1975. Transmission from photoreceptors in barnacles.

71. Pak, W. *Handbook of Genetics* **3**:703–733, 1975. Mutations affecting vision in Drosophila.

72. Peters, K. et al. *Proc. Natl. Acad. Sci.* **74**:3119–3123, 1977. Primary photochemical events in vision.

73. Rodieck, R. W. *Annu. Rev. Neurosci.* **2**:193–225, 1979. Types of ganglion cells in cat retina.

74. Rushton, W. A. H. *Sci. Am.* **232**:64–74, 1975. Visual pigments and color blindness.

75. Salvini-Plaven, L. p. 137–154 in *Visual Cells in Evolution,* Ed. J. Westfall, Raven, New York, 1982. Polyphyletic origin of photoreceptors.

76. Shaw, S. R. *Nature* **255**:481–483, 1975. Electrical resistance barriers and lateral inhibition in eye of locust.

77. Simmons, P. J. *J. Compar. Physiol.* **145**:265–276, 1981. Synaptic transmission between neurons of locust ocellus.

77a. Spudich, J. L. and R. A. Bogomolai. *Nature* **312**:509–513, 1984. Color discrimination by bacteria.

78. Starr, M. S. *Brain Res.* **151**:604–608, 1978. Taurine in retina.

79. Steinberg, R. H. and B. Oakley. *Vision Res.* **22**:767–773, 1982. *J. Neurophysiol.* **44**:897–921, 1980. Light-evoked changes in K_o in retina, origin of C-wave of ERG.

80. Stoeckenius, W. et al. *Biochem. Biophys. Acta* **505**:215–278, 1979. Properties of bacteriorhodopsin.

81. Van Buskirk, R. and J. Dowling. *Proc. Natl. Acad. Sci.* **78**:7825–7829, 1981. Transmitters of horizontal cells of carp retina.

82. Vanfleten, J. R. pp. 107–136 in *Visual Cells in Evolution,* Ed. J. Westfall, Raven, New York, 1982. Phylogeny of receptor types.

82a. Wald, G. pp. 311–345 in *Comparative Biochemistry,* Vol. 1, Ed. M. Florkin, Academic Press, New York, 1960. pp. 671–642, Vol. 1, pt 1. *Hand. Physiol. Am. Physiol. Soc.* **19**:54. Evolution of visual pigments.

82b. Walls, G. L. *The Vertebrate Eye and Its Adaptive Radiation.* Cranbrook Inst. of Sci., Bloomfield Hills, MI, 1942.

83. Waloga, G. and W. L. Pak. *J. Gen. Phys.* **71**:69–92, 1978. Ionic mechanisms of horizontal cell potentials, Nectarus.

84. Waterman, T. H. *J. Exper. Zool.* **194**:309–344, 1975. Visual sensitivity to polarized light.

85. Werblin, F. S. Ch. 12, pp. 193–211 in *Neuroscience 4th Study Program,* M. I. T. Press, Cambridge, MA, 1979. Integrative pathways of local circuits within retina.

86. Wilson, M. *J. Compar. Physiol.* **124**:297–316, 1978; **128**:347–358, 1978. Functional organization of locust ocelli.

87. Wolf, T. *Cold Spring Harbor Symposium* **3**:255–260, 1935. with S. Hecht, *J. Gen. Physiol.* **12**:727–760, 1929. Visual properties of honeybee.

88. Yau, K. W. et al. *Nature* **269**:78–80, 1977; **292**:502–505, 1981. *J. Physiol.* **288**:589–634, 1979. *Nature* **311**:661–663, 1984. Ion movements in light-sensitive retinal receptors of toad.

15

BIOELECTRIC PROPERTIES OF
CELL MEMBRANES

Analysis of animal behavior at the molecular level requires an understanding of conduction in neurons and transmission at synapses. Membranes of excitable cells are complex in molecular organization, redundant in ion conductance paths, and diverse in kinetics of ion channel activation and inactivation. Why is there so much diversity of cell membranes and how did the various modes of electrical activity evolve? An ion-selective cell membrane was essential for primitive life; electrochemical properties of membranes developed during the evolution of the earliest cells (Ch. 2). Early in prebiotic evolution several properties of cells must have become established:

1. Retention of organic molecules, many of which are anionic at physiological pHs.
2. Electrostatic balancing of the negativity that results from anion retention by elevated concentrations of cations, especially potassium.
3. Synthesis of a plasma membrane consisting of protein and phospholipid such that transmembrane resistance of some thousands of Ω cm^2 and capacitance of about 1μF/cm^2 results.
4. Maintenance of constant cell volume by active extrusion of protons and of sodium in exchange for potassium by means of specific ATPases.
5. Selective ion permeability, reducing the influx of the more common ions such as Na$^+$ and Cl$^-$.
6. Establishment of a steady state between cell contents and medium such that a transmembrane potential (negative inside) and osmotic equilibrium (Donnan equilibrium) are maintained.
7. Synthesis of specific proteins as carriers for ion exchange or active transport.
8. Influx of Na$^+$ or K$^+$ in an exchange antiport with metabolically produced H$^+$.
9. Exchange of Na$^+$ inward for Ca^{2+} outward.

10. Extrusion of Ca^{2+} by Ca^{2+} ATPase.
11. Selective channels for each of the permeating ions.

These properties are characteristic of both nerve-muscle tissue and nonexcitable cells; most are also properties of bacterial and plant cells.

The net effect of the maintenance of ionic gradients is the generation of transmembrane potential—resting potential, or E_m. Upon physiological activation, in many cells the resting potential becomes perturbed—depolarized or hyperpolarized. The perturbation is usually accompanied by a decrease in membrane resistance and an increase in conductance for one or several ions.

Bacteria have been shown by potential-sensitive dyes to be electrically negative intracellularly; the resting potential can be altered by such chemotactic stimulants as O_2 (33). *Escherichia coli* have resting membrane potential of -35 mV before and -52 mV after treatment with the nutrient substrate galactose. Some agents may alter resting potential and locomotion, but a change in membrane potential is not essential to the chemotactic response. Macrophages have a resting membrane potential of -13 mV; when stimulated by a chemical agent, such as an endotoxin, macrophages show chemotaxis and phagocytosis; the membrane becomes hyperpolarized slowly and this persists for a long time (37). In mammalian smooth muscle such agents as acetylcholine can initiate contraction in K^+-depolarized cells. Some hormones initiate protein synthesis and secretion without evoking electrical responses.

Neurons are not unique in electrical excitability and capacity to conduct electrical signals within cells or from cell to cell. Examples of intercellular electrical conduction include conduction in sheets of electrically-coupled smooth muscle, in epithelia of embryos, and in ectodern and endoderm of coelenterate animals and in algal strands. Conduction between cells occurs when electrotonic coupling is made possible by low intercellular resistance (Ch. 12). In nervous tissue, electrical signals are conducted either as all-or-none impulses (action potentials) or as graded events. Electrotonic conduction in the processes of short neurons is sufficient to transmit signals which liberate chemical transmitters at nerve endings—in the vertebrate retina, in mechanoreceptors located near the central ganglia of crustaceans, and in short neurons whose axons show no spikes. Examples of graded potentials which may trigger spikes are pacemaker potentials, sensory potentials, and postsynaptic potentials. Conduction of action potentials is by local circuits in which current flows from a depolarized region to depolarize the adjacent membrane of the advancing front, out across the membrane and back extracellularly to complete the circuit. Propagation requires the opening and closing of channels for specific ions that carry current first inward and then outward across the membrane. Inward and outward currents may be carried by several ions moving in parallel according to several time courses of channel activation. The rate of rise and duration of an impulse are partly determined by the rates of opening and closing of channels—that is, by activation and inactivation of ion conductance changes. The frequency at which spikes can be fired is determined by the time course of conductance activation and inactivation, by the rate of recharging of membrane capacitance, and by the ion pumps.

RESTING MEMBRANE POTENTIALS (E_m)

Resting membrane potentials in animal cells are the net result of several diffusion potentials plus a small contribution from an electrogenic Na^+-K^+ pump. The equilibrium potential for an ion is the potential given by the concentrations across a boundary if the ion is free to diffuse, for example,

$$E_{Cl} = \frac{RT}{zF} \ln \frac{Cl_i}{Cl_o} \text{ or } E_K = \frac{RT}{zF} \ln \frac{K_i}{K_o}$$

where E = diffusion (Nernst or equilibrium) potential, z = valence, R = gas constant, F = faraday constant, T = absolute temperature, K_i and Cl_i are K and Cl activities inside the cell, K_o and Cl_o outside. Measured resting potentials of muscle and nerve are not the same as any single equilibrium potential. Observed values close to E_{Cl} indicate that chloride is usually distributed passively according to the membrane potential. In nerve tissue measured values of E_m are near to E_K; in other tissues E_m may be more negative than E_K depending on activities of other ions. When measured at various K_o values the curve relating E_m to log K_o is nearly linear, with a slope approaching the theoretical 58 mV per log unit over the higher range of K_o; that is, in this range the membrane acts as a potassium electrode. At lower concentrations of K_o, E_m is less sensitive to K_o and other ions or a Na-K pump contribute significantly.

In some tissues (mammalian smooth muscle) E_m is maximal near physiological K_o and the membrane depolarizes in either K-free or high-K medium. Some depolarization occurs also at low temperature and when the membrane is poisoned with ouabain. It is concluded that under physiological conditions an electrogenic Na-K pump contributes some 15 mV of a total 65–70 mV resting potential in these tissues.

In some tissues, (fish skeletal muscle) permeability to Cl^- is relatively high, and chloride determines the resting potential. In bacteria, mitochondria, chloroplasts, and plant cell membranes the contribution of a proton-potassium pump establishes steep outward gradients of H^+ which may result in K^+ accumulation and transmembrane potentials of several hundred millivolts (Ch. 9).

Two measures of movement of ions across cell membranes are permeability and conductance (8). Ionic permeability is approximated by the diffusional movement driven by forces of concentration and electrical gradient as given by the Nernst-Planck equation. Fluxes across membranes measured with isotopes provide measurements of permeability constants (P); units of P are centimeters per second. The Nernst equation modified to allow for permeability to Na, K, and Cl ions is the Goldman equation:

$$E = \frac{RT}{F} \ln \frac{P_K(K)_o + P_{Na}(Na)_o + P_{Cl}(Cl)_i}{P_K(K)_i + P_{Na}(Na_i) + P_{Cl}(Cl)_o}$$

Measurements of E in systems with multiple ions which are varied in concentration (e.g., K, Na, Cl) permit calculation of relative permeabilities. Relative permeabilities for two tissues are:

	P_K	P_{Na}	P_{Cl}
At rest (48)			
Frog muscle	1	0.01	1.9
Squid axon	1	0.04	0.45
At peak activity (101)			
Frog muscle	1	20.0	
Squid axon	1	12.0	

The Goldman equation assumes a constant field within a membrane and constant P values for a given condition.

The conductance (g) of a membrane for an ion (i) is given by

$$g_i = \frac{I_i}{(V - E_i)}$$

where g_i is conductance to ion i, I_i is current carried by i, E_i is equilibrium potential for i, and V is membrane potential. Thus conductance is the reciprocal of resistance and is measured in Siemens (formerly mhos). Conductance takes account of transfer number, that is, the fraction of total current carried by an ion. The Hodgkin-Horowitz equation for a single ion (e.g., K^+) is

$$E_K = \frac{RT}{F} T_K \ln \frac{K_o}{K_i}$$

where T is the transference number. This equation assumes constant conductance (not field) within a membrane, and separate channels for conductance of different ions; it is based on an equivalent electrical circuit (Ch. 9). Conductance is obtained from voltage-current measurements, and from slope (chord) resistances in I-V curves.

Plots of resting potentials for several cell types against external K according to the Goldman equation are curvilinear, but according to the Hodgkin-Horowtiz equation they are linear, because the two equations are based on different assumptions (49a).

In simple circuits with parallel nonreactive RC (resistance-capacity) elements, the current-voltage relation is linear and conductance is given by the slope; most biological membranes are linear over a limited range of voltage. Excitable membranes are nonlinear, as shown when the *I-V* relationship is measured under voltage clamp (41). Rectification implies that membrane resistance for an ion in one direction is different from that in the opposite direction. Membranes in which nonlinearity appears on depolarization beyond zero volts or positive to some threshold voltage show outward rectification. This nonlinearity may be a disproportionate increase in outward current, normally of potassium, and is called delayed rectification (Fig. 15-1). In some membranes the outward nonlinearity is a reduction of current on depolarization, called anomalous rectification (Fig. 15-2); this occurs normally in muscle fibers, and squid axons and some eggs when K^+-conductance is blocked, as by TEA. The converse patterns of nonlinearity are for inward currents which may be either increase or decrease (Fig. 15-3) on hyperpolarization. Echinoderm eggs show inward rectification hyperpolarization. The threshold for increase in conductance and the *I-V* slope vary according to extracellular K^+ (44). Many conductances have a range near zero volts in which the *I-V* curve is relatively flat; inward rectification which is turned on in this range may result in an N-shaped *I-V* curve (Fig. 15-8); the downward slope reflects a region of negative conductance. *I-V* curves obtained with the initial current during a pulse are different from those obtained at steady state. Measurements of current-voltage relations under different ionic conditions, in the presence of channel blocking agents, and from different clamping voltages permit identification of the currents associated with activity. An example of rectification occurs in a squid giant axon, where resting inward conductance of K^+ is 160 times greater than outward conductance (48).

Voltage clamping of cell membranes by separate voltage- and current-measuring electrodes within a cell, or by a sucrose gap when cells are electrically coupled, is used for measuring membrane current I_m at different potentials, and can be used to estimate net current carried by different ions. Most of the conclusions regarding passive and active conductance in excitable cells have been derived from gross cellular voltage clamping. Recently, a technique of clamping ionic channels by means of small external electrodes closely applied to cell membranes (patch-clamping) has made possible conclusions concerning molecular events associated with the opening and closing of single channels. Many cells under resting conditions show spontaneous unitary events which are interpreted as inward currents in single channels (87). For example, in membranes of eggs of tunicates, the number of spontaneous depolarizations per second increases as the cell is depolar-

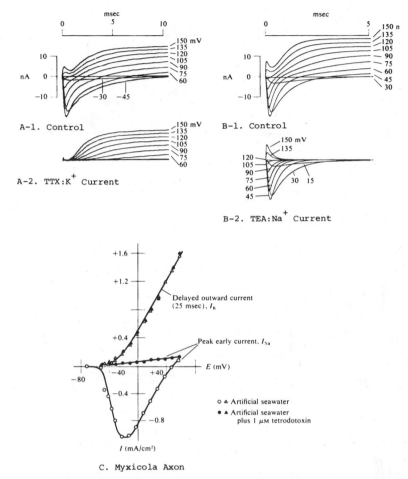

A-1. Control

A-2. TTX:K$^+$ Current

B-1. Control

B-2. TEA:Na$^+$ Current

C. Myxicola Axon

Figure 15-1. Voltage clamp records of currents from a single node of Ranvier of frog sciatic nerve fiber. Records show displacements of the membrane potential from resting level of -75 mV to the levels of depolarization indicated on the individual records. (A-1) Time course of current flow in normal Ringer solution. (A-2) Records measured in presence of TTX which blocks the Na (inward) current. Reprinted by permission from *Nature,* Vol. 210, p. 1221. Copyright © 1966, Macmillan Journals Ltd. (B-1) Control and (B-2) after blocking K current by tetraethyl ammonium (TEA). Reprinted with permission from B. Hille, *Journal of General Physiology,* Vol. 50, p. 1293. Copyright 1967, Rockefeller University Press, New York. (C) Voltage-current plot of excited *Myxicola* giant axon. Triangles designate delayed current (I_K), open symbols controls, solid symbols after TTX. Circles show early (I_{Na}) current, open symbols before and solid symbols after TTX. Reprinted with permission from L. Binstock and L. Goldman, *Journal of General Physiology,* Vol. 54, p. 736. Copyright 1969, Rockefeller University Press, New York.

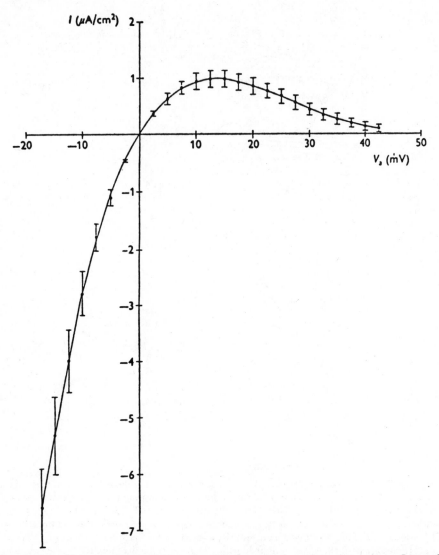

Figure 15-2. Voltage-current relation of resting frog muscle under voltage clamp. Intracellular electrodes for current and for voltage. Anomalous rectification in positive potential quadrant. Ion substitution shows that g_K is linear for hyperpolarization but at depolarizations beyond 0 mV the outward current (I_K) decreases. Reprinted with permission from R. H. Adrian and W. H. Freygang, *Journal of Physiology,* Vol. 163, p. 69. Copyright 1962, Cambridge University Press, London.

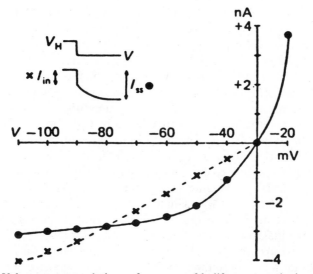

Figure 15-3. Voltage-current relations of neurons of bullfrog sympathetic ganglion. Records during hyperpolarizing steps from holding potential of -30 mV gives instantaneous ($\times$) and steady state ($\bullet$) current-voltage curves. Steady state is current level at end of a hyperpolarizing or depolarizing pulse. Note linearity of instantaneous and inward rectification of steady-state current during hyperpolarization. Reprinted with permission from P. R. Adams, D. A. Brown, and A. Constanti, *Journal of Physiology,* Vol. 330, p. 546. Copyright 1982, Cambridge University Press, London.

ized; current flow of the unitary events reverses at the equilibrium potential for Na$^+$ (35a). Patch-clamping of cultured heart muscle cells shows Ca^{2+} channels which are gated (opened) by E_m and show two open times—fast and slow. Inactivation (closing) is 20 to 30 times slower than activation; the opening time is 0.65 msec at -60 mV, 1.4 msec at -80 mV. From time and voltage dependencies it is proposed that there are closed and open states, possibly with an intermediate "activated" state (97,100).

The ions which carry the currents of depolarizing or hyperpolarizing responses—that is, of either graded or all-or-none impulses—may be ascertained by any of several measurements: (1) the effect of omitting a specific ion from the medium and substituting an impermeant ion; (2) amplitude of electrical response as a function of external ion concentration; (3) measurement of reversal potential corresponding to the potential at which ionic current changes direction; (4) effect of agents which specifically block certain ion channels; (5) measurement of current under voltage clamp; (6) determination of current-voltage curves for different times during an action potential, commonly during maximum rate of rise and at peak; and (7) ion flux measurements by means of tracers and ion-selective electrodes.

Speculation concerning the evolution of the use of different ions for the same function in various organisms and tissues is limited (47a). There is redundancy of current carriers and more than one channel in parallel for a single ion. The two following sections deal with the diverse ways in which ions are used for depolarization, repolarization, and hyperpolarization by active cells. When a cell responds electrically, there is movement of several ions and often interaction between them. Since resting potentials are negative, perturbations are usually depolarizing, sometimes hyperpolarizing. Inward depolarizing currents are mostly carried by Ca^{2+} or Na^+; outward currents of repolarization or hyperpolarization are mostly carried by K^+. Cl^- currents may be either depolarizing or hyperpolarizing according to whether E_{Cl} is negative or positive to E_m. Depolarization toward the equilibrium potential for Ca^{2+} or Na^+ is counteracted by the turning-on or activation of an outward current (usually of K^+) which brings the membrane back to resting level. Changes in membrane resistance can result in current and voltage shifts without actual movement of large amounts of ions. All cell membranes have capacitative as well as resistive components; some current serves to charge the membrane capacitance before significant voltage change occurs.

CARRIERS OF INWARD (DEPOLARIZING) CURRENTS

Calcium

Most intracellular calcium is bound to organelles or in organic complexes; in resting cells the concentration of free calcium ions is usually 10^{-8} to 10^{-7} M; during responses the concentration may reach 10^{-6} M due to Ca^{2+} influx or to mobilization of bound calcium; extracellular concentrations are millimolar. Estimates of total intracellular concentration of Ca^{2+} are inexact because of compartmentation, binding to proteins, participation of Ca^{2+} in many enzyme reactions, and interactions of Ca^{2+} with other ions. Equilibrium potentials of $+60$ to $+140$ mV have been reported (42). Inward calcium currents are blocked by ions in the series $La^{3+} > Co^{2+} > Ni^{2+} = Mn^{2+}$, and by drugs such as verapamil, nifedipine, and D-600, to different degrees according to the tissue. The organic blockers are effective on vertebrate heart and smooth muscle Ca^{2+} channels, and are much less effective on other membranes. Other divalent cations Ba^{2+}, Sr^{2+}, (but usually not Mg^{2+}) can substitute for Ca^{2+} as current carriers in some, but not in all cells; this indicates specificity as to the channels. The theoretical slope when the membrane acts as a Ca^{2+} electrode is 29 mV per decade change in Ca^{2+} concentration. Changes in cellular ionized calcium can be made visible by aequorin, a material active in luminescence of jellyfish; aequorin is very sensitive to calcium ions and gives off light transiently when Ca^{2+} concentration reaches micromolar level. A dye Arsenazo III changes in absorption correlated with intracellular concentration of Ca^{2+}; tetracycline fluoresces at low Ca^{2+}. Removal of calcium ions following an action potential is by (1) enzymatic sequestration, (2) exchange for sodium, and (3) active extrusion across plasma membrane.

Sodium

The sodium gradient is inward and the equilibrium potential is commonly about $+50$ mV. Li^+ can substitute for Na^+ as a current carrier in some cells, not in others; choline, Tris, and trimethyl amine (TMA) are commonly used as impermeant cations to replace external Na^+ but are not without side effects. The Nernstian slope for the peak overshoot of the spike is 58 mV. Numerous animal toxins are potent blockers of Na^+ channels. It is probable that these toxins evolved as paralytic agents against prey and predator animals. Most Na^+ channels can be blocked by very low concentrations of tetrodotoxin (TTX) from puffer fish or by saxitoxin (STX) from red tide flagellates (Fig. 15-1a). These agents do not block in some neurons of invertebrates and some epithelia. Batrachotoxin from a South American toad increases Na^+ currents by opening Na^+ channels and keeping these channels open. In most animal cells, Na is actively extruded by an Na-K pump, hence the passive Na^+ currents can be ascertained only if the pump is first blocked, usually by ouabain. Sodium channels probably evolved much later than Ca^{2+} channels; Na^+ currents are greater by some 40 times than Ca^{2+} currents (47a).

CARRIERS OF OUTWARD (HYPERPOLARIZING) CURRENTS

Potassium

In most cells potassium is the most abundant diffusible ion. Potassium concentration is maintained by either a Na^+-K^+ pump (animal cells) or a H^+-K^+ pump (bacteria and plant cells). The inward permeability constant of potassium is high. The K^+ gradient is outward and usually has an equilibrium potential of -70 to -90 mV. Most, but not all, potassium currents are blocked by TEA (tetraethyl ammonium) or by 4AP (4-aminopyridine) (Fig. 15-1b). Since the resting potential is fixed by concentrations and relative permeabilities of K, Na, and Cl, the resting membrane potential may coincide with or be less than E_K; if the latter, activation of K conductance uses hyperpolarization. At least four different K currents have been identified in animal cells, each in a separate channel. The principal current of repolarization is voltage-dependent; an important second K current is activated by intracellular Ca^{2+} and is slower than the voltage-activated K current; it is designated $I_K(Ca)$. Two additional K currents (A and M) have thresholds in hyperpolarizing ranges and may function to hold the cell membrane at a negative potential and thus to prevent firing (3).

Chloride

Chloride concentrations and conductances vary considerably between tissues. In most animal cells Cl_i is low and the permeability to Cl^- is much less than to potassium. The Cl^- gradient is, therefore, usually inward and E_{Cl} is usually between -70 and -90 mV. Activation of Cl^- channels, for example, by inhibitory neurotransmitters, results in hyperpolarization. Interpretation of the mechanisms of maintenance of low intracellular Cl^- is not agreed upon. Probably, maintenance of Cl^- level in animal cells is mainly passive, conforming to Donnan ratios; Cl^- pumps have been postulated (Ch. 9). In cardiac Purkinje fibers, there is four times more Cl_i than if Cl^- were in equilibrium (65). Substitutes for Cl^- are impermeant anions such as SO_4^-, methanesulfonate, acetate.

Immature oocytes of the frog *Xenopus* have relatively high Cl_i and when depolarized from rest, -70 mV to -20 mV, a transient outward current with reversal at -30 mV or E_{Cl} is activated. The outward Cl^- current requires intracellular calcium; that is, the outward current is $I_{Cl(Ca)}$ (10).

Hydrogen Ions

Hydrogen ion is not in equilibrium with E_m; the normal concentration gradient for H^+ is outward, but the electrical gradient is inward (Ch. 9). A proton pump functions at cell boundaries of bacteria, of many plant cells, and in the inner membranes of mitochondria. Proton currents may hyperpolarize these membranes. In most excitable animal cells H^+ does not function as a carrier of outward current, but in some molluscan neurons, on depolarization, the intracellular pH declines and g_{H+} increases (62). The effects of fluxes of H^+ are synergistic with those of Ca^{2+} on the activity of rhythmically bursting molluscan neurons (38a).

DISTRIBUTION OF CONDUCTANCE CHANNELS

Probably most excitable cells have several channels of inward and outward ionic currents. Clues to the adaptive functions of the parallel channels and their origins may be found by surveying different tissues and animals. Most information has come from a few types of cells large enough for penetration by two electrodes. Evidence for evolutionary trends is available only by inference.

Inward Calcium and Sodium Currents

Depolarization by inward Ca^{2+} currents appears to be more widespread and more primitive than depolarization by inward Na^+ currents. This may be related to the functions of Ca within the cell, in motility, protein synthesis, and secretion. Calcium has been referred to as a second messenger within cells in coupling membrane to intracellular events.

In starfish eggs, activation by sperm or by electrical stimulation triggers a wave of depolarization which reflects an inward current of calcium ions. Voltage clamp experiments indicate two types of Ca^{2+} currents; one current is activated on depolarization in the range of -73 to -55 mV; the other current is activated at -7 mV. Both currents depend on Ca^{2+} (Sr^{2+} or Ba^{2+} can substitute); the first current is faster, more sensitive to cobalt, and is blocked by procaine; the other current decreases if either Na^+ or Ca^{2+} is reduced, but when Ca^{2+} is absent, addition of Na^+ does not restore inward current. Apparently Na^+ modulates the Ca^{2+} current (44). The slope resistance is high near the resting potential but decreases on hyperpolarization; this anomalous rectification reflects an increase in g_K.

Many muscles show predominantly Ca^{2+} action potentials. The muscle fibers of a barnacle are large enough (> 1 mm diameter) for internal perfusion. The inward current in these large striated fibers is carried by Ca^{2+}, which is competitively inhibited by La^{3+}, Co^{2+}, and Mn^{2+}. Overshoot of zero potential is proportional to Ca_o^{2+}. Transmembrane stimulation triggers spikes, and if barnacle fibers have been injected with aequorin, a flash of light accompanies the action potential. Influx of Ca^{2+} increases from 0.11 at rest to 0.38 pm/cm^2/sec in activity, and if H_i^+ is elevated, g_{Ca} is increased (29). The Ca^{2+} efflux after a spike is Na-dependent, 3 Na_i^+ and 1 Ca_o^{2+} (78). Outward current is carried by K^+; and when TEA (tetraethyl ammonium) is applied there is only an inward current; in Na^+- and Ca^{2+}-free medium there is only an outward current (44). The muscle of a serpulid worm shows action potentials similar to the barnacle's with an overshoot proportional to Ca^{2+} (27b). Action potentials are Ca^{2+} spikes in myoepithelia of the pharynx of a polychaete worm.

In the lamella-like muscle cells of amphioxus, spikes can be blocked by TTX or by Tris replacing Na^+. After TTX block, treatment with procaine permits stimulation to elicit regenerative spikes which depend on Ca^{2+} and can be blocked by Ca^{2+} or La^{3+}. Either Na^+ or Ca^{2+} can carry inward current of spikes (43). Similarly, reduction of intracellular pH induces Ca^{2+} spiking in crayfish muscle by reducing g_{K+} (65).

In vertebrate visceral smooth muscles, spikes which trigger contractions are due to inward Ca^{2+} currents and can be blocked by Co, Mn, verapamil, and D-600. Spikes persist in Na-free medium, but if both Ca^{2+} and Na^+ are reduced so as to maintain a constant ionic ratio (Na^+/Ca^{2+}), there is less reduction in amplitude than when Ca_o^{2+} only is reduced; hence there is some synergism between Ca^{2+} and Na^+ even though the effective current is carried by Ca^{2+} (73,103).

Calcium influx appears to be necessary for transmitter liberation at the presynaptic terminals of all synapses. In frog neuromuscular junctions, when presynaptic impulses are blocked by TTX and electrical pulses are applied, focal application of calcium for as brief a time as 50 μsec before the pulse results in transmitter release as measured by postsynaptic potentials. Ca^{2+} influx occurs during depolarization of a presynaptic membrane, and Mg^{2+} blocks the presynaptic action of Ca^{2+}, possibly by competing for some protein site in the membrane (Ch. 12). In squid giant synapses treated with TTX and TEA to block Na^+ and K^+ conductances, the release of transmitter in response to depolarizing pulses is proportional to Ca_o^{2+}. Ca^{2+} entry can be shown by an aequorin flash (58). In *Aplysia* synapses, presynaptic inhibition depends on a decrease in influx of Ca^{2+}, best seen when the membrane is clamped below E_{Cl}. Presynaptic facilitation may be due to increased influx of Ca^{2+} (50).

Calcium has a stabilizing effect on cell membranes, and even where it may not serve as a significant carrier of current, calcium may modify sodium conductance. In squid giant axons in low Ca medium, excitability is enhanced and responses to a single stimulus may be multiple spikes or oscillatory local responses; in high Ca_o, threshold is raised and responses are single (6,48a). In squid axons Na^+ is the current carrier but Li^+ can replace Na^+ for spike production, not for gradient recovery by Na ion pumping.

In gastropod mollusc ganglia the large neurons are not alike in the proportion of inward current carried by Na^+ and Ca^{2+}. In a snail Helix, soma spikes of identified neurons are proportional to Ca_o^{2+} in amplitude; spikes are blocked by cobalt, in the same neurons the axon spikes are Na-dependent but are TTX insensitive. In cells of a ganglion of the snail Anisodoris, two inward currents separated by a notch in the *I-V* curve are detected, a fast inward TTX-resistant Na^+ current, and a slow current due to influx of Ca^{2+} which results in a plateau. Influx of Ca^{2+} can be shown by optical changes in Arsenazo III (a Ca-complexing dye); it is calculated that Ca_i^{2+} can reach concentration of $2 \times 10^{-7}M$; Ba can substitute for Ca^{2+} as a slow current carrier (21). In ganglia of some snails three cell types are identified, those with pure Ca^{2+} spikes, with pure Na^+ spikes, and with Na-dependent Ca^{2+} spikes. According to one report on Helix neurons (95), the Ca^{2+} and Na^+ currents have the same time course; reports on other snails indicate that Ca^{2+} currents are slower in rate of activation and inactivation than Na^+ currents. The relative amounts of the two currents may not be the same in different neurons or in soma and axon of the same neuron. In a much-studied neuron, R15 of *Apylsia,* the maximum conductance is twice the g_{Ca}; the early fast current due to Na^+ reverses at $+54$ mV and is blocked by TTX; the late slow inward current due to Ca^{2+} reverses at $+65$ and is blocked by Mn, Co, Ni; depolarization to -30 mV activates the inward Na current, depolarization to -10 mV activates $I_{Ca^{2+}}$ (1,4). In the *Aplysia* R15 neuron, three inward currents are distinguished: (1) a fast Na^+ current, activated on depolarization to -25 mV, reversing at $+54$ mV and blocked by TTX; (2) a Ca^{2+} current (for which Sr or Ba can substitute), blocked by Mn or verapamil, reversing at $+65$ mV; (3) a slow current carried by either Na^+ or Ca^{2+} (2).

Currents due to Ca^{2+} occur in dendrites of Purkinje neurons of mammalian cerebellum; the dendritic currents are blocked by Mn^{2+} or Co^{2+}; the axons of these neurons have predominantly Na$^+$ spikes (58). In *Xenopus* tadpoles calcium channels in dorsal root neurons can be blocked by cobalt. In rat hippocampal neurons the dendrites give both fast Na$^+$ spikes and slow Ca^{2+} spikes (106a).

Embryonic myoblasts and neuroblasts give Ca^{2+} action potentials; after differentiation the action potentials are mediated by Na$^+$. For example, in amphibian embryos the myoblasts first have Ca^{2+} action potentials, then Na$^+$ action potentials that are insensitive to TTX, and finally Na$^+$ action potentials which can be blocked by TTX (8a). One type of central neuron (Rohan Beard cells) in salamander larvae is electrically inexcitable at an early stage in development; later these cells respond with potentials due to increase in g_{Ca}, after metamorphosis, with sodium spikes (94).

In some tissues, for example, vertebrate smooth muscle and some cardiac muscle, in which most of the conductance change is normally due to Ca^{2+}, large depolarizing waves can occur in a Ca-free medium containing EGTA. In some smooth muscles these large flat-topped spikes occur spontaneously, in cardiac muscle they occur only when triggered. Ion substitutions show that these potentials are due to Na$^+$ entering through channels normally occupied by Ca^{2+} (73b). The potentials are TTX-insensitive but can be blocked by Mn^{2+}, Co^{2+}, or verapamil.

Giant neurons of a jellyfish *Aglantha* support both Na$^+$ and Ca^{2+} action potentials in the same axon. The Na$^+$ spikes mediate fast swimming; Ca^{2+} spikes serve slow swimming (58b).

Potassium Currents

Inward (anomalous) rectification on hyperpolarization and outward (delayed) rectification on depolarization occur in separate potassium channels. In most action potentials, potassium currents activate more slowly than sodium and calcium currents. Usually the K current begins as the inward current reaches its peak; the outward current returns the membrane to its resting potential well after the inward current has ceased. Differences in rates of inactivation of Ca^{2+} and Na$^+$ currents and rates of activation of K$^+$ current result in different durations of spikes, depolarizing plateaus, afterpotentials, and hyperpolarizing responses.

Several types of K$^+$ action currents are recognized. The principal outward current of most action potentials is I_K which is activated at a slightly more depolarized potential than the inward current. I_K reaches its peak as I_{Na} declines (Fig. 15-4). A delayed K$^+$ current (I_C) is activated not by membrane potential but by Ca^{2+} which enters during the inward calcium current. I_C can be activated by Sr^{2+} but not usually by Ba^{2+} (3,21). Ca^{2+}-activated K$^+$ current occurs in many excitable tissues; this outward current is identified by nonoccurrence in a Ca-free medium. Two types of Ca^{2+}-activated K$^+$ channels have been identified. Circular muscle of cat intestine produces

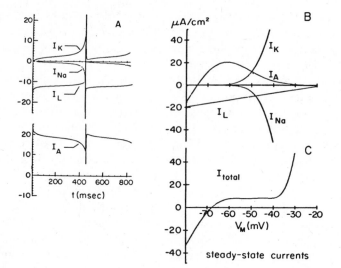

Figure 15-4. (*a*) Plot of ion currents underlying firing pattern in rhythmic neuron of Aniso-doris. (*b*) Steady-state values of individual ion currents versus voltage, I_L leak current. (*c*) Steady-state values of total membrane current versus voltage. Reproduced from J. Connor et al., in the *Biophysical Journal*, 1977, Vol. 18, pp. 92, 94 by copyright permission of the Biophysical Society.

Ca²⁺-spikes which trigger contractions. Repolarization is due to outward K⁺ current which depends on an influx of Ca²⁺ during the spike; the outward K⁺ current can be blocked by TEA. Barium can substitute for Ca²⁺ for spike production but not for activation of I_K(Ca) (103).

Several types of K⁺ channel are primed or turned on by hyperpolarization. One of these is a transient fast outward current, the A current, widely occurring in rhythmic neurons, best studied in mollusc rhythmic cells. The A current is activated by depolarization especially following a hyperpolarization. Under voltage clamp the A current is activated at −45 to −60 mV; it is inactivated by hyperpolarization more negative than E_K. In mollusc neurons the A current is maximum at 10–50 ms and in crustacean axons at 2–3 ms after stimulation (Fig. 5-4, 15-5). The A current is blocked by 4-aminopyridine, reduced by TEA. A-currents function in many rhythmic neurons and muscles to set interspike intervals; the outward K current of repolarization following a spike hyperpolarizes; the hyperpolarization primes the A channel and the A activity delays the inward Na current of the next spike (21, 22).

In some cells the resting potential is less negative than E_K; an active response may be hyperpolarization toward E_K. In the distal retina of a scallop eye a hyperpolarizing response to light is due to increase in g_K; in a K-free medium the response is larger than in a normal medium (61).

In sympathetic ganglion neurons of frog a K⁺ current called the M current is responsible for a prolonged EPSP by blocking the development of inward

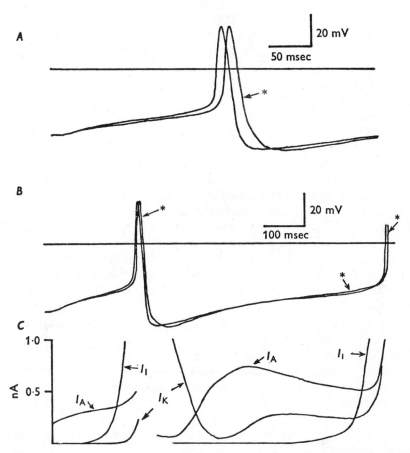

Figure 15-5. (*a*) Comparison of action potential recorded from cell in ganglia of Anisodoris with calculated (asterisk) A.P. (*b*) Action potentials at compressed time scale for comparison of different ion curves. (*c*) Membrane currents associated with voltage behavior in B. I_t total current, I_A potassium A current, I_K voltage-sensitive potassium current. Reprinted with permission from J. A. Connor and C. F. Stevens, *Journal of Physiology,* Vol. 213, p. 47. Copyright 1971, Cambridge University Press, London.

Na^+ currents. Under voltage clamp the M current is activated slowly in the range -60 to -10 mV. Muscarinic agonists (ACh), LHRH and Ba^{2+} reduce or block outward steady-state current in M channels. Inhibition of I_M decreases outward rectifying currents and prolongs the EPSP. This is an example of a transmitter acting by turning off a K^+ channel.

Reverse or negative-going spikes can result from stimulation of K^+ channels in absence of stimulation of Na^+ channels. An example is the muscle of the pharynx of the roundworm Ascaris; relaxation of the muscle closes the lumen. The cells give negative spikes due to increase in K^+ conductance when stimulated (16a). Regenerative hyperpolarizations have been reported in Paramecium and in rods of the retina of Necturus.

Other examples of currents stimulated by hyperpolarization are in sinus venosus (frog heart) and Purkinje fibers (rabbit heart). These hyperpolarization-initiated currents help to set the rhythm of the heart (see below).

Chloride Currents

In animal cells the Cl^- concentration inside, $[Cl]_i$, is much lower than the concentration outside, $[Cl]_o$, hence the passive Cl^- gradient is inward and is limited by the resting membrane conductance. In many plants Cl_i is high, especially in vacuolar fluid; action potentials result from efflux of Cl^- out across the vacuolar membrane. From *Torpedo* electroplax, extracts of protein fractions have been incorporated into planar phospholipid membranes; potential changes across the membrane cause currents carried by Cl^-.

In some muscles the resting permeability to Cl^- is so high as to contribute significantly to the resting potential. In frog striated muscles the resting membrane conductance is 68 percent due to Cl^-; at high temperatures more Cl^- passes inward and more K^+ goes outward. The resting potential of many fish muscles is dominated by Cl^-; reduction of Cl^- by substitution with an impermeant anion depolarizes the cell. In sunfish muscles Cl^- replacement causes depolarization of resting potential from -85 mV to about -35 mV (54). Similar Cl^- dependence has been described for muscles of a stingray in which the g_{Cl} at rest is 8–10 times greater than g_K (43). Voltage clamping at depolarized potentials shows inward currents carried predominantly by Na^+. Membrane resistance R_m in fish muscle is low, hence there is much shunting and spikes are not readily triggered; in muscles of cold-acclimated fish both g_K and g_{Cl} are reduced. In cold-acclimated sunfish muscle K conductance exceeds g_{Cl}, membrane resistance is high, and spiking is readily triggered (54).

The resting Cl^- conductance in striated muscle may be different in the plasma membrane from that in transverse (T) tubules. Treatment of muscle with hypertonic glycerol-Ringer disrupts the T-tubules with little or no effect on plasma membrane. In frog sartorius, the resting conductance for Cl^- in surface membrane is high, in T-tubules it is nil, but both membranes are permeable to potassium (35b):

g_{Cl} surface	219 μmhos/cm^2	g_{Cl} T-tubule	0 μmhos/cm^2
g_K surface	29 μmhos/cm^2	g_K T-tubule	55 μmhos/cm^2

Active conductance change in surface membrane and in T-tubule membrane is a depolarization due to a sodium current.

In mammalian muscles, the T-tubules may be more Cl-permeable than they are in frog muscles. In fibers of rat diaphragm some 85 percent of the resting conductance is due to Cl^-, seven times as much as that due to K^+. Disruption of the tubules reduces total g_{Cl} but not total g_K. In goats with myotonia, muscle fibers respond to single stimuli with trains of spikes; the repetitive response is associated with a decrease in g_{Cl}; glycerol treatment abolishes the delayed depolarization and the repetitive firing. Normal (non-myotonic) muscle responds repetitively in Cl-free medium (5). Crayfish striated muscles have wide invaginations of the surface membrane which lead to T-type tubules (Ch. 16) (97). Indirect evidence indicates that these intrafiber membranes have a very high resting g_{Cl}.

Depolarizing spikes resulting from an outward current of Cl_i^- may occur when the Cl_i^- exceeds Cl_o^-. In tissue cultures of chick heart triggered spikes are increased when Cl_o^- is replaced by acetate or when Cl^- is injected into the cells; hence depolarization occurs due to Cl^- efflux (35). Electroplax membranes from a skate (elasmobranch fish) give regenerative spikes in Cl^--free medium, presumably due to Cl^- efflux.

The usual role of active Cl^- conductance is hyperpolarization due to influx of Cl^-. Inhibitory postsynaptic potentials (IPSPs) in crayfish muscle are due to inward movement of Cl^-. Chloride ion fluxes are increased in low Ca_o (97). Spinal motorneurons are hyperpolarized in response to inhibitory transmitter; the reversal potential for this hyperpolarization can be shifted by injection into the neurons of anions with higher or lower permeability than Cl^-. Inhibitory transmitters Gly and GABA activate different receptors which may share the same Cl^- channel (44a). In motorneurons of cat trochlear nucleus, the IPSP reverses if Cl^- ions are injected such that the gradient is reversed; an afterhyperpolarization (to -90 mV) is due to increased g_K; the inhibitory potential (to -100 mV) is due to increase in g_{Cl}. In crayfish stretch receptors with resting potential of -64 mV, an IPSP shows reversal at -81 mV; the hyperpolarization is due to increased g_{Cl}; when Cl is iontophoresed onto the receptor the E_{IPSP} may approach 0 mV. The inhibitory hyperpolarization due to inward Cl^- current is distinct from spike afterhyperpolarization due to outward K^+ current (47a).

A few types of cells have outward currents carried by chloride. Cultured chick myoblasts respond to pulses by slow (several seconds) depolarizing responses which are due to outflux of Cl^- (35). An outward current triggered by depolarization in *Xenopus* oocytes is carried by Cl^-, is activated by Ca_i^{2+} and shows reversal at -30 mV or E_{Cl}-(10).

In *Aplysia* abdominal ganglion acetylcholine excites (depolarizes) some cells and inhibits (hyperpolarizes) others. The depolarizing responses (EPSPs) are due to increased Na^+ conductance, the hyperpolarizing responses to increased conductance to Cl^-.

CILIATE PROTOZOANS

In ciliates such as *Paramecium,* the resting potential is about -30 mV. When the anterior end of a ciliate is stimulated mechanically the animal reverses its ciliary beat and backs away (Fig. 15-6). The anterior membrane gives a graded receptor potential and a regenerative spike in which the depolarizing current is carried by calcium. Repolarization is by a potassium current. After a protozoan has been deciliated, touch does not elicit the receptor potential. Caudal stimulation initiates a hyperpolarization; the outward current is due to increased K conductance (25,31). The behavioral result of the hyperpolarization is increased forward swimming. Three currents in *Paramecium* are: (1) a transient inward current carried by Ca^{2+}, (2) a voltage-induced outward K^+ current, and (3) a prolonged K current, probably induced by Ca^{2+} (81,83) (Fig. 15-7). In the ciliate *Stylonichia* there are two Ca^{2+} channels: (1) Mg^{2+} can substitute for Ca^{2+} in carrying the current of response to mechanical stimulation, but not to electrical depolarization; (2) Ba^{2+} and Sr^{2+} can replace Ca^{2+} for the electrical depolarization (26). In Stylonichia a Ca^{2+} action potential activates ciliary beat; touch stimulation initiates a Ca^{2+} receptor potential. Sr^{2+} or Ba^{2+} can substitute for Ca^{2+} as inward current carriers (25).

Some genetic mutants of Paramecium are deficient in conductance of specific ions. The mutant "pawn" has no Ca^{2+} action potential, hence cannot reverse its swimming direction. A TEA-insensitive Paramecium mutant shows prolonged depolarization and swims backward continually (80,84). A mutant (paranoic) of Paramecium shows negative resistance in its curve, maximum at -20 mV; calcium induces a conductance, the channels of which pass Na^+ to give a plateau response (80). Chemical studies on deficient mutants may aid identification of the proteins of ion channels.

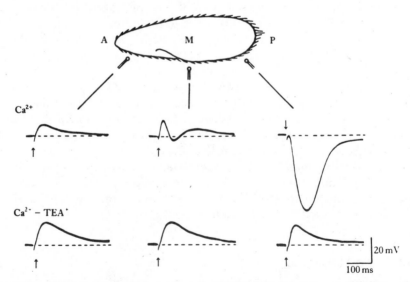

Figure 15-6. Membrane potential responses to mechanical stimulation at different regions of a *Paramecium* (pawn mutant) bathed in solution containing Ca^{2+} plus TEA (lower records). A anterior end, M middle, P posterior end. Reprinted with permission from T. Satow, A. D. Murphy, and C. Kung, *Journal of Experimental Biology,* Vol. 103, p. 256. Copyright 1983, Biochemical Society, Colchester, Essex, England.

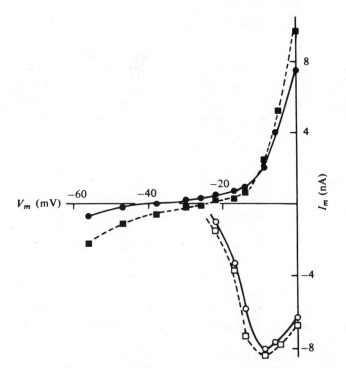

Figure 15-7.   Voltage-current curves of *Paramecium* measured under voltage clamp. Closed symbols, steady state (after 10-sec duration); open symbols, peak Ca^{2+} transients (after 5 msec). Circles are currents in medium containing K^+ and no Na^+; squares in solution containing Na^+ and no K^+. Reprinted with permission from Y. Siami and C. Kung, *Journal of Experimental Biology,* Vol. 88, p. 310. Copyright 1980, Biochemical Society, Colchester, Essex, England.

RHYTHMICITY; MOLLUSCAN NEURONS

Endogenous oscillatory activity is characteristic of many biological systems. Long-term rhythms may be annual, monthly, or circadian—in behavior, reproduction, or glandular secretion. Membrane rhythmicity in periods of seconds to milliseconds is characteristic of hearts, visceral muscles, and many central neurons. Membrane oscillations have been examined in detail in the central neurons of molluscs and crustaceans. Neuronal rhythmicity usually consists of pacemaker potentials which may or may not reach the threshold for spikes. Some neurons fire a single spike per pacemaker depolarization; others fire several spikes and are called bursters—frequently their spike pattern is parabolic (36). The ionic events in the membrane of an oscillating neuron or heart cell consist of varying proportions of passive (leak) conductance, and of active outward and inward currents.

The Aplysia neuron R15 (Ch. 11), is a parabolic burster; voltage clamp experiments show inward currents carried by Na^+ and Ca^{2+}, and three outward currents carried by K^+. An initial rapidly rising current is carried by Na^+, a slower inward current by Ca^{2+} (1). A burst consists of the following: a slow inward current, partly by Na^+ and partly by Ca^{2+}, depolarizes, and at a critical threshold fast inward currents cause spikes. The Ca_i is high during a burst and declines during the succeeding hyperpolarization. Observations made with Arsenazo III in the neuron R15 show increase in Ca_i during each pacemaker-induced burst, more increase if prolonged by TEA; the increase in g_{Ca} can be blocked by lanthanum ions (39).

The outward current is carried by potassium and has three components, each in a separate channel: (1) a voltage-sensitive I_K, activated by depolarization, (2) a voltage-insensitive Ca^{2+}-activated I_K, and (3) a transient current, the A current, which is activated at -40 mV; this is detected under voltage clamp as a hyperpolarizing current (Fig. 15-4, 15-5). After a tetanus or a burst of spikes in R15 a prolonged posttetanic hyperpolarization (PTH) is caused by outward K^+ current. PTH is absent if the preparation is in a Ca-free medium, and PTH is prolonged if Ca is intracellularly injected; the I_K responsible for the undershoot is Ca-activated K conductance. The interval between pacemaker potentials is shorter in a Ca-free medium or after injection of EGTA.

One function of the mentioned previously A currents is to set the intervals between spikes in rhythmic neurons. Pacemaker activity is due to gradual depolarization by increase in g_{Na^+} or $g_{Ca^{2+}}$. During the interspike interval, the membrane potential is dominated by the A current when g_A is activated by small depolarization (to -40 mV); this fast outward current restrains the pacemaker depolarization to the point where the slow inward and fast outward currents cross (Fig. 15-5) (21,22). Similar control of interspike intervals by the A current occurs in repetitively responding crustacean axons. In bursting neurons, the hyperpolarization between bursts is due to a Ca-induced increase in g_K; the interval becomes reduced if the medium is free of Ca or if EGTA has been injected. The interspike interval within a burst is determined by I_A. In rhythmically active neurons and muscles, the phase of inward rectification is prolonged and may be evident as a region of negative slope resistance in the I-V curve. Negative slope resistance results from the opening of additional inward channels. The reversal of the I-V curve occurs in the range of -35 to -50 mV.

Measurements with H^+-sensitive electrodes show that each Ca^{2+} ion entering the cell releases a proton and lowers pH_i (62,98). Intracellular levels of Ca^{2+} may be regulated by cyclic nucleotides. The slow inward current (I_{si}) in some neurons is produced by a Ca^{2+} influx which increases Ca_i^{2+} and H_i^+; phosphodiesterase is then activated (possibly by calmodulin). The resulting decrease in c-AMP accompanies a decrease in I_{si} and causes burst termination. Ca^{2+} is then sequestered or exchanged; phosphodiesterase activity declines and c-AMP increases, the channels for I_{si} are activated (presumably by phosphorylation), and inward Ca^{2+} current depolarizes; Ca_i increases and the cycle is repeated (38a).

In summary, as g_K declines, the inward conducting channels for Na^+ and Ca^{2+} are opened and depolarization occurs; at a critical threshold spiking begins. Elevated Ca_i activates g_K which hyperpolarizes, firing stops, I_A is activated, and the opening of inward channels is delayed. During depolarization due to passive influx of Na^+ and Ca^{2+} the inward current increases at spike threshold.

MEMBRANES OF HEART MUSCLE OF VERTEBRATES

In cardiac muscles of vertebrates several ionic currents are active in parallel. The proportions of the currents and the kinetics differ in the regions of a heart and in the corresponding regions of various vertebrate animals. The electrocardiogram recorded at a distance from the heart shows a series of events as designated by cardiologists. The P wave comes from conduction in the atria, PQ is the delay at the AV junction, QRS complex is the spread of excitation over the ventricle, and T occurs at repolarization of the ventricle. Strips of atrial muscle, ventricular muscle, and Purkinje fibers show prolonged action potentials or plateau potentials which correspond approximately to the ST period of the electrocardiogram. Rhythmicity is autonomous in sinus venosus (fish, amphibian) and in sino-auricular node (mammal). Under suitable conditions, auricular muscle and the conducting Purkinje fibers beat rhythmically.

The terminology of the several inward and outward currents is confusing because several designations have been used in different laboratories; currents differ with heart regions and species (14). Cardiac muscle shows time-independent passive currents: background inward currents of Na^+ and Ca^{2+}, background outward current of K^+, and electrogenic Na-K pump current (69,70). Time-dependent currents are: (1) a fast inward TTX-sensitive Na current (I_{Na}) with threshold at -60 to -40 mV; (2) a slow TTX-insensitive Ca current (I_{si}), blocked by verapamil, nifedipine, or Mn^{2+}, with threshold at -30 mV; (3) a time-dependent inward current (I_f or I_h) activated on hyperpolarization, more negative than -40 mV; and (4) an outward time-dependent K current activated in the range of -50 to $+10$ mV (I_K), the current of repolarization. I_K in turn has three components: (1) fast (I_{K1}), voltage-dependent, blocked by Ba; (2) slow (I_{K2}), probably Ca-activated; and (3) I_{accum} caused by accumulation of K^+ between muscle fibers. Slow activation of I_{in} and delayed activation of I_K result in the plateau, a period of high membrane resistance and low g_K (11,51,60). Fast inward I_{Na} is sensitive to TTX in Purkinje fibers, but insensitive in frog sinus venosus.

Current-voltage curves of cardiac muscle are linear for instantaneous steps (Fig. 15-8). *I-V* curves show anomalous rectification, and increased

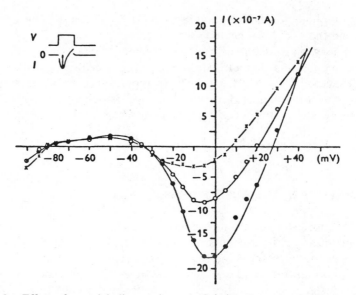

Figure 15-8. Effect of acetylcholine on inward (Ca^{2+}/Na^{+}) current in fibers of bullfrog atrial trabeculum. Preparation bathed with Ringer solution containing TTX 2.0×1^{-8}/ml, ●; current voltage relation in 3×10^{-8} M ACh, ○; current voltage relation in 1.2×10^{-7} M ACh, ×. Acetylcholine reduces negative rectification. Reprinted with permission from W. Giles and J. J. Noble, *Journal of Physiology,* Vol. 261, p. 108. Copyright 1976, Cambridge University Press, London.

conductance on hyperpolarization for steady state steps. In frog sinus venosus, inward rectification at voltages positive to -75 mV results in a negative slope in the range positive to -60 mV.

The current I_f in sinus venosus of frog has been measured during hyperpolarization under voltage clamp (15,16,49). When pacemaker fibers are clamped at -45 mV and then subjected to a hyperpolarizing step (to -60 mV), an inward current is recorded which shows inward (anomalous) rectification. I_f resembles I_{K2} in Purkinje fibers of rabbit in that both currents activate on hyperpolarization, are blocked by Cs, increase in elevated K_o and disappear in reduced Na_o (28).

The sequence of currents in pacemaker processes (Figs. 15-9, 15-10) is: as I_K decreases the conductance for potassium (g_K) declines, and in some nonspecified way the fall in membrane resistance allows background current to flow inward, primarily carried by Na^+, partly by Ca^{2+}; this gradual depolarization brings the membrane to threshold for I_{Na} and upon more depolarization to threshold for I_{Ca}. The role of I_f in pacemaking is uncertain; I_f is activated near pacemaker threshold and may set the interval in sinus venosus. Hyperpolarization and decline in g_K unmask background currents (28,38b,82). Elevated Ca_o results in steeper initial slope of the pacemaker

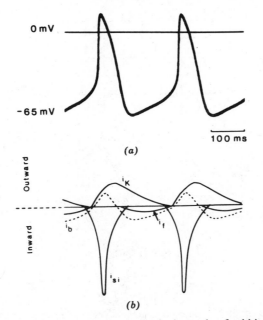

Figure 15-9. (*a*) Spontaneous activity in sinoauricular node of rabbit heart. (*b*) Diagrammatic representation of membrane currents of heart. I_K time dependent outward current, I_f inward current activated by hyperpolarization, I_b probable changes in total background current, I_{si} slow inward current. Reprinted with permission from H. F. Brown, *Physiological Review*, Vol. 62, No. 2, p. 520. Copyright 1982, American Physiological Society, Bethesda, MD.

potential. Decline in g_K during repolarization is enhanced due to the negative slope of the rectifier function of I_K.

Heart rate and amplitude of beat are normally modified by neurotransmitters. Epinephrine accelerates heart rate by shortening the duration of inward Ca^{2+} current (I_{si}) and increasing I_K, thus reducing the diastolic interval. Epinephrine has little effect if I_{si} is blocked by D600 (71). Epinephrine shortens the plateau and decreases tension. Acetylcholine slows the heart by decreasing the limits of g_{si} and increasing g_K; acetylcholine hyperpolarizes, and as a result of its effect on I_{si} the slope of the pacemaker potential is reduced and the intervals lengthened (38) (Fig. 15-8).

Relative proportions of the different currents change during development. In chick embryonic heart, sensitivity to TTX and the rate of pacemaker depolarization increase with age; there is a shift from Ca^{2+} toward Na^+ control (74).

Cardiac muscle membranes have ATPases for Na^+-K^+ and Ca^{2+}. Most of the efflux of the Ca^{2+} which enters during an action potential is by Ca^{2+}-Na^+ exchange, the energy for which is given by gradients generated by the Na-K pump. The exchange is neutral if the ratio is 2 Na^+/1 Ca^{2+}, but ex-

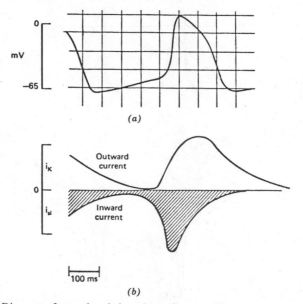

Figure 15-10.   Diagram of postulated time-dependent current changes in a pacemaker cell of heart. (*a*) Intracellular record of activity in rabbit S-A node cell, pacemaker potential precedes spikes; (*b*) Schematic diagram of inward and outward currents associated with action potential. Activation of slow inward current (i_{si}) occurs during development of pacemaker depolarization as well as in depolarization of action potential. Reprinted with permission from W. Giles et al., *Cellular Pacemakers,* Vol. 1, D. O. Carpenter, Ed., Copyright 1982. John Wiley and Sons, Inc., New York.

change is voltage-dependent and Ca^{2+} can move in either direction according to Na_o. If Ca^{2+} enters then exits by exchange, the effect is electrogenic (66). Reduction in Na_i accelerates the exchange with Ca^{2+}.

IONIC CHANNELS; GATING

Cell membranes are approximately 125 Å thick and consist of a bilayer of phospholipids bounded by and penetrated by protein molecules (Ch. 9). On a molar basis, phospholipids in membranes are ten times more abundant than proteins, but the proteins are specific in function—as transport enzymes and specifiers for ion conductances. In squid axons sodium channels are only one-tenth as abundant as sodium pumping sites (8). Previous sections of this chapter and of Chapter 12 list a variety of different channels of ion conductance. How the many specific channel proteins evolved is unknown. Channel proteins with other functions may have evolved before there were action potentials generated by given ions.

The proteins for sodium channels have been isolated and characterized by binding membrane fractions to radioactive blocking agents. Three different Na$^+$ channel proteins in brain are identified by blocking agents: STX (saxitoxin) and TTX (tetrodotoxin) bind to a water-soluble voltage-sensitive channel; these toxins block sodium activation. Batrachotoxin, veratradine, and aconitine bind to a lipid-soluble channel protein; they alter the voltage sensitivity to Na$^+$ and shift conductance toward a negative potential. Scorpion venom or a sea urchin toxin blocks the inactivation of a Na$^+$ channel, causing it to persist in open state (18). Half-saturation values (K_d) for TTX binding to its receptor are: 1.1 nM for electroplax, 4.3 nM for squid giant axon, 1–2 nM for brain synaptosomes; for STX binding, the value is 1.4 nM in muscle plasma membrane (9). In rat muscle it is calculated from STX binding that there are 30 binding sites per square micrometer. After denervation, the density of binding sites decreases (9).

The sodium channel protein is a large molecule. In Electrophorus electroplax the single polypeptide is of molecular weight 260,000 to 300,000 and contains four repeated amino acid sequences. The purified protein has 1820 amino acid residues and the cloned cDNA has 7230 base pairs, including long leader and tail sequences (7,70a). In bovine brain the principal protein is of 260,000 Da, plus two subunits of 39,000 and 37,000 Da. The three units have been incorporated into a planar bilipid layer which is voltage-sensitive; STX and TTX bind to the external surface of the bilayer (45a).

The nicotinic postsynaptic channels which are opened by the transmitter acetylcholine allow multiple ions (Na$^+$ and K$^+$) to enter. By comparison of the entry of various cations, the dimensions of the ACh channel opening are calculated to be 6.5×6.5 Å (47a).

Several electrical events occur in depolarizing excitation. A brief capacitative event charges the membrane according to the time constant given by the product of resistance and capacitance (RC). The capacitative current can be separated from active conductance by applying a hyperpolarizing pulse which charges the membrane in a nonexciting polarity, then subtracting this from the current when excitation occurs (depolarizing). The gating current or opening of an ion channel results from shifts in charges within the membrane.

Gating is quantal and results in two states of channels, open and closed. Gating currents are non-ionic. The probability of a channel being open or closed is voltage-dependent, as is the duration of the open state. The voltage dependence indicates two sensitive sites, one cis and one trans (96).

Elucidation of channel opening and closing, conductance activation and inactivation, is based on patch-clamp records and on kinetic analysis of action potentials. Steady-state reversal potential measurements of a Na$^+$ channel are interpreted as indicating four energy barriers; one ion passes at a time; the height of the energy barrier determines the zero-current potential; the depth of the energy well (binding) determines the current amplitude. The

Na$^+$ ion may undergo stepwise dehydration as its interacts with ionized carboxylic groups (47). The channel proteins may have an enzyme-like action in reducing the energy required for transmembrane transfer of an ion. The activation and inactivation processes can be described in terms of fast and slow components, each with its own time constant.

Membrane channels for K$^+$ have been classified in several categories (57): (1) ion specific K channels are responsible for delayed rectification; they can pass some other ions besides K$^+$, but with lower conductance than for K$^+$; (2) valence-selective channels open according to charge, not to specific ions (for example, the ACh-induced channel); (3) nonselective channels are large, and permit movement irrespective of charge; (4) maxi-K channels favor potassium and have higher rates of transfer of K$^+$ than other channels. The small K$^+$ channels are responsible for delayed rectification and are voltage-sensitive. TEA blocks at the K channel mouth; Cs, Na, or Li can enter, but cannot pass through this channel. Ions are pictured as passing in single file.

ELECTRICAL POTENTIALS IN PLANTS

The contribution of an electrogenic pump to resting potentials is greater in bacteria, fungi, algae, and higher plants than in animals. Proton pumps occur more frequently and probably preceded Ca^{2+} and Na$^+$-K$^+$ pumps in early evolution (Ch. 9). Potentials generated by ion pumps provide more versatility than diffusion potentials in volume regulation and acquisition of nutrients in the variable environments in which bacteria and plants grow. All transporters make use of some kind of ATPase. A plant has a more complex membrane system than an animal, with a central vacuole bounded by a tonoplast membrane and a thin cortical cytoplasm bounded on the outside by the plasmalemma. Both membranes in plant cells support potential gradients; pH of the vacuole in some plants may be as low as 3.0 and pH of the cytoplasm near 7.0; protons are pumped from cytoplasm at both membranes. Recently, methods have been devised for studying tonoplast and plasmalemma membranes in isolation. In *Chara* in the absence of ATP, the membrane potential of plasmalemma is -125 mV, near E_K, but in presence of ATP E_m is -180 mV (89). H$^+$ is pumped with a current of 20 mA/cm^2; measured under voltage clamp, pump values are 1.5–2.5 H$^+$/ATP.

Plant cells are turgid, and intracellular osmotic concentrations are so high that several atmospheres of pressure may exist across cell membranes. The effect of osmotic gradients may correlate with the active proton extrusion by a proton pump which limits accumulation of metabolically produced H$^+$. Proton extrusion couples metabolic energy to the transport of solutes, and permits accumulation of K$^+$ by a H$^+$-K$^+$ exchange. Protons reenter the cells, and the uptake of sugar and amino acids is coupled to an influx of H$^+$ down the H$^+$ electrochemical gradient (Fig. 15-11). When these substrates

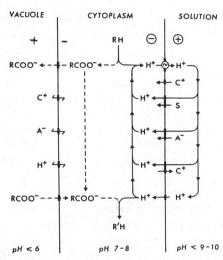

VACUOLE   CYTOPLASM   SOLUTION

pH < 6   pH 7-8   pH < 9-10

Figure 15-11. Diagram of H^+ transport between plant cells and bathing medium. H^+ efflux is active and H^+ influx is passive. H^+ influx may be coupled to influx of uncharged solutes (e.g., sugar, S), or inorganic anions (A^-) or to efflux of inorganic cations (C^+). Reproduced, with permission, from T. A. Smith and P. Raven, *Annual Review of Plant Physiology,* Vol. 30, p. 296 1979, by Annual Reviews Inc.

are taken into a cell, the pH of the medium rises and the cell is depolarized. In a few plants entry of H^+ is coupled as an antiport with Na^+ extrusion, but most plants do not require Na and in them its extrusion is independent of proton movement (Ch. 9).

A cotransport system for sugar uptake coupled with protons has been measured in *Lemna, Chlorella,* lily pollen grains, and sugarcane leaf cells. It is postulated that proton and substrate combine with a carrier molecule at the outer surface of plasmalemma, the entire complex then enters, driven by the electrochemical gradient for H^+; on the inner surface the complex dissociates and the protons are actively transported outward. The outward pumping is by an ATPase using energy from ATP hydrolysis. In several plants it has been shown that two protons are transported for every ATP hydrolyzed. The active extrusion of protons can be blocked by inhibitors of a membrane ATPase; Na^+ is not extruded actively, but by exchange for H^+. Both H^+ extrusion and Na^+/H^+ antiport are essential for accumulation of K^+ (Fig. 15-11). Similar transport occurs in bacteria. Glycolyzing cells of *Streptococcus faecalis* accumulate organic cations and potassium, down an electrochemical gradient due to extrusion of H^+ and Na^+.

The net effect of the outward proton transport is to lower the pH of the medium and to make the cytoplasm electrically negative. In oat coleoptiles and corn roots, resting potentials are -120 to -160 mV. Corn root cells have two systems; one is a H^+-K^+ ATPase which functions at low activities of K_o, < 0.02 mM K_o; the other functions at high K_o. At high K_o, block of respiration depolarizes the corn cells nearly to E_K; inhibiting ATPase has a smaller effect on potential but does reduce K^+ influx as shown by the following (19):

		K_o (mM)	0.02	4.6
Control	K influx (mM/g/h)		3.5	6.9
	Potential (mV)		-142	-100
Oligomycin	K influx (μM/g/h)		0.39	2.7
block of ATPase	Potential (mV)		-124	-92

In *Nitella* H^+ is extruded from the cytoplasm at the plasma membrane and probably also at the vacuolar membrane. In the marine alga *Acetabularia,* the resting potential of cytoplasm to medium is -174 mV in light. In dark or at 10°C in the presence of dinitrophenol, the resting potential is -70 mV. Chloride influx is 10 times as great as Na^+ influx. The potential in the dark may be a K^+ diffusion potential; in the light the potential may result from an inwardly directed Cl^- pump (40,79,92,93). Membrane conductance is high, 8 μS/cm^2. In the fungus *Neurospora* azide and dinitrophenol block K^+ uptake; at pH above 7.0 K^+ uptake decreases. The resting potential of *Neurospora* is -200 mV, greater than the equilibrium potential of any of the contained ions; this potential is proportional to cytoplasmic ATP and two protons are pumped per ATP. In high K_o, however, some depolarization occurs (88). The importance of an ATP-requiring proton pump for maintaining intracellular ions in yeast is shown by the following (49b):

Intracellular Concentration (mM)		
	Control	With DNP
K	261	150
Na	7	0
Mg	14	20
Ca	0	3

The liverwort *Riccia* depolarizes in dark, hyperpolarizes in light; in low K_o (0.1 mM) E_m in dark was -130 mV, in light -215 mV (33a).

Beet cells yield isolated vacuoles which have a potential of -55 mV in low K_o; when metabolic phosphorylation is uncoupled, the vacuoles depolarize to -35 mV (30). ATP makes the membrane negative and its ATPase can be blocked by vanadate. In most plants, the vacuole is positive by a small amount to the cytoplasm. In cells from leaves of *Kalanchoe,* a CAM plant, the potential of cytoplasm to extracellular fluid is -180 mV, the vacuole to extracellular fluid -155 mV; therefore the transtonoplast potential is $+25$ mV. In photosynthesis, malic acid accumulates in the vacuole with a gradient generated by the proton pump; the vacuole potential varies with pH; the slope is 54 mV per decade change in H^+. The potentials of cytoplasm and vacuole in cells of *Lupinus* roots and *Acer* leaves are similar to *Kalanchoe* (77). In the giant marine alga *Valonia* the vacuole is electropositive to the cytoplasm. With electrodes in both cytoplasm and vacuole, measured potentials were: cytoplasm to medium -70 mV in light, -69 mV in dark; vacuole to cytoplasm $+86$ mV in light, $+69$ mV in dark; vacuole to medium $+16$ mV. Protons are pumped from cytoplasm to vacuole and medium, and K^+ is actively transported into the vacuole; each of these actions makes the vacuole positive (24). Isolated vacuoles from mesophyll of pea accumulate sugar by an active process stimulated by ATP; the accompanying cation is unknown.

The greater importance of proton pumps for ionic balance in bacteria and plant cells than in animals may relate to the probable use of proton pumps for volume regulation and for K^+ accumulation in media of diverse osmotic and ionic concentrations. Also, animal cells are less tolerant of a wide range of pH.

Plant Action Potentials

In plants which make rapid movements by turgor changes in cells—*Mimosa* (sensitive plant) and *Drosera* (sundew)—mechanical stimuli elicit action potentials which trigger movement. In *Drosera,* conduction of action potentials is at 4.3 mm/sec down the stem, 9.9 mm/sec up the stem. A series of action potentials averaging 86 mV was recorded on a resting potential of -105 to -150 mV (105). Plasmodesmata connect cells and presumably permit intercellular conduction. The action potentials are brief spikes and plateaus lasting several seconds. In *Drosera* epidermal cells the resting potential is sensitive to K_o^+ and may be a K^+ diffusion potential; high K_o^+ reduces the action potentials. No impedance change has been observed, but the action potentials are sensitive to Mg, Ca, Cl, and K. In the sensitive plant *Mimosa*, motor cells of the pulvinus have a tannin vacuole surrounded by a central vacuole; calcium is high in the tannin vacuole and on stimulation, calcium moves from the tannin vacuole to the central vacuole; rise in Ca^{2+} in the central vacuole triggers K^+ efflux; the extracellular K^+ draws H_2O from the cell which causes a volume decrease and leaf folding. Latency for mechanical response is 70–120 msec. Changes in turgor pressure can trigger action potentials in *Chara;* in *Acetabularia* Cl^- is released when cells are subjected to increased turgor and when they give action potentials. Steady turgor is maintained by active influx of Cl^- (104).

The filamentous algae *Chara* and *Nitella* show action potentials which are conducted from cell to cell. In *Chara* the amplitude of the action potential is 170 mV at an internode, less at a node. An action potential can be initiated by an increase in internal pressure, by mechanical stimulation, or by electrical shock. Recordings from cytoplasm show two components; the second is generated at the vacuolar membrane. The first component is caused by increase in g_{Cl} outward across the plasmalemma; Ca^{2+} is necessary for the Cl^- response, and there may be some Ca^{2+} influx. Repolarization appears due to increase in g_K. Amplitude of action potential corresponds to chloride concentration and E_{Cl} is $+40$ mV (60 mV/decade change in Cl^-) (107). When *Chara* is cultured in Cl-free medium for a few days Cl_i decreases; the action potential due to outward flux of Cl^- diminishes; when Cl^- is replaced in the medium a greater than normal influx of Cl^- occurs. *Acetabularia* shows both spontaneous and elicited action potentials of some 120 mV and 30–300 sec duration (40). Several algae have outward K^+ currents through TEA-sensitive channels (34a).

In the luminescent flagellate *Noctiluca* an action potential of 55 mV is recorded on a resting vacuole potential of -155 mV, and the response is 58 mV per decade change in H^+ in the vacuole; acidification of the vacuole increases amplitude of the action potential and the potential triggers a flash of light (67b).

The mold *Neurospora* shows two kinds of endogenous electrical activity, rhythmic oscillations and fast spikes. The spikes are 40 mV depolarizations on a -180 mV resting potential. The action potentials are due to decrease in the proton pump current, possibly also to some outward Cl^- flux (88).

The nature of action potentials in plant cells is far from completely ascertained, but evidently some potentials may result from efflux of Cl^-, possibly from influx of Ca^{2+}; probably K^+ efflux accounts for repolarization. An alternative explanation is that the action potentials result from decreased activity of the proton pump.

FUNCTION OF STATIONARY POTENTIALS

Direct or stationary current (DC) potentials have been recorded between regions of multicellular organisms, particularly between regions of growth and stability and between regions of high and low metabolic activity. Asymmetry potentials across epithelia, in skin, bladder, kidney tubules, are DC potentials. Epithelial potentials are generated by differential ion transport across the two faces of a flat cell. In frog skin, sodium is actively transported toward the inner surface; by this means frogs living in fresh water actively absorb salt (Ch. 9).

One probable function of DC potentials is in development and regeneration. It has been known for many years that motile aquatic cells, such as ciliates, rhizopods, and leukocytes, orient toward the cathode in a direct-current field; this is electrotaxis. Explants of pieces of embryonic nervous systems send processes toward the cathode and suppress growth toward the anode. Growing tips of seedlings may bend in an electric field; the lower sides of corn coleoptiles which are placed horizontally, and hence are stimulated to turn upward with respect to gravity, are 80 mV positive on their lower (faster growing) side with respect to the upper side. Transected hydroid stems regenerate new heads; the end of a stem which grows a head is electropositive to the base. Growth of a new head toward the anode can be induced by placing a hydroid stem in a DC field.

Eggs become polarized early in development. Fertilization of an egg by sperm or activation by a mechanical prick or brief electrical pulse evokes a Ca^{2+} action potential which is followed by a series of cellular responses. Eggs of a polychaete worm upon fertilization are depolarized from -58 to $+40$ mV and remain at this potential for 65 min (56a). Mammalian eggs hyperpolarize on fertilization, probably by release of intracellular Ca^{2+}. During germination of fertilized eggs of the alga Fucus, depolarization occurs; at the end where cell division is most active, a small steady potential develops, and occasional pulses 100 seconds in duration occur at intervals of several minutes. Positive current of some 300 $\mu A/cm^2$ enters at the animal pole, leaves at the vegetal pole, is due to entry of Ca^{2+}, and can be blocked by La^{3+}, Mn^{2+}, or verapamil.

During regrowth of a leg amputated from a newt, currents of 10–100 $\mu A/cm^2$ are generated between the end of the stump and surrounding skin. This current is generated by the transepithelial potentials of the skin; when these electrogenic potentials are blocked by amiloride or Na-free medium, regeneration of the limb stops. It is postulated that regeneration may be driven by the current field. Regeneration across a transected region of spinal cord of a lamprey is enhanced when a DC field is applied.

How direct-current fields influence orientation and stimulate growth is unknown. In plant tips there may be enhanced transport of anions. Intracellular alignment of macromolecules or stimulation of protein synthesis have been suggested as possible mechanisms.

Behavioral effects of electric and magnetic fields have been well documented. Fishes (Ch. 13) possessing ampullar organs are excited by electric fields generated by muscle potentials of other fish. Many fishes have specialized electroreceptors which detect discharge from electric organs of electric fish. Many aquatic invertebrates orient in magnetic fields; bees and pigeons respond to very small magnetic fields. Magnetite is present in these animals but no connections to sensory nerves have been found. How magnetite is formed and whether it functions in orienting behavior are not known.

CONCLUSIONS AND ADAPTIVE INTERPRETATION

The preceding account indicates the diversity of ionic mechanisms of bioelectric properties of cell membranes. For both static and active functions in living cells, an electrical gradient regularly accompanies volume and osmotic regulation, energy metabolism, and organic anion accumulation. The usual arrangement in plasma membranes of lipid bilayers bounded by proteins is so universal that it must have appeared when cellular organization was first established. Ionic gradients and some of the membrane proteins which determine the electrical properties of membranes must have evolved very early, and changes in such membrane components as ion channels must have occurred as behavior became more complex.

Electrical balance results from both passive and active properties of membranes. Electrical properties of cell membranes were the consequence of the difference in inorganic ion composition of the cytoplasm from the medium. In marine organisms, there must have been the following concentrations within cells relative to seawater: high K_i^+, low Na_i^+, high H_i^+, very low $Mg^{2+} + SO_4^{2-}$, and low ionized Ca^{2+}. These ionic gradients persisted at the cellular level in all multicellular organisms. Attempts have been made to account for retention of some ions and exclusion of others on the basis of charge, crystal diameter, size as hydrated ions, lipid solubility, or chemical reactivity (Ch. 2). Selective permeability—high for K^+ relative to Na^+— established the passive diffusional component of negative resting potentials. The lipoproteins responsible for selective permeabilities must have been among the first to be coded by primitive nucleotides. The capacity to go from rest to activity required (1) voltage-sensitive proteins which could undergo changes in conformation or charge distribution, and (2) other specifically reactive proteins which facilitated ion movements across the energy barriers of conductance channels. Active and exchange transport of ions, particularly Na^+, K^+, and H^+, contributes electrogenically to the establishment of gradients at rest and to repolarization and hyperpolarization in activity. Every cell membrane is a mosaic of patches of ion transfer, passive and active (Ch. 9).

Membrane structure is complex—many proteins and phospholipids are arranged in an orderly fashion within a thickness of some 100 Å. Models of membrane structure are described in Chapter 9. A consequence of the structural composition and arrangement of membranes is high electrical resistance (1000 Ω cm^2) and capacitance (1 μF/cm^2). The transition between rest and activity involves changes in charge distribution which convert membrane channels from closed to open states, from rest to activation followed by inactivation of an ion conductance.

The retention of potassium as the principal cation within cells must have evolved very early for balancing organic anions and consequently for providing a basis for transmembrane diffusion potentials. Potassium participates also in many enzyme functions. Probably the first metabolic pump maintaining high intracellular K^+ was a proton exchange pump; this functions at present in bacteria, chloroplasts, and mitochondria. In plant cell membranes, a specific ATPase is used. The proton pump couples ATP synthesis with ionic gradients (Ch. 9). A later energy-requiring mode of K^+ retention, especially in animals, was the Na-K pump. It is possible that the H^+ pump was replaced by a Na^+ pump because of the narrow range of tolerance of pH change in animal cells. Channels for H^+ may have been the earliest to evolve; these became coupled to K^+ conductance. Several K^+ channels came later. From actions of blocking agents and kinetics it appears that the basic structure of K^+ channels has been conserved.

Calcium was probably regulated as cells differentiated. Calcium has many cellular functions—for example, in membrane stabilization and as cofactor for numerous enzymes. It was essential that intracellular Ca^{2+} be kept low in order not to precipitate sulphates and phosphates. Concentration of Ca^{2+} was kept low by the high binding affinity of Ca^{2+} for many proteins. Calcium pumps (Ca-ATPase) appeared in cellular organelles. Calcium could then have become a messenger between membrane potentials and enzyme reactions because of its affinity for proteins. One early function of Ca^{2+} was probably the activation of protein kinases; a more recent function is the activation of the kinase for the phosphorylation of myosin.

Calcium functioned early as the principal ion in action potentials; it is the most widely distributed inward ion carrier. Sodium became a carrier of inward current in many excitable cells of animals. In algae, action potentials result from efflux of Cl^- from cytoplasm across the plasmalemma and efflux is modulated by Ca^{2+}; repolarization is by outward K^+ current (58a). In mechanosensitive plants such as *Mimosa,* the movement of petioles is produced by water efflux triggered by extracellular accumulation of K^+ and Cl^- (87a). Many kinds of muscles and neurons use calcium and sodium in various proportions as inward current carriers. Neither calcium nor sodium action potentials have been observed in bacteria. Calcium spikes occur in ciliates. Coelenterate and ctenophore myoepithelia and neurons have both Na^+ and Ca^{2+} action potentials. Some mollusc and echinoderm nerves have Na^+ action potentials that are not TTX-sensitive. It is probable that Na and Ca^{2+} channels evolved early in the Cambrian.

Calcium modifies Na^+ conductance in several ways: (1) Ca^{2+} has a screening action on proteins and thus sets the charge on Na^+ channels; (2) Ca^{2+} may alter direct Na^+ binding to protein of specific channels; (3) Ca^{2+} may affect Na^+ binding allosterically at a distance from the binding site; (4) calcium may change Na^+ conductance by altering gating constants. Calcium is a modulator of many electrical events in excitable cells—at presynaptic membranes, in the photochemical pathways of visual excitation, and in controlling interaction of myosin with actin in muscle. Sodium as an inward current carrier in animals makes use of the most abundant extracellular cation. In some neurons and muscles, there is more than one Ca^{2+} channel. The kinetics of increase in Na^+ conductance provide for a fast depolarization, and Ca^{2+} influx for slow response. In squid axons, Na^+ currents are 40 times as large as Ca^{2+} currents (47a). During development, precursors of excitable tissues—neuroblasts, myoblasts—first use Ca^{2+} for action potentials and gradually change to Na^+. In a few aberrant situations, Na^+ can pass through the channels normally used by Ca^{2+}. A possible reason for transition from Ca^{2+} to Na^+ and Cl^- as principal carriers of inward currents is that calcium has many more intracellular functions than sodium and chloride, also that Na^+ and Cl^- are more abundant extracellularly.

Potassium is the most used ion for repolarization of cells, both animal and plant. Several different potassium currents are recognized in neurons, including a fast current which activates and inactivates at negative potentials and serves to counteract inward Na^+ current; K^+ current may be either voltage sensitive or Ca^{2+}-activated. The Ca^{2+}-activated K current may be primitive and may have been functional in cells with a Ca^{2+} action potential only. In most excitable cells, the outward K^+ current is nonlinear with respect to voltage; this is "delayed" rectification. Inward rectification during hyperpolarization (muscle, marine eggs) is by a potassium channel. Other means of repolarization in animal cells are by inward flux of Cl^- and activation of the Na-K pump. Membrane events in prokaryotes and higher plants are different from those in animals; rapid ion conductance changes are rare, and Na-K pumps are absent or different.

The general features of ionic events in neural activity are similar in all animals with nervous systems. Specializations provide for diversity of response systems and complex behavior. Capacity for rhythmicity extends the range of responses to environmental stimuli and fosters autonomous activity. Many kinds of rhythmicity have evolved. The most common are slow depolarizations which are pacemaker potentials, coupled with spike thresholds and followed by undershooting potentials. Conductance during repolarization may be activated in such a way as to delay pacemaker depolarization by Na^+ current (I_f heart) or to prolong synaptic potential (M channel in sympathetic genglia). Special K^+ channels (A current) can activate within the range of a pacemaker and thus can determine spike intervals (in molluscan neurons). Most autonomously rhythmic neurons and muscles show marked inward (anomalous) rectification in the voltage range of membrane threshold. Some neural rhythmic systems depend on circus conduction.

Electrical specialization occurs at the membranes of chemical synapses. Excitatory (depolarizing) postsynaptic potentials involve conductance increases for Na^+ and K^+ by channels with different kinetics from those in conducting membranes. Calcium is essential presynaptically, but not postsynaptically. Inhibitory (hyperpolarizing) postsynaptic potentials usually result from increased Cl^- conductance, sometimes from increased K^+ conductance when E_K is more negative than the resting potential. Postsynaptic membranes, both excitatory and inhibitory, differ in their channel proteins from those of axon or muscle conducting membranes. Evolution of the proteins was presumably coupled with the evolution of chemical transmitters.

A mosaic structure of membranes occurs in cells which have different electrical properties on two surfaces. This is most evident in flat cells, such as epithelia and electroplaques. Across the entire cell, a large asymmetry potential is developed, and one face may have ion transport or postsynaptic properties while the other face may be conductive.

Specializations determining differences in conduction velocities include differences in cell (axon) diameter, membrane time and space constants, presence of insulative sheath, of nodes or other means of saltatory conduction. Another kind of specialization permits electrical conduction from cell to cell by low-resistance nexuses (Ch. 12). The most specialized of nerve and muscle impulses are all-or-none spikes; graded potentials provide for diverse functions—sensory, pacemaker, synaptic.

The gap between the study of biophysical properties of excitable membranes and of animal behavior is wide. Complex behavior could not have evolved without diversity of electrical properties of membranes. The electrical properties of membranes evolved in primitive organisms from gradients of ions between the cell interior and the medium.

Besides the role of electrical properties of cell membranes in animal behavior, many other organismic properties are controlled or influenced by transmembrane potentials. The effects of electrical potentials on secretion, growth, and differentiation are less well understood than ion conduction in nerve and muscle. Transmembrane ionic gradients, established by ion pumps which lead to electrical potentials, provide means for moving ions into and out of cells. Such movements of ions provide cotransport for entry or exit of metabolites (e.g., sugars and amino acids). Sugars and amino acids may be transported into or across cells in association with protons or sodium, which are then actively pumped out. In multicellular organisms where there is electrical coupling (low resistance) between cells, as in filamentous algae and embryos, electrical fields provide for ion flow between cells.

Many cellular functions are correlated with transmembrane potentials. When an egg is fertilized, waves of electrical activity and increase in intracellular Ca^{2+} occur. This leads to intracellular organization, such as aggregation of cortical granules, and prevents multiple fertilization. As embryos develop they become electrically polarized. Electrical fields are present, possibly universally, in growth and regeneration; the molecular role of these fields is unknown. Electrical potentials are correlated with cellular function in glandular secretion. Direct current fields have been found in active glands; pulses of current, some of them rhythmic, are associated with secretion in glands such as in the pancreas. The multiple inward and outward ionic channels exemplify redundancy of essential functions. The primitive ionic gradients and resulting potentials have persisted and have been selected for many biological functions.

REFERENCES

1. Adams, D. J. and P. W. Gage. *J. Physiol.* **289:**115–141, 1979. Ionic current in R15 neuron. *Science* **192:**783–784, 1976.
2. Adams, D. J. et al. *Annu. Rev. Neurosci.* **3:**141–167, 1980. Ionic currents in molluscan neuron.

3. Adams, P. R. et al. *J. Physiol.* **330:**537–572, 1982. *Nature* **296:**746–749, 1982. Voltage-current analysis of neurons in bullfrog sympathetic ganglia.

4. Adams, W. B. and J. B. Levitan. *Proc. Natl. Acad. Sci.* **79:**3877–3880, 1982. Serotonin-induced increase in K conductance in Aplysia R15.

5. Adrian, R. H. and S. H. Bryant. *J. Physiol.* **240:**505–515, 1974. Muscle potentials of myotonic goats.

6. Adrian, R. H. and W. H. Freygang. *J. Physiol.* **163:**61–103, 1962. K and Cl conductance of frog muscle.

7. Agnew, W. S. et al. *Proc. Natl. Acad. Sci.* **75:**2606–2610, 1978. Properties of sodium channels from Electrophorus electroplax.

8. Aidley, D. J. *The Physiology of Excitable Cells,* 2nd ed., Cambridge University Press, New York, 1978.

8a. Baccaglini, P. I. and N. S. Spitze. *J. Physiol.* **271:**93–117, 1977. *Nature* **280:**208–214, 1979. Transition in ionic control in embryonic muscle.

9. Barchi, R. L. and J. B. Weigele. *J. Physiol.* **295:**383–396, 1979. Sodium channel of muscle membrane.

10. Barish, M. E. *J. Physiol.* **342:**309–325, 1983. Ca-dependent Cl$^-$ current in *Xenopus* oocytes.

11. Beeler, G. W. and H. Reuter. *J. Physiol.* **268:**177–210, 1977. Ionic currents in cardiac fibers.

12. Bittar, E. E., Ed. *Membrane Structure and Function,* Vol. 2, Wiley, New York, 1980.

13. Brehm, P. and R. Eckert. *Science* **202:**1203–1206, 1978. Ca activation in paramecium.

14. Brown, H. F. *Physiol. Rev.* **62:**505–528, 1982. Electrophysiology of sinoatrial node.

15. Brown, H. F. et al. *J. Physiol.* **258:**547–577, 579–613, 615–629; **271:**783–816, 1977. Membrane currents of frog sinus venosus and atrium.

16. Brown, H. F. and D. Noble. *Nature* **280:**235–236, 1979. *J. Exper. Biol.* **81:**175–204, 1979. Action of adrenalin on the heart.

16a. Byerly, L. and M. O. Masudo. *J. Physiol.* **288:**263–284, 1979. K-currents of negative spikes in ascaris muscle.

17. Cambraia, J. and T. Hodges in *Membrane Transport in Plants.* Ed. R. W. Spanswick, Elsevier, New York, 1981. ATPase of plasma membrane of oat roots.

18. Catterall, W. A. *Trends Neurosci.* **5:**303–306, 1982. Catterall, W. A. et al. *J. Biol. Chem.* **254:**11379–11387, 1979; **258:**2488–2495, 1983. *J. Gen. Physiol.* **80:**753–768, 1982. *Proc. Natl. Acad. Sci.* **77:**639–643, 1980. Characterization of Na channels in brain and muscle.

19. Cheeseman, J. M. et al. *Plant Physiol.* **64:**842–845, 1979; **65:**1139–1145, 1980. Role of ATPase on cell potentials in corn root.

20. Christoffersen, G. R. J. *Compar. Biochem. Physiol.* **46A:**371–389, 1973. Cl conductance in neurons of Helix.

21. Connor, J. A. *Biophys. J.* **18:**81–102, 1977. *Fed. Proc.* **37:**2139–2145, 1978. *J. Physiol.* **286:**41–60, 1979. Ch. 6, pp. 187–217 in *Cellular Pacemakers,* Ed. D.

O. Carpenter, Academic Press, New York, 1982. *Annu. Rev. Physiol.* **47**:17–28, 1985. Analysis of ionic currents in molluscan neurons.

22. Connor, J. A. and C. F. Stevens. *J. Physiol.* **213**:31–53, 1971. Current changes in repetitive neurons of molluscs.

23. Daniels, C. J. et al. *Proc. Natl. Acad. Sci.* **78**:5396–5400, 1981. Membrane potentials of E. coli.

24. Davis, R. F. *Plant Physiol.* **67**:825–835, 1981. Potentials across membranes of alga Valonia.

25. Deitmer, J. W. *J. Exper. Biol.* **96**:239–249, 1982. pp. 5163 in *The Physiology of Excitable Cells,* Eds. A. N. Grinnell and W. J. Moody, A. Liss, New York, 1983. Calcium and potassium currents in ciliate Stylonychia.

26. dePeyer, J. E. and H. Machemer. *J. Compar. Physiol.* **127**:225–266, 1978. *J. Exper. Biol.* **88**:73–89, 1980. Action potentials of ciliate Stylonychia.

27. dePeyer, J. E. et al. *Proc. Natl. Acad. Sci.* **79**:4207–4211, 1982. Activation of Ca and K conductance by phosphorylation in Helix neurons.

28. Di Francesco, D. *J. Physiol.* **308**:353–367, 1980. **314**:377–393, 1981; **329**:485–507, 1982. *Pflügers Arch.* **387**:83–90, 1980. Ionic currents in rabbit SA node and Purkinje fibers.

29. Dipolo, R. *Biochem. Biophys. Acta* **298**:279–283, 1973. Ca currenty in barnacle muscle fibers.

30. Doll, S. and R. Hauer. *Planta* **152**:153–158, 1981. Membrane potentials in isolated vacuoles of beet cells.

31. Eckert, R. and P. Brehm. *Annu. Rev. Biophys. Bioeng.* **8**:353–384, 1979. Ionic mechanisms of excitation in Paramecium.

32. Eckert, R. and H. D. Lux. *Brain Res.* **83**:486–489, 1974. *J. Physiol.* **254**:129–151, 1976. Conductance in rhythmic neuron of snail ganglion.

33. Eisenbach, M. et al. *Biophys. J.* **45**:463–467, 1984. *Biochemistry* **12**:6818–68256, 1982. Membrane potentials measured by voltage sensitive dyes in E. coli.

33a. Felle H. and F. W. Bentouk. *J. Membr. Biol.* **27**:153–157, 1976. Effect of light on membrane potentials in liverwort.

34. Fenwick, E. N. et al. *J. Physiol.* **331**:577–597, 1982. Patch clamp of chromaffin cells.

34a. Findlay, G. P. and H. Coleman. *J. Membr. Biol.* **75**:241–251, 1983. Ion channels in algae.

35. Fukuda, J. *Science* **185**:76–77, 1974. Chloride spikes in tissue cultured muscle cells.

35a. Fukushima, Y. *Proc. Natl. Acad. Sci.* **78**:1274–1277, 1981. Patch-clamping of tunicate eggs.

35b. Gage, P. and R. Eisenberg. *J. Gen. Physiol.* **53**:265–278, 1969. Properties of T-tubule membranes.

36. Gainer, H. *Brain Res.* **39**:403–418, 1972. Electrophysiology of rhythmic neurons in a snail.

37. Gallin, E. K. *Science* **214**:458–460, 1981. *J. Cell Biol.* **85**:160–165, 1980; **86**:653–662, 1975; **75**:277–289, 1977. Membrane potentials in macrophages.

38. Garnier, D. et al. *J. Physiol.* **247**:381–396, 1978. Action of ACh on frog atrium.

38a. Gillette, R. *J. Neurosci.* **49**:509–515, 1983. Intracellular pH affects on inward current and bursting in mollusc neuron.

38b. Giles, W. et al. Ch. 3, pp. 91–121 in *Cellular Pacemakers,* Vol. 1, Ed. D. O. Carpenter, Wiley, New York, 1982. Interpretation of cardiac action potentials.

39. Gorman, A. *J. Physiol.* **275**:357–376, 1978. Ca currents in R15 neuron.

40. Gradmann, D. *Planta* **93**:323–353, 1970. with H. Mummert pp. 333–344 in *Membrane Transport in Plants,* Ed. R. W. Spanswick, Elsevier, New York, 1981. Membrane potentials in Acetabularia.

40a. Grinnell, A. D. and W. J. Moody, Eds. *The Physiology of Excitable Cells,* A. Liss, New York, 1983.

41. Grundfest, H. *Annu. New York Acad. Sci.* **94**:405–456, 1961. Current-voltage curves and electrogenesis theory.

42. Hagiwara, S. *Adv. Biophys.* **4**:71–102, 1973. with L. Byerly. *Annu. Rev. Neurosci.* **4**:65–125, 1981. Review of Ca channels.

43. Hagiwara, S. et al. *J. Gen. Physiol.* **50**:583–601, 1967. *J. Physiol.* **190**:479–518, 1967; **238**:109–127, 1974. Ion permeability in muscle membranes of fishes.

44. Hagiwara, S. et al. *J. Gen. Physiol.* **65**:617–644, 1975, **67**:621–638, 1976. *Annu. Rev. Biophys. Biolng.* **8**:385–416, 1979. Voltage clamp on marine eggs.

44a. Hamill, A. P. et al. *Nature* **305**:805–808, 1982. Cl⁻ channel in spinal motor neurons.

45. Harold, F. M. and D. Papineau. *J. Membr. Biol.* **8**:45–62, 1972. Electrogenesis in Streptococcus.

45a. Hartshorne, R. P. et al. *Proc. Natl. Acad. Sci.* **82**:240–244, 1985. Reconstitution of Na channel in planar lipid bilayers.

46. Hermann, A. and A. L. Gorman. *J. Gen. Physiol.* **78**:87–110, 1981. Effects of TEA on I_K in molluscan neuron.

47. Hille, B. *J. Gen. Physiol.* **58**:599–619, 1971; **61**:669–686, 1973; **66**:535–560, 1975; **72**:409–442, 1978. Kinetic analysis of sodium and potassium channels in axons.

47a. Hille, B. *Ionic Channels of Excitable Membranes,* Sinauer, Sunderland, MA, 1984.

47b. Hille, B. in *New Insights into Synaptic Function,* Ed. W. Gall and W. M. Cowan, Wiley, New York, 1985. Evolutionary origins of voltage-gated channels and synaptic transmission.

48. Hodgkin, A. L. et al. *J. Physiol.* **116**:442–448, 1952; **116**:449–506, 1952. Current voltage relations in squid giant axons.

48a. Huxley, A. F. *Annu. New York Acad. Sci.* **81**:221–246, 1959. Ion movements during nerve activity.

49. Irisawa, H. *Physiol. Rev.* **58**:461–498, 1978. with K. Yanagihara. *Pflügers Arch.* **385**:11:19, 1980. Comparative physiology of cardiac pacemaker mechanisms.

49a. Jaffee, L. F. *J. Theoret. Biol.* **48**:11–18, 1974. Interpretation of voltage vs ion relations.

50. Kandel, E. R. *Cellular Basis of Behavior,* W. H. Freeman Co., San Francisco, 1976.

51. Kass, R. S. and R. W. Tsien. *J. Gen. Physiol.* **66:**169–172, 1975. Ca currents in Purkinje fibers.

52. Katz, A. M. *Physiology of the Heart,* Raven Press, New York, 1977.

53. Kerkut, G. A. and R. C. Thomas. *Compar. Biochem. Physiol.* **14:**167–183, 1965. Electrogenic sodium pump in snail neuron.

54. Klein, M. G. *J. Exper. Biol.* **144:**563–579, 581–598, 1985. Ion conductances in muscle of cold and warm acclimated sunfish.

55. Krueger, B. K. et al. *J. Membr. Biol.* **50:**287–310, 1979. *Nature* **303:**172–175, 1983. The binding of toxin to brain membrane.

56. Kung, C. et al. *Science* **188:**898–904, 1975. Membrane properties of mutant paramecium.

56a. Lansman, J. B. pp. 233–246 in *The Physiology of Excitable Cells,* Eds. A. D. Grinnell and W. J. Moody, A. Liss, New York, 1983. Fertilization responses of eggs.

57. Latorre, R. and C. Miller. *J. Membr. Biol.* **71:**11–30, 1983. Conduction and selectivity in potassium channels.

58. Llinas, R. et al. *Proc. Natl. Acad. Sci.* **73:**2918–2922, 1976. Presynaptic Ca current.

58a. Lunevsky, V. Z. et al. *J. Membr. Biol.* **72:**43–58, 1983. Action potentials of algae.

58b. Mackie, G. O. and R. W. Meech. *Nature* **313:**791–793, 1985. Na and Ca spikes in the same axon.

59. Mayer, M. L. and D. L. Westbrook. *J. Physiol.* **340:**19–45, 1983. Voltage clamp analysis of mouse spinal ganglion neuron.

60. McAllister, R. E. et al. *J. Physiol.* **251:**1–59, 1975. Reconstruction of action potential in multiple channels in cardiac muscle fibers.

61. McReynolds, J. S. and A. Gorman. *Science* **183:**658–659, 1974. Ionic basis of hyperpolarizing receptor potential in scallop eye.

62. Meech, R. W. et al. *J. Physiol.* **249:**211–239, 1975; **265:**867–879, 1977. *Annu. Rev. Biophys. Bioeng.* **7:**1–18, 1978. pp. 65–72 in *The Physiology of Excitable Cells,* Eds. A. D. Grinnell and W. J. Moody, A. Liss, New York, 1983. Effects of Ca and pH on voltage current relations in Helix neurons.

63. Merickel, M. and S. B. Kater. *J. Compar. Physiol.* **94:**195–206, 1974. Electrogenic pump in resting potential of molluscan neurons.

64. Miller, C. and M. White. *Proc. Natl. Acad. Sci.* **81:**2772–2775, 1984. Dimeric structure of Cl channels in Torpedo electroplax.

64a. Mirolli, M. *Nature* **292:**251–253, 1981. Nonspiking crustacean receptors.

65. Moody, W. *Annu. Rev. Neurosci.* **7:**257–278, 1984. Effects of protons on electrical properties of cell membranes.

66. Mullins, L. *Am. J. Physiol.* **236:**C103–C110, 1979. *J. Physiol.* **338:**295–319, 1983. Generation of currents in cardiac fibers by Na/Ca exchange.

67. Mummert, H. and D. Gradmann. *Biochem. Biophys. Acta* **443:**443–450, 1976. Voltage-dependent K fluxes in Acetabularia.

67a. Naitoh, Y. and R. Eckert. *J. Exper. Biol.* **59:**53–65, 1973; **56:**667–681, 1972. Sensory mechanisms in Paramecium.

67b. Nawata, T. and Sibaska. *J. Compar. Physiol.* **134:**137–149, 1979. Action potentials and bioluminescence in Noctiluca.

68. Neher, E. *J. Gen. Physiol.* **58:**36–53, 1971. Fast transient current in snail neuron.

69. Noble, D. and R. W. Tsien. *J. Physiol.* **200:**233–254, 1969. Repolarization in heart muscle.

70. Noble, D. *Initiation of Heartbeat,* Clarendon Press, Oxford, 1979.

70a. Noda, M. et al. *Nature* **312:**121–127, 1984. Structure of Electrphorus Na channel from cDNA sequence.

71. Noma, A. et al. *Pflügers Arch.* **388:**1–9, 1980. Epinephrine action on SA node of heart.

72. Nonner, W. et al. *Nature* **284:**360–363, 1980. Effect of pH on gating of sodium channels in frog muscle.

73. Ohmori, H. and S. Hagiwara. *Proc. Natl. Acad. Sci.* **78:**4960–4964, 1981. Single channel currents of anomalous rectification in cultured rat myotubes.

73a. Prosser, C. L. and C. Connor. *Am. J. Physiol.* **226:**1212–1218, 1974. Contribution of Na pump to resting potentials, smooth muscle.

73b. Prosser, C. L. et al. *Am. J. Physiol.* **233:**C19–C24, 1977. Prolonged potentials when Na^+ occupies Ca^{2+} channels.

74. Renaud, J. F. et al. *Proc. Natl. Acad. Sci.* **78:**5348–5354, 1981. Fast sodium channels in the embryonic heart cells.

75. Reuter, H. *Prog. Biophys. Molec. Biol.* **26:**1–43, 1973. *Nature* **301:**569–574, 1983. Calcium channel modulators in heart cells.

76. Robinson, R. B. and W. W. Sleator. *Am. J. Physiol.* **223:**H203–H210, 1977. Effects of Ca and catecholamines on guinea pig atrium.

77. Rona, J. P. et al. *J. Membr. Biol.* **57:**25–35, 1980. Transmembrane potentials in plant (Kalanchoe) cells.

78. Russell, J. M. and M. P. Blaustein. *J. Gen. Physiol.* **63:**144–167, 1974. Ca efflux from barnacle muscle fiber.

79. Saddler, H. D. W. *J. Gen. Physiol.* **55:**802–821, 1970. *J. Exper. Bot.* **21:**345–359, 1970. Potentials in Acetabularia.

80. Saimi, Y. et al. *Proc. Natl. Acad. Sci.* **80:**5112–5116, 1983. Ion channel mutant in Paramecium.

81. Saimi, Y. and C. Kung. *J. Exper. Biol.* **88:**305–325, 1980. Ca-induced Na current in Paramecium.

82. Sakmann, B. and W. Trautwein. *Nature* **303:**250–253, 1983. Muscarinic K channels of pacemaker cells of heart.

83. Satow, Y. and C. Kung. *J. Exper. Biol.* **78:**149–161, 1979; **88:**293–303, 1980. Voltage-sensitive Ca channels transient inward current Paramecium; Ca-induced K-outward current.

84. Satow, Y. and C. Kung. *J. Exper. Biol.* **84:**57–71, 1980. Membrane currents of pawn mutant of Paramecium.

85. Satow, Y., A. Murphy, and C. Kung. *J. Exper. Biol.* **103**:253–264, 1983. Depolarizing mechanoreceptor potentials in Paramecium.

86. Satow, Y. and C. Kung. *J. Exper. Biol.* **65**:51–63, 1976. *J. Neurobiol.* **7**:325–338, 1976. *J. Compar. Physiol.* **119**:99–110, 1977. *Compar. Biochem. Physiol.* **71A**:29–40, 1982. Ionic currents, Paramecium.

87. Sigworth, F. and E. Neher. *Nature* **287**:140–143, 447–449, 1980. Technique for patch planting in nerve axons.

87a. Simons, P. J. *New Phytol.* **87**:11–37, 1981. Electricity in plant movements.

87b. Skaer, H. L. *J. Exper. Biol.* **60**:351–370, 1974. Excitability of muscle in serpulid polychaete.

88. Slayman, C. L. et al. *Biochem. Biophys. Acta* **426**:732–44, 1976. pp. 179–194 in *Membrane Transport in Plants,* Ed. R. W. Spanswick, Elsevier, New York, 1981. Action potentials in Neurospora.

89. Smith, T. A. and R. Raven. *Annu. Rev. Plant Physiol.* **30**:289–311, 1979. Intracellular pH in relation to membrane potentials in plant cells.

90. Smith, T. G. and H. Gainer. *Nature* **253**:450–452, 1975. Bursting pacemakers in molluscan neurons.

91. Spanswick, R. W., Ed. *Membrane Transport in Plants,* Elsevier, New York, 1981.

92. Spanswick, R. M. *Planta* **102**:215–227, 1972. Electrogenesis in plant cells.

93. Spanswick, R. M. et al. *Plant Physiol.* **72**:837–846, 1983. *Annu. Rev. Plant Physiol.* **32**:267–289, 1981. Electrogenic ion pumps in plants.

94. Spitzer, N. C. *Annu. Rev. Neurosci.* **2**:363–397, 1979. Ion channels in developing neurons.

95. Standen, N. B. *J. Physiol.* **249**:241–268, 1975. *Nature* **293**:158–159, 1981. Action potentials in Helix neuron.

96. Stevens, C. F. *Fed. Proc.* **34**:1364–1370, 1975. with R. Tsien, *Nature* **297**:501–504, 1982. Gating of sodium and calcium channels.

97. Takeuchi, A. N. *J. Physiol.* **212**:337–351, 1971. Anion conductance in crayfish muscle.

98. Thomas, R. C. and R. W. Meech. *Nature* **299**:826–828, 1982. Proton currents in voltage clamp snail neurons.

99. Thompson, S. H. *J. Physiol.* **265**:465–488, 1977. Potassium channels in molluscan neurons.

100. Tsien, R. W. et al. *Nature* **297**:498–501, 501–504, 1982. Ionic channels in heart muscle cells. *J. Physiol.* **251**:1–59, 1975.

101. Ulbricht, W. *Annu. Rev. Biophys. Bioeng.* **6**:7–31, 1977. Gating currents in excitable membranes.

102. Wareham, A. C. et al. *Compar. Biochem. Physiol.* **48A**:765–797, 1974; **52A**:295–298, 1975. Muscle potentials of insects.

103. Weigel, R. J. et al. *Am. J. Physiol.* **237**:C247–C256, 1979. Ion currents in circular muscles of intestine.

104. Wendler, S. et al. *J. Membr. Biol.* **72**:75–84, 1983. Effect of turgor pressure on membrane potentials in Acetabularia.

105. Williams, S. E. and R. M. Spanswick. *J. Compar. Physiol.* **102:**211–223, 1976. Propagation in the plant Drosera.

106. Wilson, W. A. and H. Wachtel. *Science* **186:**932–934, 1974. Ch. 7, pp. 219–235 in *Pacemakers,* Ed. D. Carpenter, Wiley, New York, 1982. Pattern bursting neurons of invertebrates.

106a. Wong, R. K. S. et al. *Proc. Natl. Acad. Sci.* **76:**986–990, 1979. Dendritic potentials in hippocampal neurons.

107. Zimmerman, W. and F. Bekers. *Planta* **138:**173–179, 1978. Action potentials of alga Chara.

16

MOTILITY; MUSCLES

Motility is characteristic of living organisms. In the prebiotic period substances moved toward and away from, into and out of macromolecular aggregates by diffusion and by water currents. With increased size and organization of living material, other means than diffusion became essential. Movement of materials from surface to interior, from one region of a cell to another, is by protoplasmic streaming, cyclosis; actin and myosin participate in streaming. In eukaryotes the alignment of chromosomes on spindles and their movement in mitosis and meiosis is precise and is brought about by tubulin. Movements of cell division, sometimes by budding, often by cleavage, use actin and myosin.

Neurons transport compounds that they have manufactured under nuclear influence down axons to nerve endings that may be meters away from the cell body. Axoplasmic components flow at several speeds, and the mechanisms of movement remains largely unknown; neurotubules and neurofilaments probably participate.

Another kind of movement is the growth or extension of regions of a cell. Neurons are prime examples of the extension of cell processes which find their way in a collage of neural and nonneural tissue. The growing tips of nerve processes are motile as are pollen tubes and projections of epithelial cells.

In many invertebrates—sponges, coelenterates, molluscs—water is moved by ciliary action through channels or over gills. Cilia participate in the transport of fluid in most animals (except arthropods and nematodes) and in many plants. Fluid is moved by cilia in the tubes of respiratory, urinary and genital tracts in vertebrates. Cilia move by means of tubulin filaments. Transport of fluids and gases from one region to another of a multicellular organism is done by hydraulic or aerodynamic forces. Sap ascends in plants and trees by root pressure and the suction created by evaporation. Transport of gas in tracheas of insects and respiratory tubes of air-breathing vertebrates is under positive and negative forces. Flow of blood in most animals is due to pressure developed by a pump, a heart.

Plants change orientation to the sun by changes in turgor. When the sensitive plant Mimosa is stimulated mechanically, water shifts out of pulvinar cells and the petiole droops. Some unicells and metazoan embryos move by amoeboid or ciliary action.

ANIMAL LOCOMOTION

Kinomatic observation together with hydrodynamic and aerodynamic measurements provide information about the forces developed by muscle systems in various types of animal movements. Patterns of locomotion in water are: peristaltic crawling and swimming by polychaetes; undulating waves by leeches, some polychaetes, and fishes such as eels; jet propulsion by squids; fin and body movements by fishes; swimming strokes of appendages in aquatic amphibians, turtles, mammals. Locomotion on land may be performed by peristaltic crawling (earthworms), waves of leg movement (myriapods), alternation of leg movement (hexapod and decapod arthropods), walking and running (quadrupedal and bipedal reptiles, birds and mammals). Flying is by wingbeat and soaring. Examples of shifts between slow and fast locomotion are: in insects, changes in leg simultaneity; in fast fishes, shifts from rhythmic opercular breathing to jet ventilation with mouth open; in running mammals, changes of gait; in fishes and mammals, shifts from primarily aerobic muscles to primarily glycolytic muscles. A human changes from walk to run at 2.4 m/sec; a horse changes gait from walk to trot to gallop; the metabolic cost at each gait is minimal for the speed that is characteristic of the gait (62). Mammals differ in strategies for seeking prey; some fast runners chase prey, other kinds sit and pounce. Work done in locomotion is measured by the increase in metabolism above the metabolism at rest. Work is done to overcome frictional drag, to maintain buoyancy in water and body elevation in air, to move appendages.

The energetic cost of locomotion is given in calories (or other units) per body weight for a unit distance moved. Cost of locomotion is given by the metabolic rate per velocity: cost of transportation equals

$$\frac{\text{cal/g/hr}}{\text{km/hr}} = \text{cal/g/km}$$

The cost of locomotion decreases as speed increases, is minimum for some given speed, and cost increases at speeds above and below that minimum. Data for minimum energy cost for movement for swimming, flying, and running are given in Figure 6-5b (124). In general, for comparable body weights, swimmers are more efficient than flyers and flyers more efficient than runners. Each liter of oxygen used is equivalent to 20.1 kJ of energy or 4.8 kcal (at an R.Q. of 0.73). Metabolic cost (M) in calories per gram (W) per kilometer traveled is for mammals $M = 8.46\ W^{-0.40}$; for flying birds $M = W^{-0.227}$ (124). The minimal energetic cost for a flying mammal is less by 20-fold than for a walking or running mammal. Bipedal locomotion is only half as efficient as quadrupedal, as shown by comparison of a rhea or a man with a quadrupedal mammal, such as a carnivore. The cost for running in a biped (man, bird) is $M = 11.5 \ W^{-0.24}$, for a quadruped $M = 10.7 \ W^{-0.4}$, where W is mass and M is ml O_2/g/km (141). The increased metabolism given by metabolic power for running over standard metabolism is 5–9 times for the range of mass from a 0.009 kg pigmy mouse to a 500 kg horse. The ability to utilize available energy in running increases with training; this is partly the result of improved neural integration but mainly an increase in energy liberation resulting from a greater number of mitochondria per muscle cell. A 70 kg man at a brisk walk covers 1.75 m/sec, at a fast jog 3.5 m/sec. The speed of locomotion by a sloth is 0.25–0.3 km/hr. The sloth muscles contract at one-third the rate of comparable muscles of cat, and relax at one-fifth the rate of cat muscle; this is true even when correction is made for the slightly lower body temeprature of the sloth. The actin-activated myosin ATPase of sloth is only one-sixth as active as that of cat muscle (46).

Birds are more efficient in flying than mammals are in running; on long flights birds use body fat as fuel at about the same efficiency as high octane gasoline is used in an airplane (148). Some birds—hummingbirds and golden plover on very long migratory flights—have marginal energy reserves and must rely on assists by the wind. Many birds, hawks in soaring flight, for example, make use of air currents. The minimal cost of flight per unit weight calculates out to the same curve for insects and birds (Fig. 6-5a).

Speed of flight in insects ranges from 2 m/sec for houseflies to 2.5–4.0 m/sec in honeybees and 15 m/sec in sphingid moths. In insects the energy required for hovering is much greater than for flight and is least for walking; O_2 consumption in flight is 50–100 times as great as at rest. The power required is given by drag, times the cube of velocity (53). Sphingids, some butterflies, and locusts, which migrate like birds, can travel several hundred miles without a rest; and these insects, like birds, burn body fat. The aerodynamic power (i.e., thrust) in a locust is positive and large for downstroke of the wing, negative for upstroke, because on upstroke the wing is actively lifted by airflow (158). In insects aerodynamic power is low, less than 1 W/N at 3.5–4.5 m/sec flight speed; O_2 consumption in flight is 50–100 times above resting metabolism. Maximum power for sphingid moths is 27–70, for noctuid moths 4–63 W/N (53). The metabolic rate in O_2 consumed per kilogram of muscle for maintained flight is for Calliphora and for Apis 1300–2200, Drosophila 600–700, Schistocerca 400–800. Drosophila increases total O_2 consumption in flight over that at rest some 22 times, Apis 52 times, blowfly 30–50 times. These increases are much greater than those for a hummingbird, which increases its metabolism in flight over that at rest by 5.5 times. A Drosophila uses mainly sugar, a locust uses proportionately more fat. Insects have postural reflexes of orientation in flight. They steer mainly by the unequal beat of the wings on the two sides and by rotation of wings; each wing can change inclination independently. Diptera have the forewings modified as halteres, balancers; the halteres can change direction and amplitude of vibration to maintain position of the insect in rolling and yawing (158).

Fishes occur in various body forms; all fish which are active swimmers are streamlined and lose little energy by water friction. Drag results from frictional and inertial forces. A thin layer of water moves with the fish, and pressure drag is minimum when this boundary layer adheres to the body. Swimming is energetically very efficient. As with runners and fliers, energy expenditure in calories per gram per kilometer decreases with increasing mass and increases exponentially with increasing velocity (Figs. 6-5b, 16-1) (14a). For swimming a unit distance by a spermatozoan, an extrapolation of energy cost falls on the same curve as for a fish (124). Many fish are positively buoyant because of the swim bladder; other fish, such as scombrids (tuna, mackerel), are negatively buoyant and must swim constantly to maintain position. Speed increases as number of tail beats per second increases. Prolonged swimming increases to a maximum rate as temperature is raised (Fig. 7-3). The temperature for maximum velocity in fish is higher after acclimation at elevated temperature (14a). At maximum swimming rate, oxygen consumption may increase above the consumption at rest by 15-fold. For each increase in swimming speed of one body length per second, there is twofold to threefold increase in metabolism.

Some fishes show the behavioral adaptation of swimming in schools; the movement of the water mass of a school requires less individual effort than movement of an isolated fish. Many fish have physiological adaptations for increased speed, such as shift from rhythmic to ram jet (continuous) ventilation. In most fishes, at high speeds aerobic metabolism decreases and anaerobic metabolism increases. Most fish have both white (phasic) and red (tonic) muscles and some intermediate fibers. Red muscle is highly oxidative and is used in steady swimming; white muscle is glycolytic and is used for bursts of fast swimming. A carp uses red muscles in swimming at speeds of less than 0.5 body length per second; the carp recruits intermediate-type fibers at 1.1–1.5 body lengths and recruits white muscle at speeds greater than 2.0–2.5 lengths per second. The activity of energy-yielding myosin ATPase is in a ratio for red : pink : white muscles of 1 : 2 : 4 (3,68). Five histological types of muscle were observed in dogfish, three in carp, and two in brook trout. Maximum velocities of shortening in muscle lengths per second are: dogfish slow muscle 0.67, fast muscle 2.34 (67). When trout were made to swim at high swimming speed for 21 days, the red muscle hypertrophied and showed increased fat metabolism and glycolytic enzyme activity (68). The energy-yielding enzyme patterns of tonic and phasic muscles are different, and the myosins in the red fibers are different from those in the white fibers. Recruitment of white muscle occurs at a slower swimming speed in carp at low temperature than in warm water (120).

The metabolic cost of isometric contractions of isolated frog muscles increases with temperature with a Q_{10} of 3. The cost of running by a lizard is independent of temperature. Extrapolation between *in vivo* and *in vitro* ob-

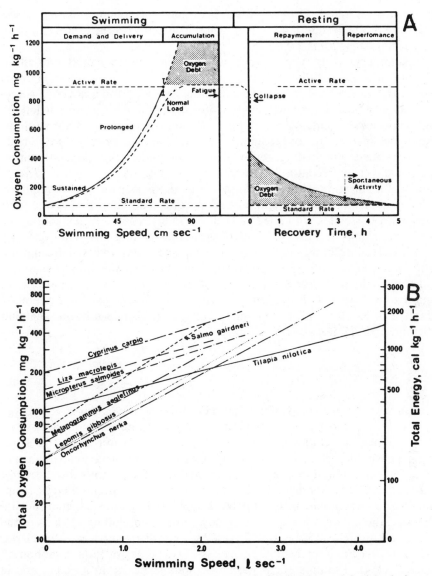

Figure 16-1. (*a*) O₂ consumption in relation to speed of swimming and O₂ debt during recovery in sockeye salmon. Reprinted with permission from J. R. Brett, *Journal of the Fisheries Research Board of Canada,* Vol. 21, p. 1221. Copyright 1964, Fisheries Research Board of Canada, Ottawa, Ontario, Canada. (*b*) O₂ consumption, energy utilization, and swimming speed in several species of fishes. Reprinted with permission from F. W. H. Beamish in *Fish Physioloy,* Vol. 7, *Locomotion.* W. S. Hoar and D. J. Randall, Eds. Copyright 1978, Academic Press, New York.

servations is limited. Except under maximum work conditions only a few fibers are active at one time *in vivo* whereas usually all fibers are activated in experiments (120) *in vitro*.

In summary, many types of locomotion of animals are adaptive to different environments and life styles. Few biological functions show such diversity at the organismic level. Specializations provide for amoeboid movement, ciliary beat, burrowing and crawling, swimming by jet propulsion, by body muscles, by appendages; flying, walking, running, and climbing. Morphological adaptations for locomotion are multiple, and provide for the body transport needed in feeding, reproducing, schooling, territory maintenance, and escape from predators. At organ and tissue levels, there are diverse modes of respiration, shifts from one fuel to another, from aerobic to anaerobic metabolism, and from one type of myosin to another. Speed of animal movement correlates with speed of nerve conduction and reflex action; the most important correlations are with properties of muscles. The following sections detail differences in histology of muscle fibers, in mechanical properties such as elasticity, viscosity, resting and active tension-length relations, and force-velocity curves; other cellular correlations are with patterns of neural activation and of myogenic activity. At the molecular level, there are adaptive differences in myosins and actins, in modulation by intracellular calcium, and in energy-yielding enzymes.

NONMUSCLE MOVEMENT

Several kinds of proteins which function in cellular movement have evolved separately. The most widely occurring are tubulins and actins-myosins. The rotational movement of bacteria is effected by proteins of the flagellar basal body. Spasmoneme is a contractile protein of ciliate protozoans; spasmoneme filaments function in cellular contractions of such animals as *Stentor* and *Vorticella*. Shortening of spasmoneme filaments is not initiated by membrane action potentials or by ATP, but is activated by increase in intracellular Ca^{2+}; the energy for contraction may come from binding of Ca^{2+} to the contractile protein (158a). Tubulins are widely distributed as intracellular microtubules. Each microtubule consists of two types of tubulin monomer, α and β (134). Tubulins occur in mitotic spindle fibers, in all cilia, and in neurons. Cilia have a circumferential ring of nine doublets of microtubules plus a central pair; the doublets are bound together by a protein, nexin; the central pair is surrounded by a sheath and connected by spokes to the outer tubules. Projecting from one of the paired tubules to the other are paired arms of dynein, which has ATPase action. Electron micrographs show that during the active phase of a ciliary stroke, the tubules of each doublet slide past each other (19). Brain tubulin occurs as single scattered microtubules. Dimers of brain tubulin are formed from α and β monomers (134). Activation of rhythmic ciliary movement requires ATP and is potentiated by Ca^{2+} at concentrations greater than $10^{-7}\,M$.

MUSCLE

Few biological systems show so much diversity of structure and biochemistry as muscles. No direct evolutionary trends are recognizable; rather there is considerable diversity, and structures such as cross-striations have evolved numerous times. Actins are widely distributed in nonmuscle cells; cytoplasmic actins differ from muscle actins. All muscles use actin and myosin for contraction, and many cellular and molecular properties are adaptive to different kinds of movements. Universal properties of muscles are: (1) myosin occurs in thick filaments and actin in thin ones, (2) contraction is by sliding of thick over thin filaments (Fig. 16-2a), (3) regulation of contraction is by increase in intracellular calcium ion concentration. No single classification of the properties of muscle is adequate, and for each category intermediate types are recognized. A single muscle may contain several fiber types, and characterization is better done with single fibers than with

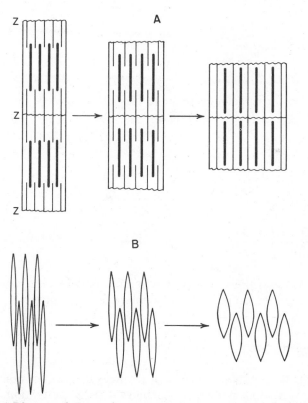

Figure 16-2. (a) Diagram of contraction sequence in cross-striated muscle. (b) Diagram of contraction of a vertebrate smooth muscle fiber. Reprinted with permission from C. L. Prosser in *Basic Biology of Muscles: A Comparative Approach,* B. M. Twarog, R. J. C. Levine, M. M. Dewey, Eds. Copyright 1982, Raven Press, New York.

Table 16-1.   Speed of Muscle Contraction and Relaxation (from 113a and 62)

Muscle	Contraction Time	Relaxation Time
Indirect flight and stridulation muscles of insects	2–4 msec	
Swim bladder (toadfish)	4–5 msec (20°C)	5–6 msec
Internal rectus (cat)	7.5–10 msec (36°C)	
Soleus (cat)	70–80 msec (36°C)	80 msec
Posterior lateral dorsi (chicken)	50 msec (25°C)	200 msec
Anterior lateral dorsi (chicken)	300–400 msec	0.75–1 sec
Sartorius (frog)	40 msec (10°C)	50 msec
Molluscan fast adductor (*Pecten*)	45 msec (15°C)	40–100 msec
Proboscis restractor (*Golfingia*)		
Twitch	85 msec (15°C)	95 msec
Tonus		2–5 sec
Spindle muscle (*Golfingia*)	1–1.5 sec	6.5–9 sec
Byssus retractor (*Mytilus*)		
Phasic	2 sec	1–7 sec
Tonic	2–3 sec	5–1,000 min
Holothurian lantern retractor (*Thyone*)	4 sec	5–7 sec
Holothurian body wall retractor (*Thyone*)	0.5–1 sec	2–8 sec
Mammalian intestine	2–7 sec (37°C)	5–10 sec
Turtle intestine	30 sec	360 sec
Coelenterate		
(*Aurelia*) phasic	0.5–1 sec	0.6–1 sec
(*Metridium*) tonic	30 sec	1–2 min
Sponge (*Microciona*)	1–3 sec	10–60 sec

whole muscles (62). Fiber types are characterized by size, myofilament organization, innervation patterns, and energetics (glycolytic or oxidative).

Speed of Contraction

One classification of muscles is by differences in speed of contraction, rate of relaxation, holding power, fatiguability, neuron activation, and autonomous rhythmicity (Table 16-1). Muscles that are attached to skeletal elements function mainly by development of tension, predominantly isometric contractions; some muscles are attached either to other muscle fibers or to epithelia, and these contract mainly by shortening, predominantly by isotonic contractions. Speed of muscular movement is measured by rate of development of tension or of shortening, time for relaxation, frequency of stimuli for mechanical fusion, facilitating or nonfacilitating contractions, velocity of shortening under different loads (force-velocity curves), passive and active tension-length relations, and redevelopment of tension after release at different times during a contraction.

Speed of contraction depends on the amount of series and parallel viscoelastic elements and on fiber arrangement and sarcomere length. Speed of contraction is related to intrinsic properties of contractile proteins, patterns of energy mobilization, rate of calcium release from intracellular stores, and kinetic activity of actin-activated myosin ATPase, which releases energy from ATP. Speed is also related to whether a muscle is activated by muscle action potentials or by junction potentials, or to whether a muscle shows endogenous activity. Isometric contractions move appendages; isotonic contractions compress fluid-filled organs or body cavities. Fast contractions permit rapid behavior: attack, escape, quick locomotion, production of sound; slow movements are adapted for visceral functions and for locomotion of hollow-bodied animals. Table 16-1 gives representative speeds of contraction within the physiological range. The fastest muscles function in sound production—bat cricothyroid muscles contract at 200/sec, sonic muscles of katydid, cicada, and toadfish contract and relax in 1.0–1.5 msec. A hummingbird hovering above a flower beats its wings at 33–45 times per second. The fastest mammalian muscles are extrinsic ocular muscles, which contract in 5–6 msec with shortening speeds of 65 μm/sec. The fastest movements of muscles are those of the wings of some flies and bees, in which beats as frequent as 500 per second have been recorded; however, these are not full activation and relaxation cycles but mechanical resonance.

Most muscles which move limbs of vertebrates and arthropods have contraction times in the range of 40–100 msec, a little faster at high temperatures than at low. Some nonstriated postural muscles of invertebrates contract nearly as fast as striated muscles of fishes and amphibians. Very slow contractions of swaying sea anemones take minutes. Molluscan adductors hold tension for minutes, even for hours. In general, visceral muscles—gastrointestinal, urogenital, and vascular—are slow, with contraction times in seconds. Cardiac muscles are intermediate in speed. No single explanation can account for the full range of speeds. Rather, there are multiple mechanisms.

Wide-Fibered and Narrow-Fibered Muscles

The classification of muscles as striated or nonstriated (smooth) is based on gross morphology; classification as wide-fibered and narrow-fibered is more functional in relating membrane events to protein contractions. A functional

Table 16-2.   Classification of Narrow-Fibered and Wide-Fibered Muscles

Narrow-Fibered Muscles	Wide-Fibered Muscles
Random thick and thin filaments or striations present transversely or diagonally	Transverse alignment of thick and thin filaments
Fiber diameters 2–10 μm in cylindrical (spindle) fibers or from top to bottom in ribbon fibers	Fiber diameters 50–2000 μm
Uninucleate	Usually multinucleate
Surface:volume ratio high	Surface:volume ratio low
No T-tubules, sparse SR; membrane caveolae in vertebrates	Membrane invaginations, large or as T-tubules; extensive SR
Direct membrane coupling to proteins	Indirect coupling to proteins
Ca^{2+} action potentials predominant, sometimes Ca^{2+} plus Na^+	Na^+ action potentials or junction potentials, rarely Ca^{2+}
Dense bodies or plaques for thin filament attchment, narrow Z-lines in the diagonally striated	Z-bands as transverse discs
Ca control by phosphorylation of myosin by Ca-dependent myosin kinase. Ca-sensitive light chain or some myosins	Tropomyosin-troponin modulation of Ca control of actin-myosin interaction. Some myosin phosphorylation of uncertain function

Activation:
   Nerve activated
   Myogenically rhythmic
	All nerve activated

Electronic coupling by nexuses in myogenically rhythmic, not in neurally activated	No electronic coupling
Series, parallel	Parallel

classification which relates fiber structure, particularly fiber diameter, to coupling between membrane events and the activation of contractile filaments is given in Table 16-2 (114). Wide-fibered muscles usually have transverse alignment of thick and thin filaments—that is, they are cross-striated (Fig. 16-2a, 16-3). Fibers of these muscles have diameters 50–200 μm and are multinucleate; they have invaginating transverse (T) tubules which conduct excitatory signals inward from cell membranes to the Ca^{2+}-releasing sarcoplasmic reticulum (SR) and thus activate contractile filaments indirectly. Narrow-fibered muscles are 2–10 μm from cell surface to contractile filaments, are spindle or ribbon-shaped, usually have a single central nucleus, have no T-tubules; coupling between cell membranes and contractile

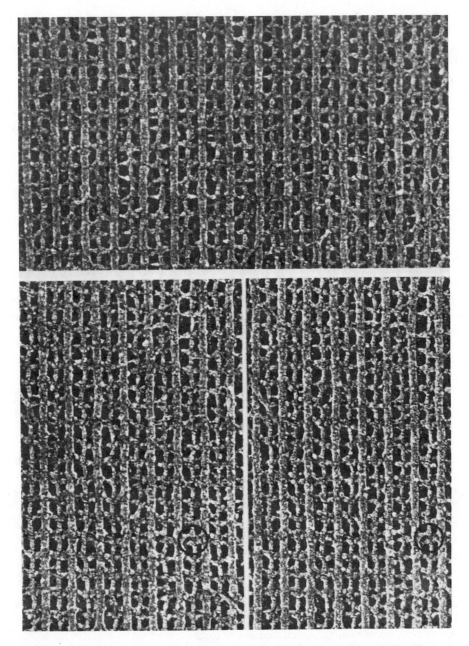

Figure 16-3. Electron micrographs of insect muscle to visualize thick and thin filaments and bridges. Myosin-actin interactions visualized by quick-freeze and deep-etch technique. From J. E. Heuser and R. Cooke, with permission from *Journal of Molecular Biology*, Vol. 169. Copyright 1983 by Academic Press, Inc. (London).

filaments is direct, either by influx of Ca^{2+} or by its release from intracellular stores (Fig. 16-2b).

Wide-fibered muscles are usually activated by Na^+ action potentials or by endplate potentials; narrow-fibered muscles have Ca^{2+} or a combination of Ca^{2+} and Na^+ action potentials. All wide-fibered muscles are activated by motor nerve impulses; many narrow-fibered muscles are nerve-activated, and some are endogenously rhythmic with neural modulation. In wide-fibered muscles tension is generated by fibers in parallel, in narrow-fibered muscles by fibers in series as well as in parallel. Some narrow-fibered muscles have ribbon-shaped fibers, relatively wide but with distance less than 5 μm from top or bottom cell membrane to contractile filaments. In some narrow-fibered muscles, the myofilaments are arranged in a spiral, hence the fibers are said to have diagonal striations (Fig. 16-4). In some muscles of this category the central region of a fiber contains nuclei, mitochondria, and other organelles, while a thin cortex contains myofilaments (e.g., ctenophore *Beroe*) (55) (Fig. 16-5). The ctenophore fiber diameter is 15–20 μm. Action potentials are due to both Ca^{2+} and Na^+ currents, repolarization is by a TEA-sensitive K^+ current (5a). In some muscles and myoepithelial cells the contractile filaments are along one side of each fiber. Transverse striations of thick and thin filaments have evolved many times, and neither nonstriated nor cross-striated muscles can be considered primitive.

Innervation; Energetics

Several classifications of muscles relate speeds of contraction to patterns of innervation, to relative activities of oxidative and glycolytic enzymes, and to concentrations of actin-activated myosin ATPase (Table 16-4). Many postural muscles of vertebrates have focal (uni- or bi-terminal innervation) with a single motor axon innervating a group of muscle fibers, thus forming a motor unit. Gradation of movement is by varying the number of motor units activated in the spinal cord and the frequency of firing of single units. Endplate potentials trigger muscle action potentials that are conducted in muscle membranes. Applied transmitter (acetylcholine) is effective only in the region of a motor endplate.

Contractions may be fast (phasic) or slow (tonic). Force-velocity curves of phasic muscles show higher speeds of shortening at small loads than do such curves of tonic muscles. Activity of the actin-activated ATPase of phasic muscles is higher than that of tonic muscles. Examples of phasic white muscles are rat extensor digitorum longus, frog sartorius, chicken posterior latissimus dorsi; examples of tonic white muscles are rat soleus and some fish "pink" muscles. Resting potentials of fast white muscles are large, often −90 mV; ratio of resting permeabilities P_{Na}/P_K is low, 0.01 to 0.04; intracellular potassium concentration is high as it is in all muscles.

Red muscles contain myoglobin (muscle Hb) and, in general, red muscles

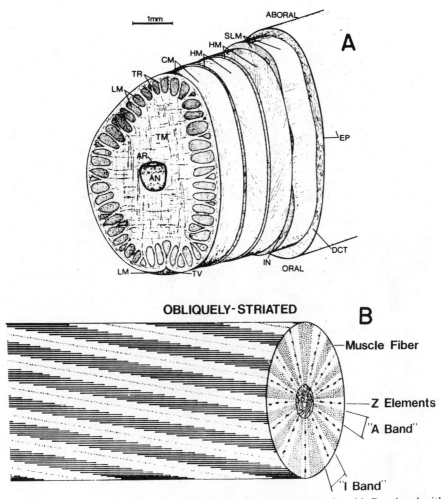

Figure 16-4.   (*a*) Functional arrangement of muscles in a tentacle of squid. Reprinted with permission from W. M. Kier, *Journal of Morphology,* Vol. 172, p. 182. Copyright 1982, Alan R. Liss, Inc., New York. (*b*) Model of diagonal striations of squid muscle. By personal communication from W. M. Kier.

fatigue less rapidly (i.e., maintain contractions longer, are slower in contraction than fast white muscles). Red muscles have high activity of oxidative enzymes, whereas white muscles are more glycolytic. Examples of red muscles are some fibers in rectus abdominus in frog, leg muscles of chicken, red muscle bands along lateral midline of fishes, and fin muscles of fishes.

One classification of mammalian muscles is: (1) white fast muscles which are relatively low in oxidative enzymes, high in glycolytic enzymes, high in myosin ATPase (e.g., the superficial fibers of quadriceps); (2) red muscles which are low in glycolytic, high in oxidative enzymes, and have low myosin

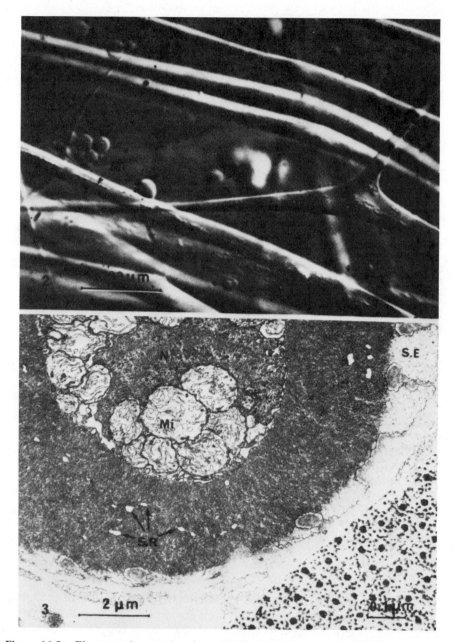

Figure 16-5. Electron micrographs of muscle fibers of ribbon-shaped diagonally striated muscle of ctenophore Beroe. Reproduced from Hernandez-Nicaise, the *Journal of General Physiology*, 1980, Vol. 75, p. 84 by copyright permission of the Rockefeller University Press.

Table 16-3.   Enzyme Activity of Quadriceps Red and White Fibers in Sedentary Persons (s) and Runners (r) (10)[a]

	White	Red
Cytochrome oxidase	167	840
(s)	339	2041
(r)		
Citrate synthase	10.3	35.5
(s)	18.5	69.6
(r)		
Carnitine palmityl-transferase	0.11	0.72
(s)	0.20	1.20
(r)		
Cytochrome c	3.2	16.5
(s)	6.3	28.4
(r)		

[a]Units of enzyme activities.

ATPase activity (e.g., deep quadriceps fibers) (Table 16-3); and (3) muscles of a third type which are intermediate in oxidative activity but low in myosin ATPase (e.g., soleus). Amphibian phasic muscles have been classified in five types (128a); three give fast contractions and two give slow contractions. ATPase is high in the three fast types, and staining for succinic dehydrogenase varies from very dark (type 3) to pale (type 1). Fiber diameters are small in the two slow types, glycogen and lipid stores are large, sarcoplasmic reticulum is slight to absent from the slow and extensive in the fast muscles (128a). Muscles of active fish (trout) have more mitochondria than those of sluggish fish (plaice) and fast muscles are more glycolytic than slow muscles (Table 16-4).

Patterns of innervation and of metabolism as shown histochemically in muscle of fishes are of several types and no evolutionary scheme can yet be given (Table 16-2) (62,67). Most fishes have two or three types of skeletal muscles, each with a different innervation pattern. Twitch fibers similar to the slow white fibers of mammals have conducted action potentials; this type occurs in some fish (e.g., salmonids). White fibers in elasmobranchs, urodeles, and some teleosts are innervated focally by two axons to the same endplate; the vesicles of the two axons differ in size (17). In dogfish *Scylliorhinus*, fibers of an outer layer of white muscle are of fast twitch type, and the inner white layer shows less SDH than the outer (17). In addition, *Scylliorhinus* has red fibers with distributed innervation showing only junction-type potentials. Cyprinid fish have multiple innervation (67). In an eel a fast muscle fiber has a single motor axon; a sculpin muscle fiber receives two to five axons. *Tilapia* has both slow facilitating muscle fibers and fast

Table 16-4.   Activities of Enzymes from Slow (S) and Fast (F) Muscle Fibers of Fish (68)[a]

	Brook Trout		Crucian Carp		Plaice	
	F	S	F	S	F	S
Hexokinase	0.6	0.3	0.7	2.0	0.06	0.4
Phosphofructokinase	14.0	11.4	4.2	1.9	29.0	17.1
Lactate dehydrogenase	345	200	237	440	242	174
Citrate synthase	0.7	4.9	0.9	9.2	0.3	7.1
Cytochrome oxidase	0.4	2.3	0.8	5.6	0.2	3.9
Mitochondrial volume (%)	9.3	31	4.6	25.5	2.0	24.6

[a]Activities in μM substrate/g_{ww}/min.

high-frequency fibers (69). White fibers of marlin develop three times the force and contract twice as fast as red fibers (68).

A general classification of two types of muscles in fishes and frogs follows (17). See also Table 16-4 (68).

	Fast	Slow
Diameter of fibers	large	small
Myoglobin	none	myoglobin usually present
Mitochondria	few	large, abundant
Enzymes (by histochemistry)	glycolytic	oxidative
ATPase	large amounts	small amounts
Storage particles	little lipid	much lipid and glycogen
Sarcoplasmic reticulum	abundant	scarce
Innervation	focal or distributed	distributed
Resting potentials	highly negative	low
Action potentials	usually propagated spikes	usually junction potentials only
Contraction time	brief	prolonged

Prolonged exercise (training) in humans and rats leads to increase in ability to oxidize pyruvate and palmitate, and to increase in cytochrome oxidase and citrate synthase in all three types of muscles. Cytochrome c doubles in the superficial white, the deep red fibers of quadriceps, and intermediate fibers of soleus (Table 16-3) (10).

Table 16-5.   Classification by Innervation Pattern and Energy Pathway

Cross striated, wide-fibered, long, unbranched fibers. Skeletal muscles.
   Uniterminal, unineuronal; muscle action potentials.
      Fast twitch, white, glycolytic, fast myosin (frog sartorius, chick PLD, rat
        EDL).
      Slow twitch, white, glycolytic, slow myosin (rat soleus, some fish muscles)
Multiterminal, unineuronal.
   White, tonic, electrical responses are graded psps.
   Red, myoglobin, indirect, slow to fatigue.
   Spiking or large psps, usually nonfacilitating
   Nonspiking, facilitating
Multiterminal, polyneuronal, mostly white or pink.
   Conduction by nerves (psp responses or spikes) (teleost white, neurogenically
      driven arthropod hearts).
   Multiple types of innervation, excitatory and inhibitory (crustacean postural,
      insect asynchronous).
   Oscillatory muscles (insect synchronous).
Endogenously rhythmic or driven by nonneural pacemakers.
   Cardiac muscles, fibers often branched, nexal connections.
Narrow-fibered, nerve activated, postural.
   Multiunitary (Sipunculid).
   Diagonally striated, shearing (many annelids, nematodes).
   Random filaments, either dual or unitary innervation.
   Large filaments, dual innervation excitatory and relaxing (paramyosin containing
      catch muscles of molluscs).
Narrow-fibered, endogenously rhythmic, nerve modulated, random filaments,
      (vertebrate unitary smooth muscle).
Myoepithelia, nerve-activated, nexal connections.
   Cross-striated or nonstriated (coelenterates).

In the prochordate amphioxus (*Branchiostoma*) muscle fibers are 1–2 μm thick, 10–30 μm wide, and 600 μm long (Table 16-2). Thick and thin filaments are aligned, there is little sarcoplasmic reticulum, spikes are sensitive to TTX and are due to increase in g_{Na}, but when Na^+ current is blocked there are Ca^{2+} spikes (49).

A second general category of vertebrate striated muscles includes those with distributed (multiterminal) innervation—sometimes polyneuronal (Table 16-5). The resting potentials of these fibers in amphibians are low, −50 to −60 mV, while P_{Na}/P_K is high (0.1–0.18) (77). These muscles give graded nonconducting potentials which are distributed endplate potentials. Examples of multiterminal innervations are: most fibers of frog rectus abdominis, chicken anterior latissimus dorsi. Intrafusal muscle fibers (sensory spindles) in frog are either uniterminal or multiterminal; in cat the intrafusal fibers are uniterminal. Fish muscles differ in innervation with kind of fish. The "pink" muscles of many teleosts are of the slow-twitch type. White muscles of "higher" teleosts give fast-twitch contractions which may be activated by either action potentials or by junction potentials according to whether the fish are actively swimming or not.

Muscles which are both multiterminal and multineuronal and which respond differently to several motor axons are found in many arthropods—crustaceans and insects. A single muscle fiber may receive as many as five axons and a single axon may branch to innervate several muscle fibers (Table 16-5). When an axon is stimulated, the electrical and mechanical responses differ in speed according to the junction's requirement for facilitation (8). The electrical responses are graded junction potentials; however, the fibers are capable of all-or-none action potentials when stimulated directly. Fast (nonfacilitating) contractions are strong and of short latency; slow contractions are facilitating, long-latency, low-amplitude responses.

In addition to the multiple excitatory innervation, most crustacean and many insect muscles receive one inhibitory axon. Inhibition may be presynaptic and/or postsynaptic. Presynaptic inhibition reduces the amount of excitatory transmitter liberated, and postsynaptic inhibition hyperpolarizes a muscle fiber and so reduces its response to excitation. Perfusion experiments show that the excitatory transmitter is glutamate, irrespective of which excitatory axon is active, and that the inhibitory transmitter is gamma aminobutyric acid (GABA) (8).

Insects have two types of striated muscles, synchronous, which pull directly on an appendage, such as a wing, and asynchronous or indirect, which pull on a skeletal lever system. Postural muscles, such as leg muscles of locust, and flight muscles of slow-flying insects, such as butterflies, moths, and dragonflies, have distributed endplates much like crustaceans. Usually these muscles receive one excitatory and one inhibitory axon. Some postural muscles of insects have two excitors.

Narrow-fibered postural muscles (mostly of invertebrates) differ considerably in their patterns of innervation and contraction. Narrow-fibered muscles that are activated by motor nerves perform locomotion of hollow-bodied animals, such as annelids, sipunculids, many molluscs and echinoderms. Muscle fibers of some species receive usually a single motor innervation; muscles of other species have two excitatory axons eliciting facilitating and nonfacilitating responses. Inhibitory innervation has not been described in narrow-fibered muscles but may occur. Some narrow-fibered muscles receive both an excitatory and a relaxing nerve fiber, each liberating its own transmitter. The postural nerve-activated muscles of annelids contract over wide ranges of length, and yield changes in shape and size of the animals. Structurally, these muscles are diverse; some have myofilaments arranged in spiral patterns. Very few vertebrate narrow-fibered muscles are nerve-activated; these are called multiunitary. A few muscles have properties of both multiunitary and unitary (endogenously active) types; examples are vas deferens and bladder.

Narrow-fibered muscles without nerve activation function mainly for contraction of visceral organs. Some of these muscles are endogenously rhythmic when no nerves are present (e.g., chick amnion). Other smooth muscles may require the presence of special cells, such as the interstitial cells of mammalian intestine. Cardiac muscle presents a special case; ventricular fibers of vertebrates are cross-striated, wide in diameter, and unique in being branched and coupled together at intercalated discs; cardiac fibers are capable of autorhythmicity, but are normally triggered by pacemakers which are themselves not contractile (sinoatrial node). Frog heart cells of sinus venosus and auricle are typical narrow-fibered nonstriated muscle fibers. All of the muscles that are endogenously active, or driven by nonneural tissue, receive regulatory innervation which can be excitatory, increasing frequency and amplitude of contraction, or inhibitory, reducing frequency and amplitude. Contractions of visceral muscles are slow compared with skeletal muscles. Properties of visceral muscles of invertebrate animals are not well known; many of them—heart-like vessels of annelids, intestinal muscles of crustaceans and insects—are nerve-driven. Possibly the intestines of molluscs and of echinoderms are myogenic. Among invertebrate hearts, molluscan ventricles are endogenously rhythmic, while crustacean hearts (striated) are neurogenic.

ORIGINS AND EVOLUTIONARY DIVERSITY OF MUSCLES

Protoplasmic streaming and amoeboid movement have much in common with muscle contraction. Strands of slime mold *Physarum* show streaming, the direction of which reverses periodically. Internal pressure is produced by contraction of the outer protoplasmic layer, and when a strand is stretched, its tension increases like muscle; when a mechanical load is applied, the amplitude of the tension wave increases. Actin and myosin of *Physarum* are similar to muscle actin and myosin and actomyosin threads have been made from *Physarum*. Ca^{2+} and ATP are required for tension development (71).

Algae such as *Nitella* and *Chara* show cyclotic streaming which rotates the chloroplasts. Cytoplasmic streaming of the internal "endoplasm" is caused by forces produced at the cell's outer surface by contractile processes in the "ectoplasm." Actomyosin in *Acanthamoeba* resembles that in muscle. Cleavage in many dividing cells occurs by molecular mechanisms similar to those of amoeboid movement. One hypothesis is that the actomyosin of muscle had its origin in proteins of cytoplasmic flow and cell division.

Myoepithelia and Related Contractile Cells

Myocytes of sponges may be considered as primitive muscle cells; thick filaments are surrounded by thin ones. Myocytes of sponges are 2–3 μm in diameter, short, spindle-shaped, and uninucleate. They contract in response to mechanical stimulation and require a divalent ion (usually calcium) in the medium (80,113).

Myoepithelia may be precursors of muscle. Some myoepithelial cells have bundles of transversely aligned actin and myosin filaments, others have the two proteins in an apparently random distribution (90).

In the diploblastic coelenterates (Cnidaria) all movement is generated by epitheliomuscular cells. In polyps all contractile myoepithelia, both endo-dermal and ectodermal, are nonstriated; in most hydromedusae circular fibers are striated and longitudinal fibers are nonstriated; tentacles are non-striated. The nonstriated radial muscles perform involution of the margin of a medusa, the striated circular muscles contract for swimming (129). Con-tractions of striated myoepithelial cells are faster than those of nonstriated ones. Sea anemones may show slow rhythmic contractions; avoidance re-sponses to alarm pheromones may take 1.5–3.0 min (90). Coelenterate myoepithelia are activated by signals conducted by nerve nets or in epithe-lium (6).

Endodermal cells of some trematodes, myoepithelia lining the esophagus of *Ascaris,* and myoepithelial cells of the heart of ascidians have myofila-ments along one side of the long cell. In the tubular hearts of the polychaete worm *Arenicola* contractions are in myoepithelia. The microvilli of intestinal epithelia of amphibians, birds, and mammals are myoepithelia with actin filaments oriented toward the terminal web of mucosal cells. Myoepithelia containing actin occur in sweat glands and salivary glands of humans. The proventriculus of a polychaete worm *Sylla* is lined with myoepithelia with aligned thick and thin filaments making up one or two sarcomeres per cell. These large, flattened myoepithelia have membrane invaginations, some sar-coplasmic tubules; the cells generate spikes with the depolarizing phase due to increased g_{Ca} and hyperpolarizing phase due to increased g_{Cl}; contraction of these cells is nerve-activated and opens the lumen of the proventriculus (5). These fibers have properties of both smooth and striated muscle.

A few myoepithelia have true T-tubules and all measure only 1–2 μm from cell surface to contractile filaments. It is relatively easy to picture myoepi-thelia as giving rise to muscles. Striations, transverse alignment of thick and thin filaments, must have evolved early in myoepithelia as adaptations for fast movements.

In summary, myoepithelia probably preceded discrete muscles in evolu-tion. Some myoepithelia show striations, others have nonstriated bundles of actin and myosin filaments. Activation of myoepithelia is either by nerves or by epithelial nexal conduction. Spikes are probably evoked largely by increased g_{Ca}. Striated cells show faster contractions than nonstriated.

Nonstriated, Nerve-Activated Muscles of Invertebrates

In nonstriated (narrow-fibered) locomotor muscles the fibers may be organized in discrete bands or may form outer and inner layers of the body wall
(Table 16-5). The single muscle fibers have maximum distance from membrane to contractile elements of 3–5 μm and transverse tubules are absent.
The postural nonstriated muscles are activated by nerves, with either one or
two axons to a muscle cell, sometimes with two transmitters; hence the muscles may be considered analogous to multiunitary smooth muscles of vertebrates. Contractions are relatively fast (40–60 msec). Acetylcholine is the
transmitter in echinoderms, acetylcholine with serotonin or dopamine in
molluscs. Muscle spikes are predominantly Ca-mediated.

Among echinoderms, the proboscis retractors of holothurians (sea cucumbers) are five long narrow nonstriated muscles innervated from radial
nerves. Each muscle consists of bundles of fibers; cholinergic synaptic endings within the bundles are abundant; there is little or no sarcoplasmic reticulum and no nexal connections between muscle fibers (85,126). In some holothurians, these muscles show spontaneous rhythmic contractions, and in
all species the muscles can be stimulated by brief stretches (116). Rhythmicity and responses to stretch are blocked by nerve blocking agents and by
tubocurarine; the responses to stretch are enhanced by physostigmine. In a
sea urchin *Parechinus* lantern retractors consist of striated and nonstriated
fibers; the muscle gives both phasic and tonic responses; tonic contractions
result from continuous firing of neurons in the hyponeural complex (27).

In the mantle of a squid, contraction of circular fibers causes ejection of
water and contraction of radial fibers causes intake of water. The circular
fibers show phasic and tonic contractions, triggered by separate nerve fibers. Tentacles and arms of squid have peripheral bundles of longitudinal
fibers and central radial and circular fibers; a large nerve runs down the
central core. Circular and radial fiber contraction extends an appendage,
longitudinal fiber contraction retracts the arm or tentacle (76) (Fig. 16-4).

In the gastropod *Aplysia* the smooth muscle fibers of the gill are excited
by each of three identifiable neurons in the pleural ganglion; two are cholinergic, the third probably glutaminergic (20a). Each muscle fiber receives at
least two axons. The radular retractor of the snail *Helix* shows both fast
nonfacilitating and slow facilitating components in an intracellularly recorded action potential. The radula protractors of *Helix* are excited by low
concentrations of acetylcholine, relaxed by serotonin; when both substances
are applied simultaneously, the muscle contracts rhythmically. Innervation
of the penis retractor of *Helix* is probably both cholinergic and dopaminergic; rhythmicity is induced by a ganglion at the base of the muscle (151a).
The radula protractor of snail *Busycon* is red with myoglobin; its nerve endings have two types of synaptic vesicles. The snail *Rabana* radula protractor
shows dual innervation, cholinergic and tryptaminergic (78).

Nonstriated fibers of the foot of the bivalve *Mytilus* are arranged to give

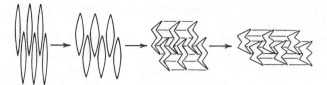

Figure 16-6.   Diagram of contraction of the sipunculid muscle with folding fibers. Reprinted with permission from C. L. Prosser in *Basic Biology of Muscles: A Comparative Approach.* B. M. Twarog, R. J. C. Levine, M. M. Dewey, Eds. Copyright 1982, Raven Press, New York.

either shortening or extension; stimulation of the cerebral ganglion elicits tonic extension, stimulation of the visceral ganglion elicits shortening; dopaminergic nerves cause shortening, tryptaminergic nerves extension.

An unusual morphological arrangement occurs in the proboscis retractors of sipunculid worms. The fibers are doubly innervated and stimulation can elicit both fast, rapidly fatiguing potentials and contractions, and slow facilitating responses. The transmitter for both responses appears to be acetylcholine. The small diameter (4–5 µm) muscle fibers are connected across the entire muscle, so that on contraction the fibers fold like an accordion and birefringent bands cross the muscle (Fig. 16-6). Tension-length curves show two optimal lengths, one when pleats are extended, the other presumably when thick and thin filaments are completely aligned. The muscle can extend or shorten over a 10-fold range of length. In sipunculid worms as in holothurians, contractions are mainly isotonic.

Diagonal striation is an adaptation for isotonic contraction of highly extensible muscles that perform shearing motion. Examples are the body wall (locomotor) muscles of many annelids, longitudinal muscles of the roundworm *Ascaris* (121), and of the prochordate amphioxus. Individual fibers of most muscles that have diagonal striations are oval or ribbon-shaped in cross section, and the distance from upper or lower membrane to contractile filaments is only 3–4 µm. The diagonally striated fibers are innervated, usually acetylcholine is the excitatory transmitter; in some annelids GABA inhibits. When diagonally striated muscles contract, the angle of the filaments with respect to the long axis of the fiber increases from between 5° and 15° when extended, to 45° when shortened. The change in angle results in shortening by shearing (Figs. 16-7a, 16-8). In light microscopy of the ribbon-shaped fibers of earthworm, the myofilaments are seen as running diagonally in one direction near the upper face and in the opposite direction in the lower face. In longitudinal sections of a polychaete muscle, electron micrographs show prominent Z-lines in diagonal rows; the distance between Z-lines of adjacent rows is less in short than in extended fibers. The measured angle is the same for a given fiber length whether the length is reached by contraction or by passive extension and release (Fig. 16-7a). Fiber diameter increases with shortening and fiber volume is maintained constant in contraction. The Z-

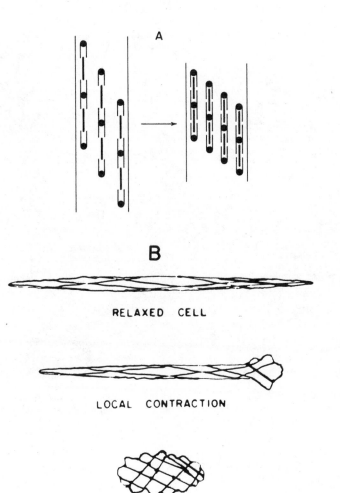

Figure 16-7. (*a*) Diagram of contraction of diagonally striated muscle showing both sliding and shearing. Reprinted with permission from C. L. Prosser in *Basic Biology of Muscles: A Comparative Approach*. B. M. Twarog, R. J. C. Levine, M. M. Dewey, Eds. Copyright 1982, Raven Press, New York. (*b*) Representation of a smooth-muscle cell showing changes in angle of filaments in local and full contraction. Dense structures represent plasma membrane dense bodies. Reprinted with permission from F. S. Fay and O. C. Delise, *Proceedings of the National Academy of Sciences* (USA), Vol. 70, p. 644. Copyright 1973, The National Academy of Sciences, Washington, D.C.

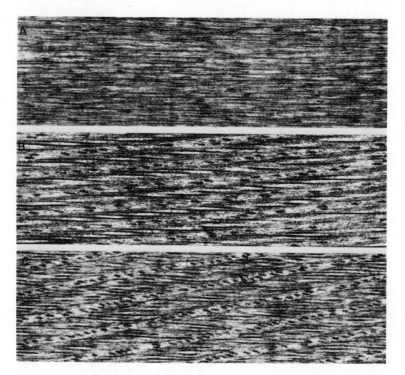

Figure 16-8.   Electron micrographs of longitudinal sections of diagonally striated muscle of a polychaete; fibers shorten in sequence from top to bottom. By personal communication from H. Iwamoto and H. Takahaski (1982).

lines are not discs, as in vertebrate cross-striated muscles, but are narrow bands (114) (Fig. 16-8).

In a muscle fiber with contractile elements in series the generated force (F) equals the force (f) of each single element. When filaments are diagonal, slanted at angle θ to the long axis, the force generated is proportional to $n \times f \times \cos \theta$, where n is number of parallel elements, f is force of each element. Less energy is needed for a given force because the parallel elements change in angle (Fig. 16-7b) (9).

In the earthworm *Lumbricus* ribbon-shaped fibers of the longitudinal layer are 4 μm thick, 45 μm wide; fibers of the circular layer are more oval. Longitudinal fibers show spontaneous miniature IPSPs due to increase in g_{Cl} and activation by GABA; they show some miniature EPSPs due to increase in g_{Ca}; these are blocked by *d*-tubocurarine (dTc), hence probably are activated by ACh (57). Some longitudinal fibers give phasic contractions of 70–80 msec contraction time, other fibers are tonic, 0.8–1.5 sec contraction time, and facilitating. In the earthworm *Pheretima* two populations of fibers are found electrically: (1) those with small resting potentials, spontaneous spikes, and (2) those with large resting potentials, no spontaneity, EPSPs due to cholinergic neural activation. The spikes of the first kind are due to g_{Na}, the EPSPs and spikes of the second kind to g_{Ca} (66).

Some molluscs and worms have "catch" muscle. Tension can be maintained for a very long time—minutes or hours—with little expenditure of energy, and only occasional activating motor impulses. "Catch" contraction occurs in adductor muscles of bivalve molluscs, byssus retractor of the clam *Mytilus,* and body wall of gorgonian worms. Contraction time is short and, once contracted, the muscle can remain in that state. There is dual innervation; one transmitter (usually ACh) elicits contraction, a second transmitter (5HT) elicits relaxation. The mechanism of "catch" is discussed later in this chapter.

Myogenically rhythmic muscles are rare among invertebrate animals. The hearts of bivalve molluscs consist of narrow-fibered nonstriated cells which can contract rhythmically in isolation. However, they receive both excitatory and inhibitory modulating innervation.

Narrow-Fibered Nonstriated Muscles of Vertebrates

Among vertebrates very few nonstriated muscles are multiunitary; examples are nictitating membrane, pilomotor muscles, sphincters. The retractor of the spiral intestine of elasmobranchs is multiunitary. Most smooth muscles of vertebrates are unitary and show myogenic rhythmicity, with neural and hormonal modulation. Contraction of iris muscle in mammals is by a central reflex. In amphibians the iris muscle contracts on illumination; no innervation is required.

Contractions in unitary smooth muscles of vertebrates are usually triggered by endogenous spikes, of which the depolarizing phase is due to increase in g_{Ca} and repolarization is due to increased g_K, largely Ca-activated (Ch. 15). Several ionic mechanisms of rhythmicity occur. One mechanism consists of pacemaker potentials which result from voltage- and time-dependent changes in Ca^{2+} conductance. These may give rise to prolonged spikes, as in amphibian and elasmobranch stomach. A less common generator of rhythmicity is an oscillating Na^+ pump, that can be blocked by ouabain or by K-free medium, and is dependent on oxidative energy (117a). A third type of rhythmicity is atropine-sensitive and results from acetylcholine liberated from plexus neurons, for example in guinea pig intestine. Nervous modulation in vertebrate unitary muscles is usually tonic; it may activate contraction or relaxation, as in sphincters. Varicosities along axons have synaptic vesicles that liberate transmitters which diffuse in extracellular space for distances of many cell diameters. Some transmitters are excitatory, some inhibitory, others have tonic actions on rhythmicity and tension; postinhibitory rebound of excitation is common. Visceral smooth muscles are sensitive to hormones, which may alter rhythmicity, spiking, and contractions, for example, in uterus. The sensitivity to several neurotransmit-

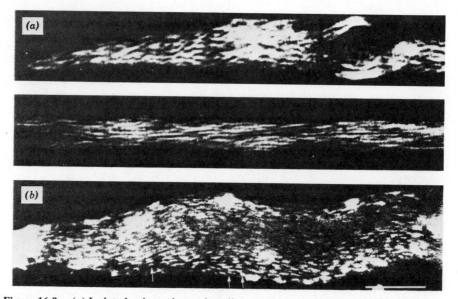

Figure 16-9. (*a*) Isolated guinea pig taenia coli smooth muscle cells viewed with polarizing optics. Above: shortened state. Below: extended cell. In shortened state the myofilaments are subtended at greater angles than in the relaxed state. From J. V. Small, reprinted by permission of *Nature*, Vol. 249, p. 326. Copyright Macmillan Journals Ltd. (*b*) Fluorescence photomicrograph of smooth muscle fiber stained with anti-α-actinin to show dense bodies aligned in rows. Reprinted with permission from F. S. Fay et al., *Journal of Cell Biology*, Vol. 96, p. 788. Copyright 1983, The Rockefeller University Press, New York.

ters and hormones suggests the presence of numerous kinds of receptors in muscle membranes.

Vertebrate smooth muscle cell membranes have pinocytotic vesicles, caveolae, many of them in longitudinal rows.' The function of the vesicles is uncertain; it could be for transfer of material in and out of the fibers (9). The surface/volume ratio in smooth muscle is much greater than in striated muscle, hence active ion pumps are needed to maintain ion gradients.

Muscles in which myofilaments are arranged diagonally contract by filament sliding together with an increase in angle. Vertebrate smooth muscles have rows of dense bodies in the cytoplasm and plaques at the cell membrane which contain α-actinin. Myofilaments attach to these dense bodies and plaques, which are analogous to Z-lines in striated muscle. Photographs by polarized light or of muscles stained with anti-α-actinin show that the spiral pitch of the filaments and their attached dense bodies increases during contraction (Fig. 16-9) (9,36a). Shearing or angle decrease adds to the tension developed by sliding.

Muscle fibers of unitary muscles of vertebrates are connected by nexuses which are regions of low electrical resistance; a strip of visceral muscle is a functional syncytium. Calcium modulation of contraction is by different intermediaries from those in striated muscle. Vertebrate visceral muscles are different from the invertebrate narrow-fibered muscles in the extent of electrical coupling, general occurrence of caveolae, myogenic rhythmicity, and neural and hormonal modulation.

In summary, nonstriated muscles are extremely diverse and no clear phylogenetic connections can be discerned. The most usual features which myoepithelia, nonstriated, and diagonally striated muscles have in common are the absence of T-tubules, the distance of 2–4 μm from cell membranes to contractile filaments, direct coupling between membrane and contractile filaments, calcium membrane potentials, and absence of troponin-tropomyosin modulation of Ca action.

Striated Muscles; Neural Control

The mode of neural activation is one aspect of muscle adaptation. The phasic vertebrate muscles described previously have unineuronal, uniterminal innervation. A second type of innervation is unineuronal, multiterminal. Some frog muscle fibers are activated by small multiterminal axons which conduct at one-fifth to one-half the speed of the motor axons to fast muscles. There is a rich branching of the axons and multiple endings on the muscle membrane. The electrical response of multiterminal muscles consists of junction potentials without spikes; the entire surface area is sensitive to the transmitter acetylcholine, and conduction is in the nerve fibers rather than in muscle. The resting membrane potential is lower than in fast fibers and the membrane is chemically excitable, but electrically inexcitable. The junctional potentials summate on repetitive activation and are graded, as is contraction, which is more prolonged than in fast fibers. Many frog muscles contain both fast (spiking) and slow (synaptic potential) fibers. In slow fibers of frog the time constant of endplate potential is three times longer than in fast fibers. In the toad *Xenopus,* early in development muscle fibers have polyneuronal innervation; in adults most fibers receive only single axons. The quantal content of transmitter in dual innervation is less than in single innervation (7). The multiple nerve endings are *en grappe* (like clusters of grapes), the single endings *en plaque* (like plates). The alignment of thick and thin filaments in slow fibers is less regular, a *Felderstruktur* (i.e., out of register); Z-bands may follow a zig-zag path and the tubule system is less regular than in fast fibers (*Fibrillenstruktur*). The equilibrium potential of an EPSP in the slow muscle response is -15 mV; resting permeability to Na^+ is higher relative to K^+ permeability than in fast muscle, and resting potentials are less sensitive to K_o than in fast twitch fibers. A few mixed muscles have been reported in mammals (e.g., tensor tympani in the ear). Rat extraocular muscles give phasic and tonic responses due to the presence of both types of muscle fibers.

A third type of striated muscle has multiple innervation by several axons, each of which evokes a different response. One muscle fiber may receive two or more motor nerve fibers with diffuse endings. Best examples are crustacean muscles, in which there may be two to five excitatory axons to one muscle fiber; one axon may innervate many fibers. One of these axons, the "fast" motor fiber, elicits electrical and mechanical responses which approach all-or-none response, and show little facilitation. Another axon, the "slow" motor fiber, elicits graded, highly facilitating responses (8). Many muscles have intermediate motor innervation as well as nonfacilitating fast and facilitating slow responses. In some muscles, a given axon may evoke a "fast" response at one end of the muscle and "slow" at the other end. Most crustacean muscle fibers are large—some as much as 1000–2000 μm in diameter—and there are wide invaginations of the plasma membrane which branch to form T-tubules inside the muscle fiber. Ion permeabilities of the invaginations are unlike those of the plasma membrane and are selective for chloride. In addition to the excitatory axons, there is usually one inhibitory axon; inhibition may be presynaptic or postsynaptic or both. Fast electrical responses are focal at the endings, and more transmitter is liberated than in slow responses. One excitatory transmitter serves for the different speeds of motor activation; probably it is glutamate, and responses to it correlate with diffuseness of innervation. The inhibitory transmitter is gamma amino butyric acid (GABA). One motor axon may innervate several muscles, possibly two antagonistic muscles. The following features make it likely that striations in arthropod muscles evolved separately from those in vertebrates: crustacean muscles have long sarcomeres, wide membrane invaginations, Cl^- permeability of the invaginated membrane, multiple innervation, myosin-linked Ca^{2+} regulation, glutamate as excitatory transmitter. Furthermore, arthropods are unrelated to vertebrates phylogenetically (115).

Many insects, especially the large, slow-flying Lepidoptera and Neuroptera, resemble crustaceans in having multiple innervation with few axons. A muscle of leg or wing may have fast and slow excitors or fast excitor plus an inhibitor. The muscles which move wings are direct flight muscles, and one spike per fiber triggers a contraction and wingbeat. Some of these insects have rapid muscle contractions; the tymbal muscle of cicada may twitch at 200/sec (69).

In fast-flying insects with indirect flight muscles, Diptera and Hymenoptera, and in Orthoptera and Homoptera with rapidly vibrating sound producing organs, a different sort of contractile system has evolved. The frequency of wingbeat or tymbal vibration depends on mechanical load, increasing when load is reduced, for example, by clipping wings, and decreasing with more load. Several contractions may follow one action potential. A blowfly flight muscle may beat at 125/sec when the action potentials are 3/sec. The fastest recorded wingbeat is in a midge at approximately 1000/sec, too fast for membrane excitation and recovery. Rhythmic contractions of this kind result from mechanical resonance. A single motor impulse triggers a contraction, the muscle shortens slightly, then the elastic skeletal element or an antagonistic muscle stretches the muscle, which remains in an active state, the muscle develops tension and shortens again and the cycle is repeated so long as the excitatory effect of the action potential remains. The property of resonance is in the contractile system (135).

Fast muscles of those insects in which contractions are synchronous with motor nerve impulses have extensive sarcoplasmic reticulum (SR); in asynchronous muscles with multiple contractions per nerve impulse much less of the fiber volume is occupied by sarcoplasmic reticulum. Stereologic measurements on electron micrographs of katydid muscles show the following (69):

	SR Thickness (μm)	Myofibrillar Area (μm²)
Synchronous muscles	0.06–0.09	0.4–0.8
Asynchronous muscles	0.016	1.1

Synchronous flight muscles of a cicada contain, by volume: 40 percent mitochondria, 20 percent myofibrils, and 30 percent sarcoplasmic reticulum. Asynchronous tymbal muscles contain 58 percent mitochondria, 48 percent myofibrils, and 2 percent SR (69). It is probable that asynchronous muscles were derived phylogenetically from synchronous ones. Asynchronous activity is more economical, uses less Ca^{2+} cycling, more fibrils, fewer spikes. Where precise control is required, as in walking, synchrony of activation and contraction are essential. Where high-frequency oscillations occur, as in indirect flight muscles, there is less need for control. Muscles of mesothorax and metathorax are similar in twitch dimensions in nymphal tettigonid crickets; however, in adults the singing mesothoracic muscles are much faster than the walking metathoracic muscles (70).

In summary, the modes of neural activation are adaptive for different speeds of movement by striated muscles. Highest frequency of movement occurs with resonant insect muscles; the fastest muscles for complete activation and relaxation are unineuronal and uniterminal; slower movements are triggered by multiterminal innervation. Muscles with multiaxonal and multiterminal endings differ from unineuronal muscles in facilitation; gradation of movement is more in the periphery than by the central nervous system.

MUSCLE PROTEINS

The genetic control of rates of contraction and relaxation, and of passive mechanical properties, is via muscle proteins. There are three kinds of muscle proteins, contractile, regulatory, and structural. Contractile proteins are the highly conserved actins and the many forms of myosins. Calcium is essential for myosin-actin interaction; three methods of Ca^{2+} regulation have evolved. In many striated muscles regulatory proteins mediate calcium control of the interaction of actin with myosin; in smooth muscles Ca^{2+} acts on an intermediate protein; in some invertebrate muscles Ca^{2+} acts directly on myosin. Structural proteins link contractile proteins together and give elasticity and limited rigidity to muscle fibers.

Actins

Actins occur in all muscles; actins also occur in motile cells that are not muscle and, as cytoplasmic proteins, in many—perhaps all—cells of eukaryotes and some prokaryotes (79). In striated muscle, actin constitutes some 21 percent of protein, in a soil amoeba *Acanthamoeba* 20–30 percent of protein, in blood platelets 20–30 percent, in liver 1–2 percent (107). The amino acid structure of actin is highly conserved. An actin molecule is composed of 374 amino acids, molecular weight approximately 45,000, isoelectric point 4.8 (36). Monomeric actins are globular (G-actin) and they polymerize to fibrillar (F-actin) under appropriate ionic conditions in the presence of Mg^{2+} plus ATP. A thin filament of muscle consists of two tropomyosin strands surrounded by a double strand of actin (Fig. 16-10). Actins constitute a family of proteins of characteristic electrophoretic patterns and amino acid sequences (151).

Tissues may have two or more actins in different proportions. The principal actin of rabbit skeletal muscle has been completely sequenced (36). The N-terminal sequence of 18 amino acids is the same in positions 7–9, 11–15, and 18 in all eukaryotes examined; these highly conserved sequences must be of critical importance. Heart actin differs from skeletal muscle actin in having a serine instead of threonine at position 357, Asp for Glu at position 2 (151,36). There are two smooth muscle actins in mammals. A vascular smooth muscle actin differs in the N-terminal ends from nonvascular actin; actin of gizzard and of gastrointestinal smooth muscle has at position 17 Cys instead of Val as in vascular muscles. Nonmuscle cytoplasmic actins (brain, thymus, kidney) are of two principal types; one differs from muscle actin in having Ileu for Cys at position 10, the other has Asp for Gly at 2 and Val for Cys at 10. Both cytoplasmic actins have Met-Cys instead of the Leu-Val found in striated muscle actin at positions 16–17 (39,151). All cytoplasmic actins differ by some 25 amino acids from muscle actins. Actin of blood platelets differs from muscle actin by 11 amino acids (88). Fibroblast actin resembles one smooth muscle actin and one cytoplasmic actin more than it does striated muscle actin (122). Comparisons of vertebrate actins show that they are more tissue-specific than species-specific (105). Actin of the slime mold *Physarum* resembles cytoplasmic actin and differs from muscle actin in 25 sequences (8 percent of total); *Acanthamoeba* actin closely resembles that of mammalian cytoplasmic actin (131). Yeast actin differs from *Physarum* actin by 39 amino acids.

Actin occurs with myosin in the cleavage furrow of dividing cells; chro-

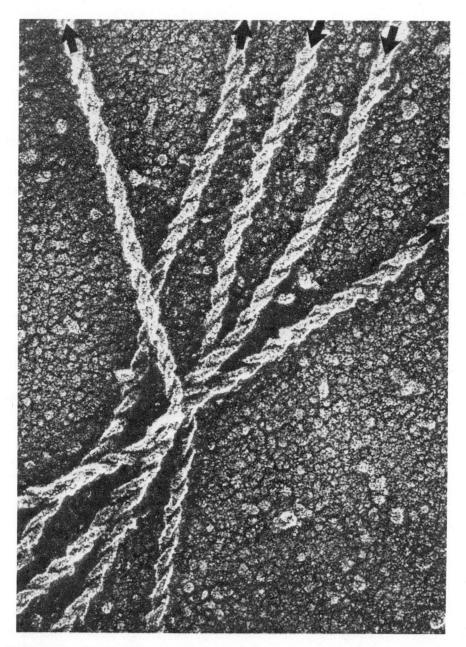

Figure 16-10. Electron micrograph of actin strands from insect muscle. From J. E. Heuser and R. Cooke with permission from *Journal of Molecular Biology,* Vol. 169, pp. 123–154. Copyright 1983 by Academic Press Inc. (London).

mosome movement uses tubulins in spindle fibers, and cleavage is by actin-myosin sliding. Actin in yeast cells has been demonstrated by its binding to mammalian myosin; yeast actin has a molecular weight of 45,000 and forms filaments. In the alga Nitella, actin filaments (5–7 nm) are associated with chloroplasts and actin plays a role in cytoplasmic streaming (104). Nitella actin can bind to myosin of rabbit muscle. Actin occurs close to the cell membranes of fibers in a vertebrate lens. Actin filaments occur in the core of intestinal epithelium microvilli and extend into the terminal web where there may also be myosin. Actin is the major protein in the acrosome of starfish sperm, and on sperm discharge a 90 nm filament is formed by polymerization. Actin in *E. coli* has a molecular weight of 45,000 (100); this bacterial actin can bind to muscle myosin (96). Actin has been found in plants—in tomatoes (150) and in tubers and pollen tubes of Amaryllis where it forms 6 nm filaments (30). Soybean seedlings have actin of 45,000 Da that reacts with antibodies to rabbit muscle actin.

The genes for actin constitute a family, often in multiple copies in a genome. In mammals there are at least six different actin genes. In mouse the same gene encodes for fetal skeletal muscle actin as for fetal cardiac actin, and this DNA hybridizes with adult cardiac actin RNA. Another gene codes for adult skeletal muscle actin (97). Drosophila has six actin genes per genome, including one structural gene 17.5 kb in length (144). Sea urchin cDNA fragments hybridize with Drosophila actin genes. The sea urchin actin gene copy number is 5–20 per haploid genome (34). Yeast actin has 374 amino acid residues coded by a single split gene. Yeast actin differs from actin of rabbit skeletal muscle by 44 amino acid residues and by 39 residues from Physarum actin (41). A plasmid containing actin cDNA 1.7 kb long from Dictyostelium has actin-encoding sequences repeated 15 times (15).

In summary, actin is one of the most widely occurring of proteins; its structure is highly conserved. There are six genes for actin in mammals. Actin may function in cytoplasmic movement by polymerization or by bonding with myosin or cytoskeletal proteins.

Myosins

Myosins are more numerous and diverse and less conserved than actins and make for differences in speed of contraction of muscles. Myosins are large—450,000–475,000 Da. Each molecule consists of two strands twisted in a double α-helix forming a rod some 140 nm long (Fig. 16-11). Cleavage of myosin by a proteolytic enzyme breaks the molecule into light meromyosin (LMM, molecular weight 140,000) and heavy meromyosin (HMM, molecular weight 340,000). Further proteolysis cleaves the HMM to a segment, S-2, some 65 nm long and of 60,000–100,000 Da, and a head, S-1, 21 nm long, of 115,000

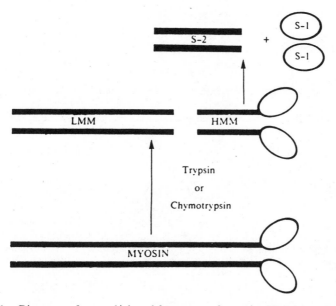

Figure 16-11. Diagrams of proteolytic subfragments of myosin. HMM heavy meromysin, LMM light meromysin. Reprinted with permission from A. Weeds and P. Wagner in *Biochemical Evolution,* H. Gutfreund, Ed. Copyright 1981, Cambridge University Press, Cambridge, England.

Da (Fig. 16-11). In vertebrate striated muscle the S-1 heads have four light chains (l.c.s). Two of the light chains, removed from striated muscle by a thiol blocking agent, dithionitrobenzoic acid (DTNB), are called regulatory l.c.s; each has a molecular weight of 18,000. The other two light chains (A_1 and A_2) can be removed by alkaline treatment. In most vertebrate fast muscles they have molecular weights of about 23,000 and 16,500 (86). In rabbit A_1 is 21,000 Da, A_2 17,000, the two regulatory l.c.s 19,000 each (133).

A cDNA recombinant plasmid is complementary to RNAs for l.c.s 1 and 3 of mouse striated muscle. The light chains l.c. 1 and 3 have the same terminal sequences at the carboxyl end—this consists of 141 amino acids—but the amino terminal of 49 sequences is different for the two light chains (118a). Multiple genes encode myosin heavy chains, and isozymes of each myosin are known. The number of genes encoding a myosin heavy chain of rat skeletal muscle is 8 to 10, cardiac myosin 2 (91). Chicken skeletal muscle light chains are encoded by a single gene. The sequences for encoding the C-terminal 141 amino acids and the 3′ untranslated region are the same for l.c. 1 and l.c. 3. The 5′ ends are different; that for l.c. 1 has 108 nucleotides, that for l.c. 3 has 78 nucleotides. The two 5′ ends are encoded in separate DNA segments of the same gene, one of 17.5 kb and the other of 8 kb; two

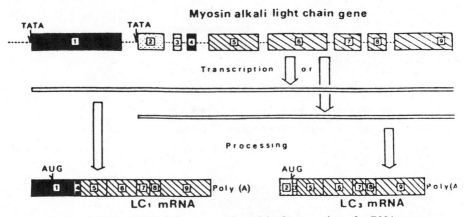

Figure 16-12. Schema of LC$_1$/LC$_3$ gene and model of processing of mRNA precursors. Hatched boxes common to both mRNAs; solid and stippled boxes unique for LC$_1$ and LC$_3$ mRNA. For explanation of non-coding regions see Chapter 3, from Nabeshima et al. Reprinted by permission from *Nature*, Vol. 308, p. 337. Copyright 1984, Macmillan Journals Ltd.

precursors are formed (Fig. 16-12). How the precursors become spliced is not known (99).

The slow myosin of tonic striated muscle and myosin of cardiac muscle of rabbits have light chains of two classes (20,000 and 17,000 Da); fast white striated myosin has three light chains (23,000, 17,400, 15,100 Da) (Fig. 16-13) (123). The heavy myosin chains of fast and slow striated and cardiac muscles differ in amino acid sequence and immunological properties. The ATPase activity of myosins of fast muscles is some four times greater than ATPase activity of slow muscle myosins.

Molecular weights of myosin fractions of cat muscle follow (157):

	Fast	Cardiac	Slow
Heavy chain	200,000	200,000	200,000
Alkaline l.c.1	25,000	27,000	27,000
Alkaline l.c.3	16,000		
DTNB l.c.2	18,000	19,000	19,000

In chicken embryos the separation of fast and slow myosin begins at the time of myoblast fusion (72). Heavy chains from chick skeletal muscle have been cloned (149). In rabbit heart, two heavy chains (α and β) have been cloned and their genes identified. In rabbit fetus, the ratio α/β is 3, at birth it is 1, and in adult 4 (127).

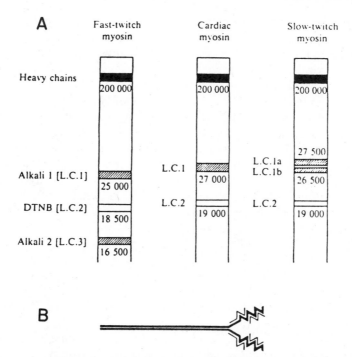

Figure 16-13. (*a*) Subunit structure of rabbit muscle myosin. Electrophoretic patterns of slow twitch and cardiac myosins. Diagonal shadings show related alkali light chains L.C.1 and 3. Regulatory (DTNB extracted) light chains L.C.2. (*b*) Schema of myosin molecule showing two pairs of light chains associated with heads of myosin molecule. Reprinted with permission from A. Weeds and P. Wagner, *Biochemical Evolution*, H. Gutfreund, Ed. Copyright 1981, Cambridge University Press, Cambridge, England.

Light chains in chicken myosin differ in size (23):

	Percentage of Light Chains		
Molecular Weight	In Fast Muscle	In Myotubes	In Fibroblasts
25,000	35	57	
20,000			55
18,000	52 (2 l.c.s)	43	
16,000	13		45

Smooth muscle myosin light chains differ in ATPase from those of striated muscle. Uterine myosin amino acid content in relation to Ser has less Pro, less Ileu, more Met and Arg (48).

Isoforms of myosin differ in auricles from those in ventricles. Three isoforms of myosin (V1, V2, and V3) have been isolated from rat and rabbit ventricles by electrophoresis and by immune reactions (24). ATPase activity of V1 is two to three times greater than of V3. Ventricle myosin of guinea pig shows a single band but reacts with antibodies to both V1 and V3 of rabbit. Cardiac myosin of Xenopus, chicken, dog, beef, and human occurs as a single band that reacts with V3 of rabbit. Genes for rabbit ventricle myosin have been cloned and designated α and β. Myosin V1 is a homodimer of heavy chain α (HCα) and V3 a dimer of HCβ (23a). In adult rabbits the α/β ratio of messenger RNA is 4:1; in fetal rabbits it is 0.3:1 (127).

Muscle hypertrophy brings about changes in proportions of isoforms of myosins. Cardiac hypertrophy can be caused by aortal constriction and can also be brought about in a few hours by injection of thyroid hormones. Cardiac atrophy occurs in hypothyroidism. The hypertrophy produced by arterial constriction is due to increase in V3 light chain; in hypertrophy by thyroid hormone V3 decreases and V1 (α chain) increases; and in hypothyroidism only the β chain occurs (127).

There are several kinds of myosins in poikilotherms. In mackerel the myosins of red and white muscles differ as follows: red muscle has two types of light chains, ATPase activity of 0.18; white muscle has three types of light chains, ATPase activity of 0.44 μM P_i/min/mg; carp muscle has three kinds of light chains (32). Myosin of the muscle of the mollusc *Pecten* has two heavy and four light chains (22). Myosin ATPase of scallop muscle has low activity—64 nM P_i/min/mg protein (22). Myosin from muscles of an ascidian has two types of light chains, and these are different from vertebrate myosin l.c.s (103). In the nematode *Caenorhabditis,* three isotypes of myosin heavy chains have been identified; type A, molecular weight 210,000 in body wall and pharynx; type C, molecular weight 206,000 in pharynx; and type B, molecular weight 203,000 in body wall (163). A 50 percent homology exists between nematode and mammalian myosin heavy chains. *Drosophila* has one gene for the heavy chain of myosin, located on the second chromosome (16).

In the soil amoeba *Acanthamoeba* there are two myosins; myosin I is located near the plasma membrane, myosin II more internally as shown by immunofluorescence (40). Amoeba myosins consist of two heavy chains and two pairs of light chains. The l.c.s of myosins I and II differ in size and immunological properties (40). *Acanthamoeba* myosin is remarkable for the absence of cysteine and for the small size of the total molecule, 180,000 Da. In *Amoeba proteus* both heavy subunits of myosin are of 225,000 Da. The myosin of slime mold *Physarum* has two chains of 225,000 Da each and two pairs of 17,000-Da light chains (75).

Astriated muscle fiber develops from many uninucleate myoblasts. These fuse to form myotubes, each with 10^2 nuclei which unite and form myofibers with 10^3–10^4 nuclei per fiber. The developmental processes from myoblasts to fibers are not well known. Which isoforms of actin or myosin are predominant changes during development.

The speed of contraction of striated muscles is, in part, determined by the type of myosin, fast or slow. The changes in myosin that occur during development of embryos differ with species. In rats the myosins of all striated muscles are "slow" at birth, and some myosins become "fast" in adults. Embryonic heavy chains differ in composition from fast and slow adult myosins; three genes are involved. A rat embryo has two of the three adult fast light chains at 16 days gestation; the primordia of EDL (extensor digitorum longus) and SOL (soleus) react with anti-fast-myosin antibodies; at 18 days reactions occur to both anti-fast and to anti-slow antibodies. The neonatal EDL has both slow and fast types. The transition from embryonic to adult myosin occurs at 18–20 days of age in rats, and is related to innervation. Rabbit fetal heart light chains are similar to embryonic striated light chains. In rabbits, the fast myosin shows three electrophoretic peaks corresponding to the three light chains. Fetal myosin also shows three electrophoretic peaks, one of which may be similar to the third peak of the adult (42,59,160). In chick embryos the time for contraction to half-maximum tension is 0.50 to 0.53 sec for both ALD (anterior latissimus dorsi) and PLD (posterior latissimus dorsi); at 18 days, contraction time is 0.48 sec for ALD and 0.17 sec for PLD. Innervation occurs at 16 days, and the differentiation of fast and slow chicken muscle occurs at that time (59). In cultured cells of chick embryos, the transformation of myosins to adult forms can be induced by electrical stimulation, as can synthesis of myosin; hence nerves may not be essential to the transformation (18). When myoblasts of quail fuse to make myotubes in culture, synthesis of myosin increases 50–100 fold. In chick myotubes, the ratio of myosin to actin is 3:1, in fibroblasts 1:1, in adult muscle 1:9. Both myotubes and fibroblasts have 2 l.c.s, adult muscle has 3 l.c.s; myotube myosin ATPase has double the activity of that in fibroblasts (23).

The actin-induced ATPase activity of myosin from fast striated muscle is four times greater than that from slow muscle; maximum velocity of shortening is proportional to ATPase activity. Values of myosin ATPase for heart muscle in nM P_i/min/mg protein are: rat 423, guinea pig 268, dog 139, rabbit 94 (33). Heart myosin ATPase activity correlates with speed of heartbeat in the series: mouse > rat > rabbit > cow (137). Specific activity of myosin ATPase of colon smooth muscle is lower than that of skeletal myosin. Myosin ATPase activity of guinea pig white striated muscle is 660, of red muscle 268 nm P_i/min/mg protein (33). Correlation between contraction time and myosin ATPase in corresponding muscles of cat and sloth is given by the following (13):

	Contraction Time (msec)	Actin-Activated Myosin ATPase (μM P_i/min/mg protein)
Cat extensor digitorum	20.8	0.56
Cat gastrocnemius	22.5	0.50
Sloth extensor digitorum	115	0.14
Sloth gastrocnemius	106	0.11

The soleus (SOL) muscle has longer contraction time; sarcomeres shorten more slowly under load than those of a fast muscle such as extensor digitorum longus (EDL) or flexor digitorum longus (FDL). Actin-activated myosin ATPase from soleus is less active, more heat-stable, more alkaline-labile, and has different electrophoretic mobility from myosin from EDL. When motor nerves to the two muscles were crossed so that the nerve to SOL went to EDL and the reverse, and the nerves were allowed to regenerate for several months, reversal in properties of the two myosins occurred—SOL became faster and EDL became slower (26,136). Reversal occurred also in soleus and flexor hallucis longus of the cat after motor nerves had been exchanged (157):

	Flexor hallucis longus		Soleus	
	Normal	Cross-Innervated	Normal	Cross-Innervated
Contraction time (msec)	61	87	87	25
Myosin ATPase (mg P_i/mg/ myosin/sec)	2.7	0.7	0.2	2.5

Complete functional reversal occurred in cross-innervated rabbit muscles in 11–12 months. Whether the neural effect on the type of myosin synthesized is due to trophic factors or to usage and tonus is debated. When ACh release was blocked for some days by botulinus toxin, slowing of contraction and relaxation of the tibialis anterior occurred, similar to postdenervation.

When motor nerves of a rabbit were stimulated at 10/sec for up to four weeks, the ATPase of myosin from two fast muscles decreased by some 30 percent and contraction and velocity of shortening slowed. Myosin light chains like those of slow muscle appeared in the fast fibers (136). Evidently continuous nerve activation influences the type of myosin synthesized.

Hypertrophy of skeletal muscles can be induced in mammals by tenotomy. In chicks, stretch applied to a growing wing induces hypertrophy and in the wing muscles brings about a relative increase of slow myosin and of oxidative enzymes (60). Increased growth of myotubes in culture can be induced by stretch. In rats forced to swim for a period daily, hypertrophy of muscles occurs and the myosin ATPase increases 20 percent above the content in controls.

Fast and slow myosins may occur in the same muscle fiber in proportions determined by the innervation, possibly by the pattern of tonic motor activation.

Speed of relaxation of different muscles correlates with rate of uptake of Ca^{2+} by the sarcoplasmic reticulum (SR). Maximum velocity of Ca^{2+} sequestration by SR of rat fast EDL was five times greater than the rate of uptake by SR of the slow soleus. Uptake of Ca^{2+} by SR of fast muscles of trout is 2.8 times faster than uptake by rat muscle, despite the 27°C lower temperature. Trout fast muscle takes up Ca^{2+} 8 times faster than frog muscle at the same temperature. Trout SR is adapted for a rapid cycle of contraction and relaxation (93).

In summary, the general structure of myosins is similar in slime molds, amoebas, and muscles; each molecule is a double helical chain with several light chains. There are differences in amino acid sequences in both heavy and light chains of different organisms. Genetic analyses show several myosin genes in a given animal; these become active at different times during development. The ATPase activity of myosin light chains is correlated with speed of muscle contraction. The synthesis of fast and slow myosins can change according to innervation and muscle activity.

FILAMENT ORGANIZATION AND THEORIES OF CONTRACTION

The organization of myofilaments may be highly geometric with six (or more) thin surrounding one thick filament. Myosin molecules are 10–15 nm in diameter and 160 nm long (Fig. 16-14). The heads point toward the two ends of a thick filament, and the bare zone in the middle forms the M-band of a striated muscle sarcomere where there are no crossbridges to actin. Thick filaments are 450–500 nm, thin filaments 8–10 nm in diameter. Thin filaments of actin are more abundant and run from Z-line to Z-line. The S-1 heads bind reversibly to actin, and also have ATPase activity which is actin-activated. The bridges between myosin and actin form a 6/2 helix of 429 nm per turn (Fig. 16-15). The DTNB light chains are not essential for ATPase or for binding to actin. Thick filaments contain several structural proteins which provide support for myosin.

In striated muscle a thin filament consists of two strands of G-actin in a helix with tropomyosin (Tm) wound around the outside, seven actins per

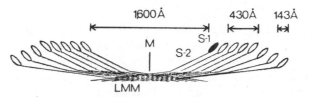

Figure 16-14. Diagram of arrangement of myosin molecules within a thick filament. From A. O. MacLachland and J. Kern. Reprinted by permission from *Nature,* Vol. 299, p. 227. Copyright 1982, Macmillan Journals Ltd.

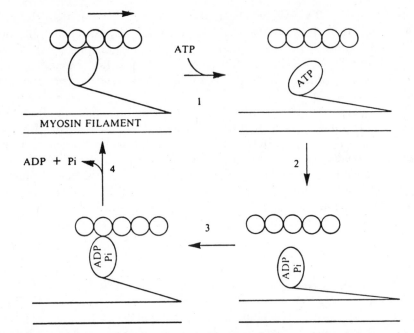

Figure 16-15.   Diagram of postulated scheme for cross-bridge cycle between myosin and actin during contraction. Reprinted with permission from A. Weeds and P. Wagner, *Biochemical Evolution,* H. Gutfreund, Ed. Copyright 1981, Cambridge University Press, Cambridge, England.

tropomyosin (Fig. 16-16). Interspersed at 40 nm intervals between the actins are troponin (Tn) complexes (106). Of the total protein, myosin constitutes 54 percent, actin 21 percent, tropomyosin 9 percent. The principal form of tropomyosin has a molecular weight of 66,000, molecular size 40 by 2 nm; tropomyosin forms two strands per thin filament and each Tm molecule contacts seven actin monomers (34a). There are three troponins: TnI, of molecular weight 20,900, which binds to actin in the presence of tropomyosin; TnC, 17,800, binds to calcium, 4 Ca^{2+} per molecule TnC; TnT, 30,500, a basic protein which links the other troponins to tropomyosin and increases Ca^{2+} sensitivity of the complex. The thin filament proteins occur in a ratio 7 actin : 1 Tm : 1 TnT : 1 TnI : 1 TnC (Fig. 16-17).

In the relaxed state, myosin and actin are not in contact, and interaction between them is prevented by the attachment of Tm and TnI to actin. When Ca^{2+} ions are released from the SR, Ca^{2+} becomes bound by TnC and this starts a chain of events. The restraint of Tm and TnI on actin is removed such that one of the myosin heads (l.c.) bridges to actin, and as ATPase becomes activated by the actin, energy is liberated (Fig. 16-18). A sarcomere shortens on contraction, but the filaments do not change in length; rather the thin filaments slide over the thick filaments. Myosin molecules probably interact with two sites on both strands of F actin (4,142).

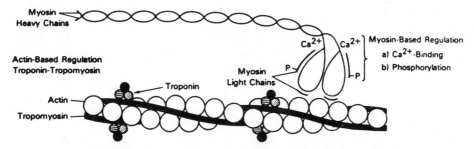

Figure 16-16. Diagrammatic representation of light chain and heavy chain with the two types of calcium regulation. Reprinted with permission from R. S. Adelstein, *Annual Review of Biochemistry*, Vol. 49, p. 925. Copyright 1980, Annual Reviews Inc., Palo Alto, California.

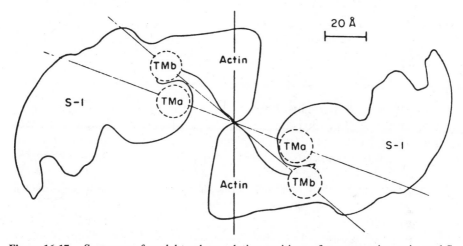

Figure 16-17. Summary of model to show relative positions of tropomyosin, actin, and S-1 of myosin in bridge. From K. A. Taylor and L. A. Amos with permission from *Journal of Molecular Biology*, Vol. 147. Copyright 1981 by Academic Press Inc. (London).

The most widely accepted hypothesis of contraction is that the myosin head makes contact with actin at a 90° angle, and at this stage the acto-myosin (AM) is at high energy; then the right angle changes to 45° as the energy is discharged, sliding occurs, and force is produced. The connecting element between the head and rod of myosin, S2, is pictured as an elastic connector 59 nm long, molecular weight 116,000. The myosin bridge gener-ates passive force at 90° and the force decreases to 0 at 45°. The myosin complex (thick filament) is bipolar so that bridges at the two ends pull thin actin filaments towards each other, and if the actin is attached to Z-lines the sarcomere shortens (63) (Fig. 16-15).

Evidence regarding the interaction between thick and thin filaments and the role of cross-bridges in contraction is based on observations on struc-

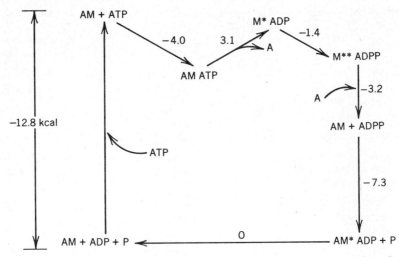

Figure 16-18. Schema for energy cycle in rabbit skeletal muscle contraction. Symbols given in text. Modified with permission from J. W. Schriver, *Trends in Biochemical Sciences*, Vol. 9, No. 7, p. 324. Copyright 1984, Elsevier Publications, Cambridge, England.

ture, mostly on geometrically regular striated muscles—frog sartorius or insect flight muscle. Tension is maximum at a length at which there is complete overlap of thick and thin filaments. When the muscle is stretched to a length at which there is no overlap, no tension is developed, and when the muscle is allowed to shorten below rest length, the filaments pile up and tension is reduced. This observation has been repeated with so many kinds of muscles that there is little doubt that thick and thin filaments slide past each other during contraction.

Electron micrographs of striated muscles fixed at different lengths and in contracted and relaxed states show that in contraction the angle of bridges from myosin filaments to actin is reduced from approximately 90° to 45° (Fig. 16-15). Pictures of insect flight muscle in which filament organization is very regular and the ratio of actin to myosin is 2 show chevron patterns of the bridges (118). Muscle has been compared in three states, relaxed, contracted, and in rigor. Rigor is a state of stiffness when ATP declines or is removed, there is no excitation-contraction coupling, and cross-bridges fail to break in the absence of ATP and Ca^{2+}.

Muscles were prepared by rapid freezing and storage under liquid nitrogen; strips were freeze-etched, replicated with platinum and carbon, and viewed by electron microscopy (56). Electron micrographs prepared in this way show features not seen with chemical fixation. Actin filaments are left-handed α helices; they show transverse bands with a 5.5 nm repeat in a pitch of 15–20° (Fig. 16-3). Insect muscle shows cross-bridges in a spiral pattern with a repeat every 38 nm. Bridges of intact muscle show two heads on the myosin filaments, and placement of the bridges appears more random than when seen under chemical fixation (56).

X-ray diffraction pictures of muscles in states of relaxation, contraction, and rigor show a meridional reflection at 14.3 nm in spirals with repeats every three units or 42.9 nm. The 14.3 nm reflection is interpreted as coming from cross-bridges. In contraction, the intensity of the 42.9 nm band decreases slightly ahead of peak tension and recovers more slowly than does tension. During contraction the 14.3 nm reflection decreases in intensity and broadens. This is interpreted as showing a disordering of cross-bridges as myosin heads go out to contact actin (52,63). According to the cross-bridge rotating-head hypothesis (63), force is generated when the angle of attachment of myosin to actin decreases from 90° to 45°. S-1 rotates while attached to actin, S-2 is stretched, and the power stroke results from retraction of the S-2 elastic component.

An alternative hypothesis for force development is a change in the helix coil configuration (50,147); helix coiling is shown in the HMM as observed by measurements of optical rotation of S-2 segments. This hypothesis pictures the myosin head (S-1) as remaining at a constant angle to the thin filament until it detaches. It is suggested that S-2 extends out from the thick filament at a HMM-LMM hinge which undergoes melting of its helix to a random coil. The helix coil transition generates force on S-2 which pulls S-1, and sliding between the filaments results.

Evidence for force generation by rotation of heads comes from observations on cytoplasmic droplets prepared from the alga *Nitella*. Chloroplasts are seen to move in the presence of HMM and ATP, Mg and actin. Cables of *Nitella* actin, five per row of chloroplasts, were treated with HMM from muscle, and were coated with 0.7 μm fluorescent beads; Mg and ATP were added and movement of the myosin along actin filaments was observed at 2.5 μm/sec; that is, a myosin molecule "walked" along a thin filament with accompanying rotation of the chloroplasts. The orientation of actin was such that its movement was opposite to the arrowhead structure of the actomyosin (125). However, no recovery occurred on dissociation.

ENERGY SOURCES FOR CONTRACTIONS

The energy for myosin-actin interaction and resulting movement comes from hydrolysis of ATP by an ATPase of the light chain of myosin when activated by actin in the presence of Mg. The amount of ATP in a resting skeletal muscle is approximately 10^{-6} mol/g. ATP must be continually regenerated and the second level of energy reserve is a phosphagen which is phosphocreatine (PC) in vertebrate muscles, phosphoarginine (PA) in muscles of many kinds of invertebrates. In annelids and sipunculids several bases replace phosphoarginine: phosphoglycocyamine, phospholombricine, phosphotaurocyamine. The occurrence of certain phosphagens in different phyla and classes may be related to their intermediary metabolism. High-energy phosphate is transferred from the intermediary phosphagen stores under action of a kinase specific for the particular base. The base, usually creatine or arginine, is rephosphorylated from the anaerobic breakdown of glycogen (sometimes of lipids), and lactate is formed glycolytically. In vertebrate muscles under aerobic conditions *in vitro,* part of the lactic acid formed is oxidized, giving energy for reconversion of the remaining lactate to glycogen. *In vivo* the lactate leaves the muscles via blood and is reconverted to glycogen in the liver; this supplies blood glucose which then replenishes the muscle glycogen. Under anaerobic conditions lactate can accumulate in muscles so long as glycogen is present, but when glycolysis is blocked by an inhibitor such as iodoacetate, glycolysis stops at the triose phosphate stage and contractions persist only so long as PC is present.

Red muscles contain myoglobin and are highly aerobic in their enzyme complement. White muscles lack myoglobin and are adapted for anaerobic glycolysis. Presence of myoglobin is not necessarily correlated with innervation and myofibrillar organization. Red muscles are often slower and are consistently more resistant to fatigue than white muscles.

Each on-off or attachment-detachment cycle of myosin with actin is accompanied by hydrolysis of one ATP per bridge. The myosin chain has ATPase activity in the presence of Mg^{2+}, and this is enhanced twofold when the myosin is attached to actin. Attachment is controlled by calcium ions. A diagram of the postulated energy sequence in contraction-relaxation is given in Figure 16-18 (125a,142). M = myosin, ATP = adenosine trisphosphate, A = actin, AM = actomyosin, ADP = adenosine diphosphate. When ATP binds to AM, detachment of a myosin head from actin occurs and relaxation results. The energy of ATP is used in separation of A from M. Association of actin with myosin-ADP is blocked by troponin-tropomyosin. The energy level of myosin is raised before actin associates with M-ADP. The initial attachment is presumed to be at a 90° angle and the power stroke is a bending toward a 45° angle. Relaxation occurs as actin dissociates from AM-ATP and calcium is resequestered in SR.

CALCIUM REGULATION OF MYOSIN-ACTIN INTERACTION

Calcium ions are requisite for contraction in all muscles. In vertebrate striated muscle, Ca^{2+} is stored mainly in cisternae (terminal enlargements) less in the tubules of sarcoplasmic reticulum (SR). Stored Ca^{2+} is in steady state with the Ca^{2+} in 30–40 Å particles of the SR tubules. On electrical activation from the T-tubules, the SR releases Ca^{2+} which binds to troponin-C in the tropomyosin-troponin complex. The nature of signal transmission from T-tubule to SR is unclear; transmission appears not to be by a chemical transmitter, but may be by a Ca^{2+} induced calcium release (31a) or by charge movement along pillars which connect T-tubules with terminal vesicles of

SR (37). The cytoplasmic side of SR tubules contains particles of high Ca^{2+}-affinity CaATPase; the tubule lumen side has calsequestrin, a protein of high binding capacity but low affinity for Ca^{2+}.

Actin does not react with myosin while actin is bound to the Tm-Tn complex; Ca^{2+}, by combining with TnC, removes the Tm-Tn inhibition on actin and permits activation of the ATPase and movement of cross-bridges. The sequence of control is: (1) calcium release from SR, (2) reaction with TnC, (3) reactions of TnC with Tm, TnT, TnI, (4) change in position of Tm and TnI relative to actin, (5) binding of myosin S-1 to actin, (6) activation of S-1 ATPase (Figs. 16-15, 16-16). Actin-myosin interaction and resulting motion is held in check in the relaxed state, and activation of bridges can be sudden and synchronous for the start of contraction. The sequence contains redundancy and safety factors. Tension increases abruptly at the critical concentration of Ca^{2+} of $10^{-5.8}$ to 10^{-6} M, relaxation occurs at 10^{-7} M, and at rest the concentration of free calcium ions is 10^{-8} M or lower. The protein parvalbumin in fast-twitch fibers may facilitate relaxation and may shuttle Ca^{2+} between TnC and the SR (21).

Release of Ca^{2+} from SR can be demonstrated if the Ca^{2+}-requiring fluorescent protein aequorin is injected into a muscle fiber. Aequorin gives a flash of light in presence of a very small amount of Ca^{2+}. Amplitude of the flash, speed of contraction, and decay of the flash, are more rapid in fast (phasic) than in slow (tonic) striated muscles.

In relaxation, the calcium uptake by SR shows a greater maximum velocity in fast than in slow muscles; uptake in rat extensor digitorum longus is five times as fast as in soleus. A Ca^{2+}-binding ATPase isolated from SR has a molecular weight of 102,000; the ATPase releases energy from ATP for the transport of Ca^{2+} from cytosol concentration 10^{-8} M to a particle concentration of 10^{-4} M. Calsequestrin has molecular weight 54,000 and binds 60–70 mol Ca^{2+}/mol protein.

Several alternative modes of Ca^{2+} regulation have evolved. The first of these to be discovered was in molluscan muscle, specifically in adductor of scallop, in which troponins are absent and regulation is by direct combination of Ca^{2+} with a regulatory light chain of myosin. Treatment of molluscan myosin with EDTA removes a Ca^{2+}-sensitive ATPase. Tropomyosin is present, but it is not essential for calcium activation and may facilitate thin-filament binding to myosin. Calcium concentration of 10^{-6} to $10^{-5}M$ is necessary for actin-myosin interaction. Calcium sequestration in the molluscan muscle is mainly in the plasma membrane, with small amounts in mitochondria and SR. Calcium spikes bring calcium ions into the muscle fibers. After the Ca^{2+}-binding light chain is removed, the myosin protein does not bind calcium, actin does not attach to myosin, and AM-MgATPase is not activated; these events require that the light chain be attached to myosin. The Ca^{2+}-binding regulatory light chain has a molecular weight of 18,000, and is distinct from another light component, a sulfhydryl-containing chain (SH l.c.) that is one-half as abundant as regulatory l.c. (22). Some interactions between regulatory l.c. and SH l.c. is probable (153). Antibodies for the regulatory l.c. and for SH l.c. are species-specific and do not cross-react with vertebrate muscle, platelet, or Physarum myosin l.c.s (153). Myosin regulation by a Ca^{2+}-sensitive light chain has been found in muscles of other invertebrate animals (e.g., in holothurians).

Muscles of most insects, crustaceans, annelids, a sipunculid, and *Ascaris* have double control—by a Ca^{2+}-sensitive light chain myosin and by a thin filament troponin. The troponins of invertebrate muscles differ from troponins of vertebrate striated muscles in that they bind less calcium than TnC. The troponin of lobster muscle binds only 1 mol of Ca^{2+} per mol (119).

A third method of calcium regulation occurs in vertebrate smooth muscle and has been most studied in chicken gizzard and in "skinned" fibers of mammalian uterus (1a,58,74). Vertebrate smooth muscle lacks troponins. A myosin light chain kinase (MLCK) is activated when Ca^{2+} is released into the cytosol. A light chain of myosin becomes phosphorylated by action of the kinase and, when phosphorylated, the myosin can interact with actin. Calcium is needed only for activation of the MLCK to phosphorylate a light chain. After ATP hydrolysis and energy transfer associated with movement, a second enzyme, a phosphatase, dephosphorylates the myosin and relaxation occurs. Phosphorylation of the kinase occurs at Ca^{2+} concentration of 10^{-6} to $10^{-7}M$, and dephosphorylation via the phosphatase occurs when the Ca^{2+} concentration drops below $10^{-7}M$ (1a).

Two substances important for Ca-activation of myosin kinase of smooth muscles are calmodulin (CM) and c-AMP. Calmodulin is a protein of wide occurrence which has four Ca^{2+}-binding sites; calmodulin molecular weight is 17,000. Calcium bound to calmodulin is more effective than Ca^{2+} alone in activating the myosin kinase. Calmodulin resembles troponin-C but is a smaller molecule; possibly calmodulin and TnC evolved from the same precursor. In dividing cells Ca^{2+}-calmodulin ($Ca^{2+}CM$) prevents disassembly of microtubules; in ciliated cells calmodulin occurs in the basal roots of cilia. Calmodulin functions in Ca^{2+} regulation in smooth muscle by buffering free Ca^{2+} concentration.

Another substance which functions in calcium regulation of contraction and relaxation in smooth muscle is cyclic AMP (c-AMP). Many chemical agents, such as norepinephrine, are capable of activating adenyl cyclase, which increases the amount of c-AMP in smooth muscle. c-AMP activates a protein kinase which phosphorylates MLCK; this weakens the Ca^{2+}-calmodulin binding to the MLCK, which then becomes inactive and the myosin light chain is no longer phosphorylated; relaxation then occurs. Inhibition of diesterase by theophylline or IBMX permits increase in cellular c-AMP. Dephosphorylation of the myosin kinase by a phosphatase restores it to the form that binds Ca^{2+} strongly and relaxation occurs (79a). The sequence can be summarized as follows: Ca^{2+}-CM activates MLCK which phosphorylates a myosin l.c. which then reacts with actin and contraction occurs. c-AMP activates a kinase which phosphorylates MLCK, resulting in decreased phosphorylation of the myosin light chain, and relaxation results. Two mechanisms of relaxation are dephosphorylation of myosin by a phosphatase and phosphorylation of myosin kinase resulting in reduced binding to actin.

Phosphorylation of the 20,000-Da light chains of gizzard smooth muscle is needed to initiate bridge cycling. The actin-activated ATPase and tension development depend on phosphorylation of both light chains of myosin (64). Calcium regulation by phosphorylation of a myosin light chain occurs in platelets, cultured kidney cells, and smooth muscle (13,83). In sarcoplasmic reticulum from cardiac muscle a protein phospholambin is 4–6 percent of the total protein. Phospholambin becomes phosphorylated in the presence of c-AMP. $CaCa^{2+}$ uptake by SR is increased by phosphorylation of phospholambin (138).

In mammalian arterial muscle, the percent of myosin light chains phosphorylated was, at rest 4.6; when stimulated by high potassium, 34.8; and when stimulated by histamine, 57.1 (98). A calcium-dependent l.c. kinase (MLCK) was essential for the phosphorylation. Phosphorylation of the 20,000 Da l.c. was maximal in 30 seconds of potassium stimulation; maximum tension was reached in 2–4 minutes. Active contraction continued long after dephosphorylation of the myosin l.c. In smooth muscle of carotid artery, the maximum shortening velocity is proportional to the phosphorylation. Phosphorylation may be needed more for force development than for force maintenance (98). In skeletal muscle myosin, the 18,000 Da l.c. (DTNB l.c.) may become phosphorylated, but this has no effect on actin-activated myosin ATPase, and the function of phosphorylation in striated muscle is unknown. Myosin of uterus becomes phosphorylated when stimulated by oxytocin, carbachol, or potassium. Flashes of light from injected aequorin have demonstrated Ca^{2+} release associated with excitation coupling in mollusc nonstriated muscle and in guinea pig stomach muscle (101).

In summary, modulation by Ca^{2+} of actin-myosin interaction is a universal requirement and at least three mechanisms of regulation have evolved.

OSCILLATORY CONTRACTIONS

Myosin filaments in asynchronous muscles of insects give mechanical resonance which allows the muscles to beat at frequencies faster than the frequencies supported by neural control and membrane depolarization-repolarization. An asynchronous muscle is attached to a skeletal element or to an antagonist muscle which exerts stretch on the contracting muscle after the muscle has applied tension to the stretching element. Less work is done on the muscle by stretching than the muscle does in shortening. Sarcoplasmic reticulum is sparse in asynchronous muscles, and isometric twitches are prolonged (i.e., active state is persistent). Asynchronous muscles have large and extensive myofibrils. The rate of twitch contraction and relaxation in-

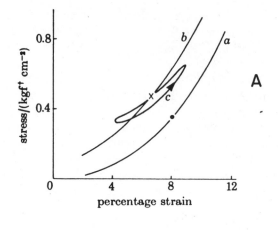

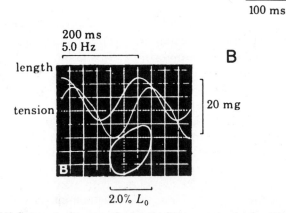

Figure 16-19. (*A*) Stress-strain curve for beetle fight muscle; (*a*) unstimulated, (*b*) stimulated with isometric or dampened isotonic recording; (*c*) oscillation when load has inertia. (*B*) Sinusoidal modulation with tension leading length and a resulting tension-length loop. Reprinted with permission from J. Pringle, *Proceedings of the Royal Society of London,* Vol. 201, pp. 110, 112. Copyright 1978, Royal Society of London, London.

crease and duration decreases as the muscle is warmed; an example is the singing muscle of a tettigonid cricket, contraction time 30.5 msec at 30°C and 25.3 msec at 35°C (69). Asynchronous muscles treated with glycerol to remove ions and soluble organic molecules can be made to oscillate under light load if supplied with ATP in the presence of appropriate amounts of Mg^{2+} and Ca^{2+}. Tension-length curves show closed patterns, circles; tension is less on stretch than on shortening (Fig. 16-19) (112). When an asynchronous muscle is stretched to a new length while in the excited state, it develops tension so that as it relaxes the tension is higher than during the contraction phase. Changes in length lead changes in active tension. There is less difference between tension during shortening and stretching when in a relaxing solution (5×10^{-9} M Ca^{2+}) than when in an activating solution (1×10^{-7} M Ca^{2+}). ATPase activity rises to a maximum at the resonant frequency.

Stretch activation such as occurs in asynchronous insect muscles has been observed in some other striated muscles, and it may be important in force-velocity relations of vertebrate heart. In summary, in insect fibrillar muscle the delayed tension or the stretch activation is responsible for rhythmic activity of some muscles of flight and stridulation. Several lines of evidence indicate the myosin filament as the location of the sensor for stretch activation; the molecular mechanism is not known (111,112).

ADAPTATION FOR MAINTAINED TENSION

An adaptation for delayed relaxation based on properties of muscle proteins is "catch" in mollusc adductors and in *Mytilus* byssus retractor. In these muscles the thick filaments are very large (several hundred nm) and myosin is wrapped around a core of noncontractile protein, paramyosin, which constitutes 90 percent of total myofibrillar protein. Between the thick filaments are actin filaments; contractions of 40–50 msec rise time occur by actin-myosin interaction. Calcium activation of A-M binding is via direct action on a Ca^{2+}-sensitive myosin chain. Myosin binds Ca^{2+}, and actin-activated ATPase depends on the Ca-bound myosin (84). Once contracted, tension can be maintained for minutes or hours with little or no energy expenditure. Maintained tensions in catch are high—some 15 kg/cm². Resistance to stretch is increased sevenfold during catch. A relaxing nerve liberates a transmitter, 5 HT, which causes relaxation.

One hypothesis for "catch" is that it results from delay in Ca^{2+} sequestration and that myosin bridges are slow to detach from actin. However, catch can be induced when Ca^{2+} is as low as 10^{-9} M. A second hypothesis based on electron micrographs is that long-range bonding or fusion occurs between thick filaments. A more likely hypothesis is that when myosin shortens the paramyosin assumes a paracrystalline state (28). It is known that the paramyosin becomes phosphorylated and that phosphorylation of paramyosin facilitates ATPase activity of actomyosin (1). The relaxing transmitter 5 HT accelerates dephosphorylation and actomyosin ATPase is inhibited during catch. Dephosphorylation is mediated by c-AMP.

Electron micrographs and X-ray diffraction pictures show that paramyosin occurs in planar ribbons which are wrapped into cylinders (Fig. 16-20a). The paramyosin molecules form diagonal patterns such that each molecule makes an axial repeat every 725 Å caused by staggering of five 145 Å units (Fig. 16-20b). A paramyosin filament is a three-dimenstional crystal with a principal period of 14.4 nm which repeats every five units (35).

That paramyosin is essential for catch is made questionable by the occur-

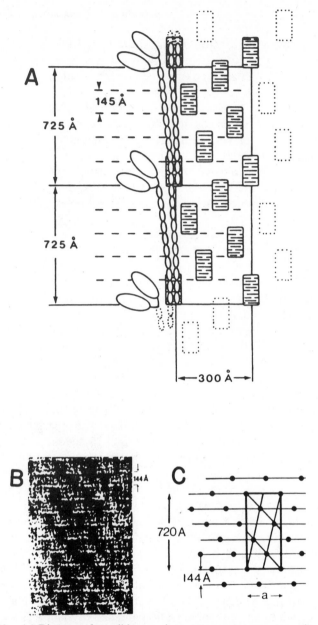

Figure 16-20. (*a*) Diagram of possible myosin arrangement on paramyosin core of a catch muscle filament. Myosin associated with one paramyosin subfragment is shown extending from the core.Reprinted with permission of C. Cohen, *Proccedings of the National Academy of Sciences* (USA), Vol. 79, p. 3177. Copyright 1982, National Academy of Sciences, Washington, D.C. (*b*) Appearance of negatively stained surface of an isolated paramyosin filament from adductor muscle of Crassostrea. Transverse stripes are at axial separations of 145 Å (filament axis is vertical). (*c*) Diagram of lattice with a 720 Å axial repeat and lateral spacings which are shown in diagram (*a*) as 300 Å. Reprinted with permission from J. Squire, *Structural Basis of Muscle Contraction*. Copyright 1981, Plenum, New York.

rence of paramyosin in many muscles that are incapable of catch (e.g., limulus striated muscle). However, in noncatch muscles the protein may provide support for the extraordinarily long sarcomeres. Paramyosin is not known is any vertebrate muscles. In these, tension is not maintained for very long times except upon repetitive stimulation.

CYTOSKELETAL PROTEINS

In addition to thick filaments (myosin) and thin filaments (actin plus tropomyosin and troponins), muscle fibers contain other filamentous proteins. Actin occurs in many nonmuscle cells (of bacteria and plants as well as animals) and may bind to other structural proteins than myosin. Cytoskeletal proteins determine cell shape, position of organelles, flow of cytoplasm.

Tubulins were mentioned previously. Cytoplasmic tubules are of 24 nm diameter and consist of protofilaments composed of alternating α and β tubulin (19). Microtubules are essential to movement of cilia, flagella, sperm tails, mitotic spindles, and tubules of neurons.

Several structural proteins have skeletal functions, providing appropriate elasticity and rigidity in muscle fibers. In striated muscle, the sarcomeres are separated by Z-lines, which are zones of attachment of thin filaments. In nonstriated muscles the "dense bodies" may serve the same function as Z-bands. Z-bands and dense bodies contain α-actinin, molecular weight 95,000, which forms a pattern of 2×30 nm filaments. α-actinin has been prepared from nonmuscle cells, for example, from Hela cells in culture and from blood platelets.

Table 16-6 gives the relative amounts and sizes of proteins in mammalian skeletal muscle (162). Striated muscle C-protein may function to hold myosin molecules together in thick filaments. Molecular weight of C-protein is 140,000 and it constitutes 2 percent of myofibrillar protein (31). Desmin may link Z and M bands to surface membranes. In smooth muscle the dense bodies contain both α-actinin and desmin (82). Titin is a large protein of molecular weight 1,000,000 which may constitute 10 percent of myofibrillar protein. In isolation, titin forms long flexible filaments. Antibodies to titin stain at AI junctions, at Z lines, and in A bands (154).

A protein of the M-line, molecular weight 170,000, is called M-protein (146). Nebulin may bind thin filaments together. Vimentin, molecular weight 52,000, has been found in skeletal, cardiac, and smooth muscle, and in fibroblasts, myoblasts, and myotubes. Vimentin filaments may serve as anchors for nuclei (82).

Proteins of the spectrin family apparently function in giving shape to cells and in some cells spectrins link actin to the plasma membrane. Spectrins are tetramers of two nonidentical subunits, molecular weights approximately 240,000 and 215,000. Spectrins occur in the terminal web of intestinal epithelial cells, in the "ruffles" of membranes of fibroblasts, in neurons, in

Table 16-6.   Muscle Proteins According to Abundance and Molecular Size

Protein	Percent (net weight)	Approximate Size (Da)
Myosin	43	45,000
Actin	22	42,000
Titin	10	1,000,000
Nebulin	5	500,000
Troponin	5	60,000
Tropomyosin	5	66,000
C-protein	2	140,000
M-protein	<1	170,000
α-actinin	2	200,000
Desmin	<1	50,000
Vimentin	<1	52,000
TNI	5	20,900
TNC	5	17,800
TNT	5	30,500
Myosin LC 1		22,000
Myosin LC 2		19,000
Myosin LC 3		15,500

muscles, and especially in red blood cells. Red cells have spectrin in the cell membrane, actin and spectrin in the cytoskeleton (143,29). Red blood cells also contain several proteins known from their position in electrophoretic patterns—band 2 and band 4, molecular weight approximately 78,000. There may be 100,000 spectrin-binding sites per red cell (162). The interaction between actin and spectrin is analogous to that of actin and myosin (157).

In addition to thick and thin filaments, muscles have intermediate filaments, 10 nm diameter (38). In epithelia, intermediate filaments are composed of a protein of the keratin family. In neurofilaments of neurons, glia and mesenchyme, the intermediate filaments are a protein vimentin, in muscle desmin.

The preceding list is a partial enumeration of cytoskeletal proteins. The physiology of cytoskeletal proteins, especially of intermediate filaments, and their evolution and relationships remain to be elucidated.

GENERAL SUMMARY

Movement is characteristic of all active organisms. Intracellular movements are: protoplasmic cyclosis and transport or flow from one region of a cell to another, chromosome translation in mitotic spindles, budding, disc formation, and cleavage in cell division. Cellular movements are: muscle contraction, amoeboid locomotion, beating of cilia and flagellae. All movements are produced by protein filaments or tubules, usually by actin-myosin or tubulin. Nonmotile protein filaments anchor contractile filaments, maintain shape of cells and the position of organelles.

In animals, actomyosin-produced movements include peristalsis in worms, jet propulsion in cephalopods, swimming, flying, walking and running by animals, all at characteristic speeds. Examples of nonlocomotor movement are: beating of hearts, contraction of blood vessels, peristalsis of visceral organs. Differences in rates of movement result from speed of muscle contraction and duration of tension maintenance (relaxation rate). Muscle speeds vary over a 10,000-fold range from the slowest to the fastest muscles, all making use of similar actomyosins. Animals depend on movement to obtain food, to escape from predators, for reproduction, and for social behavior. In general, muscles that move appendages act mainly by developing tension; muscles that move fluids contained in hollow organs or soft bodies (e.g., by peristalsis) act mainly by shortening.

Proteins of cellular movement appeared early in biotic evolution. Actins are present in bacteria (*E. coli*), in plant cells which show streaming, in most, probably all, animal cells; actins occur at high concentration in all muscles. Actins can cause movement by polymerization (G-F transformation); by reversible bridging to larger proteins, especially to myosins and spectrins; and by sliding or shearing of the bridges. Actins together with other proteins (tropomyosins and troponins) form thin filaments in muscles. Actins are highly conserved in composition; the actin of yeast is similar to that in mammalian muscle. Vertebrate actins are different from one another in skeletal, cardiac, and smooth muscle, and in nonmuscle cells. The number of genes for actins shows much variation with kind of organism. Actin genes may occur repetitively in a genome.

Myosins are large filamentous proteins with heavy chains and light chains. The light chains bridge reversibly to actin and have the enzymatic function of energy liberation from ATP. There are many myosins, each coded by a separate gene, and each myosin with properties adapted for particular movements and kind of animal. In one animal there may be several kinds of myosin. Genes for myosin light chains have been cloned. The type of myosin synthesized varies with developmental stage, with innervation, and with exercise or applied stretch. Vertebrate skeletal muscles are in early stages uninucleate myoblasts and later develop into fibers with $> 10^3$ nuclei.

Tubulins occur in most eukaryotes—in cilia, mitotic spindles, neurotubules. Movements of flagellated bacteria and ciliated protozoans are produced by special proteins different from actomyosin or tubulin.

Muscles are adaptively diverse in structure, physiology, and chemistry. Muscles in which thick and thin filaments are transversely aligned (striated) contract faster than nonstriated muscles. Cross-alignment of filaments occurs in various degrees of regularity; striations have apparently evolved numerous times. Large-diameter muscle fibers have membrane invaginations and a system of intracellular tubules for activating contractile filaments; large-fibered muscles are striated, multinucleate, activated mostly by motor nerves which evoke junction potentials with or without spikes, usually by inward Na^+ current, and with indirect coupling by T-tubules and sarcoplasmic reticulum between cell membrane and myofilaments. Narrow-diameter muscle fibers are cylindrical or ribbon-shaped, are usually uninucleate, lack transverse tubules, usually have calcium action potentials, myofilaments are random or in spirals, and the fibers have direct coupling between cell membrane and myofilaments. Narrow-fibered muscles may be either nerve-activated or endogenously activated.

The several patterns of innervation of both wide and narrow-fibered muscles are: (1) single axonal terminals eliciting graded junctional potentials and spikes, (2) multiple terminals from one axon eliciting only junction potentials (conduction in nerve rather than in muscle), (3) multiple terminals from several axons to one muscle fiber, (4) both excitatory and inhibitory axons, (5) nerves liberating transmitter which diffuses through a sheet of muscle. A nerve may activate a muscle or may modulate ongoing activity. Gradation of animal movement is by central control of the number and frequency of firing of motorneurons, or by a peripheral balance between facilitating and nonfacilitating excitation and inhibition. In oscillatory (asynchronous) muscles of insects, a muscle may give multiple contractions to one motor impulse. In holding muscles (molluscs), relaxation is very slow in a "catch" mechanism which probably depends on a structural protein paramyosin. In both striated and nonstriated muscles contraction takes place by sliding of thick over thin filaments; however, the precise mechanisms of force generation remain undetermined. Narrow-fibered muscles (smooth) with diagonal filaments shorten by change of filament angle (shearing).

In all muscles, movement of filaments is initiated by calcium ions. Ca^{2+} can be released from storage organelles (sarcoplasmic reticulum), from nonspecific storage sites, or can enter as Ca^{2+} currents. Regulation is brought about as follows: (1) by a series of proteins, tropomyosin, and three troponins which prevent actin from interacting with myosin until Ca^{2+} combines with one troponin; (2) by a direct effect of Ca^{2+} on a myosin light chain; or (3) indirectly by the effect of Ca^{2+} on a myosin kinase that phosphorylates a myosin light chain before it can bridge to actin.

Neural activation is best known for fast vertebrate striated muscles. Timing in the sequence of activation is precise and complicated. Motorneurons are activated by central nervous pathways. All-or-none motor impulses are conducted to neuromuscular endplates where a sequence of transmitter synthesis, packaging, and release occurs. At postsynaptic membranes, the transmitter elicits a graded endplate potential that may or may not initiate a muscle action potential. An electrical signal spreads intracellularly via transverse tubules to the sarcoplasmic reticulum. The SR granules liberate Ca^{2+}, which combines with troponin-C, which by interacting with other troponins and with tropomyosin removes the block on bridging between myosin and actin. Energy is liberated by the actin-activated ATPase of myosin and a cycle of breaking and making A-M bridges leads to sliding between thick and thin filaments. Ca^{2+} is then resequestered in SR to take part in the next contraction. There is limited redundancy in this complex sequence and all steps are precisely timed. The probable evolution of such a chain of events cannot be traced with present information. No muscle can be considered as more primitive than any other; narrow-fibered muscles are as complex as striated. Both striated and nonstriated myoepithelia are found in coelenterates; muscles may have evolved from myoepithelia.

Muscle contractions require energy, and more energy is used for animal locomotion than for maintenance. Metabolic adaptations of muscle are as essential as contractile properties. ATP is the proximate source of energy for all filament interactions—by actomyosins or by tubulins. AMP and ADP are recharged from intermediate phosphagens, principally phosphocreatine and phosphoarginine. These in turn are energized from metabolism, usually of carbohydrates. Many animals use muscles with oxidative pathways for low-level movements but shift to glycolysis when energy is needed for bursts of high activity. Oxidation is supported in a few kinds of muscles by an intracellular pigment, myoglobin, for O_2 transfer. Muscles rely on anaerobic means to varying degrees; visceral muscles, although slow, are mainly glycolytic. The coupling of energetics must have evolved in coordination with contractile proteins.

Besides the contractile and Ca^{2+}-regulating proteins, there are numerous structural or cytoskeletal proteins. Some of these anchor thin filaments at Z-discs or their intracellular and membrane equivalents; two of these proteins are α-actinin and desmin. Other proteins anchor actin filaments at cell membranes (e.g., spectrins). Many cytoskeletal proteins control cell shape and organelle position.

In the evolution of mechanisms for movement, morphological changes accompanied selection of contractile proteins. Intracellular movement early transferred substances from regions of synthesis to utilization, and implemented chromosome movement and cell division. Amoeboid movement and cyclosis were made possible by actin and myosin. In parallel with these there evolved tubulin systems, not so versatile as actomyosin, but retained in nearly all eukaryotes. The functions of actin and myosin in prokaryotes and plants are not well understood. The requirement for Ca^{2+} by both tubulins and actomyosins was established early in evolution. Several methods of Ca^{2+} regulation have persisted in parallel in different kinds of muscles. Different speeds of motion are adaptive for locomotion, for life in various media, and for kinds of behavior, for example, visceral and transport functions.

The evolution of biological movement well exemplifies causal connections and interrelations between holistic functions and molecular processes.

REFERENCES

1. Achazi, R. K. *Pflügers Arch.* **379**:197–201, 1979. pp. 291–304 in *Basic Biology of Muscles,* Ed. B. Twarog, et al., Raven, New York, 1982. Filament structure, phosphorylation in catch muscles.

1a. Adelstein, R. S. and E. Eisenberg. *Annu. Rev. Biochem.* **49**:921–956, 1980. Regulation of actin-myosin-ATP interaction.

2. Albrecht-Buehler, G. *Proc. Natl. Acad. Sci.* **77**:6639–6643, 1980. Autonomous movements of cytoplasmic fragments, fibroblasts, platelets.

3. Altringham, J. D. and J. A. Johnston. *J. Compar. Physiol.* **143**:123–127, 1981. Recruitment of fast muscle fibers in swimming of cod and carp.

4. Amos, L. A. et al. *Nature* **299**:467–469, 1982. Myosin heads may interact with two sites on F actin.

5. Anderson, M. *J. Exper. Biol.* **75**:113–122, 1978; **82**:227–238, 1979. pp. 309–322 in *Basic Biology of Muscles,* Ed. B. Twarog et al., Raven, New York, 1982. Structure and function of striated myoepithelium of Syllis pharynx.

5a. Anderson, P. A. *J. Compar. Physiol.* **B154**:257–268, 1984. Electrophysiology of muscle fibers from a ctenophore.

6. Anderson, P. A. V. and W. E. Schwab. *J. Morphol.* **170**:383–399, 1981. Structure of muscle and nerve in jellyfish Cyanea.

7. Angant-Petit, D. et al. *J. Physiol.* **289**:203–218, 1979. Dual innervation in amphibian muscles.

8. Atwood, H. *Am. Zool.* **13**:357–387, 1973. *Prog. Neurobiol.* **7**:291–391, 1976. Neuromuscular transmission in crustaceans.

9. Bagby, R. Ch. 1, pp. 1–84 in *Biochemistry of Smooth Muscle,* Vol. 1, Ed. N. L. Stephens, CRC Press, Boca Raton, FL, 1983. Structure of vertebrate smooth muscle.

10. Baldwin, K. M. et al. *Am. J. Physiol.* **222**:373–378, 1972. Enzymes of red, white, and intermediate muscles in relation to exercise.

11. Barany, M. et al. *Europ. J. Biochem.* **2**:156–164, 1967. Myosin ATPase from muscles of cat and sloth.

12. Barany, M. and K. Barany. *Annu. Rev. Physiol.* **42**:275–292, 1980. Phosphorylation of myofibrillar proteins.

13. Barany, M. et al. *J. Biol. Chem.* **254**:4954–4956, 1979; **255**:6238–6244, 1980. Phosphorylation of proteins in arterial muscles.

14. Bartholomew, G. N. and J. M. Casey. *J. Exper. Biol.* **76**:11–25, 1978. Oxygen consumption by insects at rest, in flight, and post-flight.

14a. Beamish, F. W. H. Ch. 2, pp. 101–189 in *Fish Physiology,* Vol. 7, Ed. W. Hoar and J. Randall, Academic Press, New York, 1978. Swimming capacity in fishes.

15. Bender, W. et al. *Cell* **15**:779–788, 1978. Genes for actin of Dictyostelium.

16. Bernstein, S. I. et al. *Nature* **302**:393–397, 1983. Myosin genes in Drosophila.

17. Bone, Q. J. *Journal of the Marine Biological Association, United Kingdom* **46**:321–349, 1966. Ch. 6, pp. 361–424 in *Fish Physiology,* Vol. 7, Ed. W. Hoar and J. Randall, Academic Press, New York, 1978. Functional types of muscle fibers in fishes.

18. Brevet, A. et al. *Science* **193**:1152–1154, 1976. Myosin synthesis in cultured skeletal muscle cells.

19. Brokaw, C. J. pp. 165–180 in *Molecules and Cell Movement,* Ed. S. Inoue and R. Stephens, Raven, New York, 1975. Sliding filaments in flagella.

20. Burridge, K. and J. Feramisco. *Nature* **294**:565–567, 1981. Nonmuscle α-actin.

20a. Carew, T. et al. *J. Neurophysiol.* **37**:1020–1040, 1974. Neuromuscular transmission to gill muscle, Aplysia.

21. Celio, M. R. et al. *Nature* **297**:504–506, 1982. Ca-binding by parvalbumin in fast striated muscle.

22. Chantler, P. D. and A. G. Szent-Györgyi. *J. Molec. Biol.* **138**:473–492, 1982. Scallop muscle light chains.

23. Chi, J. C. H. et al. *J. Cell Biol.* **67**:523–537, 1975. *Proc. Natl. Acad. Sci.* **72**:4999–5003, 1975. Myosin and actin in cultured muscle cells and fibroblasts.

23a. Chizzonite, R. A. et al. *J. Biol. Chem.* **257**:2056–2065, 1982. Iosomyosins of rabbit ventricle.

24. Clark, W. A. et al. *J. Biol. Chem.* **257**:5449–5454, 1982. Cardiac isomyosins.

25. Cleveland, D. W. *Cell* **34**:330–332, 1983. Review of tubulins.

26. Close, R. *J. Physiol.* **197**:461–477, 1968; **204**:331–346, 1969. Changes in fast and slow skeletal muscles after cross-innervation.

27. Cobb, L. S., Jr., and M. S. Laverarck. *Proc. R. Soc. London* **B164**:624–650, 1966. *Compar. Biochem. Physiol.* **24**:311–315, 1968. Lantern retractor muscles of Echinus.

28. Cohen, C. *Proc. Natl. Acad. Sci.* **79**:3176–3178, 1982. Molecules in catch mechanism.

29. Cohen, C. M. et al. *Nature* **279**:163–165, 1979. Actin and spectrin in red blood cell cytoskeleton.

30. Condeelis, J. S. *Exper. Cell Res.* **88**:435–439, 1974. *J. Cell Biol.* **74**:901–927, 1977. *Nature* **292**:161–163, 1981. Actin in plants and slime molds.

31. Craig, R. and G. Offer. *Proc. R. Soc. London* **B192**:451–461, 1976. Location of C-protein in muscle fibers.

31a. Curtis, B. and R. Eisenberg. *J. General Physiol.* **85**:383–408, 1985. Calcium influx in frog fast fibers.

32. Dabrowska, R. and A. Szpacenko. *Compar. Biochem. Physiol.* **56B**:139–142, 1977. Composition of muscle proteins of carp and rabbit muscle.

33. De Saye, C. and B. Swynghedauw. *Pflügers Arch.* **355**:39–47, 1975. Comparative study of heart myosins.

34. Durica, P. S. *Proc. Natl. Acad. Sci.* **77**:5683–5687, 1980. Actin genes in sea urchin.

740   Motility; Muscles

34a. Ebashi, S., M. Endo, and I. Ohtsuki. *Q. Rev. Biophys.* **2**:351–384, 1969. Control of muscle contraction.

35. Elliott, A. pp. 11–28 in *Basic Biology of Muscles,* Ed. B. M. Twarog et al., Raven, New York, 1982. Structure of thick filaments in catch muscles.

36. Elzinga, M. et al. *Proc. Natl. Acad. Sci.* **70**:2687–2691, 1973. *J. Biol. Chem.* **250**:5915–5920, 1975. Sequence analysis of skeletal muscle actins.

36a. Fay, F. S. and O. C. Delise. *Proc. Natl. Acad. Sci.* **70**:641–645, 1973. *J. Cell Biol.* **96**:783–795, 1983. Spiral structure in smooth muscle fibers as indicated by α-actinin.

37. Franzini-Armstrong, C. *J. Cell Biol.* **58**:630–642, 1973. *Fed. Proc.* **39**:2403–2409, 1980. with L. D. Peachey. Ch. 3, pp. 23–71 in *Handbook of Physiology,* Section 10, skeletal muscle. American Physiological Society, Bethesda, MD, 1983. Structure of Z-line and of T-tubules and SR in striated muscle.

38. Fuchs, E. and I. Hanukagen. *Cell* **34**:332–334, 1983. Review of structure of intermediate filaments.

39. Gabbiani, G. et al. *Proc. Natl. Acad. Sci.* **78**:293–302, 1981. Vimentin filaments and actin in vascular muscle cells.

40. Gadasi, H. and E. D. Korn. *Nature* **286**:452–456, 1980. Intracellular localization of myosin isoenzymes in Acanthamoeba.

41. Gallowitz, D. and I. Sures. *Proc. Natl. Acad. Sci.* **77**:2546–2550, 1980. Actin genes in yeast.

42. Gauthier, G. F. et al. *J. Cell Biol.* **92**:471–484, 1982. Development of fast and slow myosins.

43. Geiger, B. et al. *Proc. Natl. Acad. Sci.* **77**:4127–4131, 1980. Vinculin in gizzard smooth muscle.

44. Geisler, N. et al. *Proc. Natl. Acad. Sci.* **78**:4120–4123, 1981. Sequence analysis of proteins from intermediate filaments.

45. Gergely, J. *Fed. Proc.* **35**:1283–1287, 1976. Excitation-contraction coupling in cardiac myofilaments.

46. Goffart, M. *Form and Function in Sloths,* Pergamon, New York, 1971.

47. Gordon, D., E. Eisenberg, and E. Korn. *J. Biol. Chem.* **251**:4778–4786, 1976. Acanthamoeba actin.

48. Groschel-Stewart, U. *Biochem. Biophys. Acta* **229**:322–334, 1971. Composition of uterine smooth muscle.

49. Hagiwara, S. and Kidokora. *J. Physiol.* **217**:217–232. Action potentials of muscles of Amphioxus.

50. Harrington, W. T. *Proc. Natl. Acad. Sci.* **76**:5066–5070, 1979. Helix coil folding of S2 as theory of contraction.

51. Harwicke, P. M. et al. *Nature* **301**:478–482, 1983. Light chain regulation in scallop myosin.

52. Haselgrove, J. C. and C. D. Rodger. *J. Muscle Res. Cell Motility* **1**:371–390, 1980. Review of X-ray diffraction patterns of muscle.

53. Heath, J. E. and M. S. Heath. *Annu. Rev. Physiol.* **44**:133–143, 1982. Energetics of locomotion in endothermic insects.

54. Heiny, J. A. et al. *Nature* **301**:164–166, 1983. Electrical signals in transverse tubules of skeletal muscle.

55. Hernandez-Nicaise, M. L. et al. *J. Gen. Physiol.* **75:**79–105, 1980. *Biol. Bull.* **167:**210–228, 1984. pp. 513–522 in *Coelenterate Evolution and Behavior,* Ed. G. Mackie, Plenum, New York, 1976. Giant nonstriated muscle cells of ctenophore Beroe.

56. Heuser, J. E. *J. Molec. Biol.* **169:**97–122, 123–154, 1983. Ultrastructure of cross-bridge lattice in striated muscle without chemical fixatives.

57. Hidaka, T. et al. *J. Exper. Biol.* **50:**363–375, 387–403, 1969. Electrical properties of earthworm muscle.

58. Hoar, P. E., W. G. Kerrick, and P. Cassidy. *Science* **204:**503–506, 1979. Ca regulation in smooth muscle of chicken gizzard.

59. Hoh, J. F. *Biochemistry* **14:**742–747, 1975. Neural regulation of mammalian fast and slow muscle myosins.

60. Holly, R. G. et al. *Am. J. Physiol.* **238:**C62–C71, 1980. Stretch-induced growth in chicken wing muscles.

61. Hoyle, G. *Compar. Biochem. Physiol.* **66A:**57–68, 1980. Comparison of frog and arthropod muscles.

62. Hoyle, G. *Muscles and Their Neural Control,* Wiley, New York, 1983.

63. Huxley, H. E. *Proc. Natl. Acad. Sci.* **78:**2297–2301, 1981. *J. Molec. Biol.* **158:**637–648, 1982; **169:**469–506, 1983. X-ray reflections from contracting striated muscle.

64. Ikebe, N. and D. Hartshorne. *J. Biol. Chem.* **258:**14770–14773, 1983. Ca-modulation, gizzard.

65. Inoune, S. and R. E. Stephens, Eds. *Molecules and Cell Movement,* Raven, New York, 1975. Papers on tubulins, cilia.

66. Ito, L. et al. *J. Exper. Biol.* **50:**107–118, 1969. Excitatory junction potentials in earthworm muscles.

67. Johnston, I. A. Ch. 2, pp. 36–67 in *Fish Biomechanics,* Eds. P. Webb and D. Weiks, Praeger, New York, 1983. Dynamic properties of fish muscle.

68. Johnston, I. A. et al. *Compar. Biochem. Physiol.* **49B:**367–373, 1974. *J. Compar. Physiol.* **114:**203–216, 1977. *J. Physiol.* **295:**49P, 1979. *J. Exper. Biol.* **87:**177–194, 1980; **11:**171–177, 1984. *Cell Tissue Res.* **219:**93–109, 1981; **237:**253–258, 1983; **227:**179–199, 1982. Structural, physiological, and biochemical properties of red, white, and mixed muscles of fishes.

69. Josephson, R. K. pp. 20–44 in *Insect Thermoregulation,* Ed. B. Heinrich, Wiley, New York, 1981. *J. Exper. Biol.* **80:**69–81, 1979; **91:**219–237, 1981; **99:**109–125, 1982. Body temperature and effects of temperature on muscle contraction in insects.

70. Josephson, R. *J. Exper. Biol.* **108:**77–96, 1984. Contraction dynamics of cricket muscles.

71. Kamiya, N. *Annu. Rev. Plant Physiol.* **32:**205–236, 1981. Review of cytoplasmic streaming in slime molds and filamentous algae.

72. Keller, L. R. and C. P. Emerson. *Proc. Natl. Acad. Sci.* **77:**1020–1024, 1980. Synthesis of myosin by muscle cultures.

73. Kendrick Jones, J. et al. pp. 255–272 in *Basic Biology of Muscles,* Ed. B. M. Twarog et al., Raven, New York, 1982. Role of myosin light chains in Ca regulation of molluscan muscles.

74. Kerrick, G. L. et al. *J. Gen. Physiol.* **77**:177–190, 1981. Ca-regulatory mechanisms in nonstriated muscles.

75. Kessler, D. et al. *Cell Motility* **1**:63–71, 1981. Physarum myosin.

76. Kier, W. *J. Morphol.* **127**:179–192, 1982. Muscles of squid tentacles and arms.

77. Kirby, A. C. *Am. J. Physiol.* **219**:1446–1450, 1970; **225**:166–170, 1970. Membrane properties of phasic and tonic muscle fibers of frog.

78. Kobayashi, M. et al. *Compar. Biochem. Physiol.* **60C**:115–122, 1978; **65C**:73–79, 1980; **72C**:343–348, 1982. Responses of radula protractors and retractors to putative transmitters.

79. Korn, E. D. *Proc. Natl. Acad. Sci.* **75**:588–599, 1978. Biochemistry of actomyosin-dependent cell motility.

79a. Lanerolle, P. D. et al. *Science* **223**:1415–1417, 1984. Phosphorylation of myosin light chain kinase and effects of c-AMP in smooth muscle.

80. Lawn, I. D. and G. O. Mackie. *Science* **211**:1169–1171, 1981. Conduction in a sponge.

81. Lazarides, E. J. *Cell Biol.* **65**:549–561, 1975. Localization of tropomyosin in nonmuscle cells.

82. Lazarides, E. et al. *Nature* **283**:249–255, 1980. *Cell* **23**:524–532, 1981. Distribution of actin, filamin, vimentin, and desmin in chick skeletal muscle fibers.

83. Lebowitz, E. A. and R. Cooke. *J. Biochem.* **85**:1489–1494, 1979. Phosphorylation of uterine smooth muscle myosin.

84. Lehman, W. and A. G. Szent Györgyi. *J. Gen. Physiol.* **66**:1–30, 1975. Ca regulation in mollusc muscle.

85. Levine, R. J. C. et al. pp. 37–52 in *Basic Biology of Muscles,* Ed. B. Twarog et al., Raven, New York, 1982. Molecular organization of Limulus and holothunan thick filaments.

86. Lowey, S. et al. *Nature* **282**:522–524, 1979. Distribution of light chains in fast skeletal myosins.

87. Lowy, J. and F. R. Pulsen. *Nature* **299**:308–312, 1982. X-ray diffraction patterns of myosin heads of contracting unstriated muscle of Mytilus.

88. Lu, R. C. and M. Elzinga. *Biochemistry* **16**:5801–5806, 1977. Sequence homologies of actin from muscles, Acanthamoeba, and slime mold.

89. Lundholm, L. et al. Ch. 2, pp. 85–109 in *Biochemistry of Smooth Muscle,* vol. 2, Ed. N. L. Stephens, CRC Press, Boca Raton, FL, 1982. Phosphorylation sequence in smooth muscle.

90. Mackie, G. O. *J. Neurobiol.* **6**:357–378, 1975. Myoepthelia of jellyfish.

91. Mahdawi, V. et al. *Nature* **297**:659–664, 1982. *Proc. Natl. Acad. Sci.* **81**:2626–2630, 1984. Genes for myosin heavy chains for adult heart myosin.

92. Mannherz, H. G. and R. S. Goody. *Annu. Rev. Biochem.* **45**:428–464, 1974. Energy cycle in contractile systems.

93. McArdle, J. J. and I. A. Johnston. *Experimental Cell Biol.* **25**:103–107, 1981. Ca^{2+} uptake by SR from fish muscles.

94. McArdle, J. J. and E. X. Albuquerque. *J. Gen. Physiol.* **61**:1–23, 1973. Reinnervation of fast and slow mammalian muscles.

95. Meiss, D. E. and C. K. Govind. *J. Exper. Biol.* **79**:99–114, 1979. Regional differentiation in lobster muscle.

96. Minkoff, L. and R. Damidian. *J. Bacteriol.* **125**:353–365, 1976. Actin from E. coli.

97. Minty, A. J. et al. *Cell* **30**:186–192. Expression of genes for actin.

98. Murphy, R. A. et al. *Am. J. Physiol.* **240**:C222–233, 1981. Role of Ca and myosin light chain phosphorylation in arterial muscles.

99. Nabeshima, Y. et al. *Nature* **308**:333–338, 1984. Transcription splicing for mRNAs for two myosin light chains.

100. Nakamura, K. and S. Watanabe. *J. Biochem.* (Japan) **83**:1459–1470, 1978. Myosin and actin from E. coli.

101. Neering, I. R. and K. G. Morgan. *Nature* **288**:585–587, 1980. Aequorin in excitation-contraction coupling in smooth muscle.

102. Nguyen, H. T. et al. *Proc. Natl. Acad. Sci.* **79**:5230–5234, 1982. Genetic coding of myosins.

103. Obinata, T. et al. *Compar. Biochem. Physiol.* **76B**:437–442, 1983. Myosin and actin from ascidian muscle.

104. Palevitz, B. A. and P. Hepler. *Proc. Natl. Acad. Sci.* **71**:363–366, 1974. *J. Cell Biol.* **65**:29–38, 1975. Actin at endoplasm-ectoplasm interface in Nitella.

105. Pardo, J. V. et al. *Cell* **32**:1093–1103, 1983. Subcellular sorting of isoactins.

106. Phillips, G. N. et al. *Nature* **278**:413–417, 1979. *Biophys. J.* **32**:485–500, 1980. Crystal structure and molecular interactions of tropomyosin.

107. Pollard, T. D. and E. D. Korn. *J. Biol. Chem.* **248**:4682–4690, 1973. Acanthamoeba myosin.

108. Pollard, T. D. and P. R. Weiking. *C.R.C. Cortical Reviews in Biochemistry* **2**:1–65, 1974. Review of actin and myosin in cell movement.

109. Potter, J. D. *Arch. Biochem. Biophys.* **162**:436–441, 1974. Proportions of different proteins in rabbit skeletal muscle.

110. Poulsen, F. R. and J. Lowy. *Nature* **303**:146–152, 1983. X-ray scattering from myosin heads of relaxed and rigor frog muscle.

111. Pringle, J. pp. 283–329 in *Physiology of Insects,* Vol. 2, Ed. C. Rockstein, Academic Press, New York, 1965. Efficiency of flight muscles of insects.

112. Pringle, J. *Proc. R. Soc. London* **B201**:107–130, 1978. Stretch activation of muscle.

113. Prosser, C. L. *Z. vergl. Physiol.* **54**:109–120, 1967. Contractions in sponge myocytes.

113a. Prosser, C. L. Ch. 16, pp. 719–788 in *Comparative Animal Physiology,* Ed. C. L. Prosser, Saunders, Philadelphia, 1973.

114. Prosser, C. L. pp. 381–398 in *Basic Biology of Muscles,* Ed. B. Twarog et al., Raven, New York, 1982. Diversity of narrow-fibered and wide-fibered muscles.

115. Prosser, C. L. Ch. 21, pp. 635–670 in *Handbook of Physiology,* Vol. 2, Cardiovascular System, American Physiological Society, Bethesda, MD, 1980. Evolution and diversity of nonstriated muscles.

116. Prosser, C. L. and G. O. Mackie. *J. Compar. Physiol.* **136**:103–112, 1980. Contractions of holothurian muscles.

117. Prosser, C. L. and A. Mangel. *J. Exper. Biol.* **86**:237–248, 1980. Rhythms in toad stomach.

117a. Prosser, C. L. and A. Mangel. Ch. 10, pp. 273–301 in *Cellular Pacemakers,* Ed. D. Carpenter, Wiley, New York, 1982. Rhythmicity in smooth muscle.

118. Reedy, M. K. *J. Molec. Biol.* **31**:155–176, 1968. Ultrastructure of insect flight muscle.

118a. Robert, B. et al. *Proc. Natl. Acad. Sci.* **79**:2437–2441, 1982. DNA complementary to mRNA for l.c.s 1 and 3 of mouse myosin.

119. Regenstein, J. M. *Compar. Biochem. Physiol.* **56B**:239–244, 1977. Lobster striated muscle myosin.

120. Rome, L. C. et al. *Am. J. Physiol.* **247**:R272–279, 1984. Muscle fiber activity in carp as function of swimming speed and temperature. *J. Exper. Biol.* **99**:269–277, 1982; **108**:429–439, 1984. Energetics of running in lizards and jumping in frogs.

121. Rosenbluth, J. *J. Cell Biol.* **25**:494–515, 1965; **54**:566–579, 1972. Ultrastructure of diagonally striated muscles of Ascaris and Lumbricus.

122. Rubenstein, P. A. and K. A. Spudick. *Proc. Natl. Acad. Sci.* **74**:120–123, 1977. Actins in chick fibroblasts and adult smooth muscle.

123. Sarkar, S. et al. *Proc. Natl. Acad. Sci.* **68**:946–950, 1971. Myosin light chains from red, white, and cardiac muscle, molecular size.

124. Schmidt-Nielsen, K. *Science* **177**:222–228, 1972. Metabolic cost of flying, swimming, and walking for animals of different sizes.

125. Sheetz, M. P. and J. A. Spudich. *Nature* **303**:31–35, 1983. Movement of myosin beads on actin cables in vitro.

125a. Shriver, J. M. *Trends Biochem. Sci.* **9**:312–328, 1984. Energy transduction by myosin.

126. Singla, C. L. *Cell Tissue Res.* **188**:317–327, 1978. Locomotion and neuromuscular system of jellyfish Aglantha.

127. Sinha, A. M. et al. *Proc. Natl. Acad. Sci.* **79**:5847–5851, 1982. Cloning of mRNA sequences for cardiac α and β myosin heavy chains.

128. Small, J. V. and A. Sobieszek. Ch. 2, pp. 86–140 in *Biochemistry of Smooth Muscle,* Vol. 1, Ed. N. L. Stephens, CRC Press, Boca Raton, FL, 1982. Contractile and structural proteins.

128a. Smith, R. S. and W. K. Ovalle. *J. Anat.* **116**:1–24, 1973. Five types of amphibian muscle fibers.

129. Spencer, A. N. *J. Compar. Physiol.* **144**:401–407, 1981; **148**:353–363, 1982. Excitation and contraction of swimming myoepithelia of a medusoid coelenterate.

130. Spudich, J. A. *J. Biol. Chem.* **246**:4866–4871, 1971; **250**:7485–7491, 1975. Actin in skeletal muscle.

131. Spudich, J. A. *J. Biol. Chem.* **249**:6013–6020, 1974. Actin from amoebae of Dictyostelium.

132. Spudich, J. A. and M. A. Clarke. *Annu. Rev. Biochem.* **46**:797–822, 1977. Role of actin and myosin in cell motility and shape determination.

133. Squire, J. M. *Structural Basis of Muscle Contraction,* Plenum, New York, 1981. *TINS,* October 1983. Molecular mechanisms in muscular contraction.

133a. Stephens, N. L. *Biochemistry of Smooth Muscle,* 3 vols., CRC Press, Boca Raton, FL, 1983.

134. Stephens, R. E. pp. 181–206 in *Molecules and Cell Movement,* Ed. S. Inoue and R. Stephens, Raven, New York, 1975. Structure and function of tubulin filaments.

135. Stokes, D. R. et al. *J. Exper. Zool.* **193:**281–300, 1975; **194:**379–408, 1975. Neural control of fast muscles in singing katydid.

136. Streter, F. A. et al. *J. Biol. Chem.* **241:**5772–5776, 1966. *Nature* **241:**17–18, 1973. *J. Gen. Physiol.* **66:**811–821, 1975. Effects of cross-inneration between fast and slow muscles on muscle proteins.

137. Syrovy, J. *Compar. Biochem. Physiol.* **72B:**289–293, 1982. Myosin ATPase of hearts of mammals.

138. Tada, M. and A. M. Katz. *Annu. Rev. Physiol.* **44:**401–423, 1982. Phosphorylation of proteins of cardiac sarcolemma.

139. Takahashi, H. and H. Iwamoto. Personal communication.

140. Takamatsu, H. et al. *Compar. Biochem. Physiol.* **70B:**435–439, 1981. Myosin from dog colon.

141. Taylor, C. R. *Am. J. Physiol.* **219:**1104–1107, 1970. *J. Exper. Biol.* **86:**9–18, 1980. pp. 161–170 in *Companion to Animal Physiology,* Ed. C. R. Taylor et al. Cambridge University Press, New York, 1982. Energetic constants with body size in locomotion of mammals.

142. Taylor, K. A. and L. A. Amos. *J. Molec. Biol.* **147:**297–324, 1981.

143. Tilney, L. G. and P. Detmers. *J. Compar. Physiol.* **66:**508–520, 1975. Actin and spectrin from erythrocyte ghosts.

144. Tobin, S. L. *Cell* **19:**121–131, 1980. Actin genes in Drosophila.

145. Toh, B. H. et al. *Cell Tissue Res.* **199:**117–126, 1976. Localization of actin and myosin by fluorescent antibodies.

146. Trinick, J. and S. Lowey. *J. Molec. Biol.* **113:**343–368, 1977. Proteins of M-line.

147. Tsong, T., T. Karr, and W. F. Harrington. *Proc. Natl. Acad. Sci.* **76:**1109–1113, 1979. Helix-coil transitions in S-2 region of myosin.

148. Tucker, V. A. *Compar. Biochem. Physiol.* **34:**841–846, 1970. *Am. Zool.* **11:**115–124, 1971. *Am. Sci.* **63:**413–419, 1975. Energetic cost of locomotion.

149. Umeda, P. K. et al. *Proc. Natl. Acad. Sci.* **78:**2843–2847, 1981. Cloning of two fast myosin heavy chains from chicken muscle.

150. Vahey, M. *Biophys. J.* **21:**23AM-AM-G7, 1978. Myosin in tomatoes.

151. Vanderkerckhove, J. and K. Weber. *Proc. Natl. Acad. Sci.* **75:**1106–1110, 1978. *J. Molec. Biol.* **126:**783–802, 1978. *Nature* **276:**720–721, 1978. *FEBS Lett.* **102:**219–222, 1979. Amino acid sequences of actins from different tissues.

151a. Wabnitz, R. W. *Compar. Biochem. Phys.* **54C:**75–80, 1976; **55A:**253–259, 1976. Excitation-contraction coupling in Helix muscle.

152. Wagner, P. D. *J. Biol. Chem.* **256:**2493–2498, 1981. *Nature* **292:**560–562, 1981. Role of heavy chains of myosin in ATPase action; hybrids of light chains.

153. Walliman, T. and A. G. Szent Györgyi. *Biochemistry* **20:**1188–1197, 1981. *J.*

Molec. Biol. **156**:141–173, 1972. Ca regulation by specific light chain of myosin in scallop muscle.

154. Wang, K. et al. *Proc. Natl. Acad. Sci.* **81**:3685–3689, 1984. Properties of titin.

155. Water, R. D. et al. *J. Bacteriol.* **144**:1143–1151, 1980. Yeast actin.

156. Waterston, R. H. et al. *J. Molec. Biol.* **90**:285–290, 1974. Myosin mutants in nematode.

157. Weeds, A. G. et al. *Nature* **247**:135–139, 1974. Myosins from cross-innervated fast and slow muscles of cat.

158. Weis-Fogh, T. *J. Exper. Biol.* **59**:169–230, 1971. pp. 729–762 in *Swimming and Flying in Nature,* Plenum, New York, 1971. Aerodynamics of insect flight.

158a. Weis-Fogh, T. and W. Amos. *Nature* **236**:301–304, 1972. Energy for shortening of ciliate spasmonemes.

159. Westgaard, R. H. *J. Physiol.* **251**:683–697, 1975. Influence of activity on passive electrical properties of denervated muscle, rat.

160. Whalen, R. G. et al. *Proc. Natl. Acad. Sci.* **73**:2018–2022, 1976; **76**:5197–5201, 1979. *Nature* **286**:731–733, 1980; **292**:805–809, 1981. Development of myosin isozymes in rats; embryos and neonates.

161. Yates, L. D. and M. L. Greaser. *J. Molec. Biol.* **168**:123–141, 1983. Quantitative measures of protein fractions in rabbit skeletal muscle.

162. Yu, J. and S. R. Goodman. *Proc. Natl. Acad. Sci.* **76**:2340–2344, 1979. Cytoskeletal proteins.

163. Zengel, J. M. and H. F. Epstein. *Proc. Natl. Acad. Sci.* **77**:852–856, 1980. Nematode muscles.

17

GENERAL SUMMARY

The objectives of this book are (1) to summarize the evidence for functional adaptations as the basis for evolution; (2) to describe the physiology of present-day organisms in terms of their evolutionary history; (3) to describe life as a series of dynamic processes; and (4) to present a generalized rather than a specialized view of biological adaptations.

Not since the post-Darwinian period has there been so much controversy among biologists (Ch. 1). Many biologists take a reductionist position and propose that life can be understood in strictly molecular terms. Other biologists with holistic views consider that knowledge of organismic integration is the way to understand life.

According to one viewpoint, evolution occurs by gradual accumulation of small genetic changes in geographically isolated populations; other evolutionists, viewing saltatory steps in the geological record, propose that evolution is by major genetic changes that occur periodically.

The species concept has several meanings—(a) descriptive morphological taxonomy, (b) phylogenetic cladistics, (c) reproductive isolation as identified by population genetics, (d) hybridization and colonization, especially in plants, and (e) physiological variations in different ecological situations.

Another disagreement is in the interpretation of behavior. One view is that description of behavior can be reduced to membrane properties, synaptic transmission, and neural circuitry; the opposing view is that complex behavior cannot be known from the study of simple nervous systems, that behavior consists of integrative functions at different levels from neural analysis, that mind cannot be understood by the study of brain. It is my contention in this book that a functional or adaptational approach can aid in the resolution of these and other conflicts of opinion.

Several general properties of life must be considered prior to discussing specific topics of adaptational biology. These general properties are detailed in Chapter 1.

The first to be considered in this summary is the hierarchical organization of biological structure. In living systems the properties of wholes are not equal to the sum of the properties of separate parts. Macromolecules in isolation are unlike the same molecules when they are contained in organelles. Functions of cells cannot be predicted from the properties of their chemical components, and whole organisms are not equal to the sum of their cells. An aim of modern biology is to describe organisms in terms of component molecules. However, the functional principles at one level are very different from those at another level, and at each level of integration properties emerge which differ from those of less integrated components. It is difficult to describe properties at one level in terms of another.

An intact organism does not tolerate stress over as wide a range as its cells and tissues, and those cells and tissues tolerate less stress than the molecules that compose them. The more highly integrated a biological system is, the more circumscribed is its range of tolerance and function. The limits of genetically inherited functions are constrained by biological organization.

A related topic is the thermodynamic description of living organisms. Some theoreticians have attempted to treat organisms as closed systems thermodynamically. In fact, organisms are quasi-open systems with exchange between organism and environment. However, at any instant in time, input equals output, and for practical purposes boundaries must be set within which life processes are limited. Life converts energy from a closed equilibrium state to an open nonequilibrium state. As noted by several thermodynamic treatments, "life evades decay toward equilibrium"; life opposes the forces of entropy.

Life functions and information content of living organisms increase nonlinearly as evolution proceeds and as complexity increases. Nonlinearity results from various feedback and amplification reactions that complicate theoretical treatment of the origin of life.

Biological indeterminacy or quasi-randomness occurs in single components of complex living systems. Mutation is not directed, although it can vary in frequency; single neurons in an aggregation of similar cells in a brain show controlled randomness; as development proceeds equipotency of cells diminishes. In each of these and other examples, control of the system is influenced or directed by external factors. Evolution is directed by natural selection; feedback sensory and motor controls monitor neuronal activity.

It is certain that elementary physicochemical formulations are inadequate to define life.

ORIGINS

The age of the earth is estimated at some 4.6 billion years. Adaptational biology starts with the premise that much biochemical evolution, perhaps the most important part, occurred before there were organisms we would recognize as such. There is little doubt that organic molecules were formed in quantity during the prebiotic period by the action of energy from both solar and terrestrial sources, in an aquatic medium, possibly in saline pools or on clay particles. Organic acids, sugars, fatty acids, amino acids, and organophosphates that now take part in intermediary metabolism have been synthesized under simulated prebiotic conditions. The earliest Cyanobacteria, estimated from stromatolites, are 3 billion–3.5 billion years old. At that time the atmosphere changed from reducing to oxidizing.

Some of the following universally critical molecules probably became established by selection pressures which can only be speculated upon: 20 L-amino acids as building blocks for proteins; polymerization of peptides, formation of nucleotides by joining purines and/or pyrimidines with pentoses and phosphates; coupling of phosphates with sugars as energy carriers, specifically ATP; polymer formation of nucleotides; arrangement of phospholipids and proteins as bounding membranes with selective permeability; self-replication of nucleic acids, probably the first being tRNAs; selection of nucleotides in DNA with a three-letter genetic code; association of RNAs with amino acids leading to synthesis of proteins, some of which exerted feedback interactions with RNA; specific inorganic ions used as cofactors in chemical reactions and as solutes in primitive cells (most cells concentrate K^+ and have higher H^+ than the milieu); complexing of metals with porphyrin rings which became versatile electron-carrier molecules. One postulate is that the first "organisms" were hypercycles of transfer RNAs and proteins which were capable of replication by feedback interactions. Besides the capacity for replication, the first organisms must have had a membrane containing a proton pump and ion-selective channels.

As soon as organic solutes, usually anionic, accumulated in cells, some mechanism for maintenance of osmotic equilibrium became necessary. At the same time regulation of intracellular pH and of redox potential, with resulting transmembrane electric potential, were essential. Selective permeability was insufficient, and active extrusion of cations must have evolved. A proton pump is present in prokaryotes, in the mitochondrial membrane of eukaryotes and in plasma membranes of most plants. In animal cells a sodium-potassium pump serves similar functions.

Proteins are similar in general composition in all organisms. Protein composition reflects availability of amino acids and of nucleotide codon assignments. Some 300 amino acids are known in nature; why were only 20 of them used in proteins? These 20 amino acids are in functional classes: negatively charged or acidic aspartic and glutamic acids; positively charged basic lysine, arginine, histidine; polar hydrophilic molecules—glycine, serine, threonine, tyrosine, asparagine, glutamine, tyrosine; nonpolar hydrophobic alanine, valine, leucine, isoleucine, proline, methionine, phenylalanine, tryptophan; imidazole histidine; SH-bridging cysteine; proton-transferring histidine; proline at kinks in secondary folds of peptides; aromatic phenylalanine. Hydrophobic amino acids tend to be in the center of protein molecules, hydrophilic ones in the periphery.

The earliest organisms were undoubtedly heterotrophic, using energy from preexisting compounds. An energy-yielding pathway that is so universal as to have evolved very early is anaerobic glycolysis. Before glycolysis, there must have appeared transfer agents which maintained redox balance, and which persist now as coenzymes. Before the glycolytic pathway was established the individual reaction steps probably functioned separately.

Evidence for early metabolic "experimentation" in chemoautotrophy and then photoautotrophy comes from the diversity of known prokaryotes, many of which are probably different from their progenitors. Some of these, the archaebacteria, persist in extreme environments such as may have obtained in the prebiotic period—hypersaline pools, high temperatures, acid water, anaerobic media.

Some chemoautotrophic eubacteria use sulfur compounds as energy sources. The first photoautotrophic organisms were anaerobic and synthesized sugar by a single photocenter without liberating oxygen. Diversity of photosynthetic pathways in bacteria indicates considerable testing before the two-center photosynthesis of cyanobacteria and modern plants evolved. When oxygen came to be liberated from water molecules in photosynthesis, many anaerobes were poisoned and oxygenase enzymes for counteracting the toxic O_2 appeared. A reasonable postulate is that when O_2 was available as an electron acceptor more energy was released from substrates, and glycolytic steps reversed from synthesizing to energy-yielding. Cytochromes which had functioned anaerobically became electron carriers to oxygen.

After eukaryotes evolved, cellular metabolism became more specialized and complex. However, from our present perspective the most striking aspect of prebiotic evolution is the early establishment of all the fundamental processes of protein and nucleic acid synthesis and degradation, and of energy metabolism. The selective pressures for those processes that have become nearly universal for living organisms may become elucidated as techniques of simulation advance.

GENETIC REPLICATION AND PROTEIN SYNTHESIS

It was essential for evolution that organisms should have the capacity for replication and for transmission of codes for synthesis of proteins and nucleic acids. The earliest nucleic acids are postulated to have been tRNAs because these are small in size, relatively simple in structure, and function as carriers of specific amino acids. In some present-day viruses the genome consists of single strands of RNA, analogous to messenger RNA. DNA must have evolved very early; it contains one pyrimidine different from RNA and deoxyribose instead of ribose. DNA is capable of polymerizing into very long molecules. DNA carries the genetic information in all prokaryotes and eukaryotes. A DNA molecule consists of a double helix of two strands of covalently linked deoxyribose and phosphate connected by pairs of nucleotides. The nucleotides consist of four bases: a purine adenine (A) or guanine (G) bonded to a pyrimidine cytosine (C) or thymine (T). Each gene is composed of a specific series of nucleotide pairs which serve as template for encoding messenger; mRNA has the same nucleotides as DNA except for uracil instead of thymine. Every protein is encoded by its mRNA; each amino acid in a protein is represented by four nucleotide sequences of three bases each. The genetic code in DNA is transcribed to a corresponding pattern in RNA. If the sequence in DNA is known, that of mRNA can be deduced, and thus the amino acid sequence of an encoded protein. Such computed protein sequences agree well with those obtained by direct analysis of proteins.

The origin of the genetic code is obscure, but it must be very ancient, because the same code is used by all organisms (except for some mitochondrial and chloroplast DNA). It has been proposed that one- or two-letter codes that would encode 4^1 or 4^2 messages would have been inadequate and that these may have been selected against. However, a three-letter code provides for 64 possible triplets. In actuality, since there are 20 amino acids in proteins, 61 of the possible triplets are used, the remaining three serving for termination sequences. Of the three letters in each code, the third is the most degenerate (i,e., can be varied without much change in the message). The second base is highly conserved; for example, when the second letter is U (uracil) hydrophobic amino acids are encoded. The origin of the genetic code may have been by a stereochemical fit between amino acids and their nucleotide codons. There could have been autocatalytic reciprocal cycles between polynucleotides and polypeptides. Selection could have resulted if some of the synthesized proteins had catalytic properties.

Transfer RNAs specify amino acids, and each tRNA has a cloverleaf configuration. The nucleotide sequences in tRNAs vary in different organisms, but the general plan and the sequences must have evolved very early. Several different tRNAs can carry the same amino acid.

Besides tRNAs and mRNAs there are ribosomal rRNAs, which may constitute 80–90 percent of the total RNA. It is on the ribosomes that a mRNA molecule aligns the amino acids transferred to it in acylated form by a tRNA. Translation from nucleic acids to protein results from the message read via the mRNA from one site to another on the ribosome. The universality of the genetic code, the transcription sequence from DNA to mRNA, and the eventual translation to protein argue for coevolution of these constituents in prebiotic times.

The genome of all organisms is highly dynamic, and this guarantees genetic diversity. Dynamic genetics demonstrates long noncoding regions in genomes of eukaryotes and many nucleotide substitutions that are not reflected in proteins. Some of the characteristics of dynamic genomes are:

1. Most genes consist of start and stop nucleotide sequences, encoding exons, and noncoding regions or introns. Exons may be interrupted by long introns in most genes of eukaryotes, not in prokaryotes. The net

effect is longer genes in eukaryotes. Why prokaryotes lack introns is not clear.

2. Another dynamic characteristic of genomes is jumping genes, which shift from one position to another within a genome, usually on the same chromosome.

3. Deletions and splicing, mostly in noncoding regions, result in changes in the order of nucleotide pairs and of entire genes.

4. Many genes recur several times (i.e., are repetitive). For example, the gene for actin is copied 5–20 times per haploid genome in numerous animals.

5. Many genomes have regions of satellite DNA for which there is no complementary mRNA.

6. Pseudogenes are noncoding regions which have been referred to as "dead" genes.

Noncoding regions of genomes have been described as regulatory to the coding exons. The nature of such regulation is obscure. The long noncoding sequences may represent reserve DNA, and may be very critical for evolutionary change. Another view is that much noncoding DNA is residual "junk." Genetic diversity is, therefore, much greater than would be concluded from breeding experiments and protein composition. Genomes are considered by some biologists as self-perpetuating, with the rest of the organism as a "host."

Genes for secreted proteins encode large precursor pro-proteins and pre-pro-proteins which are cleaved posttranslationally into two or more proteins. How genes evolved for proteins that later are cleaved into small fractions of the original precursor is not known. Feedback controls from proteins to genetic coding must have been selected; probably some precursor or product proteins now have different functions from their original ones. The role of proteolysis in protein formation has not been much considered. Several instances are now known in which a precursor mRNA is encoded, and this is then processed to a second-level mRNA for specific proteins.

Examples of the differences from DNA to mRNA to final processed protein are numerous. The stomach hormone gastrin contains 17 amino acids, progastrin has 104 amino acids. The cDNA for gastrin contains 602 base pairs (bps) of which 312 are encoding exons; the mRNA has 61 base pairs in its 5' end, 56 in the 3' end, and 86 in a polyadenylate tail, in addition to the approximately 100 for the progastrin.

An example of stepwise processing is insulin. In the laboratory rat there are two insulins, I and II, each of which has A and B chains. Pre-pro-insulin is a chain of 110 amino acids. Posttranslational processing consists of elimination of the first 24 amino acids in the pre-region. The remaining pro-insulin then folds and splits to A and B chains; disulfide bridges form and by proteolysis a C-peptide is cleaved away. In rat the gene for insulin I has one intron; gene II has two introns, one of 119 bps, the other of 499 bps. The smaller intron in each encodes the 5' noncoding region of mRNA. The two pre-pro-insulin genes may be products of a recent gene duplication. From amino acid sequences it is estimated that the rat genes I and II diverged some 20 million years ago.

Most genes become expressed only at certain times in development, and the means of control of timing has not been fully elucidated. A striking example is the silk gland of the moth Bombyx, in which the cells become polyploid during a few days in late larval development. Each fibroin gene leads to manufacture of 10⁴ molecules of mRNA, and each mRNA codes the formation of 10⁵ molecules of silk protein.

One of the most complex examples of posttranslational processing is the precursor protein in mammals of pro-opiomelanocortin. The gene contains 1091 base pairs, the precursor protein has 132 amino acids. A leader nucleotide sequence is followed by the sequence for gamma MSH (melanocyte-stimulating hormone), then comes the sequence for ACTH (adrenocorticotropin hormone), a 39 amino acid protein which contains the sequence for α MSH of 13 amino acids (positions 82–99 in the gene), then a protein CLIP (corticotropin-like intermediate peptide) of 21 amino acids, then β LTP (lipotropin) of 91 amino acids (nucleotides 42–132) which includes γ LTP of 58 amino acids (sequences 42–101), then β endorphin of 30 amino acids (sequences 104–134) and a termination sequence. Thus, the entire pro-opiomelanocortin gene encodes precursors for two types of MSH, β endorphin, β and γ LTP, ACTH plus leader sequences. The sequence for encoding enkephalin is nearly as complex and differs in hypothalamus from adrenal.

The gene for skeletal muscle myosin is repeated 8–10 times and a precursor mRNA is first encoded; this is then processed to the final mRNA which encodes myosin heavy chain (Ch. 16).

BIOLOGICAL VARIATION

Variation is characteristic of life. Whether variation refers to polymorphism between individuals or to diversity of populations, subspecies, species, higher taxonomic levels or clones of genetic strains, diverse organisms provide the raw material for natural selection. Experimentalists attempt to minimize individual variation by controlling heredity and environment, by inbreeding, and by factoring out variability. However, variation has broad biological meaning and experimentalists could well give more attention to it.

Taxonomic species are useful for identification of organisms; criteria are mainly morphological. Structural variations may be obviously adaptive or they may be neutral. The functions of structures of fossils can only be inferred. Taxonomic species need not correspond with phylogenetic relationships. Functional characters may supplement identification by systematists. Experimentalists working with wild organisms in different geographic areas may not be aware that they are using taxonomically different groups.

Biological species are those between which gene exchange does not occur, even when they are sympatric. Reproductive isolation is difficult to ascertain for organisms that hybridize extensively or that reproduce asexually. For both population ecologists and physiologists, organisms represent a continuum of variation, with boundaries between taxa set more or less arbitrarily by systematists. As finer criteria of variation are applied, the amount of known variation becomes enormous. Much evolutionary literature is concerned with the origin of species. It is my conviction that more attention should be given to the origin of biological diversity. Since no two species can occupy the same ecological niche and geographic range throughout their life histories, descriptions of adaptations to the total environment characterize the functional units, the "physiological species" on which natural selection operates.

Every organism is determined in its morphology and chemistry by three influences: genetic, developmental, and environmental. The genotype is the product of a long evolutionary history and it sets the limits within which phenotypic variation can occur. The simple hypothesis of one gene for one protein is no longer tenable. Feedback controls from proteins to genetic encoding must have been selected.

One view of genetic change is that most mutations are negative or neutral. There is much genetic "noise" and not all nucleotide substitutions are demonstrably adaptive. There is far more variation in genomes than in mRNAs, and more variation in mRNAs than in synthesized proteins. Many changes in amino acid sequences are trivial. Nonadaptive changes can be carried as neutral characters by linkage with adaptive genes. A physiologist argues for selection on the grounds that when subtle kinetic measurements are made on proteins, differences are found that have advantages for the ecology and life history of plants and animals. Simple comparisons of survival, growth rate, reproduction, or gross metabolism (as O_2 consumption) may not show adaptive properties of small changes in amino acid substitutions. Sequence differences (primary structure) may be less important for evolution than higher-order conformation of proteins.

Inherited Adaptations

The following paragraphs list a few examples of biological variation related to genetic determination. Since genes are mainly expressed by the proteins they encode, comparison of protein composition allows estimates of relatedness of organisms. One method used by population biologists to assess relatedness is to compare corresponding proteins by their electrophoretic patterns or by amino acid sequencing. Isozymes (encoded by separate genes) and allozymes (mutant forms of one gene) can be identified by bands on an electrophoretic gel. Polymorphism within a population is assessed by comparing positions, numbers, and intensities of these bands in many individuals. In practice, protein patterns are comparable to morphological descriptions and can be measured quantitataively. Unfortunately, many proteins which are useful for population analysis because they stain well on a gel are not well known as to function (e.g., esterases, phosphatases, proteases).

A second level of protein analysis is the sequencing of amino acids and determination of primary structure, and from it of secondary and tertiary structure. Measurements of protein composition by amino acid sequences and by corresponding DNA sequences have shown that there is much more variation within populations than had been previously supposed on morphological grounds; there are vast reserves of genetic material. Some sibling species, morphologically similar, are distinguishable by their protein patterns.

Families of related proteins show various degrees of homology between members; proteins are useful for assessing relationships between groups of organisms and between species within a group. Comparison of differences in proteins with paleontological data leads to construction of phylogenetic trees. The physiological significance of changes over evolutionary time is inferred from protein function and from the ecology of present-day forms. Some proteins are little changed throughout evolution. The time in millions of years computed for 1 percent change in sequence is called the unit evolutionary period. This period for histone (H_4) is 400 million years. Histones provide the structural skeleton of chromosomes, and this has not changed in 10^9 years. Some special features of histone structure have not been explained; genes for sea urchin histones contain more noncoding nucleotides than there are in histone genes of other known organisms. Histones of the ciliate Tetrahymena are so different from those of other eukaryotes as to indicate early divergence. Intermediate rates of evolution seem probable for other proteins; the unit evolutionary period for LDH A_4 is 19 million years (myr), for cytochrome c 15 myr. Hemoglobins have evolved fast among vertebrates, with unit evolutionary period of 3.7 myr. The fastest evolving proteins known are vertebrate immunoglobulins, with unit evolutionary period of 0.7 myr.

The best known example of protein evolution correlated with function is vertebrate hemoglobin (Hb). Hemoglobins occur in unrelated invertebrate animals—many annelids, a few pelecypods, holothurians, microcrustaceans, and fly larvae—but these hemoglobins are diverse and no connection to vertebrate hemoglobin is known. Hemoglobin occurs in root nodules of nitrogen-fixing legumes, but this may represent a secondary insertion. The first chordate hemoglobin was probably a monomer similar to the Hb of modern cyclostomes and to the myoglobin of other vertebrates. The time for transition from a monomer to a tetramer is uncertain; this change permitted cooperativity between the four chains, allowed a sigmoid oxygen dissociation curve, and expanded the possible physiology considerably. Perhaps the first such HB was a homotetramer α2α2. A gene duplication in early jawed fishes at about 500 myr ago resulted in the formation of two Hb lines, α-like and β-like; these differ in the amino acid sequences at the carboxyl terminal. There are 30 sites for contact between the Hb chains, of which 27 are in similar positions on both chains. In adult human Hb each α chain contains 141 amino acids, each β chain 146 amino acids; there are 75 differences and 65 similarities in sequences between the α and β chains.

The evolution of both α and β chains was rapid in fishes; evolution of positions for critical functions was four times as fast as for structural positions. The rate of evolution of Hb slowed in early tetrapod divergence. Evolution was faster in marsupials than in other mammals. From mammalian ancestors to primates, a period 90 to 65 myr ago, there was a fourfold to fivefold acceleration. From a primate ancestor to humans (35 myr to present) the rate decelerated and the α substitutions were at the rate of only two nucleotide changes per 100 codons per 10^9 years (Ch. 4). In early mammals a duplication of the β locus occurred with appearance of the γ Hb gene. The γ Hb is similar to β protein in 73 percent of sequences; γ Hb is used in fetal Hb such that HbF is α2 γ2 instead of adult α2 β2. Fetal hemoglobin has higher affinity for O_2 and lower potentiation of O_2 binding by diphosphoglycerate than adult Hb. Early human embryos have another β substitution, a mutant called ε. The principal human adult Hb is HbA$_2$, which consists of α2 β2; a second adult form, HbA$_2$, consists of α2 δ2 and constitutes 2–3 percent of circulating Hb molecules. Many allotypes are known; for example, in human sickle cell Hb the β chain is altered by substitution of a valine for a glutamate. Two allotypes of γ Hb are G (glycine) and A (alanine).

Fishes have many more isotypic and allotypic Hbs than mammals. In fish hemoglobins more than one type of α chain and β chain associate in different proportions as hybrid tetramers.

Functional properties of the amino acid substitutions in allotypes and isotypes are adaptive. In many Hbs the affinity for O_2 is reduced when the pH of the milieu is lowered. This enhances unloading of O_2 in tissues and the converse effect facilitates loading in the respiratory surface. O_2 affinity is also modulated by organophosphates. The positions affected by pH and by organophosphate are on the β chains. Imidazole groups of C-terminal histidines of β chains and differences in terminal amino groups of β chains with a phenylalanine substituted for the terminal histidyl residue show no or reverse CO_2 effect. In some catastomid fishes the β chain of the cathodal Hb has a carboxy terminal of -Tyr-Phe while the anodal β chain ends in -Tyr-His. The anodal Hbs are sensitive to both pH and ATP, while the cathodal Hb is insensitive to pH and is negligibly affected by ATP. The fishes of one subgenus catastomid occur in fast streams; the cathodal Hb has high O_2 affinity, no pH effect, and is probably used for O_2 transport when the fish are highly active. The subspecies with all Hbs anodal are sensitive to pH; they live in pools and sluggish water.

The family of cytochromes c is an example of extreme biochemical conservatism in that the protein structure and functions are similar in primitive bacteria and in advanced eukaryotes. From similarities in the fold into which heme binds, from some invariant amino acid sequences, and from functions in electron transport it is concluded that cytochromes c originated very early, probably in ancestors of anaerobic photosynthesizing bacteria. Anaerobic sulfur bacteria, green or purple, use cytochromes in photosynthesis. There are 27 invariant amino acid residues in eukaryotic cytochromes c in fungi, fish, plants, and mammals. Structural differences in interactions of cytochrome c with cyt-oxidase and cyt-reductase indicate functional adaptations. Cytochrome c of heart has two binding sites for cyt-oxidase, one of high affinity and the other of low affinity. The high-affinity site is sensitive to ions and is inhibited by ATP.

In addition to functional measurements with respect to differences in amino acid sequences, studies of kinetic properties of enzymes are a subtle way to demonstrate the role of different phenotypes of proteins. Kinetic measurements are K_m, k_{cat}, which measure affinities for substrates and coenzymes, and K_d, which measures binding of inhibitors. Other kinetic measures are activation energies ΔG measured by temperature effects, values of entropy ΔH and enthalpy ΔS constants, volume responses to pressure and temperature of denaturation (Ch. 5). An example is lactate dehydrogenase, which catalyzes the reactions between lactate and pyruvate and uses NAD^+ as a cofactor. LDH is a tetramer of varying amounts of two isozymes A and B. The B enzyme (B_4) is predominant in tissues which are highly aerobic (heart, brain), and B_4 has high affinity for NAD^+. The A isozyme (A_4) is predominant in tissues low in O_2-white muscle-tissues which produce but do not utilize lactate. The A_4 isozyme has high affinity for NADH. LDH-B is inhibited by lactate and pyruvate in lower concentrations than LDH-A. K_m of mammalian B_4 for pyruvate is lower than the K_m of A_4. Species differences in proportions of the two forms of LDH are considerable and have not been fully explained. A third isozyme, LDH-C, occurs in spermatocytes of birds and mammals. In the retina and brain of many fishes another isozyme, LDH-E, has been found.

Temperature is an environmental parameter for which many molecular adaptations have been described. A number of allozymes, especially of LDH-B, occur in proportions that correlate with habitat. In the killifish Fundulus the allozyme B^a is more abundant in populations from cold water (Halifax) and B^b in warm water populations (Florida). Concentrations of ATP in red blood cells of Fundulus are higher in fish with the B^b allotype. At high temperatures the ratio ATP:Hb increases in correlation with LDH-B^b. Fish with B^aB^a hatch sooner than B^bB^b fish, and those with B^bB^b swim greater distances when stimulated. The two allozymes occur in a cline from north to south.

For many enzymes, activation energies (E_a) are lower for enzymes that normally function over a wide range of temperature (in poikilotherms), and in animals living at low temperatures, than for enzymes of high-temperature animals. Enzymes from cold-living species are less temperature-sensitive and have lower Q_{10} values than from warm-living animals.

In all chemically catalyzed reactions the energy barrier is climbed by increase in enthalpy (heat content) or by decrease in entropy (decreased randomness). Endothermic animals have higher heat content than ectothermic. For a constant change in activation energy (ΔG), changes in enthalpy (ΔS) are relatively greater in enzymes of ectotherms, whereas in endotherms changes in entropy (ΔH) are greater. Low enthalpy reduces temperature dependence. Energy of activation ΔG is less in ectotherms. For example, at normal cell temperatures ΔH for LDH-B$_4$ of rabbit is 12.5 Cal/mol, for tuna 8.7, whereas ΔS for rabbit is 2.5 Cal/mol and for tuna 13.4. For myosin ATPases of fish from cold water or from tropical regions ΔH and ΔS change compensatorily. These represent genetic differences.

Thermal stability of enzymes is correlated with the temperature at which plants or animals normally function. Enzymes from Antarctic fishes are inactivated by warming to temperatures a few degrees above zero; corresponding enzymes of tropical fishes are cold-inactivated at about 30°C.

Another kinetic property in which corresponding enzymes differ is change in volume (ΔV) as measured by effects of hydrostatic pressure. Enzymes of fishes that live at pressure of one atmosphere are much more sensitive to applied pressure than the corresponding enzymes of deep-water fishes.

In conclusion, most of the correlations made in the past of amino acid composition with protein function for enzymes are based on maximum velocity (V_{max}) measurements. More subtle measurements reveal the adaptive significance of small differences in protein structure. Kinetic and structural properties of proteins are adaptive for certain life patterns, and many genetic changes which are thought to be neutral are likely to prove to have been selected when subtle properties of the proteins are measured.

Adaptive Development

Another determinant of variation is development. Unless an organism passes through each requisite stage in the sequence of development, it will not become the normal phenotype. Many genes are expressed for only short periods; some genes are specific for only certain tissues. Classical embryology recognized the differences between indeterminate and determinate cleavage, the equipotency or fixed pattern of cell organization during development. The importance of development "canalization" was recognized by Waddington.

Examples of adaptive developmental properties of proteins are the isotypes in different tissues. Hemoglobins occur in several isotypes—early embryonic, fetal, and adult. Muscle develops from myoblasts to myotubes to muscle fibers, and the light chains of myosin differ at each stage. Myosins and actins of skeletal muscle change in amount and composition with exercise; these proteins differ in striated, smooth, and cardiac muscle. The number of dendritic trees in pyramidal neurons of cerebral cortex increases with enriched experience during the animal's development. Neurons of the visual cortex fail to develop unless young mammals see objective patterns at critical stages in development. The ways in which genes are turned on and off at certain stages in specific tissues, how neuronal processes find their way during development, which of many neurons degenerate during development, are virtually unknown.

Adaptations to the Environment

Environmental factors modify all organisms within the limits set by the genotype and development. All organisms are somewhat plastic, capable of change according to environmental experience. Organisms are adapted to their total environment. Modifications by acclimatization to a natural environment may be different from acclimation-modifications in the laboratory where a single parameter may be varied. Every organism is played upon by multiple factors—physical factors, such as temperature, salinity, ions, oxygen, hydrostatic pressure, nutrients, and biotic factors, such as predators, prey, conspecific and heterospecific competitors or cooperators. Changes in organisms can occur over three kinds of time spans. Long-term changes resulting in genetic alterations over many generations were mentioned previously. Over periods of days, months, seasons there may be biochemical reorganization, acclimatization which is adaptive to the altered environment. Short-term effects or direct responses are adaptive to immediate environmental change. All three time courses are genetically coded.

One class of direct response and of intermediate time change is the internal state of the organism as a function of the environment. The internal state may be the same as the external. Body temperature of poikilotherms conforms to that of the environment; many marine organisms have osmoconcentration of body fluids equal to that of the environment (sea water). The level of environmental oxygen may cause metabolism to rise or fall. A general term for internal state in direct proportion to the environment is *conformity*.

The contrasting mode of internal state is *regulation* or homeostasis. Homeotherms maintain relatively constant body temperature over a widely fluctuating environmental range; all organisms, including bacteria, have their internal ion composition different from the medium. All fresh-water and most estuarine organisms maintain hyperosmoticity to the medium. Many animals regulate metabolism at a constant level down to some critical partial pressure of oxygen below which regulation fails. Within one organism there can be either conformity or regulation corresponding to different ranges of an environmental parameter. Some estuarine animals conform osmotically in high concentrations of the medium and regulate if it is dilute. Many marine animals are osmotic conformers but ionic regulators. Internal state of all organisms is controlled by both enzymatic and membrane adjustments for which the genetic basis has been only partly elucidated.

Another type of plasticity of organisms has to do with activity metabolism in a broad context and is called *capacity adaptation*. In metabolic conformers, energy liberation rises and falls with internal state in an adaptive way. One response to cold is to reduce metabolism in conformity with lower activity. Lowered metabolism is an opportunistic response particularly useful during seasonal change, tidal cycles, and in drought. Examples are animals that hibernate in winter or estivate in summer. The converse is regulation of energy liberation such that activity is maintained at a relatively constant level when the internal state varies or when an organism is stressed by environmental changes. Regulation of metabolism and constancy of energy liberation is called *homeokinesis*.

The principles of *capacity regulation* of metabolism can be applied to environmental parameters such as salinity, oxygen, or nutrients, but have been most studied for temperature. When oxygen consumption or activity of an energy-yielding reaction is measured at an intermediate temperature and the temperature is then lowered, the rate first declines in a direct response. The magnitude of the response may be adaptive in that the Q_{10} is usually greater for constant-temperature organisms than for variable-temperature ones (Ch. 7). After a period of days, metabolism may rise to a level approaching that before the initial drop in temperature. At a new steady state, compensation for cold has occurred. Patterns of acclimation are: no change, undercompensation or paradoxical response, partial compensation, complete compensation, overcompensation. When temperature is raised above the initial median level, reactions which are the converse to those in the cold occur. Many variations in these patterns are known, mostly with respect to the temperature range within which compensation occurs. Animals that become lethargic in the cold may compensate in a midrange but may show inverse or no compensation at extremes. The net effect of compensatory acclimation, whether to temperature, oxygen, or salinity is to maintain constancy of energy production, or homeokinesis. An animal may be a conformer in its internal state (i.e., not show homeostasis) yet show energy regulation, homeokinesis.

The mechanisms of metabolic acclimation, homeokinesis, are summarized as follows. One mechanism is shift in metabolic pathway, for example, between glycolysis and oxidation, between fat and carbohydrate metabo-

lism. Some tissues, such as white muscle and liver, are predominantly glycolytic; others, such as heart and brain, are oxidative. In exercise, shifts
from oxidative to glycolytic metabolism occur. Lipid metabolism increases
during cold acclimation and in hibernators. Lipid is the principal fuel of migrating birds.

Another mechanism of temperature compensation is the shift in proportions of isozymal and allotypal forms. The relative amounts of isotypes of
LDH and of G6PDH vary with tissue and with metabolic stress. During compensatory acclimation to low temperatures the activity of energy-yielding
enzymes increases. This increase is brought about partly by changes in absolute amounts of critical enzymes, partly by changes in the lipid composition of membranes to which enzymes are bound, and partly by shifts in
amounts of isotypes or allotypes of enzymes. Fatty acids laid down at low
temperatures are more unsaturated than those laid down at high temperatures. Membrane-bound enzymes, such as succinic dehydrogenase of
mitochondria, are altered according to the lipoproteins of the organelle
membranes.

The molecular biology of changes in metabolic pathways and in enzyme
properties associated with acclimatization involves feedback controls to protein-synthesizing systems. The details of the signals, how they are detected
and how they act are virtually unexplored. Primary cultures of fish liver cells
maintained at high or low temperatures for several weeks show acclimatory
compensation of enzymes similar to the changes *in vivo*. Differences in general protein synthesis between acclimated cultured cells and intact fish suggest hormonal effects *in vivo*.

Related to the metabolic adaptations are changes in cell membranes, both
plasma membranes and those of intracellular organelles such as mitochondria. Changes in lipid composition make for relative compensation in membrane fluidity, but whether these control permeability to ions is uncertain.
Changes in active transport of ions are adaptive for life at different salinities.
Fish that migrate between sea water and fresh water are faced with reverse
osmotic problems; their gills transport sodium and chloride outward in the
sea and actively absorb ions in fresh water. Changes in amounts of Na-K
ATPase in specialized cells of the gills occur on transfer from one medium
to the other.

Capacity adaptation by behavior is used by most animals. Animals tend
to select appropriate temperatures when in a gradient; they bask in the sun
or seek shade. They avoid extremes of salinity or of low oxygen. The selection of environment can be altered by acclimation experience. Selective
adaptive behaviors tend to put the animals into environments which are favorable for life processes.

Resistance adaptations are those changes, environmentally induced
within genetically set limits, which permit activity beyond the environmental
limits usually encountered. When organisms—bacteria, plants, or animals—
are severely stressed, one of the first signs of damage is increase in ion

permeability. Cells leak potassium and gain sodium. Leaky cell membranes precede inactivation of enzymes. In fact, most organisms die well before denaturation of metabolic proteins occurs. Lipids change in structure and hence membranes change in fluidity. In animals the first system to fail under stress—cold, heat, hypoxia, toxins, and so on—is the nervous system. Some synapses are more sensitive than others, and axons are generally less sensitive than synapses. Ion channels differ in temperature sensitivity, and tissue differences in conductances for ions indicate protein specificities. Heat shock proteins are synthesized when tissues are exposed to moderate heat, and later these tissues are more resistant to extreme heat than untreated tissues. Synthesis of other proteins virtually stops while heat shock proteins are made; how they protect is unknown.

A few examples of biological variation presented in preceding chapters give a glimpse of the wide panorama in living organisms of variations determined genetically, developmentally, and environmentally. Correlations of biological variations with ecological distribution and phylogenetic relationships are so pervasive that causations seem incontrovertible. Fine-scale measurements of nucleotide and amino acid sequences, kinetic properties of enzyme proteins, and experimental modifications of these properties give functional explanations of distribution and phylogenetic relations not possible from gross morphological and metabolic measurements.

MEMBRANES AND BEHAVIOR

The necessity for selective cell membranes bounding the earliest cells was mentioned in Chapter 9. As organic solutes, mostly anionic, were formed within cells, membranes with active transport properties evolved and osmotic and electrical steady states were maintained. Probably at the same time concentrations of hydrogen and potassium ions became higher and such ions as sodium, magnesium, and sulfate were excluded from cells. Cell membranes today keep essential substances in and noxious substances out. The most primitive and general of ion pumps move protons and are found in bacteria, mitochondria, and many plant plasma membranes. The net effect of a proton pump is to maintain the pH of the cell, to rid the cell of metabolically produced protons, to maintain intracellular electronegativity, and, by exchange, to raise intracellular potassium. Ion pumps for calcium are widely distributed in plasma membranes and intracellular organelles. In most animal cells a Na-K ATPase actively extrudes sodium and takes up potassium, often electrogenically. Ion pumps function also as carriers for amino acids and sugars. Active transport of chloride and of magnesium is probable.

Transport at the organismic level is particularly important in aquatic animals and plants. By specialized cells fresh-water organisms actively absorb essential ions and sea-water organisms extrude ions.

The evolution of cell membranes paralleled the evolution of nucleotides and proteins. Cell membranes contain proteins and phospholipids in varying proportions and quantities. Probably the high specificity of selective permeability and of active transport is determined by the proteins. How the membrane lipids act other than in a supportive fashion is unclear.

Cell membranes also mediate responses to environmental transients. Bacteria show negative and positive chemotaxis to nutrients such as amino acids; bacterial membranes are highly specific in sensitivity. Many cell membranes respond to mechanical deformation—for example, the pulvinar cells of sensitive plants, stretch and touch receptors of animals, and the ultrasensitive hair cells within vertebrate ears. In ciliate protozoa mechanical stimuli evoke one electrical response at the anterior end of the cell and a different response at the posterior end. Some cell membranes are activated by very small electric currents, strikingly so the electroreceptors of electric fishes.

The best known responses to stimulation—electrical, mechanical, or chemical—are electrical. Some responses are graded with the strength of the stimulus: sensory potentials, postsynaptic responses, pacemaker potentials. Other responses are of constant or near-constant size and are usually self-propagating by local electrical currents. In some plant cells depolarization results from efflux of Cl^-. Hyperpolarizing responses of synapses in animals are due to either an influx of Cl^- or to increased efflux of K^+. Probably the most widepsread depolarizing responses in animals are by influx of Ca^{2+}, but many muscle and nerve membrane action potentials are produced by inward currents of Na^+. Calcium action potentials are probably more primitive than sodium ones. In both Ca and Na action potentials, the repolarizing phase represents outward K^+ current, which can be activated either electrically or by Ca^{2+}. Specific channels are used for the different ions, and blocking agents are known for each of these. More kinds of channels occur for K^+ than for other ions.

Information transfer from one cell to another may be electrical—between embryonic cells, between epithelial cells, and from neuron to neuron where speed is essential. Most synaptic transmission in nervous systems is by chemical agents. The most primitive transmitters are probably the amino acids that are attractive or repellent to bacteria and to some invertebrate animals. Only a few amino acids have neurotransmitter functions—glutamic acid as an excitor, glycine and gamma-amino butyric acid as inhibitors. An evolutionary transition was probable between amino acids and catecholamines and monoamines. Why some of these, such as the series derived from tyrosine, have been selected while others, such as homarine, betaine, and so on, occur widely but are not neurotransmitters is unknown. Choline is a constituent of most membrane phospholipids and the acetylated form is the most widely occurring neurotransmitter. The nicotinic acetylcholine receptor is a tetrameric protein with anionic and esteratic sites.

A group of neurotransmitters of great variety are the neuropeptides. Some mammalian neuropeptides were discovered as gastrointestinal hormones and are now known to function also in regions of the brain as transmitters or modulators. Several neuropeptides have in common some amino acid se-

quences. Many neuropeptides of invertebrates resemble those of vertebrates, others are totally different. The distinction between transmitters,
neuromodulators, and hormones breaks down when neuropeptides are considered in detail.

Some transmitters and sensory signals, for example, light stimuli to photoreceptors, trigger a series of intracellular signals, second messengers such
as calcium and cyclic AMP.

Other kinds of chemical signals are pheromones, such as attractants between male and female insects, and repellents between polyps of neighboring encroaching corals. Some plants produce substances in small amounts
that repel most insects; but a few kinds of insects have mutated so that they
are attracted by the compounds. Selection of plant food is based on attractant compounds; the glycosides of cabbage attract cabbage butterflies.

NEURAL BASES OF ANIMAL BEHAVIOR

The origin of nervous systems is obscure. Nerve nets of coelenterates have
integrative properties similar to those of more complex animals. Nervous
systems of leeches, gastropods, crustaceans, and insects have general properties similar to central nervous systems in vertebrates. General aspects of
central nervous systems are: fixed action patterns; endogenous rhythmic
activity of nerve cells and of nerve networks; modulation by sensory inputs;
hierarchical controls of behavior; command neurons and networks; sequential activity which coordinates rhythmic movements; modification of responses by repetition (e.g., posttetanic potentiation, heterosynaptic facilitation, and habituation); persistent modification; memory. However, it is
uncertain how far simple nervous systems can be used as models for functions of mammalian nervous systems.

Animal behavior provides many examples of physiological adaptation.
Many species of animals are isolated by reproductive behavior. Examples
are found in the seasonal timing of behavior, its circadian and tidal rhythms,
courtship and mating sequential events, and signals between sexes, especially by sound, visual displays, and pheromones. Much of animal behavior
is critical for survival, growth, and reproduction: feeding patterns, social
interactions, communication between parent and offspring, between mates,
between members of communities. The neural basis of complex animal behavior is barely known, certainly not at molecular levels of analysis. Yet
these behaviors have much adaptive significance in animal evolution.

Most reproductive behaviors are highly specific and stereotyped. They
are clearly programmed in the nervous system and genetically transmitted.
Analysis of fixed action patterns and of stereotyped but complex behaviors
in terms of nervous circuits has not been fully successful. The general locus
of some components of behavior is clear, for example, centers for hearing
of calls in frogs and crickets, but the exact nature of genetic control of the

central nervous patterns of species-specific stereotyped behavior remains one of the important questions of adaptational biology. In vertebrates, especially mammals and birds, neural centers and endocrines interact, for example, hypothalamus and gonad.

Genetic coding for some ion channels and transmitters is known and the genes for several of them have been cloned. Genetic coding for many specific neurons and neural networks remains to be eludicated. Mutants for behavior have been recognized. Behavioral mutants in Drosophila show altered patterns of wing beat, locomotion, and activity. Mutants of nematodes (Coenorhabditis) have abnormal movement patterns coded in a nervous system containing very few neurons. Domestic animals, especially dogs, have been selected for behavioral patterns that are genetically determined. Genetic analysis of nervous systems is much more difficult than of other systems. Rat brain has six to eight more kinds of mRNA and translated proteins than organs such as kidney or liver. This means that the expressed part of the genome is greater for brain.

Behaviors can be arranged in order of increasing complexity, from taxes and kineses through stereotyped reflexes and fixed action patterns to social interactions. The relationship between brain and mind is often interpreted in anthropomorphic terms. Several aspects of complex behavior can be described in terms of neural function, others only as psychological concepts. To what extent can single neurons process sensory input and motor output, and to what extent are neural nets requisite? The critical parts of nervous systems for behavior are interneurons, which determine routes for messages to follow. Critical properties of brain include neuronal redundancy, hierarchical organization, convergence and divergence of pathways, information content, and number of interacting cells. In the most highly specialized brains, in cephalopods and vertebrates, there is an abundance of small neurons. The information content of the brain, like that of a computer, increases exponentially with the number of elements. Information processing can be rigidly specified or it can be a function of cells which have some randomness and duplication of function.

A number of adaptive properties of nervous systems are described in holistic terms. An example is perception contrasted with sensation. An animal may give direct motor responses to spectral colors or to tones of certain wave frequencies according to sensitivities of sense organs. However, sensory reception is only a first step in perception. Interneurons in the association cortex of awake monkeys respond to pattern and movement of light and to faces. The neural cell responses may occur only when the monkey is motivated to the visual stimulus. These interneurons have the properties expected of decision-making cells.

A difficult ethological concept is self-awareness or consciousness; this is usually considered in anthropomorphic terms. Objective measures of self-awareness are possible only with humans.

A related problem is intelligence. Whether intelligence is attributable to a honeybee in search for nectar and communication within a hive or to a bird that constructs a species-specific type of nest is a matter of definition. Ability to perform maze-learning is not an adequate measure of intelligence. Perhaps the best we can do at present is to attempt a description of behavior which is adaptive to the total natural environment of an animal.

All nervous systems are somewhat plastic. At the cellular level this is shown by growth patterns, by functional changes due to repetition, habituation, facilitation, and long-term inhibition. The number of synapses on cortical neurons can be modified according to whether experience has been in enriched or impoverished environments. Modification of excitability which resembles learning is accompanied by biochemical changes in synaptic membranes, for example, in the mollusc Aplysia. It is questionable how far cellular models can go in elucidating learning.

An important component of behavior is communication. This can be stereotyped and genetically programmed, as in the dance language of honeybees. Communication can make use of visual, auditory, and chemical signals. Much communicatory ability is genetically programmed. Some communication is imprinted, for example bird song. Another type of communication is the use of signs as symbols by advanced primates. One use of language is in cognition. The speech centers of human cortex are so large and complex as to make for qualitative differences between humans and the most closely related primates. Attribution of cognition to apes is anthropomorphic.

Finally and of utmost importance is the question of applicability of adaptive neurobiology to humans. The transmission of cultural characters within societies is by very different means from genetic transmission, and widens the range of possible functioning of individual humans. That all humans are one biological species and have certain characters, transmitted socially rather than genetically are unifying principles. The basic brain structures of humans and apes are similar, but the quantitative differences are extensive. There has been coevolution of genetically and culturally transmitted characters. The symbiosis of the two types of transmission gives humans their uniqueness.

INDEX

Abstract thinking, 435
Acclimation, 2
  changes:
    energy metabolism, 15
    enzyme activity, 15
    organ size, 298
    phospholipid, 299
    protein synthesis, 15
    synaptic membrane, 16
  enzyme activity, 295
    and fatty acid composition, 15
    mechanisms, 293
    and protein synthesis, 297
Acetylcholine:
  effect on heart rate, 663
  hydrolysis, 529
  methyl group, 530
Acetylcholine esterase, 15, 123, 527
Acetylcholine receptors, 528
  binding rate, 531
  monoclonal antibodies for, 531
  muscarinic, 529
  nicotinic, 529
  subunits, 531
Acid-base balance, 301
Acid phosphatase, 100
Acrosome, 714
Actin, 329, 688, 712
  amino acid sequence, 712
  binding to myosin, 714
  genes for, 714
  molecular weight, 714
  non-muscle, 689
Actinomycin D, 101, 503
Action potential, 273
Activation energy, 173
Actomyocin, 701
Adaptation:
  determinants, 1

developmental, 1
and environment, 1
genetically specified, 1
as immediate response to change, 14
inherited, 754
limitations, 3
and natural selection, 3
for osmotic balance, 346
physiological, 1
  classification, 6
  resistance adaptation, 12
seasonal, 100
to temperature, 262
time course of, 14
Adaptational genetics, 66
Adaptive behavior, animal, 433
Adaptive diversity, 2, 21
Adaptive evolution, 2
Adenosine triphosphatase, 34
  heat inactivation, 276
  ionophores on, 364
  myofibrillar, 279
Adenosine triphosphate, 32, 539
  hydrolysis, 725
  as neurotransmitter, 540
Adenosine triphosphate sulfurylase, 324
Adenosine triphosphate synthetase, 365
Adenylates, 33
Adenylsulfate reductase, 324
Adenyl sulfite, 42
Adipose tissue, 222
α-Adrenergic receptors, 536
Aerobic scope, 221
Agonistic actions, animal behavior, 401
Aggression, in animals, 402
  nervous system control, 445
  visual display in, 403
Alanine-aspartate transferase, 140
Albumin, immunological distance, 132

767